Xenopu

Xenopus

From Basic Biology to Disease Models in the Genomic Era

Edited by
Abraham Fainsod and Sally A. Moody

CRC Press
Taylor & Francis Group
Boca Raton London

CRC Press is an imprint of the
Taylor & Francis Group, an **informa** business

First edition published 2022
by CRC Press
6000 Broken Sound Parkway NW, Suite 300, Boca Raton, FL 33487–2742

and by CRC Press
2 Park Square, Milton Park, Abingdon, Oxon, OX14 4RN

CRC Press is an imprint of Taylor & Francis Group, LLC

Library of Congress Cataloging-in-Publication Data
Names: Moody, Sally A., editor. | Fainsod, Abraham, editor.
Title: Xenopus : from basic biology to disease models in the genomic era / edit by Sally Moody and Abraham Fainsod.
Description: Boca Raton : CRC Press, 2022. | Includes bibliographical references and index. | Summary: "Xenopus is unique among the model animals used in the biology. Several books of protocols used Xenopus. Missing is a book taking an historical perspective documenting cell and developmental discoveries and illustrating how Xenopus contributes to the understanding of genes. These topics will be covered in the proposed book"—Provided by publisher.
Identifiers: LCCN 2021038920 (print) | LCCN 2021038921 (ebook) | ISBN 9780367505271 (hardback) | ISBN 9780367505349 (paperback) | ISBN 9781003050230 (ebook)
Subjects: LCSH: Xenopus. | Xenopus—Molecular aspects. | Animal models in research.
Classification: LCC QL668.E265 X45 2022 (print) | LCC QL668.E265 (ebook) | DDC 597.8/654—dc23
LC record available at https://lccn.loc.gov/2021038920
LC ebook record available at https://lccn.loc.gov/2021038921

ISBN: 978-0-367-50527-1 (hbk)
ISBN: 978-0-367-50534-9 (pbk)
ISBN: 978-1-003-05023-0 (ebk)

DOI: 10.1201/9781003050230

Typeset in Times
by Apex CoVantage, LLC

Contents

Preface vii
Sally A. Moody and Abraham Fainsod

Contributors ix

SECTION I Contributions to Cell, Developmental, and Molecular Biology

Chapter 1 A Quick History of *Xenopus*: "The Humble Batrachian" 3
John B. Wallingford

Chapter 2 The Study of Cell Division Control and DNA Replication in *Xenopus* Egg Extracts 13
Allison M. Jevitt and Susannah Rankin

Chapter 3 Maternal mRNAs and the Making of Cell Lineages in the Early Xenopus Embryo 25
Douglas W. Houston

Chapter 4 Signaling Components in Dorsal-Ventral Patterning and the Organizer in *Xenopus* 43
Edward M. De Robertis and Nydia Tejeda-Muñoz

Chapter 5 Signaling Pathways in Anterior-Posterior Patterning 51
Chenbei Chang

Chapter 6 Wnt Signaling in Tissue Differentiation and Morphogenesis 65
Stefan Hoppler and Michael Kühl

Chapter 7 Multiple Functions of Notch Signaling during Early Embryogenesis 77
Silvia L. López

Chapter 8 The Development and Evolution of the Vertebrate Neural Crest: Insights from *Xenopus* 125
Joshua R. York and Carole LaBonne

Chapter 9 The Use of *Xenopus* Oocytes to Study the Biophysics and Pharmacological Properties of Receptors and Channels 143
Ataúlfo Martínez-Torres and Elizabeth Pereida-Jaramillo

SECTION II Systems Biology and the Genomic Era

Chapter 10 The Continuing Evolution of the *Xenopus* Genome 155
Mariko Kondo and Masanori Taira

Chapter 11 Dynamics of Chromatin Remodeling during Embryonic Development ... 173
Gert Jan C. Veenstra

Chapter 12 Gene Regulatory Networks Controlling *Xenopus* Embryogenesis ... 185
Ken W.Y. Cho and Ira L. Blitz

Chapter 13 The Development of High-Resolution Proteomic Analyses in *Xenopus* ... 197
Elizabeth Van Itallie and Leonid Peshkin

Chapter 14 Advances in Genome Editing Tools ... 207
Marko E. Horb, Anita Abu-Daya, Marcin Wlizla, Anna Noble, and Matt Guille

SECTION III From Basic Biological Insights to Human Disease

Chapter 15 Formation of the Left-Right Axis; Insights from the *Xenopus* Model ... 225
Axel Schweickert and Tim Ott

Chapter 16 Discovering the Function of Congenital Heart Disease Genes ... 233
Delfina Pearledith González and Mustafa K. Khokha

Chapter 17 Craniofacial Development and Disorders—Contributions of *Xenopus* ... 245
Ashwin Lokapally and Hazel Sive

Chapter 18 Modeling Digestive and Respiratory System Development and Disease in *Xenopus* ... 259
Scott A. Rankin and Aaron M. Zorn

Chapter 19 Functional Neurobiology in *Xenopus* Provides Insights into Health and Disease ... 277
Clayton Gordy, Michael Forsthofer, Parthena Soupiadou, Suzan Özugur, and Hans Straka

Chapter 20 Leaping toward Understanding of Spinal Cord Regeneration ... 289
Paula G Slater, Gabriela Edwards-Faret, and Juan Larraín

Chapter 21 Studying Tumor Formation and Regulation in *Xenopus* ... 301
Dieter Tulkens and Kris Vleminckx

Chapter 22 *Xenopus*: A Model to Study Natural Genetic Variation and Its Disease Implications ... 313
Avi Leibovich, Sally A. Moody, Steven L. Klein, and Abraham Fainsod

Chapter 23 Using *Xenopus* to Understand Pluripotency and to Reprogram Cells for Therapeutic Use ... 325
Meghana S. Oak and Eva Hörmanseder

Index ... 337

Preface

The story of *Xenopus* as an experimental model system seems like a version of the Grimm Brothers' "The Frog Prince." Here, a frog, *Xenopus laevis*, also came from humble beginnings, and with a little "tender loving care," it became a prince, or, as many would prefer, a "princess," for many fields of biomedical research and basic biology. Once upon a time, the founders of experimental embryology used many different species of amphibians, including a variety of newts and frogs. For half a century, pioneers, including Wilhelm Roux, Hans Spemann, Thomas Hunt Morgan, Oscar Hertwig, and others, used these species to decipher many of nature's secrets, including basic principles in cell biology, neurobiology, morphogenetic movements, cell communication, and the contributions of the genome. Some of these breakthroughs were afforded the highest accolades, such as the awarding of the 1935 Nobel prize to Hans Spemann for the discovery of embryonic induction. In parallel, some of these experimental models provided scientific tools to study the effects of the environment on biological processes in early toxicological and teratogenic studies. However, the inability to maintain or breed these animals in the laboratory environment limited many of these studies.

In the 1930s, a new frog animal model caught the interest of the medical and scientific community. Initially introduced to the clinic for performing pregnancy tests, *Xenopus laevis*, also known as the African clawed frog, proved very resilient in the laboratory environment. One of the consequences of using *Xenopus* in pregnancy tests was its wide distribution around the world, and colonies were established at multiple universities and hospitals. The maintenance of these colonies revealed that *Xenopus* husbandry is very simple, and eggs and embryos can be readily obtained year-round with minimal effort. This frog species thrived in the laboratory, efficiently breeding to produce large numbers of eggs after injection of readily available reproductive hormones. Since then, many of the classical experiments previously performed in other amphibian models have been repeated and expanded using *Xenopus*.

Each experimental model system has advantages and disadvantages, and the chapters in this book describe some of the particular strengths of *Xenopus* as a vertebrate experimental model system to study basic biological principles and human disease. Because amphibians are tetrapods, *Xenopus* is closer evolutionarily to mammals, and in particular humans, than fish. More importantly, most of the biological and biomedical principles identified and characterized in *Xenopus* are generally applicable to all vertebrates. The *Xenopus laevis* oocyte is relatively large in size (~1.4 mm in diameter), and a female can lay from several hundred to several thousand eggs in a single day. This large clutch size provides an excellent source of biochemical materials and has been used to study the cell cycle, DNA replication, and chromatin structure. Beyond the clear advantages for biochemical studies, the large clutch size allows performing rather complex experiments composed of multiple samples from a similar genetic background and large enough sample sizes for statistical significance of the results. The large size of the *Xenopus* oocyte also allows the easy use of glass needles to inject the oocytes with mRNAs that can undergo translation and subsequent post-translational processing into proteins that can be properly folded, chemically modified, and either retained intracellularly, integrated into membranes, or secreted. Oocytes can also be injected with chemicals, antibodies, DNA molecules, and other reagents to study many biological processes.

An important direction in which *Xenopus laevis* became a cornerstone was in the analysis of embryonic development. The ease with which hundreds of embryos can be obtained in a single clutch, allowing large samples, combined with the external development that allows analysis of all developmental stages, were essential for characterizing basic principles of embryogenesis in vertebrates. Already from the early studies, it was clear that amphibian embryos were amenable to microsurgical manipulation and transplantations; *Xenopus* embryos proved no different and even particularly resilient to this type of experiment. The egg's large size and its rapid development into an embryo became an asset for the production and analysis of gene products following mRNA injection. The detailed fate maps of individual cells made it possible to target microinjections to particular regions, sometimes on one side of the embryo, thus minimizing off-target effects and providing an internal control in each embryo. Thus, many basic principles of vertebrate embryogenesis were elucidated using *Xenopus* embryos, and numerous aspects in the characterization of signaling pathways were worked out using *Xenopus* embryos. Over the years, studies based on *Xenopus* have received worldwide recognition and prizes, including the Nobel prize in 2012 to Sir John Gurdon for the discovery of genetic reprogramming of mature cells to pluripotency.

The allotetraploid genetic composition of *Xenopus laevis* and its relatively long generation time (1–1.5 years) had initially posed a challenge for loss-of-function studies based on mutants. These apparent "deal breakers" as a model system drove research in directions that proved extremely fruitful for our understanding of vertebrate embryogenesis. First, the *Xenopus* community focused on the identification and cloning of numerous novel genes central to almost every developmental and regeneration process. In many instances, this massive cloning effort helped identify and elucidate signaling pathways, developmental processes, cellular morphogenesis, biochemical interactions, and many more processes. This focused cloning effort eventually resulted in a massive collection of cDNA and EST clones that drove forward research in *Xenopus* and in multiple instances paved the way for studies in other experimental model systems.

Second, the apparent lack of genetic tools for a couple of decades drove the analysis of gene function in *Xenopus laevis* to rely on gene-specific antisense oligonucleotides, overexpressed dominant-negative protein variants, antimorphic protein constructs, and pharmacological manipulation of protein function.

The complex genome composition of *Xenopus laevis* necessitated the search for strategies to enable genetic approaches. One of the solutions to this genetic quest culminated in the introduction of *Xenopus tropicalis*, a true diploid *Xenopus* species with a short generation time (4–6 months) similar to mouse and zebrafish. *X. tropicalis* research benefited from the wealth of methods already developed for *X. laevis*, which were rapidly adapted for the newcomer. Also, due to the high similarity in the genome sequence, gene discovery in *X. tropicalis* surged forward based on the clones available from *X. laevis*. The story of these two truly complementary models is a story of real cross-pollination. As a true diploid, the *X. tropicalis* genome project was completed in parallel to mammals. The difficulties of sequencing the allotetraploid *X. laevis* genome could finally be solved using the high sequence homology to *X. tropicalis* as a reference genome. In parallel, methods were developed in both *Xenopus* species to generate transgenic strains as well as TALEN and CRISPR/Cas9 genome editing. Even with the longer generation time, genetically modified *X. laevis* lines are becoming mainstream in analyzing normal developmental processes and generating disease models.

Close to a century and a half of amphibian-based research led to an extensive convergence to *Xenopus* as the main model system. This is not to say that it is exclusive, as other *Xenopus* species are incorporated for evolutionary studies, other frog species are used for neurophysiological research, and axolotl plays a major role in regeneration studies. With the extensive genome information for *laevis* and *tropicalis* and the genome editing technologies efficiently implemented to generate mutations, both species have surged forward as excellent model systems in biomedical research. More importantly, *Xenopus* is an excellent model system to recapitulate many aspects of human disease and the in-depth study of its etiology. The ease of pharmacological manipulation in *Xenopus* has also allowed the study of chemically induced birth defects and disease. The large clutch size together with the efficient implementation of genome editing technologies allow the performance of relatively rapid disease studies and screens. These studies initially analyze founder animals, overcoming the need to establish genetically modified lines and allowing faster analysis of disease-causing changes.

One aspect that significantly contributed to advancing *Xenopus* as a major model system was a collaborative and interactive community. In 1984, two leaders in the field of developmental biology who utilized *Xenopus*, Igor Dawid (NIH) and John Gurdon (Cambridge University), brought together a group of about 20 international investigators to discuss their research interests and deliberate on the advantages of having a regular meeting focused on *Xenopus* developmental genetics. Thus was born the International *Xenopus* Conference, which has been held every two years since then. Over the years, the conference has expanded to encompass cell biology, neurobiology, regeneration, and disease models. In addition, NIH-supported workshops in the 1990s led to the development of critical research resources for the community, and similar community-organized meetings have continued on a biannual basis as the *Xenopus* Resources and Emerging Technologies meeting. These truly international efforts of several hundred laboratories have significantly contributed to the many advances that are summarized in the chapters presented in this book.

Thus, this humble frog was transformed into a prince, or princess, by the dedication and persistence of many researchers, and by an interactive and supportive community. The chapters of this book summarize some of the advantages of working with *Xenopus* in biomedical research and some of the major contributions of *Xenopus* to our biological knowledge. This collection shows how and why *Xenopus*-based research is not only poised but has already made major contributions to our understanding of human biology and disease.

Abraham Fainsod and Sally A. Moody

Contributors

Anita Abu-Daya
European *Xenopus* Resource Centre and School of Biological Sciences
University of Portsmouth
Portsmouth, UK

Ira L. Blitz
Department of Developmental and Cell Biology
University of California, Irvine
Irvine, California, USA

Chenbei Chang
Department of Biological Chemistry
Department of Cell, Developmental and Integrative Biology
University of Alabama at Birmingham
Birmingham, Alabama, USA

Ken W.Y. Cho
Department of Developmental and Cell Biology
University of California, Irvine
Irvine, California, USA

Edward M. De Robertis
Department of Biological Chemistry
David Geffen School of Medicine, University of California
Los Angeles, California, USA

Abraham Fainsod
Department of Developmental Biology and Cancer Research
Institute for Medical Research Israel-Canada, Faculty of Medicine, The Hebrew University of Jerusalem
Jerusalem, Israel

Gabriela Edwards-Faret
Life & Medical Sciences Institute
University of Bonn
Bonn, Germany

Michael Forsthofer
Department Biology II
Ludwig-Maximilians-University Munich
Planegg, Germany

Delfina P. González
Department of Pediatrics
Yale University School of Medicine
New Haven, Connecticut, USA

Clayton Gordy
Department Biology II
Ludwig-Maximilians-University Munich
Planegg, Germany

Matt Guille
European *Xenopus* Resource Centre and School of Biological Sciences
University of Portsmouth
Portsmouth, UK

Stefan Hoppler
Aberdeen Developmental Biology Group and Aberdeen Cardiovascular and Diabetes Centre
Institute of Medical Sciences, University of Aberdeen
Aberdeen, Scotland, UK

Marko E. Horb
Marine Biological Laboratory
National *Xenopus* Resource and Eugene Bell Center for Regenerative Biology and Tissue Engineering
Woods Hole, Massachusetts, USA

Eva Hörmanseder
Helmholtz-Zentrum München
Institute for Epigenetics and Stem Cells
Munich, Germany

Douglas W. Houston
Department of Biology
The University of Iowa
Iowa City, Iowa, USA

Allison M. Jevitt
Program in Cell Cycle and Cancer Biology
Oklahoma Medical Research Foundation
Oklahoma City, Oklahoma, USA

Mustafa K. Khokha
Department of Pediatrics
Yale University School of Medicine
New Haven, Connecticut, USA

Steven L. Klein
Department of Anatomy and Cell Biology
George Washington University School of Medicine and Health Sciences
Washington, DC, USA

Mariko Kondo
Laboratory of Aquatic Molecular Biology and Biotechnology
Graduate School of Agricultural and Life Sciences, University of Tokyo
Yayoi, Bunkyo, Tokyo, Japan

Michael Kühl
Institute of Biochemistry and Molecular Biology
Ulm University
Ulm, Germany

Carole LaBonne
NSF-Simons Center for Quantitative Biology
Northwestern University
Evanston, Illinois, USA

Juan Larraín
Department of Cell and Molecular Biology
Pontifical Catholic University of Chile
Santiago, Chile

Avi Leibovich
Department of Developmental Biology and Cancer Research
Institute for Medical Research Israel-Canada, Faculty of Medicine, The Hebrew University of Jerusalem
Jerusalem, Israel

Ashwin Lokapally
Department of Biology
Northeastern University
Boston, Massachusetts, USA

Silvia L. López
Faculty of Medicine, Department of Cell Biology and Histology
University of Buenos Aires
Buenos Aires, Argentina

Ataúlfo Martínez-Torres
Laboratorio de Neurobiología Molecular y Celular
Instituto de Neurobiología
Universidad Nacional Autónoma de México
Querétaro, México

Sally A. Moody
Department of Anatomy and Cell Biology
George Washington University School of Medicine and Health Sciences
Washington, DC, USA

Anna Noble
European *Xenopus* Resource Centre and School of Biological Sciences
University of Portsmouth
Portsmouth, UK

Meghana S. Oak
Helmholtz-Zentrum München
Institute for Epigenetics and Stem Cells
Munich, Germany

Tim Ott
Institute of Biology, Fg. Zoology
University of Hohenheim
Stuttgart, Germany

Suzan Özugur
Department Biology II
Ludwig-Maximilians-University Munich
Planegg, Germany

Elizabeth Pereida-Jaramillo
Laboratorio de Neurobiología Molecular y Celular
Instituto de Neurobiología
Universidad Nacional Autónoma de México
Querétaro, México

Leonid Peshkin
Marine Biological Laboratory
Eugene Bell Center
Woods Hole, Massachusetts, USA

Scott A. Rankin
Center for Stem Cell and Organoid Medicine (CuSTOM), Division of Developmental Biology, Perinatal Institute
Cincinnati Children's Hospital Medical Center
Cincinnati, Ohio, USA

Susannah Rankin
Cell Biology Department
University of Oklahoma Health Sciences Center
Oklahoma City, Oklahoma, USA

Axel Schweickert
Institute of Biology, Fg. Zoology
University of Hohenheim
Stuttgart, Germany

Hazel Sive
Department of Biology
Northeastern University
Boston, Massachusetts, USA

Paula G. Slater
Department of Cell and Molecular Biology
Pontifical Catholic University of Chile
Santiago, Chile

Parthena Soupiadou
Department Biology II
Ludwig-Maximilians-University Munich
Planegg, Germany

Hans Straka
Department Biology II
Ludwig-Maximilians-University Munich
Planegg, Germany

Masanori Taira
Department of Biological Sciences, Faculty of Science and Engineering
Chuo University
Kasuga, Bunkyo-ku, Tokyo, Japan

Nydia Tejeda-Muñoz
Department of Biological Chemistry
David Geffen School of Medicine, University of California
Los Angeles, California, USA

Dieter Tulkens
Department of Biomedical Molecular Biology
Ghent University
Ghent, Belgium

Elizabeth Van Itallie
Department of Systems Biology
Harvard Medical School
Boston, Massachusetts, USA

Gert Jan C. Veenstra
Faculty of Science, Department of Molecular Developmental Biology
Radboud University
Nijmegen, The Netherlands

Kris Vleminckx
Department of Biomedical Molecular Biology
Ghent University
Ghent, Belgium

John B. Wallingford
Department of Molecular Biosciences
Patterson Labs
Austin, Texas, USA

Marcin Wlizla
Marine Biological Laboratory
National *Xenopus* Resource and Eugene Bell Center for Regenerative Biology and Tissue Engineering
Woods Hole, Massachusetts, USA

Joshua R. York
Department of Molecular Biosciences
Northwestern University
Evanston, Illinois, USA

Aaron M. Zorn
Department of Pediatrics, College of Medicine
University of Cincinnati
Cincinnati, Ohio, USA

Section I

Contributions to Cell, Developmental, and Molecular Biology

1 A Quick History of *Xenopus*
"The Humble Batrachian"

John B. Wallingford

CONTENTS

1.1. Introduction 3
1.2. What's in a Name? 3
1.3. Pregnancy and Prominence: The Rise of Xenopus as a Laboratory Animal 6
1.4. Some Unsung Heroes of Early Developmental Biology Research With *Xenopus* 8
1.5. *The Ninth Day of Creation: Xenopus* at the Dawn of Molecular Biology 9
1.6. Where History Stops and "the Literature" Begins 9
Acknowledgments 10
References 10

> Might I suggest that a little credit might be given to the humble batrachian, which seems to give an invariably correct diagnosis.
>
> **—J.W.C. Gunn (1939)**

> When a developmental biologist says "in mammals" he probably means "in the mouse"; by "in birds" he almost certainly means "in the chick". When he generalizes more widely than this about mechanisms of differentiation in "all developing cells" there is a danger that he is referring to work carried out exclusively on Xenopus.
>
> **—Elizabeth Deuchar (1972)**

1.1. INTRODUCTION

Frogs have played so central a role in biological research that many people's only memory of actual biological study involves dissecting one in high school. It's no surprise, then, that frogs have been central to scientific discovery for centuries. Marcello Malpighi had at least an inkling of the concept we now call "model organisms." He repeatedly extolled the frog as an outstanding system for study, and it was in a frog that he first discovered capillaries of the circulatory system in 1661 (Holmes, 1993; West, 2013). He even wrote to his friend Giovanni Borelli that "indeed, things show up much more clearly in frogs" (Boorstin, 1985). Likewise, the entire field of electrophysiology is frequently considered to have originated with Galvani's experiments on frog legs in 1791 (Piccolino, 1997).

The frog's external development is another boon, one that was exploited by embryologists at least since 1758, when Johann Rösel von Rosenhof engraved the first chronological depictions of eggs developing into tadpoles and then into frogs (Wellmann, 2017). In 1886, nuclear transplantation in frogs and toads would first be attempted in an exploration of the hereditary control of development (Rauber, 1886), presaging Nobel prize-winning work on nuclear totipotency nearly another century later. In the late 19th century, newts and salamanders became the favored amphibian for embryologists (Beetschen, 2004), but beginning in the mid-20th century, a curious foray into endocrinology in South Africa led to the rise of *Xenopus* frogs as the dominant amphibian for laboratory studies of biology across the world.

Since that time, *Xenopus* was used for discoveries as varied as the first description of nuclear pores, to the first isolation of a eukaryotic gene, to the demonstration of the totipotence of nuclei, to the invention of *in situ* hybridization. Several excellent historical memoirs of research on *Xenopus* have been published previously, but each has a more specific focus on discrete elements of our frog's success (Blow and Laskey, 2016; Brown, 2004; Deuchar, 1975b; Gurdon and Hopwood, 2000; Maller, 2012). My goal here is to provide a more general history of research with *Xenopus*. I will cover the period spanning the first description of the genus in 1803 through about 1980, when work with *Xenopus* exploded, establishing the vibrant model organism that we use today. I hope the chapter will provide an entertaining journey back in time for the *Xenopus* community, my scientific family for over 30 years.

1.2. WHAT'S IN A NAME?

As has been frequently described, the first scientific description of *Xenopus* frogs comes from the French naturalist François Marie Daudin in 1803. But who was this man? As it happens, he was a tragic figure: having lost the use of his legs (and also his mother) while still a child, he died of tuberculosis at the young age of 27, shortly after writing his description of what he called *Bufo laevis* or *Crapaud lisse* ("smooth toad"). Despite his physical handicap and

DOI: 10.1201/9781003050230-2

his untimely death, Daudin is credited with the description of dozens of new species of animals, working in many cases from preserved specimens in collections around Paris (Bour, 2011).

Among these is the first description of the genus now called *Xenopus*, though exactly which species seems unclear. Working from a single preserved specimen in the *Museum d'Histoire Naturelle* in Paris, Daudin succinctly described the smooth body, the dorsally set eyes, and the "longitudinal rows of small prickles" along the back (Daudin, 1803), an anatomical landmark that even the newest student injecting hormones to induce egg laying will recognize immediately. Daudin's book also includes the first scientific illustration of *Xenopus* (Figure 1.1). As it happens, his wife Adélaïde was a talented artist, and she provided many illustrations for his work, though the image of the smooth toad is attributed instead to an artist named Prévost. In a further tragedy, Adélaïde also died of tuberculosis, a few days before her husband (Bour, 2011).

Surprisingly, neither Daudin's description nor the accompanying illustration makes note of the claws on what would later come to be known as the African clawed frog. Thus, when the great French naturalist Georges Cuvier assembled his magnum opus *Le Règne Animal* in 1829, he chided Daudin for his omission of the claws, and indeed, it does seem like quite the blunder. However, in the first focused and comprehensive description of our frog in 1864, J.E. Gray notes that the animals' claws are "deciduous in spirits" (Gray, 1864), suggesting the possibility that the claws on Daudin's preserved specimen had dissolved before he made his observations. Regardless, Cuvier renamed the genus *Dactylethra*, a rendering of "finger sheath" or "thimble" (Cuvier, 1829).

The genus name *Dactylethra* was commonly used by 19th-century writers, including the first description and images of tadpoles of the species (Gray, 1864). The name *Xenopus* only became fashionable in the 1890s, which is curious, since it was actually first suggested in 1827. Even more curious is the forum in which that name was first bestowed: a footnote appended to a letter written by Heinrich Boie and published in Lorenz Oken's journal, *Isis* (Wagler, 1827). Previous authors have attributed this footnote to "Wagler, H.," but thanks to the digitized archives available from the *Biodiversity Heritage Library* (www.biodiversitylibrary.org/), I have been able to learn a bit more about this important figure in the history of *Xenopus*.

As it happens, the footnote proposing the genus name "*Xenopus*" has no first initial indicated; it is signed simply: "Wagler" (Figure 1.2). This almost certainly refers to Johann Georg Wagler, a noteworthy German naturalist, who in 1830 and 1831 published two monographs on the Amphibia. In fact, in his *Natürliches System der Amphibien* of 1830, he cites the earlier footnote in *Isis*, and moreover makes frequent mention of *Xenopus*, making a special note of its clawed toes (Wagler, 1830). Moreover, in his original footnote, Wagler even lets Daudin off the hook for missing the claws, suggesting that he must have observed a decayed specimen (Wagler, 1827). In an unsettling coincidence, Wagler, like Daudin, died tragically, in this case at 32 years old from an accidental, self-inflicted gunshot wound (Hess, 1896).

Wagler's naming of the genus is the one that finally stuck, but not until long after his death. *Dactylethra* was used in several important 19th-century studies of the animal. Just the same, text searches of the *Biodiversity Heritage Library* revealed that Wagler's naming of *Xenopus* had been translated into French by 1829 (Boie and Wagler, 1829), and even as early as 1831, an English synopsis of Cuvier's book acknowledges Wagler and includes *Xenopus* as a synonym for *Dactylethra* (Griffith, 1831). Though my survey was in no way systematic, the earliest instance I found of *Xenopus* being used as the primary name (with *Dactylethra* as a junior synonym) was in a *Handbook of Zoology* from 1858 (van der Hoeven, 1858).

The name *Xenopus* became increasingly common in the 1890s, starting with a description of *Xenopus* breeding in the wild (Leslie, 1890), and shortly thereafter, with the first description of *Xenopus* breeding in captivity by Frank Beddard, working with animals at the London Zoo (Beddard, 1894). Beddard included some quite nice illustrations of the tadpole, but he laments that the "intervening Sunday prevented me from examining the early stages" (Beddard, 1894). His account is filled with other rich details, including that the tadpoles he observed had been deposited on Saturday May 27th. Such details may also be his undoing, however, as he also reports that the frogs he studied were collected by a Mr. Finn in Zanzibar, the home not of *X. laevis* but of *X. mulleri*. These and other interesting minutiae of early *Xenopus* research were the subject of a fun series of blog posts by Malcolm Peaker in his *Zoology Jottings* (https://zoologyweblog.blogspot.com/).

Xenopus may be said to have formally entered the realm of developmental biology with a sprawling and impressive paper on their early development from Edward Bles in 1905. He was a noted naturalist, though "for the shackles of departmental teaching and organization, Bles had some distaste" (Hopkins, 1926). Lucky for him, he also had tremendous personal wealth. After several efforts as a traditional academic, he decided that independent, entirely self-funded science was best and pursued that course for over 20 years (Hopkins, 1926) While working at the University of Glasgow, Bles set about to breed and raise *Xenopus* frogs. Of course, work with tropical frogs in the cold of Glasgow required new apparatus, and a tropical aquarium devised by Samuel Budgett (of the now-popular Budgett's frog [Amin et al., 2015]) provided just the thing (Bles, 1905).

With this apparatus, Bles was able to provide a remarkably comprehensive description of the early development of *Xenopus laevis*, and moreover, the work was accompanied by a gorgeous series of illustrations by A.K. Maxwell (Bles, 1905). These images of *Xenopus* would launch Maxwell's decades-long career as a scientific illustrator of the first order, with work ranging from medical illustration of wounds during WWI, to updating the illustrations in *Gray's Anatomy*, to rendering Max Perutz's first model of hemoglobin (De

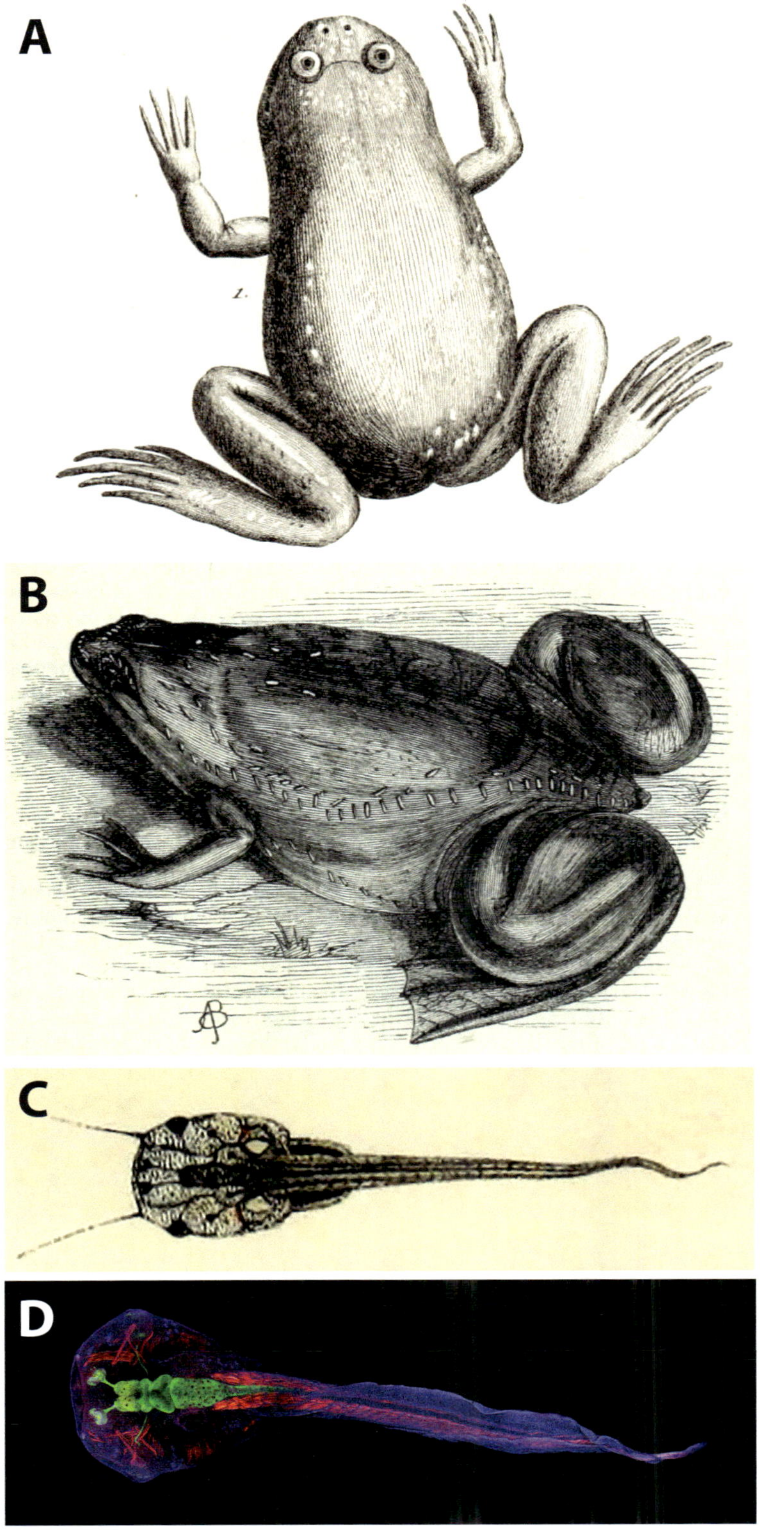

FIGURE 1.1 Images of Xenopus, then and now. (A) An illustration from Daudin's first description of *Xenopus* in 1803. Note the absence of claws. (B) An illustration, showing the claws, from Gray's 1864 report. (C) A *Xenopus* tadpole as illustrated by Pierre Jacques Smit in Beddard's report on the first captive breeding in 1894. (D) A modern image of a *Xenopus* tadpole with immunostaining for neural and muscle tissues.

Source: Images in A–C are used with permission from the Biodiversity Heritage Library. D courtesy of Helen Willsey.

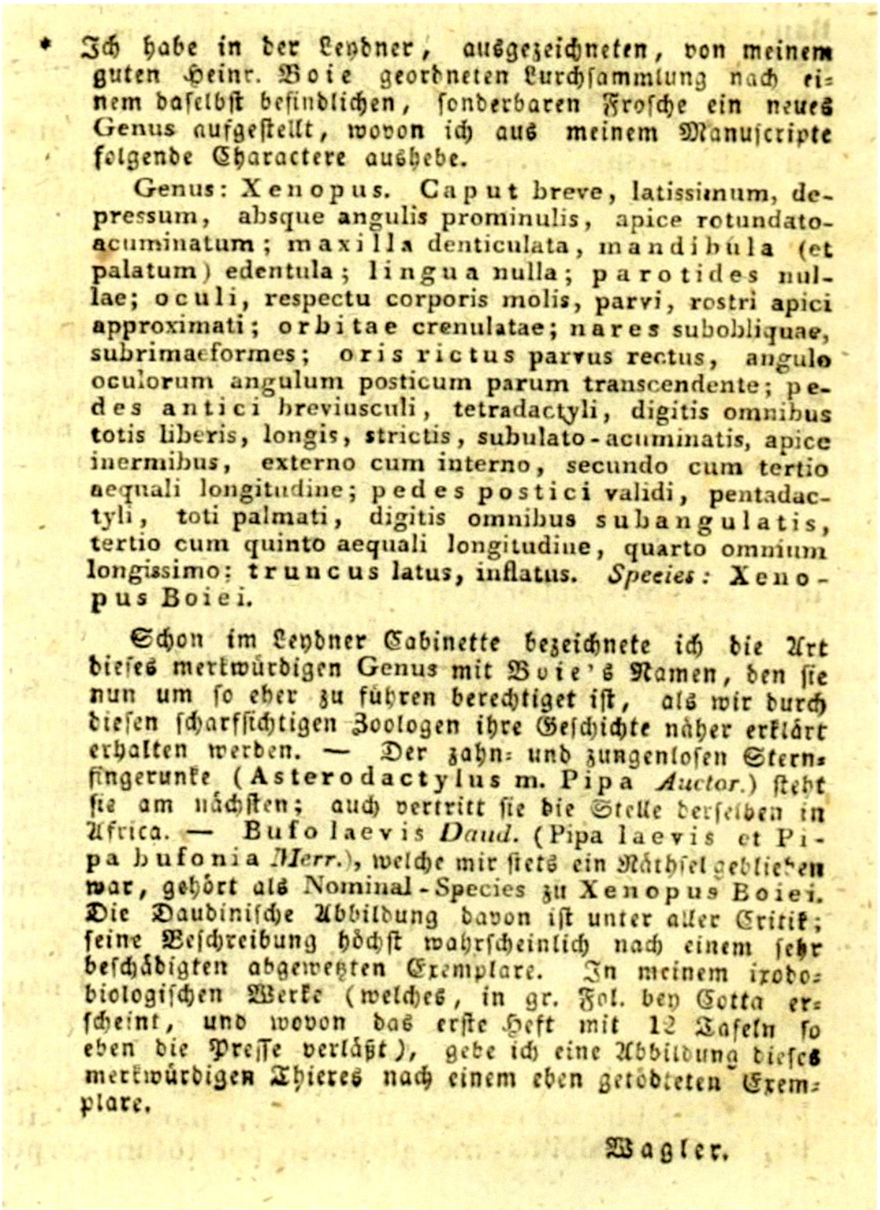

* Ich habe in der Leydner, ausgezeichneten, von meinem guten Heinr. Boie geordneten Lurchsammlung nach einem daselbst befindlichen, sonderbaren Frosche ein neues Genus aufgestellt, wovon ich aus meinem Manuscripte folgende Charactere aushebe.

Genus: Xenopus. Caput breve, latissimum, depressum, absque angulis prominulis, apice rotundato-acuminatum; maxilla denticulata, mandibula (et palatum) edentula; lingua nulla; parotides nullae; oculi, respectu corporis molis, parvi, rostri apici approximati; orbitae crenulatae; nares subobliquae, subrimaeformes; oris rictus parvus rectus, angulo oculorum angulum posticum parum transcendente; pedes antici breviusculi, tetradactyli, digitis omnibus totis liberis, longis, strictis, subulato-acuminatis, apice inermibus, externo cum interno, secundo cum tertio aequali longitudine; pedes postici validi, pentadactyli, toti palmati, digitis omnibus subangulatis, tertio cum quinto aequali longitudine, quarto omnium longissimo; truncus latus, inflatus. *Species:* Xenopus Boiei.

Schon im Leydner Cabinette bezeichnete ich die Art dieses merkwürdigen Genus mit Boie's Namen, den sie nun um so eher zu führen berechtiget ist, als wir durch diesen scharfsichtigen Zoologen ihre Geschichte näher erklärt erhalten werden. — Der zahn- und zungenlosen Sternfingerunke (Asterodactylus m. Pipa *Auctor.*) steht sie am nächsten; auch vertritt sie die Stelle derselben in Africa. — Bufo laevis *Daud.* (Pipa laevis et Pipa bufonia *Merr.*), welche mir stets ein Räthsel geblieben war, gehört als Nominal-Species zu Xenopus Boiei. Die Daudinische Abbildung davon ist unter aller Critik; seine Beschreibung höchst wahrscheinlich nach einem sehr beschädigten abgewetzten Exemplare. In meinem zoobiologischen Werke (welches, in gr. Fol. bey Cotta erscheint, und wovon das erste Heft mit 12 Tafeln so eben die Presse verläßt), gebe ich eine Abbildung dieses merkwürdigen Thieres nach einem eben getödteten Exemplare.

Wagler.

FIGURE 1.2 Wagler's footnote to a letter from Heinrich Boie in *Isis*, containing the original suggestion of the name "*Xenopus*" in 1827.

Chadarevian, 2002; Elliott, 1999). Maxwell's gorgeous illustrations are worth a look by any *Xenopus* researcher today, even rivaling the also fantastic images recently generated by Natalya Zahn (Zahn et al., 2017).

1.3. PREGNANCY AND PROMINENCE: THE RISE OF XENOPUS AS A LABORATORY ANIMAL

How *Xenopus* rose to worldwide prominence as a model organism is a fascinating story that has been told quite well in both academic and popular writing (Deuchar, 1975b; Gurdon and Hopwood, 2000; Yong, 2017). One reason the story resonates is, as Ed Yong put it in *The Atlantic*, because of the "fantastic name of Lancelot Hogben" (Yong, 2017). Hogben is a well-known personality from the early 20th century, not just for his scientific contributions (e.g. as an eminent physiologist and geneticist, a co-founder of the *British Society for Experimental Zoology*) but also for his compelling story (an ardent atheist and humanist, he was jailed as a conscientious objector during the First World War and was something of a gadfly for his entire career) (Erlingsson, 2016; Hogben, 1998; Sarkar, 1996). Born in England and wanting "to be a biologist long before I was twelve" (Hogben, 1998, p. 2), Hogben had a wildly peripatetic early career in academic biology, working as faculty at the University of Edinburgh, at McGill University in Canada, and by the late 1920s, at the University of Cape Town in South Africa.

A major research thrust for Hogben at the time, which had begun with Julien Huxley (Huxley and Hogben, 1922), was hormonal control of color changes in amphibia. He adopted *Xenopus* immediately upon his arrival in Cape Town, calling the animal a "godsend" (Hogben, 1998) (p. 101). Not only were the frogs plentiful, but they were "endowed with a very striking capacity for colour change" (Hogben et al., 1931). Moreover, the animals could routinely survive precise removal of the pituitary, allowing him to ask important questions. In a brief report to the *Transactions of the Royal Society of South Africa* in 1930, and in a more thorough study the following year in the *Journal of Experimental Zoology*, he reported that not only did removal of the pituitary disrupt color change but also elicited a profound "retrogression" of the ovaries (Hogben, 1930, 1998). This result was important because while the pituitary was implicated in the control of ovarian function, this direct experiment had not been possible in mammals. Perhaps even more striking, however, was the finding that extracts of the pituitary, when injected into *Xenopus*, were sufficient to induce ovulation (Hogben et al., 1931).

It's worth noting here that the *J. Exp. Zool.* paper was co-authored together with David Slome and also Hogben's wife, Enid Charles, who was pursuing her PhD in Cape Town. Enid was every bit as compelling a character as her husband, and she, too, was both an ardent activist and an exceptional scientist. She worked first on endocrinology and reproductive biology but later emerged as a pioneering demographer and worked for the World Health Organization (Wargon, 2005); she also once risked her life in South Africa to smuggle two Bantu men in the trunk of her car past vigilante checkpoints looking to lynch the men (Hogben, 1998, pp. 114–115). With the rise of Apartheid, Lancelot and Enid left South Africa abruptly in the 1930s, with Hogben becoming the chair of social biology at the London School of Economics, bringing *Xenopus* with him (Hogben, 1998, p. 121). Curiously, though their experiments in South Africa clearly demonstrated that, in principle, *Xenopus* could be used as a pregnancy test, this obvious application was never mentioned in those first two papers.

Exactly what happened next in this story has been much debated, but Hillel Shapiro and Harry Zwarenstein in Cape Town pursued the use of *Xenopus* as a pregnancy test, reporting their findings first in the *Proceedings of the Royal Society of South Africa* in 1933 (see (Gurdon and Hopwood, 2000) and then in *Nature* in 1934 (Shapiro and Zwarenstein, 1934). In London, Charles Bellerby was simultaneously pursuing the same goal (Bellerby, 1934). Though Shapiro and Zwarenstein reported their results first, the test came to be known as the "Hogben Pregnancy Test" after an article by the influential head of the Pregnancy Diagnosis Laboratory in Edinburgh, Frank Crew (1939). This led to a lively argument in the pages of the *British Medical Journal*, which sadly was never resolved (see Gurdon and Hopwood, 2000). I'll not re-litigate the issue and instead will simply concur with J.W.C. Gunn, who—serendipitously presaging a famous quote by Viktor Hamburger—suggested that "a little credit be given to the humble batrachian, which seems to give an invariably correct diagnosis" (Gunn, 1939). Over the next 20 years or so, the *Xenopus* pregnancy test became the worldwide state of the art. Thus, by the mid-20th century, substantial colonies of *Xenopus* could be found in several universities (Van Sittert and Measey, 2016).

Among the most significant for this story was the colony at the Utrecht Laboratories in the Netherlands, where a young Pieter Nieuwkoop pursued his PhD studies under Nazi occupation during World War II (Gerhart, 1987). Nieuwkoop described *Xenopus* as an "important acquisition" for embryologists, noting its robust and rapid development as well as its amenability to microsurgery (Nieuwkoop and Van De Kamer, 1946). Shortly after, Nieuwkoop and Faber began the tedious but essential task of creating an intimately detailed normal table for *Xenopus* (1956), a sourcebook whose wealth of information is still critical to the day-to-day work of *Xenopus* researchers and is available in a more recent reprinting (Nieuwkoop and Faber, 1994).

Of course, amphibians had already played a central role in experimental embryology for half a century, but researchers had been restricted to the breeding seasons of their local species (Beetschen, 2004). Now, *Xenopus* and the pregnancy test provided not just an amphibian that could be induced to lay eggs year-round but also one that was already kept in laboratories across the world. *Xenopus* then quickly evolved into a "model organism," a concept just coming into focus in the mid-20th century (Leonelli and Ankeny, 2013). Nieuwkoop, of course, went on to use *Xenopus* to make a wide range of seminal contributions to our understanding of early vertebrate development (Gerhart, 1997, 1999).

The Pregnancy Diagnosis Center in Edinburgh also remained an important source of *Xenopus* for developmental biologists, especially C.H. Waddington and his legions of trainees (Slack, 2002). Among these was Michaïl Fischberg, who joined Waddington's group in 1948. When Fischberg left to establish his own group at Oxford, he took *Xenopus* with him (Gurdon and Hopwood, 2000). Fischberg then had the foresight to isolate mutants of *Xenopus* that lack nucleoli (Elsdale et al., 1958), a resource that would have a profound impact on both developmental and molecular biology. Fischberg, too, would introduce *Xenopus* to its most important advocate, a young PhD student named John Gurdon.

A middling student of biology in school, Gurdon was admitted to the biology program at Oxford essentially by accident (Gurdon, 2008). However, during his PhD, Fischberg put him onto the knotty problem of nuclear totipotency, and the rest is history. The concept of nuclear transplantation originated with Rauber's unsuccessful experiments with frogs and had been attempted several times over the decades (Beetschen and Fischer, 2004; Rauber, 1886). Fischberg and Gurdon knew that work done in *Rana* by Briggs and King suggested that some transplanted nuclei from somatic cells could sustain development, but this capacity seemed to be lost at later stages (Briggs and King, 1952, 1953; King and Briggs, 1954). In what must be among the most remarkable PhD theses in modern biology, Gurdon used genetically marked anucleolate mutants in a series of papers to convincingly demonstrate the totipotency of somatic nuclei (Fischberg et al., 1958; Gurdon, 1960; Gurdon, 1962; Gurdon et al., 1958; Laskey and Gurdon, 1970). These experiments laid the foundation of our modern understanding of nuclear reprogramming (Gurdon, 2017), and, together with Shinya Yamanaka, Gurdon was honored with the Nobel Prize in 2012.

Accordingly, much has now been written about Gurdon and his work, so I won't add the story here (Yamada et al., 2015; Yamanaka and Blau, 2010). I will, however, direct the reader to some delightful retrospectives by Sir John himself (Gurdon, 2013a, 2013b). When reading these "memoirs" from 2013, one should bear in mind that in the same year, Gurdon continued his work on nuclear reprogramming (Miyamoto et al., 2013) and produced an authoritative review of the current state of the field (Halley-Stott et al., 2013).

1.4. SOME UNSUNG HEROES OF EARLY DEVELOPMENTAL BIOLOGY RESEARCH WITH *XENOPUS*

Gurdon's work had an outsized impact on developmental biology, and I'd argue it also had a big impact on the history of *Xenopus*. In terms of biology, those high-profile findings obviously helped to spur the frog to widespread use as a model organism. From the standpoint of history, however, the glare of those bright discoveries may also have obscured other work using *Xenopus* in the mid-20th century, as little is now written of several other notable discoveries.

For example, nuclear pores were discovered in *Xenopus* (Beck and Hurt, 2017). Harold "Mick" Callan, who had built radar equipment during WWII, worked under Waddington in Edinburgh in the beginning of his career in biology (Gall, 2003). He would become far more well known for his work on lampbrush chromosomes, but in 1949 and 1950, with the help of S.G. Tomlin at King's College London, he became the first to examine the nuclear membrane with "the" electron microscope (Callan et al., 1949; Callan and Tomlin, 1950). Using both *Xenopus* and *Triturus* oocyte nuclei, they described the double layers of nuclear membranes as well as the nuclear pores for the first time. They did, however, mistakenly conclude that the pores traversed only the outer nuclear membrane. It would take decades for the idea of nuclear pores to become commonly accepted, but Callan's method of exploiting large oocyte nuclei would be crucial to that effort (Beck and Hurt, 2017; Gall, 2003).

Xenopus also played a key role in our understanding of primordial germ cells, still a murky area in the 1950s. Working at times with Fischberg, Antoine Blackler developed methods for the transplantation of germ cells in *Xenopus*, again using the anucleolate mutant as a marker, and thereby provided the first direct experimental demonstration that germ plasm-containing cells in the very early embryo colonized the gonad and were responsible for producing the gametes (Blackler and Fischberg, 1961; Blackler, 1958, 1960). A modified version of this germ cell transplantation approach was recently developed for isolating CRISPR-based mutations in essential genes (Blitz et al., 2016). There are, of course, numerous other discoveries from this period, but from the large cast of characters using *Xenopus* in the middle 20th century, I'll discuss two in more detail that I feel deserve more attention from modern practitioners.

The first is Elizabeth Deuchar. A PhD student with C.H. Waddington in Edinburgh, she would publish dozens of papers from her independent lab and write a handful of books. Her work included not only of-their-time microsurgical experiments on embryonic induction and somite segmentation (Deuchar and Burgess, 1967; Waddington and Deuchar, 1953) but also more forward-looking biochemical studies in embryos (Deuchar, 1956, 1961), as well as early studies of regeneration in *Xenopus* (Deuchar, 1975a). Later, she would make the move to mammalian embryos, describing a decades-ahead-of-its-time method for time-lapse imaging of gastrula stage rat embryos (Deuchar and Parker, 1972).

Accounts of the time report that she was quiet and shy (Bellairs, 1980; Fellows of St Hugh's College, 1980), though it's clear that she wasn't easily intimidated. Her hilariously scathing letter to *Nature* taking issue with a paper by Francis Crick on diffusion in embryos should be required reading (Crick, 1970; Deuchar, 1970). In the end, time has proven Crick largely right, but the issue continues to be studied a half-century later (e.g. Müller et al., 2013), and we should all admire her lively debating style! Sadly, Elizabeth Deuchar passed away from cancer in 1979, at the young age of 52.

Luckily for us, she completed a book in 1975 that provides a remarkably comprehensive accounting of *Xenopus* research during the mid-20th century. In the book's rather touching preface, she paid tribute to *Xenopus* and to the embryos she obtained from tests at the Pregnancy Diagnosis Center in Edinburgh. She laments, however, that many of her embryos "alas!—perished in the cold and vibration as I bicycled with them . . . over the cobbled streets" back to her lab (Deuchar, 1975b).

Another now-underappreciated pioneer of *Xenopus* research is Osamu Nakamura. Nakamura trained under Yo Kaname Okada, who together with Katsuma Dan established Japan as a powerhouse of experimental embryology in the first half of the 20th century (Okada, 1994). In the 1920s and 1930s, Okada and others like Tsuneo Yamada brought cutting-edge techniques in amphibian embryology back from France and Germany, and Nakamura's early successes included improving the methods for vital dye fate mapping and correcting earlier errors in urodele fate maps, taking advantage of the fact that "Kyoto is a newt paradise" (Asashima, 2002). In the 1960s, Nakamura published an important series of lectures advocating the use of the new methods of molecular biology for studying the embryo, though of course this idea was ahead of its time and would not become a reality for two decades more.

By the 1970s, Nakamura had adopted *Xenopus* as his primary research material. In 1971, he used vital dye staining to produce the very first fate map of the blastula-stage *Xenopus* embryo (Nakamura and Kishiyama, 1971). Later fate maps using more advanced methods are now commonly used by the community to target microinjections (Dale and Slack, 1987; Moody, 1987), but these—and also a largely forgotten fate map by Nakamura's colleague Hiroko Takasaki (1987)—proved Nakamura's early conclusions mostly correct.

Nakamura also carried out a remarkable series of studies on the effects of removing specific blastomeres from the early embryo. These studies helped to precisely define the inductive interactions that pattern the early embryo (Nakamura et al., 1970, 1971) and later influenced some the earliest molecular studies of vertebrate axial patterning (e.g. Gurdon et al., 1984; Rosa et al., 1988; Yisraeli et al., 1990). Finally, in a little-noticed paper from 1971, he used electron microscopy to explore the developmental biology of nucleoli in the early embryos (Nakamura and Yamada, 1971), extending prior work done by Elizabeth Hay and John Gurdon (Hay and Gurdon, 1967). Professor Nakamura went on to become president of two universities and was widely celebrated in Japan.

1.5. *THE NINTH DAY OF CREATION: XENOPUS* AT THE DAWN OF MOLECULAR BIOLOGY

As the 20th century passed into its second half, the advent of molecular analysis revolutionized biology. It may seem hard to imagine now, but before the invention of recombinant DNA technologies, isolating genes was the domain of biochemists (Birnstiel, 2002; Brown, 1994). Given the massive amounts of material that can be obtained from *Xenopus*, the frog played a crucial role. In fact, the sheer scale of discovery using *Xenopus* precludes a comprehensive discussion here, so I direct the reader to a great review by one of the principal figures (Brown, 2004). A brief summary of the key landmarks might look like this:

Owing to its very high CG content, ribosomal DNA (rDNA) could be separated on cesium chloride gradients, and thus the rDNA of *Xenopus* became the first eukaryotic gene ever isolated, by Max Birnstiel (Birnstiel et al., 1966). He further demonstrated that the anucleolate mutants of *Xenopus* isolated by Fischberg lacked rDNA, providing a glimpse of the future in which genetics and biochemistry would work hand in hand (Wallace and Birnstiel, 1966). Shortly thereafter, Don Brown isolated the second family of genes: that encoding the 5S rRNA in *Xenopus* (Brown et al., 1971). As a result, the *Xenopus* 5S rDNA was the first eukaryotic gene ever to be fully sequenced (Fedoroff and Brown, 1978; Miller et al., 1978). (For modern molecular biologists hoping to grasp the massive effort required here, note that the sequencing of this one gene warranted back-to-back papers in *Cell*.)

As if that weren't enough, this string of exceptional studies was paralleled by several other contemporaneous breakthroughs in other realms of molecular biology, starting with a paper titled simply "Deoxyribonucleic Acid in Amphibian Eggs," in which Igor Dawid discovered mitochondrial DNA and demonstrated its maternal inheritance in *Xenopus* (Dawid, 1965, 1966). Moreover, studying lampbrush chromosomes in *Xenopus*, Joe Gall demonstrated that chromosomes of higher eukaryotes (with their much larger genomes) consisted of single DNA strands (Gall, 1963), as Meselson and Stahl had shown for prokaryotes. Together with Mary Lou Pardue, Joe Gall also used *Xenopus* for the invention of *in situ* hybridization (Gall and Pardue, 1969).

Xenopus also played a role in early explorations of the central dogma, providing an exceptional system for studies of transcription and translation. The rabbit *globin* gene became the first eukaryotic mRNA to be isolated in 1969, and mammalian cell-free lysates were used to translate this into Globin protein shortly thereafter, but only vanishingly small amounts of protein could be made (Lockard and Lingrel, 1969). In 1971, John Gurdon (again) showed that the *Xenopus* oocyte could produce enormous amounts of protein when injected with mRNA (Gurdon et al., 1971), leading to its widespread use for this purpose. Around the same time, Bob Roeder discovered that *Xenopus* oocytes were loaded with RNA polymerases (Roeder, 1974), and Gurdon (yet again) showed that purified DNA could be transcribed when injected in *Xenopus* oocytes (Mertz and Gurdon, 1977), making this system an effective platform for studies of transcription. Indeed, the *Xenopus* 5S gene would have yet another star turn, when Roeder and colleagues isolated the first eukaryotic transcription factor, TFIIIA, and showed that it bound internal control regions in this gene (Engelke et al., 1980; Sakonju et al., 1981).

This golden age of molecular biology with *Xenopus* led directly to its modern use in cell and developmental biology. Ultimately, Brown, Gall, and Roeder would each receive a Lasker Award for their discoveries made in *Xenopus*, and each has written entertaining retrospectives on these discoveries (Brown, 2012; Gall, 2006; Roeder, 2019).

1.6. WHERE HISTORY STOPS AND "THE LITERATURE" BEGINS

This chapter has covered roughly 180 years of the history of research with *Xenopus*, and obviously, there is a great deal more to tell. The 1980s was a decade of explosive growth for *Xenopus* research, and the frog would be used for seminal contributions spanning the biochemical analysis of the cell cycle, to fundamental analyses of replication and transcription, to the hunt for molecular regulators of development. For example, Manfred Lohka and Jim Maller would develop the *Xenopus* egg extract system (1985), building on Lohka's work with Lasker Award winner Yoshio Masui with extracts from *Rana* (Lohka and Masui, 1983). Such extracts would prove invaluable for a wide range of studies of the cell cycle, DNA replication, and the cytoskeleton (Blow and Laskey, 2016; Maller, 2012; Masui, 2001). At this same time, molecular analysis of development in *Xenopus* also exploded, a story that has been insouciantly told by Jonathan Slack in his book *Egg & Ego* (Slack, 1999). On a personal note, the close of the 1980s would see me squeeze my first frogs and look at the tadpoles with a microscope.

That experience changed my life forever. I have been inspired by "the humble batrachian" and the people who study it for over 30 years now. I can't wait to see what the next 30 years will bring.

ACKNOWLEDGMENTS

This chapter is dedicated to Mike Danilchik, Betsy Brown, Rob Savage, Jim Denegre, and Jean Bertman; they taught me to squeeze frogs. This work would have been impossible without the resources of the Biodiversity Heritage Library. I thank Frédéric and Marie-Noëlle Bour for providing the manuscript by their late father, Roger Bour; Anne-Helene Monsoro-Burq for French translations; Dan Dickinson for German translations; and John Gurdon for sharing the citation to Rauber's work from 1886; and Helen Willsey for sharing the image for Figure 1.1D.

REFERENCES

Amin, N.M., M. Womble, C. Ledon-Rettig, M. Hull, A. Dickinson, and N. Nascone-Yoder. 2015. Budgett's frog (*Lepidobatrachus laevis*): A new amphibian embryo for developmental biology. *Developmental Biology*. 405:291–303.

Asashima, M. 2002. A view of amphibian embryology research in Japan through the scientific biography of Professor Osamu Nakamura. *International Journal of Developmental Biology*. 38:155–165.

Beck, M., and E. Hurt. 2017. The nuclear pore complex: Understanding its function through structural insight. *Nat Rev Mol Cell Biol*. 18:73–89.

Beddard, F.E. 1894. Notes upon the tadpole of *Xenopus laevis*. *Proc. Zool. Soc. London*. 1894:101–107.

Beetschen, J.-C. 2004. How did urodele embryos come into prominence as a model system? *International Journal of Developmental Biology*. 40:629–636.

Beetschen, J.C., and J.L. Fischer. 2004. Yves Delage (1854–1920) as a forerunner of modern nuclear transfer experiments. *Int J Dev Biol*. 48:607–612.

Bellairs, R. 1980. Dr Elizabeth Marion Deuchar (1927–1979). *Journal of Embryology and Experimental Morphology*. 57:1–2.

Bellerby, C. 1934. A rapid test for the diagnosis of pregnancy. *Nature*. 133:494–495.

Birnstiel, M.L. 2002. The dawn of gene isolation. *Gene*. 300:3–11.

Birnstiel, M.L., H. Wallace, J.L. Sirlin, and M. Fischberg. 1966. Localization of the ribosomal DNA complements in the nucleolar organizer region of *Xenopus laevis*. *Natl Cancer Inst Monogr*. 23:431–447.

Blackler, A.W. 1958. Contribution to the study of germ-cells in the anura. *J Embryol Exp Morphol*. 6:491–503.

Blackler, A.W. 1960. Transfer of germ-cells in *Xenopus laevis*. *Nature*. 185:859–860.

Blackler, A.W., and M. Fischberg. 1961. Transfer of primordial germ-cells in *Xenopus laevis*. *J. Embryol. Exp. Morph*. 9: 634–641.

Bles, E.J. 1905. The life-history of *Xenopus icevis*, Daud. *Trans. Roy. Soc. Edin*. 41:789–821.

Blitz, I.L., M.B. Fish, and K.W. Cho. 2016. Leapfrogging: Primordial germ cell transplantation permits recovery of CRISPR/Cas9-induced mutations in essential genes. *Development*. 143: 2868–2875.

Blow, J.J., and R.A. Laskey. 2016. Xenopus cell-free extracts and their contribution to the study of DNA replication and other complex biological processes. *Int J Dev Biol*. 60:201–207.

Boie, H., and J.G. Wagler. 1829. Lettre de H. Boié a M. Wagler sur quelques oiseaux et reptiles de i'ile de java. *Bulletin Des Sciences Naturelles et de Géologie*. 16:107.

Boorstin, D.J. 1985. *The discoverers: A history of man's search to know world and himself*. Vintage, New York.

Bour, R. 2011. François Marie Daudin (29 août 1776–30 novembre 1803), auteur de l'Histoire naturelle, générale et particulière, des Reptiles. *Alytes, International Journal of Batrachology*. 28:1–76.

Briggs, R., and T.J. King. 1952. Transplantation of living nuclei from blastula cells into enucleated frogs' eggs. *Proceedings of the National Academy of Sciences*. 38:455–463.

Briggs, R., and T.J. King. 1953. Factors affecting the transplantability of nuclei of frog embryonic cells. *Journal of Experimental Zoology*. 122:485–505.

Brown, D.D. 1994. Some genes were isolated and their structure studied before the recombinant DNA era. *Bioessays*. 16:139–143.

Brown, D.D. 2004. A tribute to the *Xenopus laevis* oocyte and egg. *J Biol Chem*. 279:45291–45299.

Brown, D.D. 2012. Developmental biology using purified genes. *Nat Med*. 18:1496–1498.

Brown, D.D., P.C. Wensink, and E. Jordan. 1971. Purification and some characteristics of 5S DNA from *Xenopus laevis*. *Proc Natl Acad Sci U S A*. 68:3175–3179.

Callan, H.G., J. Randall, and S. Tomlin. 1949. An electron microscope study of the nuclear membrane. *Nature*. 163: 280–280.

Callan, H.G., and S. Tomlin. 1950. Experimental studies on amphibian oocyte nuclei I. Investigation of the structure of the nuclear membrane by means of the electron microscope. *Proceedings of the Royal Society of London: Series B-Biological Sciences*. 137:367–378.

Crew, F. 1939. Biological pregnancy diagnosis tests: A comparison of the rabbit, the mouse, and the "clawed toad" (*Xenopus laevis*) as the experimental animal. *The British Medical Journal*. 1:766–770.

Crick, F. 1970. Diffusion in embryogenesis. *Nature*. 225:420–422.

Cuvier, G. 1829. *Le règne animal distribué d'après son organisation, pour servir de base à l'histoire naturelle des animaux et d'introduction à l'anatomie comparée*. Chez Déterville, Paris.

Dale, L., and J.M. Slack. 1987. Fate map for the 32-cell stage of *Xenopus laevis*. *Development*. 99:527–551.

Daudin, F.M. 1803. *Histoire naturelle des reinettes, des grenouilles et des crapauds*. Levrault, Paris.

Dawid, I.B. 1965. Deoxyribonucleic acid in amphibian eggs. *J Mol Biol*. 12:581–599.

Dawid, I.B. 1966. Evidence for the mitochondrial origin of frog egg cytoplasmic DNA. *Proc Natl Acad Sci U S A*. 56:269–276.

De Chadarevian, S. 2002. *Designs for life: Molecular biology after World War II*. Cambridge University Press, Cambridge.

Deuchar, E.M. 1956. Amino acids in developing tissues of *Xenopus laevis*. *Development*. 4:327–346.

Deuchar, E.M. 1961. Amino-acid activation in tissues of early embryos. *Nature*. 191:1006–1007.

Deuchar, E.M. 1970. Diffusion in embryogenesis. *Nature*. 225:671.

Deuchar, E.M. 1975a. Regeneration of the tail bud in Xenopus embryos. *Journal of Experimental Zoology*. 192:381–389.

Deuchar, E.M. 1975b. *Xenopus: The South African clawed frog*. New York, Wiley.

Deuchar, E.M., and A. Burgess. 1967. Somite segmentation in amphibian embryos: Is there a transmitted control mechanism? *Development*. 17:349–358.

Deuchar, E.M., and F. Parker. 1972. A method for time-lapse cinematography of primitive streak stage rat embryos in culture. *Experientia*. 28:374–376.

Elliott, P. 1999. Notable names in medical illustration: A. Kirkpatrick Maxwell. *Journal of Audiovisual Media in Medicine*. 22: 130–131.

Elsdale, T.R., M. Fischberg, and S. Smith. 1958. A mutation that reduces nucleolar number in *Xenopus laevis*. *Exp Cell Res*. 14:642–643.

Engelke, D.R., S.Y. Ng, B.S. Shastry, and R.G. Roeder. 1980. Specific interaction of a purified transcription factor with an internal control region of 5S RNA genes. *Cell*. 19: 717–728.

Erlingsson, S.J. 2016. "Enfant terrible": Lancelot Hogben's life and work in the 1920s. *Journal of the History of Biology*. 49:495–526.

Fedoroff, N.V., and D.D. Brown. 1978. The nucleotide sequence of oocyte 5S DNA in *Xenopus laevis*. I: The AT-rich spacer. *Cell*. 13:701–716.

Fellows of St Hugh's College, O. 1980. Elizabeth Marion Baldock (née Deuchar). *St. Hugh's College Chronicle*. 52:35–36.

Fischberg, M., J.B. Gurdon, and T.R. Elsdale. 1958. Nuclear transplantation in *Xenopus laevis*. *Nature*. 181:424–424.

Gall, J.G. 1963. Kinetics of deoxyribonuclease action on chromosomes. *Nature*. 198:36–38.

Gall, J.G. 2003. Harold Garnet Callan. *Biogr. Mems Fell. R. Soc. Lond*. 49:107–118.

Gall, J.G. 2006. Through the looking glass. *Nat Med*. 12:1142–1145.

Gall, J.G., and M.L. Pardue. 1969. Formation and detection of RNA-DNA hybrid molecules in cytological preparations. *Proc Natl Acad Sci U S A*. 63:378–383.

Gerhart, J. 1987. The epigenetic nature of vertebrate development: An interview of Pieter D. Nieuwkoop on the occasion of his 70th birthday. *Development*. 101:653–657.

Gerhart, J. 1997. In memoriam Pieter D. Nieuwkoop (1917–1996). *Dev Biol*. 182:1–4.

Gerhart, J. 1999. Pieter Nieuwkoop's contributions to the understanding of meso-endoderm induction and neural induction in chordate development. *Int J Dev Biol*. 43:605–613.

Gray, J.E. 1864. Notes on the clawed toad (Dactylethra) of Africa. *Proc. Zool. Soc. London*. 1864:458–464.

Griffith, E. 1831. *A classified index and synopsis of the Animal Kingdom Arranged in conformity with its organization by Baron G. Cuvier with supplementary additions to each order*. Whittaker & Co., London.

Gunn, J. 1939. Xenopus test for pregnancy. *British Medical Journal*. 2:580.

Gurdon, J.B. 1960. The developmental capacity of nuclei taken from differentiating endoderm cells of *Xenopus laevis*. *J Embryol Exp Morphol*. 8:505–526.

Gurdon, J.B. 1962. The developmental capacity of nuclei taken from intestinal epithelium cells of feeding tadpoles. *J Embryol Exp Morphol*. 10:622–640.

Gurdon, J.B. 2008. Sir John Gurdon: Godfather of cloning. Interviewed by Ruth Williams. *J Cell Biol*. 181:178–179.

Gurdon, J.B. 2013a. The cloning of a frog. *Development*. 140:2446–2448.

Gurdon, J.B. 2013b. The egg and the nucleus: A battle for supremacy. *Development*. 140:2449–2456.

Gurdon, J.B. 2017. Nuclear transplantation, the conservation of the genome, and prospects for cell replacement. *Febs J*. 284:211–217.

Gurdon, J.B., S. Brennan, S. Fairman, and T. Mohun. 1984. Transcription of muscle-specific actin genes in early Xenopus development: Nuclear transplantation and cell dissociation. *Cell*. 38:691–700.

Gurdon, J.B., T.R. Elsdale, and M. Fischberg. 1958. Sexually mature individuals of *Xenopus laevis* from the transplantation of single somatic nuclei. *Nature*. 182:64–65.

Gurdon, J.B., and N. Hopwood. 2000. The introduction of *Xenopus laevis* into developmental biology: Of empire, pregnancy testing and ribosomal genes. *Int J Dev Biol*. 44:43–50.

Gurdon, J.B., C.D. Lane, H.R. Woodland, and G. Marbaix. 1971. Use of frog eggs and oocytes for the study of messenger RNA and its translation in living cells. *Nature*. 233: 177–182.

Halley-Stott, R.P., V. Pasque, and J. Gurdon. 2013. Nuclear reprogramming. *Development*. 140:2468–2471.

Hay, E.D., and J.B. Gurdon. 1967. Fine structure of the nucleolus in normal and mutant Xenopus embryos. *Journal of Cell Science*. 2:151–162.

Hess, W. 1896. Wagler, Johann Georg. *Allgemeine Deutsche Biographie*. 41:776, Accessible here: https://de.wikisource.org/wiki/ADB:Wagler,_Johann_Georg.

Hogben, L.T. 1930. Some remarks on the relation of the pituitary gland to ovulation and skin secretion in *Xenopus laevis*. *Proc. R. Soc. S. Africa, March*. 19:1930.

Hogben, L.T. 1998. *Lancelot Hogben, scientific humanist: An unauthorised autobiography*. Merlin Press, Suffolk, UK.

Hogben, L.T., E. Charles, and D. Slome. 1931. Studies on the pituitary: VIII. The relation of the pituitary gland to calcium metabolism and ovarian function in Xenopus. *Journal of Experimental Biology*. 8:345–354.

Holmes, F.L. 1993. The old martyr of science: The frog in experimental physiology. *Journal of the History of Biology*. 26:311–328.

Hopkins, F.G. 1926. Obituary: Dr. Edward J. Bles. *Nature*. 118:90–91.

Huxley, J.S., and L.T. Hogben. 1922. Experiments on amphibian metamorphosis and pigment responses in relation to internal secretions. *Proceedings of the Royal Society of London: Series B, Containing Papers of a Biological Character*. 93:36–53.

King, T.J., and R. Briggs. 1954. Transplantation of living nuclei of late gastrulae into enucleated eggs of *Rana pipiens*. *Development*. 2:73–80.

Laskey, R., and J. Gurdon. 1970. Genetic content of adult somatic cells tested by nuclear transplantation from cultured cells. *Nature*. 228:1332–1334.

Leonelli, S., and R.A. Ankeny. 2013. What makes a model organism? *Endeavour*. 37:209–212.

Leslie, J.M. 1890. Note on the habits and oviposition of *Xenopus laevis*. *Proc. Zool. Soc. London*. 1890:69–71.

Lockard, R.E., and J.B. Lingrel. 1969. The synthesis of mouse hemoglobin beta-chains in a rabbit reticulocyte cell-free system programmed with mouse reticulocyte 9S RNA. *Biochem Biophys Res Commun*. 37:204–212.

Lohka, M.J., and J.L. Maller. 1985. Induction of nuclear envelope breakdown, chromosome condensation, and spindle formation in cell-free extracts. *J Cell Biol*. 101:518–523.

Lohka, M.J., and Y. Masui. 1983. Formation in vitro of sperm pronuclei and mitotic chromosomes induced by amphibian ooplasmic components. *Science*. 220:719–721.

Maller, J.L. 2012. Pioneering the Xenopus oocyte and egg extract system. *J Biol Chem*. 287:21640–21653.

Masui, Y. 2001. From oocyte maturation to the in vitro cell cycle: The history of discoveries of maturation-promoting factor (MPF) and cytostatic factor (CSF). *Differentiation*. 69:1–17.

Mertz, J.E., and J.B. Gurdon. 1977. Purified DNAs are transcribed after microinjection into Xenopus oocytes. *Proc Natl Acad Sci U S A*. 74:1502–1506.

Miller, J.R., E.M. Cartwright, G.G. Brownlee, N.V. Fedoroff, and D.D. Brown. 1978. The nucleotide sequence of oocyte 5S DNA in *Xenopus laevis*. II: The GC-rich region. *Cell*. 13:717–725.

Miyamoto, K., M. Teperek, K. Yusa, G.E. Allen, C.R. Bradshaw, and J. Gurdon. 2013. Nuclear Wave1 is required for reprogramming transcription in oocytes and for normal development. *Science*. 341:1002–1005.

Moody, S.A. 1987. Fates of the blastomeres of the 32-cell-stage Xenopus embryo. *Dev Biol*. 122:300–319.

Müller, P., K.W. Rogers, R.Y. Shuizi, M. Brand, and A.F. Schier. 2013. Morphogen transport. *Development*. 140:1621–1638.

Nakamura, O., and K. Kishiyama. 1971. Prospective fates of blastomeres at the 32 cell stage of *Xenopus laevis* embryos. *Proceedings of the Japan Academy*. 47:407–412.

Nakamura, O., H. Takasaki, and T. Mizohata. 1970. Differentiation during cleavage in *Xenopus laevis*. I: Acquisition of self-differentiation capacity of the dorsal marginal zone. *Proceedings of the Japan Academy*. 46:694–699.

Nakamura, O., H. Takasaki, T. Okumoto, and H. Iida. 1971. Differentiation during cleavage in *Xenopus laevis*. II: Development of inductive activity of the organizer. *Proceedings of the Japan Academy*. 47:203–208.

Nakamura, O., and K. Yamada. 1971. Differences in the fine structure and chemical constitution of nucleolar bodies and the true nucleolus in *Xenopus laevis* embryos. *Development, Growth & Differentiation*. 13:303–322.

Nieuwkoop, P.D., and J. Faber. 1956. *Normal table of Xenopus laevis (Daudin)*. North-Holland Publishing Company, Amsterdam.

Nieuwkoop, P.D., and J. Faber. 1994. *Normal table of Xenopus laevis (Daudin): A systematical and chronological survey of the development from the fertilized egg till the end of metamorphosis*. Garland Pub, New York.

Nieuwkoop, P.D., and J.C. Van De Kamer. 1946. *Xenopus laevis* as experimental object. In *Experimental embryology in the Netherlans, 1940–1945*. Vol. 10. M.W. Woerdeman and C.P. Raven, editors. Elsevier, Amsterdam. 115–118.

Okada, T.S. 1994. Experimental embryology in Japan, 1930–1960: A historical background of developmental biology in Japan. *The International Journal of Developmental Biology*. 38:135–154.

Piccolino, M. 1997. Luigi Galvani and animal electricity: Two centuries after the foundation of electrophysiology. *Trends in Neurosciences*. 20:443–448.

Rauber, A. 1886. Personaltheil und germinaltheil des individuum. *Zool. Anz*. 9:166–171.

Roeder, R.G. 1974. Multiple forms of deoxyribonucleic acid-dependent ribonucleic acid polymerase in *Xenopus laevis*: Levels of activity during oocyte and embryonic development. *J Biol Chem*. 249:249–256.

Roeder, R.G. 2019. 50+ years of eukaryotic transcription: An expanding universe of factors and mechanisms. *Nat Struct Mol Biol*. 26:783–791.

Rosa, F., A.B. Roberts, D. Danielpour, L.L. Dart, M.B. Sporn, and I.B. Dawid. 1988. Mesoderm induction in amphibians: The role of TGF-beta 2-like factors. *Science*. 239:783–785.

Sakonju, S., D.D. Brown, D. Engelke, S.Y. Ng, B.S. Shastry, and R.G. Roeder. 1981. The binding of a transcription factor to deletion mutants of a 5S ribosomal RNA gene. *Cell*. 23:665–669.

Sarkar, S. 1996. Lancelot Hogben, 1895–1975. *Genetics*. 142:655.

Shapiro, H., and H. Zwarenstein. 1934. A rapid test for pregnancy on *Xenopus laevis*. *Nature*. 133:762–762.

Slack, J.M.W. 1999. *Egg and ego: An almost true story of life in the biology lab*. Springer Science & Business Media, New York.

Slack, J.M.W. 2002. Conrad Hal Waddington: The last Renaissance biologist? *Nat Rev Genet*. 3:889–895.

Takasaki, H. 1987. Fates and roles of the presumptive organizer region in the 32-cell embryo in normal development of *Xenopus laevis*. *Development, Growth & Differentiation*. 29:141–152.

van der Hoeven, J. 1858. *Handbook of zoology*. Longman, Brown, Green, Longmans and Roberts, London.

Van Sittert, L., and G.J. Measey. 2016. Historical perspectives on global exports and research of African clawed frogs (*Xenopus laevis*). *Transactions of the Royal Society of South Africa*. 71:157–166.

Waddington, C., and E.M. Deuchar. 1953. Studies on the mechanism of meristic segmentation. I: The dimensions of somites. *Development*. 1:349–356.

Wagler, J.G. 1827. Footnote to Letter from H. Boie. *Isis*. 20:726.

Wagler, J.G. 1830. *Natürliches System der Amphibien: mit vorangehender Classification der Säugethiere und Vögel: ein Beitrag zur vergleichenden Zoologie*. J.G. Cotta'schen Buchhandlung, Munich.

Wallace, H., and M.L. Birnstiel. 1966. Ribosomal cistrons and the nucleolar organizer. *Biochim Biophys Acta*. 114:296–310.

Wargon, S. 2005. Legacy of Enid Charles, 1894–1972. *Canadian Studies in Population [ARCHIVES]*:137–153.

Wellmann, J. 2017. *The form of becoming: Embryology and the epistemology of rhythm, 1760–1830*. MIT Press, Cambridge, MA.

West, J.B. 2013. Marcello Malpighi and the discovery of the pulmonary capillaries and alveoli. *Am J Physiol Lung Cell Mol Physiol*. 304:L383–390.

Yamada, M., J. Byrne, and D. Egli. 2015. From cloned frogs to patient matched stem cells: Induced pluripotency or somatic cell nuclear transfer? *Current Opinion in Genetics & Development*. 34:29–34.

Yamanaka, S., and H.M. Blau. 2010. Nuclear reprogramming to a pluripotent state by three approaches. *Nature*. 465:704–712.

Yisraeli, J.K., S. Sokol, and D.A. Melton. 1990. A two-step model for the localization of maternal mRNA in Xenopus oocytes: Involvement of microtubules and microfilaments in the translocation and anchoring of Vg1 mRNA. *Development*. 108:289–298.

Yong, E. 2017. How a frog became the first mainstream pregnancy test. *The Atlantic*. https://www.theatlantic.com/science/archive/2017/05/how-a-frog-became-the-first-mainstream-pregnancy-test/525285/

Zahn, N., M. Levin, and D.S. Adams. 2017. The Zahn drawings: New illustrations of Xenopus embryo and tadpole stages for studies of craniofacial development. *Development*. 144:2708–2713.

2 The Study of Cell Division Control and DNA Replication in *Xenopus* Egg Extracts

Allison M. Jevitt and Susannah Rankin

CONTENTS

2.1. Introduction 13
2.2. *In Vitro* Recapitulation of Cell Cycle Events 13
2.3. Protein Degradation Drives the Cell Cycle 15
2.3.1. Pausing the Cell Cycle 16
2.4. Systems Biology of the Cell Cycle 16
2.5. DNA Replication Control 17
2.6. Ordering Events in DNA Replication 18
2.7. Origins and Timing 20
2.8. Summary and Future Directions 20
Acknowledgments 21
References 21

2.1. INTRODUCTION

The eggs of the African clawed frog, *Xenopus laevis*, have played critical roles in our understanding of the regulation of nuclear events and cell cycle control. Some of the earliest studies involved the use of *Xenopus* eggs to reprogram nuclei from an adult frog. Following injection of somatic nuclei into the enucleated egg, a considerable portion of the eggs would go on to cleave and some even go on to form adult frogs (Gurdon, 1962). These striking results challenged the prevailing notion that cells could not "dedifferentiate," that is, go from a differentiated cell type to a thoroughly pluripotent embryonic state. The second conclusion that could be drawn from this experiment was that the nuclear changes that occurred upon differentiation were largely epigenetic; factors were present in the egg that could reprogram the nuclei, leading to changes in gene expression and thus cell fate. In this extreme case, nuclei from somatic epithelial cells were reprogrammed simply through incubation in egg cytoplasm. This notion had fundamental implications for several fields of study and led to significant efforts to optimize the preparation of cytoplasmic extracts for further *in vitro* experimentation.

By the mid 1970s, it was becoming clear that the individual steps involved in cell division, including chromosome condensation, spindle assembly, nuclear envelope dynamics, and DNA replication, were somehow interdependent. Genetic experiments from Leland Hartwell and colleagues demonstrated the existence of the "cell division cycle," a regulatory system that controls the events of cell division (Hartwell et al., 1974). They identified a set of temperature-sensitive yeast mutants that arrested with the characteristic morphology of each step of cell division. This implied that an intrinsic regulatory system ensured timely, ordered completion of all the steps of cell division. The genetics were clear, but the mechanisms remained to be elucidated.

2.2. *IN VITRO* RECAPITULATION OF CELL CYCLE EVENTS

In parallel with genetic analyses in fungal models, investigators worked to develop biochemically tractable methods to characterize the stages of cell division. In groundbreaking experiments, Rao and Johnson showed that the cytoplasm from one cell could impose cell cycle control on nuclei from a cell fusion partner (Rao and Johnson, 1970). This observation inspired efforts to develop new experimental systems to monitor and manipulate the steps of cell division *in vitro*. Previous experiments suggested that amphibian eggs might provide an excellent source of material from which to build such a system: frog eggs are relatively large; laid in abundance; moderately soft and easy to lyse; and stockpiled with material required for multiple, rapid rounds of cell division in the absence of new transcription.

Foundational work by Gurdon and colleagues showed that *Xenopus* egg cytoplasm had the capacity to induce DNA replication in somatic nuclei that were micro-injected into *Xenopus* eggs (Graham et al., 1966). The induction of DNA replication in nuclei that had been isolated from post-mitotic tissues suggested that egg cytoplasm contains critical regulators of DNA replication and, interestingly, that the species from which the nuclei were isolated was relatively unimportant: it worked as well with mouse nuclei as frog nuclei. The Gurdon micro-injection experiment was essentially moved into the test tube by Benbow and Ford in 1975. They successfully prepared a concentrated cytosolic

DOI: 10.1201/9781003050230-3

extract from *Xenopus* eggs and showed that somatic frog nuclei added to it could be induced to undergo DNA replication. This experiment confirmed that nuclei from cells in the quiescent state (liver) could be induced to enter S phase by egg cytosol and that DNA replication would occur by recruitment of necessary components from the extract. These experiments supported the notion that the major drivers of cell cycle progression were in fact cytoplasmic factors and not nuclear ones.

A critical breakthrough in egg extract preparation and use was made by Masui and Lohka in 1983. Using concentrated cytoplasmic extract prepared from *Rana pipiens* eggs and sperm nuclei from *Xenopus laevis*, they documented morphological and biochemical changes associated with progression through the cell cycle. These events included nuclear assembly, DNA replication, and after some time had elapsed, nuclear envelope breakdown and chromosome condensation. Importantly, in this experiment, the nuclei had been pre-treated with detergent to remove membranes and many nuclear-associated proteins. The results suggested that the nuclei had assembled and progressed through the cell cycle through the recruitment of components from the extract. Through fractionation of the extract, they were further able to show that both the soluble *and* particulate components of the extract were required for DNA replication to occur: the particulate component contained membrane vesicles that could fuse to form a nuclear envelope, and the soluble fraction contained proteins and other factors, such as nucleotides, critical for nuclear assembly and function.

In experiments related to those using egg extracts, Gerhart, Kirschner, and colleagues characterized an activity which they named maturation promoting factor (MPF) using *Xenopus* eggs and oocytes. MPF was defined as a cytoplasmic activity that could be transferred from a mature egg into a late-staged oocyte, causing the recipient oocyte to mature into an egg (Masui and Markert, 1971; Smith and Ecker, 1971). The fact that during maturation, the meiotic cell cycle is driven from meiosis I to arrest into meiosis II suggested that MPF, a cytoplasmic factor, might itself be a driver of the cell cycle. MPF had striking and unusual properties: it caused oocyte maturation even when significantly diluted (100-fold!) prior to injection into the recipient oocyte, and the level of MPF decreased upon egg fertilization and cyclically reappeared with a period like the cleavage cycles of the early embryo, as shown in Figure 2.1 (Gerhart et al., 1984).

The characterization of MPF and the extract experiments of Lohka and Masui were consistent with a model in which cytoplasmic activities drove cell cycle progression. But what exactly were these activities? In a remarkable and exciting convergence of work from different experimental models, MPF was proven to be a kinase whose activity was controlled by a regulatory subunit called Cyclin (Kirschner, 2020). Cyclin had been identified previously as a protein whose level fluctuated with cleavage divisions in clams and sea urchins (Evans et al., 1983). Murray and Kirschner generated *Xenopus* egg extracts capable of multiple autonomous cell cycles *in vitro*, which could be monitored by changes in nuclear morphology. Using this kind of extract, they proved

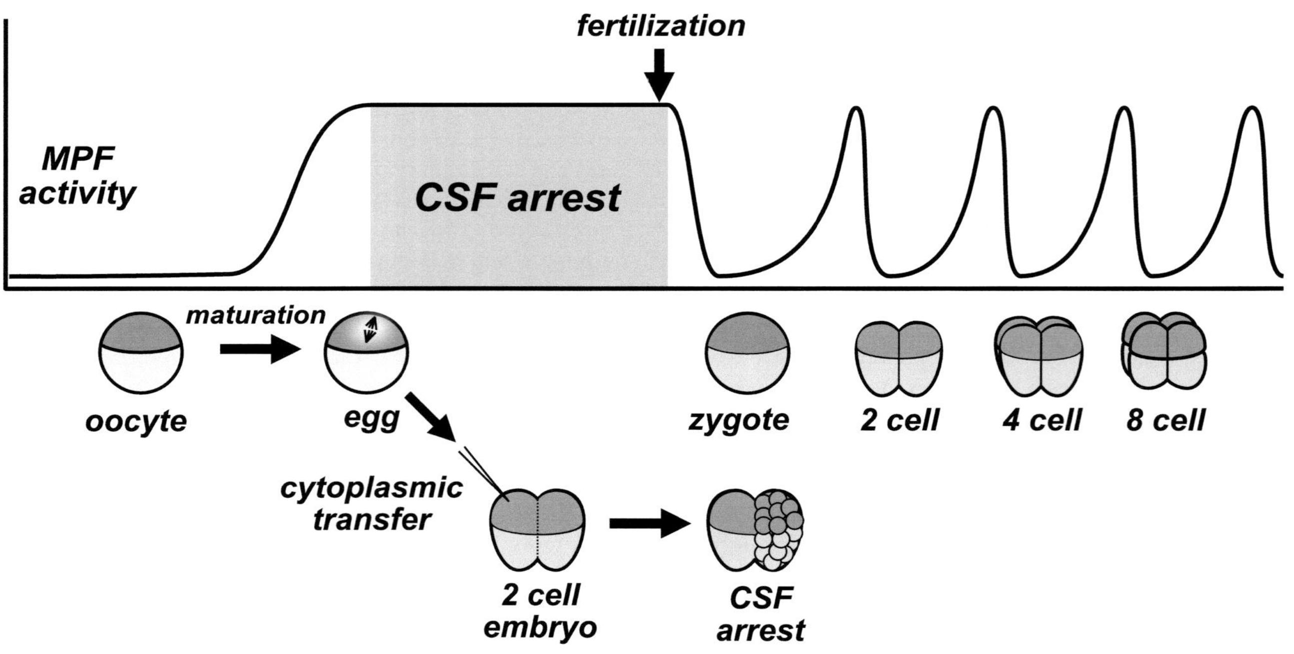

FIGURE 2.1 Early development and the cell cycle. When oocytes mature into an egg, the level of maturation promoting factor (MPF) kinase activity rises and is sustained during meiosis II meiotic arrest by the activity of cytostatic factor (CSF). Following fertilization, MPF levels fall and subsequently rise and fall in synchrony with the cleavage stage divisions. CSF can be assayed by transferring cytoplasm from an egg into one blastomere of a fertilized embryo, which causes that cell to cease dividing, while the rest of the embryo continues to cleave (bottom). MPF can similarly be assayed by transfer of cytoplasm from an M phase cell into an oocyte, which causes it to mature into an egg (not shown). These principles guided our earliest understanding of the biochemistry of cell division control. See text for details.

that *cyclin* mRNA (in this case from sea urchins) was necessary and sufficient to drive oscillations in MPF activity (Murray and Kirschner, 1989). To do this, they destroyed all endogenous RNA in the extract by nuclease treatment and then added back synthetic *cyclin* mRNA (in the presence of RNase inhibitor). This resulted in multiple cell cycles *in vitro*. In this cycling extract, the Cyclin protein accumulated during interphase and was degraded at exit from M phase. This elegant experiment showed that everything required to drive the oscillation between interphase and mitosis was present in the egg cytosol and, importantly, that the level of Cyclin protein oscillated with MPF activity. The highly conserved kinase activated by Cyclin had previously been identified as one of the cell cycle genes mutated in Hartwell's screen in budding yeast (*cdc2*) and a similar screen done in fission yeast (*cdc28*) (Hartwell et al., 1974; Lörincz and Reed, 1984). This kinase was indeed so conserved that the human gene was able to complement the temperature-sensitive allele in fission yeast, confirming the universal nature of cell cycle control in eukaryotes (Lee and Nurse, 1987).

The observation that cytoplasmic factors drive cell cycle progression was also consistent with a striking observation made previously by Hara and Kirschner (Hara et al., 1980). They had found that enucleated *Xenopus* eggs that were stimulated to exit meiotic arrest underwent a series of coordinated movements called surface contraction waves. When the eggs were examined from the side, these movements caused the eggs to appear to "bounce" up and down with the same periodicity as the cell cycle. The fact that these events occurred in the absence of nuclei was consistent with the notion that the cell cycle machinery represented an autonomous, cytoplasmic oscillator to which cytoskeletal components would respond. In the following years, we learned that cycling extracts and "bouncing eggs" are detectable in frog eggs because these cells are insensitive to external inputs: checkpoints that become active later in development are not yet fully active, so the cytoplasmic oscillatory machine trundles along, unaffected by feedback from DNA replication, a mitotic spindle, or cell cleavage. In the context of the *Xenopus* embryo, the ability to undergo more than a dozen cleavage divisions without new gene expression forms the basis for its remarkable usefulness as an *in vitro* model for studying cell division.

2.3. PROTEIN DEGRADATION DRIVES THE CELL CYCLE

In their extract experiments, Murray and Kirschner established that a particular deletion mutant of Cyclin (Δ90) was able to drive entry into mitotic (M) phase but was not destroyed like wild-type Cyclin, preventing M phase exit (Murray et al., 1989). These results suggested that regulated proteolysis of Cyclin was necessary for exit from M phase. To better understand how Cyclin destruction was regulated, the Kirschner group purified the activity leading to Cyclin degradation from *Xenopus* egg extract. Prior to degradation, Cyclin had been seen to accumulate high molecular weight derivatives, which were ultimately shown to be ubiquitin conjugates (Glotzer et al., 1991). The activity eventually purified was later identified as an E3 ubiquitin ligase, which was named the anaphase promoting complex (APC) (King et al., 1995). Satisfyingly, this large (20S) protein complex was shown to contain subunits homologous to yeast proteins that had previously been shown to control Cyclin degradation (Irniger et al., 1995). While the APC was purified from frog egg extract, the analogous complex was purified from clam embryos by Hershko and colleagues and named the Cyclosome (Sudakin et al., 1995). Thus, again, yeast genetics and extract biochemistry converged on a conserved mechanism of cell cycle control in all eukaryotes. The cell cycle resulted from the accumulation of mitotic Cyclin protein, which in turn led to activation of MPF, which in turn led to activation of the anaphase promoting complex/Cyclosome (APC/C), which in turn led to destruction of Cyclin, resetting the cycle.

It soon became evident that the APC/C had other substrates in addition to Cyclin. As mentioned previously, Murray and Kirschner had found that deletion of the N terminus of mitotic Cyclin (Cyclin Δ90) resulted in mitotic arrest. The nature of the arrest was interesting and unique: although MPF levels remained high due to the persistence of Cyclin, the APC/C remained active, creating an artificial "early anaphase"-like state *in vitro*. This discovery provided an ideal system with which to identify additional APC/C substrates. By incubating small pools of *in vitro* translated and radiolabeled proteins in egg extracts with active APC, it was possible to identify additional substrates based on their instability in Cyclin Δ90-treated extract, compared to interphase controls. In this manner, several key effectors of cell cycle control were identified (King et al., 1997; McGarry and Kirschner, 1998; Stukenberg et al., 1997; Zou et al., 1999). In time it would be shown that the APC/C in somatic cells was activated by two different substrate specificity factors: Cdc20 protein upon mitotic exit and Cdh1 in the Gap1 or Growth1 (G1) phase of the cell cycle (Fang et al., 1998a). Although Cdh1 activity is undetectable in *Xenopus* egg extracts, the addition of recombinant Cdh1 protein to interphase extract results in an artificial G1-like state in which Cdh1-dependent substrates of the APC/C were degraded. Based on these observations, additional small pool screens were performed in egg extracts to identify G1 substrates of the APC/C (Ayad et al., 2005, 2003; Rankin et al., 2005). Similar screens were performed to identify mitotic phosphoproteins, based on their shift in electrophoretic mobility in mitotic versus interphase extracts (Lustig et al., 1997; Stukenberg et al., 1997)

Xenopus egg extracts were also used to investigate the mechanisms of APC-dependent degradation. Because degradation of radiolabeled APC substrates is readily detectable in egg extracts, it was relatively straightforward to identify specific sequences in substrate proteins that promote their recognition and degradation by screening for mutations that disrupted degradation. Cdc20 substrates were shown to contain a specific sequence named the destruction- or d-box required for their turnover (Glotzer et al., 1991). Cdh1-dependent substrates were found to have a different degradation-ensuring sequence motif, or "degron," called the KEN box, named for the Lysine (K)–Glutamic Acid (E)–Asparagine (N) sequence required to signal for degradation (Pfleger and Kirschner, 2000). The number of identified APC substrates has continued to grow, suggesting involvement of the APC in diverse intracellular pathways of both actively dividing and post-mitotic cells (reviewed in (Davey and Morgan, 2016)). Characterization of the APC/Cyclosome is ongoing and continues to shape how we think about the cell cycle and its regulation.

2.3.1. Pausing the Cell Cycle

While the cell cycle is remarkably processive in egg extracts, there are distinct pause points that have been documented and used to better understand cell cycle control. In mature *Xenopus* eggs, the cell cycle is arrested in metaphase of meiosis II. This arrest is dependent on a factor called cytostatic factor (CSF), an activity that proved very challenging to define at the molecular level but was ultimately identified using *Xenopus* models. CSF, originally identified in *Rana* eggs, is an activity that results in cell cycle arrest following transfer of small amounts of egg cytoplasm into blastomeres of a cleavage stage embryo (Masui and Markert, 1971). Cytological examination of the blastomeres of similarly treated *Xenopus* embryos showed fully developed spindles and condensed chromosomes in the arrested cells, consistent with M phase arrest (Moses and Masui, 1989). The cytoplasmic transfer experiment proved that the activity of CSF is dominant, causing cycling blastomeres to arrest with high MPF levels at the next M phase following injection. The fact that CSF could arrest post-fertilization embryos suggested that CSF, while normally restricted to meiotic cells, was able to arrest the mitotic cell cycle equally well. During CSF arrest, the APC/C is inhibited, preventing degradation of Cyclin and thus exit from M phase. Experiments over a number of years ultimately showed that CSF arrest requires both the MOS-MAPK signaling that occurs during oocyte maturation (reviewed in Tunquist and Maller, 2003) and a protein called Emi2/xErp1 (Tung et al., 2005). Emi2 binds to and inhibits the APC/C and is degraded upon egg fertilization. CSF transfer can both arrest cleavage-stage embryos and be used *in vitro* to arrest egg extracts in M phase. Simply adding CSF arrested extract (prepared from unfertilized eggs) to cycling extract results in CSF arrest of the recipient extract. This approach has practical applications such as in the study of mitotic chromosomes and spindles following DNA replication *in vitro* (Silva and Rankin, 2018; Song et al., 2012).

In somatic cells, the cell cycle can be paused or arrested during mitotic divisions, and *Xenopus* models have helped elucidate these mechanisms. The spindle checkpoint is a surveillance mechanism that prevents activation of the APC/C (and therefore mitotic exit) in the presence of unattached chromosomes. This mechanism was first characterized in budding yeast (Li and Murray, 1991), but critical experiments in *Xenopus* showed that the mechanism was conserved in vertebrates. Chen et al. showed that the checkpoint can be triggered in egg extracts by the presence of high numbers of chromosomes and the microtubule poison nocodazole, which results in checkpoint signaling from unattached kinetochores (Chen et al., 1996). They also confirmed that the frog ortholog of the yeast Mad2 protein was critical to the arrest and that the Mad2 protein accumulated at the unattached kinetochores. Mad2 activity was further characterized in egg extracts and proved to be an inhibitor of the APC/C (Fang et al., 1998b; Li et al., 1997). The logic of the pathway is clear: unattached kinetochores prevent exit from mitosis by preventing Cyclin degradation. Once chromosome attachments are made, the signaling pathway is turned off, and mitotic exit can occur, ensuring accurate chromosome segregation.

There are also checkpoints that arrest the cell cycle in interphase in response to DNA damage or incomplete DNA replication. These mechanisms, which prevent mitotic *entry* through signaling cascades that inhibit activation of MPF, have been elucidated using *Xenopus* egg extracts and are nicely reviewed elsewhere (Cupello et al., 2016; Garner and Costanzo, 2009; Hoogenboom et al., 2017; Lin et al., 2019; Lupardus et al., 2007; Peng et al., 2008; Smythe and Newport, 1992). Interestingly, certain checkpoints appear not to be fully active in *Xenopus* development until the early cleavage cycles are over. The weakness of these checkpoints in eggs and cleavage-stage embryos results in a strikingly robust cell cycle prior to this time. The accumulation of nuclei and the resulting change in nuclear-cytoplasmic ratio that occur following cell cleavage divisions can be replicated *in vitro* through titration experiments. For example, signaling from a single nucleus in a microliter of extract (or an egg) is insufficient to prevent mitotic entry when DNA replication is blocked, but titration of nuclei into the same extract can generate a significant signal and prevent M phase entry. Similarly, the spindle checkpoint from a low concentration of unattached kinetochores does not cause M phase arrest in egg extract, but addition of a large number of nuclei with spindle attachment problems generates a robust mitotic arrest (Chen et al., 1996). Thus, though the checkpoints are not generally active until later in development, the components required are present in the egg and can be stimulated to act under the certain conditions.

2.4. SYSTEMS BIOLOGY OF THE CELL CYCLE

Xenopus eggs and embryos have not only provided a system to identify components that make up the cell cycle engine but have also been instrumental in the development of mathematical models for how cell cycle transitions are controlled and the feedback mechanisms that ensure switch-like transitions from interphase to M phase and back (Kim and Ferrell, 2007; Novak and Tyson, 1993; Pomerening et al., 2005; Solomon et al., 1990). The ideas proposed by this modeling have been tested and validated using egg extract (Kim and Ferrell, 2007; Pomerening et al., 2005; Sha et al., 2003; Solomon et al., 1990). The experiments ultimately confirmed the model that cell cycle transitions in extract, which occur independently of nuclear activity, are bistable: only interphase and M phase are stable states, and both positive and negative feedback loops ensure that transitions between the two are rapid and concerted.

Having worked out the nature of the transitions between S phase and M phase, another important systems-level question that was addressed using egg extracts is: How are cell cycle transitions propagated within the cytoplasm? This is of particular interest in large cells such as the amphibian egg. The surface contraction waves in the egg had been shown to represent a wave of cell cycle progression from the top (animal pole) of the egg to the bottom (Pérez-Mongiovi et al., 1998; Rankin and Kirschner, 1997). This wave model was subsequently explored *in vitro* using egg extract to elucidate the nature of the cell cycle wave (Chang and Jr, 2013). To do this, egg extract was supplemented with a recombinant fluorescent nuclear protein and loaded into capillary tubes. Cell cycle progression within the tubes was then analyzed by time-lapse analysis of nuclear accumulation of the fluorescent protein, which occurs only during interphase, when the nuclear envelope is intact. In this system, the accumulation and loss of nuclear signal serves as a surrogate for cell cycle progression. The results indicate that the cell cycle progresses as a trigger wave, expanding from sites of initiation and radiating out. Similar approaches were subsequently used to analyze propagation of apoptosis in egg extract, and recently a previously unknown impact of nuclei on cell cycle progression was analyzed using a related approach (Afanzar et al., 2020; Cheng and Ferrell, 2018).

The rapid oscillatory nature of the cell cycle in the early frog embryo has clearly enabled important studies on the nature of the cell cycle, but it is also a limitation. Because the cell cycle proceeds independently of transcription, early *Xenopus* models are generally not appropriate for the study of cell cycle entry or exit, which are transcription dependent and emerge later in development. In addition, the absence of Gap/Growth phases in the early embryo has somewhat limited the utility of egg extracts in the study of events that unfold during G1 and G2 in somatic cells.

2.5. DNA REPLICATION CONTROL

The ability to control cell cycle transitions *in vitro* using egg extracts opened the door for the study of cellular events downstream of the cell cycle. Perhaps the field that most benefited from *in vitro* control of the cell cycle was the study of DNA replication. Low-speed *Xenopus* egg extracts have several distinct advantages for DNA replication research. First, because eggs are stockpiled with material to support the first 12 divisions without new RNA synthesis, they contain all the components to assemble and replicate ~4000 nuclei per egg. Second, the low-speed extracts are rich in the membranous component that can dock and fuse to form nuclear envelopes at exit from M phase. Additionally, the extracts are essentially devoid of nuclei, allowing the addition of model replication substrates, typically *Xenopus* sperm nuclei that have been isolated and demembranated by detergent treatment. Finally, the absence of plasma membrane makes trivial the addition of labeled nucleotides, drugs, and other interventions important to understand DNA replication. As in many systems, the early embryonic divisions in *Xenopus* embryos are much more rapid than those that occur later in development. In the cleavage-stage *Xenopus* embryo, cleavages occurred in about 30 minutes, with complete genome duplication between each cleavage. Although replication occurs somewhat more slowly in egg extracts, it is largely complete within 45–60 minutes and very synchronous. This makes *Xenopus* egg extract a particularly powerful and tractable tool for studying DNA replication *in vitro*. What can take 6–20 hours in somatic cells takes only 2 hours from start to finish in the egg extract. Here we describe the contributions that were made to the understanding of DNA replication control through the use of *Xenopus* egg extracts.

The method of extract preparation would prove to be critical to the study of DNA replication. Using extract prepared by generating a clarified supernatant following homogenization of eggs, Méchali and Harland demonstrated efficient and complete replication of single-stranded DNA (1982). However, they were unable to detect replication of double-stranded DNA, limiting the usefulness of this approach. Blow and Laskey generated a low-speed extract from *Xenopus* eggs and showed that it was capable of initiating DNA replication of both sperm nuclei and plasmid DNA. The extract, following the technique established by Lohka and Masui, was prepared by centrifugation of packed eggs, which led to stratification of the egg components. Critically, the membrane-rich cytosolic layer could be collected without clarification. They showed that replication of sperm nuclei in this extract was quite efficient, while replication of plasmid DNA occurred but was less efficient. They also made the critical observation that in both cases nuclear envelopes would form around the substrates during incubation in the extract. This suggested that the nuclear envelope played a critical role in DNA replication and helped to explain why double-stranded DNA replication failed in the clarified extract of Méchali and Harland.

One critical observation made using the turbid cytosolic extracts was that replication seemed to be well controlled: it happened once and only once in nuclei added to the extract. The concept of *replication licensing* was developed to explain this phenomenon (Blow, 1993; Blow and Laskey, 1988; Coverley et al., 1993). The licensing concept suggested that there were two essential phases of DNA replication control that were mutually exclusive. Licensing of replication origins was initially proposed to be dependent on exposure of chromatin to cytosolic components in the extract ("licensing factors") that were then excluded by the nuclear envelope after it had formed (Blow and Laskey, 1988). Thus, replication origins, which were licensed by progression through M phase when the nuclear envelope was broken down, would subsequently initiate DNA replication in interphase but be unable to replicate again until progression through the next M phase.

The mechanisms that ensure a single round of DNA replication per cell cycle, and prevent re-replication, were ultimately discovered to be complex and include some redundancy. Licensing itself was defined as recruitment of the Mini chromosome maintenance 2–7 (MCM2–7) protein complex to replication origins, which were previously bound by the Origin recognition complex (ORC) (Kubota et al., 1997, 1995). The MCM hexamer is a DNA helicase that ultimately unwinds the DNA helix allowing replication fork progression. The Geminin protein, which had been identified originally as an APC/C substrate destroyed in anaphase, was shown to prevent MCM loading (McGarry and Kirschner, 1998). These studies suggested that destruction of Geminin is in fact a part of the replication licensing system.

Although it was clear that Geminin was able to prevent MCM loading, the mechanism was unknown. This mystery was eventually solved with the discovery that Geminin binds tightly to an essential replication licensing factor called Cdt1 (Maiorano et al., 2000; Wohlschlegel et al., 2000). Using the *Xenopus* extract system, it was shown that degradation of Geminin releases Cdt1, which can then bind to origins and promote MCM binding. This in part explained the role of the nuclear envelope: Geminin helped to retain Cdt1 in the cytosol. Following degradation of Geminin, Cdt1 is released to enter the nucleus, bind origins, and recruit the MCM complex. Unloading of MCM proteins following DNA replication helps to prevent re-replication.

Additional mechanisms were found to prevent DNA over-replication. The Cdt1 licensing factor itself was found to be degraded during DNA replication, which led to the discovery of a previously unknown mechanism of protein turnover (Arias and Walter, 2006, 2005). Cdt1 degradation was shown to result from its direct interaction with proliferating cell nuclear antigen (PCNA), an essential replication protein. PCNA acts as a sliding clamp to ensure the processivity of DNA replication. Cdt1 interacts with PCNA through a PCNA-interacting protein (PIP) box, a conserved PCNA interaction motif. The interaction between chromatin-bound PCNA and Cdt1 through its PIP box was found to result in ubiquitination of Cdt1 by the Cul4Cdt2 ubiquitin ligase, resulting in degradation of Cdt1 by the proteasome. The PIP box of Cdt1 was ultimately shown to have unique properties that ensure its modification by Cul4, and thus it was called a "PIP degron" (Havens and Walter, 2009). Several other proteins were found to have PIP degrons, ensuring their degradation during DNA replication, while most PCNA-interacting proteins are not affected by this mechanism. The destruction of Cdt1 during DNA replication helps to ensure that origins do not become relicensed, helping to prevent over-replication of DNA.

2.6. ORDERING EVENTS IN DNA REPLICATION

The use of *Xenopus* egg extracts has been essential to our understanding of the events that ensure controlled DNA replication in vertebrates. The synchrony and speed with which nuclei added to egg extracts undergo regulated DNA replication has been a great advantage in assigning orders of action of the various replication proteins. In a typical experiment, antibodies against a particular protein of interest are generated and used to immune-deplete the protein from the extract. The extract is then tested for replication by the addition of nuclei or other DNA substrates. Rescue experiments can easily be performed using recombinant proteins expressed in bacterial, insect, or human cells. This can be done with wild-type and mutant protein derivatives, allowing mapping of essential functions to protein domains or residues. At any time during progression through DNA replication, chromatin can be isolated from the extract and the proteins associated with the chromatin assessed, often by immunoblot. It is also possible to program the extract to express proteins from synthetic RNA added to the extract or to add radiolabeled *in vitro* translated proteins to the extract. These kinds of approaches have been used repeatedly to understand the order of binding and the interdependencies among a large number of replication proteins and their regulators.

In addition to the protein recruitment steps that lead to replication initiation, replication is also dependent on the activities of several kinases. As with other replication initiation steps, mechanisms requiring kinase activities were identified using *Xenopus* egg extracts and supported the findings of parallel studies in other systems. Some kinases were found to control the activity of replication proteins directly; others were shown to provide feedback between the cell cycle machinery and DNA replication control.

The change in MPF kinase activity during cell cycle transitions helps to orchestrate DNA replication. The binding of the ORC to replication origins is inhibited by MPF, ensuring that ORC binding occurs only after exit from M phase (Carpenter et al., 1996; Romanowski et al., 1996; Rowles et al., 1996). Additionally, activation of the APC ubiquitin ligase by MPF results in Cyclin destruction and exit from M phase (and thus ORC binding) and also causes the destruction of the replication licensing inhibitor Geminin (McGarry and Kirschner, 1998). Cell cycle-dependent destruction of Geminin ensures that licensing occurs only once during the cell cycle, during exit from M phase.

The initiation of DNA replication following replication licensing is tightly controlled by two additional kinases: Cdc7 and Cdk2. The Cdc7 kinase, previously identified in fungal models, was shown to be essential for DNA replication. In *Xenopus*, Cdc7 interacts with one of two activating subunits: Drf1 in extracts and the early embryo and Dbf4 later in development (Jares et al., 2004; Silva et al., 2006; Takahashi and Walter, 2005). Cdc7, also called Dbf4-dependent kinase (DDK) phosphorylates the MCM complex, which in turn leads to recruitment of a second kinase called Cdc45 to the replication origin (Jares and Blow, 2000). Cdc45 is localized to the site of local DNA unwinding with the MCM complex during DNA replication and is essential for loading of the GINS (go-ichi-ni-san) complex (Kubota et al., 2003; Mimura et al., 2000; Pacek et al., 2006). The GINS complex is required for both DNA replication initiation and elongation. Several other proteins required for DNA replication have been identified and/or characterized in *Xenopus* egg extracts, including Treslin/TICRR and its interacting partner Mdm2 binding protein (MTBP) (Kumagai et al., 2010; Kumagai and Dunphy, 2017).

A second type of Cyclin/Cdk complex was identified and shown to be critical during interphase for DNA replication (Blow and Nurse, 1990). Like MPF, this kinase contains an activating Cyclin subunit and a Cdk kinase. The catalytic subunit of the S phase Cdk was identified as Cdk2, a protein with sequence similar to Cdc2/Cdk1 (Elledge and Spottswood, 1991; Paris et al., 1991). Cdk2 interacts with two different Cyclin subunits, Cyclin E and Cyclin A. Cdk2 bound to Cyclin E was shown to be the principal driver of S phase entry (Chevalier et al., 1996; Strausfeld et al., 1996). The ability of Cdk2 bound to Cyclin E to drive entry into S phase is conserved in many models; Cyclin A drives activation of both Cdk1 and Cdk2 kinases and therefore plays roles in both late S phase and M phase (Coverley et al., 1993; Hwang and Clurman, 2005; Möröy and Geisen, 2004; Ohtsubo and Roberts, 1993; Strausfeld et al., 1996; Teixeira and Reed, 2018). Phosphorylation of replication proteins by Cyclin E/Cdk2 is important for replication initiation. In somatic cells, Cyclin E also drives transcription of replication proteins.

Prior to the onset of DNA replication, Cdk2 bound to Cyclin E was found to be inhibited by a small protein called Xic1 (Su et al., 1995). Xic1 is a member of a large class of proteins called cyclin-dependent kinase inhibitors (CKIs), which function by binding to Cyclin/Cdk complexes, preventing their catalytic activity. Proteolysis of Xic1 was shown to be the real trigger for S phase entry (Yew and Kirschner, 1997; You et al., 2007). Xic1 is modified by the Skp1-cullin-F-box protein (SCF) ubiquitin ligase, leading to its destruction by the proteasome. Interestingly, using *Xenopus* egg extract, Xic1 degradation was later shown to be linked directly to the assembly of replication proteins on chromatin (Furstenthal et al., 2001a, 2001b; You et al., 2002).

Over the years, there have been important technological breakthroughs that have increased the utility of *Xenopus* egg extracts for the study of DNA replication. A critical contribution was the development of nucleoplasmic extract (NPE) (Walter et al., 1998). NPE is prepared by assembling a large number of nuclei in low speed egg extract. The nuclei, which become quite large over time through the recruitment of karyophilic proteins from the extract, are then collected by flotation following a low-speed spin (Figure 2.2). After collection of the nuclei, the nuclear contents are released and clarified by high-speed spin. At this point, the extract can be frozen for future use. Importantly, NPE allows complete *in vitro* replication of DNA in the absence of a nuclear envelope. To accomplish this, the DNA substrate is first incubated in clarified membrane-free egg cytosol, which leads to replication licensing. Then, the reaction is supplemented with an appropriate volume of NPE. While normally the nuclear envelope is required in order to concentrate factors required for efficient DNA replication, these factors are pre-concentrated in NPE. The use of NPE allows the replication of even small DNA substrates, which are inefficiently replicated in low-speed extract in part because they do not assemble a proper nuclear envelope. It is likely that concentration of multiple proteins, including active Cdk2, ensures the robust activity of NPE.

Although a detailed discussion of all of the proteins required for DNA replication and how their roles were elucidated using *Xenopus* egg extract is beyond the scope of this review, it is important to note that *Xenopus* egg extracts continue to provide a uniquely tractable system with which to study vertebrate DNA replication. In recent years, for example, model substrates with engineered DNA adducts or crosslinks have been used to study how the replication machinery responds to replication barriers (Amunugama et al., 2018; Douwel et al., 2017; Hodskinson et al., 2020; Kose et al., 2019; Larsen et al., 2019). The use of synthetic or modified DNA substrates is facilitated by the development of NPE, which allows DNA replication in the absence of a nuclear envelope. Modified substrates can be added directly

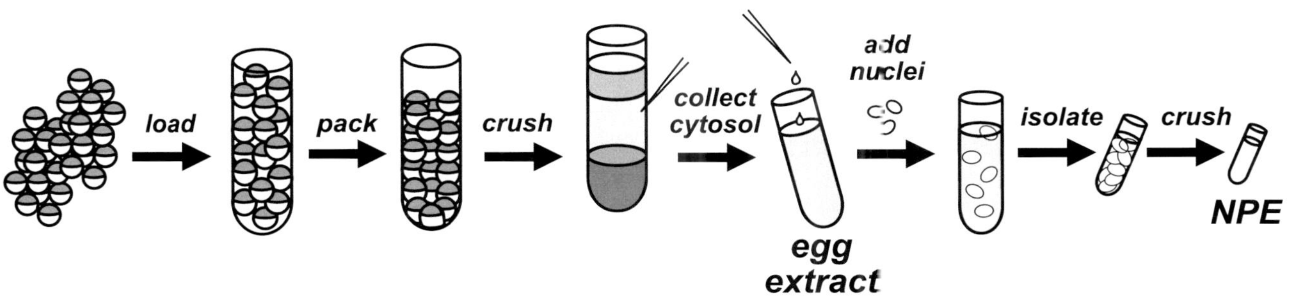

FIGURE 2.2 Preparation of extracts from *Xenopus* eggs. Following the removal of their jelly coats, eggs are packed tightly in test tubes by gentle centrifugation. Excluded buffer is removed from above the eggs, and the eggs are crushed by a high-speed spin, which causes stratification of egg contents. The middle, membrane-rich, cytosolic layer is collected and used directly or frozen for later use. To make nucleoplasmic extract (NPE), sperm nuclei are added to interphase extract, and after the nuclei have swelled and accumulated nuclear contents, they are floated by gentle centrifugation to the top of the extract. The nuclei are collected and their contents harvested following high-speed centrifugation.

to the extract, and replication does not require nuclear envelope assembly.

Another recent development includes using single-molecule approaches in *Xenopus* egg extracts in the study replication dynamics (Yardimci et al., 2012). In these elegant experiments, DNA templates are tethered to a substrate, and the incorporation of nucleotides or the behavior of replication proteins can be monitored, often in real time, with fluorescent nucleotides or proteins and fluorescence microscopy. Using microfluidics, it is possible to block new origin firing by flowing in inhibitors, and the sequential addition of different nucleotides or labeled proteins can be accomplished. In one example, fork progression was monitored in real time by analyzing the movements of a fluorescent-PCNA binding protein on the immobilized template (Loveland et al., 2012). In another, the impact of DNA replication on histone dynamics was examined (Gruszka et al., 2020). Similar approaches have been used to analyze the interaction of the replication apparatus with chromatin-bound proteins, such as the cohesin complex, or with DNA protein crosslinks (Kanke et al., 2016; Sparks et al., 2019). These and other single-molecule approaches to study chromosome biology in *Xenopus* egg extracts have recently been well reviewed elsewhere (Cameron and Yardimci, 2021).

Experiments in *Xenopus* egg extract have helped elucidate the mechanisms that link the completion of DNA replication to M phase entry. As Cyclin B accumulates in the extract, MPF (Cdk1-Cyclin B) is prevented from activation by a kinase called Myt1, a member of the Wee1 family of kinases. Wee1, which was first identified in fission yeast, places inhibitory phosphorylations on Cdk1. The removal of these phosphorylations by Cdc25 ensures mitotic entry (Gautier et al., 1991; Kumagai and Dunphy, 1991). Cdc25 itself is controlled by periodic phosphorylation (Izumi et al., 1992). Incomplete DNA replication, stimulated by the addition of DNA polymerase inhibitors, prevents activation of Cdc25, ensuring that MPF is not activated until DNA replication is complete (Gabrielli et al., 1992). These kinds of experiments have, in turn, led to many ground-breaking experiments in the study of DNA damage signaling and repair, reviewed elsewhere (Hoogenboom et al., 2017).

2.7. ORIGINS AND TIMING

Work in bacterial and yeast models in the late 1970s suggested that DNA replication originates at specific DNA sequences. In higher eukaryotes, the picture was less clear. Several groups showed that DNA injected into frog eggs or egg extract would spontaneously be replicated, even if the template contained no eukaryotic DNA sequences (Harland and Laskey, 1980; Mahbubani et al., 1992; Méchali and Kearsey, 1984). This led to the model, which persists to this day, that replication origins were not sequence specified in vertebrates.

Xenopus egg extracts provide a uniquely tractable system with which to study the nature and distribution of vertebrate replication origins. It was noted that the number of nuclei added to extract could impact the rate of completion of DNA replication, suggesting that titration of some soluble factor in the extract could result in increased replicon size (Walter and Newport, 1997). DNA combing, another sort of single-molecule experiment, was used to analyze the distribution of active origins and helped to explain the efficiency and speed of genome duplication in the early embryo. In this technique, DNA in egg extract is labeled by the addition of modified nucleotides, which are then incorporated into nascent DNA. The DNA is then isolated from the extract and spread on glass slides where it can be probed for labeled nucleotides (Blow et al., 2001). The addition of a second labeled nucleotide to the replication reaction at a later time point allows calculation of the distance between replication origins and the rate of DNA replication (Marheineke et al., 2009; Marheineke and Hyrien, 2004). It also led to the observation that fork density, that is, the number of replication forks per unit length of DNA, increased during replication progression through activation of later firing origins (Herrick et al., 2000). *Xenopus* egg extracts were also exploited to understand the nature of replication origins in nuclei from other sources. For example, the addition of Chinese hamster ovary (CHO) cell nuclei to egg extract was used to demonstrate the time in the cell cycle at which the somatic (CHO) cell origins were specified (Dimitrova and Gilbert, 1999; Li et al., 2003). These experiments ultimately led to the development of the concept of *replication timing* (Pope et al., 2013). Replication timing refers to the phenomenon in which certain regions of the genome reproducibly replicate earlier than others. Replication timing is still an area of active investigation in many model systems.

The ability to manipulate the steps that lead to replication origin firing have made *Xenopus* egg extracts uniquely useful for studying events that depend on replication. For example, the Cohesin complex, which establishes connections between sister chromatids during DNA replication, is regulated by direct interaction of Cohesin regulators with the replication machinery. Work in our lab and others has shown dependencies and often direct interaction between replication proteins and essential Cohesin regulators. The Cohesin loader Scc2 binds to DDK (Cdc7-Dbf4), and the Cohesin regulator Esco2 binds to chromatin in a licensing-dependent manner and interacts with the PCNA sliding clamp (Higashi et al., 2012; Lafont et al., 2010; Rankin et al., 2005; Song et al., 2012; Takahashi et al., 2008, 2004). The cohesion phenotypes of mitotic chromosomes can also be analyzed by driving extract into M phase following DNA replication (Silva and Rankin, 2018).

2.8. SUMMARY AND FUTURE DIRECTIONS

The eggs and embryos of *Xenopus laevis* have been used for decades in foundational investigations of DNA replication and cell division. Through the collective work of many dedicated researchers, we have a clearer understanding of basic mechanisms driving the proliferation of cells. With the ability to quickly and easily replicate DNA *in vitro*, there is a promising future for research using *Xenopus* egg extracts. Now that extracts can be compartmentalized using microfluidic systems, new adaptations to the biochemically tractable cell-free egg extract system are being developed to study other vital cellular processes. Innovations like light-inducible systems for initiating cell cycle progression are being developed and hold great promise for even more finely tuned control of extract dynamics (Bischt et al., 2019).

ACKNOWLEDGMENTS

We are grateful to all investigators who have, over the last 40 years, helped reveal the awesome power of *Xenopus* egg extracts for the study of complex vertebrate mechanisms. We apologize in advance to the investigators whose work in other systems informed the work done in *Xenopus* models; many citations were left out for brevity, as the *Xenopus* model was the focus of this review. We are grateful to animal care staff, whose fine work ensures the continued viability of *Xenopus* as a model system, to our colleagues from around the world who are generous with their expertise and reagents, and of course to the frogs themselves. This work was supported by NIH grant R01GM101250 to SR.

REFERENCES

Afanzar, O., Buss, G.K., Stearns, T., Ferrell, J.E., 2020. The nucleus serves as the pacemaker for the cell cycle. *Elife* 9, e59989. https://doi.org/10.7554/elife.59989

Amunugama, R., Willcox, S., Wu, R.A., Abdullah, U.B., El-Sagheer, A.H., Brown, T., McHugh, P.J., Griffith, J.D., Walter, J.C., 2018. Replication fork reversal during DNA interstrand crosslink repair requires CMG unloading. *Cell Reports* 23, 3419–3428. https://doi.org/10.1016/j.celrep.2018.05.061

Arias, E.E., Walter, J.C., 2005. Replication-dependent destruction of Cdt1 limits DNA replication to a single round per cell cycle in Xenopus egg extracts. *Genes & Development* 19, 114–126. https://doi.org/10.1101/gad.1255805

Arias, E.E., Walter, J.C., 2006. PCNA functions as a molecular platform to trigger Cdt1 destruction and prevent re-replication. *Nature Cell Biology* 8, 84–90. https://doi.org/10.1038/ncb1346

Ayad, N.G., Rankin, S., Murakami, M., Jebanathirajah, J., Gygi, S., Kirschner, M.W., 2003. Tome-1, a trigger of mitotic entry, is degraded during G1 via the APC. *Cell* 113, 101–113.

Ayad, N.G., Rankin, S., Ooi, D., Rape, M., Kirschner, M.W., 2005. Identification of ubiquitin ligase substrates by in vitro expression cloning. *Methods in Enzymology* 399, 404–414. https://doi.org/10.1016/s0076-6879(05)99028-9

Benbow, R.M., Ford, C.C., 1975. Cytoplasmic control of nuclear DNA synthesis during early development of *Xenopus laevis*: A cell-free assay. *Proc National Acad Sci* 72, 2437–2441. https://doi.org/10.1073/pnas.72.6.2437

Bischt, J., LeValley, P., Noren, B., McBride, R., Kharkar, P., Kloxin, A., Gatlin, J., Oakey, J., 2019. Light-inducible activation of cell cycle progression in *Xenopus* egg extracts under micro-fluidic confinement. *Lab Chip* 19, 3499–3511. http://doi.org/10/1039/C9LC00569B

Blow, J., 1993. Preventing re-replication of DNA in a single cell cycle: Evidence for a replication licensing factor. *J Cell Biology* 122, 993–1002. https://doi.org/10.1083/jcb.122.5.993

Blow, J.J., Gillespie, P.J., Francis, D., Jackson, D.A., 2001. Replication origins in Xenopus egg extract are 5–15 kilobases apart and are activated in clusters that fire at different times. *J Cell Biology* 152, 15–26. https://doi.org/10.1083/jcb.152.1.15

Blow, J.J., Laskey, R.A., 1988. A role for the nuclear envelope in controlling DNA replication within the cell cycle. *Nature* 332, 546–548. https://doi.org/10.1038/332546a0

Blow, J.J., Nurse, P., 1990. A cdc2-like protein is involved in the initiation of DNA replication in Xenopus egg extracts. *Cell* 62, 855–862. https://doi.org/10.1016/0092-8674(90)90261-c

Cameron, G., Yardimci, H., 2021. Studying chromosome biology with single-molecule resolution in *Xenopus laevis* egg extracts. *Essays Biochem*. https://doi.org/10.1042/ebc20200026

Carpenter, P.B., Carpenter, P.B., Mueller, P.R., Mueller, P.R., Dunphy, W.G., Dunphy, W.G., 1996. Role for a Xenopus Orc2-related protein in controlling DNA replication. *Nature* 379, 357–360. https://doi.org/10.1038/379357a0

Chang, J.B., Jr, J.E.F., 2013. Mitotic trigger waves and the spatial coordination of the Xenopus cell cycle. *Nature* 500, 603–607. https://doi.org/10.1038/nature12321

Chen, R.-H., Waters, J.C., Salmon, E.D., Murray, A.W., 1996. Association of spindle assembly checkpoint component XMAD2 with unattached kinetochores. *Science* 274, 242–246. https://doi.org/10.1126/science.274.5285.242

Cheng, X., Ferrell, J.E., 2018. Apoptosis propagates through the cytoplasm as trigger waves. *Science* 361, 607–612. https://doi.org/10.1126/science.aah4065

Chevalier, S., Couturier, A., Chartrain, I., Guellec, R.L., Beckhelling, C., Guellec, K.L., Philippe, M., Ford, C.C., 1996. Xenopus cyclin E, a nuclear phosphoprotein, accumulates when oocytes gain the ability to initiate DNA replication. *J Cell Sci* 109 (Pt 6), 1173–1184.

Coverley, D., Downes, C., Romanowski, P., Laskey, R., 1993. Reversible effects of nuclear membrane permeabilization on DNA replication: Evidence for a positive licensing factor. *J Cell Biology* 122, 985–992. https://doi.org/10.1083/jcb.122.5.985

Cupello, S., Richardson, C., Yan, S., 2016. Cell-free Xenopus egg extracts for studying DNA damage response pathways. *Int J Dev Biol* 60, 229–236. https://doi.org/10.1387/ijdb.160113sy

Davey, N.E., Morgan, D.O., 2016. Building a regulatory network with short linear sequence motifs: Lessons from the degrons of the anaphase-promoting complex. *Molecular Cell* 64, 12–23. https://doi.org/10.1016/j.molcel.2016.09.006

Dimitrova, D.S., Gilbert, D.M., 1999. The spatial position and replication timing of chromosomal domains are both established in early G1 phase. *Molecular Cell* 4, 983–993.

Douwel, D.K., Hoogenboom, W.S., Boonen, R.A., Knipscheer, P., 2017. Recruitment and positioning determine the specific role of the XPF-ERCC1 endonuclease in interstrand cross-link repair. *Embo J* 36, 2034–2046. https://doi.org/10.15252/embj.201695223

Elledge, S.J., Spottswood, M.R., 1991. A new human p34 protein kinase, CDK2, identified by complementation of a cdc28 mutation in *Saccharomyces cerevisiae*, is a homolog of Xenopus Eg1. *Embo J* 10, 2653–2659. https://doi.org/10.1002/j.1460-2075.1991.tb07808.x

Evans, T., Rosenthal, E.T., Youngblom, J., Distel, D., Hunt, T., 1983. Cyclin: A protein specified by maternal mRNA in sea urchin eggs that is destroyed at each cleavage division. *Cell* 33, 389–396. https://doi.org/10.1016/0092-8674(83)90420-8

Fang, G., Yu, H., Kirschner, M., 1998a. Direct binding of CDC20 protein family members activates the anaphase-promoting complex in mitosis and G1. *Molecular Cell* 2, 163–171.

Fang, G., Yu, H., Kirschner, M., 1998b. The checkpoint protein MAD2 and the mitotic regulator CDC20 form a ternary complex with the anaphase-promoting complex to control anaphase initiation. *Gene Dev* 12, 1871–1883. https://doi.org/10.1101/gad.12.12.1871

Furstenthal, L., Kaiser, B.K., Swanson, C., Jackson, P.K., 2001a. Cyclin E uses Cdc6 as a chromatin-associated receptor required for DNA replication. *J Cell Biology* 152, 1267–1278. https://doi.org/10.1083/jcb.152.6.1267

Furstenthal, L., Swanson, C., Kaiser, B.K., Eldridge, A.G., Jackson, P.K., 2001b. Triggering ubiquitination of a CDK inhibitor at origins of DNA replication. *Nat Cell Biol* 3, 715–722. https://doi.org/10.1038/35087026

Gabrielli, B.G., Lee, M.S., Walker, D.H., Piwnica-Worms, H., Maller, J.L., 1992. Cdc25 regulates the phosphorylation and activity of the Xenopus cdk2 protein kinase complex. *J Biol Chem* 267, 18040–18046. https://doi.org/10.1016/s0021-9258(19)37149-2

Garner, E., Costanzo, V., 2009. Studying the DNA damage response using in vitro model systems. *Dna Repair* 8, 1025–1037. https://doi.org/10.1016/j.dnarep.2009.04.015

Gautier, J., Solomon, M.J., Booher, R.N., Bazan, J.F., Kirschner, M.W., 1991. Cdc25 is a specific tyrosine phosphatase that directly activates p34cdc2. *Cell* 67, 197–211. https://doi.org/10.1016/0092-8674(91)90583-k

Gerhart, J., Wu, M., Kirschner, M., 1984. Cell cycle dynamics of an M-phase-specific cytoplasmic factor in *Xenopus laevis* oocytes and eggs. *J Cell Biology* 98, 1247–1255. https://doi.org/10.1083/jcb.98.4.1247

Glotzer, M., Murray, A., Kirschner, M., 1991. Cyclin is degraded by the ubiquitin pathway. *Nature* 349, 132–138.

Graham, C.F., Arms, K., Gurdon, J.B., 1966. The induction of DNA synthesis by frog egg cytoplasm. *Dev Biol* 14, 349–381. https://doi.org/10.1016/0012-1606(66)90020-0

Gruszka, D.T., Xie, S., Kimura, H., Yardimci, H., 2020. Single-molecule imaging reveals control of parental histone recycling by free histones during DNA replication. *Sci Adv* 6, eabc0330. https://doi.org/10.1126/sciadv.abc0330

Gurdon, J.B., 1962. Adult frogs derived from the nuclei of single somatic cells. *Developmental Biology* 4, 256–273.

Hara, K., Tydeman, P., Kirschner, M., 1980. A cytoplasmic clock with the same period as the division cycle in Xenopus eggs. *Proc National Acad Sci* 77, 462–466. https://doi.org/10.1073/pnas.77.1.462

Harland, R.M., Laskey, R.A., 1980. Regulated replication of DNA microinjected into eggs of *Xenopus laevis*. *Cell* 21, 761–771. https://doi.org/10.1016/0092-8674(80)90439-0

Hartwell, L.H., Culotti, J., Pringle, J.R., Reid, B.J., 1974. Genetic control of the cell division cycle in yeast. *Science* 183, 46–51. https://doi.org/10.1126/science.183.4120.46

Havens, C.G., Walter, J.C., 2009. Docking of a specialized PIP Box onto chromatin-bound PCNA creates a degron for the ubiquitin ligase CRL4Cdt2. *Molecular Cell* 35, 93–104. https://doi.org/10.1016/j.molcel.2009.05.012

Herrick, J., Stanislawski, P., Hyrien, O., Bensimon, A., 2000. Replication fork density increases during DNA synthesis in *X. laevis* egg extracts. *J Mol Biol* 300, 1133–1142. https://doi.org/10.1006/jmbi.2000.3930

Higashi, T.L., Ikeda, M., Tanaka, H., Nakagawa, T., Bando, M., Shirahige, K., Kubota, Y., Takisawa, H., Masukata, H., Takahashi, T.S., 2012. The prereplication complex recruits XEco2 to chromatin to promote cohesin acetylation in Xenopus egg extracts. *Current Biology: CB* 22, 977–988. https://doi.org/10.1016/j.cub.2012.04.013

Hodskinson, M.R., Bolner, A., Sato, K., Kamimae-Lanning, A.N., Rooijers, K., Witte, M., Mahesh, M., Silhan, J., Petek, M., Williams, D.M., Kind, J., Chin, J.W., Patel, K.J., Knipscheer, P., 2020. Alcohol-derived DNA crosslinks are repaired by two distinct mechanisms. *Nature* 579, 603–608. https://doi.org/10.1038/s41586-020-2059-5

Hoogenboom, W.S., Douwel, D.K., Knipscheer, P., 2017. Xenopus egg extract: A powerful tool to study genome maintenance mechanisms. *Dev Biol* 428, 300–309. https://doi.org/10.1016/j.ydbio.2017.03.033

Hwang, H.C., Clurman, B.E., 2005. Cyclin E in normal and neoplastic cell cycles. *Oncogene* 24, 2776–2786. https://doi.org/10.1038/sj.onc.1208613

Irniger, S., Piatti, S., Michaelis, C., Nasmyth, K., 1995. Genes involved in sister chromatid separation are needed for b-type cyclin proteolysis in budding yeast. *Cell* 81, 269–277. https://doi.org/10.1016/0092-8674(95)90337-2

Izumi, T., Walker, D.H., Maller, J.L., 1992. Periodic changes in phosphorylation of the Xenopus cdc25 phosphatase regulate its activity. *Mol Biol Cell* 3, 927–939. https://doi.org/10.1091/mbc.3.8.927

Jares, P., Blow, J.J., 2000. Xenopus cdc7 function is dependent on licensing but not on XORC, XCdc6, or CDK activity and is required for XCdc45 loading. *Genes & Development* 14, 1528–1540.

Jares, P., Luciani, M.G., Blow, J.J., 2004. A Xenopus Dbf4 homolog is required for Cdc7 chromatin binding and DNA replication. *Bmc Mol Biol* 5, 5. https://doi.org/10.1186/1471-2199-5-5

Kanke, M., Tahara, E., Veld, P.J.H. in't, Nishiyama, T., 2016. Cohesin acetylation and Wapl-Pds5 oppositely regulate translocation of cohesin along DNA. *The EMBO Journal* 35, 2686–2698. https://doi.org/10.15252/embj.201695756

Kim, S.Y., Ferrell, J.E., 2007. Substrate competition as a source of ultrasensitivity in the inactivation of Wee1. *Cell* 128, 1133–1145. https://doi.org/10.1016/j.cell.2007.01.039

King, R., Lustig, K., Stukenberg, P., McGarry, T., Kirschner, M., 1997. Expression cloning in the test tube. *Science* 277, 973–974.

King, R., Peters, J., Tugendreich, S., Rolfe, M., Hieter, P., Kirschner, M., 1995. A 20S complex containing CDC27 and CDC16 catalyzes the mitosis-specific conjugation of ubiquitin to cyclin B. *Cell* 81, 279–288.

Kirschner, M., 2020. What makes the cell cycle tick? a celebration of the awesome power of biochemistry and the frog egg. *Mol Biol Cell* 31, 2874–2878. https://doi.org/10.1091/mbc.e20-10-0626

Kose, H.B., Larsen, N.B., Duxin, J.P., Yardimci, H., 2019. Dynamics of the eukaryotic replicative helicase at lagging-strand protein barriers support the steric exclusion model. *Cell Reports* 26, 2113–2125.e6. https://doi.org/10.1016/j.celrep.2019.01.086

Kubota, Y., Mimura, S., Nishimoto, S., Masuda, T., Nojima, H., Takisawa, H., 1997. Licensing of DNA replication by a multi-protein complex of MCM/P1 proteins in Xenopus eggs. *Embo J* 16, 3320–3331. https://doi.org/10.1093/emboj/16.11.3320

Kubota, Y., Mimura, S., Nishimoto, S., Takisawa, H., Nojima, H., 1995. Identification of the yeast MCM3-related protein as a component of Xenopus DNA replication licensing factor. *Cell* 81, 601–609. https://doi.org/10.1016/0092-8674(95)90081-0

Kubota, Y., Takase, Y., Komori, Y., Hashimoto, Y., Arata, T., Kamimura, Y., Araki, H., Takisawa, H., 2003. A novel ring-like complex of Xenopus proteins essential for the initiation of DNA replication. *Genes & Development* 17, 1141–1152. https://doi.org/10.1101/gad.1070003

Kumagai, A., Dunphy, W.G., 1991. The cdc25 protein controls tyrosine dephosphorylation of the cdc2 protein in a cell-free system. *Cell* 64, 903–914. https://doi.org/10.1016/0092-8674(91)90315-p

Kumagai, A., Dunphy, W.G., 2017. MTBP, the partner of treslin, contains a novel DNA-binding domain, that is essential for proper initiation of DNA replication. *Molecular Biology of the Cell* 28, 2998–3012. https://doi.org/10.1091/mbc.e17-07-0448

Kumagai, A., Shevchenko, A., Shevchenko, A., Dunphy, W.G., 2010. Treslin collaborates with TopBP1 in triggering the initiation of DNA replication. *Cell* 140, 349–359. https://doi.org/10.1016/j.cell.2009.12.049

Lafont, A.L., Song, J., Rankin, S., 2010. Sororin cooperates with the acetyltransferase Eco2 to ensure DNA replication-dependent sister chromatid cohesion. *Proceedings of the National Academy of Sciences of the United States of America* 107, 20364–20369. https://doi.org/10.1073/pnas.1011069107

Larsen, N.B., Gao, A.O., Sparks, J.L., Gallina, I., Wu, R.A., Mann, M., Räschle, M., Walter, J.C., Duxin, J.P., 2019. Replication-coupled DNA-protein crosslink repair by SPRTN and the proteasome in Xenopus egg extracts. *Mol Cell* 73, 574–588.e7. https://doi.org/10.1016/j.molcel.2018.11.024

Lee, M.G., Nurse, P., 1987. Complementation used to clone a human homologue of the fission yeast cell cycle control gene cdc2. *Nature* 327, 31–35. https://doi.org/10.1038/327031a0

Li, F., Chen, J., Solessio, E., Gilbert, D.M., 2003. Spatial distribution and specification of mammalian replication origins during G1 phase. *J Cell Biology* 161, 257–266. https://doi.org/10.1083/jcb.200211127

Li, R., Murray, A.W., 1991. Feedback control of mitosis in budding yeast. *Cell* 66, 519–531. https://doi.org/10.1016/0092-8674(81)90015-5

Li, Y., Gorbea, C., Mahaffey, D., Rechsteiner, M., Benezra, R., 1997. MAD2 associates with the cyclosome/anaphase-promoting complex and inhibits its activity. *Proc National Acad Sci* 94, 12431–12436. https://doi.org/10.1073/pnas.94.23.12431

Lin, Y., Ha, A., Yan, S., 2019. DNA repair, methods and protocols. *Methods Mol Biology* 1999, 161–172. https://doi.org/10.1007/978-1-4939-9500-4_9

Lörincz, A.T., Reed, S.I., 1984. Primary structure homology between the product of yeast cell division control gene CDC28 and vertebrate oncogenes. *Nature* 307, 183–185. https://doi.org/10.1038/307183a0

Loveland, A.B., Habuchi, S., Walter, J.C., Oijen, A.M. van, 2012. A general approach to break the concentration barrier in single-molecule imaging. *Nat Methods* 9, 987–992. https://doi.org/10.1038/nmeth.2174

Lupardus, P.J., Van, C., Cimprich, K.A., 2007. Analyzing the ATR-mediated checkpoint using Xenopus egg extracts. *Methods (San Diego, Calif)* 41, 222–231. https://doi.org/10.1016/j.ymeth.2006.07.024

Lustig, K., Stukenberg, P., McGarry, T., King, R., Cryns, V., Mead, P., Zon, L., Yuan, J., Kirschner, M., 1997. Small pool expression screening: Identification of genes involved in cell cycle control, apoptosis, and early development. *Methods Enzymol* 283, 83–99.

Mahbubani, H.M., Paull, T., Eider, J.K., Blow, J.J., 1992. DNA replication initiates at multiple sites on plasmid DNA in Xenopus egg extracts. *Nucleic Acids Res* 20, 1457–1462. https://doi.org/10.1093/nar/20.7.1457

Maiorano, D., Moreau, J., Méchali, M., 2000. XCDT1 is required for the assembly of pre-replicative complexes in *Xenopus laevis*. *Nature* 404, 622–625. https://doi.org/10.1038/35007104

Marheineke, K., Goldar, A., Krude, T., Hyrien, O., 2009. Use of DNA combing to study DNA replication in Xenopus and human cell-free systems. *Methods in Molecular Biology (Clifton, NJ)* 521, 575–603. https://doi.org/10.1007/978-1-60327-815-7_33

Marheineke, K., Hyrien, O., 2004. Control of replication origin density and firing time in Xenopus egg extracts: Role of a caffeine-sensitive, ATR-dependent checkpoint. *The Journal of Biological Chemistry* 279, 28071–28081. https://doi.org/10.1074/jbc.m401574200

Masui, Y., Markert, C.L., 1971. Cytoplasmic control of nuclear behavior during meiotic maturation of frog oocytes. *J Exp Zool* 177, 129–145. https://doi.org/10.1002/jez.1401770202

McGarry, T.J., Kirschner, M.W., 1998. Geminin, an inhibitor of DNA replication, is degraded during mitosis. *Cell* 93, 1043–1053. https://doi.org/10.1016/s0092-8674(00)81209-x

Méchali, M., Harland, R.M., 1982. DNA synthesis in a cell-free system from Xenopus eggs: Priming and elongation on single-stranded DNA in vitro. *Cell* 30, 93–101. https://doi.org/10.1016/0092-8674(82)90015-0

Méchali, M., Kearsey, S., 1984. Lack of specific sequence requirement for DNA replication in Xenopus eggs compared with high sequence specificity in yeast. *Cell* 38, 55–64. https://doi.org/10.1016/0092-8674(84)90526-9

Mimura, S., Masuda, T., Matsui, T., Takisawa, H., 2000. Central role for Cdc45 in establishing an initiation complex of DNA replication in Xenopus egg extracts. *Genes Cells* 5, 439–452. https://doi.org/10.1046/j.1365-2443.2000.00340.x

Möröy, T., Geisen, C., 2004. Cyclin E. *Int J Biochem Cell Biology* 36, 1424–1439. https://doi.org/10.1016/j.biocel.2003.12.005

Moses, R.M., Masui, Y., 1989. Cytostatic factor (CSF) in the eggs of *Xenopus laevis*. *Exp Cell Res* 185, 271–276. https://doi.org/10.1016/0014-4827(89)90055-4

Murray, A., Kirschner, M., 1989. Cyclin synthesis drives the early embryonic cell cycle. *Nature* 339, 275–280.

Murray, A., Solomon, M., Kirschner, M., 1989. The role of cyclin synthesis and degradation in the control of maturation promoting factor activity. *Nature* 339, 280–286.

Novak, B., Tyson, J.J., 1993. Numerical analysis of a comprehensive model of M-phase control in Xenopus oocyte extracts and intact embryos. *J Cell Sci* 106 (Pt 4), 1153–1168.

Ohtsubo, M., Roberts, J., 1993. Cyclin-dependent regulation of G1 in mammalian fibroblasts. *Science* 259, 1908–1912. https://doi.org/10.1126/science.8384376

Pacek, M., Tutter, A.V., Kubota, Y., Takisawa, H., Walter, J.C., 2006. Localization of MCM2-7, Cdc45, and GINS to the site of DNA unwinding during eukaryotic DNA replication. *Molecular Cell* 21, 581–587. https://doi.org/10.1016/j.molcel.2006.01.030

Paris, J., Guellec, R.L., Couturier, A., Guellec, K.L., Omilli, F., Camonis, J., MacNeill, S., Philippe, M., 1991. Cloning by differential screening of a Xenopus cDNA coding for a protein highly homologous to cdc2. *Proc National Acad Sci* 88, 1039–1043. https://doi.org/10.1073/pnas.88.3.1039

Peng, A., Lewellyn, A.L., Maller, J.L., 2008. DNA damage signaling in early Xenopus embryos. *Cell Cycle* 7, 3–6. https://doi.org/10.4161/cc.7.1.5157

Pérez-Mongiovi, D., Chang, P., Houliston, E., 1998. A propagated wave of MPF activation accompanies surface contraction waves at first mitosis in Xenopus. *J Cell Sci* 111 (Pt 3), 385–393.

Pfleger, C., Kirschner, M., 2000. The KEN box: An APC recognition signal distinct from the D box targeted by Cdh1. *Genes & Development* 14, 655–665.

Pomerening, J.R., Kim, S.Y., Ferrell, J.E., 2005. Systems-level dissection of the cell-cycle oscillator: Bypassing positive feedback produces damped oscillations. *Cell* 122, 565–578. https://doi.org/10.1016/j.cell.2005.06.016

Pope, B.D., Aparicio, O.M., Gilbert, D.M., 2013. SnapShot: Replication timing. *Cell* 152, 1390–1390.e1. https://doi.org/10.1016/j.cell.2013.02.038

Rankin, S., Ayad, N.G., Kirschner, M.W., 2005. Sororin, a substrate of the anaphase-promoting complex, is required for sister chromatid cohesion in vertebrates. *Molecular Cell* 18, 185–200. https://doi.org/10.1016/j.molcel.2005.03.017

Rankin, S., Kirschner, M.W., 1997. The surface contraction waves of Xenopus eggs reflect the metachronous cell-cycle state of the cytoplasm. *Curr Biol* 7, 451–454. https://doi.org/10.1016/s0960-9822(06)00192-8

Rao, P.N., Johnson, R.T., 1970. Mammalian cell fusion: Studies on the regulation of DNA synthesis and mitosis. *Nature* 225, 159–164. https://doi.org/10.1038/225159a0

Romanowski, P., Madine, M.A., Rowles, A., Blow, J.J., Laskey, R.A., 1996. The Xenopus origin recognition complex is essential for DNA replication and MCM binding to chromatin. *Curr Biol* 6, 1416–1425. https://doi.org/10.1016/s0960-9822(96)00746-4

Rowles, A., Chong, J.P.J., Brown, L., Howell, M., Evan, G.I., Blow, J.J., 1996. Interaction between the origin recognition complex and the replication licensing system in Xenopus. *Cell* 87, 287–296. https://doi.org/10.1016/s0092-8674(00)81346-x

Sha, W., Moore, J., Chen, K., Lassaletta, A.D., Yi, C.-S., Tyson, J.J., Sible, J.C., 2003. Hysteresis drives cell-cycle transitions in *Xenopus laevis* egg extracts. *Proc National Acad Sci* 100, 975–980. https://doi.org/10.1073/pnas.0235349100

Silva, E.M.L. da, Rankin, S., 2018. Chromosome cohesion and condensation in Xenopus egg extracts. *Cold Spring Harbor Protocols* 1–9. https://doi.org/10.1101/pdb.prot097121

Silva, T., Bradley, R.H., Gao, Y., Coue, M., 2006. Xenopus CDC7/DRF1 complex is required for the initiation of DNA replication. *J Biol Chem* 281, 11569–11576. https://doi.org/10.1074/jbc.m510278200

Smith, L.D., Ecker, R.E., 1971. The interaction of steroids with *Rana pipiens* oocytes in the induction of maturation. *Dev Biol* 25, 232–247. https://doi.org/10.1016/0012-1606(71)90029-7

Smythe, C., Newport, J.W., 1992. Coupling of mitosis to the completion of S phase in Xenopus occurs via modulation of the tyrosine kinase that phosphorylates p34cdc2. *Cell* 68, 787–797. https://doi.org/10.1016/0092-8674(92)90153-4

Solomon, M.J., Glotzer, M., Lee, T.H., Philippe, M., Kirschner, M.W., 1990. Cyclin activation of p34cdc2. *Cell* 63, 1013–1024. https://doi.org/10.1016/0092-8674(90)90504-8

Song, J., Lafont, A., Chen, J., Wu, F.M., Shirahige, K., Rankin, S., 2012. Cohesin acetylation promotes sister chromatid cohesion only in association with the replication machinery. *Journal of Biological Chemistry* 287, 34325–34336. https://doi.org/10.1074/jbc.m112.400192

Sparks, J.L., Chistol, G., Gao, A.O., Räschle, M., Larsen, N.B., Mann, M., Duxin, J.P., Walter, J.C., 2019. The CMG helicase bypasses DNA-protein cross-links to facilitate their repair. *Cell* 176, 167–181.e21. https://doi.org/10.1016/j.cell.2018.10.053

Strausfeld, U.P., Howell, M., Descombes, P., Chevalier, S., Rempel, R.E., Adamczewski, J., Maller, J.L., Hunt, T., Blow, J.J., 1996. Both cyclin A and cyclin E have S-phase promoting (SPF) activity in Xenopus egg extracts. *J Cell Sci* 109 (Pt 6), 1555–1563.

Stukenberg, P., Lustig, K., McGarry, T., King, R., Kuang, J., Kirschner, M., 1997. Systematic identification of mitotic phosphoproteins. *Current Biology: CB* 7, 338–348.

Su, J.Y., Rempel, R.E., Erikson, E., Maller, J.L., 1995. Cloning and characterization of the Xenopus cyclin-dependent kinase inhibitor p27XIC1. *Proc National Acad Sci* 92, 10187–10191. https://doi.org/10.1073/pnas.92.22.10187

Sudakin, V., Ganoth, D., Dahan, A., Heller, H., Hershko, J., Luca, F., Ruderman, J., Hershko, A., 1995. The cyclosome, a large complex containing cyclin-selective ubiquitin ligase activity, targets cyclins for destruction at the end of mitosis. *Mol Biol Cell* 6, 185–197. https://doi.org/10.1091/mbc.6.2.185

Takahashi, T.S., Basu, A., Bermudez, V., Hurwitz, J., Walter, J.C., 2008. Cdc7-Drf1 kinase links chromosome cohesion to the initiation of DNA replication in Xenopus egg extracts. *Genes & Dev*. 22, 1894–1905. https://doi.org/10.1101/gad.1683308

Takahashi, T.S., Walter, J.C., 2005. Cdc7: Drf1 is a developmentally regulated protein kinase required for the initiation of vertebrate DNA replication. *Gene Dev* 19, 2295–2300. https://doi.org/10.1101/gad.1339805

Takahashi, T.S., Yiu, P., Chou, M.F., Gygi, S., Walter, J.C., 2004. Recruitment of Xenopus Scc2 and cohesin to chromatin requires the pre-replication complex. *Nature Cell Biology* 6, 991–996. https://doi.org/10.1038/ncb1177

Teixeira, L.K., Reed, S.I., 2018. DNA replication, from old principles to new discoveries. *Adv Exp Med Biol* 1042, 527–547. https://doi.org/10.1007/978-981-10-6955-0_22

Tung, J.J., Hansen, D.V., Ban, K.H., Loktev, A.V., Summers, M.K., Adler, J.R., Jackson, P.K., 2005. A role for the anaphase-promoting complex inhibitor Emi2/XErp1, a homolog of early mitotic inhibitor 1, in cytostatic factor arrest of Xenopus eggs. *P Natl Acad Sci Usa* 102, 4318–4323. https://doi.org/10.1073/pnas.0501108102

Tunquist, B.J., Maller, J.L., 2003. Under arrest: Cytostatic factor (CSF)-mediated metaphase arrest in vertebrate eggs. *Genes & Development* 17, 683–710. https://doi.org/10.1101/gad.1071303

Walter, J., Newport, J.W., 1997. Regulation of replicon size in Xenopus egg extracts. *Science* 275, 993–995. https://doi.org/10.1126/science.275.5302.993

Walter, J., Sun, L., Newport, J., 1998. Regulated chromosomal DNA replication in the absence of a nucleus. *Mol Cell* 1, 519–529. https://doi.org/10.1016/s1097-2765(00)80052-0

Wohlschlegel, J.A., Dwyer, B.T., Dhar, S.K., Cvetic, C., Walter, J.C., Dutta, A., 2000. Inhibition of eukaryotic DNA replication by geminin binding to Cdt1. *Science* 290, 2309–2312. https://doi.org/10.1126/science.290.5500.2309

Yardimci, H., Loveland, A.B., Oijen, A.M. van, Walter, J.C., 2012. Single-molecule analysis of DNA replication in Xenopus egg extracts. *Methods* 57, 179–186. https://doi.org/10.1016/j.ymeth.2012.03.033

Yew, P.R., Kirschner, M.W., 1997. Proteolysis and DNA replication: The CDC34 requirement in the Xenopus egg cell cycle. *Science* 277, 1672–1676. https://doi.org/10.1126/science.277.5332.1672

You, Z., Bailis, J.M., Johnson, S.A., Dilworth, S.M., Hunter, T., 2007. Rapid activation of ATM on DNA flanking double-strand breaks. *Nature Cell Biology* 9, 1311–1318. https://doi.org/10.1038/ncb1651

You, Z., Harvey, K., Kong, L., Newport, J., 2002. Xic1 degradation in Xenopus egg extracts is coupled to initiation of DNA replication. *Genes & Development* 16, 1182–1194. https://doi.org/10.1101/gad.985302

Zou, H., McGarry, T.J., Bernal, T., Kirschner, M.W., 1999. Identification of a vertebrate sister-chromatid separation inhibitor involved in transformation and tumorigenesis. *Science* 285, 418–422. https://doi.org/10.1126/science.285.5426.418

3 Maternal mRNAs and the Making of Cell Lineages in the Early Xenopus Embryo

Douglas W. Houston

CONTENTS

3.1. Historical Perspectives: The Localization Problem in Amphibians 25
3.2. Molecular Characterization of Maternal and Localized RNAs 26
3.2.1. Localized Maternal mRNAs 26
3.2.2. mRNA Localization Mechanisms 27
3.3. Analysis of Maternal Gene Function in *Xenopus* Development 28
3.4. Maternal Control of Germ Layer Induction and Patterning 29
3.4.1. Maternal Control of Endoderm and Mesoderm by Vegt 29
3.4.2. Maternal Secreted Molecules in Germ Layer Induction 30
3.4.3. Ectoderm Specification 30
3.4.4. β-Catenin in Dorsal Gene Activation 31
3.4.5. Cytoplasmic Activation of Dorsal β-Catenin 31
3.4.6. Secreted Ligand Activation of β-Catenin 32
3.5. Maternal Control of Primordial Germ Cell Formation 32
3.5.1. The Germ Plasm 32
3.5.2. Germ Plasm mRNAs in PGC Specification 33
3.5.3. Assembly of Maternal Germ Plasm 33
3.6. Concluding Remarks 34
Acknowledgments 35
References 35

3.1. HISTORICAL PERSPECTIVES: THE LOCALIZATION PROBLEM IN AMPHIBIANS

For over 200 years, embryologists have speculated on the extent that the pattern and organization of the body is determined by that of the egg. The study of amphibian development is somewhat unique among model organisms in that many modern research problems follow the questions and traditions of what perhaps were the very first embryological studies, which were aimed at addressing this very issue. The idea that the frog embryo body is organized into germ layers and along axes of polarity was introduced by the early embryologists von Baer (1834) and Remak (1855), who noted the regional origins of the germ layers, and by Newport (1854, 1851) and Roux (1888, 1887), who described relationships between the site of sperm entry and the alignment of the body axis. Roux also described what came to be called the grey crescent, a pale section of the frog egg opposite the sperm entry point (Figure 3.1) that formed in the precise position of the future dorsal lip (reviewed in Wilson, 1928). The dorsal lip was identified by Spemann as the "organizer" of axial pattern in the amphibian embryo (Spemann, 1938, 1921; Spemann and Mangold, 1924; and see this volume, Chapter 4).

Spemann's egg constriction experiments (Figure 3.1) showed that often only one-half of the embryo would form a dorsal lip (having inherited grey crescent material) and develop into a normally proportioned larva, whereas the other half would develop ventral derivatives of all three germ layers, failing to form axial tissues (Spemann, 1903, 1902, 1901). Conversely, in cases in which each half-embryo formed part of the dorsal lip, two normally proportioned embryos would arise. These experiments in amphibians paralleled contemporaneous ones in aquatic invertebrates, eventually leading to the realization that so-called "mosaic" versus "regulative" development (i.e. cell autonomous versus normal development of isolated blastomeres) was merely a function of whether maternal determinants were asymmetrically or symmetrically distributed in the early embryo (reviewed in Davidson, 1986; and in Wilson, 1928). Cytoplasmic determinants were later hypothesized to determine the fates of largely equivalent nuclei in the embryo.

Three prominent examples of cytoplasmic determinants in amphibians emerged, first in earlier frog models and salamanders and later in *Xenopus*. These were: (1) the relative dorsal movement of the egg cortex in axis formation ("rotation of symmetrization," Ancel and Vintemberger, 1948; later termed "cortical rotation," Gerhart et al., 1989), (2) the role

DOI: 10.1201/9781003050230-4

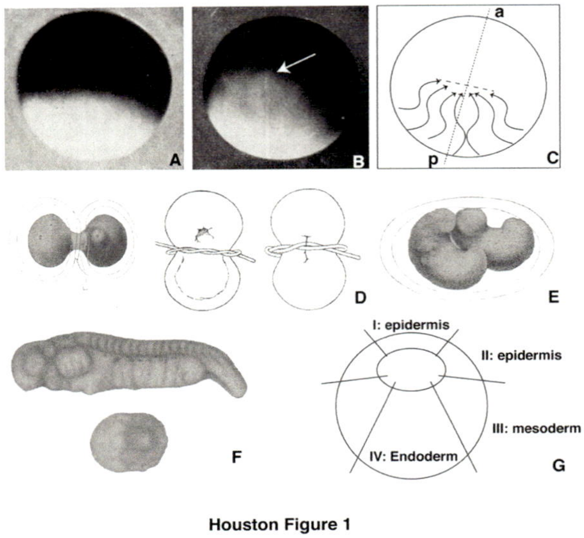

FIGURE 3.1 The organization of the amphibian egg and embryonic patterning. (A) A frog egg at fertilization; the pigmented animal pole is towards the top; the pale vegetal pole is towards the bottom. (B) The egg at 20 minutes post-fertilization; the appearance of the grey crescent is indicated by the arrow. (C) Model of cortical rotation (dorsal view of the egg); microtubules are shown as arrows. The dotted line indicates the future midline (a, anterior; p, posterior). (D) Spemann's egg ligation experiments; the left panel shows a "strongly constricted" egg at the early two cell stage; to the right are the two main outcomes—isolation of the dorsal lip to one half (middle) or bisection of the dorsal lip (right). (E) Twinned embryos result from the right-hand case in D. (F) One normal embryo and one *Bauchstück* (a ventralized "belly piece") resulting from the middle case in D. (G) Model of regional specification of the *Xenopus* blastula. Nieuwkoop recombined zones I and II with zone IV to demonstrate mesoderm induction in *Xenopus* (after Sudarwati and Nieuwkoop 1971).

Source: Panels (A–B) are reproduced from Rugh (1951); panels (D–F) reproduced from Spemann (1924).

of this dorsal-ventral regionalization of the vegetal hemisphere in mesoderm and organizer induction (*Ambystoma*: Boterenbrood and Nieuwkoop, 1973; Nieuwkoop, 1969; *Xenopus*: Sudarwati and Nieuwkoop, 1971; Figure 3.1), and (3) the presence of "germ plasm" in the vegetal cortex of the frog egg (*cytoplasme germinale*; Bounoure, 1934, 1931, related to insect "pole plasm" (*polares Plasma*, Kahle, 1908) in germline specification (see Section 5).

This chapter reviews past work on the patterning of the *Xenopus* embryo by maternal gene products, including the discovery of localized RNAs in *Xenopus* oocytes (where much of the initial progress was made); the mechanisms of localization of these RNAs; and functional studies on localized and non-localized molecules in germ layer specification, axis induction, and germline formation. These ideas and their impact on their respective fields have been reviewed separately or more comprehensively in the context of overall vertebrate development and oocyte polarity, so I will not present another broad comparative review. Rather, my goal is to provide a general background to the main biological questions relating to the maternal control of *Xenopus* development. I will also review the principal approaches and findings, following early work to current state of the art. The sections can be read independently in any order; any omissions of material and references are unintentional and reflect limited space and my own view of the field.

3.2. MOLECULAR CHARACTERIZATION OF MATERNAL AND LOCALIZED RNAS

Early discoveries in molecular biology identified the central role of RNA in interpreting information encoded in the DNA, with the definitive discovery of mRNA being reported in 1961 (Brenner et al., 1961; Gros et al., 1961; Hayashi and Spiegelman, 1961; Jacob and Monod, 1961). The critical nature of maternal mRNA in development was inferred through experiments in sea urchins and in frogs that showed the importance of new protein synthesis rather than new transcription in driving early development (Hultin, 1961; Smith and Ecker, 1965; Tyler, 1965). More direct evidence for maternal mRNA was obtained by cell-free translation assays, showing that a minor proportion of *Xenopus* oocyte total RNA could drive protein synthesis in cell-free translation assays (Davidson et al. 1966).

3.2.1. Localized Maternal mRNAs

Despite these demonstrations, the compelling notion that specific mRNAs might become specifically localized and direct the specification of different cell lineages remained speculative (Davidson and Britten, 1971; Kalthoff, 1979). However, the discovery that mRNAs are covalently modified at the three-prime end with "polyadenylic acids" was both a milestone in understanding gene expression regulation and an advance in the detection and isolation of mRNAs (Darnell et al., 1971; Edmonds et al., 1971; Lee et al., 1971). Analytical hybridization of poly(A)+ RNAs with labeled cDNAs and poly(U) *in situ* hybridization identified asymmetric accumulation of (putative) mRNA in eggs of many species, including *Xenopus* (Capco and Jeffery, 1981, 1979; Jeffery and Capco, 1978). Additional analysis of *Xenopus* vegetally enriched cDNAs indicated that a minority of poly(A) RNA sequences (about 3–5%) were enriched up to 20-fold at the vegetal pole (Carpenter and Klein, 1982), likely representing localized mRNAs. Importantly, cell-free translation of *Xenopus* oocyte vegetal pole mRNAs identified unique patterns of protein synthesis (King and Barklis, 1985), implying the presence of maternally localized transcripts with the ability to create functional protein asymmetry in the embryo.

Advances in molecular cloning technology made it possible to isolate and identify individual localized mRNA sequences. Differential screening of an oocyte cDNA library identified one vegetally localized clone (Rebagliati et al., 1985), designated *Vg1* (now *gdf1*). This molecule encodes a member of the Transforming growth factor β (Tgfβ) family and undergoes specific localization to the vegetal cortex of full-grown oocytes, becoming inherited by vegetal blastomeres (Weeks and Melton, 1987). Tangential efforts to clone maternally expressed *Wingless-type MMTV integration site* (*Wnt*) genes fortuitously identified an additional vegetally localized mRNA, *Xwnt-11* (now *wnt11b*; Ku and Melton, 1993). Both the Tgfβ and Wnt proteins had been concomitantly implicated in mesoderm induction and in axis formation (see Section 4), making the presence of these proteins in the set of vegetally localized transcripts especially intriguing.

Further fractionation of the oocyte using biochemical or physical methods facilitated the isolation of additional localized mRNAs (Elinson et al., 1993; Pondel and King, 1988) and led to the demonstration of two mechanisms of mRNA localization in the oocyte: (1) a late pathway initiated after the elaboration of animal-vegetal polarity in mid-oogenesis and (2) an early mechanism in pre-vitellogenic oocytes involving the mitochondrial cloud (Forristall et al., 1995; Kloc and Etkin, 1995). The early pattern exactly matched the known distribution of the germ plasm in *Xenopus* (Czołowska, 1972, 1969; Heasman et al., 1984; Savage and Danilchik, 1993). An intermediate pattern also became evident following the characterization of *plin2* (*née fatvg*)

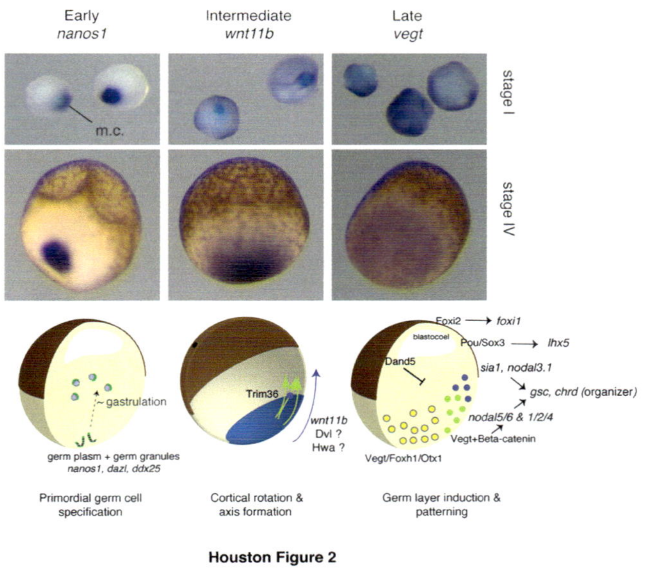

FIGURE 3.2 Localized mRNAs and their roles in early Xenopus development. Top panels show in situ hybridization patterns of representative mRNAs of early-, intermediate- and late-localizing mRNAs (*nanos1*, *wnt11b*, and *vegt*, respectively). In stage I oocytes (top row), *nanos1* is tightly restricted to the mitochondrial cloud/Balbiani body (m.c.), whereas *wnt11b* localizes to the cytoplasm and the mitochondrial cloud. *vegt* is not localized. By stage IV (middle row), *nanos1* is localized to the germ plasm at the vegetal apex, *wnt11b* is less restricted, and *vegt* is broadly localized in the vegetal hemisphere. Pigmented oocytes are shown; animal pole toward the top. Bottom panels, models for the three roles of these localized mRNA classes.

mRNA localization, which displayed aspects of both pathways (Chan et al., 1999). Representative examples of these different patterns are shown in Figure 3.2.

3.2.2. mRNA Localization Mechanisms

Additional studies on the mechanisms of localization using injected transcripts in oocytes identified both shared and distinct molecular mechanisms of localization for early and late pathway mRNAs. Notably, early pathway localization appeared to be independent of the cytoskeleton, likely involving a diffusion/entrapment mechanism (Chang et al.,2004; Kloc et al., 1996). By contrast, the late pathway required microtubule polarization and transport occurring as a consequence of general vegetal localization of organelles and the mitochondrial cloud remnants and anchoring by different cytoskeletal components (reviewed in King et al., 2005; Medioni et al., 2012; Houston, 2013).

Structure-function mutagenesis identified minimal localization elements (LEs) containing clustered repeats of motifs in the 3'UTRs of several localized transcripts as well as cognate RNA-binding proteins that bound these motifs (Claussen et al., 2004; Kloc et al., 1996; Mowry and Melton, 1992; Zhou and King, 1996a, 1996b). Counterintuitively, similar localization motifs were identified in early and late pathway mRNAs, consisting of repeated clustered UUCAC and UUUCU motifs and recognized by RNA-binding proteins Igf2bp3 and Ptbp1, respectively (reviewed in Cabral and Mowry, 2020; Houston, 2013; Oh and Houston, 2017a). In light of the fact that other localized mRNAs lack these motifs, (Chan et al., 1999; Claussen and Pieler, 2004; Horvay et al., 2006), a consensus localization element has yet to be identified. Multimers composed solely of localization elements do not localize (Lewis et al., 2004), also suggesting that a context-dependent organization in the 3'UTR is needed for proper localization.

A computational approach to localized mRNA prediction identified clusters of CAC-rich motifs in validated localization elements across species (Betley et al., 2002), but these motifs could not explain all mRNA localization. In the last decade, global transcriptomic analyses in oocytes and early embryos have identified extensive, and largely comparable, sets of maternally localized mRNAs in *Xenopus* spp. (see Table 3.1). Sindelka et al. (2018) performed a bioinformatic analysis that identified a number of putative motifs associated with vegetally localized mRNAs and some animally enriched mRNAs, including some CAC-rich sequences, but none of these mRNAs were experimentally validated. The extent to which these analyses can distinguish between distinct vegetal localization patterns still remains unclear, but newer machine-learning approaches might be usefully applied in this context.

Table 3.1 lists selected mRNAs with previously described vegetal localization in Xenopus oocytes. References cite the primary description of the localization pattern and

TABLE 3.1
Selected Vegetally Localized mRNAs in *Xenopus*

Late Pathway mRNAs	Gene Symbol	References
growth differentiation factor 1 (alias vg1)	***gdf1***	Rebagliati et al., 1985; Weeks et al., 1987; Birsoy et al., 2006
vegt	***vegt***	Zhang and King, 1996; Zhang et al., 1998
acyl-CoA synthetase long-chain 1	***acsl1b***	King et al., 2005; Wang et al., 2012
bicaudal C homolog 1	***bicc1***	Wessely and Robertis, 2000; Park et al., 2016
low density lipoprotein receptor adaptor protein 1	***ldlrap1***	Zhou et al., 2004
orthodenticle homolog 1	***otx1***	Pannese et al., 2000; Owens et al., 2017; Paraiso et al., 2019
zinc finger protein 36-like 2 (C3H-3)	***zfp36l2***	Betley et al., 2002
beta-transducin repeat containing E3 ubiquitin protein ligase	***btrc***	Hudson et al., 1996
ephrin-b1	***efnb1***	Betley et al., 2002; Owens et al., 2017
sox7	***sox7***	Claussen et al. 2015; De Domenico et al. 2015;
Early Pathway mRNAs		
nanos homolog 1 (xcat2)	***nanos1***	Mosquera et al., 1993; Lai et al., 2012
deleted in azoospermia-like	***dazl***	Houston et al., 1998; Houston and King, 2000b
ddx25 (deadsouth)	***ddx25***	MacArthur et al., 2000; Yamaguchi et al., 2013
germes	***germes***	Berekelya et al., 2003
dead (Asp-Glu-Ala-Asp) box polypeptide 59 (centroid)	***ddx59***	Kloc and Chan, 2007
xpat/pgat	***pgat***	Hudson and Woodland, 1998
xsirt 13.2	***xsirts***	Kloc et al., 1993
proprotein convertase subtilisin/kexin type 6 (pace4)	***pcsk6***	Birsoy et al., 2005
syntabulin	***sybu***	Colozza and Robertis, 2014; Oh and Houston, 2017b
RAS related 2	***rras2***	Owens et al., 2017
Intermediate/Dual Pathway mRNAs		
tripartite motif-containing protein 36	***trim36***	Cuykendall and Houston, 2009
wnt11b	***wnt11***	Ku and Melton, 1993; Kloc et al., 1998; Tao et al., 2005
perilipin 2 (alias fatvg)	***plin2***	Chan et al., 1999, 2007
DND microRNA-mediated repression inhibitor 1	***dnd1***	Weidinger et al., 2003; Mei et al., 2013
RNA binding protein with multiple splicing (alias hermes)	***rbpms***	Zearfoss et al., 2004
glutamate receptor interacting protein 2	***grip2***	Kaneshiro et al. 2007; Tarbashevich et al., 2007
low molecular weight neuronal intermediate filament	***nif***	Claussen et al., 2004
vegetally-localized 1	***velo1***	Claussen and Pieler, 2004; Nijjar and Woodland, 2013b
Strategies for Identifying Maternally Localized mRNAs	**Database**	**References**
Animal/Vegetal Half RNA sequencing (8-cell stage)	**GSE118024**	Paraiso et al., 2019
Differential RNA sequencing/Proteomics	**GSE104848**	Sindelka et al., 2018
Differential RNA sequencing	**GSE80971**	Owens et al., 2017
Differential RNA sequencing	**GSE58420**	Claussen et al., 2015
Single blastomere RNA sequencing(8-cell stage)	**N/A**	De Domenico et al., 2015
Animal/Vegetal halves/Affymetrix microarray (8-cell stage)	**GSE48659**	Grant et al., 2014
Vegetal cortex isolation/Affymetrix microarray	**GSE17713**	Cuykendall and Houston, 2010
Differential hybridization to cDNA arrays	**N/A**	Horvay et al., 2006
Differential hybridization to cDNA arrays	**N/A**	Kataoka et al., 2005
Computational analysis of 3'UTRs	**N/A**	Betley et al., 2002
Differential display PCR	**N/A**	Hudson et al., 1996
Differential cDNA library screening	**N/A**	Zhang and King, 1996
Differential cDNA library screening	**N/A**	Rebagliati et al., 1985

loss of function, if available. Intermediate pathway RNAs are defined as having mainly a late localization pattern but with additional localization to the mitochondrial cloud and to the germ plasm in embryos. Studies using different molecular or "genomics" strategies to identify vegetally localized mRNAs have increased in frequency since 2014.

3.3. ANALYSIS OF MATERNAL GENE FUNCTION IN *XENOPUS* DEVELOPMENT

The genetic assessment of maternally supplied gene products in early development requires eggs derived from a female lacking functional copies of the gene: a "maternal effect" mutation. These analyses involve either the generation of viable homozygous mutant females or of germline mosaics, a challenging prospect in amphibians. *Xenopus* nevertheless offers several advantages relative to other vertebrates for maternal gene analysis, including (1) the ability to readily culture and manipulate oocytes *in vitro*, (2) the effectiveness of DNA-based antisense-mediated mRNA degradation in oocytes, and (3) the use of oocyte/egg transfer procedures to ultimately fertilize these oocytes. In addition, large-scale new zygotic mRNA synthesis does not occur until the mid-blastula transition (MBT; 4000-cell stage/stage 8), alongside other changes in cell behavior (Newport and Kirschner, 1982), representing the main maternal-to-zygotic transition in *Xenopus* (MZT; Vastenhouw et al., 2019). Thus, depleted maternal mRNAs are unlikely to be replaced prior to early cell-fate decisions being made. Together, these properties enabled the development of powerful methods to use *Xenopus* to examine vertebrate maternal gene functions ("oocyte host-transfer"; Heasman et al., 1991; Houston, 2019, 2018; Mir and Heasman, 2008; intracytoplasmic sperm injection: Miyamoto et al., 2013, 2015a).

By the early 1970s, use of *Xenopus* oocytes became widespread owing to their use in the expression of heterologous mRNAs (Gurdon et al., 1971). Previous work on oviductal transport and species-specific fertilization of amphibian eggs also established methods for the transfer of coelomic eggs between ovulating females (Rugh, 1935). These transplantation methods were later adapted to work with cultured oocytes that were stimulated to mature *in vitro* as a test of their developmental potential (*Lithobates [née Rana] pipiens*: Smith et al., 1968; *Xenopus:* Brun, 1975). Vegetal irradiation of cultured oocytes prior to host-transfer was used to show the presence of UV-sensitive molecules important for axis and germline formation (Elinson and Pasceri, 1989; Holwill et al., 1987), establishing a paradigm for the pre-fertilization manipulation of development.

Xenopus oocytes were also an important test system for antisense DNA oligonucleotide (oligo) technology in the 1980s. Oligonucleotides injected into full-grown oocytes hybridize to complementary mRNAs and cleave the RNA strand via endogenous RNase H (Cazenave et al., 1987; Dash et al., 1987; Shuttleworth and Colman, 1988). However, the same oligo types, even when modified, were more toxic and short lived when injected into fertilized eggs (Shuttleworth et al., 1988; Woolf et al., 1990). This toxicity, along with inherent limitations in the specificity of RNase H-dependent antisense DNA oligos (Woolf et al., 1992), discouraged the widespread use of antisense technology in post-fertilization *Xenopus* embryos until the commercial availability of phosphorodiamidate morpholino oligos ("Morpholinos"; Heasman et al., 2000; Summerton and Weller, 1997).

Janet Heasman and Chris Wylie first coupled antisense mRNA depletion in oocytes with fertilization through the host transfer procedure (Heasman et al., 1991). They inaugurated this method by depleting maternal *cytokeratin* (*krt8.1*) mRNA, demonstrating that this gene was required for gastrulation and wound healing in early embryos (Torpey et al., 1992). Subsequent studies revealed a requirement for β-Catenin in dorsal axis formation (Heasman et al., 1994). Because parallel experiments in *Drosophila* demonstrated that stabilization of Armadillo/β-Catenin was a main output of Wnt signaling (Peifer et al., 1994), these *Xenopus* antisense mRNA depletion experiments strongly implicated endogenous maternal Wnt/β-catenin activation in axis formation (Section 4).

This maternal mRNA depletion approach has been beneficial for understanding many aspects of early *Xenopus* development because, in almost all cases, Morpholino injection after fertilization at best only partially inhibits maternally regulated processes and cannot affect processes initiated around the time of injection. Another benefit of the host-transfer method is that it can also be used to over- or ectopically express proteins before fertilization, including those representing genome editing reagents for F0 mutagenesis in *Xenopus* embryos (Aslan et al., 2017; Miyamoto et al., 2015b; Nakajima and Yaoita, 2015; Ratzan et al., 2017).

3.4. MATERNAL CONTROL OF GERM LAYER INDUCTION AND PATTERNING

Extensive work in *Xenopus* and other organisms has culminated in a largely unified model for germ layer formation in vertebrates (reviewed in Houston, 2017). This basic model suggests that spatiotemporal gradients of Nodal signaling (patterned by auto-regulation and by Wnt signals) induce dorsal mesendoderm/organizer at high/early doses and ventrolateral mesendoderm at low/later doses. Organizer induction by Nodal occurs in synergy with early Wnt signaling (maternal in *Xenopus*) and sets up self-regulating gradients of BMP and later Wnt activity (zygotic in *Xenopus*) to pattern the dorsoventral and anteroposterior axes.

3.4.1. Maternal Control of Endoderm and Mesoderm by Vegt

Nieuwkoop initially proposed that mesoderm (germ layer) induction would involve new mRNA synthesis (Nieuwkoop, 1969), although some early experiments on heterochronic blastomere recombinations and transplantations (Dale and Slack, 1987; Jones and Woodland, 1987) suggested a maternal mesoderm inducer. The zygotic nature of germ layer induction was demonstrated by Wylie and Heasman, who showed that vegetal masses only induced mesoderm after the onset of zygotic transcription (Wylie et al., 1996). Several groups subsequently and by different approaches identified a likely maternal transcription factor candidate, the T-domain transcription factor, Vegt, encoded by a vegetal cortex-localized mRNA (Horb and Thomsen, 1997; Lustig et al., 1996; Stennard et al., 1996; Zhang and King, 1996).

The maternal role of *vegt* mRNA was assessed using antisense oligos in host-transfer experiments (Zhang et al., 1998). These embryos largely lacked mesoderm and endoderm, and vegetal cells gained expression of ectoderm markers. Vegetal explants from *vegt*-depleted embryos failed to induce mesoderm in animal caps (Zhang et al.,

1998), and subsequent work showed that Vegt controls mesendoderm induction by directly activating a number of *nodal* and *nodal-related* genes (*nodal1, nodal2, nodal4*, and the early expressed paralogs *nodal5* and *nodal6*; Agius et al., 2000; Clements et al., 1999; Hyde and Old, 2000; Kofron et al., 1999; Lee et al., 2001; Takahashi et al., 2000). Vegt likely functions mainly as a transcriptional activator, at least maternally, although later or context-dependent repressive roles have not been excluded. Consistent with this idea, Vegt has been shown to mediate activating histone acetylation modifications (Gao et al., 2016). In addition to these transcriptional roles, depletion of the *vegt* transcript leads to other localized mRNAs becoming delocalized (Heasman et al., 2001) and to disorganization of intermediate filaments (Kloc, 2009), indicating that *vegt* has additional but still unclear roles as a structural or regulatory RNA.

Several observations indicated that maternal regulation of *nodal* expression was especially relevant for mesoderm induction. First, a highly specific Nodal antagonist (CerS) blocked mesendoderm induction *in vivo* and in Nieuwkoop assays (Agius et al., 2000). Only inhibition of Nodal by CerS mimicked pan-Tgfβ inhibition (Agius et al., 2000) and other Tgfβ molecules but not FGFs could rescue *vegt*-depletion (e.g. *derrière/gdf3*, Kofron et al., 1999). Furthermore, an early/elevated aspect of *nodal* homologue expression was dependent on maternal β-Catenin, which could also synergize with Vegt to drive higher levels of *nodal* expression and activity (Agius et al., 2000; Lee et al., 2001; Rex et al., 2002). Last, this temporal aspect and the absence of a unique (non-Nodal) dorsal signal was shown in heterochronic, modified-Nieuwkoop assays using conjugates of equatorial explants with β-Catenin-depleted vegetal masses, with activity in late-stage β-Catenin-depleted vegetal explants equivalent to early-stage control explants (Xanthos et al., 2002).

Vegt and Vegt-induced Nodal signals also regulate transcription factors important for endoderm, notably the *mix-related*, *gata4–6*, and *sox17a/b* genes (Casey et al., 1999; Clements and Woodland, 2003; Taverner et al., 2005; Xanthos et al., 2001). It has been problematic to determine the exact regulatory relationships among these genes, owing to possible overlapping roles of maternal and zygotic Vegt and multiple feedback and cross-regulatory interactions. Recent advances in genomic approaches implicated Vegt acting with vegetally localized Otx1 and Foxh1 in the establishment and function of pre-zygotic enhancer complexes at endodermal loci (Paraiso et al., 2019; see Chapters 12 and 18).

3.4.2. Maternal Secreted Molecules in Germ Layer Induction

The roles of maternal secreted molecules in germ layer induction remain relatively unclear. The identification of *gdf1* as encoding a Tgfβ family growth factor (Weeks and Melton, 1987) was suggestive, but initial antisense and dominant-negative loss of function experiments for *gdf1* were equivocal (Joseph and Melton, 1998; Kessler and Melton, 1995; Rebagliati and Melton, 1987; Woolf et al., 1990). It also became problematic to envision a strong role for Gdf1 in mesendoderm induction given the absence of Tgfβ/Nodal activity in *vegt*-depleted embryos. A revisitation of Gdf1 function using maternal mRNA depletion revealed a requirement for this molecule in the timing of Nodal versus BMP signaling and ultimately in anterior patterning (Birsoy et al., 2006). One hypothesis is that Gdf1 may affect Nodal signal propagation or sensitivity, as was shown in studies of mouse embryos, *Xenopus* explants, and embryonic stem cells (Fuerer et al., 2014; Tanaka et al., 2007). Thus, endogenous Gdf1 may have no function in the absence of properly regulated Nodal signals.

Tdgf1.3 (*née Cripto/XCr-1/FRL1*) is maternally supplied but unlocalized and thought to function as a secreted Nodal co-receptor and trafficking factor (Constam, 2009; Shen and Schier, 2000). In *Xenopus*, maternal *tdgf1.3* mRNA is translated primarily in the animal hemisphere (Zhang et al., 2013) but may also interact with maternal Wnt11b in dorsal signaling (Tao et al., 2005; see subsequently). A limited set of additional secreted signaling molecules are expressed maternally in a non-localized fashion and are involved in modulating Nodal, BMP, FGF, and other signaling pathways in the early embryo (Ism1, Tsku, Ndp/Norrin, Grem1; Hsu et al., 1998; Morris et al., 2007; Pera et al., 2002; Xu et al., 2012).

3.4.3. Ectoderm Specification

In contrast to the detailed pathways known for mesendoderm differentiation, much less is known about the initial specification of ectoderm in *Xenopus*. Ectoderm has been thought of as a maternally programmed "default state" (Weinstein and Hemmati-Brivanlou, 1999). Both inhibition of Tgfβ and depletion of *vegt* allow vegetal cells to express ectodermal genes and to adopt ectodermal cell adhesion and cell sorting properties (Houston and Wylie, 2003; Zhang et al., 1998). Because mesoderm forms out of prospective ectoderm there does not appear to be an "animalizing" gradient as in sea urchins (Dale et al., 1985; Nieuwkoop and Ubbels, 1972), and thus maternal transcription factors must ultimately specify ectoderm fate.

Two *zygotic* transcription factors were initially identified as candidate targets of this maternal ectoderm specifying pathway, *lhx5* and *foxi1* (Houston and Wylie, 2003; Mir et al., 2007; Suri et al., 2005). Lhx5 is necessary and sufficient for normal ectoderm adhesion and cell sorting properties but does not activate ectodermal gene expression or inhibit Tgfβ signaling, suggesting that Lhx5 primarily regulates cell behavior (Houston and Wylie, 2003). *foxi1* is mosaically and dynamically expressed and tightly regulated in the ectoderm, beginning dorsally and shifting ventrally during gastrulation (Mir et al., 2008, 2007; Suri et al., 2005). Foxi1 is required both for ectoderm adhesion and for repression of Tgfβ signaling and mesoderm (Mir et al., 2007; Suri et al., 2005). The mesoderm-inhibiting function of Foxi1 is indirect, mediated through activation of Tbx2 (a transcriptional repressor) in the ectoderm (Teegala et al., 2018).

Foxi2, encoded by a transcript enriched animally in the oocyte, was identified as a maternal regulator of *foxi1* through upstream enhancer analysis (Cha et al., 2012). Maternal mRNA depletion of *foxi2* showed dysregulation of *foxi1* expression and gastrulation/axis defects but no major loss of ectoderm derivatives or adhesion. Other animally enriched maternal transcriptional regulators include Sox3, which is thought to inhibit *nodal5* expression (Zhang et al., 2003; Zhang and Klymkowsky, 2007); Trim33, a putative inhibitor of Smad4-dependent Tgfβ signaling (Dupont et al., 2005); and Znf585b, a Tp53 antagonist (Sasai et al., 2008). A number of other molecules that can antagonize mesendoderm are also enriched animally (i.e. *dand5*, Bates et al., 2013). These proteins might act to protect early ectoderm fate (Reich and Weinstein, 2019).

Ectoderm development also involves homologues of the mammalian pluripotency-related transcription factor Pou5f1, including a maternal paralogue (*pou5f3.3* [*née xlpou60/oct60*]). Morpholino-based depletion of these genes (singly or together) in *Xenopus* have implicated this family in the maintenance of pluripotent character in the animal cap (Morrison and Brickman, 2006) and in repression of Nodal/Tgfβ activity (Cao et al., 2006; Snir et al., 2006). Additionally, Pou5f1 paralogs are required for ectodermal adhesion and gastrulation, likely through the activation of certain conserved Pou5f target genes, notably *lhx5* (see previously), *sall1*, and *cdx1* (Livigni et al., 2013). Maternal Pou5f3 (along with Sox3) has been shown to establish competent chromatin prior to the onset of zygotic transcription (Gentsch et al., 2019), suggesting the possibility that the maternally programmed early chromatin state may form the basis for the ectoderm "default state."

3.4.4. β-Catenin in Dorsal Gene Activation

In addition to regulating the timing of *nodal* expression, β-Catenin regulates a number of dorsal genes directly, including *sia1*, *nodal3* paralogs, *noggin*, and *chordin* (these last two are induced by β-Catenin but stabilized by Nodal), that act in multiple aspects of organizer function. β-Catenin mainly regulates gene expression by de-repressing TCF proteins and recruiting coactivators in the context of Wnt-regulated enhancer complexes (Gammons and Bienz, 2018). Both activities are supported by maternal loss-of-function experiments in *Xenopus* (*tcf7l1:* Houston et al., 2002; *pygo2* and *bcl9:* Belenkaya et al., 2002; Kennedy et al., 2010). Other maternal TCF proteins, Tcf7 and Tcf7l2, are also involved, with context-dependent activating and repressing roles on β-Catenin target genes (Roel et al., 2003; Standley et al., 2006). Genomic studies have begun to identify more complex interactions of β-Catenin with many other transcription factors (e.g. Nishita et al., 2000; Sinner et al., 2004; Zorn et al., 1999; reviewed in Abu-Remaileh et al., 2010).

More recent evidence suggests that maternal β-Catenin regulates dorsal gene expression prior to major MZT in part through an epigenetic "poising" mechanism, marking organizer-specific genes in dorsal morula nuclei for later expression by the recruitment of the histone arginine methyltransferase Prmt2 (Blythe et al., 2010). Similarly, maternal Foxh1 is also involved in presetting chromatin for later expression, largely targeting subsequent Nodal/Smad2-regulated genes (Afouda et al., 2020; Charney et al., 2017; Chiu et al., 2014; Gupta et al., 2014). Foxh1 overlaps β-Catenin targets to a significant extent in these cases, potentially acting as a "pioneer" factor to activate the opening of the chromatin conformation (Zaret and Carroll, 2011).

3.4.5. Cytoplasmic Activation of Dorsal β-Catenin

Wnt/β-Catenin activation is dependent on cortical rotation and is most sensitive to induced stimulation or inhibition around the 16–32-cell stage (Kao et al., 1986; Yang et al., 2002). Elinson, Gerhart, Kirschner, and others began reinvestigating the cortical rotation phenomenon in the 1980s, ultimately leading to the characterization of dorsally directed parallel microtubule array assembly in the vegetal sub-cortical cytoplasm and kinesin-based translocation as the underlying mechanism of cortical rotation (Gerhart et al., 1989; Houston, 2017, 2012; Weaver and Kimelman, 2004).

The molecular control of cortical rotation is not well understood. In addition to an overall increase in microtubule nucleation and assembly following fertilization, likely related to the cell cycle (Elinson, 1985; Olson et al., 2015), the ubiquitin ligase activity of Trim36, encoded by a vegetally localized mRNA is essential for normal parallel microtubule array assembly and cortical rotation (Cuykendall and Houston, 2009). Another localized mRNA product, the Dnd1 RNA-binding protein (typically involved in germline specification), is required to anchor *trim36* in the cortex (Mei et al., 2013), likely enriching Trim36 at the site of cortical rotation. The targets of Trim36 ubiquitylation remain unknown. mRNA depletion of vegetally localized *plin2* also leads to defects in cortical microtubule assembly during cortical rotation (Chan et al., 2007), although it is unclear if the protein product (Perilipin 2 is localized to lipid droplets) or the mRNA itself are the relevant moiety (Kloc, 2009).

Cytoplasmic ablation/transplantation experiments have generally supported the idea that cortical rotation dorsally displaces a potent axis-inducing activity (Holowacz and Elinson, 1993; Kageura, 1997). The identity of this activity remains unclear. Data suggest the vegetal cortical cytoplasm mimics Wnt activation and acts most similarly to overexpressed Adenomatous Polyposis Coli (APC) protein (Marikawa and Elinson, 1999). One caveat of these experiments is that the large size of the injected *apc* transcript tends produce a variety of truncated/degradation products, which may include dominant-inhibitory species (Vleminckx et al., 1997).

Dishevelled and Frat1 (GBP) proteins are also candidates for the cytoplasmic dorsalizing activity, based on the visualization of "puncta" produced by injection of GFP fusion constructs of these proteins and their potential for dorsal translocation (Miller et al., 1999; Weaver et al., 2003).

However, the localization of the endogenous proteins is not known. Furthermore, maternal Dvl2/3 depletion results in dorsalized embryos, opposite to expectation (Tadjuidje et al., 2011), and Frat1, although required for axis formation in *Xenopus* (Yost et al., 1998), may not actually function generally in Wnt/β-Catenin regulation (Amerongen et al., 2010).

Recent work on a spontaneous ventralized zebrafish maternal-effect mutation *huluwa* (*hwa*; Yan et al., 2018) has shown that the novel Hwa protein may act to promote the dorsal degradation of Axin1, a key negative regulator of β-Catenin. Hwa is encoded by a localized maternal mRNA in both fish and frogs and is a novel membrane protein that accumulates dorsally in the blastula (Yan et al., 2018). Dorsal down-regulation of Axin1 had been inferred from experiments showing the reduced ability of dorsally injected *axin1* mRNA to rescue maternal *axin1* depletion compared to ventrally injected *axin1* (Kofron et al., 2001).

3.4.6. Secreted Ligand Activation of β-Catenin

In contrast to the idea of strict intracellular activation of dorsal Wnt/β-Catenin by cytoplasmic determinants outlined previously, current models suggest that maternal Wnt11b ligand, encoded by a vegetally localized mRNA, is the main determinant for axis specification in *Xenopus*. The observation that overexpressed extracellular Wnt antagonists fail to inhibit endogenous Wnt/β-Catenin activity or axis determination tends to support the intracellular model (Hoppler et al., 1996; Leyns et al., 1997; Wang et al., 1997; Yan et al., 2018). However, antisense depletion experiments also demonstrate that inhibition of maternal *wnt11b*, encoding a secreted ligand, does result in ventralization and in reduced dorsal β-Catenin activity (Tao et al., 2005).

Although Wnt11b is commonly classified as a "non-canonical" Wnt ligand (non-β-Catenin-activating), there is a body of evidence pointing toward receptor/co-receptor context-dependent regulation of β-Catenin-dependent and -independent pathways by Wnt11 proteins and other Wnts (e.g. He et al., 1995). The timing of Wnt11b action is not known, but it must act prior to the 16-cell stage (see previously). The extent and mechanisms by which Wnt11b activity would become enriched dorsally by cortical rotation is also unclear, and recent transcriptomic data from single cleavage-stage blastomeres (*X. laevis*: Flachsova et al., 2013; *X. tropicalis*: Collart et al., 2014; Domenico et al., 2015) suggest little to no dorsal enrichment of *wnt11b* mRNA or *wnt11b* polyadenylation.

Maternal mRNA depletion studies have indirectly implicated translational regulation, potentially targeting *wnt11b*, in dorsal signaling. The RNA-binding protein Bicc1 is encoded by a localized maternal mRNA in *Xenopus* (Wessely and Robertis, 2000) and can bind directly to the 3'UTRs of *dand5*, *tdgf1.3*, and *wnt11b* and repress their translation in reporter assays (Park et al., 2016; Zhang et al., 2013). Depletion of maternal *bicc1* leads to dorso-anteriorized embryos, suggesting that translational repression, mediated by RNA-binding KH domains (Dowdle et al., 2019; Park et al., 2016), is necessary to restrict the activities of one or more of these or other molecules. Bicc1 was first identified in a screen for mRNAs with reduced polyadenylation in the embryos ventralized by UV-irradiation (Wessely and Robertis, 2000). One implication of this result is that *bicc1* polyadenylation, and hence Bicc1 protein, might be higher dorsally. However, this situation would predict that Bicc1 should normally inhibit *wnt11b* translation dorsally. In the absence of Bicc1, overactive Wnt11b signals might be involved in dorsalizing the embryo, but it remains unclear whether this mechanism controls endogenous dorsal β-Catenin accumulation.

One proposed synthesis of these ideas is that Wnt signaling may be potentiated by the endocytosis of activated Wnt receptor-coreceptor complexes ("Lrp6 signalosomes") in association with Dvl (Bilic et al., 2007). Cortical rotation might thus enrich Wnt signalosomes on the dorsal side (Dobrowolski and Robertis, 2012), activating dorsal signaling through the perpetuation of an earlier signaling event or by possibly sequestering β-Catenin degradation machinery (e.g. Gsk3b) in multi-vesicular bodies (Taelman et al., 2010). However, recent data from cell-line experiments suggest that endocytosis of Wnt receptor-co-receptor complexes is not required for subsequent signal transduction (Rim et al., 2020), and the extent that signalosomes differentially accumulate dorsally remains unknown.

While genetic studies in mice have shown that β-Catenin signaling is required for anterior visceral endoderm (AVE) formation (the key step in mammalian axial patterning), it has become clear that secreted Wnt activity is dispensable in this regard (reviewed in Houston, 2017). Similar to the case in *Xenopus*, Tdgf1-mediated signals are required upstream of β-Catenin stabilization in the AVE (Morkel et al., 2003), but the extent to which this similarity represents a coincidental convergence of signals or a deeper level of conservation in mechanisms for establishing bilateral symmetry remains unclear.

3.5. MATERNAL CONTROL OF PRIMORDIAL GERM CELL FORMATION

The idea of a separate germline and its specification by localized maternal determinants arose from classic studies in insects, following the observation of a conspicuous budding of primordial germ cells from the posterior pole of the early blastoderm (classical literature reviewed in Hegner, 1914). Additional evidence for what came to be called "germ plasm" was subsequently revealed in many other organisms (Beams and Kessel, 1974; Eddy, 1975; Extavour and Akam, 2003).

3.5.1. The Germ Plasm

The identification of germ plasm in a vertebrate embryo (in *Rana* frogs, Bounoure, 1934, 1931) suggested that cytoplasmic inheritance of germline fate might represent a general mechanism. Other evidence from salamanders, mammals, and other organisms however suggested that primordial germ cells (PGCs) could also be induced by signals acting on pluripotent cells, which may be the ancestral mode (Extavour and Akam, 2003; Houston and King, 2000a). Regardless of whether PGCs are specified by maternal germ plasm, a material similar to germ plasm has since been identified in the differentiating adult germ cells of all animals (typically termed *nuage*, Fr. "cloud"; André and Rouiller, 1956). Thus, germ plasm/*nuage* is likely a fundamental feature of the germline (Eddy, 1975, 1974). Recent experiments on the reprogramming of somatic cells to PGCs have shown that *nuage* does indeed form in reprogrammed cells (Bucay et al., 2009), further suggesting that *nuage* is tightly connected to, and likely a product of, germ-cell differentiation.

Both descriptive and experimental studies in *Rana/Lithobates* and in *Xenopus* eggs established that the vegetally localized germ plasm is essential for germline formation in Anuran amphibians (Houston and King, 2000a). The classical embryological evidence showed that UV-irradiation of the vegetal pole severely reduced PGC formation (e.g. Bounoure et al., 1954; Züst and Dixon, 1975). Unlike the case in *Drosophila* (Ephrussi and Lehmann, 1992; Illmensee and Mahowald, 1974), initial cytoplasmic transfer experiments in *Xenopus* failed to show a similar strong determinative role for germ plasm (Wakahara, 1977; Wylie et al., 1985). Recent experiments, however, indicate that germ plasm transplantation into animal pole cells can create functional PGCs in genetically marked *Xenopus* (Tada et al., 2012), although discrepancies with the previous studies remain unresolved.

3.5.2. Germ Plasm mRNAs in PGC Specification

The first germ-plasm mRNA to be studied functionally in *Xenopus* was *dazl*, identified contemporaneously with *vegt* (Houston et al., 1998). The Dazl family (Deleted in AZoospermia-like) comprises a conserved group of RRM domain RNA-binding proteins involved in many different aspects of germ cell development in animals and in human fertility (Fu et al., 2015; Kee et al., 2009). In *Xenopus*, *dazl* exhibits a mitochondrial cloud-dependent mRNA localization pattern similar to *nanos1* and remains detectable in early PGCs through the tailbud stages (Houston et al., 1998; Sekizaki et al., 2004). Maternal mRNA depletion studies identified a role for Dazl in PGC development in *Xenopus*, with *dazl*-depleted embryos exhibiting PGC migration defects and remaining abnormally clustered within the posterior endoderm (Houston and King, 2000b). Dazl is thus thought to control competence for migration in PGCs either directly or indirectly through an earlier step in establishing PGC specification. Work in other systems has suggested that Dazl likely functions in polyadenylation (Haston et al., 2009; Smorag et al., 2014).

A number of additional genes localized to the germ plasm have been identified (Table 3.1). Several of these genes have roles in PGC migration as assessed by Morpholino knockdown or antisense mRNA depletion. Notably, depletion of mRNAs encoding RNA-binding proteins *dnd1*, *ddx25*, *nanos1*, and *dazl* have similar effects (Horvay et al., 2006; Houston and King, 2000b; Lai et al., 2012; Yamaguchi, 2013). The nature of these genes and mechanistic studies of their effects on PGCs strongly suggest diverse roles in RNA metabolism, with germ plasm components interacting with and regulating each other to ultimately control PGC fate and/or migration. As an example, Dnd1 has been implicated in the protection of target mRNAs (including itself) from microRNAs (Kedde et al., 2007), in regulating Nanos1 translation (Aguero et al., 2018, 2017), and in the anchoring of localized *trim36* mRNA to the vegetal cortex (Mei et al., 2013). Recent experiments have also suggested that additional pathways are involved in restricting the action of Dnd1 (and possibly other germ plasm RNAs), including the important role of the ubiquitin-independent proteasome pathway, which is animally localized (Hwang et al., 2019). Other germ plasm localized mRNAs, including *sox7* and *efnb1*, also have roles in PGC migration (Owens et al., 2017; Butler et al., 2018).

In addition to controlling overall PGC specification, a number of mRNAs localized to the germ plasm appear to also act in morphogenesis of the germ plasm itself. *germes* encodes a *Xenopus*-specific leucine zipper/EF-hand protein that interacts with dynein light chain (Berekelya et al., 2003, 2007) and may control germ plasm morphology. Similarly, *syntabulin* (*sybu*), encoding a motor adaptor protein implicated in mitochondrial transport in neurons and axis formation, is involved in aggregation of germ plasm and in perinuclear germ plasm relocalization in *Xenopus*, resulting in PGC deficiency (Colozza and Robertis, 2014; Oh and Houston, 2017b). Similarly, Grip2 is another vesicle transport-related protein encoded by a germ plasm mRNA that is involved in PGC development, although its role in germ plasm morphogenesis has not been examined (Kaneshiro et al., 2007; Kirilenko et al., 2008; Tarbashevich et al., 2007).

3.5.3. Assembly of Maternal Germ Plasm

The assembly of germ plasm in vertebrates is not well understood. In zebrafish, *buckyball* is necessary and sufficient for germ plasm and mitochondrial cloud assembly (Bontems et al., 2009). The homologous *Xenopus* gene, *velo1*, encodes a vegetally localized mRNA that forms a major protein component of the mitochondrial cloud in fish and frog (Boke et al., 2016; Claussen and Pieler, 2004; Heim et al., 2014; Nijjar and Woodland, 2013a, 2013b). Recent data suggest that the structural role of Velo1 in the mitochondrial cloud is linked to its ability to form amyloid fibrils via an N-terminal prion-like domain (PLD; Boke et al., 2016).

Velo1 is also a highly disordered protein and contains a domain at the C-terminus involved in non-specific RNA binding. Disordered low-complexity protein domains are found in numerous RNA-binding proteins, typically mediating the formation of liquid hydrogel droplets (Kato and McKnight, 2016). Velo1 protein-protein interactions with Dazl and Rbpms2 (and interactions among the cognate RNAs and proteins) are also required for zebrafish mitochondrial cloud formation (Heim et al., 2014). Additional evidence from fish suggests that this self-assembling process is triggered by the formation of the "chromosome bouquet," a polarized cluster of telomeres (Elkouby et al., 2016). Genetically, this event is upstream of mitochondrial cloud formation (Elkouby et al., 2016; Escobar-Aguirre et al., 2017). Experiments in *Xenopus* show that Velo1 transitions from a mitochondrial cloud/amyloid state to a non-amyloid hydrogel-like state during formation of the germ plasm domain of the cloud, possibly based on phosphorylation of Velo1 (Boke et al., 2016; Nijjar and Woodland, 2013b). This more mobile state may be a prerequisite for germ plasm morphogenesis during development, in contrast to the more stable "architectural" form present in the oocyte.

3.6. CONCLUDING REMARKS

The general picture of early *Xenopus* development is well appreciated but is now being investigated with increasing sophistication at the cellular, molecular, and genetic levels. Thus far, a relatively small number of maternal mRNAs, both non-localized and localized to different extents, have been found to exert strong effects on development. But there clearly is greater underlying complexity than is immediately evident. These mRNAs have predominant roles in primordial germ cell specification, axis formation, and germ layer induction and patterning (Figure 3.2).

A broad view of early *Xenopus* development suggests that non-localized (or animally enriched) maternal RNAs would initiate general ectodermal identity throughout the embryo (in addition to housekeeping functions), within which is embedded the competence to respond to various inducers to alter germ layer identity. Maternal transcription factors Pou5f3.3 and Sox3 may function in this regard as an adjunct to their potential role in contributing to the MZT, but the distinction (or lack thereof) between the two activities is unclear. Also, maternal Foxh1 may act to mark some genes for later activation/repression by Nodal/Smad2 or other signals, but other roles in ectoderm differentiation are unknown.

mRNAs localized widely throughout the vegetal hemisphere, such as *vegt* and *otx1*, are thought to establish endodermal identity in vegetal cells and indirectly activate the expression of mesendoderm-inducing Nodal proteins. Maternally expressed signaling ligands such as Gdf1 may "prime" this mesendoderm induction or, more likely, control the range, activity, or specificity of Nodal/Nodal-related proteins. Other localized mRNAs of the more "intermediate" pattern may be dedicated in some way to dorsal-ventral patterning, with *trim36* and *dnd1* functioning to stimulate microtubule polymerization in vegetal subcortical cytoplasm for cortical rotation, controlling the distribution and activity of maternal β-Catenin, possibly through Wnt11b. Dorsal β-Catenin would then interact with ubiquitous maternal TCFs to de-repress/activate Nodal, initiating its spatiotemporal regulation, and to regulate genes important for organizer function. Last, mRNAs restricted to the germ plasm generate proteins that generally inhibit somatic fate and specify cellular properties, such as global transcriptional activity and cell migration, of a small subset of cells destined to become the PGCs.

The use of the *Xenopus* embryo has been extraordinarily successful in characterizing genes involved in early vertebrate development, and part of this success of *Xenopus* has been the ease of "tinkering" with development—trying out various hypotheses and rejecting erroneous ideas quickly. It is unlikely that many of the discoveries described in this chapter would have been made as readily without the ability to test the functions of genes at the biochemical and cellular levels almost at a whim. There are several areas in which studying the maternal control of early development in *Xenopus* holds good potential for further uncovering new knowledge regarding embryonic development.

First, although there are many efforts to map gene regulatory network models for different aspects of cell differentiation, in *Xenopus*, this analysis can be easily extended to include the contributions of maternally provided genes. In *Xenopus*, it should be possible to construct a complete description of development, beginning with contributions of maternally localized mRNAs. Also, although there has been tremendous progress in modeling embryonic development using stem cells, many of these approaches use growth factor stimulation of cells to initiate self-organizing processes. These approaches were pioneered by the late *Xenopus* investigator Yoshiki Sasai, focusing on brain and retinal development (Eiraku et al., 2011, 2008). Methods have recently been developed to approximate post-implantation-like states in mouse embryos and mouse and human pluripotent cell aggregates (Zhu and Zernicka-Goetz, 2020). These experiments are possible because of the robust ability of differentiating cells to undergo self-organization, a well-noted but not well-understood phenomenon. Developmental biologists familiar with the work of Spemann and the Holtfreters should not be surprised that self-organization seems to be the norm. Because studies in early *Xenopus* embryos do not require artificial stimulation or culture conditions, Wnt/β-Catenin activation and other signaling mechanisms can thus be investigated under endogenous activation conditions.

Second, the realization that many types of ribonucleoprotein granules are formed through liquid-like phase transitions to a hydrogel state has allowed the formulation of a "solid-state" model of information transfer via spatially distinct iterations of these states along the nucleocytoplasmic transport route (Kato and McKnight, 2016). These studies are still in the early stages, and we still do not know the full extent to which mRNA localization depends on hydrogel states (Neil et al., 2021). This intriguing possibility has recently been suggested by experiments showing that intermediate filament head domains also form hydrogels that can coaggregate with low-complexity domain-containing RNA binding proteins (Zhou et al., 2021). Such an observation could also help explain the structural roles of *vegt* and other localized mRNAs in addition to germ plasm morphogenesis. These questions could all be profitably addressed using *Xenopus* oocytes.

Last, embryonic development is being described with ever-increasing biological detail at multiple levels of analysis. The abundant protein and nucleic acids in each embryo and the ability to easily monitor cellular and intracellular behaviors make *Xenopus* an ideal organism in which to integrate many of these levels into a complex multi-scale model of development. Given the current trend towards the development of sophisticated statistical and machine-learning computational tools to explore and work with vast amounts of data, a "systems biology" approach to rapidly enabling accurate modeling of complex spatio-temporal interactions of biological molecules, genes, and cell behavior into gene regulatory networks is ideally suited to *Xenopus* embryos.

With its rich history and abundant experimental advantages across many scales of organization, the early *Xenopus* embryo holds promise for rigorously providing a united view of cellular (and subcellular) morphogenesis and the molecular structure of genetic information, with the cell as a (proper) frame of reference. Such an effort may well occupy the next 200 years.

ACKNOWLEDGMENTS

The author would like to thank Maurine Neiman for critical reading and editing of the manuscript, Malgorzata Kloc for discussion on structural RNAs, and the University of Iowa for support.

REFERENCES

Abu-Remaileh, M., Gerson, A., Farago, M., Nathan, G., Alkalay, I., Rousso, S.Z., Gur, M., Fainsod, A., Bergman, Y., 2010. Oct-3/4 regulates stem cell identity and cell fate decisions by modulating Wnt/β-catenin signalling. *EMBO J* 29, 3236–3248.

Afouda, B.A., Nakamura, Y., Shaw, S., Charney, R.M., Paraiso, K.D., Blitz, I.L., Cho, K.W.Y., Hoppler, S., 2020. Foxh1/Nodal defines context-specific direct maternal Wnt/β-catenin target gene regulation in early development. *Iscience* 23, 101314.

Agius, E., Wessely, O., Kemp, C., Robertis, E.M.D., 2000. Endodermal nodal-related signals and mesoderm induction in Xenopus. *Development* 127, 1173–1183.

Aguero, T., Jin, Z., Chorghade, S., Kalsotra, A., King, M.L., Yang, J., 2017. Maternal dead-end 1 promotes translation of nanos1 through binding the eIF3 complex. *Development* 144, dev.152611.

Aguero, T., Jin, Z., Owens, D., Malhotra, A., Newman, K., Yang, J., King, M.L., 2018. Combined functions of two RRMs in Dead-end1 mimic helicase activity to promote nanos1 translation in the germline. *Mol Reprod Dev* 85, 896–908.

Amerongen, R.V., Nawijn, M.C., Lambooij, J.-P., Proost, N., Jonkers, J., Berns, A., 2010. Frat oncoproteins act at the crossroad of canonical and noncanonical Wnt-signaling pathways. *Oncogene* 29, 93–104.

Ancel, P., Vintemberger, P., 1948. Recherches sur le déterminisme de la symétrie bilatérale dans l'œuf des amphibiens. *Bull. Biol. Fr. Belg. Suppl* 31, 1–182.

André, J., Rouiller, C., 1956. L'ultrastructure de la membrane nucléaire des ovocytes del l'araignée (Tegenaria domestica Clark). In: *Proc. European Conf. Electron Microscopy*. New York: Stokholm Acad. Press. pp. 162–164.

Aslan, Y., Tadjuidje, E., Zorn, A.M., Cha, S.-W., 2017. High-efficiency non-mosaic CRISPR-mediated knock-in and indel mutation in F0 Xenopus. *Development* 144, 2852–2858.

Baer, K. von, 1834. Die Metamorphose des Eies der Batrachier vor der Erscheinung des Embryo und Folgerungen aus ihr für die Theorie der Erzeugung. *Archiv für Anatomie, Physiologie und wissenschaftliche Medizin* 481–509.

Bates, T.J.D., Vonica, A., Heasman, J., Brivanlou, A.H., Bell, E., 2013. Coco regulates dorsoventral specification of germ layers via inhibition of TGFβ signalling. *Development* 140, 4177–4181.

Beams, H.W., Kessel, R.G., 1974. The problem of germ cell determinants. *Int Rev Cytol* 39, 413–479.

Belenkaya, T.Y., Han, C., Standley, H.J., Lin, Xinda, Houston, D.W., Heasman, J., Lin, Xinhua, 2002. Pygopus encodes a nuclear protein essential for wingless/Wnt signaling. *Development* 129, 4089–4101.

Berekelya, L.A., Mikryukov, A.A., Luchinskaya, N.N., Ponomarev, M.B., Woodland, H.R., Belyavsky, A.V., 2007. The protein encoded by the germ plasm RNA germes associates with dynein light chains and functions in Xenopus germline development. *Differentiation* 75, 546–558.

Berekelya, L.A., Ponomarev, M., Luchinskaya, N., Belyavsky, A., 2003. Xenopus germes encodes a novel germ plasm-associated transcript. *Gene Expression Patterns: GEP* 3, 521–524.

Betley, J.N., Frith, M.C., Graber, J.H., Choo, S., Deshler, J.O., 2002. A ubiquitous and conserved signal for RNA localization in chordates. *Current Biology* 12, 1756–1761.

Bilic, J., Huang, Y.-L., Davidson, G., Zimmermann, T., Cruciat, C.-M., Bienz, M., Niehrs, C., 2007. Wnt induces LRP6 signalosomes and promotes dishevelled-dependent LRP6 phosphorylation. *Science* 316, 1619–1622.

Birsoy, B., Berg, L., Williams, P.H., Smith, J.C., Wylie, C.C., Christian, J.L., Heasman, J., 2005. XPACE4 is a localized pro-protein convertase required for mesoderm induction and the cleavage of specific TGFβ proteins in Xenopus development. *Development* 132, 591–602.

Birsoy, B., Kofron, M., Schaible, K., Wylie, C.C., Heasman, J., 2006. Vg 1 is an essential signaling molecule in Xenopus development. *Development* 133, 15–20.

Blythe, S.A., Cha, S.-W., Tadjuidje, E., Heasman, J., Klein, P.S., 2010. Beta-catenin primes organizer gene expression by recruiting a histone H3 arginine 8 methyltransferase, Prmt2. *Developmental Cell* 19, 220–231.

Boke, E., Ruer, M., Wühr, M., Coughlin, M., Lemaitre, R., Gygi, S.P., Alberti, S., Drechsel, D., Hyman, A.A., Mitchison, T.J., 2016. Amyloid-like self-assembly of a cellular compartment. *Cell* 166, 637–650.

Bontems, F., Stein, A., Marlow, F., Lyautey, J., Gupta, T., Mullins, M.C., Dosch, R., 2009. Bucky ball organizes germ plasm assembly in zebrafish. *Current Biology* 19, 414–422.

Boterenbrood, E.C., Nieuwkoop, P.D., 1973. The formation of the mesoderm in urodelean amphibians. *Wilhelm Roux' Archiv für Entwicklungsmechanik der Organismen* 173, 319–332.

Bounoure, L., 1934. Recherches sur la lignée germinale chez la grenouille rousse aux premiers stades du développment. *Ann Sci Nat* 17, 67–248.

Bounoure, L., 1931. Sur l'existence d'un déterminant germinal dans l'oeuf indivis de la Grenouille rousse. *CR Acad Sci Paris* 193, 402.

Bounoure, L., Aubry, R., Huck, M.L., 1954. Nouvelles recherches experimentales sur les origines de la lignee reproductrice chez la Grenouille rousse. *The Journal of Embryology and Experimental Morphology* 2, 245–263.

Brenner, S., Jacob, F., Meselson, M., 1961. An unstable intermediate carrying information from genes to ribosomes for protein synthesis. *Nature* 190, 576–581.

Brun, R.B., 1975. Oocyte maturation in vitro: Contribution of the oviduct to total maturation in *Xenopus laevis*. *Experientia* 31, 1275–1276.

Bucay, N., Yebra, M., Cirulli, V., Afrikanova, I., Kaido, T., Hayek, A., Montgomery, A.M.P., 2009. A novel approach for the derivation of putative primordial germ cells and sertoli cells from human embryonic stem cells. *Stem Cells* 27, 68–77.

Butler, A.M., Owens, D.A., Wang, L., King, M.L., 2018. A novel role for sox7 in Xenopus early primordial germ cell development: Mining the PGC transcriptome. *Development* 145, dev.155978.

Cabral, S.E., Mowry, K.L., 2020. Chapter four organizing the oocyte: RNA localization meets phase separation. *Curr Top Dev Biol* 140, 87–118.

Cao, Y., Siegel, D., Knöchel, W., 2006. Xenopus POU factors of subclass V inhibit activin/nodal signaling during gastrulation. *Mechanisms of Development* 123, 614–625.

Capco, D.G., Jeffery, W.R., 1981. Regional accumulation of vegetal pole poly(A)+ RNA injected into fertilized Xenopus eggs. *Nature* 294, 255–257.

Capco, D.G., Jeffery, W.R., 1979. Origin and spatial distribution of maternal messenger RNA during oogenesis of an insect, *Oncopeltus fasciatus*. *J Cell Sci* 39, 63–76.

Carpenter, C.D., Klein, W.H., 1982. A gradient of poly(A)+ RNA sequences in *Xenopus laevis* eggs and embryos. *Dev Biol* 91, 43–49.

Casey, E., Tada, M., Fairclough, L., Wylie, C.C., Heasman, J., Smith, J., 1999. Bix4 is activated directly by VegT and mediates endoderm formation in Xenopus development. *Development* 126, 4193–4200.

Cazenave, C., Chevrier, M., Nguyen, T.T., Hélène, C., 1987. Rate of degradation of [alpha]- and [beta]-oligodeoxynucleotides in Xenopus oocytes: Implications for anti-messenger strategies. *Nucleic Acids Research* 15, 10507–10521.

Cha, S.-W., McAdams, M., Kormish, J., Wylie, C., Kofron, M., 2012. Foxi2 is an animally localized maternal mRNA in Xenopus, and an activator of the zygotic ectoderm activator foxi1e. *PLoS One* 7, e41782.

Chan, A.P., Kloc, M., Etkin, L.D., 1999. Fatvg encodes a new localized RNA that uses a 25-nucleotide element (FVLE1) to localize to the vegetal cortex of Xenopus oocytes. *Development* 126, 4943–4953.

Chan, A.P., Kloc, M., Larabell, C.A., LeGros, M., Etkin, L.D., 2007. The maternally localized RNA fatvg is required for cortical rotation and germ cell formation. *Mechanisms of Development* 124, 350–363.

Chang, P., Torres, J., Lewis, R.A., Mowry, K.L., Houliston, E., King, M.L., 2004. Localization of RNAs to the mitochondrial cloud in Xenopus oocytes through entrapment and association with endoplasmic reticulum. *Molecular Biology of the Cell* 15, 4669–4681.

Charney, R.M., Forouzmand, E., Cho, J.S., Cheung, J., Paraiso, K.D., Yasuoka, Y., Takahashi, S., Taira, M., Blitz, I.L., Xie, X., Cho, K.W.Y., 2017. Foxh1 occupies cis-regulatory modules prior to dynamic transcription factor interactions controlling the mesendoderm gene program. *Dev Cell* 40, 595–607.e4.

Chiu, W.T., Le, R.C., Blitz, I.L., Fish, M.B., Li, Y., Biesinger, J., Xie, X., Cho, K.W.Y., 2014. Genome-wide view of TGFβ/Foxh1 regulation of the early mesendoderm program. *Development* 141, 4537–4547.

Claussen, M., Horvay, K., Pieler, T., 2004. Evidence for overlapping, but not identical, protein machineries operating in vegetal RNA localization along early and late pathways in Xenopus oocytes. *Development* 131, 4263–4273.

Claussen, M., Lingner, T., Pommerenke, C., Opitz, L., Salinas, G., Pieler, T., 2015. Global analysis of asymmetric RNA enrichment in oocytes reveals low conservation between closely related Xenopus species. *Molecular Biology of the Cell* 26, 3777–3787.

Claussen, M., Pieler, T., 2004. Xvelo1 uses a novel 75-nucleotide signal sequence that drives vegetal localization along the late pathway in Xenopus oocytes. *Developmental Biology* 266, 270–284.

Clements, D., Friday, R.V., Woodland, H.R., 1999. Mode of action of VegT in mesoderm and endoderm formation. *Dev Camb Engl* 126, 4903–4911.

Clements, D., Woodland, H.R., 2003. VegT induces endoderm by a self-limiting mechanism and by changing the competence of cells to respond to TGF-β signals. *Dev Biol* 258, 454–463.

Collart, C., Owens, N.D.L., Bhaw-Rosun, L., Cooper, B., Domenico, E.D., Patrushev, I., Sesay, A.K., Smith, J.N., Smith, J.C., Gilchrist, M.J., 2014. High-resolution analysis of gene activity during the Xenopus mid-blastula transition. *Development* 141, 1927–1939.

Colozza, G., Robertis, E.M.D., 2014. Maternal syntabulin is required for dorsal axis formation and is a germ plasm component in Xenopus. *Differentiation* 88, 17–26.

Constam, D.B., 2009. Riding shotgun: A dual role for the epidermal growth factor-cripto/FRL-1/cryptic protein cripto in nodal trafficking. *Traffic* 10, 783–791.

Cuykendall, T. N. & Houston, D. W. 2010. Identification of germ plasm-associated transcripts by microarray analysis of Xenopus vegetal cortex RNA. *Developmental dynamics* 239, 1838–1848.

Cuykendall, T.N., Houston, D.W., 2009. Vegetally localized Xenopus trim36 regulates cortical rotation and dorsal axis formation. *Development* 136, 3057–3065.

Czołowska, R., 1972. The fine structure of the "germinal cytoplasm" in the egg of *Xenopus laevis*. *Development Genes and Evolution* 169, 335–344.

Czołowska, R., 1969. Observations on the origin of the germinal cytoplasm in *Xenopus laevis*. *Development* 22, 229–251.

Dale, L., Slack, J., 1987. Regional specification within the mesoderm of early embryos of *Xenopus laevis*. *Development* 100, 279–295.

Dale, L., Smith, J.C., Slack, J.M., 1985. Mesoderm induction in *Xenopus laevis*: A quantitative study using a cell lineage label and tissue-specific antibodies. *Journal of Embryology and Experimental Morphology* 89, 289–312.

Darnell, J.E., Wall, R., Tushinski, R.J., 1971. An adenylic acid-rich sequence in messenger RNA of HeLa cells and its possible relationship to reiterated sites in DNA. *Proc National Acad Sci* 68, 1321–1325.

Dash, P., Lotan, I., Knapp, M., Kandel, E.R., Goelet, P., 1987. Selective elimination of mRNAs in vivo: Complementary oligodeoxynucleotides promote RNA degradation by an RNase H-like activity. *Proceedings of the National Academy of Sciences of the United States of America* 84, 7896–7900.

Davidson, E.H., 1986. *Gene Activity in Early Development* (Third Edition). San Diego: Academic Press.

Davidson, E.H., Britten, R.J., 1971. Note on the control of gene expression during development. *J Theor Biol* 32, 123–130.

Davidson, E.H., Crippa, M., Kramer, F.R., Mirsky, A.E., 1966. Genomic function during the lampbrush chromosome stage of amphibian oögenesis. *Proc National Acad Sci* 56, 856–863.

De Domenico, E., Owens, N.D., Grant, I.M., Gomes-Faria, R., Gilchrist, M.J. Molecular asymmetry in the 8-cell stage Xenopus tropicalis embryo described by single blastomere transcript sequencing. *Dev Biol*. 2015 Dec 15;408(2):252–68.

Dobrowolski, R., Robertis, E.M.D., 2012. Endocytic control of growth factor signalling: Multivesicular bodies as signalling organelles. *Nature Reviews Molecular Cell Biology* 13, 53–60.

Dowdle, M.E., Park, S., Imboden, S.B., Fox, C.A., Houston, D.W., Sheets, M.D., 2019. A single KH domain in Bicaudal-C links mRNA binding and translational repression functions to maternal development. *Development* 146, dev172486.

Dupont, S., Zacchigna, L., Cordenonsi, M., Soligo, S., Adorno, M., Rugge, M., Piccolo, S., 2005. Germ-layer specification and control of cell growth by Ectodermin, a Smad4 ubiquitin ligase. *Cell* 121, 87–99.

Eddy, E.M., 1975. Germ plasm and the differentiation of the germ cell line. *Int Rev Cytol* 43, 229–280.

Eddy, E.M., 1974. Fine structural observations on the form and distribution of nuage in germ cells of the rat. *The Anatomical Record* 178, 731–757.

Edmonds, M., Vaughan, M.H., Nakazato, H., 1971. Polyadenylic acid sequences in the heterogeneous nuclear RNA and rapidly-labeled polyribosomal RNA of HeLa cells: Possible evidence for a precursor relationship. *Proc National Acad Sci* 68, 1336–1340.

Eiraku, M., Takata, N., Ishibashi, H., Kawada, M., Sakakura, E., Okuda, S., Sekiguchi, K., Adachi, T., Sasai, Y., 2011. Self-organizing optic-cup morphogenesis in three-dimensional culture. *Nature* 472, 51–56.

Eiraku, M., Watanabe, K., Matsuo-Takasaki, M., Kawada, M., Yonemura, S., Matsumura, M., Wataya, T., Nishiyama, A., Muguruma, K., Sasai, Y., 2008. Self-organized formation of polarized cortical tissues from ESCs and its active manipulation by extrinsic signals. *Cell Stem Cell* 3, 519–532.

Elinson, R.P., 1985. Changes in levels of polymeric tubulin associated with activation and dorsoventral polarization of the frog egg. *Developmental Biology* 109, 224–233.

Elinson, R.P., King, M.L., Forristall, C., 1993. Isolated vegetal cortex from Xenopus oocytes selectively retains localized mRNAs. *Developmental Biology* 160, 554–562.

Elinson, R.P., Pasceri, P., 1989. Two UV-sensitive targets in dorsoanterior specification of frog embryos. *Development* 106, 511–518.

Elkouby, Y.M., Jamieson-Lucy, A., Mullins, M.C., 2016. Oocyte polarization is coupled to the chromosomal bouquet, a conserved polarized nuclear configuration in meiosis. *PLoS Biology* 14, e1002335.

Ephrussi, A., Lehmann, R., 1992. Induction of germ cell formation by oskar. *Nature* 358, 387–392.

Escobar-Aguirre, M., Elkouby, Y.M., Mullins, M.C., 2017. Localization in oogenesis of maternal regulators of embryonic development. *Advances in Experimental Medicine and Biology* 953, 173–207.

Extavour, C.G., Akam, M., 2003. Mechanisms of germ cell specification across the metazoans: Epigenesis and preformation. *Development* 130, 5869–5884.

Flachsova, M., Sindelka, R., Kubista, M., 2013. Single blastomere expression profiling of *Xenopus laevis* embryos of 8 to 32-cells reveals developmental asymmetry. *Scientific Reports* 3.

Forristall, C., Pondel, M., Chen, L., King, M., 1995. Patterns of localization and cytoskeletal association of two vegetally localized RNAs, Vg1 and Xcat-2. *Development* 121, 201–208.

Fu, X.-F., Cheng, S.-F., Wang, L.-Q., Yin, S., Felici, M.D., Shen, W., 2015. DAZ family proteins, key players for germ cell development. *International Journal of Biological Sciences* 11, 1226–1235.

Fuerer, C., Nostro, M.C. Constam, D.B., 2014. Nodal·Gdf1 heterodimers with bound prodomains enable serum-independent nodal signaling and endoderm differentiation. *J Biol Chem* 289, 17854–17871.

Gammons, M., Bienz, M., 2018. Multiprotein complexes governing Wnt signal transduction. *Curr Opin Cell Biol* 51, 42–49.

Gao, L., Zhu, X., Chen, G., Ma, X., Zhang, Y., Khand, A.A., Shi, H., Gu, F., Lin, H., Chen, Y., Zhang, H., He, L., Tao, Q., 2016. A novel role for Ascl1 in the regulation of mesendoderm formation via HDAC-dependent antagonism of VegT. *Development* 143, 492–503.

Gentsch, G.E., Spruce, T., Owens, N.D.L., Smith, J.C., 2019. Maternal pluripotency factors initiate extensive chromatin remodelling to predefine first response to inductive signals. *Nat Commun* 10, 4269.

Gerhart, J., Danilchik, M., Doniach, T., Roberts, S., Rowning, B., Stewart, R., 1989. Cortical rotation of the Xenopus egg: Consequences for the anteroposterior pattern of embryonic dorsal development. *Development* 107 Suppl, 37–51.

Grant, P.A., Yan, B., Johnson, M.A., Johnson, D.L.E., Moody, S.A., 2014. Novel animal pole-enriched maternal mRNAs are preferentially expressed in neural ectoderm. *Developmental Dynamics* 243, 478–496.

Gros, F., Hiatt, H., Gilbert, W., Kurland, C.G., Risebrough, R.W., Watson, J.D., 1961. Unstable ribonucleic acid revealed by pulse labelling of *Escherichia coli*. *Nature* 190, 581–585.

Gupta, R., Wills, A., Ucar, D., Baker, J., 2014. Developmental enhancers are marked independently of zygotic Nodal signals in Xenopus. *Dev Biol* 395, 38–49.

Gurdon, J.B., Lane, C.D., Woodland, H.R., Marbaix, G., 1971. Use of frog eggs and oocytes for the study of messenger RNA and its translation in living cells. *Nature* 233, 177–182.

Haston, K.M., Tung, J.Y., Pera, R.A.R., 2009. Dazl functions in maintenance of pluripotency and genetic and epigenetic programs of differentiation in mouse primordial germ cells in vivo and in vitro. *PLoS One* 4, e5654.

Hayashi, M., Spiegelman, S., 1961. The selective synthesis of informational RNA in bacteria. *Proc National Acad Sci* 47, 1564–1580.

He, X., Saint-Jeannet, J.F., Woodgett, J.R., Varmus, H.E., Dawid, I.B., 1995. Glycogen synthase kinase-3 and dorsoventral patterning in Xenopus embryos. *Nature* 374, 617–622.

Heasman, J., Crawford, A., Goldstone, K., Garner-Hamrick, P., Gumbiner, B., McCrea, P., Kintner, C., Noro, C.Y., Wylie, C.C., 1994. Overexpression of cadherins and underexpression of beta-catenin inhibit dorsal mesoderm induction in early Xenopus embryos. *Cell* 79, 791–803.

Heasman, J., Holwill, S., Wylie, C.C., 1991. Fertilization of cultured Xenopus oocytes and use in studies of maternally inherited molecules. *Methods Mol. Biol.* 36, 213–230.

Heasman, J., Kofron, M., Wylie, C.C., 2000. Beta-catenin signaling activity dissected in the early Xenopus embryo: A novel antisense approach. *Developmental Biology* 222, 124–134.

Heasman, J., Quarmby, J., Wylie, C.C., 1984. The mitochondrial cloud of Xenopus oocytes: The source of germinal granule material. *Developmental Biology* 105, 458–469.

Heasman, J., Wessely, O., Langland, R., Craig, E.J., Kessler, D.S., 2001. Vegetal localization of maternal mRNAs is disrupted by VegT depletion. *Developmental Biology* 240, 377–386.

Hegner, R.W., 1914. Studies on germ cells. I: The history of the germ cells in insects with special reference to the

Keimbahn-determinants. II: The origin and significance of the . . . *Journal of Morphology* 25, 375–509.

Heim, A.E., Hartung, O., Rothhämel, S., Ferreira, E., Jenny, A., Marlow, F.L., 2014. Oocyte polarity requires a buckyball-dependent feedback amplification loop. *Development* 141, 842–854.

Holowacz, T., Elinson, R.P., 1993. Cortical cytoplasm, which induces dorsal axis formation in Xenopus, is inactivated by UV irradiation of the oocyte. *Development* 119, 277–285.

Holwill, S., Heasman, J., Crawley, C., Wylie, C., 1987. Axis and germ line deficiencies caused by u.v irradiation of Xenopus oocytes cultured in vitro. *Development* 100, 735–743.

Hoppler, S., Brown, J.D., Moon, R.T., 1996. Expression of a dominant-negative Wnt blocks induction of MyoD in Xenopus embryos. *Gene Dev* 10, 2805–2817.

Horb, M.E., Thomsen, G.H., 1997. A vegetally localized T-box transcription factor in Xenopus eggs specifies mesoderm and endoderm and is essential for embryonic mesoderm formation. *Dev Camb Engl* 124, 1689–1698.

Horvay, K., Claussen, M., Katzer, M., Landgrebe, J., Pieler, T., 2006. Xenopus dead end mRNA is a localized maternal determinant that serves a conserved function in germ cell development. *Developmental Biology* 291, 1–11.

Houston, D.W., 2019. Culture and host transfer of Xenopus oocytes for maternal mRNA depletion and genome editing experiments. In: Pelegri, F. (Ed.), *Vertebrate Embryogenesis: Embryological, Cellular, and Genetic Methods*. New York: Humana. pp. 1–16.

Houston, D.W., 2018. Oocyte Host-Transfer and Maternal mRNA Depletion Experiments in Xenopus. *Cold Spring Harbor Protocols* 10, pdb-prot096982.

Houston, D.W., 2017. Vertebrate axial patterning: From egg to asymmetry. In: Pelegri, F., Sutherland, A., Danilchik, M. (Eds.), *Vertebrate Development, Maternal to Zygotic Control, Advances in Experimental Medicine and Biology*. Cham: Springer. pp. 209–306.

Houston, D.W., 2013. Regulation of cell polarity and RNA localization in vertebrate oocytes. *International Review of Cell and Molecular Biology* 306, 127–185.

Houston, D.W., 2012. Cortical rotation and messenger RNA localization in Xenopus axis formation. *Wiley Interdisciplinary Reviews: Developmental Biology* 1, 371–388.

Houston, D.W., King, M.L., 2000a. Germ plasm and molecular determinants of germ cell fate. *Curr Top Dev Biol* 50, 155-IN2.

Houston, D.W., King, M.L., 2000b. A critical role for Xdazl, a germ plasm-localized RNA, in the differentiation of primordial germ cells in Xenopus. *Development* 127, 447–456.

Houston, D.W., Kofron, M., Resnik, E., Langland, R., Destrée, O., Wylie, C.C., Heasman, J., 2002. Repression of organizer genes in dorsal and ventral Xenopus cells mediated by maternal XTcf3. *Development* 129, 4015–4025.

Houston, D.W., Wylie, C., 2003. The Xenopus LIM-homeodomain protein Xlim5 regulates the differential adhesion properties of early ectoderm cells. *Development* 130, 2695–2704.

Houston, D.W., Zhang, J., Maines, J.Z., Wasserman, S.A., King, M.L., 1998. A Xenopus DAZ-like gene encodes an RNA component of germ plasm and is a functional homologue of Drosophila boule. *Development* 125, 171–180.

Hsu, D., Economides, A., Wang, X., Eimon, P., Harland, R., 1998. The Xenopus dorsalizing factor Gremlin identifies a novel family of secreted proteins that antagonize BMP activities. *Molecular Cell* 1, 673–683.

Hudson, C., Woodland, H.R., 1998. Xpat, a gene expressed specifically in germ plasm and primordial germ cells of *Xenopus laevis*. *Mech Develop* 73, 159–168.

Hudson, J.W., Alarcón, V.B., Elinson, R.P., 1996. Identification of new localized RNAs in the Xenopus oocyte by differential display PCR. *Developmental Genetics* 19, 190–198.

Hultin, T., 1961. The effect of puromycin on protein metabolism and cell division in fertilized sea urchin eggs. *Experientia* 17, 410–411.

Hwang, H., Jin, Z., Krishnamurthy, V.V., Saha, A., Klein, P.S., Garcia, B., Mei, W., King, M.L., Zhang, K., Yang, J., 2019. Novel functions of the ubiquitin-independent proteasome system in regulating Xenopus germline development. *Development* 146, dev.172700.

Hyde, C.E., Old, R.W., 2000. Regulation of the early expression of the Xenopus nodal-related 1 gene, Xnr1. *Dev Camb Engl* 127, 1221–1229.

Illmensee, K., Mahowald, A.P., 1974. Transplantation of posterior polar plasm in Drosophila: Induction of germ cells at the anterior pole of the egg. *Proceedings of the National Academy of Sciences of the United States of America* 71, 1016–1020.

Jacob, F., Monod, J., 1961. Genetic regulatory mechanisms in the synthesis of proteins. *J Mol Biol* 3, 318–356.

Jeffery, W.R., Capco, D.G., 1978. Differential accumulation and localization of maternal poly(A)-containing RNA during early development of the ascidian, Styela. *Developmental Biology* 67, 152–166.

Jones, E., Woodland, H., 1987. The development of animal cap cells in Xenopus: A measure of the start of animal cap competence to form mesoderm. *Development* 101, 557–563.

Joseph, E.M., Melton, D.A., 1998. Mutant Vg1 ligands disrupt endoderm and mesoderm formation in Xenopus embryos. *Dev Camb Engl* 125, 2677–2685.

Kageura, H., 1997. Activation of dorsal development by contact between the cortical dorsal determinant and the equatorial core cytoplasm in eggs of *Xenopus laevis*. *Development* 124, 1543–1551.

Kahle, W., 1908. *Die Paedogenesis der Decidomyiden*. Stuttgart: E. Schweizerbart.

Kalthoff, K., 1979. Determinants of spatial organization. *Cytoplasmic Localization Early Dev* 97–126.

Kaneshiro, K., Miyauchi, M., Tanigawa, Y., Ikenishi, K., Komiya, T., 2007. The mRNA coding for Xenopus glutamate receptor interacting protein 2 (XGRIP2) is maternally transcribed, transported through the late pathway and localized to the germ plasm. *Biochemical and Biophysical Research Communications* 355, 902–906.

Kao, K.R., Masui, Y., Elinson, R.P., 1986. Lithium-induced respecification of pattern in *Xenopus laevis* embryos. *Nature* 322, 371–373.

Kataoka, K., Tazaki, A., Kitayama, A., Ueno, N., Watanabe, K., Mochii, M., 2005. Identification of asymmetrically localized transcripts along the animal-vegetal axis of the Xenopus egg. *Develop Growth Differ* 47, 511–521.

Kato, M., McKnight, S.L., 2016. A solid-state conceptualization of information transfer from gene to message to protein. *Annu Rev Biochem* 87, 1–40.

Kedde, M., Strasser, M.J., Boldajipour, B., Vrielink, J.A.F.O., Slanchev, K., Sage, C. le, Nagel, R., Voorhoeve, P.M., Duijse, J. van, Ørom, U.A., Lund, A.H., Perrakis, A., Raz, E., Agami, R., 2007. RNA-binding protein Dnd1 inhibits microRNA access to target mRNA. *Cell* 131, 1273–1286.

Kee, K., Angeles, V.T., Flores, M., Nguyen, H.N., Pera, R.A.R., 2009. Human DAZL, DAZ and BOULE genes modulate primordial germ-cell and haploid gamete formation. *Nature* 462, 222–225.

Kennedy, M.W., Cha, S.-W., Tadjuidje, E., Andrews, P.G., Heasman, J., Kao, K.R., 2010. A co-dependent requirement of xBcl9 and Pygopus for embryonic body axis development in Xenopus. *Developmental Dynamics* 239, 271–283.

Kessler, D.S., Melton, D.A., 1995. Induction of dorsal mesoderm by soluble, mature Vg1 protein. *Dev Camb Engl* 121, 2155–2164.

King, M.L., Barklis, E., 1985. Regional distribution of maternal messenger RNA in the amphibian oocyte. *Developmental Biology* 112, 203–212.

King, M.L., Messitt, T.J., Mowry, K.L., 2005. Putting RNAs in the right place at the right time: RNA localization in the frog oocyte. *Biology of the Cell* 97, 19–33.

Kirilenko, P., Weierud, F.K., Zorn, A.M., Woodland, H.R., 2008. The efficiency of Xenopus primordial germ cell migration depends on the germplasm mRNA encoding the PDZ domain protein Grip2. *Differentiation* 76, 392–403.

Kloc, M., 2009. Teachings from the egg: New and unexpected functions of RNAs. *Mol Reprod Dev* 76, 922–932.

Kloc, M., Chan, A.P., 2007. Centroid, a novel putative DEAD-box RNA helicase maternal mRNA, is localized in the mitochondrial cloud in *Xenopus laevis* oocytes. *The International Journal of Developmental Biology* 51, 701–706.

Kloc, M., Etkin, L.D., 1995. Two distinct pathways for the localization of RNAs at the vegetal cortex in Xenopus oocytes. *Development* 121, 287–297.

Kloc, M., Larabell, C., Chan, A.P., Etkin, L.D., 1998. Contribution of METRO pathway localized molecules to the organization of the germ cell lineage. *Mechanisms of Development* 75, 81–93.

Kloc, M., Larabell, C., Etkin, L.D., 1996. Elaboration of the messenger transport organizer pathway for localization of RNA to the vegetal cortex of Xenopus oocytes. *Developmental Biology* 180, 119–130.

Kloc, M., Spohr, G., Etkin, L.D., 1993. Translocation of repetitive RNA sequences with the germ plasm in Xenopus oocytes. *Science* 262, 1712–1714.

Kofron, M., Klein, P., Zhang, F., Houston, D.W., Schaible, K., Wylie, C.C., Heasman, J., 2001. The role of maternal axin in patterning the Xenopus embryo. *Developmental Biology* 237, 183–201.

Kofron, M., Xanthos, J.B., Sun, B., Sive, H., Osada, S., Wright, C.V.E., Wylie, C.C., Heasman, J., 1999. Mesoderm induction in Xenopus is a zygotic event regulated by maternal VegT via TGFbeta growth factors. *Development* 126, 5759–5770.

Ku, M., Melton, D.A., 1993. Xwnt-11: A maternally expressed Xenopus wnt gene. *Development* 119, 1161–1173.

Lai, F., Singh, A., King, M.L., 2012. Xenopus Nanos1 is required to prevent endoderm gene expression and apoptosis in primordial germ cells. *Development* 139, 1476–1486.

Lee, M.A., Heasman, J., Whitman, M., 2001. Timing of endogenous activin-like signals and regional specification of the Xenopus embryo. *Development* 128, 2939–2952.

Lee, S.Y., Mendecki, J., Brawerman, G., 1971. A polynucleotide segment rich in adenylic acid in the rapidly-labeled polyribosomal RNA component of mouse sarcoma 180 ascites cells. *Proc National Acad Sci* 68, 1331–1335.

Lewis, R.A., Kress, T.L., Cote, C.A., Gautreau, D., Rokop, M.E., Mowry, K.L., 2004. Conserved and clustered RNA recognition sequences are a critical feature of signals directing RNA localization in Xenopus oocytes. *Mechanisms of Development* 121, 101–109.

Leyns, L., Bouwmeester, T., Kim, S.H., Piccolo, S., Robertis, E.M.D., 1997. Frzb-1 is a secreted antagonist of Wnt signaling expressed in the Spemann organizer. *Cell* 88, 747–756.

Livigni, A., Peradziryi, H., Sharov, A.A., Chia, G., Hammachi, F., Migueles, R.P., Sukparangsi, W., Pernagallo, S., Bradley, M., Nichols, J., Ko, M.S.H., Brickman, J.M., 2013. A conserved Oct4/POUV-dependent network links adhesion and migration to progenitor maintenance. *Current Biology* 23, 2233–2244.

Lustig, K.D., Kroll, K.L., Sun, E.E., Kirschner, M.W., 1996. Expression cloning of a Xenopus T-related gene (Xombi) involved in mesodermal patterning and blastopore lip formation. *Development* 122, 4001–4012.

MacArthur, H., Houston, D.W., Bubunenko, M., Mosquera, L., King, M.L., 2000. DEADSouth is a germ plasm specific DEAD-box RNA helicase in Xenopus related to eIF4A. *Mech Develop* 95, 291–295.

Marikawa, Y., Elinson, R.P., 1999. Relationship of vegetal cortical dorsal factors in the Xenopus egg with the Wnt/beta-catenin signaling pathway. *Mechanisms of Development* 89, 93–102.

Medioni, C., Mowry, K., Besse, F., 2012. Principles and roles of mRNA localization in animal development. *Development* 139, 3263–3276.

Mei, W., Jin, Z., Lai, F., Schwend, T., Houston, D.W., King, M.L., Yang, J., 2013. Maternal Dead-End1 is required for vegetal cortical microtubule assembly during Xenopus axis specification. *Development* 140, 2334–2344.

Miller, J.R., Rowning, B.A., Larabell, C.A., Yang-Snyder, J.A., Bates, R.L., Moon, R.T., 1999. Establishment of the dorsal: ventral axis in Xenopus embryos coincides with the dorsal enrichment of dishevelled that is dependent on cortical rotation. *The Journal of Cell Biology* 146, 427–438.

Mir, A., Heasman, J., 2008. How the mother can help: Studying maternal Wnt signaling by anti-sense-mediated depletion of maternal mRNAs and the host transfer technique. *Methods Mol. Biol.* 469, 417–429.

Mir, A., Kofron, M., Heasman, J., Lang, S., Birsoy, B., Wylie, C.C., 2008. Long- and short-range signals control the dynamic expression of an animal hemisphere-specific gene in Xenopus. *Developmental Biology* 315, 161–172.

Mir, A., Kofron, M., Zorn, A.M., Bajzer, M., Haque, M., Heasman, J., Wylie, C.C., 2007. FoxI1e activates ectoderm formation and controls cell position in the Xenopus blastula. *Development* 134, 779–788.

Miyamoto, K., Simpson, D., Gurdon, J.B., 2015a. Manipulation and in vitro maturation of *Xenopus laevis* oocytes, followed by intracytoplasmic sperm injection, to study embryonic development. *J Vis Exp* e52496.

Miyamoto, K., Suzuki, K.-I.T., Suzuki, M., Sakane, Y., Sakuma, T., Herberg, S., Simeone, A., Simpson, D., Jullien, J., Yamamoto, T., Gurdon, J.B., 2015b. The expression of TALEN before fertilization provides a rapid knock-out phenotype in *Xenopus laevis* founder embryos. *PLoS One* 10, e0142946.

Miyamoto, K., Teperek, M., Yusa, K., Allen, G.E., Bradshaw, C.R., Gurdon, J.B., 2013. Nuclear Wave1 is required for reprogramming transcription in oocytes and for normal development. *Science* 341, 1002–1005.

Morkel, M., Huelsken, J., Wakamiya, M., Ding, J., Wetering, M. van de, Clevers, H., Taketo, M.M., Behringer, R.R., Shen, M.M., Birchmeier, W., 2003. Beta-catenin regulates Cripto- and Wnt3-dependent gene expression programs in mouse axis and mesoderm formation. *Development* 130, 6283–6294.

Morris, S.A., Almeida, A.D., Tanaka, H., Ohta, K., Ohnuma, S., 2007. Tsukushi modulates Xnr2, FGF and BMP signaling: Regulation of Xenopus germ layer formation. *PLoS One* 2, e1004.

Morrison, G.M., Brickman, J.M., 2006. Conserved roles for Oct4 homologues in maintaining multipotency during early vertebrate development. *Development* 133, 2011–2022.

Mosquera, L., Forristall, C., Zhou, Y., King, M.L., 1993. A mRNA localized to the vegetal cortex of Xenopus oocytes encodes a protein with a nanos-like zinc finger domain. *Development Camb Engl* 117, 377–386.

Mowry, K.L., Melton, D.A., 1992. Vegetal messenger RNA localization directed by a 340-nt RNA sequence element in Xenopus oocytes. *Science* 255, 991–994.

Nakajima, K., Yaoita, Y., 2015. Highly efficient gene knockout by injection of TALEN mRNAs into oocytes and host transfer in *Xenopus laevis*. *Biology Open* 4, 180–185.

Neil, C.R., Jeschonek, S.P., Cabral, S.E., O'Connell, L.C., Powrie, E.A., Otis, J.P., Wood, T.R., Mowry, K.L., 2021. L-bodies are RNA–protein condensates driving RNA localization in Xenopus oocytes. *Mol Biol Cell* 32, ar37.

Newport, G., 1854. Researches on the impregnation of the ovum in the amphibia: And on the early stages of development of the embryo. (Third Series): Philosophical transactions of the royal society of London Series B. *Biological Sciences* 144, 229–244.

Newport, G., 1851. On the impregnation of the ovum in the amphibia. (First Series): Philosophical transactions of the royal society of London Series B. *Biological Sciences* 141, 169–242.

Newport, J., Kirschner, M., 1982. A major developmental transition in early Xenopus embryos. I: Characterization and timing of cellular changes at the midblastula stage. *Cell* 30, 675–686.

Nieuwkoop, P.D., 1969. The formation of the mesoderm in urodelean amphibians. *Wilhelm Roux' Archiv Für Entwicklungsmechanik Der Org* 163, 298–315.

Nieuwkoop, P.D., Ubbels, G.A., 1972. The formation of the mesoderm in urodelean amphibians. *Wilhelm Roux' Archiv für Entwicklungsmechanik der Organismen* 169, 185–199.

Nijjar, S., Woodland, H.R., 2013a. Localisation of RNAs into the germ plasm of vitellogenic Xenopus oocytes. *PLoS One* 8, e61847.

Nijjar, S., Woodland, H.R., 2013b. Protein interactions in Xenopus germ plasm RNP particles. *PLoS One* 8, e80077.

Nishita, M., Hashimoto, M.K., Ogata, S., Laurent, M.N., Ueno, N., Shibuya, H., Cho, K.W.Y., 2000. Interaction between Wnt and TGF-|[beta]| signalling pathways during formation of Spemann's organizer. *Nature* 403, 781–785.

Oh, D., Houston, D.W., 2017a. RNA localization in the vertebrate oocyte: Establishment of oocyte polarity and localized mRNA assemblages. *Results and Problems in Cell Differentiation* 63, 189–208.

Oh, D., Houston, D.W., 2017b. Role of maternal Xenopus syntabulin in germ plasm aggregation and primordial germ cell specification. *Developmental Biology* 432, 237–247.

Olson, D.J., Oh, D., Houston, D.W., 2015. The dynamics of plus end polarization and microtubule assembly during Xenopus cortical rotation. *Developmental Biology* 401, 249–263.

Owens, D.A., Butler, A.M., Aguero, T.H., Newman, K.M., Booven, D.V., King, M.L., 2017. High-throughput analysis reveals novel maternal germline RNAs crucial for primordial germ cell preservation and proper migration. *Development* 144, 292–304.

Pannese, M., Cagliani, R., Pardini, C.L., Boncinelli, E., 2000. Xotx1 maternal transcripts are vegetally localized in *Xenopus laevis* oocytes. *Mechanisms of Development* 90, 111–114.

Paraiso, K.D., Blitz, I.L., Coley, M., Cheung, J., Sudou, N., Taira, M., Cho, K.W.Y., 2019. Endodermal maternal transcription factors establish super-enhancers during zygotic genome activation. *Cell Reports* 27, 2962–2977.e5.

Park, S., Blaser, S., Marchal, M.A., Houston, D.W., Sheets, M.D., 2016. A gradient of maternal Bicaudal-C controls vertebrate embryogenesis via translational repression of mRNAs encoding cell fate regulators. *Development* 143, 864–871.

Peifer, M., Sweeton, D., Casey, M., Wieschaus, E., 1994. Wingless signal and Zeste-white 3 kinase trigger opposing changes in the intracellular distribution of Armadillo. *Dev Camb Engl* 120, 369–380.

Pera, E.M., Kim, J.I., Martinez, S.L., Brechner, M., Li, S.Y., Wessely, O., Robertis, E.M.D., 2002. Isthmin is a novel secreted protein expressed as part of the Fgf-8 synexpression group in the Xenopus midbrain-hindbrain organizer. *Mechanisms of Development* 116, 169–172.

Pondel, M.D., King, M.L., 1988. Localized maternal mRNA related to transforming growth factor beta mRNA is concentrated in a cytokeratin-enriched fraction from Xenopus oocytes. *Proceedings of the National Academy of Sciences of the United States of America* 85, 7612–7616.

Ratzan, W., Falco, R., Salanga, C., Salanga, M., Horb, M.E., 2017. Generation of a *Xenopus laevis* F1 albino J strain by genome editing and oocyte host-transfer. *Developmental Biology* 426, 188–193.

Rebagliati, M.R., Melton, D.A., 1987. Antisense RNA injections in fertilized frog eggs reveal an RNA duplex unwinding activity. *Cell* 48, 599–605.

Rebagliati, M.R., Weeks, D.L., Harvey, R.P., Melton, D.A., 1985. Identification and cloning of localized maternal RNAs from xenopus eggs. *Cell* 42, 769–777.

Reich, S., Weinstein, D.C., 2019. Repression of inappropriate gene expression in the vertebrate embryonic ectoderm. *Genes-Basel* 10, 895.

Remak, R., 1855. *Untersuchungen über die Entwickelung der Wirbelthiere*. Berlin: Walter De Gruyter Incorporated.

Rex, M., Hilton, E., Old, R., 2002. Multiple interactions between maternally-activated signalling pathways control Xenopus nodal-related genes. *The International Journal of Developmental Biology* 46, 217–226.

Rim, E.Y., Kinney, L.K., Nusse, R., 2020. β-catenin-mediated Wnt signal transduction proceeds through an endocytosis-independent mechanism. *Mol Biol Cell* 31, 1425–1436.

Roel, G., Broek, O. van den, Spieker, N., Peterson-Maduro, J., Destree, O., 2003. Tcf-1 expression during Xenopus development. *Gene Expression Patterns: GEP* 3, 123–126.

Roux, W., 1888. Beitrage zur Entwickelungsmechanik des Embryo. *V. Virchows Arch Path Anat* 114, 113–153.

Roux, W., 1887. Beiträge zur Entwickelungsmechanik des Embryo. *Archiv für Mikroskopische Anatomie* 29, 157–211.

Rugh, R., 1951. *The Frog, Its Reproduction and Development*. Philadelphia: The Blakiston Co.

Rugh, R., 1935. Ovulation in the frog. II: Follicular rupture to fertilization. *J Exp Zool* 71, 163–193.

Sasai, N., Yakura, R., Kamiya, D., Nakazawa, Y., Sasai, Y., 2008. Ectodermal factor restricts mesoderm differentiation by inhibiting p53. *Cell* 133, 878–890.

Savage, R.M., Danilchik, M.V., 1993. Dynamics of germ plasm localization and its inhibition by ultraviolet irradiation in early cleavage Xenopus embryos. *Developmental Biology* 157, 371–382.

Sekizaki, H., Takahashi, S., Tanegashima, K., Onuma, Y., Haramoto, Y., Asashima, M., 2004. Tracing of *Xenopus tropicalis* germ plasm and presumptive primordial germ cells with the *Xenopus tropicalis* DAZ-like gene. *Developmental Dynamics* 229, 367–372.

Shen, M.M., Schier, A.F., 2000. The EGF-CFC gene family in vertebrate development. *Trends Genet* 16, 303–309.

Shuttleworth, J., Colman, A., 1988. Antisense oligonucleotide-directed cleavage of mRNA in Xenopus oocytes and eggs. *The EMBO Journal* 7, 427–434.

Shuttleworth, J., Matthews, G., Dale, L., Baker, C., Colman, A., 1988. Antisense oligodeoxyribonucleotide-directed cleavage of maternal mRNA in Xenopus oocytes and embryos. *Gene* 72, 267–275.

Sindelka, R., Abaffy, P., Qu, Y., Tomankova, S., Sidova, M., Naraine, R., Kolar, M., Peuchen, E., Sun, L., Dovichi, N., Kubista, M., 2018. Asymmetric distribution of biomolecules of maternal origin in the *Xenopus laevis* egg and their impact on the developmental plan. *Scientific Reports* 8, 8315.

Sinner, D., Rankin, S., Lee, M., Zorn, A.M., 2004. Sox17 and β-catenin cooperate to regulate the transcription of endodermal genes. *Development* 131, 3069–3080.

Smith, L.D., Ecker, R.E., 1965. Protein synthesis in enucleated eggs of *Rana pipiens*. *Science* 150, 777–779.

Smith, L.D., Ecker, R.E., Subtelny, S., 1968. In vitro induction of physiological maturation in *Rana pipiens* oocytes removed from their ovarian follicles. *Developmental Biology* 17, 627–643.

Smorag, L., Xu, X., Engel, W., Pantakani, D.V.K., 2014. The roles of DAZL in RNA biology and development. *Wiley Interdiscip Rev Rna* 5, 527–535.

Snir, M., Ofir, R., Elias, S., Frank, D., 2006. *Xenopus laevis* POU91 protein, an Oct3/4 homologue, regulates competence transitions from mesoderm to neural cell fates. *The EMBO Journal* 25, 3664–3674.

Spemann, H., 1938. *Embryo Development and Induction*. New Haven: Yale University Press.

Spemann, H., 1924. Vererbung und Entwicklungsmechanik. *Naturwissenschaften* 12, 65–79.

Spemann, H., 1921. Die Erzeugung tierischer Chimären durch heteroplastische embryonale Transplantation zwischen Triton cristatus und taeniatus. *Wilhelm Roux' Archiv für Entwicklungsmechanik der Organismen* 48, 533–570.

Spemann, H., 1903. Entwickelungsphysiologische Studien am Triton-Ei. III. *Archiv Für Entwicklungsmechanik Der Org* 16, 551–631.

Spemann, H., 1902. Entwickelungsphysiologische Studien am Triton-Ei. II. *Wilhelm Roux'Archiv für Entwicklungsmechanik der Organismen* 15, 448–534.

Spemann, H., 1901. Entwickelungsphysiologische Studien am Triton-Ei. I. *Wilhelm Roux'Archiv für Entwicklungsmechanik der Organismen* 12, 224–264.

Spemann, H., Mangold, 1924. Uber Induktion von Embryonenanlagen durch Implantation artfremder Organisatoren. *Wilhelm Roux' Arch. EntwMech. Org*. 100, 599–638.

Standley, H.J., Destree, O., Kofron, M., Wylie, C.C., Heasman, J., 2006. Maternal XTcf1 and XTcf4 have distinct roles in regulating Wnt target genes. *Developmental Biology* 289, 318–328.

Stennard, F., Carnac, G., Gurdon, J.B., 1996. The Xenopus T-box gene, Antipodean, encodes a vegetally localised maternal mRNA and can trigger mesoderm formation. *Development* 122, 4179–4188.

Sudarwati, S., Nieuwkoop, P.D., 1971. Mesoderm formation in the anuran *Xenopus laevis* (Daudin). *Development Genes and Evolution* 166, 189–204.

Summerton, J., Weller, D., 1997. Morpholino antisense oligomers: Design, preparation, and properties. *Antisense Nucleic Acid Drug Dev* 7, 187–195.

Suri, C., Haremaki, T., Weinstein, D.C., 2005. Xema, a foxi-class gene expressed in the gastrula stage Xenopus ectoderm, is required for the suppression of mesendoderm. *Development* 132, 2733–2742.

Tada, H., Mochii, M., Orii, H., Watanabe, K., 2012. Ectopic formation of primordial germ cells by transplantation of the germ plasm: Direct evidence for germ cell determinant in Xenopus. *Developmental Biology* 371, 86–93.

Tadjuidje, E., Cha, S.-W., Louza, M., Wylie, C.C., Heasman, J., 2011. The functions of maternal Dishevelled 2 and 3 in the early Xenopus embryo. *Developmental Dynamics* 240, 1727–1736.

Taelman, V.F., Dobrowolski, R., Plouhinec, J.-L., Fuentealba, L.C., Vorwald, P.P., Gumper, I., Sabatini, D.D., Robertis, E.M.D., 2010. Wnt signaling requires sequestration of glycogen synthase kinase 3 inside multivesicular endosomes. *Cell* 143, 1136–1148.

Takahashi, S., Yokota, C., Takano, K., Tanegashima, K., Onuma, Y., Goto, J., Asashima, M., 2000. Two novel nodal-related genes initiate early inductive events in Xenopus Nieuwkoop center. *Development* 127, 5319–5329.

Tanaka, C., Sakuma, R., Nakamura, T., Hamada, H., Saijoh, Y., 2007. Long-range action of Nodal requires interaction with GDF1. *Genes & Development* 21, 3272–3282.

Tao, Q., Yokota, C., Puck, H., Kofron, M., Birsoy, B., Yan, D., Asashima, M., Wylie, C.C., Lin, X., Heasman, J., 2005. Maternal Wnt11 activates the canonical Wnt signaling pathway required for axis formation in Xenopus embryos. *Cell* 120, 857–871.

Tarbashevich, K., Koebernick, K., Pieler, T., 2007. XGRIP2.1 is encoded by a vegetally localizing, maternal mRNA and functions in germ cell development and anteroposterior PGC positioning in *Xenopus laevis*. *Developmental Biology* 311, 554–565.

Taverner, N.V., Kofron, M., Shin, Y., Kabitschke, C., Gilchrist, M.J., Wylie, C., Cho, K.W.Y., Heasman, J., Smith, J.C., 2005. Microarray-based identification of VegT targets in Xenopus. *Mech Develop* 122, 333–354.

Teegala, S., Chauhan, R., Lei, E., Weinstein, D.C., 2018. Tbx2 is required for the suppression of mesendoderm during early Xenopus development. *Dev Dynam* 247, 903–913.

Torpey, N., Wylie, C.C., Heasman, J., 1992. Function of maternal cytokeratin in Xenopus development. *Nature* 357, 413–415.

Tyler, A., 1965. The biology and chemistry of fertilization. *Am Nat* 99, 309–334.

Vastenhouw, N.L., Cao, W.X., Lipshitz, H.D., 2019. The maternal-to-zygotic transition revisited. *Development* 146, dev161471.

Vleminckx, K., Wong, E., Guger, K., Rubinfeld, B., Polakis, P., Gumbiner, B.M., 1997. Adenomatous polyposis coli tumor suppressor protein has signaling activity in *Xenopus laevis* embryos resulting in the induction of an ectopic dorsoanterior axis. *The Journal of Cell Biology* 136, 411–420.

Wakahara, M., 1977. Partial characterization of "primordial germ cell-forming activity" localized in vegetal pole cytoplasm in anuran eggs. *Development* 39, 221–233.

Wang, H., Fang, J., Kuang, X., Miao, L., Wang, C., Xia, G., King, M.L., Zhang, J., 2012. Activity of long-chain acyl-CoA synthetase is required for maintaining meiotic arrest in *Xenopus laevis*. *Biology of Reproduction* 87, 74.

Wang, S., Krinks, M., Lin, K., Luyten, F., Moos, M., 1997. Frzb, a secreted protein expressed in the Spemann organizer, binds and inhibits Wnt-8. *Cell* 88, 757–766.

Weaver, C., Farr, G.H., Pan, W., Rowning, B.A., Wang, J., Mao, J., Wu, D., Li, L., Larabell, C.A., Kimelman, D., 2003. GBP binds kinesin light chain and translocates during cortical rotation in Xenopus eggs. *Development* 130, 5425–5436.

Weaver, C., Kimelman, D., 2004. Move it or lose it: Axis specification in Xenopus. *Development* 131, 3491–3499.

Weeks, D.L., Melton, D.A., 1987. A maternal mRNA localized to the vegetal hemisphere in Xenopus eggs codes for a growth factor related to TGF-beta. *Cell* 51, 861–867.

Weidinger, G., Stebler, J., Slanchev, K., Dumstrei, K., Wise, C., Lovell-Badge, R., Thisse, C., Thisse, B., Raz, E., 2003. Dead end, a novel vertebrate germ plasm component, is required for zebrafish primordial germ cell migration and survival. *Current Biology* 13, 1429–1434.

Weinstein, D.C., Hemmati-Brivanlou, A., 1999. Neural induction. *Annu Rev Cell Dev Bi* 15, 411–433.

Wessely, O., Robertis, E.M.D., 2000. The Xenopus homologue of Bicaudal-C is a localized maternal mRNA that can induce endoderm formation. *Development* 127, 2053–2062.

Wilson, E., 1928. *The Cell in Development and Heredity* (Third Edition, with corrections. ed., The Macmillian Company). New York: The Macmillian Company.

Woolf, T.M., Jennings, C.G., Rebagliati, M., Melton, D.A., 1990. The stability, toxicity and effectiveness of unmodified and phosphorothioate antisense oligodeoxynucleotides in Xenopus oocytes and embryos. *Nucleic Acids Research* 18, 1763–1769.

Woolf, T.M., Melton, D.A., Jennings, C.G., 1992. Specificity of antisense oligonucleotides in vivo. *Proceedings of the National Academy of Sciences of the United States of America* 89, 7305–7309.

Wylie, C.C., Holwill, S., O'Driscoll, M., Snape, A., Heasman, J., 1985. Germ plasm and germ cell determination in *Xenopus laevis* as studied by cell transplantation analysis. *Cold Spring Harbor Symposia on Quantitative Biology* 50, 37–43.

Wylie, C.C., Kofron, M., Payne, C., Anderson, R., Hosobuchi, M., Joseph, E., Heasman, J., 1996. Maternal beta-catenin establishes a "dorsal signal" in early Xenopus embryos. *Development* 122, 2987–2996.

Xanthos, J.B., Kofron, M., Tao, Q., Schaible, K., Wylie, C.C., Heasman, J., 2002. The roles of three signaling pathways in the formation and function of the Spemann organizer. *Development* 129, 4027–4043.

Xanthos, J.B., Kofron, M., Wylie, C.C., Heasman, J., 2001. Maternal VegT is the initiator of a molecular network specifying endoderm in *Xenopus laevis*. *Development* 128, 167–180.

Xu, S., Cheng, F., Liang, J., Wu, W., Zhang, J., 2012. Maternal xNorrin, a canonical Wnt signaling agonist and TGF-β antagonist, controls early neuroectoderm specification in Xenopus. *PLoS Biology* 10, e1001286.

Yamaguchi, T., Taguchi, A., Watanabe, K., Orii, H. 2013. DEAD-South protein localizes to germ plasm and is required for the development of primordial germ cells in *Xenopus laevis*. *Biology Open* 2, 191.

Yan, L., Chen, J., Zhu, X., Sun, J., Wu, X., Shen, W., Zhang, W., Tao, Q., Meng, A. 2018. Maternal Huluwa dictates the embryonic body axis through β-catenin in vertebrates. *Science* 362, eaat1045.

Yang, J., Tan, C., Darken, R.S., Wilson, P.A., Klein, P.S. 2002. Beta-catenin/Tcf-regulated transcription prior to the mid-blastula transition. *Development* 129, 5743–5752.

Yost, C., Farr, G.H., Pierce, S.B., Ferkey, D.M., Chen, M.M., Kimelman, D., 1998. GBP, an inhibitor of GSK-3, is implicated in Xenopus development and oncogenesis. *Cell* 93, 1031–1041.

Zaret, K.S., Carroll, J.S., 2011. Pioneer transcription factors: Establishing competence for gene expression. *Gene Dev* 25, 2227–2241.

Zearfoss, N.R., Chan, A.P., Wu, C.F., Kloc, M., Etkin, L.D., 2004. Hermes is a localized factor regulating cleavage of vegetal blastomeres in *Xenopus laevis*. *Developmental Biology* 267, 60–71.

Zhang, C., Basta, T., Jensen, E.D., Klymkowsky, M.W., 2003. The beta-catenin/VegT-regulated early zygotic gene Xnr5 is a direct target of SOX3 regulation. *Development* 130, 5609–5624.

Zhang, C., Klymkowsky, M.W., 2007. The Sox axis, Nodal signaling, and germ layer specification. *Differentiation* 75, 536–545.

Zhang, J., Houston, D.W., King, M.L., Payne, C., Wylie, C.C., Heasman, J., 1998. The role of maternal VegT in establishing the primary germ layers in Xenopus embryos. *Cell* 94, 515–524.

Zhang, J., King, M.L., 1996. Xenopus VegT RNA is localized to the vegetal cortex during oogenesis and encodes a novel T-box transcription factor involved in mesodermal patterning. *Development* 122, 4119–4129.

Zhang, Y., Cooke, A., Park, S., Dewey, C.N., Wickens, M., Sheets, M.D., 2013. Bicaudal-C spatially controls translation of vertebrate maternal mRNAs. *RNA* 19, 1575–1582.

Zhou, X., Lin, Y., Kato, M., Mori, E., Liszczak, G., Sutherland, L., Sysoev, V.O., Murray, D.T., Tycko, R., McKnight, S.L., 2021. Transiently structured head domains control intermediate filament assembly. *Proc National Acad Sci* 118, e2022121118.

Zhou, Y., King, M.L., 1996a. Localization of Xcat-2 RNA, a putative germ plasm component, to the mitochondrial cloud in Xenopus stage I oocytes. *Development* 122, 2947–2953.

Zhou, Y., King, M.L., 1996b. RNA transport to the vegetal cortex of Xenopus oocytes. *Developmental Biology* 179, 173–183.

Zhou, Y., Zhang, J., King, M.L., 2004. Polarized distribution of mRNAs encoding a putative LDL receptor adaptor protein, xARH (autosomal recessive hypercholesterolemia) in Xenopus oocytes. *Mechanisms of Development* 121, 1249–1258.

Zhu, M., Zernicka-Goetz, M., 2020. Principles of self-organization of the mammalian embryo. *Cell* 183, 1467–1478.

Zorn, A.M., Barish, G.D., Williams, B.O., Lavender, P., Klymkowsky, M.W., Varmus, H.E., 1999. Regulation of Wnt signaling by Sox proteins XSox17α/β and XSox3 physically interact with β-catenin. *Mol Cell* 4, 487–498.

Züst, B., Dixon, K., 1975. The effect of u.v. irradiation of the vegetal pole of *Xenopus laevis* eggs on the presumptive primordial germ cells. *The Journal of Embryology and Experimental Morphology* 34, 209–220.

4 Signaling Components in Dorsal-Ventral Patterning and the Organizer in *Xenopus*

Edward M. De Robertis and Nydia Tejeda-Muñoz

CONTENTS

4.1. Historical Background 43
4.2. The Search for Spemann Organizer Molecules in *Xenopus* 44
4.2.1. Goosecoid 44
4.3. The Ventral Signaling Center 45
4.4. BMP Antagonists and Tissue Differentiation 45
4.5. Neural Induction by the Organizer 47
4.6. The Chordin/Tolloid/Twisted Gastrulation/Crossveinless-2/BMP Ancestrally Conserved D-V Patterning System 47
4.7. Self-Regulation by Spemann Organizer Relocalization 47
4.8. Future Avenues of Research 48
References 49

"Les théories passent, la grenouille reste."
"Theories pass, but the frog remains."
Jean Rostand—Le carnet d'un biologiste—1959

4.1. HISTORICAL BACKGROUND

Watching a fertilized animal egg develop into an embryo with many tissues is fascinating. How could something as complex happen? Embryology became the forefront of biological research when researchers realized that rather than a descriptive approach, experimental challenges were required to unravel the mechanisms of development. The frog embryo led the way. In 1883, Wilhelm Roux killed one of the two blastomeres of a frog embryo with a hot needle and found that the surviving cell gave rise to only a half embryo. However, in 1891, Hans Driesch separated the first two blastomeres of a sea urchin embryo and found that each cell could self-organize and give rise to a complete, albeit smaller, embryo (reviewed in Spemann, 1938). The discordant results were clarified by Thomas Hunt Morgan (who before becoming a geneticist was an experimental embryologist), who repeated Roux's experiment and found that if the dead blastomere were gently pipetted out of the frog embryo, frogs also could self-regulate and generate a complete tadpole from half an egg (Morgan, 1895). Using baby hair loops to slowly constrict newt eggs at the one- or two-cell stage, Hans Spemann later obtained twins from the same amphibian egg (Spemann, 1938).

One can imagine that understanding the mechanisms leading to making two out of one would be next to impossible. Yet the way forward was pointed by an experiment carried out by a graduate student at Freiburg University, Hilde Mangold. Under the direction of Spemann, who had found that the dorsal lip of the blastopore was the first region of the embryo to become determined, she transplanted dorsal lips into the ventral side of host embryos from newt species that differed in their degree of pigmentation. In experiments based largely on two embryos with secondary axes—no statistical significance analyses were required back then—she described in wonderful camera lucida drawings of histological sections that the transplanted dorsal organizer gave rise mostly to notochord, while the neighboring cells were induced to form a Siamese twin containing dorsal tissues such as somites and central nervous system. This was the most famous experiment in embryology (Spemann and Mangold, 1924). Tragically, Hilde Mangold, who had a small baby, died before her paper was published. Spemann received the Nobel Prize for Physiology or Medicine in 1935 for the discovery of embryonic induction of histotypic differentiation.

The Spemann-Mangold experiment marked the apogee of experimental embryology. A flurry of experiments attempted to identify the chemical substance that was able to induce the central nervous system (CNS), also called the primary inducer. However, given the methods available at the time, all efforts failed, and experimental embryology gradually faded (Hurtado and De Robertis, 2007). The genetics founded by Morgan with the use of the *Drosophila* fruit fly became the pre-eminent biological discipline for most of the 20th century.

By the 1970s, it was common to hear famous professors say that Spemann had set back developmental biology by 50 years. Experimental embryology was forgotten. That all changed in the 1990s, when molecular biology became practical and new genes could be readily isolated. A wonderful memoir about

DOI: 10.1201/9781003050230-5

graduate student days in the Spemann lab also played a role in the renaissance of experimental embryology (Hamburger, 1988). This short book, written when Viktor Hamburger was 88 years old, inspired many of us in the *Xenopus* community to take a second look. What followed was a revolution.

4.2. THE SEARCH FOR SPEMANN ORGANIZER MOLECULES IN *XENOPUS*

In our laboratory, we used manually dissected *Xenopus* dorsal lips to prepare cDNA libraries from which the first organizer-specific gene, *goosecoid*, was isolated (Cho et al., 1991). Other laboratories followed some months later, treating *Xenopus* animal cap explants with Activin, which induces dorsal mesoderm and resulted in the isolation of other organizer transcription factors such as *lim-1* and *forkhead-1* (Taira et al., 1992; Dirksen and Jamrich, 1992). Using a very productive functional assay of microinjecting pools of synthetic mRNAs which were then sib-selected, Richard Harland succeeded in isolating Noggin, the first secreted protein of the Spemann organizer (Smith and Harland, 1992). Douglas Melton found that Follistatin was also secreted by the organizer (Hemmati-Brivanlou et al., 1994).

The Spemann organizer proved a very productive fishing ground for novel signaling molecules. At early stages of development, cells are engaged in exchanging signals specifying their positional information before tissue differentiation takes place. The *Xenopus* gastrula has been the subject of saturating screens for dorsal and ventral molecules (Figure 4.1A). This is analogous to the case of *Drosophila*, in which, using genetic screens, the signaling networks involved in cuticle patterning have been saturating as well. In our lab, we used various types of dorsal-ventral (D-V) probes to isolate *xnot-2* (Gont et al., 1993), *chordin* (*chd*) (Sasai et al., 1994), *cerberus* (Bouwmeester et al., 1996), and *frzb-1* (Leyns et al., 1997). Unbiased screens for *Xenopus* proteins secreted by cultured mammalian cells identified IGFBP-5 (Insulin-like growth factor binding protein 5), the Crescent and sFRP2 Wnt inhibitors, and many ventrally secreted proteins (Pera et al., 2000).

With the sequencing of the *Xenopus laevis* genome (Session et al., 2016), a new era started, and high-throughput RNA-seq became possible. This produced an exhaustive and quantitative catalog of transcripts expressed in the ventral and dorsal sides of the *Xenopus* gastrula (Ding et al., 2017a). A list of 44,000 transcripts arranged according to dorsal to ventral expression is available in Table S1 of Ding et al. (2017a). This study identified secreted Pkdcc (protein kinase domain containing, cytoplasmic) also known as Vlk (vertebrate lonesome kinase) as a Wnt inhibitor (Ding et al., 2017a). RNA-seq identified an early dorsal β-Catenin gene signature of 123 genes, a plethora of ventral genes (Ding et al., 2017b), and Angiopoietin-like 4 (Angptl4) and Bighead as Wnt inhibitors that promote LRP6 receptor endocytosis (Kirsch et al., 2017; Ding et al., 2018).

4.2.1. Goosecoid

Goosecoid was the first gene identified in Spemann's organizer and has proved a very reliable marker of organizer tissue (Cho et al., 1991). For example, the location of the

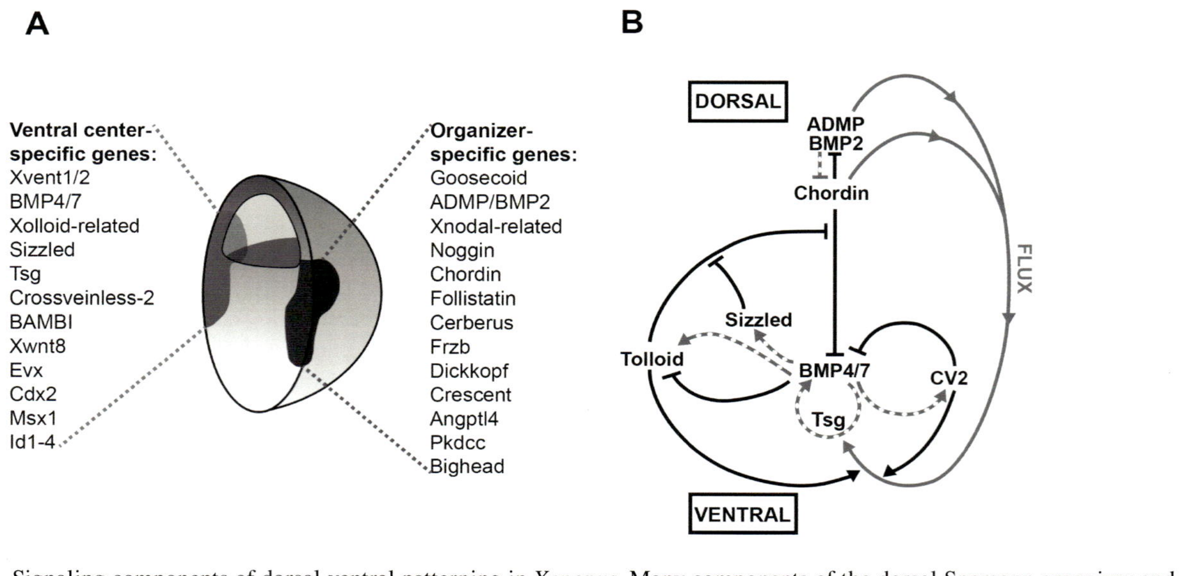

FIGURE 4.1 Signaling components of dorsal-ventral patterning in *Xenopus*. Many components of the dorsal Spemann organizer and the ventral center at the opposite pole of the embryo have been isolated in saturating molecular and functional screens carried out in many laboratories. The *Xenopus* gastrula has been a rich source of new molecules and developmental mechanisms. (A) Many of the novel genes identified encoded secreted antagonists of the BMP and Wnt pathways. (B) Components of the extracellular Chordin/Tolloid/Tsg/CV2/BMP pathway. Direct protein-protein interactions demonstrated biochemically are shown by solid black lines, transcriptional regulation by stippled lines, and flux of Chordin/BMP/Tsg complexes towards the ventral center by gray lines. The rate-limiting step is the proteolytic cleavage of Chordin by Tolloid metalloproteinase. BMP signaling is maximal in the ventral and lowest in the dorsal side, setting up the transcriptional control of this self-regulating morphogen gradient system conserved throughout the animal kingdom.

mouse organizer was not known. When I presented the first *in situ* hybridizations of mouse *goosecoid* in 1992 at the first vertebrate molecular embryology meeting in Les Diablerets, Switzerland, great consternation was caused. Our collaborator Stephen Gaunt had found *goosecoid* expression in the anterior primitive streak. Mouse embryologists almost unanimously rose to counter that the organizer was located in the node that forms posterior to the notochord at a later stage of development. Fortunately, on the train down from the mountain, I sat, dejected, next to Azim Surani, who suggested that if the mouse organizer were where we thought it was, transplantation into the *Xenopus* blastula cavity might reveal its inductive activity. We did the experiment with Martin Blum and Herbert Steinbeisser, and that was the case; *goosecoid* has pinpointed the location of the organizer in many vertebrate embryos since (De Robertis, 2004).

4.3. THE VENTRAL SIGNALING CENTER

The dorsal organizer has been the center of attention, but it is emerging that the ventral side of the gastrula is equally important. For each action on the dorsal side, there is a reaction on the ventral side. Ventral genes are turned on by BMP4 signaling, while dorsal genes are transcribed when BMP signaling levels are low (Karaulanov et al., 2004; Reversade and De Robertis, 2005). This transcriptional seesaw explains to a large degree the self-regulation and resilience of the *Xenopus* embryo. The reason the ventral center was ignored for a long time was that when transplanted, ventral tissue becomes incorporated into the host site instead of inducing changes in the neighboring tissues (Spemann, 1938). However, when BMP2/4/7 are depleted simultaneously so that no epidermis is formed, transplantation of wild-type ventral tissue can induce epidermal differentiation (high BMP) at a great distance (Reversade and De Robertis, 2005).

Xenopus BMP4 was found to be expressed in the ventral region and to induce ventralization (Fainsod et al., 1994). However, the realization that the ventral side serves to antagonize the effects of the organizer was not formulated until Christof Niehrs discovered *Xvent-1* and *Xvent-2*, two homeobox target genes of BMP4 with similarities to *Drosophila bar* (Gawantka et al., 1995). It was later found that these genes mediate the effects of BMP4 (Ladher et al., 1996; Onichtchouk et al., 1996). There is a whole panoply of genes part of the BMP4 synexpression group (Niehrs and Pollet, 1999) that are transcriptionally activated by BMP4 signaling, such as Id1–4 (inhibitor of differentiation 1–4) (Karaulanov et al., 2004) and BAMBI (BMP and Activin Membrane Bound Inhibitor), a transmembrane pseudoreceptor lacking the cytosolic Serine-Threonine kinase domain (Onichtchouk et al., 1999).

The function of Xvent1/2 as repressors of the Spemann organizer has been investigated using antisense morpholino oligonucleotides (MOs). This method of obtaining loss of function can be very effective and specific in *Xenopus* embryos. In zebrafish, on occasion, MOs may have toxic effects, and this has greatly confused the field, as zebrafish researchers now demand genetic mutations while negating the many favorable properties of MOs in *Xenopus* and many other organisms, for which we have advocated elsewhere (Blum et al., 2015). Xvent1 and 2 have redundant functions, but when both are depleted, embryos are strongly dorsalized with expanded heads and short trunks (Sander et al., 2007). Notably, the expression of *goosecoid* is greatly expanded by loss of Xvent1/2. Depletion of *Xenopus* Goosecoid with MO resulted in cyclopic embryos with small heads and enlarged ventral tissues. Unexpectedly, triple depletion of Xvent1, Xvent2, and Goosecoid rescued almost completely normal D-V and anterior-posterior (A-P) development in a variety of assays (Sander et al., 2007). Thus, it is as if these three genes are dispensable for embryogenesis and exist to balance deviations from the norm of their counterparts. Xvent1/2 and Goosecoid in *Xenopus* mediate a remarkable self-adjusting mechanism to ensure a perfect tadpole is formed time after time.

The most abundant ventral center transcript in the ventral center is Sizzled (Szl), a divergent sFRP (secreted frizzled-related protein) that lost its ability to bind Wnts (Collavin and Kirschner, 2003; Ding et al., 2017a). Sizzled, like Xolloid-related protease (Piccolo et al., 1997), Twisted gastrulation (Tsg) (Oelgeschläger et al., 2000), and Crossveinless-2 (CV-2) (Ambrosio et al., 2008) are ventral center-secreted proteins that function in the Chordin morphogenetic pathway.

4.4. BMP ANTAGONISTS AND TISSUE DIFFERENTIATION

The Spemann organizer directs the differentiation of dorsal tissues. *Xenopus* is ideal for the analysis of D-V histotypic cell differentiation because the embryo undergoes a cortical rotation at the one-cell stage that displaces the maternal pigment towards the ventral side (which is the sperm entry side). If one pays close attention, the rotation continues at the two- and four-cell stage. The result is that if one selects symmetrically dividing embryos, at the four-cell stage, the two blastomeres containing the less pigmented dorsal crescent will reliably mark the formation of the dorsal blastopore lip and the embryonic midline (Klein, 1987). This is a very powerful tool that allows embryologists to direct D-V microinjections, lineage tracing, and transplantations from the earliest stages of development. Surprisingly, this useful tool was disputed by lineage tracing random embryos, which resulted in the publication of a paper (Danilchik and Black, 1988), but, fortunately, it did not stop further lineage-tracing research in *Xenopus*. It is good that today some journals are starting to publish papers confirming, not only negating, previous observations.

At the 16-cell stage, the *Xenopus* embryo has a predictable cell lineage (Moody, 1987). At this stage, four segments can be distinguished (S1–S4) on each side, which can be marked individually by four microinjections of fluorescent lineage tracers (Moriyama and De Robertis, 2018). In the beautiful embryo shown in Figure 4.2, the descendants of the organizer can be followed in red and progressively more ventral tissues in green, blue, and finally orange. This tour-de-force by Yuki Moriyama reveals that the dorsal organizer gives rise not only to notochord and floor plate but also to the medial somite that lies next to the notochord. Lineage tracing at the 32-cell stage has delineated in detail the origin of the dorsal lip of the circular blastopore at early neurula (Bauer et al., 1994; Vodicka and Gerhart, 1995). When lineage traced at late gastrula/early neurula, the dorsal lip organizer continues its gastrulation movements inside the tailbud embryo all the way to the tip of the tail, where it gives rise to the stem cells of the chordoneural hinge (Gont et al., 1993).

The main finding of the Spemann organizer molecular studies was that this tissue is a source of secreted growth factor antagonists, many of which were novel proteins at the time (Figure 4.1A). Thus, Chordin, Noggin, and Follistatin were BMP antagonists (Piccolo et al., 1996; Zimmerman et al., 1996; Fainsod et al., 1997). The head inducer Cerberus is a secreted inhibitor of Nodal, BMP, and Wnt (Piccolo et al., 1999). Dickkopf-1 (Dkk1) (Glinka et al., 1998), Frzb-1, Crescent, Angptl4, Pkdcc, and Bighead are all secreted Wnt antagonists (Figure 4.1A).

In the case of BMP antagonists, the simultaneous depletion of all three is required for the loss of all dorsal structures (Khokha et al., 2005). The depletion of Chordin leads to a partial loss of dorsal structures and the complete loss of inductive activity of transplanted organizers (Oelgeschläger et al., 2003). Activin, a TGF-β superfamily growth factor that induces dorsal mesoderm in animal cap ectodermal explants, is only able to induce ventral mesoderm when Chordin is

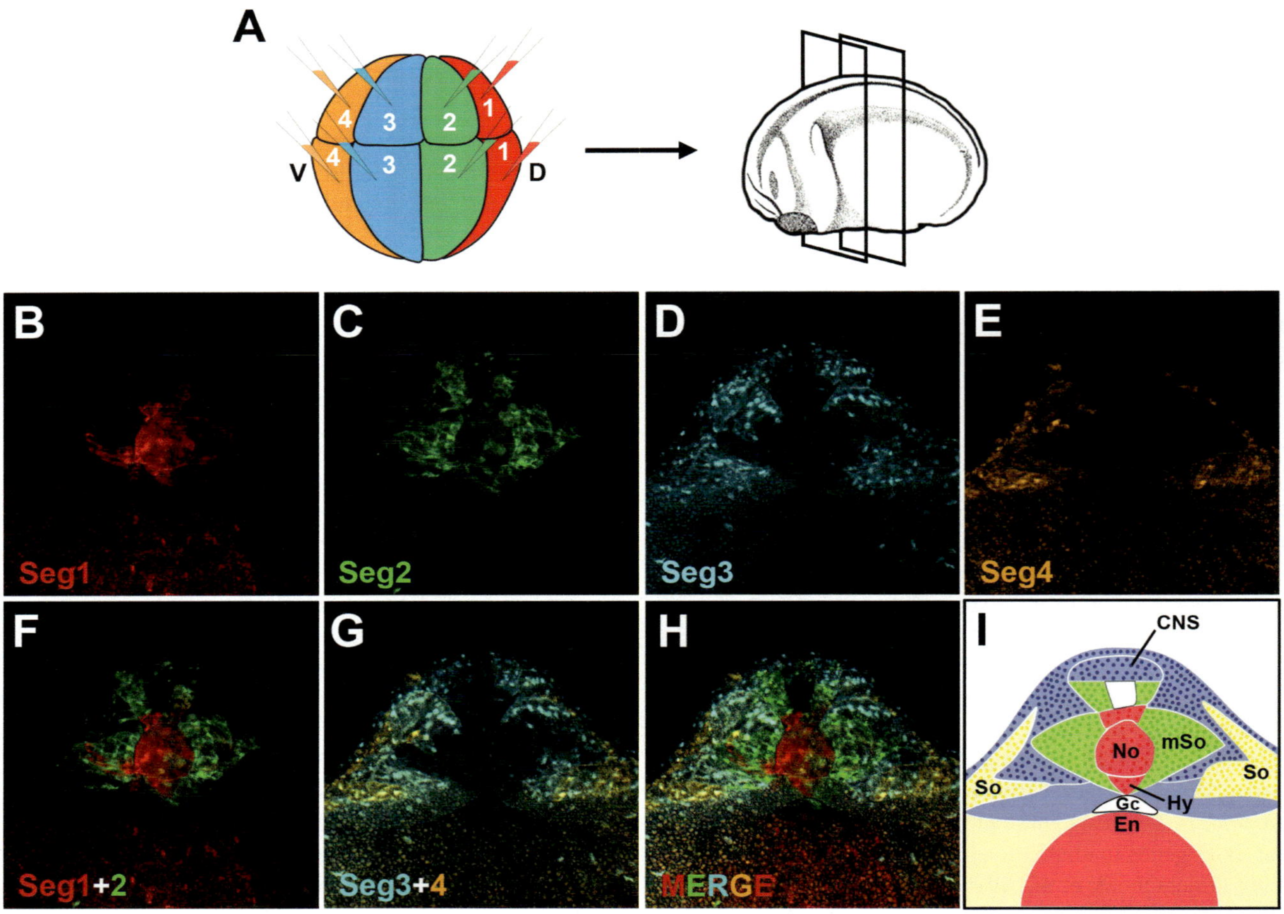

FIGURE 4.2 Inductive signals from the Spemann organizer govern the highly stereotypical histotypic development of the *Xenopus* embryo. In symmetrically cleaving embryos, the cell lineage can be followed by labeling the four segments of the *Xenopus* blastula at the16-cell stage with red, green, blue, and orange conjugated Dextran amines. (A) Experimental diagram; embryo received 16 injections, and Vibratome sections were prepared at tailbud stage. (B) Segment 1 (red Fl568-DA) gives rise to notochord, hypochord, dorsal endoderm, and ventral-most CNS, with a weaker contribution to the medial somite. (C) Segment 2 (green, Fluorescein-Dextran amine, F-DA) gives rise to most of the medial somite and spinal cord. (D) Segment 3 (Cascade Blue-Dextran amine, CsBl-DA) gives rise to most of the lateral somite, dorsal CNS and epidermis. (E) Segment 4 gives rise to the outermost parts of the somite, intermediate mesoderm, lateral plate, and ventral epidermis. (F) Progeny of the dorsal segments 1 and 2. (G) Progeny of ventral segments 3 and 4. (H) Merged image of this beautiful four-channel confocal image. (I) Diagram summarizing the origins of D-V tissues in *Xenopus*, which shares the same stereotypical D-V differentiation with all vertebrate embryos. CNS, central nervous system; En, endoderm; Gc, gut cavity; Hy, hypochord; mSo, medial somite; No, notochord; So, somite.

Source: Modified from supplementary information of Moriyama and De Robertis, 2018; reproduced with permission from the Proceedings of the National Academy of Sciences USA.

depleted, indicating an essential role of Chordin in the dorsalization of mesoderm (Oelgeschläger et al., 2003).

4.5. NEURAL INDUCTION BY THE ORGANIZER

Xenopus blastula ectodermal explants, also called animal caps, provide an ideal system to study the induction of the CNS, which has fascinated biologists since Spemann (De Robertis and Kuroda, 2004). Animal cap explants develop into epidermis and contain high levels of BMP signaling (Wilson and Hemmati-Brivanlou, 1995). Treatment of animal cap cells with Noggin protein induced anterior brain tissue (Lamb et al., 1993). Similarly, microinjection of mRNAs encoding Chordin, Noggin, or Follistatin also induced anterior neural tissue (Sasai et al., 1995; Kuroda et al., 2004).

Neural tissue is the default state of the animal cap, and when cells are dissociated in low calcium, they become neural due to sustained activation of the Ras/MAPK pathway (Kuroda et al., 2005). MAPK phosphorylation primes the BMP transcriptional regulators Smad1/5/8 for phosphorylation by GSK3 and subsequent degradation, resulting in reduced BMP signaling (Fuentealba et al., 2007). Neural induction requires the inhibition of both the BMP (Smad1) and TGF-β (Smad2) pathways (Chang and Harland, 2007). This can be achieved through polyubiquitinylation triggered by FGF/MAPK/GSK3 phosphorylations of the shared Smad4 subunit (which is not a substrate for Serine-Threonine kinase receptors) (Demagny et al., 2014). Modern studies of the Spemann organizer have greatly helped formulate current models of the regulation of CNS induction.

4.6. THE CHORDIN/TOLLOID/TWISTED GASTRULATION/CROSSVEINLESS-2/ BMP ANCESTRALLY CONSERVED D-V PATTERNING SYSTEM

Chordin is at the center of D-V patterning and is a key component of a biochemical pathway of interacting extracellular proteins in *Xenopus* (Figure 4.1B). Chordin facilitates the diffusion (or flux) of dorsal BMPs towards the ventral side of the embryo, where they are released by a specific chordinase called Tolloid (Xolloid-related in *Xenopus*) (Piccolo et al., 1997). At the highest levels of BMP signaling, Sizzled is secreted at levels comparable to those of Chordin (Lee et al., 2006). Sizzled functions as a competitive inhibitor of Tolloid, as it is bound by the active site of the protease but cannot be cut (Lee et al., 2006). However, Tolloid enzyme activity is also non-competitively inhibited by the binding of BMP to the so-called CUB domains of this metalloproteinase (Lee et al., 2009). Chordin and Sizzled are the highest enriched transcripts both at the dorsal and ventral poles of the embryo (Ding et al., 2017a).

The Chordin/Tolloid system is self-regulating and adjustable to size-dependent scaling (Ben-Zvi et al., 2008; Inomata et al., 2013). Diffusion of Chordin/BMP driven by copious degradation by Tolloid takes place in the narrow region of extracellular matrix that separates ectoderm from mesoderm, called Brachet's cleft in *Xenopus* (Plouhinec et al., 2013). Tsg secreted ventrally forms a ternary complex with BMP and Chordin, facilitating the transfer of BMP to its cell-surface receptors after Tolloid cleavage (Oelgeschläger et al., 2000; Zinski et al., 2018). CV2 does not diffuse but serves as a binding site to attract Chd/Tsg/BMP to the ventral side for cleavage by Tolloid and release of BMP/Tsg (Ambrosio et al., 2008) (Figure 4.1B). This D-V patterning pathway is strongly supported by loss-of-function mutations in zebrafish (Little and Mullins, 2006; Zinski et al., 2018).

Chordin is the homologue of *Drosophila* Short gastrulation (Sog) (Holley et al., 1995). Tsg, Tolloid, and CV2, but not Sizzled, have *Drosophila* homologues as well. In zebrafish, Sizzled is called Ogon/Mercedes (Little and Mullins, 2006). As reviewed elsewhere, the remarkable Chd/Tolloid/Tsg/CV2/BMP biochemical pathway is ancestral to bilateral animals and is even conserved in the sea anemone *Nematostella* (Bier and De Robertis, 2015; De Robertis et al., 2017).

4.7. SELF-REGULATION BY SPEMANN ORGANIZER RELOCALIZATION

Spemann had been able to obtain twins by hair-loop constrictions of the fertilized egg (Spemann, 1938). Many years later, we realized that identical twins could be obtained in *Xenopus* by cutting blastula embryos sagittally with a scalpel blade, but this occurred only at very low frequencies. More recently, the frequency of twinning was greatly improved by bisecting blastula embryos with an eyelash knife and improving culture conditions. With frequencies of twinning of 50% or more, it became possible to follow the molecular changes that take place when the entire missing half of the embryo is regenerated (Moriyama and De Robertis, 2018).

Figure 4.3 shows the results of sagittal or dorsal/ventral bisections. We unexpectedly found that, in twins that healed properly, the maternal egg pigmentation was routinely asymmetric on the left or right sides (Figure 4.3B'' and C''), prompting a series of in-depth lineage tracing investigations. The half embryo heals the large wound left by bisection within 60 minutes, bringing the most dorsal segment 1 fated to become the Spemann organizer in direct contact with the ventral-most cells of segment 4. From this opposition, the formation of the dorsal organizer, which is not yet established at mid blastula, is displaced by 90° (Figure 4.3B' and C'). This explains the pigment asymmetry, for the most pigmented epidermis arises from the ventral-most segment 4 after the mesoderm involutes forming the left and right side. Studies using Chordin and phospho-Smad1/5/8 confirmed that a new D-V gradient is formed by repositioning the D-V axis, explaining self-organization of twins after bisection (Moriyama and De Robertis, 2018). When bisection is performed perpendicularly to the sagittal plane, the dorsal half can scale the size of the gradient into an approximately normal embryo (Figure 4.3D–D''), but the ventral half forms only ventral mesoderm, as it lacks a Spemann

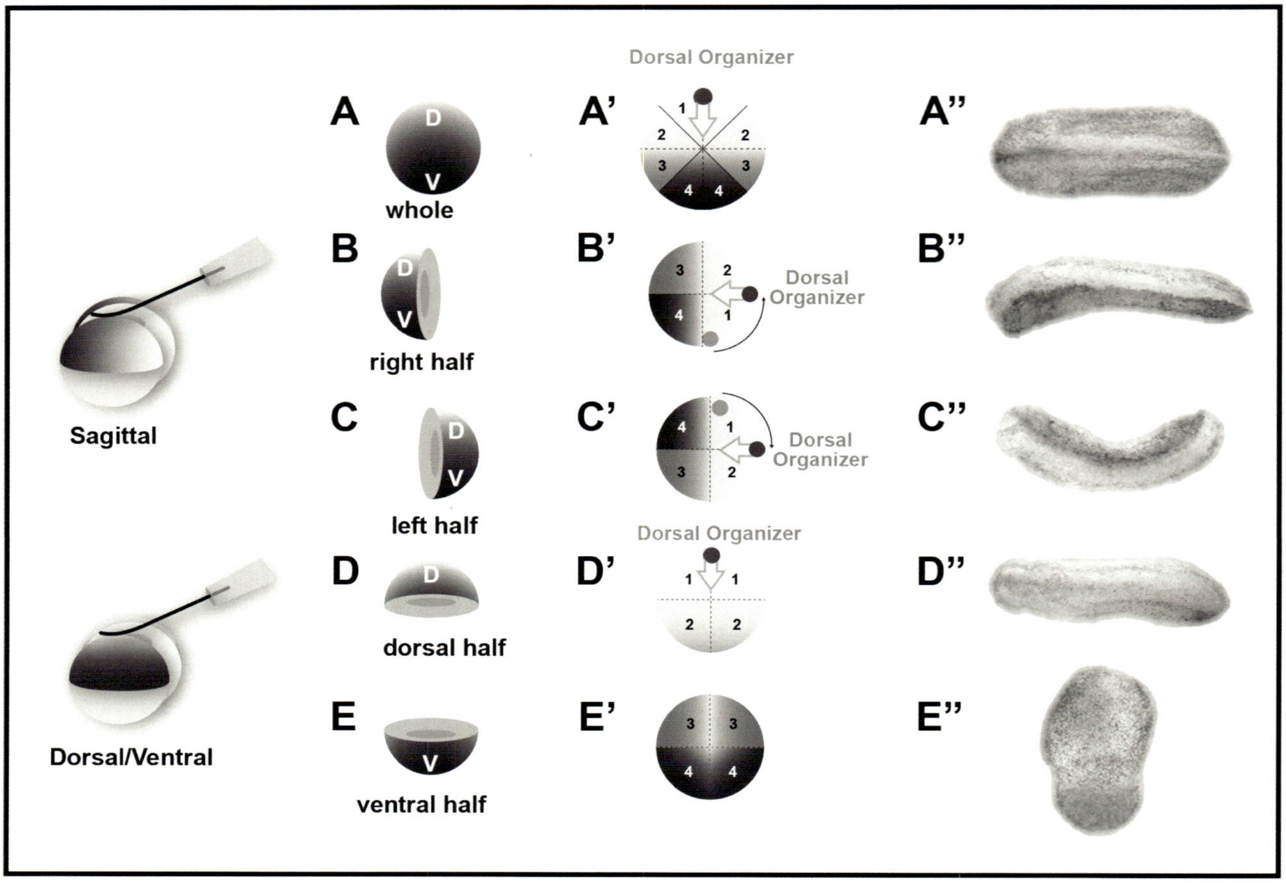

FIGURE 4.3 *Xenopus laevis* half embryos have powerful self-organizing properties that are revealed by bisection. On the left, drawing of blastula embryos being bisected in the sagittal or dorsal/ventral directions with an eyelash knife. (A) Uncut whole embryo. (B) Right half embryo cut sagittally; the maternal pigment is darker on the left side of the tailbud tadpole (while it is uniform in the whole embryo). (C) Left half from the same blastula; pigment asymmetry at tailbud is concentrated on the right side. (D) Dorsal half embryo at tailbud showing almost perfect scaling of the half embryo along the antero-posterior axis to form a well-proportioned tadpole. (E) Ventral half embryo; since it lacks the Spemann organizer, it develops into a belly-piece consisting of ventral tissues, such as blood and lateral plate, without any dorsal axis. All embryos were from the same experimental batch. The diagrams indicate how displacement by 90° of the Spemann organizer explains the tissue regeneration and pigment asymmetry observed in twinned embryos (Moriyama and De Robertis, 2018). Numbers indicate the four segments of the 16-cell embryo that were lineage-traced. After sagittal bisection, the dorsal-most segment 1 becomes juxtaposed to the ventral-most segment 4, which has high BMP and Wnt expression potential. The organizer is not yet formed at midblastula when embryos are bisected, but by early gastrula, the new organizer (indicated by the red dot) forms 90° away from its original D-V location.

Source: Embryo images from Moriyama and De Robertis, 2018; reproduced with permission from the *Proceedings of the National Academy of Sciences* USA.

organizer (such fragments were called belly-pieces by Spemann) (Figure 4.3E–E'').

4.8. FUTURE AVENUES OF RESEARCH

The embryos of the frog *Xenopus laevis* provide a marvelous biological material. The advances derived from studying the signaling components that control dorsal-ventral patterning and their remarkable regeneration properties after experimental manipulations have been profound. Many new molecules have been discovered and the nature of a morphogenetic gradient dissected. As we approach the centennial of the Spemann-Mangold experiments in 2024, many unknowns remain. The nature of the ventral center genes and their communication with dorsal signals over long distances has been incompletely explored. The regulatory mechanisms by which the D-V and A-P axis are entwined remain largely unknown. We do not know whether other morphogen gradient-fields will be regulated by opposing poles of high and low signaling by growth factor pathways. How a multitude of extracellular signals such as Wnt, FGF, and BMP signals are integrated at the level of hard-wired intracellular protein phosphorylations to generate simple cell differentiation decisions is only starting to emerge. More cell biological aspects, such as the relation between embryonic induction and membrane trafficking, are at their infancy. The questions change, but the frog remains.

REFERENCES

Ambrosio, A.L., Taelman, V.F., Lee, H.X., Metzinger, C.A., Coffinier, C. and De Robertis, E.M. 2008. Crossveinless-2 is a BMP feedback inhibitor that binds Chordin/BMP to regulate *Xenopus* embryonic patterning. *Dev. Cell* 15, 248–260.

Bauer, D.V., Huang, S. and Moody, S.A. 1994. The cleavage stage origin of Spemann's organizer: Analysis of the movements of blastomere clones before and during gastrulation in Xenopus. *Development* 120, 1179–1189.

Ben-Zvi, D., Shilo, B., Fainsod, A. and Barkai, N. 2008. Scaling of the BMP activation gradient in Xenopus embryos. *Nature* 453, 1205–1211.

Bier, E. and De Robertis, E.M. 2015. BMP gradients: A paradigm for morphogen-mediated developmental patterning. *Science* 348, 1443 and aaa5838.

Blum, M., De Robertis, E.M., Wallingford, J.B. and Niehrs, C. 2015. Morpholinos: Antisense and sensibility. *Dev. Cell.* 35, 145–149.

Bouwmeester, T., Kim, S.H., Sasai, Y., Lu, B. and De Robertis, E.M. 1996. Cerberus is a head-inducing secreted factor expressed in the anterior endoderm of Spemann's organizer. *Nature* 382, 595–601.

Chang, C. and Harland, R.M. 2007. Neural induction requires continued suppression of both Smad1 and Smad2 signals during gastrulation. *Development* 134, 3861–3872.

Cho, K.W.Y., Blumberg, B., Steinbeisser, H. and De Robertis, E.M. 1991. Molecular nature of Spemann's organizer: The role of the *Xenopus* homeobox gene *goosecoid*. *Cell* 67, 1111–1120.

Collavin, L. and Kirschner, M.W. 2003. The secreted Frizzled-related protein Sizzled functions as a negative feedback regulator of extreme ventral mesoderm. *Development* 130, 805–816.

Danilchik, M. and Black, S. 1988. The first cleavage plane and the embryonic axis are determined by separate mechanisms in *Xenopus laevis*. *Dev. Biol.* 128, 58–64.

Demagny, H., Araki, T. and De Robertis, E.M. 2014. The tumor suppressor Smad4/DCP4 is at the crossroads of Wnt, FGF and TGF-β signaling. *Cell Rep.* 9, 688–700.

De Robertis, E.M. 2004. Goosecoid. In: *Gastrulation*, C. Stern, Ed., Cold Spring Harbor Laboratory Press, New York, pp. 581–589.

De Robertis, E.M. and Kuroda, H. 2004. Dorsal-ventral patterning and neural induction in *Xenopus* embryos. *Annu. Rev. Cell Dev. Biol.* 20, 285–308.

De Robertis, E.M., Moriyama, Y. and Colozza, G. 2017. Generation of animal form by the Chordin/Tolloid/BMP gradient: 100 years after D'Arcy Thompson. *Dev. Growth Differ.* 59, 580–592.

Ding, Y., Colozza, G., Sosa, E.A., Moriyama, Y., Rundle, S., Salwinski, L. and De Robertis, E.M. 2018. Bighead is a novel Wnt antagonist secreted by the *Xenopus* Spemann organizer that promotes Lrp6 endocytosis. *Proc. Natl. Acad. Sci. USA* 115, E9135–E9144.

Ding, Y., Colozza, G., Zhang, K., Moriyama, Y., Ploper, D., Sosa, E.A., Benitez, M.D.J. and De Robertis, E.M. 2017a. Genome-wide analysis of dorsal and ventral transcriptomes of the *Xenopus laevis* gastrula. *Dev. Biol.* 426, 176–178.

Ding, Y., Ploper, D., Sosa, E.A., Colozza, G., Moriyama, Y., Benitez, M.D.J., Zhang, K., Merkurjev, D. and De Robertis, E.M. 2017b. Spemann organizer transcriptome induction by early b-Catenin, Wnt, Nodal and Siamois signals in *Xenopus laevis*. *Proc. Natl. Acad. Sci. USA* 114, E3081–E3090.

Dirksen, M.L. and Jamrich, M., 1992. A novel, activin-inducible, blastopore lip-specific gene of *Xenopus laevis* contains a *fork head* DNA-binding domain. *Genes Dev.* 6, 599–608.

Fainsod, A., Deissler, K., Yelin, R., Marom, K., Epstein, M., Pillemer, G., Steinbeisser, H. and Blum, M. 1997. The dorsalizing and neural inducing gene follistatin is an antagonist of BMP-4. *Mech. Dev.* 63, 39–50.

Fainsod, A., Steinbeisser, H. and De Robertis, E.M. 1994. On the function of BMP-4 in patterning the marginal zone of the *Xenopus* embryo. *EMBO J.* 13, 5015–5025.

Fuentealba, L.C., Eivers, E., Ikeda, A., Hurtado, C., Kuroda, H., Pera, E.M., and De Robertis, E.M. 2007. Integrating patterning signals: Wnt/GSK3 regulates the duration of the BMP/Smad1 signal. *Cell* 131, 980–993.

Gawantka, V., Delius, H., Hirschfeld, K., Blumenstock, C. and Niehrs, C. 1995. Antagonizing the Spemann organizer: The role of the homeobox gene Xvent-1. *EMBO J* 14, 6268–6279.

Glinka, A., Wu, W., Delius, H., Monaghan, A.P., Blumenstock, C. and Niehrs, C. 1998. Dickkopf-1 is a member of a new family of secreted proteins and functions in head induction. *Nature* 391, 357–362.

Gont, L.K., Steinbeisser, H., Blumberg, B. and De Robertis, E.M. 1993. Tail formation as a continuation of gastrulation: The multiple cell populations of the *Xenopus* tailbud derive from the late blastopore lip. *Development* 119, 991–1004.

Hamburger, V., 1988. *The Heritage of Experimental Embryology: Hans Spemann and the Organizer*. Oxford University Press, Oxford.

Hemmati-Brivanlou, A., Kelly, O.G. and Melton, D.A. 1994. Follistatin, an antagonist of activin, is expressed in the Spemann organizer and displays direct neuralizing activity. *Cell* 77, 283–295.

Holley, S.A., Jackson, P.D., Sasai, Y., Lu, B., De Robertis, E.M., Hoffman, F.M. and Ferguson, E.L. 1995. A conserved system for dorsal-ventral patterning in insects and vertebrates involving short gastrulation and chordin. *Nature* 376, 249–253.

Hurtado, C. and De Robertis, E.M. 2007. Neural induction in the absence of organizer in salamanders is mediated by MAPK. *Dev. Biol.* 307, 282–289.

Inomata, H., Shibata, T., Haraguchi, T. and Sasai, Y. 2013. Scaling of dorsal-ventral patterning by embryo size-dependent degradation of Spemann's organizer signals. *Cell* 153, 1296–1311.

Karaulanov, E., Knöchel, W. and Niehrs, C. 2004. Transcriptional regulation of BMP4 synexpression in transgenic *Xenopus*. *EMBO J* 23, 844–856.

Khokha, M.K., Yeh, J., Grammer, T.C. and Harland, R.M., 2005. Depletion of three BMP antagonists from Spemann's organizer leads to a catastrophic loss of dorsal structures. *Dev. Cell* 8, 401–411.

Kirsch, N., Chang, L.S., Koch, S., Glinka, A., Dolde, C., Colozza, G., Benitez, M.D.J., De Robertis, E.M. and Niehrs, C. 2017. Angiopoietin-like 4 is a Wnt signaling antagonist that promotes LRP6 turnover. *Dev. Cell* 43, 71–82.

Klein, S.L. 1987. The first cleavage furrow demarcates the dorsal-ventral axis in Xenopus embryos. *Dev. Biol.* 120, 299–304.

Kuroda, H., Fuentealba, L., Ikeda, A., Reversade, B. and De Robertis, E.M. 2005. Default neural induction: Neuralization of dissociated *Xenopus* cells is mediated by Ras/MAPK activation. *Genes Dev.* 19, 1022–1027.

Kuroda, H., Wessely, O. and De Robertis, E.M. 2004. Neural induction in *Xenopus*: Requirement for ectodermal and endomesoderm signals via Chordin, Noggin, β-Catenin and Cerberus. *PLoS Biology* 2, 623–633.

Ladher, R., Mohun, T.J., Smith, J.C. and Snape, A.M. 1996. *Xom*: A *Xenopus* homeobox gene that mediates the early effects of BMP-4. *Development* 122, 2385–2394.

Lamb, T.M., Knecht, A.K., Smith, W.C., Stachel, S.E., Economides, A.N., Stahl, N., Yancopolous, G.D. and Harland, R.M. 1993. Neural induction by the secreted polypeptide noggin. *Science* 262, 713–718.

Lee, H.X., Ambrosio, A.L., Reversade, B. and De Robertis, E.M. 2006. Embryonic dorsal-ventral signaling: Secreted Frizzled-related proteins as inhibitors of Tolloid proteinases. *Cell* 124, 147–159.

Lee, H.X., Mendes, F.A., Plouhinec, J.L. and De Robertis, E.M. 2009. Enzymatic regulation of pattern: BMP4 binds CUB domains of Tolloids and inhibits proteinase activity. *Genes Dev*. 23, 2551–2562.

Leyns, L., Bouwmeester, T., Kim, S.H., Piccolo, S. and De Robertis, E.M. 1997. Frzb-1 is a secreted antagonist of Wnt signaling expressed in the Spemann organizer. *Cell* 88, 747–756.

Little, S.C. and Mullins, M.C. 2006. Extracellular modulation of BMP activity in patterning the dorsoventral axis. *Birth Def. Res*. 78, 224–242. Moody, S.A. 1987. Fates of the blastomeres of the 16-cell stage *Xenopus* embryo. *Dev. Biol*. 119, 560–578.

Moriyama, Y. and De Robertis, E.M. 2018. Embryonic regeneration by relocalization of the Spemann organizer during twinning in *Xenopus*. *Proc. Natl. Acad. Sci. USA* 115, E4815–E4822.

Morgan, T. H. 1895. Half embryos and whole embryos from one of the first two blastomeres. *Anat. Anz*. 10, 623–638.

Niehrs, C. and Pollet, N. 1999. Synexpression groups in eukaryotes. *Nature* 402, 483–487.

Oelgeschläger, M., Kuroda, H., Reversade, B. and De Robertis, E.M. 2003. Chordin is required for the Spemann organizer transplantation phenomenon in *Xenopus* embryos. *Dev. Cell* 4, 219–230.

Oelgeschläger, M., Larraín, J., Geissert, D. and De Robertis, E.M. 2000. The evolutionary conserved BMP-binding protein Twisted Gastrulation promotes BMP signalling. *Nature* 405, 757–763.

Onichtchouk, D., Chen, Y.G., Dosch, R., Gawantka, V., Delius, H., Massagué, J. and Niehrs, C. 1999. Silencing of TGF-beta signalling by the pseudoreceptor BAMBI. *Nature* 401, 480–485.

Onichtchouk, D., Gawantka, V., Dosch, R., Delius, H., Hirschfeld, K., Blumenstock, C. and Niehrs, C. 1996. The *Xvent-2* homeobox gene is part of the *BMP-4* signalling pathway controlling dorsoventral patterning of *Xenopus* mesoderm. *Development* 122, 3045–3053.

Pera, E. and De Robertis, E.M. 2000. A direct screen for secreted proteins in *Xenopus* embryos identifies distinct activities for the Wnt antagonists Crescent and Frzb-1. *Mech. Dev*. 96, 183–195.

Piccolo, S., Agius, E., Leyns, L., Battacharyya, S., Grunz, H., Bouwmeester, T. and De Robertis, E.M. 1999. The head inducer Cerberus is a multifunctional antagonist of Nodal, BMP and Wnt signals. *Nature* 397, 707–710.

Piccolo, S., Agius, E., Lu, B., Goodman, S., Dale, L. and De Robertis, E.M. 1997. Cleavage of Chordin by the Xolloid metalloprotease suggests a role for proteolytic processing in the regulation of Spemann organizer activity. Cell *91*, 407–416.

Piccolo, S., Sasai, Y., Lu, B. and De Robertis, E.M. 1996. Dorsoventral patterning in Xenopus: Inhibition of ventral signals by direct binding of Chordin to BMP-4. *Cell* 86, 589–598.

Plouhinec, J.L., Zakin, L., Moriyama, Y. and De Robertis, E.M. 2013. Chordin forms a self-organizing morphogen gradient in the extracellular space between ectoderm and mesoderm in the *Xenopus* embryo. *Proc. Natl. Acad. Sci. U.S.A*. 110, 20372–20379.

Reversade, B. and De Robertis, E.M. 2005. Regulation of ADMP and BMP2/4/7 at opposite embryonic poles generates a self-regulating morphogen field. *Cell* 123, 1147–1160.

Sander, V., Reversade, B. and De Robertis, E.M. 2007. The opposing homebox genes *Goosecoid* and *Ventl/2* self-regulate *Xenopus* patterning. *EMBO J*. 26, 2955–2965.

Sasai, Y., Lu, B., Steinbeisser, H. and De Robertis, E.M. 1995. Regulation of neural induction by the Chd and BMP-4 antagonistic patterning signals in *Xenopus*. *Nature* 376, 333–336.

Sasai, Y., Lu, B., Steinbeisser, H., Geissert, D., Gont, L.K. and De Robertis, E.M. 1994. Xenopus *chordin*: A novel dorsalizing factor activated by organizer-specific homeobox genes. *Cell* 79, 779–790.

Session, A.M., et al. 2016. Genome evolution in the allotetraploid frog *Xenopus laevis*. *Nature* 538, 336–343.

Smith, W.C. and Harland, R.M. 1992. Expression cloning of Noggin, a new dorsalizing factor localized to the Spemann organizer in Xenopus embryos. *Cell* 70, 829–840.

Spemann, H. 1938. *Embryonic Development and Induction*. Yale University Press, New Haven, reprinted by Hafner Publishing Company, 1962.

Spemann, H. and Mangold, H. 1924. Induction of embryonic primordial by implantation of organizers from a different species. *Roux's Arch. Entw. Mech*. 100, 599–638. Reprinted and translated in *Int. J. Dev. Biol*. 2001, 45, 13–31.

Taira, M., Jamrich, M., Good, P.J. and Dawid, I.B. 1992. The LIM domain-containing homeo box gene *Xlim-1* is expressed specifically in the organizer region of *Xenopus* gastrula embryos. *Genes Dev* 6, 356–366.

Vodicka, M.A. and Gerhart, J.C. 1995. Blastomere derivation and domains of gene expression in the Spemann organizer of *Xenopus laevis*. *Development* 121, 3505–3518.

Wilson, P. A. and Hemmati-Brivanlou, A. 1995. Induction of epidermis and inhibition of neural fate by Bmp-4. *Nature* 376, 331–333.

Zimmerman, L.B., De Jesús-Escobar, J.M. and Harland, R.M. 1996. The Spemann organizer signal Noggin binds and inactivates bone morphogenetic protein 4. *Cell* 86, 599–606.

Zinski, J., Tajer, B. and Mullins, M.C. 2018. TGF-β family signaling in early vertebrate development. *Cold Spring Harb. Perspect. Biol*. 10, a033274.

5 Signaling Pathways in Anterior-Posterior Patterning

Chenbei Chang

CONTENTS

5.1. Introduction 51
5.2. Classical Experimental Embryology Studies Brought Forth Different Models on Anterior-Posterior Neural Patterning 51
5.3. Insight into Signaling Pathways Controlling Anterior-Posterior Embryonic Patterning from Molecular Biological Studies 54
5.3.1. Retinoic Acid Signaling 54
5.3.2. Fibroblast Growth Factor Signaling 55
5.3.3. Wnt/β-Catenin Signaling Pathway 56
5.3.4. Multiple Pathways and Signal Integration 57
5.4. Synthesis of the Classical and Modern Studies and Some Unresolved Issues 58
Acknowledgments 60
References 60

5.1. INTRODUCTION

The work of Hans Spemann and Hilde Mangold on induction of a secondary dorsal axis by a piece of transplanted dorsal marginal tissue was one of the most impactful in the field of embryology (Spemann and Mangold, 1924; Spemann and Mangold, 2001). The experiments demonstrated not only developmental plasticity of early amphibian embryos but also the phenomenon of cell fate re-specification in response to inductive tissue interactions, which provide strong support for cell-cell communication based on the "organizer" idea (reviewed in Chapter 4). In the same work, Spemann and Mangold also noted that the anterior portion of the secondary neural tube and the optic vesicles were missing. They discussed the possibility of deficiency in certain parts of the implanted organizer that would be necessary for the induction of anterior neural plate with eye primordia (Spemann and Mangold, 1924; Spemann and Mangold, 2001). The issue of the organizer with regionalized head- and trunk-inducing ability was further taken up by Spemann in later experiments that used dorsal lips from early or older gastrula embryos, which showed that early organizers could induce a full secondary axis, including a head, whereas older organizers induced a partial ectopic axis with only trunk and tail regions (Hamburger, 1988; Spemann, 1936). Hence, it was understood early on that neural induction and specification of regional characteristics (patterning) of the induced neural tissues (head or trunk) were interlinked processes, and the source of the inducers, the timing of induction, and the properties of the responding tissues might all influence the outcome. The issue of patterned neural induction along the embryonic axis was actively pursued in subsequent studies. Spemann's students as well as scientists from other institutions investigated inductive interactions that gave rise to distinct brain structures or spinal cord, and different models were put forward to account for the signals involved in the process (Doniach, 1993; Slack and Tannahill, 1992). In the absence of the identities of the inducing substances, the quest for the anterior (cranial) and posterior (trunk and tail) neural inducers persisted to modern days. The advent of technologies for examining gene expression and manipulating gene functions in recent decades greatly enhanced our ability to interrogate this classical question with new rigor, and important insights have been gained into molecular control of both neural induction and neural patterning. This is especially pertinent using *Xenopus*, currently the most popular amphibian model of early development. This chapter aims to review key aspects of classical works that inspired different models on anterior-posterior (AP) neural patterning and discuss several signaling pathways identified using molecular biology approaches that regulate AP neural specification. Considerations on a number of remaining issues regarding AP patterning are also discussed briefly.

5.2. CLASSICAL EXPERIMENTAL EMBRYOLOGY STUDIES BROUGHT FORTH DIFFERENT MODELS ON ANTERIOR-POSTERIOR NEURAL PATTERNING

Development of structures along the head-to-tail (a.k.a. anterior-posterior or rostral-caudal) axis in vertebrate

DOI: 10.1201/9781003050230-6

embryos was best studied using amphibian animals in the first half of the 20th century due to practical advantages. *In vitro* growth of amphibian embryos in simple salt solutions not only allowed observation of embryogenesis at all stages but more importantly made embryos accessible to microsurgery. Once the composition of culture media was rendered optimal for embryo growth and infection could be minimized by careful operations, microsurgery became the key tool in studies of developmental trajectories of tissues. Isolated pieces cultured *in vitro* (explants), as well as embryonic parts transplanted into a host embryo, were examined in detail to gain information about tissue determination and plasticity. Stage- and region-specific microsurgeries were performed to examine progressive changes with time and at different embryonic positions in terms of both inducing and responsive capacity of the manipulated tissues. Gross morphological and histological analyses were employed to assess developmental outcomes of the operations. These experimental embryological approaches were used to explore tissue interactions involved in AP axis development. The patterning of the nervous system was especially investigated in detail due to the ease of identifying morphological and histological features of the brain regions and the spinal cord, and a range of amphibian species, including salamanders, newts, and frogs (urodeles/pleurodeles and anurans), were surveyed. These studies cumulated in several models that aimed to elucidate general principles underlying AP neural induction and patterning in amphibians.

Early cell fate mapping studies showed that mesendodermal and ectodermal precursor cells in blastula and early gastrula embryos had opposite AP polarity on the surface of the embryos (Slack and Tannahill, 1992; Vogt, 1929). The prospective anterior neural cells were localized toward the animal pole whereas the anterior mesendodermal precursors were close to the vegetal side. Internationalization and morphogenetic movements of mesendoderm during gastrulation led to reversal of mesendodermal AP polarity and subsequent registration of the AP axis between the germ layers (Figure 5.1A). The fate map studies raised questions about when the AP tissue identities were developed and how mesendodermal and neural AP development was coordinated.

In the neural tissue, morphological and histological features in the amphibian tadpoles, such as the position of sensory organs and characteristic patterns of ventricular thickness and axonal tracks in the brain, made it easy to distinguish between the forebrain, midbrain, hindbrain, and spinal cord (Slack and Tannahill, 1992). However, there were no definitive features to subdivide the prechordal mesendoderm, and both the notochord and the somites looked morphologically similar along the trunk. The AP distinction of the mesendoderm was hence limited only to the prechordal and the trunk regions in induced tissues. Because of this, the analysis of AP development was done mainly using the neural tissues. Distinct research approaches, including culture of explants, conjugation of different embryonic parts, and various types of tissue transplantations, were used by many groups to address specific issues concerning AP neural development.

To explore how AP neural features were acquired during embryogenesis, tissue conjugates were made with the presumptive inducing materials taken from the dorsal lip of gastrula embryos, tissues from different AP positions along the archenteron roof, or pieces of the neural plate at distinct AP levels. These isolates were sandwiched between one or two pieces of early gastrula-stage animal caps, and the resulting structures were scored for the appearance of brain or spinal cord traits (Doniach, 1993; Slack and Tannahill, 1992). The studies showed that early dorsal lips induced only anterior structures, whereas more posterior structures also formed when late dorsal lips were used (Okada and Takaya, 1942). In addition, anterior pieces of archenteron roof or neural plate induced only anterior structures in the animal caps, such as brain and eyes, whereas posterior archenteron roof or neural plate induced more posterior structures (Doniach, 1993; Slack and Tannahill, 1992). Though some differences were seen in inductive activities of archenteron roof and neural plate, the results from the sandwich studies demonstrated that AP neural characters could be induced in both temporal- and spatial-dependent manners. Forebrain-like structures were induced by early-stage organizer and anterior tissues, whereas spinal cord was induced by late-stage organizer and posterior tissues.

The idea that mesoderm at different AP positions induced neural tissues of corresponding AP characters was in fact suggested by an earlier experiment from Otto Mangold (1933) using a different approach. He inserted stripes of dorsal mesoderm taken from underneath the neural plate at different AP positions of early neurula embryos into the blastocoel cavity of early gastrula-stage embryos (the Einsteck experiment). Gastrulation movements would push the implanted tissues against the ventral wall of the blastocoel and allow inductive interaction between the implants and the host. Otto Mangold (1933) showed that when the implants were taken from the anterior regions, only anterior structures were induced in the ectopic position. However, when donor tissues were taken from the posterior regions, ectopic trunk structures appeared. He thus proposed that different mesodermal regions contained different neural inducers that could induce neural tissues with corresponding AP characteristics (Figure 5.1B).

A very different model of AP patterning was suggested by Pieter Nieuwkoop based on his tissue transplant experiments (Nieuwkoop, 1952a, 1952b, 1952c). Instead of using *in vitro* tissue explants or transplantation in gastrula embryos, he devised a distinctive way of inserting folded flaps of competent ectoderm into the neural plate of neurula stage embryos at different axial levels. He then used histological landmarks to score for formation of different structures in the folds at later stages. He observed that neural tissues were induced in the folds along the entire AP axis. Decreased

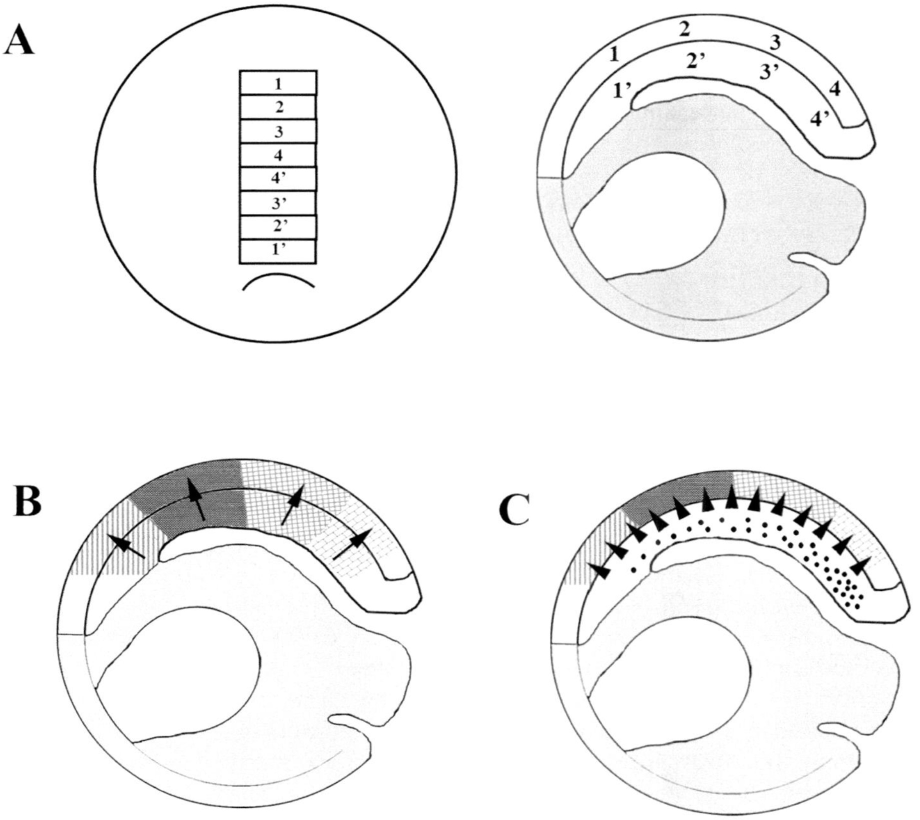

FIGURE 5.1 Distinct models for anterior-posterior neural induction. (A) Fate map of amphibian embryos reveals that the orientation of the anterior-posterior neural tissues (1–4) is opposite to that of the mesendodermal tissues (4'–1') prior to gastrulation. Internalization and movements of mesoderm and endoderm during gastrulation results in alignment of AP axis between the neural and the mesendoderm tissues. Left panel: dorsal view of early gastrula with anterior to the top and blastopore lip indicated. Right panel: mid-sagittal section of late gastrula with anterior to the left. Shaded areas indicate endoderm and ventral non-neural ectoderm. (B) Mangold's model of distinct regional neural inducers (arrows) that specify different neural characters, indicated by different patterns, along the anterior-posterior body axis. (C) Nieuwkoop's activation-transformation model proposes that a common neural inducer expressed along the entire anterior-posterior axis (arrowheads) can induce only forebrain-like neural tissues, whereas a separate transforming agent (dots), which cannot induce neural tissues on its own, can transform forebrain-like structures into more posterior neural tissues due to its graded distribution with a higher concentration at the posterior end or a longer exposure to the transforming agent by the posterior tissues.

neural induction was observed in more posterior territories. Within the folds, more posterior neural characteristics were found proximal to the junctions between the implants and the host, whereas more anterior neural characteristics could be identified distal to the implant-host connection. Nieuwkoop proposed that two distinct inducers were present in the embryo: an activator that was expressed throughout the axial mesoderm and induced forebrain-like structures, and a transformer that had higher concentrations at the posterior end and could transform forebrain into more posterior neural characters (Figure 5.1C). This model was supported by studies of other investigators, such as Eyal-Giladi and Yamada (Eyal-Giladi, 1954; Yamada, 1990). Yamada proposed that two events were required for formation of an organized neural system: (1) ectodermal dorsalization that was responsible for neural and neural crest (mesectoderm) differentiation and (2) caudalization that was responsible for the expression of posterior structures (Yamada, 1990). Like Nieuwkoop, Yamada suggested that caudalization functioned as a gradient. Besides the activation-transformation model, a modified two-signal hypothesis was also articulated by Saxén and Toivonen. They postulated a two-gradient model with "neuralizing" and "mesodermalizing" events that induced forebrain-like neural and mesodermal derivatives, respectively. They emphasized that the ratio of induced neural and mesodermal cells in the responding tissues determined the AP characteristics of the neural tissue, whereas secondary interactions among induced neural and mesodermal cells were important for organizing the specific AP neural structures (Saxen and Toivonen, 1961; Saxen et al., 1964).

The studies of classical experimental embryology using a variety of amphibian species thus brought forth several models of AP embryonic patterning of the neural tissues. While the activation-transformation model was gaining acceptance, active discussions remained about the source and the nature of the inducers, the competence of the ectoderm, and the timing of induction. The conversations about these questions would be clarified when the molecules responsible for neural induction and transformation were identified.

5.3. INSIGHT INTO SIGNALING PATHWAYS CONTROLLING ANTERIOR-POSTERIOR EMBRYONIC PATTERNING FROM MOLECULAR BIOLOGICAL STUDIES

Experimental embryology studies relied on morphological and histological features to assess the AP characteristics of the neural structures. The analyses were therefore performed long after the initial AP patterning event occurred. A high degree of expertise was required to distinguish structural differences in the neural and the associated tissues, especially since the induced structures were often disorganized. To overcome these issues, tissues from different species were used, and inductive experiments were conducted in explants and implants. These experimental limitations restricted further precision and quantitative understanding of the mechanisms controlling AP neural patterning.

The application of molecular biological techniques to embryological studies from the 1980s greatly facilitated investigation of developmental principles on all fronts, including issues concerning AP specification. By then, *Xenopus laevis* had become the most widely used amphibian model (Gurdon and Hopwood, 2000). *Xenopus* not only possesses all the advantages of the other amphibian species, such as accessibility to microsurgery, but also the adult *Xenopus* frogs are responsive to gonadotrophin hormone so that they can be induced to spawn throughout the year. The number of eggs can range from hundreds to thousands from a single female in a day, and the rate of embryonic development is rapid, with the embryos reaching the tadpole stage in three to four days. Molecular studies were thus implemented heavily in this species, and enormous progress was made on the molecular control of embryonic patterning in a relatively short time.

One important aspect that aided molecular analyses of AP neural induction was the cloning of molecular markers that not only revealed neural identity but also distinguished anterior from posterior neural tissues. Gene expression that specifically marked cement gland, forebrain, midbrain, hindbrain, and spinal cord made it possible to uncover events involved in early AP neural patterning prior to tissue differentiation and formation of specific structures (Blitz and Cho, 1995; Bradley et al., 1993; Brivanlou and Harland, 1989; Papalopulu et al., 1991a; Saha and Grainger, 1992; Sharpe et al., 1987; Sharpe and Gurdon, 1990; Sive et al., 1989). Using these markers, some of the classical experiments performed in other amphibian species were re-examined using *Xenopus*. Examples include the discovery of regional differences of the organizer (dorsal lip) that had distinct ability to induce different AP neural markers (Stewart and Gerhart, 1990; Vodicka and Gerhart, 1995; Zoltewicz and Gerhart, 1997), progressive conversion of anterior to more posterior cell fates in the nascent neural plate underlaid by post-involuted mesoderm (Sive et al., 1989), and regional differences in dorsal mesoderm in inducing AP neural markers at gastrula and neurula stages (Sharpe and Gurdon, 1990). These position-specific markers also allowed examination of different signaling pathways that control AP neural patterning. Several pathways have been identified that have the ability to convert anterior to more posterior neural tissues, with their antagonists expressed in the anterior region to promote development of the anterior structures.

5.3.1. Retinoic Acid Signaling

Retinoic acid (RA) was known for its teratogenic effects on mammals and could induce limb defects and microcephaly (Conlon, 1995). The impact of RA on head reduction was particularly intriguing and enticed several groups to investigate its action on nervous system development. Using exogenously supplied RA on whole embryos, early neural explants, or conjugates of competent ectoderm with dorsal mesoderm, it was demonstrated that RA reduced expression of genes characteristic of anterior structures, such as cement gland, olfactory pits, eyes, forebrain, and midbrain, and increased expression or anteriorly shifted the expression of posterior neural genes (Durston et al., 1989; Lloret-Vilaspasa et al., 2010; Papalopulu et al., 1991b; Ruiz i Altaba and Jessell, 1991b; Sive et al., 1990). The response to RA peaked at gastrula and gradually diminished during early neurula stages. RA not only affected general AP neural patterning but also influenced the differentiation of neurons at different AP positions (Papalopulu and Kintner, 1996). Measurement of endogenous RA in *Xenopus* embryos using HPLC or an RA-responsive reporter construct indicated that RA was present at gastrula and neurula stages when neural tissues were patterned along the AP axis, and a posterior-to-anterior concentration gradient of RA existed at early neurula stages (Chen et al., 1994; Durston et al., 1989; Yelin et al., 2005). RA did not seem to affect neural induction but could transform anterior to more posterior neural tissues. Mesodermal tissues could be altered similarly to express more posterior characters (Ruiz i Altaba and Jessell, 1991a). These results supported the notion that RA might potentially function as a posteriorizing factor in the "activation-transformation" model proposed by Nieuwkoop.

To complement the gain-of-function experiments applying ectopic RA, loss-of-function studies were performed using dominantly interfering receptor constructs for the RA receptors, RAR or RXR, both of which act as RA-dependent transcription activators. In addition to using explant experiments, dominant-negative (DN) receptors were often introduced into one side of the early *Xenopus* embryos so that the expression of a battery of AP neural markers could be compared with that on the control, uninjected side in the same embryo. These studies confirmed that blocking RA signaling directly impacted the level and/or the AP position of the region-specific neural markers (Blumberg et al.,

1997; Koide et al., 2001; Kolm et al., 1997; van der Wees et al., 1998). The hindbrain, but not the forebrain or posterior spinal cord, was particularly sensitive to manipulated RA signaling levels (Godsave et al., 1998; Kolm et al., 1997; van der Wees et al., 1998). However, the interpretation of the results might be complicated by the discovery that the unliganded RA receptors were not inactive but could bind to their targets and recruit co-repressors to act as a transcription repressor in the absence of RA (Koide et al., 2001; Weston et al., 2003). Examination of the distribution of RAR and RXR receptors (Ellinger-Ziegelbauer and Dreyer, 1991; Pfeffer and De Robertis, 1994) revealed that they were not only expressed in the posterior mesodermal and neural cells but also in the anterior mesendoderm (RARγ) or posterior to the hindbrain regions (RXRα). Thus, the pattern of RA receptor distribution was consistent with its role in regulating hindbrain development and posterior patterning.

The level of available RA in cells is controlled by multiple enzymes and RA binding proteins that influence the synthesis, degradation, and activity of RA. Several of the RA biogenesis enzymes, such as Aldh1a2/Raldh2, Rdh10, Sdr16C5/Rdhe2, and Cyp26c1, were analyzed for their functions in AP neural specification. Consistent with the studies manipulating RA or RA receptor activities, increased levels of *aldh1a2/raldh2*, *rdh10*, or *sdr16c5/rdhe2*, which promote RA biosynthesis, caused posteriorization of neural domains (Belyaeva et al., 2012; Chen et al., 2001; Strate et al., 2009). In contrast, increased levels of *cyp26c1*, which is involved in RA neutralization, anteriorized hindbrain and rescued defects induced by RA (Hollemann et al., 1998; Tanibe et al., 2008). *aldh1a2/raldh2* and *cyp26c1* are expressed in non-overlapping domains in *Xenopus* gastrula- and neurula-stage embryos and have antagonistic functions in regulating AP neural gene expression. In addition, ectopic expression of a cellular retinoic acid binding protein (Crabp2) induced embryonic AP defects and enhanced expression posterior *hox* genes, an effect similar to providing exogenous RA (Dekker et al., 1994). Taken together, the data strongly support a role of RA signaling in transforming anterior to more posterior neural tissues of *Xenopus* embryos, especially in the hindbrain region.

5.3.2. Fibroblast Growth Factor Signaling

Interest in fibroblast growth factor (FGF) signaling in early *Xenopus* development was stimulated in the late 1980s by the discovery that FGF could induce mesoderm formation in naïve ectodermal explants, referred to as animal caps (Kimelman et al., 1988; Slack et al., 1989, 1987, 1988). Molecular cloning of mammalian FGF homologs in *Xenopus* revealed that various FGF ligands and their receptors were expressed in early embryos. *In situ* hybridization on embryonic sections or whole embryos showed that the mRNAs encoding many FGF ligands, including *fgf4/efgf*, *fgf3/int-2*, and *fgf8*, were expressed in the marginal region above the blastopore at the gastrula stages and had dynamic distribution patterns including a posterior domain during neurula and tailbud stages (Christen and Slack, 1997; Isaacs et al., 1995, 1992; Lea et al., 2009; Lombardo et al., 1998; Tannahill et al., 1992). The seemingly graded expression of the FGF ligands and one of the FGF receptors, *fgfr1* (Friesel and Dawid, 1991; Golub et al., 2000; Lea et al., 2009), in the posterior region of late gastrula- and neurula-stage embryos was particularly interesting, as it implied a function of FGF signaling in posterior development.

A role of the FGF pathway in forming posterior structures was supported directly by both gain- and loss-of-function studies. Ectopic expression of FGF ligands, such as *fgf3*, *fgf4/efgf*, *fgf8*, or *fgf9* caused head reduction in the manipulated tadpoles (Christen and Slack, 1997; Lombardo et al., 1998; Pownall et al., 1996; Song and Slack, 1996), whereas expression of a dominant negative FGF receptor 1 (DN-Fgfr1 or XFD) led to trunk and tail truncation (Amaya et al., 1991; Godsave and Durston, 1997). As FGF was shown to regulate both mesodermal formation and posterior development, the direct posteriorization activity of FGF signaling was tested using explants. Neuralized animal caps or explants from anterior neural plate cultured with FGF could express hindbrain (*egr2/krox20*) and spinal cord (*hoxb9*) markers, whereas control explants lacked expression of these posterior neural genes (Cox and Hemmati-Brivanlou, 1995; Fletcher et al., 2006). The induction of posterior neural gene expression occurred in the absence of mesodermal markers, suggesting that FGF signaling could transform anterior into posterior neural tissues, supporting the transformation or caudalization hypothesis. When XFD or DN-Ras was expressed in neuralized animal caps treated with Fgf2/bFGF or conjugated with dorsal mesoderm, posterior, but not anterior, neural marker expression was blocked (Cox and Hemmati-Brivanlou, 1995; Holowacz and Sokol, 1999; Ribisi et al., 2000).

Although the experiments described above point to a crucial role of FGF signaling in caudalization of the nervous system, interpretation of the results may be complicated by several issues. One issue, as mentioned, is that FGF signaling is involved in both mesodermal induction and posterior neural development. It can be debated whether impairment of posterior structures when the pathway is inhibited *in vivo* is due to the activities of FGF on early mesoderm induction and not subsequent patterning of the neural tissue. Several approaches were used to demonstrate that FGF could caudalize neural tissues directly. The explant experiments described before took advantage of the ease of providing the Fgf2/bFGF protein at the desired temporal points of development and demonstrated that FGF signaling at gastrula to early neurula stages could induce posterior neural genes without concurrent induction of mesodermal markers (Cox and Hemmati-Brivanlou, 1995; Fletcher et al., 2006). FGF-soaked beads were also used for implantation into the neural plate directly, hence avoiding an early effect of stimulating FGF signaling (Lombardo and Slack, 1998; Pownall et al., 1996). In addition, a synthetic Fgfr1 receptor that can be activated by a dimerizing agent but not FGF ligands was used in embryos to induce FGF signaling with the drug at different time points. Stimulation of FGF signaling during mid-gastrulation could still lead to induction of posterior markers (Pownall et al., 2003). Furthermore, transgenic approaches have been used to block FGF signaling during gastrulation and neurulation, with the results supporting a function of FGF in posterior development (Pownall et al., 1998). A spliced isoform of fgf8, fgf8a, was further found to have minimal mesodermal inducing ability but was fully active in inducing posterior neural markers. Knockdown of this isoform did not affect mesodermal marker expression at gastrula stages but reduced expression of posterior neural genes in neurula stage embryos (Christen and Slack, 1997; Fletcher et al., 2006). Taken together, the data demonstrate that FGF signaling can act at two temporal phases to induce mesoderm and posteriorize neural tissues, respectively.

Another issue concerning FGF as a caudalizing factor is that while XFD blocked posterior *hox* neural gene induction by FGF in explants neuralized with BMP inhibitors, it had variable effects in preventing the induction of these *hox* genes in animal caps conjugated with the wild-type organizer (Curran and Grainger, 2000; Holowacz and Sokol, 1999). This implied that redundant endogenous signals could operate in the absence of FGF to posteriorize neural tissues. As we now know, other signals indeed work in parallel and cooperate with FGF to caudalize neural tissues.

A perhaps more serious issue is centered around whether FGF can directly induce posterior neural markers. Based on the model proposed by Mangold, there should be separate inducers for anterior and posterior neural tissues (Doniach, 1993; Mangold, 1933; Slack and Tannahill, 1992). However, if the activation-transformation model proposed by Nieuwkoop is in operation, the transforming agent cannot activate the neural program alone and should only change the characteristics of the neural tissues already induced by an activator present along the entire AP axis (Nieuwkoop, 1952a, 1952b, 1952c). Several groups reported that FGF induced both pan-neural markers and posterior neural genes in animal cap explants in the absence of mesoderm (Kengaku and Okamoto, 1995; Lamb and Harland, 1995). However, these studies seemed to have used explants that were either dissociated or kept partially open by a low calcium medium prior to the treatment with FGF. It is well documented that explants raised under these conditions experience decreased BMP signaling (Grunz and Tacke, 1989; Sato and Sargent, 1989). Therefore, these explants might have already been primed toward a neural state, which could then be changed into the posterior neural characters by FGF. Loss-of-function experiments using XFD/DN-Fgfrl or DN-Ras revealed that it did not block pan-neural markers, but another DN FGF receptor, DN-Fgfr4a, was shown to reduce neural induction by organizer or cell dissociation (Holowacz and Sokol, 1999; Hongo et al., 1999; Ribisi et al., 2000). The data imply that distinct FGF ligands and receptors may be involved in neural induction or maintenance in particular embryonic regions, whereas other FGF ligand/receptor pairs may play more crucial roles in neural caudalization.

The idea that the FGF pathway does not simply act as a graded posteriorizing signal is also suggested by the observation that interference of the pathway not only led to posterior truncation but also induced specific defects in sensory organs and/or head organization. The malformation was consistent with the expression patterns of FGF ligands and receptors, with many of them found in specific domains in the head regions from neurula stages onward (Lea et al., 2009). Hence, although FGF signaling is essential for posterior development, the specific ligand and receptor complexes expressed in different spatial positions and different developmental times influence the outcome of experiments that address AP neural specification.

5.3.3. Wnt/β-Catenin Signaling Pathway

The intensive exploration of growth factor signaling in early vertebrate development before the turn of the 21st century revealed that Wnt/β-Catenin signaling pathway was crucial for dorsal cell fate determination and could induce a secondary axis when activated in the ventral tissues of cleavage-stage embryos (Christian et al., 1991; McMahon and Moon, 1989; Smith and Harland, 1991; Sokol et al., 1991). Interestingly, it was reported that the opposite effects on embryonic development could be obtained depending on the timing of Wnt signal stimulation. Ectopic expression of Wnt/β-Catenin pathway components from mRNAs, which could be translated into proteins soon after injection, led to a dorsalized phenotype, whereas expression of the same signaling components from plasmids, which were transcribed and translated after the mid-blastula transition (MBT), resulted in head truncation (Christian et al., 1991; Christian and Moon, 1993; Darken and Wilson, 2001; Fredieu et al., 1997). These data pointed to the distinct functions of Wnt signaling during embryogenesis, with an early phase involved in dorsal-ventral patterning and a late phase in head suppression. A role of Wnt signaling in AP patterning was further supported both by expression patterns of Wnt pathway components and by functional manipulation of Wnt signaling levels. Ectopic expression, knockdown, or dominantly interfering approaches in post-MBT embryos or in neuralized animal caps and neural plate explants showed that changing the levels or the activities of Wnt ligands, signal transducers (e.g. Dishevelled), or nuclear transcription factors (e.g. β-catenin) all led to alterations in AP neural marker expression or defects in head or trunk structures (Darken and Wilson, 2001; Domingos et al., 2001; Fredieu et al., 1997; Itoh and Sokol, 1997; McGrew et al., 1997; McGrew et al., 1995; Wheeler et al., 2000). These results suggest that, similarly to RA and FGF signals, the Wnt/β-Catenin pathway can act as a transforming factor to specify posterior neural development.

5.3.3.1. Wnt Pathway in Head Formation

Though AP patterning of the neural tissues is often at the center of the investigation, key insights about Wnt/β-Catenin signaling in AP patterning of the whole embryo, which includes mesendodermal tissues, came from studies of organizer-enriched molecules, including the secreted factors Cerberus and the Frizzled-like gene Frzb-1 (Bouwmeester et al., 1996; Leyns et al., 1997). Their expression domains were complementary to that of zygotic *wnt8a*, and both genes, *cer1* and *frzb1*, were shown to promote head formation in *Xenopus* embryos. While Frzb-1 was demonstrated to inhibit Wnt/β-Catenin signaling by direct binding to the ligands (Leyns et al., 1997; Lin et al., 1997), the action of Cerberus initially was unclear. The breakthrough came when it was discovered that induction of a complete secondary axis on the ventral side required not only inhibition of BMP signaling but also simultaneous inhibition of the Wnt pathway. BMP inhibitors induced only a secondary trunk, whereas inhibition of both BMP and Wnt signals produced a secondary axis with both the head and the trunk (Glinka et al., 1997). Functional cDNA expression library screening for endogenous molecules to synergize with BMP inhibitors to induce a complete secondary axis identified Dickkopf (Dkk1), a secreted molecule that was required for head formation during *Xenopus* development (Glinka et al., 1998). Dkk1 turned out to be a Wnt inhibitor via binding to the Wnt co-receptor Lrp5/Lrp6 to prevent Wnt signaling (Mao et al., 2001; Semenov et al., 2001; Yamamoto et al., 2008). Further investigations demonstrated that Cerberus also inhibited both Wnt and BMP signals by direct binding to ligands of both families to promote head formation (Glinka et al., 1997; Piccolo et al., 1999; Silva et al., 2003). This series of studies helped revise our understanding on how AP patterning is achieved. Instead of the two-signal model of an activator and a transformer in neural AP patterning, signal antagonists are required to actively repress posteriorizing influence of the Wnt signal to ensure the normal development of anterior structures. In these cases, the Wnt antagonists may regulate anterior mesendoderm as well as neural tissues, hence complicating the interpretation of whether the antagonists act directly to influence AP neural patterning.

Once it was known that inhibition of Wnt signaling is crucial for head development, the assays for genes to either directly regulate head formation or to cooperate with BMP inhibitors to induce a complete secondary axis were used to identify other molecules that may regulate Wnt signal transduction. The expression and the endogenous functions of these molecules in head-trunk formation in *Xenopus* embryos were then analyzed. These studies revealed a surprisingly large number of regulators of Wnt signaling that have relevant expression domains for patterning the AP axis of the embryo and act to modulate Wnt signaling to affect head or trunk formation. Moreover, novel regulatory mechanisms were uncovered that show how the array of the new genes modulate Wnt/β-Catenin signaling at different subcellular levels. For example, the type I transmembrane receptors Kremen1 and 2 were shown to bind to Dkk1 and Lrp5/6 to form a ternary complex that enhances endocytosis and removal of Lrp5/6 from the plasma membrane to inhibit Wnt signaling (Davidson et al., 2002; Mao et al., 2002). Knockdown of Kremen1/2 leads to head truncation. Interestingly, in the absence of Dkk1, Kremen2 associates with Lrp6 to enhance Wnt/β-Catenin signaling (Hassler et al., 2007). Kremen therefore seems to modulate Wnt signaling levels in a context-dependent manner to regulate early *Xenopus* development. Regulation of the Wnt receptors on the plasma membrane can also be achieved by other molecules. The ER-localized protein Shisa is expressed in the organizer, can prevent maturation and surface expression of the Wnt receptor Frizzled, and functions to promote head formation in *Xenopus* embryos (Yamamoto et al., 2005). The organizer-expressed protein tyrosine phosphatase receptor-type kappa (PTPRK) suppresses Wnt signaling by regulating surface levels of both Lrp6 and Frizzled via the transmembrane E3 ubiquitin ligase ZNRF3 to control head formation (Chang et al., 2020). The ER transmembrane protein TMEM79 interacts with and inhibits the deubiquitinase, USP8, to facilitate degradation of Frizzled receptor (Chen et al., 2020). Bighead, a secreted organizer-specific protein, binds to Lrp6 to remove it from the cell surface. Gain- and loss-of-function assays reveal that Bighead is required for head formation (Ding et al., 2018). Besides the receptors, the availability of the Wnt ligands is modulated by a variety of factors. Tiki1 is a transmembrane protein that cleaves Wnt to promote oxidation and oligomerization of Wnt ligands and prevents them from binding to the Wnt receptors (Zhang et al., 2012). The secreted factor Notum is a Wnt deacylase and promotes deacylated Wnt ligands to form oxidized oligomers to prevent Wnt signaling (Zhang et al., 2015). Both Tiki1 and Notum are required for head formation, though they may act in different embryonic regions, with Tiki1 in the organizer and Notum broadly in the ectoderm and weakly in the mesoderm. Cytoplasmic and nuclear signaling molecules of the Wnt pathway are also regulated by different factors that influence head-trunk development. The phosphatase Pgam5 interacts with and dephosphorylates Dishevelled to prevent Wnt signaling (Rauschenberger et al., 2017). The GPCR proteins Flop1 and 2 promote β-Catenin degradation (Miyagi et al., 2015). March2, a membrane-associated E3 ubiquitin ligase, promotes degradation of Dishevelled via the adaptor protein Dapper to regulate head formation (Lee et al., 2018). Idax, a Dvl binding protein, prevents interaction between Dvl and Axin to inhibit the pathway (Michiue et al., 2004). NF2/Merlin, a FERM-domain containing protein, also inhibits Wnt signaling upstream of β-Catenin (Zhu et al., 2015). Custos binds to and controls cytoplasmic to nuclear shuttling of β-Catenin to inhibit Wnt signaling (Komiya et al., 2014). The studies of head-trunk formation in *Xenopus* therefore prove a fruitful ground for exciting discoveries of novel Wnt regulators and detailed mechanisms of Wnt regulation in development.

5.3.4. Multiple Pathways and Signal Integration

Although RA, FGF, and Wnt are the main signals being investigated in depth for their roles in AP patterning of the neural plate and the body axis, other signals also regulate

AP axis formation. Both the Nodal and the BMP branches of the TGF-β pathway have been shown to modulate AP patterning (Piccolo et al., 1999; Polevoy et al., 2019). Ectopic stimulation of Nodal or BMP signaling leads to head truncation of *Xenopus* embryos, whereas Cerberus inhibits both Nodal and BMP in addition to Wnt. Crescent, another organizer-specific secreted factor, belongs to the soluble Frizzled-related protein (sFRP) family and also regulates both Wnt and BMP signals (Pera and De Robertis, 2000; Ploper et al., 2011; Shibata et al., 2005; Shibata et al., 2000). The Nodal-like ligand Derriere is also involved in posterior development. Derriere is expressed in the marginal zone with high dorsal levels during gastrulation and is detected subsequently in the posterior region of neurula embryos. Ectopic expression of Derriere leads to the reduction of head structures (Sun et al., 1999). In addition, insulin-like growth factor (IGF) signaling regulates head formation via inhibition of the Wnt pathway (Richard-Parpaillon et al., 2002). Therefore, many pathways participate in controlling AP patterning during *Xenopus* embryogenesis.

As expected, the signals do not work independently and often cooperate with each other. Activation of posterior markers in neuralized animal caps or embryos exhibits a degree of co-dependence between the Wnt, RA, and FGF signaling pathways so that in the absence of one, the induction of hindbrain or spinal cord markers by another signal is often compromised (Domingos et al., 2001; McGrew et al., 1997; Roche et al., 2009; Shiotsugu et al., 2004). Redundancy of multiple signals *in vivo* may make detection of signal cooperation more complex in whole embryos. Quite often a regulatory protein for AP patterning can simultaneously modulate different pathways: Cerberus antagonizes Nodal, BMP, and Wnt signals (Piccolo et al., 1999), and Crescent regulates Wnt as well as BMP pathways (Ploper et al., 2011). The transmembrane protein Shisa controls maturation and cell surface localization of both FGFR and Frizzled receptors (Yamamoto et al., 2005). Cytoplasmic kinases and phosphatases can also modulate phosphorylation of an array of proteins involved in distinct signaling pathways. On the other hand, closely related protein members of the same family can have different regulatory activities for distinct pathways. Examples include Noggin 4, which blocks Wnt instead of BMP signaling (Eroshkin et al., 2016); Noggin 2, which inhibits Nodal and Wnt in addition to BMP (Bayramov et al., 2011); and Dkk3, which regulates TGF-β and FGF pathways instead of Wnt pathways (Pinho and Niehrs, 2007).

In addition to direct interaction of components of different pathways, crosstalk between signals can happen at the transcription level. Expression of Fgfrl and Fgfr4 depends on RA signaling (Shiotsugu et al., 2004), whereas Wnt3a activates transcription of FGF3 and FGF8 via the transcription factor Meis3 (Gutkovich et al., 2010), and FGF and Wnt ligands can be co-regulated by the same transcription factors involved in AP patterning (e.g. Tbx6 activates expression of *fgf8* and *wnt8a*, whereas JunB stimulates expression of both *fgf3* and *wnt8a* (Lou et al., 2006; Yoshida et al., 2016). Distinct signals can also converge directly in the gene regulatory networks that control expression of caudalizing factors. DNA binding motifs for Ets and TCF, the transcription factors downstream of FGF and Wnt signals, respectively, are both present and often positioned in close proximity in the regulatory region of *cdx4* and *cdx1*, encoding caudal-like homeodomain proteins involved in posterior development (Haremaki et al., 2003; Isaacs et al., 1998; Kjolby and Harland, 2017; Kjolby et al., 2019). FGF, Wnt, and RA have all be shown to regulate *hox* gene expression (Dekker et al., 1992; Durston, 2019; In der Rieden et al., 2010; Janssens et al., 2010) and may potentially cooperate at the regulatory regions of the *hox* genes via binding by respective effector transcription factors of these signals, although the exact mechanisms of regulatory input have not been investigated in detail for all the signals.

Control of AP patterning by multiple pathways begs the questions whether all signals function similarly and the redundancy simply exists to ensure robustness of the patterning system, or different signals have non-overlapping activities to influence specific aspects of AP development. Most of the molecular studies on AP neural formation focused on a limited set of markers, hence restricting a distinction between the different effects by various signals. Conclusions drawn from explant assays are frequently considered to equate to those using whole embryos, but the expression levels of the markers are often analyzed in explants and the positions of the marker domains assessed in manipulated embryos. Results from explant and embryo studies are therefore not always congruent, especially in the spinal cord region. For example, although posterior *hox* genes can be blocked efficiently by functional reduction of a caudalizing factor in explants, the expression domains of the same *hox* genes can remain the same in manipulated embryos (Curran and Grainger, 2000). The embryonic regions most sensitive to altered signaling levels are in the fore-, mid- and hindbrain, with their respective markers shifting anteriorly or posteriorly relative to the control side when the signal levels are modified. The comparative insensitivity of the spinal cord to signal manipulation may reflect functional redundancy of multiple signals, but it can also indicate that the signals have limited AP patterning capacity in the trunk region of the embryos. Simultaneous modulation of several pathways may help to resolve the issue.

5.4. SYNTHESIS OF THE CLASSICAL AND MODERN STUDIES AND SOME UNRESOLVED ISSUES

Classical experimental embryologists used microsurgery, explant, or transplant studies to address several crucial questions concerning AP embryonic patterning in amphibians. What regions of early embryonic tissues give rise to anterior or posterior structures at later stages? When do the tissues acquire AP characteristics? How can naïve tissues be specified along a particular developmental path; can this developmental trajectory be altered; and, if so, how? How do tissue interactions between mesendoderm and ectoderm or neural and ectoderm influence AP neural specification? What is the physicochemical nature of the inducers that control AP development? Does the induction always occur between tissues of different germ layers (vertical induction), or can it happen within cells of the same germ layer (planar or homeogenetic induction)? How many signals are employed to provide positional information along the AP axis? Once molecular biology emerged on the scene, modern developmental biologists took advantage of molecular tools to identify, monitor the expression of, and manipulate genes in *Xenopus* to tackle these same key questions at the molecular levels. What signals can affect expression patterns of AP markers in explants or embryos? When are the signals required? How many signals work in parallel or sequentially? What is the regulatory function of each signal for AP development? Discoveries made using molecular approaches helped to provide explanations to some of the classical questions and lent support to or refuted the models put forward decades ago.

One main question that occupied the attention of both experimental and molecular biologists is how AP neural patterning is executed by post-involuted mesendoderm as it advances toward the future anterior end of the embryo during gastrulation. Based on tissue explant and transplant experiments, distinct models were proposed that favored either the presence of multiple inducers along the AP axis or the existence of two main signals with a universal inducer of anterior neural character and a separate graded transformer responsible for posterior neural development (Figure 5.1) (Mangold, 1933; Nieuwkoop, 1952a, 1952b, 1952c). The evaluation of the two models in the modern era was facilitated by the identification of organizer-localized secreted BMP inhibitors as direct neural inducers. All these inducers, including Noggin, Chordin, and Follistatin, which are expressed in midline dorsal mesoderm along the entire notochord during neurulation, can only induce anterior neural markers in animal cap explants (Hemmati-Brivanlou et al., 1994; Lamb et al., 1993; Sasai et al., 1995, 1994; Smith and Harland, 1992). When combined with activation of the signaling pathways discussed previously, posterior neural genes can be expressed as well. These results, together with gain- and loss-of-function studies *in vivo*, accelerated the acceptance of the activation-transformation model of AP specification by the field. However, the story seems to be more complex than expected. FGF signaling has been implicated in inducing posterior neural markers directly without any "activators," but only particular pairs of FGF ligands (Fgf8a) and receptors (Fgfr4) may be involved in direct induction. This implies that the multiple-inducer model of Mangold may be compatible with some results. In addition, as discussed previously, interference of caudalizing signals often impairs expression of markers in the hindbrain, with less effect on markers of the spinal cord. It is possible that these results reflect redundancy of signals in the trunk, but it is also conceivable that distinct ratios of various active signals at different AP positions serve to divide the neural plate into separate signaling zones, with each specifying a particular AP cell fate. This would be a hybrid paradigm between the multi-inducer and activation-transformation models in that instead of the presence of multiple position-specific inducers, a universal neural inducer works within multiple transformation zones to specify AP neural tissues. Combinations of varying gradients of growth factor signals would define the zones and encode AP positional information. The dynamic expression patterns of signaling molecules and the presence of a myriad of inhibitors for all three major caudalizing pathways are consistent with this model. One additional consideration is the timing of AP specification. Some conclusions from the classical experiments were based on the studies using neurula-stage embryos when the AP axis had already been well laid out. The signals at these stages are not exactly the same as those at mid- to late gastrula stages when AP tissues are being specified. Dynamic temporal expression of signaling molecules is often observed during *Xenopus* embryogenesis so that neurula stage signals may play more important roles in maintenance or elaboration of posterior tissue development rather than inducing AP characteristics. Stage-specific manipulations of gene functions and dynamic temporal analyses of marker expression are both required to address this issue.

While molecular biological studies have contributed novel insights and proposed detailed mechanisms concerning AP patterning, several issues remain unresolved. The patterned expression of signaling molecules and their modulators in distinct domains of the Spemann organizer precedes future AP specification. However, little is known about molecular circuitry controlling organizer subdivision. The competence of neuralized animal caps or neural plate explants to respond to caudalizing signals is lost during mid-neurulation, yet the molecular machinery responsible for this loss is not understood. The activation-transformation model proposed by Nieuwkoop may operate via gradient distribution of a substance with an apex level at the posterior end or different durations that competent ectoderm is exposed to a constant transforming agent at different AP positions. Most molecular studies so far tend to focus on the chemical gradients, but the temporal gradient of signal exposure has not been investigated in depth. Though multiple signals are known to specify posterior tissues, how the signals are transduced and converge to control downstream target gene expression has only been explored slightly. Several transcription factors, such as Cdx members, Meis1/3, and Hox proteins, are required for hindbrain or posterior neural development (Dibner et al., 2001, 2004; Elkouby et al., 2010; Epstein et al., 1997; Faas and Isaacs, 2009; Gutkovich et al., 2010; Pillemer et al., 1998; Schyr et al., 2012). The bidirectional interplay between these transcription factors and the caudalizing signals is not understood in detail. The eventual development of AP neural tissues requires not only instructions from the underlying mesoderm but also further refinement within the neural plate via self-organization. This process is not well appreciated at the molecular level. Most of the work so far centers on AP patterning of the neural tissues, with only limited studies on AP specification of mesodermal and endodermal tissues (Deimling and Drysdale, 2009, 2011; McLin et al., 2007; Rankin et al., 2018). AP patterning not only involves cell fates but also morphogenesis, with posterior tissues undergoing elongation. How caudalizing signals modulate expression or activity of regulators of cell behaviors is largely unclear (Janesick et al., 2014). These issues await further examination. With microsurgery, molecular biology, and genomic tools at hand, we expect to obtain a more detailed picture and deeper understanding of AP embryonic patterning in the future.

ACKNOWLEDGMENTS

This work is supported by NIH grants R01GM127371 and R01HD102015. I apologize to the investigators whose works are not cited here due to the space limitation.

REFERENCES

Amaya, E., T.J. Musci, and M.W. Kirschner. 1991. Expression of a dominant negative mutant of the FGF receptor disrupts mesoderm formation in Xenopus embryos. *Cell*. 66:257–270.

Bayramov, A.V., F.M. Eroshkin, N.Y. Martynova, G.V. Ermakova, E.A. Solovieva, and A.G. Zaraisky. 2011. Novel functions of Noggin proteins: Inhibition of Activin/Nodal and Wnt signaling. *Development*. 138:5345–5356.

Belyaeva, O.V., S.A. Lee, M.K. Adams, C. Chang, and N.Y. Kedishvili. 2012. Short chain dehydrogenase/reductase rdhe2 is a novel retinol dehydrogenase essential for frog embryonic development. *J Biol Chem*. 287:9061–9071.

Blitz, I.L., and K.W. Cho. 1995. Anterior neurectoderm is progressively induced during gastrulation: The role of the Xenopus homeobox gene orthodenticle. *Development*. 121:993–1004.

Blumberg, B., J. Bolado, Jr., T.A. Moreno, C. Kintner, R.M. Evans, and N. Papalopulu. 1997. An essential role for retinoid signaling in anteroposterior neural patterning. *Development*. 124:373–379.

Bouwmeester, T., S. Kim, Y. Sasai, B. Lu, and E.M. De Robertis. 1996. Cerberus is a head-inducing secreted factor expressed in the anterior endoderm of Spemann's organizer. *Nature*. 382:595–601.

Bradley, L.C., A. Snape, S. Bhatt, and D.G. Wilkinson. 1993. The structure and expression of the Xenopus Krox-20 gene: Conserved and divergent patterns of expression in rhombomeres and neural crest. *Mech Dev*. 40:73–84.

Brivanlou, A.H., and R.M. Harland. 1989. Expression of an engrailed-related protein is induced in the anterior neural ectoderm of early Xenopus embryos. *Development*. 106:611–617.

Chang, L.S., M. Kim, A. Glinka, C. Reinhard, and C. Niehrs. 2020. The tumor suppressor PTPRK promotes ZNRF3 internalization and is required for Wnt inhibition in the Spemann organizer. *Elife*. 9.

Chen, M., N. Amado, J. Tan, A. Reis, M. Ge, J.G. Abreu, and X. He. 2020. TMEM79/MATTRIN defines a pathway for Frizzled regulation and is required for Xenopus embryogenesis. *Elife*. 9.

Chen, Y., L. Huang, and M. Solursh. 1994. A concentration gradient of retinoids in the early *Xenopus laevis* embryo. *Dev Biol*. 161:70–76.

Chen, Y., N. Pollet, C. Niehrs, and T. Pieler. 2001. Increased XRALDH2 activity has a posteriorizing effect on the central nervous system of Xenopus embryos. *Mech Dev*. 101:91–103.

Christen, B., and J.M. Slack. 1997. FGF-8 is associated with anteroposterior patterning and limb regeneration in Xenopus. *Dev Biol*. 192:455–466.

Christian, J.L., J.A. McMahon, A.P. McMahon, and R.T. Moon. 1991. Xwnt-8, a Xenopus Wnt-1/int-1-related gene responsive to mesoderm-inducing growth factors, may play a role in ventral mesodermal patterning during embryogenesis. *Development*. 111:1045–1055.

Christian, J.L., and R.T. Moon. 1993. Interactions between Xwnt-8 and Spemann organizer signaling pathways generate dorsoventral pattern in the embryonic mesoderm of Xenopus. *Genes Dev*. 7:13–28.

Conlon, R.A. 1995. Retinoic acid and pattern formation in vertebrates. *Trends Genet*. 11:314–319.

Cox, W.G., and A. Hemmati-Brivanlou. 1995. Caudalization of neural fate by tissue recombination and bFGF. *Development*. 121:4349–4358.

Curran, K.L., and R.M. Grainger. 2000. Expression of activated MAP kinase in *Xenopus laevis* embryos: Evaluating the roles of FGF and other signaling pathways in early induction and patterning. *Dev Biol*. 228:41–56.

Darken, R.S., and P.A. Wilson. 2001. Axis induction by Wnt signaling: Target promoter responsiveness regulates competence. *Dev Biol*. 234:42–54.

Davidson, G., B. Mao, I. del Barco Barrantes, and C. Niehrs. 2002. Kremen proteins interact with Dickkopf1 to regulate anteroposterior CNS patterning. *Development*. 129:5587–5596.

Deimling, S.J., and T.A. Drysdale. 2009. Retinoic acid regulates anterior-posterior patterning within the lateral plate mesoderm of Xenopus. *Mech Dev*. 126:913–923.

Deimling, S.J., and T.A. Drysdale. 2011. FGF is required to regulate anterior-posterior patterning in the Xenopus lateral plate mesoderm. *Mech Dev*. 128:327–341.

Dekker, E.J., M. Pannese, E. Houtzager, A. Timmermans, E. Boncinelli, and A. Durston. 1992. Xenopus Hox-2 genes are expressed sequentially after the onset of gastrulation and are differentially inducible by retinoic acid. *Dev Suppl*:195–202.

Dekker, E.J., M.J. Vaessen, C. van den Berg, A. Timmermans, S. Godsave, T. Holling, P. Nieuwkoop, A. Geurts van Kessel, and A. Durston. 1994. Overexpression of a cellular retinoic acid binding protein (xCRABP) causes anteroposterior defects in developing Xenopus embryos. *Development*. 120:973–985.

Dibner, C., S. Elias, and D. Frank. 2001. XMeis3 protein activity is required for proper hindbrain patterning in *Xenopus laevis* embryos. *Development*. 128:3415–3426.

Dibner, C., S. Elias, R. Ofir, J. Souopgui, P.J. Kolm, H. Sive, T. Pieler, and D. Frank. 2004. The Meis3 protein and retinoid signaling interact to pattern the Xenopus hindbrain. *Dev Biol*. 271:75–86.

Ding, Y., G. Colozza, E.A. Sosa, Y. Moriyama, S. Rundle, L. Salwinski, and E.M. De Robertis. 2018. Bighead is a Wnt antagonist secreted by the Xenopus Spemann organizer that promotes Lrp6 endocytosis. *Proc Natl Acad Sci U S A*. 115:E9135–E9144.

Domingos, P.M., N. Itasaki, C.M. Jones, S. Mercurio, M.G. Sargent, J.C. Smith, and R. Krumlauf. 2001. The Wnt/beta-catenin pathway posteriorizes neural tissue in Xenopus by an indirect mechanism requiring FGF signalling. *Dev Biol*. 239: 148–160.

Doniach, T. 1993. Planar and vertical induction of anteroposterior pattern during the development of the amphibian central nervous system. *J. Neurobio*. 24:1256–1275.

Durston, A.J. 2019. What are the roles of retinoids, other morphogens, and Hox genes in setting up the vertebrate body axis? *Genesis*. 57:e23296.

Durston, A.J., J.P. Timmermans, W.J. Hage, H.F. Hendriks, N.J. de Vries, M. Heideveld, and P.D. Nieuwkoop. 1989. Retinoic acid causes an anteroposterior transformation in the developing central nervous system. *Nature*. 340:140–144.

Elkouby, Y.M., S. Elias, E.S. Casey, S.A. Blythe, N. Tsabar, P.S. Klein, H. Root, K.J. Liu, and D. Frank. 2010. Mesodermal Wnt signaling organizes the neural plate via Meis3. *Development*. 137:1531–1541.

Ellinger-Ziegelbauer, H., and C. Dreyer. 1991. A retinoic acid receptor expressed in the early development of *Xenopus laevis*. *Genes Dev*. 5:94–104.

Epstein, M., G. Pillemer, R. Yelin, J.K. Yisraeli, and A. Fainsod. 1997. Patterning of the embryo along the anterior-posterior axis: The role of the caudal genes. *Development*. 124:3805–3814.

Eroshkin, F.M., A.M. Nesterenko, A.V. Borodulin, N.Y. Martynova, G.V. Ermakova, F.K. Gyoeva, E.E. Orlov, A.A. Belogurov, K.A. Lukyanov, A.V. Bayramov, and A.G. Zaraisky. 2016. Noggin4 is a long-range inhibitor of Wnt8 signalling that regulates head development in *Xenopus laevis*. *Sci Rep*. 6:23049.

Eyal-Giladi, H. 1954. Dynamic aspects of neural induction in amphibia. *Arch Biol (Liege)*. 65:179–259.

Faas, L., and H.V. Isaacs. 2009. Overlapping functions of Cdx1, Cdx2, and Cdx4 in the development of the amphibian *Xenopus tropicalis*. *Dev Dyn*. 238:835–852.

Fletcher, R.B., J.C. Baker, and R.M. Harland. 2006. FGF8 spliceforms mediate early mesoderm and posterior neural tissue formation in Xenopus. *Development*. 133:1703–1714.

Fredieu, J.R., Y. Cui, D. Maier, M.V. Danilchik, and J.L. Christian. 1997. Xwnt-8 and lithium can act upon either dorsal mesodermal or neurectodermal cells to cause a loss of forebrain in Xenopus embryos. *Dev Biol*. 186:100–114.

Friesel, R., and I.B. Dawid. 1991. cDNA cloning and developmental expression of fibroblast growth factor receptors from *Xenopus laevis*. *Mol Cell Biol*. 11:2481–2488.

Glinka, A., W. Wu, H. Delius, A.P. Monaghan, C. Blumenstock, and C. Niehrs. 1998. Dickkopf-1 is a member of a new family of secreted proteins and functions in head induction. *Nature*. 391:357–362.

Glinka, A., W. Wu, D. Onichtchouk, C. Blumenstock, and C. Niehrs. 1997. Head induction by simultaneous repression of Bmp and Wnt signalling in Xenopus. *Nature*. 389:517–519.

Godsave, S.F., and A.J. Durston. 1997. Neural induction and patterning in embryos deficient in FGF signaling. *Int J Dev Biol*. 41:57–65.

Godsave, S.F., C.H. Koster, A. Getahun, M. Mathu, M. Hooiveld, J. van der Wees, J. Hendriks, and A.J. Durston. 1998. Graded retinoid responses in the developing hindbrain. *Dev Dyn*. 213:39–49.

Golub, R., Z. Adelman, J. Clementi, R. Weiss, J. Bonasera, and M. Servetnick. 2000. Evolutionarily conserved and divergent expression of members of the FGF receptor family among vertebrate embryos, as revealed by FGFR expression patterns in Xenopus. *Dev Genes Evol*. 210:345–357.

Grunz, H., and L. Tacke. 1989. Neural differentiation of *Xenopus laevis* ectoderm takes place after disaggregation and delayed reaggregation without inducer. *Cell Differ Dev*. 28:211–217.

Gurdon, J.B., and N. Hopwood. 2000. The introduction of *Xenopus laevis* into developmental biology: Of empire, pregnancy testing and ribosomal genes. *Int J Dev Biol*. 44:43–50.

Gutkovich, Y.E., R. Ofir, Y.M. Elkouby, C. Dibner, A. Gefen, S. Elias, and D. Frank. 2010. Xenopus Meis3 protein lies at a nexus downstream to Zic1 and Pax3 proteins, regulating multiple cell-fates during early nervous system development. *Dev Biol*. 338:50–62.

Hamburger, V. 1988. *The Heritage of Experimental Embryology: Hans Spemann and the Organizer*. Oxford University Press, Oxford.

Haremaki, T., Y. Tanaka, I. Hongo, M. Yuge, and H. Okamoto. 2003. Integration of multiple signal transducing pathways on Fgf response elements of the Xenopus caudal homologue Xcad3. *Development*. 130:4907–4917.

Hassler, C., C.M. Cruciat, Y.L. Huang, S. Kuriyama, R. Mayor, and C. Niehrs. 2007. Kremen is required for neural crest induction in Xenopus and promotes LRP6-mediated Wnt signaling. *Development*. 134:4255–4263.

Hemmati-Brivanlou, A., O.G. Kelly, and D.A. Melton. 1994. Follistatin, an antagonist of activin, is expressed in the Spemann organizer and displays direct neuralizing activity. *Cell*. 77:283–295.

Hollemann, T., Y. Chen, H. Grunz, and T. Pieler. 1998. Regionalized metabolic activity establishes boundaries of retinoic acid signalling. *EMBO J*. 17:7361–7372.

Holowacz, T., and S. Sokol. 1999. FGF is required for posterior neural patterning but not for neural induction. *Dev Biol*. 205:296–308.

Hongo, I., M. Kengaku, and H. Okamoto. 1999. FGF signaling and the anterior neural induction in Xenopus. *Dev Biol*. 216:561–581.

In der Rieden, P.M., F.L. Vilaspasa, and A.J. Durston. 2010. Xwnt8 directly initiates expression of labial Hox genes. *Dev Dyn*. 239:126–139.

Isaacs, H.V., M.E. Pownall, and J.M. Slack. 1995. eFGF is expressed in the dorsal midline of *Xenopus laevis*. *Int J Dev Biol*. 39:575–579.

Isaacs, H.V., M.E. Pownall, and J.M. Slack. 1998. Regulation of Hox gene expression and posterior development by the Xenopus caudal homologue Xcad3. *EMBO J*. 17:3413–3427.

Isaacs, H.V., D. Tannahill, and J.M. Slack. 1992. Expression of a novel FGF in the Xenopus embryo: A new candidate inducing factor for mesoderm formation and anteroposterior specification. *Development*. 114:711–720.

Itoh, K., and S.Y. Sokol. 1997. Graded amounts of Xenopus dishevelled specify discrete anteroposterior cell fates in prospective ectoderm. *Mech Dev*. 61:113–125.

Janesick, A., T.T. Nguyen, K. Aisaki, K. Igarashi, S. Kitajima, R.A. Chandraratna, J. Kanno, and B. Blumberg. 2014. Active repression by RARgamma signaling is required for vertebrate axial elongation. *Development*. 141:2260–2270.

Janssens, S., T. Denayer, T. Deroo, F. Van Roy, and K. Vleminckx. 2010. Direct control of Hoxd1 and Irx3 expression by Wnt/beta-catenin signaling during anteroposterior patterning of the neural axis in Xenopus. *Int J Dev Biol*. 54:1435–1442.

Kengaku, M., and H. Okamoto. 1995. bFGF as a possible morphogen for the anteroposterior axis of the central nervous system in Xenopus. *Development*. 121:3121–3130.

Kimelman, D., J.A. Abraham, T. Haaparanta, T.M. Palisi, and M.W. Kirschner. 1988. The presence of fibroblast growth factor in the frog egg: Its role as a natural mesoderm inducer. *Science*. 242:1053–1056.

Kjolby, R.A.S., and R.M. Harland. 2017. Genome-wide identification of Wnt/beta-catenin transcriptional targets during Xenopus gastrulation. *Dev Biol*. 426:165–175.

Kjolby, R.A.S., M. Truchado-Garcia, S. Iruvanti, and R.M. Harland. 2019. Integration of Wnt and FGF signaling in the Xenopus gastrula at TCF and Ets binding sites shows the importance of short-range repression by TCF in patterning the marginal zone. *Development*. 146.

Koide, T., M. Downes, R.A. Chandraratna, B. Blumberg, and K. Umesono. 2001. Active repression of RAR signaling is required for head formation. *Genes Dev*. 15:2111–2121.

Kolm, P.J., V. Apekin, and H. Sive. 1997. Xenopus hindbrain patterning requires retinoid signaling. *Dev Biol*. 192:1–16.

Komiya, Y., N. Mandrekar, A. Sato, I.B. Dawid, and R. Habas. 2014. Custos controls beta-catenin to regulate head development during vertebrate embryogenesis. *Proc Natl Acad Sci U S A*. 111:13099–13104.

Lamb, T.M., and R.M. Harland. 1995. Fibroblast growth factor is a direct neural inducer, which combined with noggin generates anterior-posterior neural pattern. *Development*. 121:3627–3636.

Lamb, T.M., A.K. Knecht, W.C. Smith, S.E. Stachel, A.N. Economides, N. Stahl, G.D. Yancopolous, and R.M. Harland. 1993. Neural induction by the secreted polypeptide noggin. *Science*. 262:713–718.

Lea, R., N. Papalopulu, E. Amaya, and K. Dorey. 2009. Temporal and spatial expression of FGF ligands and receptors during Xenopus development. *Dev Dyn*. 238:1467–1479.

Lee, H., S.M. Cheong, W. Han, Y. Koo, S.B. Jo, G.S. Cho, J.S. Yang, S. Kim, and J.K. Han. 2018. Head formation requires Dishevelled degradation that is mediated by March2 in concert with Dapper1. *Development*. 145.

Leyns, L., T. Bouwmeester, S.H. Kim, S. Piccolo, and E.M. De Robertis. 1997. Frzb-1 is a secreted antagonist of Wnt signaling expressed in the Spemann organizer. *Cell*. 88:747–756.

Lin, K., S. Wang, M.A. Julius, J. Kitajewski, M. Moos, Jr., and F.P. Luyten. 1997. The cysteine-rich frizzled domain of Frzb-1 is required and sufficient for modulation of Wnt signaling. *Proc Natl Acad Sci U S A*. 94:11196–11200.

Lloret-Vilaspasa, F., H.J. Jansen, K. de Roos, R.A. Chandraratna, M.H. Zile, C.D. Stern, and A.J. Durston. 2010. Retinoid signalling is required for information transfer from mesoderm to neuroectoderm during gastrulation. *Int J Dev Biol*. 54:599–608.

Lombardo, A., H.V. Isaacs, and J.M. Slack. 1998. Expression and functions of FGF-3 in Xenopus development. *Int J Dev Biol*. 42:1101–1107.

Lombardo, A., and J.M. Slack. 1998. Postgastrulation effects of fibroblast growth factor on Xenopus development. *Dev Dyn*. 212:75–85.

Lou, X., P. Fang, S. Li, R.Y. Hu, K.M. Kuerner, H. Steinbeisser, and X. Ding. 2006. Xenopus Tbx6 mediates posterior patterning via activation of Wnt and FGF signalling. *Cell Res*. 16:771–779.

Mangold, O. 1933. Uber die Induckionsfahigkeit der verschiedenen Bezirke der Neurula von Urodelen. *Naturwissenshafien*. 43:761–766.

Mao, B., W. Wu, G. Davidson, J. Marhold, M. Li, B.M. Mechler, H. Delius, D. Hoppe, P. Stannek, C. Walter, A. Glinka, and C. Niehrs. 2002. Kremen proteins are Dickkopf receptors that regulate Wnt/beta-catenin signalling. *Nature*. 417:664–667.

Mao, B., W. Wu, Y. Li, D. Hoppe, P. Stannek, A. Glinka, and C. Niehrs. 2001. LDL-receptor-related protein 6 is a receptor for Dickkopf proteins. *Nature*. 411:321–325.

McGrew, L.L., S. Hoppler, and R.T. Moon. 1997. Wnt and FGF pathways cooperatively pattern anteroposterior neural ectoderm in Xenopus. *Mech Dev*. 69:105–114.

McGrew, L.L., C.J. Lai, and R.T. Moon. 1995. Specification of the anteroposterior neural axis through synergistic interaction of the Wnt signaling cascade with noggin and follistatin. *Dev Biol*. 172:337–342.

McLin, V.A., S.A. Rankin, and A.M. Zorn. 2007. Repression of Wnt/beta-catenin signaling in the anterior endoderm is essential for liver and pancreas development. *Development*. 134:2207–2217.

McMahon, A.P., and R.T. Moon. 1989. Ectopic expression of the proto-oncogene int-1 in Xenopus embryos leads to duplication of the embryonic axis. *Cell*. 58:1075–1084.

Michiue, T., A. Fukui, A. Yukita, K. Sakurai, H. Danno, A. Kikuchi, and M. Asashima. 2004. XIdax, an inhibitor of the canonical Wnt pathway, is required for anterior neural structure formation in Xenopus. *Dev Dyn*. 230:79–90.

Miyagi, A., T. Negishi, T.S. Yamamoto, and N. Ueno. 2015. G protein-coupled receptors Flop1 and Flop2 inhibit Wnt/beta-catenin signaling and are essential for head formation in Xenopus. *Dev Biol*. 407:131–144.

Nieuwkoop, P.D. 1952a. Activation and organization of the central nervous system in amphibians. Part I: Induction and activation. *J. Exp. Zool*. 120:1–31.

Nieuwkoop, P.D. 1952b. Activation and organization of the central nervous system in amphibians. Part II: Differentiation and organization. *J. Exp. Zool*. 120:33–81.

Nieuwkoop, P.D. 1952c. Activation and organization of the central nervous system in amphibians. Part III: Synthesis of a working hypothesis. *J. Exp. Zool*. 120:83–108.

Okada, Y.K., and H. Takaya. 1942. Experimental investigation of regional differences in the inductive capacity of the organizer. *Proc. Imp. Acad. (Tokyo)*. 18:505–513.

Papalopulu, N., J.D. Clarke, L. Bradley, D. Wilkinson, R. Krumlauf, and N. Holder. 1991a. Retinoic acid causes abnormal development and segmental patterning of the anterior hindbrain in Xenopus embryos. *Development*. 113:1145–1158.

Papalopulu, N., and C. Kintner. 1996. A posteriorising factor, retinoic acid, reveals that anteroposterior patterning controls the timing of neuronal differentiation in Xenopus neuroectoderm. *Development*. 122:3409–3418.

Papalopulu, N., R. Lovell-Badge, and R. Krumlauf. 1991b. The expression of murine Hox-2 genes is dependent on the differentiation pathway and displays a collinear sensitivity to retinoic acid in F9 cells and Xenopus embryos. *Nucleic Acids Res*. 19:5497–5506.

Pera, E.M., and E.M. De Robertis. 2000. A direct screen for secreted proteins in Xenopus embryos identifies distinct activities for the Wnt antagonists Crescent and Frzb-1. *Mech Dev*. 96:183–195.

Pfeffer, P.L., and E.M. De Robertis. 1994. Regional specificity of RAR gamma isoforms in Xenopus development. *Mech Dev*. 45:147–153.

Piccolo, S., E. Agius, L. Leyns, S. Bhattacharyya, H. Grunz, T. Bouwmeester, and E.M. De Robertis. 1999. The head inducer Cerberus is a multifunctional antagonist of Nodal, BMP and Wnt signals. *Nature*. 397:707–710.

Pillemer, G., M. Epstein, B. Blumberg, J.K. Yisraeli, E.M. De Robertis, H. Steinbeisser, and A. Fainsod. 1998. Nested expression and sequential downregulation of the Xenopus caudal genes along the anterior-posterior axis. *Mech Dev*. 71:193–196.

Pinho, S., and C. Niehrs. 2007. Dkk3 is required for TGF-beta signaling during Xenopus mesoderm induction. *Differentiation*. 75:957–967.

Ploper, D., H.X. Lee, and E.M. De Robertis. 2011. Dorsal-ventral patterning: Crescent is a dorsally secreted Frizzled-related protein that competitively inhibits Tolloid proteases. *Dev Biol*. 352:317–328.

Polevoy, H., Y.E. Gutkovich, A. Michaelov, Y. Volovik, Y.M. Elkouby, and D. Frank. 2019. New roles for Wnt and BMP signaling in neural anteroposterior patterning. *EMBO Rep*. 20.

Pownall, M.E., H.V. Isaacs, and J.M. Slack. 1998. Two phases of Hox gene regulation during early Xenopus development. *Curr Biol*. 8:673–676.

Pownall, M.E., A.S. Tucker, J.M. Slack, and H.V. Isaacs. 1996. eFGF, Xcad3 and Hox genes form a molecular pathway that establishes the anteroposterior axis in Xenopus. *Development*. 122:3881–3892.

Pownall, M.E., B.E. Welm, K.W. Freeman, D.M. Spencer, J.M. Rosen, and H.V. Isaacs. 2003. An inducible system for the study of FGF signalling in early amphibian development. *Developmental Biology*. 256:90–100.

Rankin, S.A., K.W. McCracken, D.M. Luedeke, L. Han, J.M. Wells, J.M. Shannon, and A.M. Zorn. 2018. Timing is everything: Reiterative Wnt, BMP and RA signaling regulate developmental competence during endoderm organogenesis. *Dev Biol*. 434:121–132.

Rauschenberger, V., D.B. Bernkopf, S. Krenn, K. Jalal, J. Heller, J. Behrens, M. Gentzel, and A. Schambony. 2017. The phosphatase Pgam5 antagonizes Wnt/beta-Catenin signaling in embryonic anterior-posterior axis patterning. *Development*. 144:2234–2247.

Ribisi, S., Jr., F.V. Mariani, E. Aamar, T.M. Lamb, D. Frank, and R.M. Harland. 2000. Ras-mediated FGF signaling is required for the formation of posterior but not anterior neural tissue in *Xenopus laevis*. *Dev Biol*. 227:183–196.

Richard-Parpaillon, L., C. Heligon, F. Chesnel, D. Boujard, and A. Philpott. 2002. The IGF pathway regulates head formation by inhibiting Wnt signaling in Xenopus. *Dev Biol*. 244:407–417.

Roche, D.D., K.J. Liu, R.M. Harland, and A.H. Monsoro-Burq. 2009. Dazap2 is required for FGF-mediated posterior neural patterning, independent of Wnt and Cdx function. *Dev Biol*. 333:26–36.

Ruiz i Altaba, A., and T. Jessell. 1991a. Retinoic acid modifies mesodermal patterning in early Xenopus embryos. *Genes Dev*. 5:175–187.

Ruiz i Altaba, A., and T.M. Jessell. 1991b. Retinoic acid modifies the pattern of cell differentiation in the central nervous system of neurula stage Xenopus embryos. *Development*. 112:945–958.

Saha, M.S., and R.M. Grainger. 1992. A labile period in the determination of the anterior-posterior axis during early neural development in Xenopus. *Neuron*. 8:1003–1014.

Sasai, Y., B. Lu, H. Steinbeisser, and E.M. De Robertis. 1995. Regulation of neural induction by the Chd and Bmp-4 antagonistic patterning signals in Xenopus. *Nature*. 376:333–336.

Sasai, Y., B. Lu, H. Steinbeisser, D. Geissert, L.K. Gont, and E.M. De Robertis. 1994. Xenopus chordin: A novel dorsalizing factor activated by organizer-specific homeobox genes. *Cell*. 79:779–790.

Sato, S.M., and T.D. Sargent. 1989. Development of neural inducing capacity in dissociated Xenopus embryos. *Dev Biol*. 134:263–266.

Saxen, L., and S. Toivonen. 1961. The two-gradient hypothesis in primary induction: The combined effect of two types of inductors mixed in different ratios. *J Embryol Exp Morphol*. 9:514–533.

Saxen, L., S. Toivonen, and T. Vainio. 1964. Initial stimulus and subsequent interactions in embryonic induction. *J Embryol Exp Morphol*. 12:333–338.

Schyr, R.B., Y. Shabtai, C.S. Shashikant, and A. Fainsod. 2012. Cdx1 is essential for the initiation of HoxC8 expression during early embryogenesis. *FASEB J*. 26:2674–2684.

Semenov, M.V., K. Tamai, B.K. Brott, M. Kuhl, S. Sokol, and X. He. 2001. Head inducer Dickkopf-1 is a ligand for Wnt coreceptor LRP6. *Curr Biol*. 11:951–961.

Sharpe, C.R., A. Fritz, E.M. De Robertis, and J.B. Gurdon. 1987. A homeobox-containing marker of posterior neural differentiation shows the importance of predetermination in neural induction. *Cell*. 50:749–758.

Sharpe, C.R., and J.B. Gurdon. 1990. The induction of anterior and posterior neural genes in *Xenopus laevis*. *Development*. 109:765–774.

Shibata, M., M. Itoh, H. Hikasa, S. Taira, and M. Taira. 2005. Role of crescent in convergent extension movements by modulating Wnt signaling in early Xenopus embryogenesis. *Mech Dev*. 122:1322–1339.

Shibata, M., H. Ono, H. Hikasa, J. Shinga, and M. Taira. 2000. Xenopus crescent encoding a Frizzled-like domain is expressed in the Spemann organizer and pronephros. *Mech Dev*. 96:243–246.

Shiotsugu, J., Y. Katsuyama, K. Arima, A. Baxter, T. Koide, J. Song, R.A. Chandraratna, and B. Blumberg. 2004. Multiple points of interaction between retinoic acid and FGF signaling during embryonic axis formation. *Development*. 131:2653–2667.

Silva, A.C., M. Filipe, K.M. Kuerner, H. Steinbeisser, and J.A. Belo. 2003. Endogenous Cerberus activity is required for anterior head specification in Xenopus. *Development*. 130:4943–4953.

Sive, H.L., B.W. Draper, R.M. Harland, and H. Weintraub. 1990. Identification of a retinoic acid-sensitive period during primary axis formation in *Xenopus laevis*. *Genes Dev*. 4:932–942.

Sive, H.L., K. Hattori, and H. Weintraub. 1989. Progressive determination during formation of the anteroposterior axis in *Xenopus laevis*. *Cell*. 58:171–180.

Slack, J.M., B.G. Darlington, L.L. Gillespie, S.F. Godsave, H.V. Isaacs, and G.D. Paterno. 1989. The role of fibroblast growth factor in early Xenopus development. *Development*. 107 Suppl:141–148.

Slack, J.M., B.G. Darlington, J.K. Heath, and S.F. Godsave. 1987. Mesoderm induction in early Xenopus embryos by heparin-binding growth factors. *Nature*. 326:197–200.

Slack, J.M., H.V. Isaacs, and B.G. Darlington. 1988. Inductive effects of fibroblast growth factor and lithium ion on Xenopus blastula ectoderm. *Development*. 103:581–590.

Slack, J.M., and D. Tannahill. 1992. Mechanism of anteroposterior axis specification in vertebrates: Lessons from the amphibians. *Development*. 114:285–302.

Smith, W.C., and R.M. Harland. 1991. Injected Xwnt-8 RNA acts early in Xenopus embryos to promote formation of a vegetal dorsalizing center. *Cell*. 67:753–765.

Smith, W.C., and R.M. Harland. 1992. Expression cloning of noggin, a new dorsalizing factor localized to the Spemann organizer in Xenopus embryos. *Cell*. 70:829–840.

Sokol, S., J.L. Christian, R.T. Moon, and D.A. Melton. 1991. Injected Wnt RNA induces a complete body axis in Xenopus embryos. *Cell*. 67:741–752.

Song, J., and J.M. Slack. 1996. XFGF-9: A new fibroblast growth factor from Xenopus embryos. *Dev Dyn*. 206:427–436.

Spemann, H. 1936. *Experimentelle Beitrage zu einer Theorie der Entwicklung*. Verlag Julius Springer, Berlin.

Spemann, H., and H. Mangold. 1924. Über Induktion von Embryonalanlagen durch Implantation Artfremder Organisatoren. *Roux'Arch. Entw. Mech.* 100:599–638.

Spemann, H., and H. Mangold. 2001. Induction of embryonic primordia by implantation of organizers from a different species. 1923. *Int J Dev Biol.* 45:13–38.

Stewart, R.M., and J.C. Gerhart. 1990. The anterior extent of dorsal development of the Xenopus embryonic axis depends on the quantity of organizer in the late blastula. *Development.* 109:363–372.

Strate, I., T.H. Min, D. Iliev, and E.M. Pera. 2009. Retinol dehydrogenase 10 is a feedback regulator of retinoic acid signalling during axis formation and patterning of the central nervous system. *Development.* 136:461–472.

Sun, B.I., S.M. Bush, L.A. Collins-Racie, E.R. LaVallie, E.A. DiBlasio-Smith, N.M. Wolfman, J.M. McCoy, and H.L. Sive. 1999. derriere: A TGF-beta family member required for posterior development in Xenopus. *Development.* 126:1467–1482.

Tanibe, M., T. Michiue, A. Yukita, H. Danno, M. Ikuzawa, S. Ishiura, and M. Asashima. 2008. Retinoic acid metabolizing factor xCyp26c is specifically expressed in neuroectoderm and regulates anterior neural patterning in *Xenopus laevis*. *Int J Dev Biol.* 52:893–901.

Tannahill, D., H.V. Isaacs, M.J. Close, G. Peters, and J.M. Slack. 1992. Developmental expression of the Xenopus int-2 (FGF-3) gene: Activation by mesodermal and neural induction. *Development.* 115:695–702.

van der Wees, J., J.G. Schilthuis, C.H. Koster, H. Diesveld-Schipper, G.E. Folkers, P.T. van der Saag, M.I. Dawson, K. Shudo, B. van der Burg, and A.J. Durston. 1998. Inhibition of retinoic acid receptor-mediated signalling alters positional identity in the developing hindbrain. *Development.* 125:545–556.

Vodicka, M.A., and J.C. Gerhart. 1995. Blastomere derivation and domains of gene expression in the Spemann organizer of *Xenopus laevis*. *Development.* 121:3505–3518.

Vogt, W. 1929. Gestaltungsanalyse am Amphibienkeim mit Ortlicher Vitalfarbung: II. Teil Gastrulation und Mesodermbildung bei Urodelen und Anuren. *Wilhelm Roux Arch Entwickl Mech Org.* 120:384–706.

Weston, A.D., B. Blumberg, and T.M. Underhill. 2003. Active repression by unliganded retinoid receptors in development: Less is sometimes more. *J Cell Biol.* 161:223–228.

Wheeler, G.N., F.S. Hamilton, and S. Hoppler. 2000. Inducible gene expression in transgenic Xenopus embryos. *Curr Biol.* 10:849–852.

Yamada, T. 1990. Regulations in the induction of the organized neural system in amphibian embryos. *Development.* 110:653–659.

Yamamoto, A., T. Nagano, S. Takehara, M. Hibi, and S. Aizawa. 2005. Shisa promotes head formation through the inhibition of receptor protein maturation for the caudalizing factors, Wnt and FGF. *Cell.* 120:223–235.

Yamamoto, H., H. Sakane, H. Yamamoto, T. Michiue, and A. Kikuchi. 2008. Wnt3a and Dkk1 regulate distinct internalization pathways of LRP6 to tune the activation of beta-catenin signaling. *Dev Cell.* 15:37–48.

Yelin, R., R.B. Schyr, H. Kot, S. Zins, A. Frumkin, G. Pillemer, and A. Fainsod. 2005. Ethanol exposure affects gene expression in the embryonic organizer and reduces retinoic acid levels. *Dev Biol.* 279:193–204.

Yoshida, H., M. Okada, K. Takebayashi-Suzuki, N. Ueno, and A. Suzuki. 2016. Involvement of JunB proto-oncogene in tail formation during early Xenopus embryogenesis. *Zoolog Sci.* 33:282–289.

Zhang, X., J.G. Abreu, C. Yokota, B.T. MacDonald, S. Singh, K.L. Coburn, S.M. Cheong, M.M. Zhang, Q.Z. Ye, H.C. Hang, H. Steen, and X. He. 2012. Tiki1 is required for head formation via Wnt cleavage-oxidation and inactivation. *Cell.* 149:1565–1577.

Zhang, X., S.M. Cheong, N.G. Amado, A.H. Reis, B.T. MacDonald, M. Zebisch, E.Y. Jones, J.G. Abreu, and X. He. 2015. Notum is required for neural and head induction via Wnt deacylation, oxidation, and inactivation. *Dev Cell.* 32:719–730.

Zhu, X., Z. Min, R. Tan, and Q. Tao. 2015. NF2/Merlin is required for the axial pattern formation in the *Xenopus laevis* embryo. *Mech Dev.* 138(Pt 3):305–312.

Zoltewicz, J.S., and J.C. Gerhart. 1997. The Spemann organizer of Xenopus is patterned along its anteroposterior axis at the earliest gastrula stage. *Dev Biol.* 192:482–491.

6 Wnt Signaling in Tissue Differentiation and Morphogenesis

Stefan Hoppler and Michael Kühl

CONTENTS

6.1. Historical Background 65
6.2. Characterizing the Wnt Signaling Pathways 65
 6.2.1. Canonical Wnt Signaling 66
 6.2.2. Non-Canonical Wnt Signaling 68
 6.2.3. Integration of Canonical and Non-Canonical Wnt Signaling 68
 6.2.4. Integrated Understanding of Wnt Signaling in Vertebrate Embryonic Development 69
6.3. Recent Advances 69
 6.3.1. Wnt Signaling in Cardiac Organogenesis 69
 6.3.2. Modeling Wnt Signaling 69
 6.3.3. Context-Specific Wnt Signaling 70
6.4. Future Directions and Important Questions 71
 6.4.1. Wnt Interactions with Other Signaling Pathways 71
 6.4.2. Regulation of Gene Expression by Non-Canonical Wnt Signaling 71
 6.4.3. Integration of Wnt Signaling into the Embryonic Signaling and Regulatory Environment 71
 6.4.4. Xenopus as a Model for Human Disease 71
Acknowledgments 72
References 72

6.1. HISTORICAL BACKGROUND

The *Xenopus* experimental model system has been instrumental for key discoveries about Wnt signaling in vertebrates, both biochemical mechanisms and developmental functions. Study of Wnt function in dorsal axis establishment, morphogenetic movements during gastrulation, and neural development provided insights into Wnt pathway mechanisms inside and between cells. Wnt signaling research continues to make important discoveries using the *Xenopus* system (see Sections 3 and 4).

The success of *Xenopus* as a model system is explained by it being at the crossroads of different disciplines, as indicated by different techniques and complementary experimental approaches, such as: stem-cell/organoid-like embryonic explants or the ability to perform transplantation experiments (for embryology), microscopy analysis (for cell biology), or extracting lysates in large amounts (for biochemistry). These unique experimental advantages of the *Xenopus* model system have served Wnt signaling research immensely and widely.

6.2. CHARACTERIZING THE WNT SIGNALING PATHWAYS

Wnt glycoproteins are secreted extracellular signal proteins that initiate intracellular signal transduction by binding to receptors primarily of the Frizzled family of seven transmembrane receptors. Co-receptors, such as LRP5/6, are also involved in some signaling contexts (Hoppler and Nakamura, 2014). Several members of tyrosine kinase receptors also have been linked to Wnt signaling such as Ryk, Ror, Ptk7, and MuSK (Roy et al., 2018).

Following the independent co-discovery of Wnt genes in mouse (*int-1*, Nusse and Varmus, 1982) and *Drosophila* (*wingless*, Baker, 1987), injection of mRNA coding for Int-1 into early *Xenopus* embryos demonstrated that Wnt proteins also function in vertebrate development; this caused a dramatic axis duplication (McMahon and Moon, 1989). This fundamental experiment became established as an important assay for investigating Wnt signaling. Wnt pathway activity was subsequently shown to be required for normal axis development (e.g. Heasman et al., 2000), and again it was *Xenopus* that was used as the model system of choice. However, the originally proposed requirement for endogenous Wnt ligand and receptor function (Kofron et al., 2007; Tao et al., 2005) has recently been challenged (Yan et al., 2018).

Using the axis duplication assay, Wnt signals were subdivided into different classes. Members of the Wnt1 class resulted in formation of secondary axis upon mRNA injection into early *Xenopus* embryos. In addition, they triggered transformation of C57MG mouse mammary

DOI: 10.1201/9781003050230-7

epithelial cells and could later be linked to canonical Wnt signaling. In contrast, members of the Wnt5a class failed in the secondary axis induction assay but rather inhibited the activity of Wnt1 class members. They affected Cadherin mediated cell adhesion (Torres et al., 1996), cell migration (Tada and Smith, 2000), and tumor metastasis (Dissanayake et al., 2007). Those members were linked to non-canonical Wnt signaling.

6.2.1. Canonical Wnt Signaling

6.2.1.1. Canonical Wnt Signaling in the Cytoplasm

Xenopus dorsal axis development served as a powerful experimental assay to investigate mechanisms and functions of **canonical Wnt signaling**. Analysis of *Xenopus* dorsal axis development served to confirm in vertebrates important components of the canonical signal transduction pathway in the cytoplasm, such as GSK3 (Yost et al., 1996), APC (Vleminckx et al., 1997), Axin (Zeng et al., 1997), and Casein Kinase I (CKI) (Peters et al., 1999). Crucially, the hallmark of canonical Wnt pathway, β-Catenin was also confirmed as the important downstream component causing *Xenopus* axis induction (Funayama et al., 1995).

Xenopus research was particularly powerful at integrating the molecular functions at the core of the canonical Wnt signal transduction pathway in the "β-Catenin destruction complex" (Hedgepeth et al., 1999; Itoh et al., 1998; Xing et al., 2003), including the N-terminal phosphorylation of β-Catenin protein by GSK3 (Yost et al., 1996) and CK1epsilon (Liu et al., 2002), which is further controlled by Protein Phosphatase 2A (Li et al., 2001), leading to β-TrCP-mediated degradation of the β-Catenin protein (Liu et al., 1999; Marikawa and Elinson, 1998).

These discoveries led to the concept of the "futile β-Catenin protein turnover cycle" (reviewed by Chen et al., 2014). In the absence of canonical Wnt signaling, β-Catenin protein is synthesized and then promptly degraded by the β-Catenin destruction complex, whereas active upstream Wnt signaling disrupts the β-Catenin destruction complex. This results in stabilized β-Catenin protein that then functions to induce axis development and by extension all the other functions of canonical Wnt signaling in other tissues at other stages and of course in other organisms.

Xenopus researchers were able to re-constitute the β-Catenin destruction complex in egg extracts (Hyde et al., 2016; Salic et al., 2000) to study the biochemistry in detail, which revealed that, in early *Xenopus* development, at least, Axin protein is clearly the limiting component (Lee et al., 2003) with likely implications for regulation of β-Catenin destruction complex assembly (see subsection 2.1.3.).

6.2.1.2. Canonical Wnt Signaling in the Nucleus

Xenopus embryos clearly illustrate that activated canonical Wnt signaling allows transport of β-Catenin protein into the nucleus (Figure 6.1) (Schneider et al., 1996; Yost et al., 1996), and again *Xenopus* dorsal axis development served in the discovery of how β-Catenin regulates developmental change in the nucleus. β-Catenin does not make direct contact with DNA. *Xenopus* experiments were crucial in demonstrating functional interaction with the Lymphoid Enhancer Factor/T-Cell Factor (LEF/TCF) family of DNA-binding proteins (Behrens et al., 1996; Molenaar et al., 1996) and the very first identification of some of the direct Wnt target genes (*sia*, Brannon et al., 1997; *nodal3*, McKendry et al., 1997; *twin*, a.k.a. *sia2*, Laurent et al., 1997; *fibronectin*, Gradl et al., 1999; *engrailed2*, McGrew et al., 1999; and *nodal5* and *nodal6*, Yang et al., 2002).

Xenopus experiments dissected what has subsequently been named the "transcriptional switch" (reviewed by Ramakrishnan et al., 2018). In the presence of nuclear β-Catenin, when the canonical Wnt pathway is active, LEF/TCF proteins establish a transcriptional activation complex together with other transcriptional co-activator proteins such as pCBP (Takemaru and Moon, 2000). In the absence of nuclear β-Catenin, when the canonical Wnt pathway is inactive, LEF/TCF proteins establish a transcriptional repression complex together with transcriptional co-repressors such as TLE/Groucho (Roose et al., 1998). The structure of LEF/TCF proteins reveals functional domains that mediate interaction with nuclear β-Catenin, and thus transcriptional activation, with TLE/Groucho, and thus transcriptional repression, and an HMG DNA binding domain (reviewed by Hoppler and Kavanagh, 2007; Hoppler and Waterman, 2014). The large protein complexes assembled by LEF/TCF proteins on regulatory DNA sequences have since been named the "Wnt enhanceosome" (Gammons and Bienz, 2018) (Figure 6.2).

6.2.1.3. Canonical Wnt Signaling at the Membrane

Xenopus embryos were instrumental in confirming in vertebrates that Frizzled proteins function as Wnt receptors (Yang-Snyder et al., 1996) and LRP5/6 (Tamai et al., 2000) as important co-receptors in canonical Wnt signaling. The interaction between the *Xenopus* Wnt8a ligand and Frizzled8 receptor has since been described in structural detail (Janda et al., 2012). *Xenopus* experiments contributed critically to the concept that in canonical signaling Wnt, Frizzled, and LRP form a complex at the membrane (Holmen et al., 2002; Tamai et al., 2000), with the intracellular domain of LRP being important for the recruitment of Axin (Davidson et al., 2005; Zeng et al., 2005) in a process that involves Disheveled to form a mega-protein-complex that has subsequently been named the "Wnt signalosome" (Bilic et al., 2007; Gammons and Bienz, 2018).

Experiments with *Xenopus* embryos were particularly fertile at identifying and studying extracellular partners modifying Wnt signaling at the membrane. The Dickkopf (DKK) family of proteins (Glinka et al., 1998) and Sclerostin (SOST) (Semenov et al., 2005) bind to LRP5/6 co-receptors (Mao et al., 2001). Secreted Frizzled related proteins (sFrp), including Frzb (a.k.a. Sfrp3) (Leyns et al., 1997; Wang et al., 1997) and at later stages Sfrp1 (Gibb et al., 2013), initially were described as Wnt inhibitors, but they may modulate Wnt

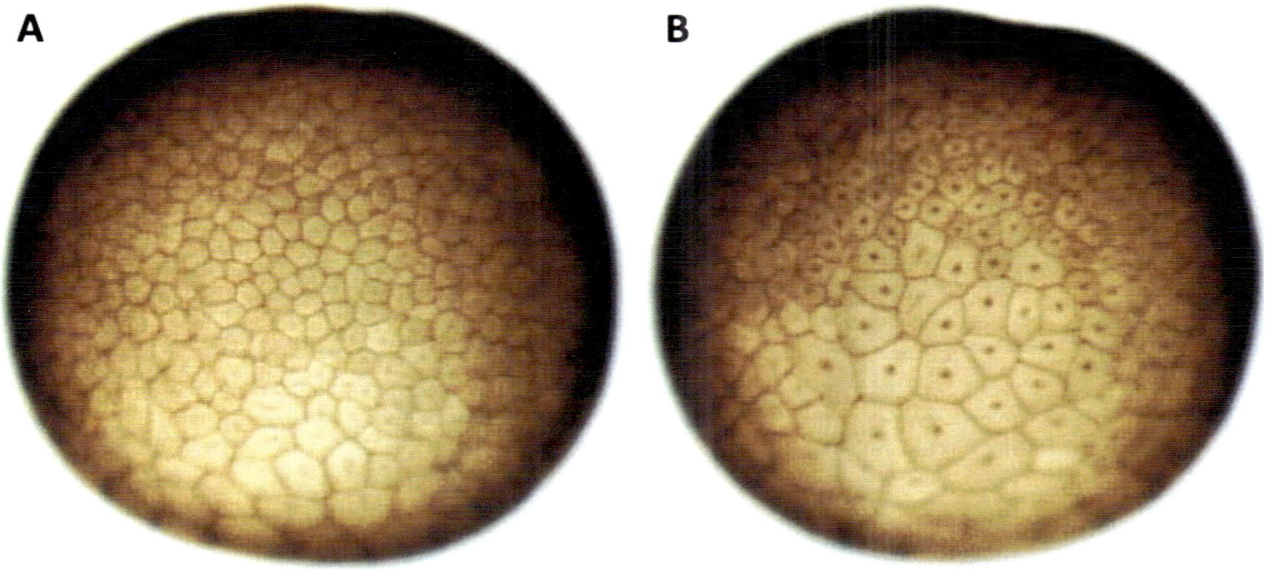

FIGURE 6.1 Nuclear localization of β-Catenin protein in dorsal blastomeres. Embryology, cell biology, and molecular biology come together in this image of β-Catenin localization in *Xenopus laevis* blastula (Schneider et al., 1996). (A) view of the prospective ventral side of blastula (stage 8.5 embryo). (B) view of the prospective dorsal side of the same blastula-stage embryo. Note that β-Catenin localization traces the outlines of the blastomere cells (in A and B, through its molecular association with adherence junctions); additionally, but only on the prospective dorsal side (B), β-Catenin also localizes to cell nuclei. Detection of nuclear β-Catenin is now established as the hallmark of activated canonical Wnt signaling.

Source: Images reproduced with permission from Elsevier publishers [Copyright Clearance Center (RightsLink) License Number 4961291291588)].

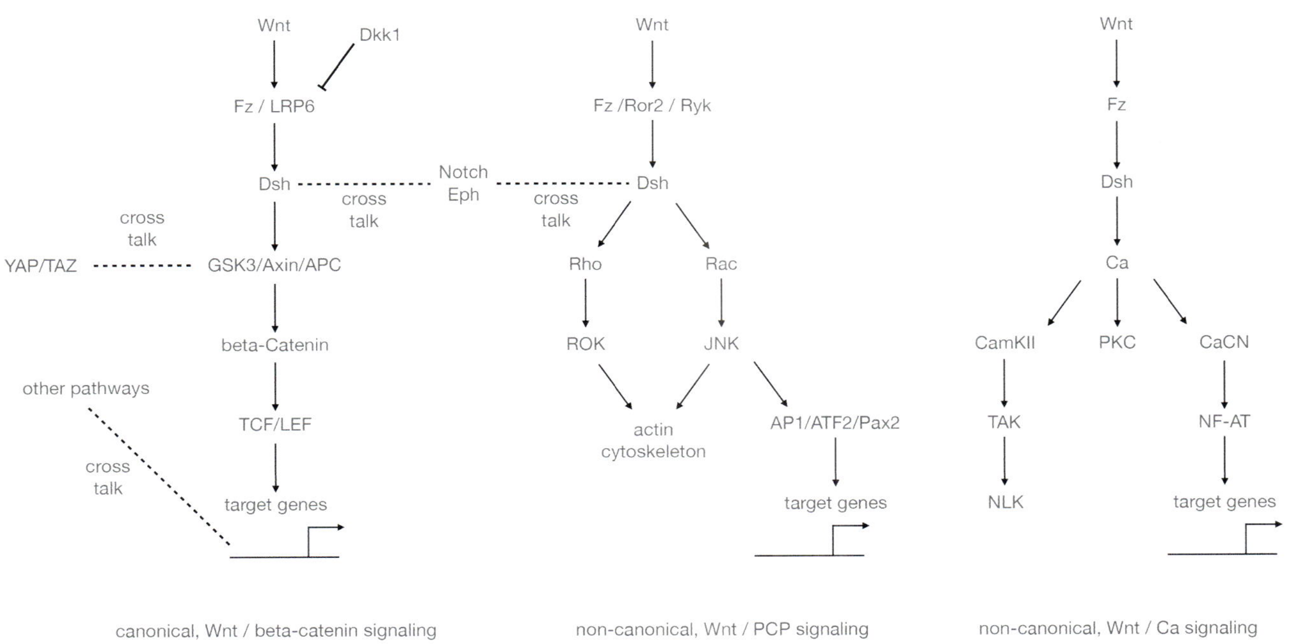

FIGURE 6.2 Different Wnt signaling pathways as discussed in the text. Left: Canonical, Wnt/b-catenin signaling. Middle: Non-canonical, Wnt/PCP signaling. Right: Non-canonical, Wnt/Ca signaling. Dashed lines indicate selected cross talks of Wnt signaling with other pathways.

signaling in more complex ways (discussed in Bovolenta et al., 2014). An image emerges with potentially several kinds of extracellular ligands (Wnt, Sfrp, Dkk, etc.) influencing Wnt signaling at the membrane.

6.2.2. Non-Canonical Wnt Signaling

Wnt proteins that failed to induce a secondary axis were originally classified as Wnt5a class members (Du et al., 1995; Torres et al., 1996). So-called **non-canonical Wnt signaling** was defined by its independence from β-Catenin (Figure 6.2). Again, *Xenopus* served as a powerful model system to dissect the function of those Wnt members and the molecular nature of the activated signaling pathways. It turned out, however, that non-canonical Wnt signaling involves more than just one signaling pathway. According to their main molecular effectors or cellular effect, they were named Wnt/PCP (planar cell polarity), Wnt/JNK (jun-N-terminal Kinase), Wnt/calcium, or Wnt/Ror signaling pathways. For historical reasons, we deal here with these pathways independently, but these originally described different pathways overlap heavily to represent a Wnt signaling network.

6.2.2.1. Wnt/PCP/JNK Signaling

Cell migration is thought to be regulated by **Wnt/PCP signaling,** often also named Wnt/JNK signaling, such as during vertebrate gastrulation (Djiane et al., 2000; Wallingford et al., 2000) or migration of neural crest cells (De Calisto et al., 2005). *Xenopus* assays including dorsal marginal zone explants as well as Activin-induced animal caps served to reveal the mechanisms and functions of this non-canonical Wnt signaling pathway during gastrulation, neural crest cell migration, and morphogenesis. This process is mediated by directed and regulated cell movements called convergent extension. This involves polarization of mesodermal cells and migration of cells toward the dorsal midline, resulting in medio-lateral narrowing and anterior-posterior elongation of the embryo. Work in *Xenopus* showed that Wnt11b is required for polarization of mesodermal cells (Yamanaka and Nishida, 2007), whereas Wnt5a regulates cell migration toward the dorsal midline (Schambony and Wedlich, 2007). Use of such *Xenopus* explants helped to further characterize these pathways and to identify components of this pathway, including Prickle (Veeman et al., 2003), Ror2 and JNK (Schambony and Wedlich, 2007), PTK7 (Lu et al., 2004; Podleschny et al., 2015), Cdc42 (Penzo-Mendez et al., 2003), Rac and Rho (Habas et al., 2003), ROK (Kim and Han, 2005), WGEF (Tanegashima et al., 2008), and Daam1 (Habas et al., 2001).

Wnt/PCP signaling links to cilia formation (Wallingford and Mitchell, 2011), which regulates left/right patterning. In particular, Wnt11b is required for the polarization of the gastrocoel roof plate and subsequently for the cilia-driven leftward flow (Walentek et al., 2013). Explants of the *Xenopus* animal cap were essential for these findings.

6.2.2.2. Wnt/Calcium Signaling

The unique feature of *Xenopus* also helped to elucidate **Wnt/calcium signaling** (Figure 6.2). Groundbreaking work by Randall Moon and colleagues indicated that certain Wnts trigger an intracellular calcium release (Slusarski et al., 1997a, 1997b). Work in other model systems later showed that this calcium release occurs within seconds and thus is very likely a direct response of Wnt signaling (Dejmek et al., 2006; Jenei et al., 2009). The identification of calcium effectors in this pathway again made use of the unique features of the *Xenopus* system. Activation of PKC through calcium is accompanied by a translocation of the enzyme toward the membrane, which can be monitored in animal cap explants taking advantage of fluorescence microscopy (Sheldahl et al., 1999). Activation of Calcium-Calmodulin Dependent Kinase II (CaMKII) was studied in cytoplasmic lysates of *Xenopus* embryos before the onset of zygotic gene transcription at the midblastula transition (Kuhl et al., 2000). Also, the calcium-dependent phosphatase, Calcineurin, and its transcriptional regulator Nuclear Factor of Activated T cells (NFAT) were shown to be regulated by Wnts using *Xenopus* as a model system (Saneyoshi et al., 2002). Wnt/calcium signaling was shown to be involved in dorso-ventral patterning using *Xenopus* embryos (Kuhl et al., 2000).

6.2.3. Integration of Canonical and Non-Canonical Wnt Signaling

While the concept of "canonical" versus "non-canonical" had been useful, cracks in this wall separating canonical from non-canonical Wnt signaling appeared from the beginning; these have subsequently only grown larger and larger. Canonical Wnt signaling may be justifiably defined in the sense of involving components such as LRP co-receptors and, of course, β-Catenin and its partners in the nucleus. Non-canonical Wnt signaling—it now appears—is really only defined by not involving β-Catenin and is a collective term for every other form of Wnt signaling. Cracks in the wall separating canonical from non-canonical Wnt signaling particularly appear where they share pathway components as shown by work in *Xenopus*, such as Disheveled (Rothbacher et al., 2000; Sokol, 1996; Wallingford et al., 2000) or Frizzled 7 functioning at the crossroads of different Wnt pathways (Medina et al., 2000; Medina and Steinbeisser, 2000).

Furthermore, some components of one pathway inhibit the activity of the other. NFAT, for example, is a calcium-sensitive transcription factor activated by Wnt/calcium signaling yet also inhibits canonical β-Catenin dependent signaling in several models (Huang et al., 2011; Saneyoshi et al., 2002; Wang et al., 2013). Nemo-like Kinase is activated by Wnt/calcium signaling involving CamKII, and this finally results in a downregulation of canonical Wnt/β-Catenin signaling (Ishitani et al., 2003). Taken together, these and other cross-regulatory effects between canonical and non-canonical Wnt signaling resulted in the idea of a Wnt signaling network (Kestler and Kuhl, 2008).

6.2.4. Integrated Understanding of Wnt Signaling in Vertebrate Embryonic Development

While axis establishment and gastrulation assays have served to study fundamental pathway mechanisms in gene regulation and control of morphogenesis, at the same time, *Xenopus* experiments have contributed to our general understanding of wider functional roles for Wnt signaling in vertebrate embryonic development, such as: (1) mesoderm induction (e.g. Liu et al., 2005; Schohl and Fagotto, 2003) and patterning (e.g. Hoppler et al., 1996); (2) patterning of the neural plate (Domingos et al., 2001; McGrew et al., 1997), including neural crest induction (LaBonne and Bronner-Fraser, 1998) and migration (De Calisto et al., 2005), as well as midbrain development (Kunz et al., 2004); and (3) organogenesis of the eye (Maurus et al., 2005; Rasmussen et al., 2001), kidney (Lyons et al., 2004; Saulnier et al., 2002), and heart (Marvin et al., 2001; Schneider and Mercola, 2001) (see section IIIA). These investigations have led to a comprehensive and integrated understanding of Wnt signaling in embryonic development of a typical model vertebrate.

6.3. RECENT ADVANCES

Xenopus research continues to make contributions at the very forefront of new and important discoveries about Wnt signaling. For example, it has been a mystery how Wnt proteins with hydrophobic protein structures are able to establish a signaling gradient in the extracellular space through tissues. The importance of proteoglycans was demonstrated with *Xenopus* embryos, particularly the role of extracellular sulfate and acetylate modifications in regulating dispersal through tissues of Wnt signals, as well as Sfrp proteins (recently reviewed by Mii and Takada, 2020). At the membrane, the regulation of receptor turnover and the intricate role of R-spondins in Wnt signaling has become better understood (Chang et al., 2020; Chen et al., 2020; Ding et al., 2018; Szenker-Ravi et al., 2018). Recent genomics studies (e.g. Nakamura et al., 2016) suggest functions for β-Catenin beyond its association with LEF/TCF proteins. For example, a functional interplay between Wnt pathway-regulated β-Catenin and Sox17 was recently carefully dissected in *Xenopus* endoderm patterning (Mukherjee et al., 2020). *Xenopus* further contributes toward a more detailed understanding of non-canonical Wnt signaling pathways, particularly in regulating morphogenetic movements during gastrulation and neurulation (e.g. Butler and Wallingford, 2018; Shindo et al., 2019). Recent fundamental discoveries about Wnt signaling in neural crest induction and patterning are groundbreaking far beyond the *Xenopus* model system, including pluripotency of neural crest (Buitrago-Delgado et al., 2015), surprising functions of Dkk2 (Devotta et al., 2018), and a role for non-canonical Wnt signaling in neural crest cells (Ossipova et al., 2018). Space constraints unfortunately prevent listing all of the important and excellent research advances that have been facilitated by the *Xenopus* model; in the following sections, three areas are explored.

6.3.1. Wnt Signaling in Cardiac Organogenesis

Work in *Xenopus* was instrumental for deciphering roles for Wnt signaling during heart development (reviewed by Hoppler and Conlon, 2020; Hoppler et al., 2014). Whereas initial work focused on the role of inhibitors of canonical signaling (Marvin et al., 2001; Schneider and Mercola, 2001), our two laboratories were the first to identify and confirm a role for non-canonical Wnt signaling during this process (Afouda et al., 2008; Pandur et al., 2002). Wnt11b as an extracellular ligand together with JNK and PKC as intracellular signaling mediators were shown to be involved in this process (Pandur et al., 2002). Our work also showed how Wnt11b is tied into a network of GATA transcription factors during this early phase of cardiac development (Afouda and Hoppler, 2011; Afouda et al., 2008). This initial finding in *Xenopus* was later confirmed in other biological models including murine and human embryonic stem cells (Mazzotta et al., 2016).

Later during cardiac development, Wnt11a is important for terminal differentiation (Gessert et al., 2008; Hempel et al., 2017), ventricular trabeculation, and outflow tract formation both in *Xenopus* (Hempel et al., 2017) and mice (Nagy et al., 2010; Zhou et al., 2007). Of interest is the finding that the cell adhesion molecule Alcam turned out to be critically involved in this process (Gessert et al., 2008; Hempel et al., 2017). During the normal cardiac differentiation program, Wnt11a is regulated by the canonical Wnt signaling inhibitor Dkk1 (Guo et al., 2019). It remains unclear at the moment how inhibition of canonical Wnt signaling regulates Wnt11a expression on a molecular level.

The precise regulation of both canonical and non-canonical Wnt signaling seems to be crucial for cardiogenesis. Inhibitors of canonical Wnt signaling, such as Dkk1, Crescent, FrzB, and Sizzled, were shown to trigger cardiogenesis in *Xenopus* in overexpression experiments (Schneider and Mercola, 2001). In contrast, loss of Dkk1 experiments indicated its role during cardiac differentiation (Guo et al., 2019). Also, the canonical Wnt inhibitor Sfrp1 turned out to be important to regulate the size of the heart muscle during development via regulation of Wnt6 signaling (Gibb et al., 2013), as explained in the next section.

6.3.2. Modeling Wnt Signaling

Work in *Xenopus* was also instrumental for work on mathematical modeling of Wnt signaling. The group of Marc Kirschner (Lee et al., 2003) built a mathematical model of the pathway, focused on the main intracellular components of canonical Wnt signaling with a set of ordinary differential equations (ODEs). This first model supported a rigorous analysis of this part of Wnt signaling *in silico* allowing the prediction of embryology experiments. Based on experiments performed with egg extracts, the critical role of Axin in regulating pathway activity was deciphered. The ODE-based model was later extended by including Dkk1 and Axin2 feedback showing that the pathway may show oscillatory behavior under some conditions (Wawra et al., 2007).

A more complex mathematical model describes a Wnt signaling network with negative cross-regulation of canonical and non-canonical Wnt signaling during dorso-ventral axis formation in *Xenopus*. Wnt11b was earlier shown to activate CamkII and to be required for ventral development in *Xenopus* (Kuhl et al., 2000). However, Wnt11b was also shown to be required for dorsal development in *Xenopus* (Tao et al., 2005). This contradiction was resolved by the observation that Wnts can activate different signaling branches in a concentration-dependent manner; high concentrations of Wnt3a favored canonical, whereas low concentrations favored non-canonical, Wnt signaling (Nalesso et al., 2011). Together with the cross-inhibition of different Wnt signaling branches, this implicated a switch-like behavior of the Wnt signaling network (Kestler and Kuhl, 2011). Indeed, the function of Wnt11b during dorso-ventral axis formation could be recapitulated in a mathematical model that also reflects the findings of gain- and loss-of-function studies in *Xenopus* (Strang et al., 2017). Taken together, this work showed that mathematical models in combination with experiments in *Xenopus* embryos can be used to gain new insights into the mechanisms underlying Wnt-mediated signal transduction.

Mathematical models also proved helpful for analyzing gene regulatory network architecture involved in embryonic pattern formation. For example, during early heart development, Wnt6 signaling is involved in patterning the lateral plate, cardiogenic mesoderm into heart muscle (myocardium, toward the inside), and non-muscular heart tissue (pericardium, toward the outside) (Lavery et al., 2008). Mathematical modeling alerted us that we were missing an additional important component, which led to study of the role of Sfrp1 in this context. We demonstrated that Sfrp1 interacts with Wnt6 in an intricate gene regulatory network (Gibb et al., 2013): Wnt6 signaling from the ectoderm adjacent to the cardiogenic mesoderm promotes nearby pericardium differentiation and confines *sfrp1* expression and myocardium differentiation toward the inside of the cardiogenic mesoderm. While *sfrp1* is initially only expressed in a few cells beyond the reach of strong Wnt6-mediated repression, after being secreted, Sfrp1 in turn inhibits Wnt6 signaling and thus reduces the reach of this Wnt6-mediated repression (and pericardium differentiation) and thereby increases the myocardium domain.

Mathematical modeling opened our eyes. While previously we had been constrained by our Wnt6-centered perspective, modeling provided us with an alternative outlook: *wnt6* expression just provides positional information about where and how far away the edge of the cardiogenic mesoderm is located. *sfrp1* expression is really in charge by using this positional information to carve out an appropriately sized and positioned heart muscle.

6.3.3. Context-Specific Wnt Signaling

Wnt signaling is one of remarkably few molecular cell-to-cell signaling pathways that are used repeatedly during embryonic development, both in different embryonic tissues and at different stages of development. The early *Xenopus* embryo provided us with an experimentally accessible model system to investigate mechanisms determining context-specific Wnt signaling function.

In one example, before zygotic gene activation (ZGA), maternal Wnt signaling promotes subsequent dorsal embryonic cell fate (illustrated in the famous axis duplication essay discussed previously), yet after ZGA, zygotic Wnt8a functions to promote essentially the opposite, subsequent ventral (and lateral) mesodermal cell fate (reviewed by Zylkiewicz et al., 2014). Dorsal-promoting maternal and ventral-promoting zygotic Wnt8a signaling are both mediated by canonical Wnt/β-Catenin pathway mechanisms (Hamilton et al., 2001); thus, the relevant context-specific Wnt signaling mechanisms are to be found downstream of β-Catenin in the regulation of presumably two different classes of direct Wnt target genes: direct maternal Wnt/β-Catenin target genes (normally expressed early in prospective dorsal cells) and direct zygotic Wnt8a/β-Catenin target genes (normally expressed later in prospective ventrolateral mesoderm). Since nuclear β-Catenin is considered the hallmark of active canonical Wnt signaling (see previously), genome-wide transcriptome analysis (RNA-seq) was combined with mapping of physical β-Catenin protein association to gene loci on chromosomes (ChIP-seq) to identify direct Wnt/β-Catenin target genes comprehensively in these two contexts, before and after ZGA (Afouda et al., 2020; Nakamura et al., 2016). Wnt signaling-regulated physical β-Catenin protein association to gene loci on chromosomes is, surprisingly, not sufficient for transcriptional regulation. Wnt signaling initiates β-Catenin association to many gene loci, with additional context-specific mechanisms combining with Wnt signaling to determine which of these β-Catenin associated genes are expressed for the correct context-specific Wnt target gene response (Nakamura and Hoppler, 2017).

Two different classes were originally expected in early *Xenopus* development: dorsal/maternal and ventral/zygotic direct Wnt target genes. Our analysis suggests not two but a useful definition of about five different classes of direct Wnt target genes. This includes two classes of maternal dorsal Wnt target genes (both co-regulated by Nodal signaling but with genes in the slightly later expressed class additionally regulated by products of the slightly earlier expressed class in a feed-forward regulatory loop, Afouda et al., 2020); a few universal Wnt target genes directly regulated by both dorsal maternal and ventral zygotic Wnt signaling (including *axin2* and *sp5*); and two classes of specific zygotic Wnt8a/β-Catenin target genes (Nakamura et al., 2016), one class co-regulated by BMP signaling (see also Hoppler and Moon, 1998) and one class co-regulated by FGF signaling (Haremaki et al., 2003).

This reveals quite a remarkable complexity of Wnt target genes for just the early stages of *Xenopus* embryonic development.

6.4. FUTURE DIRECTIONS AND IMPORTANT QUESTIONS

6.4.1. Wnt Interactions with Other Signaling Pathways

Historically, the canonical Wnt signaling pathway was described, and later β-Catenin-independent pathways were added that were thought to act independently of each other. As described previously, multiple pieces of evidence indicated that these different branches of Wnt signaling are in fact highly connected and that Wnt signaling should be represented as a signaling network. Recent evidence suggests that, in addition, Wnt signaling interacts with other signaling pathways. First, YAP and TAZ—two components of the Hippo signaling pathway—were found to associate with the destruction complex (Azzolin et al., 2014), as well as other Wnt signaling components (reviewed in Piccolo et al., 2014). Second, Disheveled recently was shown to interact with Ephrin signaling, specifically with Sipa1L3, an interactor of Epha4 (Rothe et al., 2017). This interaction was shown to be crucial for proper eye development in *Xenopus* embryos by balancing Wnt signaling. This study also raised the possibility that Disheveled may interact with other signaling pathways such as Notch, a notion supported by earlier *Xenopus* experiments that demonstrated an inhibitory cross-regulation between Wnt and Notch signaling (Collu et al., 2012). Taken together, these examples illustrate that in-depth analyses of the molecular mechanisms underlying the cross-talk of Wnt signaling with other signal pathways, such as Hippo, Ephrin, and Notch, warrant further investigation.

6.4.2. Regulation of Gene Expression by Non-Canonical Wnt Signaling

Non-canonical Wnt signaling is narrowly considered by many researchers to be the Wnt/PCP pathway regulating cell polarity and cell migration by modulating the cytoskeleton or the polarized sub-cellular distribution of its components. However, since some components of the non-canonical Wnt signaling pathway, such as JNK, CamKII, or Calcineurin, are also known to be involved in gene regulation, does non-canonical Wnt signaling directly regulate gene expression? The first genes thought to be regulated at the transcriptional level by non-canonical Wnt, via JNK, were PAPC (Feike et al., 2010; Schambony and Wedlich, 2007) and EAF 2 (Maurus et al., 2005) in *Xenopus* and TGFβ2 (Zhou et al., 2007) in mice. Since JNK regulates phosphorylation of the transcription factor ATF2, an ATF2-luciferase reporter was used to map non-canonical Wnt signaling in *Xenopus* embryos (Ohkawara and Niehrs, 2011). Further, using dorsal marginal zone explants deficient of either Wnt5a or Wnt11, Gradl and colleagues recently identified *pbk* as a Wnt5a target gene, whereas *rab11fip5* was shown to be a specific Wnt11b target gene (Wallkamm et al., 2016).

The likely best-described target gene of non-canonical Wnt signaling is given by the cell adhesion molecule ALCAM. First identified in a screen for non-canonical Wnt target genes (Prieve and Moon, 2003), it was later found to be regulated by Wnt11a during cardiogenesis in *Xenopus* (Gessert et al., 2008). This finding was subsequently confirmed in the zebrafish (Choudhry and Trede, 2013). In follow-up studies, the promotor of *alcam* was isolated in the *Xenopus* system. A Frizzled3 responsive element was identified by the use of reporter gene assays. This element is regulated through ATF2 and Pax2 in the *Xenopus* pronephros and chromatin IP experiments confirmed *in vivo* binding in a non-canonical Wnt-dependent manner (Cizelsky et al., 2014). In the *Xenopus* eye, this gene also was found to be regulated through non-canonical Wnt (Seigfried et al., 2017).

These studies raise the important question: Are there specific non-canonical Wnt cis-regulatory responsive elements on DNA level? It will be an important and fundamental research question for the future to determine to what extent non-canonical Wnt signaling regulates gene expression and to identify unifying underlying mechanisms. Recent loss-of-function experiments in *Xenopus* have shown that Wnt5a or Wnt11b regulate fewer genes than canonical Wnt signaling (Wallkamm et al., 2016), demonstrating in principle that *Xenopus* is a very suitable model in which to answer these questions.

6.4.3. Integration of Wnt Signaling into the Embryonic Signaling and Regulatory Environment

The recent identification of direct Wnt target genes (e.g. Afouda et al., 2020; Nakamura et al., 2016) highlighted the importance of the co-regulatory environment in the embryo to ensure that Wnt target genes are expressed at the right time and in the right place. Recent articles have demonstrated that *Xenopus* is a leading experimental model for investigating the embryonic function and molecular mechanisms of such combinatorial regulation of Wnt target gene expression, such as with FGF signaling (Kjolby et al., 2019) or with BMP signaling (Polevoy et al., 2019) but also to dissect the role of chromatin modification (Hontelez et al., 2015) and, most likely related to that, chromatin accessibility (Esmaeili et al., 2020).

6.4.4. Xenopus as a Model for Human Disease

CRISP/R-mediated gene editing technology has clearly transformed the potential for *Xenopus* research to model human disease, including some involving Wnt pathway mechanisms. This approach was pioneered with TALEN-mediated gene editing of the *apc* gene in *Xenopus tropicalis* to create an animal model that successfully phenocopied Familial Adenomatous Polyposis (Van Nieuwenhuysen et al., 2015). *Xenopus* experiments discovered the molecular connection between EMC1 variants in patients and Wnt signaling and neural crest development consistent with the observed diverse birth defects (Marquez et al., 2020). Several heterotaxy-correlated birth-defect candidate genes have also recently been linked to Wnt signaling by *Xenopus* experiments, such as *RAPGEF5* (Griffin et al., 2018) and *AGMO* (Duncan et al., 2019). These examples clearly demonstrate the enormous potential for *Xenopus* to model human disease related to the Wnt signaling network.

ACKNOWLEDGMENTS

The authors would like to thank their lab members (past and present) and colleagues of the *Xenopus* and Wnt signaling research fields from around the world for many stimulating discussions and scientific encouragement.

REFERENCES

Afouda, B.A., Hoppler, S., 2011. Different requirements for GATA factors in cardiogenesis are mediated by non-canonical Wnt signaling. *Dev Dyn* 240, 649–662.

Afouda, B.A., Martin, J., Liu, F., Ciau-Uitz, A., Patient, R., Hoppler, S., 2008. GATA transcription factors integrate Wnt signalling during heart development. *Development* 135, 3185–3190.

Afouda, B.A., Nakamura, Y., Shaw, S., Charney, R.M., Paraiso, K.D., Blitz, I.L., Cho, K.W.Y., Hoppler, S., 2020. Foxh1/nodal defines context-specific direct maternal Wnt/beta-catenin target gene regulation in early development. *iScience* 23, 101314.

Azzolin, L., Panciera, T., Soligo, S., Enzo, E., Bicciato, S., Dupont, S., Bresolin, S., Frasson, C., Basso, G., Guzzardo, V., Fassina, A., Cordenonsi, M., Piccolo, S., 2014. YAP/TAZ incorporation in the beta-catenin destruction complex orchestrates the Wnt response. *Cell* 158, 157–170.

Baker, N.E., 1987. Molecular cloning of sequences from wingless, a segment polarity gene in Drosophila: The spatial distribution of a transcript in embryos. *EMBO J* 6, 1765–1773.

Behrens, J., von Kries, J.P., Kuhl, M., Bruhn, L., Wedlich, D., Grosschedl, R., Birchmeier, W., 1996. Functional interaction of beta-catenin with the transcription factor LEF-1. *Nature* 382, 638–642.

Bilic, J., Huang, Y.L., Davidson, G., Zimmermann, T., Cruciat, C.M., Bienz, M., Niehrs, C., 2007. Wnt induces LRP6 signalosomes and promotes dishevelled-dependent LRP6 phosphorylation. *Science* 316, 1619–1622.

Bovolenta, B., Gorny, A.-K., Esteve, P., Steinbeisser, H., 2014. Secreted Wnt inhibitors or modulators. In: Hoppler, S., Moon, R.T. (Eds.), *Wnt Signaling in Development and Disease: Molecular Mechanisms and Biological Functions*. John Wiley & Sons, Ltd, Hoboken, NJ, pp. 179–193.

Brannon, M., Gomperts, M., Sumoy, L., Moon, R.T., Kimelman, D., 1997. A beta-catenin/XTcf-3 complex binds to the siamois promoter to regulate dorsal axis specification in Xenopus. *Genes Dev* 11, 2359–2370.

Buitrago-Delgado, E., Nordin, K., Rao, A., Geary, L., LaBonne, C., 2015. Neurodevelopment: Shared regulatory programs suggest retention of blastula-stage potential in neural crest cells. *Science* 348, 1332–1335.

Butler, M.T., Wallingford, J.B., 2018. Spatial and temporal analysis of PCP protein dynamics during neural tube closure. *Elife* 7.

Chang, L.S., Kim, M., Glinka, A., Reinhard, C., Niehrs, C., 2020. The tumor suppressor PTPRK promotes ZNRF3 internalization and is required for Wnt inhibition in the Spemann organizer. *Elife* 9.

Chen, M., Amado, N., Tan, J., Reis, A., Ge, M., Abreu, J.G., He, X., 2020. TMEM79/MATTRIN defines a pathway for Frizzled regulation and is required for Xenopus embryogenesis. *Elife* 9.

Chen, T., Wallace, H., Ahmed, Y., Lee, E., 2014. Wnt signal transduction in the cytoplasm: An introduction to the destruction complex. In: Hoppler, S., Moon, R.T. (Eds.), *Wnt Signaling in Development and Disease: Molecular Mechanisms and Biological Functions*. John Wiley & Sons, Ltd, Hoboken, NJ, pp. 33–49.

Choudhry, P., Trede, N.S., 2013. DiGeorge syndrome gene tbx1 functions through wnt11r to regulate heart looping and differentiation. *PLoS One* 8, e58145.

Cizelsky, W., Tata, A., Kuhl, M., Kuhl, S.J., 2014. The Wnt/JNK signaling target gene alcam is required for embryonic kidney development. *Development* 141, 2064–2074.

Collu, G.M., Hidalgo-Sastre, A., Acar, A., Bayston, L., Gildea, C., Leverentz, M.K., Mills, C.G., Owens, T.W., Meurette, O., Dorey, K., Brennan, K., 2012. Dishevelled limits Notch signalling through inhibition of CSL. *Development* 139, 4405–4415.

Davidson, G., Wu, W., Shen, J., Bilic, J., Fenger, U., Stannek, P., Glinka, A., Niehrs, C., 2005. Casein kinase 1 gamma couples Wnt receptor activation to cytoplasmic signal transduction. *Nature* 438, 867–872.

De Calisto, J., Araya, C., Marchant, L., Riaz, C.F., Mayor, R., 2005. Essential role of non-canonical Wnt signalling in neural crest migration. *Development* 132, 2587–2597.

Dejmek, J., Safholm, A., Kamp Nielsen, C., Andersson, T., Leandersson, K., 2006. Wnt-5a/Ca2+-induced NFAT activity is counteracted by Wnt-5a/Yes-Cdc42-casein kinase 1alpha signaling in human mammary epithelial cells. *Mol Cell Biol* 26, 6024–6036.

Devotta, A., Hong, C.S., Saint-Jeannet, J.P., 2018. Dkk2 promotes neural crest specification by activating Wnt/beta-catenin signaling in a GSK3beta independent manner. *Elife* 7.

Ding, Y., Colozza, G., Sosa, E.A., Moriyama, Y., Rundle, S., Salwinski, L., De Robertis, E.M., 2018. Bighead is a Wnt antagonist secreted by the Xenopus Spemann organizer that promotes Lrp6 endocytosis. *Proc Natl Acad Sci U S A* 115, E9135–E9144.

Dissanayake, S.K., Wade, M., Johnson, C.E., O'Connell, M.P., Leotlela, P.D., French, A.D., Shah, K.V., Hewitt, K.J., Rosenthal, D.T., Indig, F.E., Jiang, Y., Nickoloff, B.J., Taub, D.D., Trent, J.M., Moon, R.T., Bittner, M., Weeraratna, A.T., 2007. The Wnt5A/protein kinase C pathway mediates motility in melanoma cells via the inhibition of metastasis suppressors and initiation of an epithelial to mesenchymal transition. *J Biol Chem* 282, 17259–17271.

Djiane, A., Riou, J., Umbhauer, M., Boucaut, J., Shi, D., 2000. Role of frizzled 7 in the regulation of convergent extension movements during gastrulation in *Xenopus laevis*. *Development* 127, 3091–3100.

Domingos, P.M., Itasaki, N., Jones, C.M., Mercurio, S., Sargent, M.G., Smith, J.C., Krumlauf, R., 2001. The Wnt/beta-catenin pathway posteriorizes neural tissue in Xenopus by an indirect mechanism requiring FGF signalling. *Dev Biol* 239, 148–160.

Du, S.J., Purcell, S.M., Christian, J.L., McGrew, L.L., Moon, R.T., 1995. Identification of distinct classes and functional domains of Wnts through expression of wild-type and chimeric proteins in Xenopus embryos. *Mol Cell Biol* 15, 2625–2634.

Duncan, A.R., Gonzalez, D.P., Del Viso, F., Robson, A., Khokha, M.K., Griffin, J.N., 2019. Alkylglycerol monooxygenase, a heterotaxy candidate gene, regulates left-right patterning via Wnt signaling. *Dev Biol* 456, 1–7.

Esmaeili, M., Blythe, S.A., Tobias, J.W., Zhang, K., Yang, J., Klein, P.S., 2020. Chromatin accessibility and histone acetylation in the regulation of competence in early development. *Dev Biol* 462, 20–35.

Feike, A.C., Rachor, K., Gentzel, M., Schambony, A., 2010. Wnt5a/Ror2-induced upregulation of xPAPC requires xShcA. *Biochem Biophys Res Commun* 400, 500–506.

Funayama, N., Fagotto, F., McCrea, P., Gumbiner, B.M., 1995. Embryonic axis induction by the armadillo repeat domain of beta-catenin: Evidence for intracellular signaling. *J Cell Biol* 128, 959–968.

Gammons, M., Bienz, M., 2018. Multiprotein complexes governing Wnt signal transduction. *Curr Opin Cell Biol* 51, 42–49.

Gessert, S., Maurus, D., Brade, T., Walther, P., Pandur, P., Kuhl, M., 2008. DM-GRASP/ALCAM/CD166 is required for cardiac morphogenesis and maintenance of cardiac identity in first heart field derived cells. *Dev Biol* 321, 150–161.

Gibb, N., Lavery, D.L., Hoppler, S., 2013. Sfrp1 promotes cardiomyocyte differentiation in Xenopus via negative-feedback regulation of Wnt signalling. *Development* 140, 1537–1549.

Glinka, A., Wu, W., Delius, H., Monaghan, A.P., Blumenstock, C., Niehrs, C., 1998. Dickkopf-1 is a member of a new family of secreted proteins and functions in head induction. *Nature* 391, 357–362.

Gradl, D., Kuhl, M., Wedlich, D., 1999. The Wnt/Wg signal transducer beta-catenin controls fibronectin expression. *Mol Cell Biol* 19, 5576–5587.

Griffin, J.N., Del Viso, F., Duncan, A.R., Robson, A., Hwang, W., Kulkarni, S., Liu, K.J., Khokha, M.K., 2018. RAPGEF5 regulates nuclear translocation of beta-catenin. *Dev Cell* 44, 248–260 e244.

Guo, Y., Dorn, T., Kuhl, S.J., Linnemann, A., Rothe, M., Pfister, A.S., Vainio, S., Laugwitz, K.L., Moretti, A., Kuhl, M., 2019. The Wnt inhibitor Dkk1 is required for maintaining the normal cardiac differentiation program in *Xenopus laevis*. *Dev Biol* 449, 1–13.

Habas, R., Dawid, I.B., He, X., 2003. Coactivation of Rac and Rho by Wnt/Frizzled signaling is required for vertebrate gastrulation. *Genes Dev* 17, 295–309.

Habas, R., Kato, Y., He, X., 2001. Wnt/Frizzled activation of Rho regulates vertebrate gastrulation and requires a novel Formin homology protein Daam1. *Cell* 107, 843–854.

Hamilton, F.S., Wheeler, G.N., Hoppler, S., 2001. Difference in XTcf-3 dependency accounts for change in response to beta-catenin-mediated Wnt signalling in Xenopus blastula. *Development* 128, 2063–2073.

Haremaki, T., Tanaka, Y., Hongo, I., Yuge, M., Okamoto, H., 2003. Integration of multiple signal transducing pathways on Fgf response elements of the Xenopus caudal homologue Xcad3. *Development* 130, 4907–4917.

Heasman, J., Kofron, M., Wylie, C., 2000. Beta-catenin signaling activity dissected in the early Xenopus embryo: A novel antisense approach. *Dev Biol* 222, 124–134.

Hedgepeth, C.M., Deardorff, M.A., Rankin, K., Klein, P.S., 1999. Regulation of glycogen synthase kinase 3beta and downstream Wnt signaling by axin. *Mol Cell Biol* 19, 7147–7157.

Hempel, A., Kuhl, S.J., Rothe, M., Rao Tata, P., Sirbu, I.O., Vainio, S.J., Kuhl, M., 2017. The CapZ interacting protein Rcsd1 is required for cardiogenesis downstream of Wnt11a in *Xenopus laevis*. *Dev Biol* 424, 28–39.

Holmen, S.L., Salic, A., Zylstra, C.R., Kirschner, M.W., Williams, B.O., 2002. A novel set of Wnt-Frizzled fusion proteins identifies receptor components that activate beta-catenin-dependent signaling. *J Biol Chem* 277, 34727–34735.

Hontelez, S., van Kruijsbergen, I., Georgiou, G., van Heeringen, S.J., Bogdanovic, O., Lister, R., Veenstra, G.J.C., 2015. Embryonic transcription is controlled by maternally defined chromatin state. *Nat Commun* 6, 10148.

Hoppler, S., Brown, J.D., Moon, R.T., 1996. Expression of a dominant-negative Wnt blocks induction of MyoD in Xenopus embryos. *Genes Dev* 10, 2805–2817.

Hoppler, S., Conlon, F.L., 2020. Xenopus: Experimental access to cardiovascular development, regeneration discovery, and cardiovascular heart-defect modeling. *Cold Spring Harb Perspect Biol* 12.

Hoppler, S., Kavanagh, C.L., 2007. Wnt signalling: Variety at the core. *J Cell Sci* 120, 385–393.

Hoppler, S., Mazzotta, S., Kuhl, M., 2014. Wnt signaling in heart development. In: Hoppler, S., Moon, R.T. (Eds.), *Wnt Signaling in Development and Disease*. John Wiley & Sons, Hoboken.

Hoppler, S., Moon, R.T., 1998. BMP-2/-4 and Wnt-8 cooperatively pattern the Xenopus mesoderm. *Mech Dev* 71, 119–129.

Hoppler, S., Nakamura, J., 2014. *Cell-to-Cell Signalling in Development: Wnt Signalling, eLS*. Wiley Online Library, Hoboken, NJ.

Hoppler, S., Waterman, M.L., 2014. Evolutionary diversification of vertebrate TCF/LEF structure, function, and regulation. In: Hoppler, S., Moon, R.T. (Eds.), *Wnt Signaling in Development and Disease: Molecular Mechanisms and Biological Functions*. John Wiley & Sons, Ltd, Hoboken, NJ, pp. 225–237.

Huang, T., Xie, Z., Wang, J., Li, M., Jing, N., Li, L., 2011. Nuclear factor of activated T cells (NFAT) proteins repress canonical Wnt signaling via its interaction with Dishevelled (Dvl) protein and participate in regulating neural progenitor cell proliferation and differentiation. *J Biol Chem* 286, 37399–37405.

Hyde, A.S., Hang, B.I., Lee, E., 2016. Reconstitution of the cytoplasmic regulation of the Wnt signaling pathway using Xenopus egg extracts. *Methods Mol Biol* 1481, 101–109.

Ishitani, T., Kishida, S., Hyodo-Miura, J., Ueno, N., Yasuda, J., Waterman, M., Shibuya, H., Moon, R.T., Ninomiya-Tsuji, J., Matsumoto, K., 2003. The TAK1-NLK mitogen-activated protein kinase cascade functions in the Wnt-5a/Ca(2+) pathway to antagonize Wnt/beta-catenin signaling. *Mol Cell Biol* 23, 131–139.

Itoh, K., Krupnik, V.E., Sokol, S.Y., 1998. Axis determination in Xenopus involves biochemical interactions of axin, glycogen synthase kinase 3 and beta-catenin. *Curr Biol* 8, 591–594.

Janda, C.Y., Waghray, D., Levin, A.M., Thomas, C., Garcia, K.C., 2012. Structural basis of Wnt recognition by Frizzled. *Science* 337, 59–64.

Jenei, V., Sherwood, V., Howlin, J., Linnskog, R., Safholm, A., Axelsson, L., Andersson, T., 2009. A t-butyloxycarbonyl-modified Wnt5a-derived hexapeptide functions as a potent antagonist of Wnt5a-dependent melanoma cell invasion. *Proc Natl Acad Sci U S A* 106, 19473–19478.

Kestler, H.A., Kuhl, M., 2008. From individual Wnt pathways towards a Wnt signalling network. *Philos Trans R Soc Lond B Biol Sci* 363, 1333–1347.

Kestler, H.A., Kuhl, M., 2011. Generating a Wnt switch: It's all about the right dosage. *J Cell Biol* 193, 431–433.

Kim, G.H., Han, J.K., 2005. JNK and ROKalpha function in the noncanonical Wnt/RhoA signaling pathway to regulate Xenopus convergent extension movements. *Dev Dyn* 232, 958–968.

Kjolby, R.A.S., Truchado-Garcia, M., Iruvanti, S., Harland, R.M., 2019. Integration of Wnt and FGF signaling in the Xenopus gastrula at TCF and Ets binding sites shows the importance of short-range repression by TCF in patterning the marginal zone. *Development* 146.

Kofron, M., Birsoy, B., Houston, D., Tao, Q., Wylie, C., Heasman, J., 2007. Wnt11/beta-catenin signaling in both oocytes and early embryos acts through LRP6-mediated regulation of axin. *Development* 134, 503–513.

Kuhl, M., Sheldahl, L.C., Malbon, C.C., Moon, R.T., 2000. Ca(2+)/calmodulin-dependent protein kinase II is stimulated by Wnt and Frizzled homologs and promotes ventral cell fates in Xenopus. *J Biol Chem* 275, 12701–12711.

Kunz, M., Herrmann, M., Wedlich, D., Gradl, D., 2004. Autoregulation of canonical Wnt signaling controls midbrain development. *Dev Biol* 273, 390–401.

LaBonne, C., Bronner-Fraser, M., 1998. Neural crest induction in Xenopus: Evidence for a two-signal model. *Development* 125, 2403–2414.

Laurent, M.N., Blitz, I.L., Hashimoto, C., Rothbacher, U., Cho, K.W., 1997. The Xenopus homeobox gene twin mediates Wnt induction of goosecoid in establishment of Spemann's organizer. *Development* 124, 4905–4916.

Lavery, D.L., Martin, J., Turnbull, Y.D., Hoppler, S., 2008. Wnt6 signaling regulates heart muscle development during organogenesis. *Dev Biol* 323, 177–188.

Lee, E., Salic, A., Kruger, R., Heinrich, R., Kirschner, M.W., 2003. The roles of APC and Axin derived from experimental and theoretical analysis of the Wnt pathway. *PLoS Biol* 1, E10.

Leyns, L., Bouwmeester, T., Kim, S.H., Piccolo, S., De Robertis, E.M., 1997. Frzb-1 is a secreted antagonist of Wnt signaling expressed in the Spemann organizer. *Cell* 88, 747–756.

Li, X., Yost, H.J., Virshup, D.M., Seeling, J.M., 2001. Protein phosphatase 2A and its B56 regulatory subunit inhibit Wnt signaling in Xenopus. *EMBO J* 20, 4122–4131.

Liu, C., Kato, Y., Zhang, Z., Do, V.M., Yankner, B.A., He, X., 1999. beta-Trcp couples beta-catenin phosphorylation-degradation and regulates Xenopus axis formation. *Proc Natl Acad Sci U S A* 96, 6273–6278.

Liu, C., Li, Y., Semenov, M., Han, C., Baeg, G.H., Tan, Y., Zhang, Z., Lin, X., He, X., 2002. Control of beta-catenin phosphorylation/degradation by a dual-kinase mechanism. *Cell* 108, 837–847.

Liu, F., van den Broek, O., Destree, O., Hoppler, S., 2005. Distinct roles for Xenopus Tcf/Lef genes in mediating specific responses to Wnt/beta-catenin signalling in mesoderm development. *Development* 132, 5375–5385.

Lu, X., Borchers, A.G., Jolicoeur, C., Rayburn, H., Baker, J.C., Tessier-Lavigne, M., 2004. PTK7/CCK-4 is a novel regulator of planar cell polarity in vertebrates. *Nature* 430, 93–98.

Lyons, J.P., Mueller, U.W., Ji, H., Everett, C., Fang, X., Hsieh, J.C., Barth, A.M., McCrea, P.D., 2004. Wnt-4 activates the canonical beta-catenin-mediated Wnt pathway and binds Frizzled-6 CRD: Functional implications of Wnt/beta-catenin activity in kidney epithelial cells. *Exp Cell Res* 298, 369–387.

Mao, B., Wu, W., Li, Y., Hoppe, D., Stannek, P., Glinka, A., Niehrs, C., 2001. LDL-receptor-related protein 6 is a receptor for Dickkopf proteins. *Nature* 411, 321–325.

Marikawa, Y., Elinson, R.P., 1998. beta-TrCP is a negative regulator of Wnt/beta-catenin signaling pathway and dorsal axis formation in Xenopus embryos. *Mech Dev* 77, 75–80.

Marquez, J., Criscione, J., Charney, R.M., Prasad, M.S., Hwang, W.Y., Mis, E.K., Garcia-Castro, M.I., Khokha, M.K., 2020. Disrupted ER membrane protein complex-mediated topogenesis drives congenital neural crest defects. *J Clin Invest* 130, 813–826.

Marvin, M.J., Di Rocco, G., Gardiner, A., Bush, S.M., Lassar, A.B., 2001. Inhibition of Wnt activity induces heart formation from posterior mesoderm. *Genes Dev* 15, 316–327.

Maurus, D., Heligon, C., Burger-Schwarzler, A., Brandli, A.W., Kuhl, M., 2005. Noncanonical Wnt-4 signaling and EAF2 are required for eye development in *Xenopus laevis*. *EMBO J* 24, 1181–1191.

Mazzotta, S., Neves, C., Bonner, R.J., Bernardo, A.S., Docherty, K., Hoppler, S., 2016. Distinctive roles of canonical and non-canonical Wnt signaling in human embryonic cardiomyocyte development. *Stem Cell Reports* 7, 764–776.

McGrew, L.L., Hoppler, S., Moon, R.T., 1997. Wnt and FGF pathways cooperatively pattern anteroposterior neural ectoderm in Xenopus. *Mech Dev* 69, 105–114.

McGrew, L.L., Takemaru, K., Bates, R., Moon, R.T., 1999. Direct regulation of the Xenopus engrailed-2 promoter by the Wnt signaling pathway, and a molecular screen for Wnt-responsive genes, confirm a role for Wnt signaling during neural patterning in Xenopus. *Mech Dev* 87, 21–32.

McKendry, R., Hsu, S.C., Harland, R.M., Grosschedl, R., 1997. LEF-1/TCF proteins mediate wnt-inducible transcription from the Xenopus nodal-related 3 promoter. *Dev Biol* 192, 420–431.

McMahon, A.P., Moon, R.T., 1989. Ectopic expression of the proto-oncogene int-1 in Xenopus embryos leads to duplication of the embryonic axis. *Cell* 58, 1075–1084.

Medina, A., Reintsch, W., Steinbeisser, H., 2000. Xenopus frizzled 7 can act in canonical and non-canonical Wnt signaling pathways: Implications on early patterning and morphogenesis. *Mech Dev* 92, 227–237.

Medina, A., Steinbeisser, H., 2000. Interaction of Frizzled 7 and Dishevelled in Xenopus. *Dev Dyn* 218, 671–680.

Mii, Y., Takada, S., 2020. Heparan sulfate proteoglycan clustering in Wnt signaling and dispersal. *Front Cell Dev Biol* 8, 631.

Molenaar, M., van de Wetering, M., Oosterwegel, M., Peterson-Maduro, J., Godsave, S., Korinek, V., Roose, J., Destree, O., Clevers, H., 1996. XTcf-3 transcription factor mediates beta-catenin-induced axis formation in Xenopus embryos. *Cell* 86, 391–399.

Mukherjee, S., Chaturvedi, P., Rankin, S.A., Fish, M.B., Wlizla, M., Paraiso, K.D., MacDonald, M., Chen, X., Weirauch, M.T., Blitz, I.L., Cho, K.W., Zorn, A.M., 2020. Sox17 and beta-catenin co-occupy Wnt-responsive enhancers to govern the endoderm gene regulatory network. *Elife* 9.

Nagy, II, Railo, A., Rapila, R., Hast, T., Sormunen, R., Tavi, P., Rasanen, J., Vainio, S.J., 2010. Wnt-11 signalling controls ventricular myocardium development by patterning N-cadherin and beta-catenin expression. *Cardiovasc Res* 85, 100–109.

Nakamura, Y., de Paiva Alves, E., Veenstra, G.J., Hoppler, S., 2016. Tissue- and stage-specific Wnt target gene expression is controlled subsequent to beta-catenin recruitment to cis-regulatory modules. *Development* 143, 1914–1925.

Nakamura, Y., Hoppler, S., 2017. Genome-wide analysis of canonical Wnt target gene regulation in *Xenopus tropicalis* challenges beta-catenin paradigm. *Genesis* 55.

Nalesso, G., Sherwood, J., Bertrand, J., Pap, T., Ramachandran, M., De Bari, C., Pitzalis, C., Dell'accio, F., 2011. WNT-3A modulates articular chondrocyte phenotype by activating both canonical and noncanonical pathways. *J Cell Biol* 193, 551–564.

Nusse, R., Varmus, H.E., 1982. Many tumors induced by the mouse mammary tumor virus contain a provirus integrated in the same region of the host genome. *Cell* 31, 99–109.

Ohkawara, B., Niehrs, C., 2011. An ATF2-based luciferase reporter to monitor non-canonical Wnt signaling in Xenopus embryos. *Dev Dyn* 240, 188–194.

Ossipova, O., Kerney, R., Saint-Jeannet, J.P., Sokol, S.Y., 2018. Regulation of neural crest development by the formin family protein Daam1. *Genesis* 56, e23108.

Pandur, P., Lasche, M., Eisenberg, L.M., Kuhl, M., 2002. Wnt-11 activation of a non-canonical Wnt signalling pathway is required for cardiogenesis. *Nature* 418, 636–641.

Penzo-Mendez, A., Umbhauer, M., Djiane, A., Boucaut, J.C., Riou, J.F., 2003. Activation of Gbetagamma signaling downstream of Wnt-11/Xfz7 regulates Cdc42 activity during Xenopus gastrulation. *Dev Biol* 257, 302–314.

Peters, J.M., McKay, R.M., McKay, J.P., Graff, J.M., 1999. Casein kinase I transduces Wnt signals. *Nature* 401, 345–350.

Piccolo, S., Dupont, S., Cordenonsi, M., 2014. The biology of YAP/TAZ: Hippo signaling and beyond. *Physiol Rev* 94, 1287–1312.

Podleschny, M., Grund, A., Berger, H., Rollwitz, E., Borchers, A., 2015. A PTK7/Ror2 Co-receptor complex affects Xenopus neural crest migration. *PLoS One* 10, e0145169.

Polevoy, H., Gutkovich, Y.E., Michaelov, A., Volovik, Y., Elkouby, Y.M., Frank, D., 2019. New roles for Wnt and BMP signaling in neural anteroposterior patterning. *EMBO Rep* 20.

Prieve, M.G., Moon, R.T., 2003. Stromelysin-1 and mesothelin are differentially regulated by Wnt-5a and Wnt-1 in C57mg mouse mammary epithelial cells. *BMC Dev Biol* 3, 2.

Ramakrishnan, A.-B., Sinha, A., Fan, V.B., Cadigan, K.M., 2018. The Wnt transcriptional switch: TLE removal or inactivation? *BioEssays* 40.

Rasmussen, J.T., Deardorff, M.A., Tan, C., Rao, M.S., Klein, P.S., Vetter, M.L., 2001. Regulation of eye development by frizzled signaling in Xenopus. *Proc Natl Acad Sci U S A* 98, 3861–3866.

Roose, J., Molenaar, M., Peterson, J., Hurenkamp, J., Brantjes, H., Moerer, P., van de Wetering, M., Destree, O., Clevers, H., 1998. The Xenopus Wnt effector XTcf-3 interacts with Groucho-related transcriptional repressors. *Nature* 395, 608–612.

Rothbacher, U., Laurent, M.N., Deardorff, M.A., Klein, P.S., Cho, K.W., Fraser, S.E., 2000. Dishevelled phosphorylation, subcellular localization and multimerization regulate its role in early embryogenesis. *EMBO J* 19, 1010–1022.

Rothe, M., Kanwal, N., Dietmann, P., Seigfried, F.A., Hempel, A., Schutz, D., Reim, D., Engels, R., Linnemann, A., Schmeisser, M.J., Bockmann, J., Kuhl, M., Boeckers, T.M., Kuhl, S.J., 2017. An Epha4/Sipa1l3/Wnt pathway regulates eye development and lens maturation. *Development* 144, 321–333.

Roy, J.P., Halford, M.M., Stacker, S.A., 2018. The biochemistry, signalling and disease relevance of RYK and other WNT-binding receptor tyrosine kinases. *Growth Factors* 36, 15–40.

Salic, A., Lee, E., Mayer, L., Kirschner, M.W., 2000. Control of beta-catenin stability: Reconstitution of the cytoplasmic steps of the wnt pathway in Xenopus egg extracts. *Mol Cell* 5, 523–532.

Saneyoshi, T., Kume, S., Amasaki, Y., Mikoshiba, K., 2002. The Wnt/calcium pathway activates NF-AT and promotes ventral cell fate in Xenopus embryos. *Nature* 417, 295–299.

Saulnier, D.M., Ghanbari, H., Brandli, A.W., 2002. Essential function of Wnt-4 for tubulogenesis in the Xenopus pronephric kidney. *Dev Biol* 248, 13–28.

Schambony, A., Wedlich, D., 2007. Wnt-5A/Ror2 regulate expression of XPAPC through an alternative noncanonical signaling pathway. *Dev Cell* 12, 779–792.

Schneider, S., Steinbeisser, H., Warga, R.M., Hausen, P., 1996. Beta-catenin translocation into nuclei demarcates the dorsalizing centers in frog and fish embryos. *Mech Dev* 57, 191–198.

Schneider, V.A., Mercola, M., 2001. Wnt antagonism initiates cardiogenesis in *Xenopus laevis*. *Genes Dev* 15, 304–315.

Schohl, A., Fagotto, F., 2003. A role for maternal beta-catenin in early mesoderm induction in Xenopus. *EMBO J* 22, 3303–3313.

Seigfried, F.A., Cizelsky, W., Pfister, A.S., Dietmann, P., Walther, P., Kuhl, M., Kuhl, S.J., 2017. Frizzled 3 acts upstream of Alcam during embryonic eye development. *Dev Biol* 426, 69–83.

Semenov, M., Tamai, K., He, X., 2005. SOST is a ligand for LRP5/LRP6 and a Wnt signaling inhibitor. *J Biol Chem* 280, 26770–26775.

Sheldahl, L.C., Park, M., Malbon, C.C., Moon, R.T., 1999. Protein kinase C is differentially stimulated by Wnt and Frizzled homologs in a G-protein-dependent manner. *Curr Biol* 9, 695–698.

Shindo, A., Inoue, Y., Kinoshita, M., Wallingford, J.B., 2019. PCP-dependent transcellular regulation of actomyosin oscillation facilitates convergent extension of vertebrate tissue. *Dev Biol* 446, 159–167.

Slusarski, D.C., Yang-Snyder, J., Busa, W.B., Moon, R.T., 1997a. Modulation of embryonic intracellular calcium signaling by Wnt5A. *Dev Biol* 182, 114–120.

Slusarski, D.C., Corces, V.G., Moon, R.T., 1997b. Interaction of Wnt and a Frizzled homologue triggers G-protein-linked phosphatidylinositol signalling. *Nature* 390, 410–413.

Sokol, S.Y., 1996. Analysis of Dishevelled signalling pathways during Xenopus development. *Curr Biol* 6, 1456–1467.

Strang, J.E., Schuler, R., Kuhl, M., Kestler, H.A., 2017. Switch-like behavior enables Wnt11 concentration specific response during dorso-ventral axis formation in *Xenopus laevis*. *J Theor Biol* 429, 82–94.

Szenker-Ravi, E., Altunoglu, U., Leushacke, M., Bosso-Lefevre, C., Khatoo, M., Thi Tran, H., Naert, T., Noelanders, R., Hajamohideen, A., Beneteau, C., de Sousa, S.B., Karaman, B., Latypova, X., Basaran, S., Yucel, E.B., Tan, T.T., Vlaminck, L., Nayak, S.S., Shukla, A., Girisha, K.M., Le Caignec, C., Soshnikova, N., Uyguner, Z.O., Vleminckx, K., Barker, N., Kayserili, H., Reversade, B., 2018. RSPO2 inhibition of RNF43 and ZNRF3 governs limb development independently of LGR4/5/6. *Nature* 557, 564–569.

Tada, M., Smith, J.C., 2000. Xwnt11 is a target of Xenopus Brachyury: Regulation of gastrulation movements via Dishevelled, but not through the canonical Wnt pathway. *Development* 127, 2227–2238.

Takemaru, K.I., Moon, R.T., 2000. The transcriptional coactivator CBP interacts with beta-catenin to activate gene expression. *J Cell Biol* 149, 249–254.

Tamai, K., Semenov, M., Kato, Y., Spokony, R., Liu, C., Katsuyama, Y., Hess, F., Saint-Jeannet, J.P., He, X., 2000. LDL-receptor-related proteins in Wnt signal transduction. *Nature* 407, 530–535.

Tanegashima, K., Zhao, H., Dawid, I.B., 2008. WGEF activates Rho in the Wnt-PCP pathway and controls convergent extension in Xenopus gastrulation. *EMBO J* 27, 606–617.

Tao, Q., Yokota, C., Puck, H., Kofron, M., Birsoy, B., Yan, D., Asashima, M., Wylie, C.C., Lin, X., Heasman, J., 2005. Maternal wnt11 activates the canonical Wnt signaling pathway required for axis formation in Xenopus embryos. *Cell* 120, 857–871.

Torres, M.A., Yang-Snyder, J.A., Purcell, S.M., DeMarais, A.A., McGrew, L.L., Moon, R.T., 1996. Activities of the Wnt-1 class of secreted signaling factors are antagonized by the Wnt-5A class and by a dominant negative cadherin in early Xenopus development. *J Cell Biol* 133, 1123–1137.

Van Nieuwenhuysen, T., Naert, T., Tran, H.T., Van Imschoot, G., Geurs, S., Sanders, E., Creytens, D., Van Roy, F., Vleminckx, K., 2015. TALEN-mediated apc mutation in *Xenopus tropicalis* phenocopies familial adenomatous polyposis. *Oncoscience* 2, 555–566.

Veeman, M.T., Slusarski, D.C., Kaykas, A., Louie, S.H., Moon, R.T., 2003. Zebrafish prickle, a modulator of noncanonical

Wnt/Fz signaling, regulates gastrulation movements. *Curr Biol* 13, 680–685.

Vleminckx, K., Wong, E., Guger, K., Rubinfeld, B., Polakis, P., Gumbiner, B.M., 1997. Adenomatous polyposis coli tumor suppressor protein has signaling activity in *Xenopus laevis* embryos resulting in the induction of an ectopic dorsoanterior axis. *J Cell Biol* 136, 411–420.

Walentek, P., Schneider, I., Schweickert, A., Blum, M., 2013. Wnt11b is involved in cilia-mediated symmetry breakage during Xenopus left-right development. *PLoS One* 8, e73646.

Wallingford, J.B., Mitchell, B., 2011. Strange as it may seem: The many links between Wnt signaling, planar cell polarity, and cilia. *Genes Dev* 25, 201–213.

Wallingford, J.B., Rowning, B.A., Vogeli, K.M., Rothbacher, U., Fraser, S.E., Harland, R.M., 2000. Dishevelled controls cell polarity during Xenopus gastrulation. *Nature* 405, 81–85.

Wallkamm, V., Rahm, K., Schmoll, J., Kaufmann, L.T., Brinkmann, E., Schunk, J., Kraft, B., Wedlich, D., Gradl, D., 2016. Regulation of distinct branches of the non-canonical Wnt-signaling network in Xenopus dorsal marginal zone explants. *BMC Biol* 14, 55.

Wang, Q., Zhou, Y., Rychahou, P., Liu, C., Weiss, H.L., Evers, B.M., 2013. NFAT5 represses canonical Wnt signaling via inhibition of beta-catenin acetylation and participates in regulating intestinal cell differentiation. *Cell Death Dis* 4, e671.

Wang, S., Krinks, M., Lin, K., Luyten, F.P., Moos, M., Jr., 1997. Frzb, a secreted protein expressed in the Spemann organizer, binds and inhibits Wnt-8. *Cell* 88, 757–766.

Wawra, C., Kuhl, M., Kestler, H.A., 2007. Extended analyses of the Wnt/beta-catenin pathway: Robustness and oscillatory behaviour. *FEBS Lett* 581, 4043–4048.

Xing, Y., Clements, W.K., Kimelman, D., Xu, W., 2003. Crystal structure of a beta-catenin/axin complex suggests a mechanism for the beta-catenin destruction complex. *Genes Dev* 17, 2753–2764.

Yamanaka, H., Nishida, E., 2007. Wnt11 stimulation induces polarized accumulation of Dishevelled at apical adherens junctions through Frizzled7. *Genes Cells* 12, 961–967.

Yan, L., Chen, J., Zhu, X., Sun, J., Wu, X., Shen, W., Zhang, W., Tao, Q., Meng, A., 2018. Maternal Huluwa dictates the embryonic body axis through beta-catenin in vertebrates. *Science* 362.

Yang, J., Tan, C., Darken, R.S., Wilson, P.A., Klein, P.S., 2002. Beta-catenin/Tcf-regulated transcription prior to the mid-blastula transition. *Development* 129, 5743–5752.

Yang-Snyder, J., Miller, J.R., Brown, J.D., Lai, C.J., Moon, R.T., 1996. A frizzled homolog functions in a vertebrate Wnt signaling pathway. *Curr Biol* 6, 1302–1306.

Yost, C., Torres, M., Miller, J.R., Huang, E., Kimelman, D., Moon, R.T., 1996. The axis-inducing activity, stability, and subcellular distribution of beta-catenin is regulated in Xenopus embryos by glycogen synthase kinase 3. *Genes Dev* 10, 1443–1454.

Zeng, L., Fagotto, F., Zhang, T., Hsu, W., Vasicek, T.J., Perry, W.L., 3rd, Lee, J.J., Tilghman, S.M., Gumbiner, B.M., Costantini, F., 1997. The mouse Fused locus encodes Axin, an inhibitor of the Wnt signaling pathway that regulates embryonic axis formation. *Cell* 90, 181–192.

Zeng, X., Tamai, K., Doble, B., Li, S., Huang, H., Habas, R., Okamura, H., Woodgett, J., He, X., 2005. A dual-kinase mechanism for Wnt co-receptor phosphorylation and activation. *Nature* 438, 873–877.

Zhou, W., Lin, L., Majumdar, A., Li, X., Zhang, X., Liu, W., Etheridge, L., Shi, Y., Martin, J., Van de Ven, W., Kaartinen, V., Wynshaw-Boris, A., McMahon, A.P., Rosenfeld, M.G., Evans, S.M., 2007. Modulation of morphogenesis by non-canonical Wnt signaling requires ATF/CREB family-mediated transcriptional activation of TGFbeta2. *Nat Genet* 39, 1225–1234.

Zylkiewicz, E., Sokol, S.Y., Hoppler, S., 2014. Wnt signaling in early vertebrate development: From fertilization to gastrulation. In: Hoppler, S., Moon, R.T. (Eds.), *Wnt Signaling in Development and Disease: Molecular Mechanisms and Biological Functions*. John Wiley & Sons, Ltd, Hoboken, NJ, pp. 253–266.

7 Multiple Functions of Notch Signaling during Early Embryogenesis

Silvia L. López

CONTENTS

7.1. Historical Background 77
7.2. The Notch Pathway 77
7.2.1. Canonical Notch Signaling 78
7.2.2. Non-Canonical Notch Signaling 78
7.3. Notch Signaling during *Xenopus* Embryogenesis 78
7.3.1. Establishing the Dorsal-Ventral Axis 90
7.3.2. Germ Layer Formation 90
7.3.3. Dorsal Midline Tissues 98
7.3.4. Primary Neurogenesis 99
7.3.5. Neural Plate Border and Midbrain-Hindbrain Boundary 102
7.3.6. Somitogenesis 107
7.4. Future Directions 116
References 117

7.1. HISTORICAL BACKGROUND

The Notch pathway is a key cell-cell communication mechanism utilized during metazoan development. Its outcome depends on cell context: it can inhibit or promote cell fates, cell proliferation, or cell death through ligand-receptor signaling between neighboring cells (Kopan and Ilagan, 2009). The story of Notch began when John S. Dexter, in Thomas Hunt Morgan's laboratory, found a mutant phenotype in *Drosophila* with characteristic serrations at the wings' ends, which he called *Perfect Notched* (Dexter, 1914; Bridges and Morgan, 1916). This was caused by the disruption of a dominant sex-linked gene resulting in male lethality, which received the name *Notch* in subsequent publications (Bridges and Morgan, 1916; Morgan, 1917; Mohr, 1919). In the 1930s, Donald Poulson studied the lethal phenotype and noticed aberrant germ layer development (Poulson, 1937). This was later interpreted as a switch in ectodermal cell fate from dermoblast to neuroblast, since different mutant alleles of *Notch, Delta, mastermind, neuralized, Enhancer of split, almondex*, and *big brain* resulted in nervous system hypertrophy at the expense of the epidermis (Lehmann et al., 1983). These so-called "neurogenic" genes are all involved in the Notch pathway and have vertebrate counterparts (flybase.org; Lehmann et al., 1983; Thurmond et al., 2019).

Seven decades after Dexter's discovery, the fly *Notch* gene was cloned (Artavanis-Tsakonas et al., 1983), and the first vertebrate homologue, *notch1*, was isolated from *Xenopus laevis* (Coffman et al., 1990). Frog experiments using a construct lacking the extracellular domain provided the first clues that the Notch intracellular domain (NICD) mediates signal transduction (Coffman et al., 1993). This truncation resulted in a gain-of-function phenotype that affected germ layer development. Cloning the *Xenopus* gene encoding a ligand, Delta-like-1 (Dll1), demonstrated that Delta/Notch signaling plays a neurogenic role in vertebrates through lateral inhibition, as previously defined in *Drosophila* (Chitnis et al., 1995; Campos-Ortega, 1985; Lewis, 1998). Because of its relative simplicity, primary neurogenesis in *Xenopus* provided an ideal paradigm for Notch pathway research and for unraveling the molecular and cellular bases of vertebrate neural development. Since these ground-breaking studies, the accessibility of *Xenopus* embryos has made them an outstanding model for revealing the role of the Notch pathway in multiple developmental processes and for testing heterologous molecules from different species such as mouse and human, wild-type and mutant forms of pathway components, and to study their function and biochemical modulation *in vivo* (Ali et al., 2014; Hein et al., 2015; Oswald et al., 2016).

7.2. THE NOTCH PATHWAY

Most of what is known about Notch signaling can be categorized in either canonical or non-canonical pathways.

DOI: 10.1201/9781003050230-8

7.2.1. Canonical Notch Signaling

Excellent reviews describe the typical mechanism by which Notch signals, which is summarized in Figure 7.1A (Davis and Turner, 2001; Lai, 2004; Fortini, 2009; Kopan and Ilagan, 2009; Jorissen and De Strooper, 2010; Kovall and Blacklow, 2010; Tanigaki and Honjo, 2010; Groot and Vooijs, 2012; Bray, 2016). In the absence of signaling, a complex containing the DNA-binding protein RBPJ and co-repressors occupies the enhancers of Notch targets to silence them by recruiting histone deacetylases (HDACs) or other chromatin-modifying enzymes. When the mature Notch receptor is bound by a ligand, Delta (Dll), or Jagged (Jag), presented by the sending cell undergoes a conformational change that exposes a cleavage site in its extracellular domain, which is then cleaved by a membrane-tethered ADAM metalloprotease. This renders a membrane-tethered Notch intermediate (Notch extracellular truncation, NEXT), which is cleaved at the transmembrane domain by a γ-Secretase enzyme complex, whose active subunit is Presenilin (Psen). This releases the NICD, which enters the cell nucleus and forms a complex with RBPJ that recruits the co-activator Mastermind-like (MAML), displacing the RBPJ repressor complex and activating Notch-targets. Typically, Notch targets are members of the *hes/hey* gene families encoding bHLH-Orange (bHLH-O) transcriptional repressors.

Hes/Hey bHLH-O transcription factors (TFs) bind to their target DNA sequence through the basic domain, and also achieve transcriptional repression via: (1) a C-terminal tetrapeptide motif WRPW/Y that recruits transcriptional co-repressors of the TLE/Groucho family and (2) the Orange domain, located just C-terminal to the bHLH domain, that controls selection of the bHLH partner for heterodimerization. Hes proteins form homo- or heterodimers with Hey proteins and repress transcription actively or passively. Active repression involves DNA binding to the N box (CACNAG) or the class C site [CACG(C/A)G] and recruitment of TLE/Groucho co-repressors. Passive repression involves heterodimerization with other bHLH factors like E47. The WRPW motif is also necessary for polyubiquitylation, which confers short half-lives to Hes proteins by proteosome degradation, a key feature for their oscillatory expression during somitogenesis (see Section 3.6) (Davis and Turner, 2001; Bertrand et al., 2002; Huang et al., 2014; Kageyama et al., 2007; Imayoshi and Kageyama, 2014). Notably, Hey1 lacks the WRPW motif and does not bind TLE/Groucho (Pichon et al., 2004).

The traditional description of how the Notch pathway is used during development includes the following (Figure 7.1B). (1) In lateral inhibition, Notch prompts binary cell fate choices in cell populations of equal developmental potential. The ligand-sending cell signals to its neighbors, which in response repress ligand expression, "ligand-sending cell fate," and acquire an alternative fate or remain as uncommitted precursors. This usually results in salt-and-pepper patterns of cells of different fates, with roughly regular spacing between specific cell types. (2) In lateral induction, the ligand-sending cell induces ligand expression in its neighbors and instructs them to adopt the same fate. This propagates a cascade of Notch activation through a field of adjacent cells. Also, some cases of boundary formation between two cell populations involve lateral induction (Lewis, 1998; Gazave et al., 2009; Sjöqvist and Andersson, 2019). These models sometimes are insufficient to explain the complex mechanisms controlled by Notch (Favarolo and López, 2018).

7.2.2. Non-Canonical Notch Signaling

Several core components of the canonical Notch pathway also function in what are collectively known as non-canonical pathways; these are likely part of ancestral mechanisms for regulating cell differentiation in metazoans since the canonical pathway did not appear until the bilaterian lineage (Layden and Martindale, 2014). Non-canonical pathways have been described in different cell contexts and a variety of animal models, including (1) activation of Notch targets through NICD without RBPJ participation; (2) activation of Notch targets without NICD participation, with or without RBPJ mediation (Sanalkumar et al., 2010; Tanigaki and Honjo, 2010); (3) interaction with atypical ligands, atypical nuclear cofactors, and other signaling pathways (D'Souza et al., 2010; Heitzler, 2010); and (4) non-nuclear Notch activities, independent of typical ligand interaction and RBPJ-mediated transcription, involving the cytoplasmic tyrosine kinase Abl (Heitzler, 2010) or β-Catenin destabilization (Hayward et al., 2005; Hayward et al., 2008; Sanders et al., 2009; Muñoz-Descalzo et al., 2010; Acosta et al., 2011; Kwon et al., 2011). The latter was first described in *Xenopus* in the context of axis formation (see Section 3.1).

7.3. NOTCH SIGNALING DURING *XENOPUS* EMBRYOGENESIS

The *Xenopus laevis* and *tropicalis* genomes contain four Notch receptor genes (*notch1–4*), three Delta-like ligand genes (*dll1, dlc, dll4*), and two Jagged ligand genes (*jag1, jag2*) (Michiue et al., 2017; Karimi et al., 2018) (Table 7.1) (Figure 7.1). Members of two groups of *hes/hey* genes (*hey1/hey2/hey-L* and *hes1–7*) (Figure 7.2) are regulated by the canonical Notch pathway (Davis and Turner, 2001; Zhou et al., 2012) (Table 7.2). Most of them are up-regulated by Notch, but atypical responses were described for a few *hes1–7* genes (Tables 7.2, 7.3). In addition, cross-regulation between *Xenopus hes/hey* genes was described (Table 7.4).

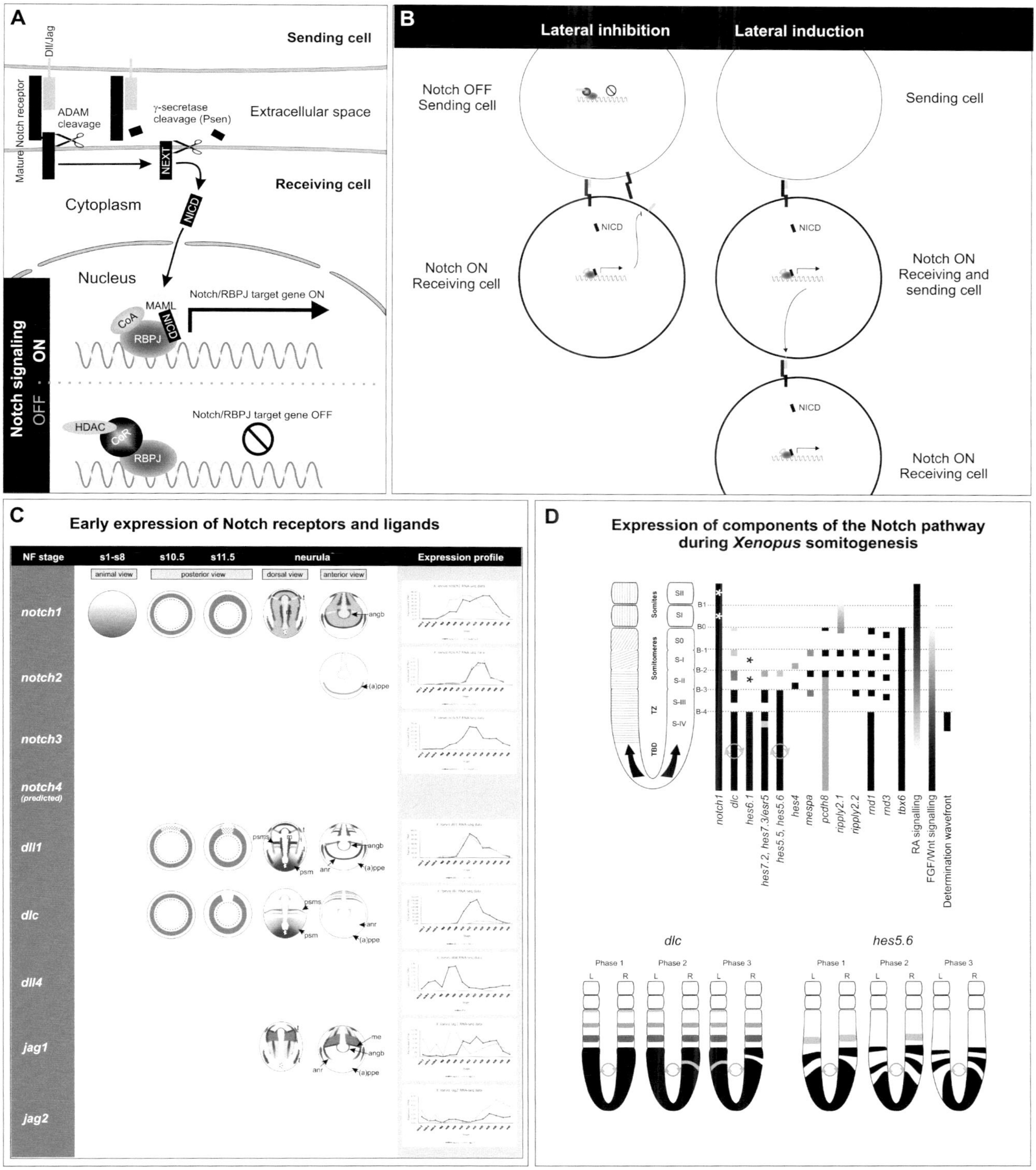

FIGURE 7.1 The canoncial Notch signaling device, expression of Notch receptors and ligands during early embryogenesis, and components of the Notch pathway during somitogenesis in *Xenopus*. (A) Simplified scheme showing the canonical Notch signaling pathway in vertebrates. CoA, co-activators; CoR, co-repressors; see additional abbreviations in the main text. (B) Modes of action of Notch signaling. (C) Early expression patterns of Notch receptors and ligands in *Xenopus*. References for building the expression domain's diagrams (left) are listed in Table 7.1. Quantitative expression profiles from *Xenopus laevis* (average TPM values from Taira and Ueno samples) were plotted with RefSeq data extracted from (Session et al., 2016) (right column). Abbreviations for Figure 7.1C and Figure 7.2: a, anterior; An, animal; anb, anterior neural border; anf, anterior neural fold; ang, anterior neural groove; angb, anterior neural groove border (prospective ventral forebrain and ventral midbrain) (Lahaye et al., 2002); anp, anterior neural plate; anr, anterior neural ridge; (d), deep layer; dbl, dorsal blastopore lip; dmz, dorsal marginal zone; ee, epidermal ectoderm; ego, early gastrula organizer; fp, floor plate; i, intermediate domain of primary neurogenesis; imz, involuting marginal zone; l, lateral domain of primary neurogenesis;

FIGURE 7.1 (Continued)

(l), lateral; lgo, late gastrula organizer; m, domain of primary neurogenesis adjacent to the midline; me, future mesencephalon; mhb, midbrain/hindbrain boundary; ncc, neural crest cells; ncb, neural crest boundaries; nb, neural border; nimz, non-involuting marginal zone; np, neural plate; npe, neural plate edge; (p), posterior; pm, prechordal mesoderm; psms, presomitic mesoderm stripe (indicative of somitogenesis); ppe, pre-placodal ectoderm; psm, pre-somitic mesoderm; prhc, prospective hypochord; prngb, prospective neural groove border; (s), superficial layer; Veg, vegetal; t, trigeminal ganglion. D. Expression of Notch pathway genes during *Xenopus* somitogenesis. Upper diagram: summary of the somitogenesis domains (left) and the expression of Notch pathway genes compared with RA and FGF/Wnt opposite gradients and other relevant segmentation genes discussed in the text (right) (adapted from Sparrow, 2008, with additional information from references listed in Table 7.12 and Kondow et al., 2007; Hitachi et al., 2008; Hitachi et al., 2009; Goda et al., 2009). Before segmentation, *Xenopus* myotomal cells form a parallel array that lies perpendicular to the embryonic long axis and undergoes a 90-degree rotation associated with the appearance of the intersegmental furrow during segmentation (left). A consistent nomenclature for somitogenesis domains and the distinct phases of cyclic gene expression in the presomitic mesoderm (PSM) was conventionally adopted for all vertebrate species (Pourquié and Tam, 2001) (right). In the already segmented paraxial mesoderm, somites are numbered with Roman numerals, with SI the most recently formed one. S0 is the most anterior presumptive somite in the PSM, which is about to be segmented, followed caudally by prospective somites sequentially numbered with negative Roman numerals. Borders between prospective somites (B) are numbered with negative Arabic numerals, with B0 the intersegmental fissure between SI and the PSM (Pourquié and Tam, 2001). The PSM is divided into three regions, according to gene expression patterns. The region encompassing S0 to S-II is known as the somitomere region; S-III and S-IV make up the transition zone (TZ); caudal to the TZ is the tailbud domain (TBD), populated by immature paraxial mesoderm cells (Moreno and Kintner, 2004; Sparrow, 2008). As the embryo elongates caudally, new paraxial cells are produced at the caudal tip of the TBD and are displaced anteriorly (arrows), gradually occupying the TZ, then forming somitomeres (S-III through S0) and finally segregating as a mature somite (S1) from the anterior end of the PSM. Gene expression domains are shown with black/gray bars. Known oscillating behavior is indicated by a circle with double arrows. When the expression of a gene was not studied with enough detail to assign a precise location, asterisks indicate their approximate expression. See Table 7.12 for more details. Lower diagrams: typical expression phases of the oscillatory genes *dlc* and *hes5.6* in the *Xenopus* PSM.

Source: Adapted from Durston et al. (2018), Kirby et al. (2003).

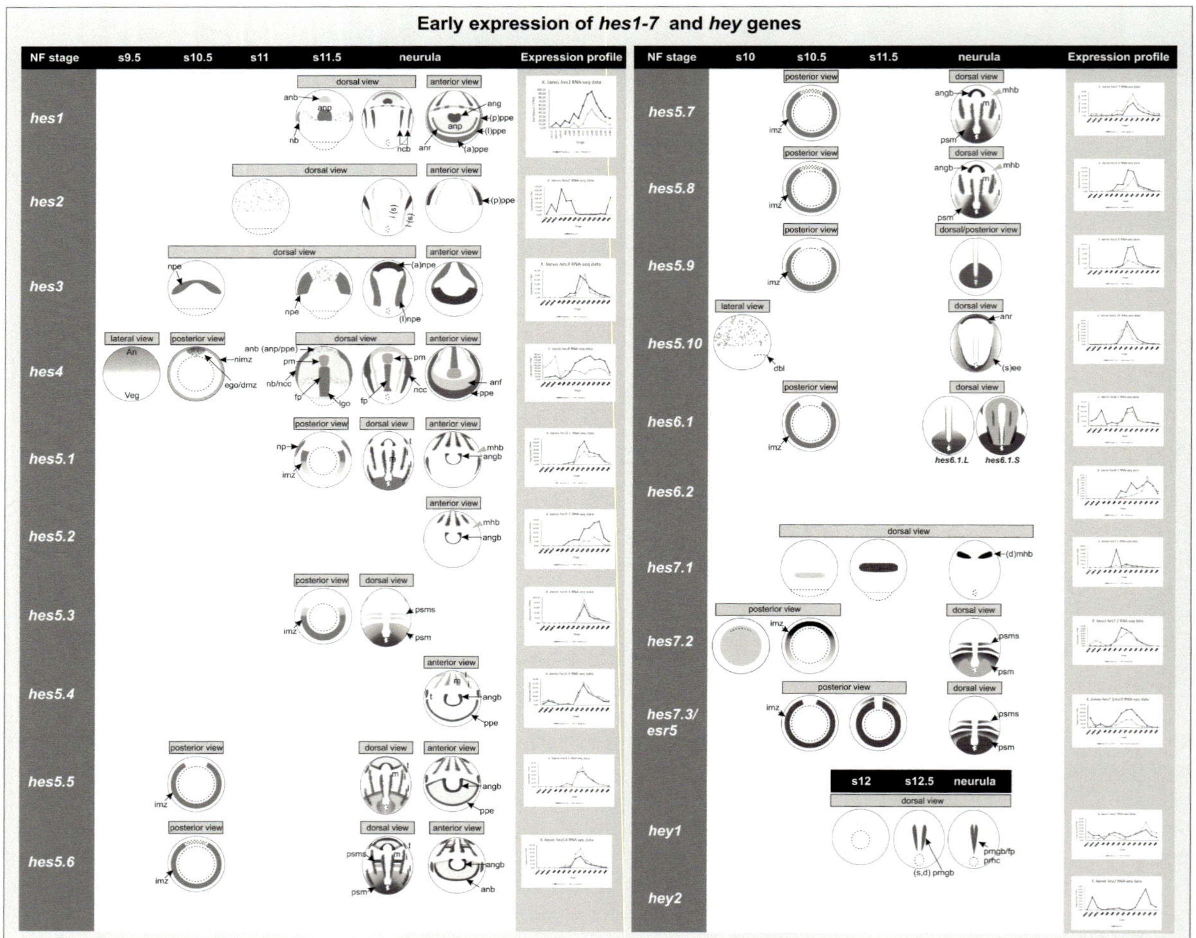

FIGURE 7.2 Early expression patterns of *hes4–7* and *hey* genes in *Xenopus*. References for building the expression domain diagrams (left) are listed in Table 7.2. Quantitative expression profiles from *Xenopus laevis* (average TPM values from Taira and Ueno samples) were plotted with RefSeq data extracted from (Session et al., 2016) (right column). See abbreviations in Figure 7.1 legend.

TABLE 7.1
Main Components of the Notch Pathway in *Xenopus*

Components of the Notch pathway present in *Xenopus tropicalis* and *Xenopus laevis* with current nomenclature were obtained from Michiue et al. (2017) and Xenbase (www.xenbase.org/, RRID:SCR_003280; Karimi et al., 2018). Until very recently, it was believed that only three of the four mammalian orthologues encoding Notch receptors were present in the *Xenopus* genome (Michiue et al., 2017). However, a gene model for *notch4* recently appeared in Xenbase. Besides, the *X. laevis dll4* gene was wrongly identified as a singleton in the transcriptomic analysis of Session et al. (2016; Michiue et al., 2017). Two *rbpj* isoforms were identified in *X. laevis*. Originally, they were termed *XSu(H)1* and *XSu(H)2* (Wettstein et al., 1997; Ito et al., 2007a; Ito et al., 2007b). Transcripts differ in their 5'UTRs, but their predicted protein sequences are almost identical, except for a 20 residue length difference at their N-termini (Wettstein et al., 1997; Ito et al., 2007a). With the availability of the *X. laevis* genome, now it is possible to predict that both are variants from the *RBPJ.S* homeolog. *XSu(H)2* is referred to as *rbpj.S-v2* to distinguish it from the *XSu(H)1* variant (referred to as *rbpj.S-v1*) that was studied elsewhere since Wettstein et al. (1997); it is now clear that *X. laevis* also has an *rbpj.L* homeolog (Michiue et al., 2017). The first (S1 cleavage) of Notch occurs in the secretory pathway, where a furin-like convertase processes the Notch full-length polypeptide. This generates the mature receptor, consisting of a NECD-NTMIC heterodimer (Notch extracellular domain-Notch transmembrane and intracellular domain), with both polypeptides bound by non-covalent interactions. ADAM10 is the best confirmed candidate in cleaving the mature receptor at the S2 site in the extracellular domain, as a consequence of ligand binding (Groot and Vooijs, 2012). Other core components of the pathway are discussed in more detail in the text.

Main Components of the Notch Pathway		*X. tropicalis*	*X. laevis* (homeologs)	Synonyms	Expression (References for Figure 7.1)
Receptors	Notch	*notch1*	*notch1.L* *notch1.S*	*L: LOC108698191* *S: notch, xotch, xnotch, notch-1, xnotch1, x-notch-1*	(Chitnis et al., 1995; Andreazzoli et al., 2003; López et al., 2003; Yan and Moody, 2007; Miazga and McLaughlin, 2009; Castro Colabianchi et al., 2018)
		notch2	*notch2.L* *notch2.S*	*ags2*	(Ogino et al., 2008)
		notch3	*notch3.L* *notch3.S*	*casil, cadasil*	
		notch4	*notch4.L* *notch4.S*	*L: loc108700568* *S: loc100488695*	
Ligands	Delta-like	*dll1*	*dll1.L* *dll1.S*	*X-delta-1, delta1, delta-1, Xdelta-1, XDelta1, x-delta, Delta-1, Xdelta1*	(Chitnis et al., 1995; Beck and Slack, 1998; Howell et al., 2002; López et al., 2005; Nichane et al., 2008b; Ogino et al., 2008)
		dlc	*dlc.L* *dlc.S*	*x-delta-2, delta-2, delta2, X-Delta, dll3*	(Jen et al., 1997; Peres and Durston, 2006; Peres et al., 2006; Ogino et al., 2008; Onai et al., 2015)
		dll4	*dll4.L* *dll4.S*	*delta 2*	
	Jagged	*jag1*	*jag1.L* *jag1.S*	*X-Serrate-1, serrate-1, serrate, jagged1*	(Kiyota et al., 2001)
		jag2	*jag2.L* *jag2.S*		
Processing	Furin (S1 cleavage, secretory pathway)	*furin*	*furin.L* *furin.S*	*PACE, spc1, xfurin*	
	ADAM-secretase (S2 cleavage)	*adam10*	*adam10.L* *adam10.S*	*xadam10, kuz, ad10, madm, cd156c, kuzbanian*	
	Presenilin (the active subunit of the γ-secretase complex, S3 cleavage)	*psen1*	*psen1.L* *psen1.S*	*L: presenilin-alfa, X-PS-alpha*	
		psen2	*psen2.L* *psen2.S*	*S: presenilin-beta, X-PS-beta*	
Transcription factor	RBPJ	*rbpj*	*rbpj.L* *rbpj.S*	*X-Su(H), XSu(H), Su(H), suppressor of hairless, rbpsuh, CBF1, csl, lag-1, CBF-1* *L: LOC108698058; S: X-Su(H)1 and 2*	
Transcriptional co-activator	Mastermind-like (MAML)	*maml1*	*maml1.L* *maml1.S*	*XMam1, Mastermind1, mam1, mam-1, Mastermind*	
		maml2	*maml2.L* *maml2.S*		
		maml3	*maml3.L* *maml3.S*		

TABLE 7.2
Genes of the *hes1–7* and *hey* Groups in the *Xenopus tropicalis* and *Xenopus laevis* Genomes and Their Responsiveness to Notch Signaling

Data for building the list of genes were obtained from Watanabe et al. (2017) and Xenbase (xenbase.org, RRID:SCR_003280; Karimi et al., 2018). The nomenclature proposed by Watanabe et al. (2017) was based on phylogenetic and syntenic analyses that revealed that the old names were misleading. The current nomenclature in Xenbase coincides with that proposed by Watanabe et al. (2017), except for *hes7.3*, which is still named *esr5* in Xenbase. Therefore, this gene is referred to as *hes7.3/esr5* throughout the chapter. Whenever possible, the correspondences of L and S homeologs with old synonyms were checked according to RefSeqs and are indicated. See Table 7.3 for details of experimental evidence of Notch responsiveness. *esr9b* (accession no. AB211547) was considered in Takada et al. (2005) as a possible *esr9* pseudoallele and was called thereafter as *esr9* for simplicity in that publication, leading to confusion with a different gene, *hes5.6.L*, which was also previously called *esr9*. The sequence AB211547 corresponds to Xenbase:XB-GENE-6253435 or *hes5.7.L*. The *X. laevis* gene formerly known as *hes9.1.S* is indeed on the L chromosome and is currently named *hes5.8.L* (Watanabe et al., 2017).

X. tropicalis	X. laevis	Synonyms	Notch Responsiveness	Expression (References for Figure 7.2)
hes1	*hes1.L* *hes1.S*	*hairy1, Xhairy1, hes-1* *L: hes1-a; S:hes1-b*	Positive	(Andreazzoli et al., 2003; Vega-López et al., 2015; Hardwick and Philpott, 2019)
hes2	*hes2.L*	*Xhes2*	Positive	(Sölter et al., 2006; Riddiford and Schlosser, 2016)
hes3	*hes3.L* *hes3.S*		Positive	(Hong and Saint-Jeannet, 2018)
hes4	*hes4.L* *hes4.S*	*Xhairy2, hairy2, H2* *L: hairy2b, Xhairy2b, hes4b, hes4-b; S: hairy2a, XHairy2a, hes4a, hes4-a*	Positive	(Turner and Weintraub, 1994; Tsuji et al., 2003; López et al., 2005; Murato et al., 2007; Nichane et al., 2008b; Nichane et al., 2008a; Murato and Hashimoto, 2009; Aguirre et al., 2013; Vega-López et al., 2015)
hes5.1	*hes5.1.L* *hes5.1.S*	*esr1, esr-1, XESR-1* *L: hes5-like; S: ESR1b*	Positive	(Lamar et al., 2001; Takada et al., 2005; Kuriyama et al., 2006; Blewitt, 2009; Maguire et al., 2012)
hes5.2	*hes5.2.L* *hes5.2.S*	*bHLHb38, esr3. L: esr7, esr-7, hes5.2-a; S: ESR3/7b, hes5.2-b*	Positive	(Nieber et al., 2013; Heeg-Truesdell and LaBonne, 2006)
hes5.3	*hes5.3.L* *hes5.3.S*	*L: esr2, hes3.3*	Positive	(Hayata et al., 2009; Blewitt, 2009; Maguire et al., 2012)
hes5.4	*hes5.4.L* *hes5.4.S*	*hes8*	Positive	(Riddiford and Schlosser, 2016)
hes5.5	*hes5.5.L* *hes5.5.S*	*L: HES-5-like, hes10.L; S: esr10.S, esr10xb, 11A10*	Positive	(Gawantka et al., 1998; Miazga and McLaughlin, 2009; Nieber et al., 2013)
hes5.6	*hes5.6.L* *hes5.6.S*	*L: hes9.1.L, esr9, 8C9; S: hes-5 like*	Positive	(Gawantka et al., 1998; Li et al., 2003; Miazga and McLaughlin, 2009; Riddiford and Schlosser, 2016)
hes5.7	*hes5.7.L* *hes5.7.S*	*esr9, hes9.1-b. L: ESR9b, hes9.1.S; S: HES-5-like vX1*	Positive	(Takada et al., 2005; Taverner et al., 2005; Xenbase community submitted by Nicolas Pollet; Karimi et al., 2018; www.xenbase.org/, RRID:SCR_003280)
hes5.8	*hes5.8.L* *hes5.8.S*	*Xtr: loc100495414* *Xla: L: loc108696616; S: loc108697696, hes5_x2*	Unknown	(Pollet et al., 2005; Xenbase community submitted by Nicolas Pollet; Karimi et al., 2018; www.xenbase.org/, RRID:SCR_003280)
hes5.9	*hes5.9.L* *hes5.9.S*	*Xtr: loc733709* *Xla: L: loc108696614; S: loc108697697, hes5_x1*	Unknown	(Kjolby and Harland, 2017)
hes5.10	*hes5.10.L* *hes5.10.S*	*esr6e, esr-6e. L: hes3.1.L; S: hes3.1.S*	Positive Atypical response to RBPJ	(Chalmers et al., 2002; Chen et al., 2005; Xenbase community submitted by Naoto Ueno; Karimi et al., 2018; www.xenbase.org/, RRID:SCR_003280)
hes6.1	*hes6.1.L* *hes6.1.S*	*XHes6, Xhes-6. L: clone 29B3–2; S: clone 10C6*	Negative	(Koyano-Nakagawa et al., 2000; Cossins et al., 2002; Hufton et al., 2006; Murai et al., 2007; Murai et al., 2011; Kjolby and Harland, 2017)
hes6.2	*hes6.2.L* *hes6.2.S*		Negative, Positive	
hes7.1	*hes7.1.L* *hes7.1.S*	*HES-related 1, XHR1*	Positive	(Shinga et al., 2001; Takada et al., 2005)
hes7.2	*hes7.2.L* *hes7.2.S*	*esr4, ESR-4, enhancer-of-split-related 4, ESR 4*	Positive	(Gawantka et al., 1998; Taverner et al., 2005; Peres et al., 2006; Xenbase community submitted by Naoto Ueno; Karimi et al., 2018; www.xenbase.org/, RRID:SCR_003280)
hes7.3/esr5	*hes7.3.L/esr5.L* *hes7.3.S/esr5.S*	*L: esr5, x-esr5, Xesr5, ESR 5*	Negative, Positive	(Taverner et al., 2005; Blewitt, 2009; Kinoshita et al., 2011; Kjolby and Harland, 2017; Janesick et al., 2017)
hey1	*hey1.L* *hey1.S*	*hrt1, XHRT1, chf2, hrt-1, hesr1, herp2, oaf1, bc8*	Positive	(Pichon et al., 2002)
hey2	*hey2.L*	*hesr2, gridlock*	Unknown	

TABLE 7.3
Experimental Evidence for the Responsiveness of *hes1–7* and *hey* Genes to the Notch Pathway in *Xenopus*

No references were found for *hes5.8, hes5.9*, and *hey2* regulation by the Notch pathway in *Xenopus*, and therefore they are not listed in this table. Abbreviations for Tables 7.3 to 7.12: AP, anterior-posterior; aPM, anterior prechordal mesoderm; ANB, anterior neural border; bHLH, basic helix-loop-helix; ChIP, chromatin immunoprecipitation; CHX, cycloheximide (an inhibitor of protein synthesis); Dex, dexamethasone; DIMZ, dorsal involuting marginal zone; *dll1STU*, antimorph of *dll1*, lacking the intracellular domain (Chitnis et al., 1995); DML, dorsal midline; DMZ, dorsal marginal zone; DN, dominant-negative; DBM, DNA binding mutant; DNIMZ, dorsal non-involuting marginal zone; E1A constructs, protein of interest fused to the activation domain of the human type 5 adenovirus E1a protein; EnR constructs, protein of interest fused to the *Drosophila* Engrailed repressor domain; ER-constructs, hormone-inducible forms of proteins with nuclear functions under the control of the ligand-binding domain of the human estrogen receptor. These recombinant fusion proteins translocate to the cell nucleus when estradiol (E2) is added to the culture medium at the desired NF stage; FP, floor plate (ventral midline of the neural tube); GO, gastrula organizer; GR-constructs, hormone-inducible forms of proteins with nuclear functions under the control of the ligand-binding domain of the human glucocorticoid receptor. These recombinant fusion proteins translocate to the cell nucleus when Dex is added to the culture medium at the desired NF stage; GR-*NICD1*–22: hormone-inducible form of the *X. laevis* Notch1 intracellular domain (Wettstein et al., 1997); HDAC, histone deacetylase; HUA, hydroxyurea and aphidicolin (strong inhibitors of cell proliferation); ICD, intracellular domain; IMZ, involuting marginal zone; ISH, *in situ* hybridization; MHB, midbrain/hindbrain boundary; MO, antisense morpholino oligonucleotide; NB, neural border; NCCs, neural crest cells; NF, Nieuwkoop and Faber stage of development in *Xenopus*; NICD, Notch intracellular domain (NICD constructs are constitutively active); NIMZ, non-involuting marginal zone; *notch1-ΔE, notch1* construct without most of the extracellular domain, constitutively active; pH3, phosphorylated form of histone H3 (marker of cell proliferation); PN, primary neurogenesis; PPE, pre-placodal ectoderm; pPM, posterior prechordal mesoderm; PSM, presomitic mesoderm; RA, retinoic acid; RAR, retinoic acid receptor; *RBPJ*, recombination signal binding protein for immunoglobulin kappa J region, transcription factor that mediates canonical Notch signaling; *RBPJ-Ank*: constitutively active form of the transcription factor RBPJ, consisting of the *X. laevis* Notch1 ankyrin repeats fused to the C-terminus of *X. laevis* RBPJ (Wettstein et al., 1997); *RBPJDBM*, DNA binding mutant form of *X. laevis* RBPJ that binds to NICD but lacks the ability to bind target sites in the DNA and acts as a dominant-negative protein by forming non-functional complexes (Wettstein et al., 1997); sq, semiquantitative; TBD, tailbud region of the presomitic mesoderm; TF, transcription factor; TZ, transition zone of the presomitic mesoderm; VMZ, ventral marginal zone; Δ, deletion; ↑up-regulation, expansion, or increase; ↓down-regulation, reduction, or decrease.

Gene	Regulated by	Regulation/Details	Gain-of-Function	Loss-of-Function (Dominant-Negative/Antagonist/Morpholino)
hes1	*dll1/ notch/ RBPJ*	Positive in pronephros and neural plate.	GR-*RBPJ-Ank*, Dex NF18: ↑*hes1* in pronephros (Taelman et al., 2006). GR-*RBPJ-Ank*, Dex unstated stage: ↑*hes1* neural plate domains (Vega-López et al., 2015).	*RBPJDBM*:↓ *hes1* in pronephros (Taelman et al., 2006). *dll1STU*:↓*hes1* in neural plate domains (Yan et al., 2009; Vega-López et al., 2015).
hes2	*notch1/ RBPJ*	Positive in NB.	*NICD1* or *RBPJ-Ank*: ectopic *hes2* restricted to the lateral border of the neural plate (Sölter et al., 2006).	
hes3	*notch1*	Positive.	*NICD1*: ↑*hes3* (NF18, microarray) (Vasiliu et al., 2015).	
hes4	*dll1/ notch1/ RBPJ*	Positive in NB/NCC.	*NICD1*: ↑NB *hes4* domain (gastrulation, ISH) (López et al., 2005). *GR-RBPJ-Ank* or *GR-NICD1–22:* Dex, unstated stage: ↑ NCC *hes4* domain (neural plate stage) (Vega-López et al., 2015). Dex NF12: ↑ NCC *hes4* domain (neural fold stage) (Glavic et al., 2004).	*GR-RBPJDBM* (Dex NF12): ↓*hes4* NCC domain at neural fold stage (Glavic et al., 2004). *dll1STU*: ↓*hes4* in NCC (neural plate stage) (Vega-López et al., 2015).
	notch/ RBPJ	No regulation in NCC.	*GR-RBPJ-Ank* (Dex NF11): did not affect the NCC *hes4* domain at neural plate stage (Nichane et al., 2008a).	*GR-RBPJDBM* (Dex NF11): did not affect the NCC *hes4* domain at neural plate stage (Nichane et al., 2008a).

(Continued)

TABLE 7.3 (Continued)

Experimental Evidence for the Responsiveness of *hes1–7* and *hey* Genes to the Notch Pathway in *Xenopus*

Gene	Regulated by	Regulation/Details	Gain-of-Function	Loss-of-Function (Dominant-Negative/Antagonist/Morpholino)
	dlc/notch/ RBPJ	Positive in somitogenesis. *dlc* and *hes4:* complementary expression patterns in somitomeres, consistent with an inductive role of *dlc* on *hes4* through Notch activation in neighboring cells (Jen et al., 1997).		*dlc-tr* or *RBPJDBM:* suppressed the segmental *hes4* prepattern in somitomeres (Jen et al., 1997). *dlc* MO: ↓*hes4* (ISH, tailbud stage) (Peres et al., 2006).
	notch/ RBPJ *hes4* 3'UTR	Positive. A paired RBPJ motif in the proximal promoter and a minimal 25 bp sequence in the 3'UTR, are necessary for *hes4* expression in neural tissue, pronephros, and anterior PSM, but are insufficient for *hes4* expression in the FP. 3′ UTR confers global instability to *hes4* mRNA, except in the anterior PSM. *In vivo* gene reporter, tailbud stage (Davis et al., 2001). Notch1 is associated with the RBPJ site of the *hes4* genomic loci (ChIP assay, NF25) (Sakano et al., 2010).		
	dll1/ notch1/ RBPJ	Positive in DML precursors during gastrulation.	*NICD1:* ↑*hes4*+ population of FP precursors in the GO and *hes4* FP domain in neurulae (López et al., 2005). *dll1* overexpression:↑*hes4*+ population of FP precursors in the GO (López et al., 2005).	*RBPJ*DBM: ↓*hes4*+ population of FP precursors in the GO and *hes4* FP in neurulae (López et al., 2005).
	dll1/ notch1	Positive in ectoderm.	*NICD1:* ↑*hes4* in animal caps (RT-PCR, NF20) and cell-autonomously in ectoderm (ISH, NF20) (Cui, 2005).	*dll1STU:* prevented the induction of *hes4* by *neurog2* in animal caps (RT-PCR, NF20) (Cui, 2005).

hes5.1	*dll1/ notch1/ RBPJ/ maml1*	Positive. An HDAC inhibitor enhanced *hes5.1* response to induction by *dll1* in neuralized animal caps, supporting the hypothesis that activation by Dll1/Notch disrupts the formation of the repressor complex containing RBPJ and HDAC-1 that maintains Notch-target genes repressed in the absence of Notch signaling (Kao et al., 1998). Paired RBPJ binding site proximal to the TATA box, necessary but not sufficient for neural expression *in vivo* (Lamar and Kintner, 2005).	*NICD1*: ↑*hes5.1* in animal caps (NF10.5, sqRT-PCR) (Kinoshita et al., 2011) and whole embryos (NF18, 28, microarray) (Vasiliu et al., 2015). (RT-qPCR, sometime between NF9.5–10) (Mir et al., 2008). Induced ectopic *hes5.1* in neural and non-neural ectoderm (neural plate stage) (Deblandre et al., 1999). *dll1* overexpression, *RBPJ-Ank* or *NICD1* but not *RBPJ*: ↑*hes5.1* in neuralized or naive animal caps (neurula stage) (Wettstein et al., 1997; Lamar et al., 2001; Lahaye et al., 2002; Pichon et al., 2002). The RAM23 domain and ankyrin repeats (Ank) of NICD1 were essential for this induction, with some contribution of sequences downstream of the Ank repeats (Wettstein et al., 1997). *hGR-NICD1(22)* or GR-*RBPJ*-Ank (Dex 1 hr.) +/− CHX: ↑*hes5.1* in neuralized animal caps (Wettstein et al., 1997). *hGR-NICD1(22)* or *hGR-RBPJ-Ank*, Dex 30 min: ↑*hes5.1* in neuralized animal caps (Lahaye et al., 2002). *hGR-RBPJ-Ank*, Dex 1 hr +/− CHX: ↑*hes5.1* in naive animal caps (Lahaye et al., 2002). *RBPJ.S v2-Ank* ↑ *hes5.1* in naive animal caps (RT-sqPCR) (Ito et al., 2007b). *RBPJ* overexpression: inhibited *hes5.1* induction by *dll1* in neuralized animal caps (similar to *RBPJDBM*) (Wettstein et al., 1997).	*RBPJDBM:* suppressed *NICD1*'s ability to induce *hes5.1* in naive animal caps (neurula stage) (Lahaye et al., 2002; Pichon et al., 2002) and *dll1*'s or *neurog2*'s ability to induce *hes5.1* in neuralized animal caps (Wettstein et al., 1997; Lamar et al., 2001). *hes5.1* was downregulated in embryos derived from *RBPJDBM*-injected oocytes (RT-qPCR, sometime between NF9.5–10) (Mir et al., 2008). *RBPJ.S v2DBM:* ↓*hes5.1* in naive animal caps (RT-sqPCR) (Ito et al., 2007b). *DN-maml1:* ↓*hes5.1* in domains of PN (Katada and Kinoshita, 2003).
	jag1	Positive during PN.	*jag1* overexpression but not *jag1*ICD: ↑ *hes5.1* in neuralized animal caps (Kiyota and Kinoshita, 2004).	
hes5.2	*notch1/ RBPJ*	Positive in the ectoderm/neuroectoderm. Paired RBPJ binding site proximal to the TATA box (Lamar and Kintner, 2005).	*NICD1:* induced ectopic *hes5.2* in neural and non-neural ectoderm (Deblandre et al., 1999): ↑*hes5.2* in neuralized animal caps (NF14–15) (Sölter et al., 2006) and whole embryos (NF18, NF28, microarray) (Vasiliu et al., 2015).	
hes5.3	*notch1*	Positive.	*NICD1:* ↑*hes5.3* (NF18, NF28, microarray) (Vasiliu et al., 2015).	
hes5.4	*notch1/ RBPJ*	Positive in placodes.	*NICD1:* ↑*hes5.4* in placodes and adjacent non-neural ectoderm, even in the absence of function of the PPE genes *six1/eya1* (ISH, neural plate stage) (Riddiford and Schlosser, 2017).	*RBPJDBM:* ↓*hes5.4* in neural plate and placodes (ISH, neural plate stage) (Riddiford and Schlosser, 2017).

(*Continued*)

TABLE 7.3 (Continued)
Experimental Evidence for the Responsiveness of *hes1–7* and *hey* Genes to the Notch Pathway in *Xenopus*

Gene	Regulated by	Regulation/Details	Gain-of-Function	Loss-of-Function (Dominant-Negative/Antagonist/Morpholino)
hes5.5	*notch1*	Positive.	*NICD1:* ↑*hes5.5* (NF18, 28; microarray) (Vasiliu et al., 2015).	
	RBPJ	Positive in the nervous system. A paired RBPJ binding site proximal to the TATA box is necessary but not sufficient for neural expression *in vivo* (Lamar and Kintner, 2005).		
	notch1/ RBPJ	Positive in the IMZ.	*GR-RBPJ-VP16* or *GR-NICD1* (Dex NF10): ↑IMZ *hes5.5* domain (gastrula, ISH) (Miazga and McLaughlin, 2009).	*GR-RBPJ-EnR* (Dex NF10): ↓IMZ *hes5.5* domain (gastrula, ISH) (Miazga and McLaughlin, 2009).
hes5.6	*notch1*	Positive.	*NICD1:* ↑*hes5.6* (NF18, 28; microarray) (Vasiliu et al., 2015).	
	notch1/ RBPJ	Positive during gastrulation/IMZ.	*GR-RBPJ-VP16* or *GR-NICD1* (Dex NF10): ↑IMZ *hes5.6* domain (gastrula, ISH) (Miazga and McLaughlin, 2009). *NICD1:* ↑*hes5.6* (mid-gastrula, RT-qPCR, NF11) (Castro Colabianchi et al., 2018).	*GR-RBPJ-EnR* (Dex NF10): ↓IMZ *hes5.6* domain (gastrula, ISH) (Miazga and McLaughlin, 2009). *notch1* MO: ↓*hes5.6*; rescued by *NICD1* (RT-qPCR, NF11) (Castro Colabianchi et al., 2018)
hes5.7	*notch1*	Positive.	*NICD1:* ↑*hes5.7* (NF28, microarray) (Vasiliu et al., 2015).	
hes5.10	*notch1/* atypical response to *RBPJ*.	Positive in the non-neural ectoderm.	*NICD1:* ↑*hes5.10* throughout the non-neural ectoderm (NF14) (Deblandre et al., 1999); ↑*hes5.10* (NF18, NF28, microarray) (Vasiliu et al., 2015).	*RBPJDBM:* ↑*hes5.10* in the inner cells of isolated ectoderm (NF11, RNAse protection) (Deblandre et al., 1999).
hes6.1	*notch1/ RBPJ*	Negative in the neural plate.	*NICD1:* ↓*hes6.1* (a positive regulator of neurogenesis) in the neural plate and did not appear to affect expression in PSM at neurula stage (Koyano-Nakagawa et al., 2000), although some expansion in the PSM staining in Figure 3A of this work is noticed.	*RBPJDBM:* ↑*hes6.1*+ cell population in the neural plate [data not shown in (Koyano-Nakagawa et al., 2000)].
hes7.1	*notch1*	Negative regulation in MHB establishment (non-canonical pathway?).	*NICD1:* ↓*hes7.1* at the MHB (NF13) (Takada et al., 2005).	*RBPJDBM:* did not affect *hes7.1* expression at the MHB (Takada et al., 2005).
	notch1	Positive.	*NICD1:* ↑*hes7.1* (NF28, microarray) (Vasiliu et al., 2015).	
hes7.2	*dlc/ notch/ RBPJ*	Positive in somitogenesis.	*dlc* overexpression or *RBPJ-Ank:* expanded *hes7.2* into the gaps of the somitomeric region (ISH late neurula/early tailbud) (Jen et al., 1999). *dlc* overexpression or *NICD1*: loss of the stripped *hes7.2* pattern in the somitomeric region (ISH, neurula) (Peres et al., 2006).	*RBPJDBM:* ↓*hes7.2* in the somitomeric, TZ, and TBD regions (ISH late neurula/early tailbud) (Jen et al., 1999; Peres et al., 2006). *dlc* MO: ↓ *hes7.2* (ISH, neurula) (Peres et al., 2006).
	notch1	Positive.	*NICD1:* ↑*hes7.2* (NF28, microarray) (Vasiliu et al., 2015).	

hes7.3/esr5	*dlc/notch/RBPJ*	Positive in somitogenesis.	*dlc* overexpression or *RBPJ-Ank:* expanded *hes7.3/esr5* expression into the gaps of the somitomeric region (ISH late neurula/early tailbud) (Jen et al., 1999).	*RBPJDBM:* ↓*hes7.3/esr5* in the somitomeric region and the TZ but not in the TBD (ISH late neurula/early tailbud) (Jen et al., 1999).
	notch1	Positive.	*NICD1:* ↑*hes7.3/esr5* (NF28, microarray analysis) (Vasiliu et al., 2015).	
	notch1/hes5.1	Negative during mesoderm induction.	*NICD1:* unable to induce *hes7.3/esr5* in naive animal caps. *NICD1* or *hes5.1:* inhibited the induction of *hes7.3/esr5* by the mesodermal inducer *nodal2* in animal caps (NF10.5, sqRT-PCR). *hes5.1* overexpression*:* ↓*hes7.3/esr5* in the IMZ (NF10.5, ISH) (Kinoshita et al., 2011).	*hes5.1 ΔWRPW:* ↑*hes7.3/esr5* in animal caps mesodermalized by *nodal2* (NF10.5, sqRT-PCR) (Kinoshita et al., 2011).
hey1	*notch1/RBPJ*	Positive.	*RBPJ-Ank* or *NICD1*: ectopic *hey1* in whole embryos and ectodermal explants (neurula) (Pichon et al., 2002).	*RBPJDBM:* ↓*NICD1*'s ability to induce ectopic *hey1* in whole embryos and ectodermal explants (Pichon et al., 2002).
	notch1/RBPJ	Positive (head, somites, pronephros).	GR-*NICD1* or GR-*RBPJ-VP16*, Dex NF11–13 or NF15–19, ISH NF24–30: ↑*hey1* in all domains, except in the pronephros with the early Dex treatment (↓*hey1*) (Rones et al., 2002).	GR-*RBPJDBM*, Dex NF11–13 or NF15–19 ISH NF24–30: ↓*hey1* (Rones et al., 2002).
	notch1	Positive.	*NICD1:* ↑*hey1* (NF18, NF28, microarray analysis) (Vasiliu et al., 2015).	

TABLE 7.4
Cross-Regulation between *hes* and *hey* Genes in *Xenopus*
See abbreviations in Table 7.3 legend.

Gene	Regulated by	Regulation/Details	Gain-of-function	Loss-of-function (Dominant-negative/Antagonist/Morpholino)
hes1	*hes4*	Negative in ectoderm (Koyano-Nakagawa et al., 2000).	*hes4* overexpression: ↓*hes1* in whole embryos and animal caps (neural plate stage). Effect in animal caps reversed by co-injection of *hes6.1* mRNA (Koyano-Nakagawa et al., 2000).	*hes4-ΔWRPW-Gal4:* ↑*hes1* in animal caps (Koyano-Nakagawa et al., 2000)
	hes6.1	Inhibits *hes1* post-transcriptionally and induces it indirectly (Koyano-Nakagawa et al., 2000). Hes6.1 protein binds Hes1 *in vitro* and in embryos, antagonizing Hes1 ability to suppress PN in a TLE/Groucho-independent way (Koyano-Nakagawa et al., 2000; Murai et al., 2011).	*hes6.1* overexpression: ↑*hes1* in animal caps and whole embryos (neural plate stage). Inhibited *hes4's* ability to repress *hes1* in animal caps. The induction of *hes1* is explained by the blockade of the *hes1/hes4* auto-regulatory negative feedback loop (Koyano-Nakagawa et al., 2000).	*hes6.1*DBM: ↑*hes1* in whole embryos (neural plate stage) like *hes6.1* overexpression. Therefore, *hes6.1* does not need to bind DNA to induce *hes1*. (Koyano-Nakagawa et al., 2000).
	hey1	Hey1 and Hes1 heterodimerize in embryos, enhancing the binding of Hey1 to a class B E-box oligonucleotide (Taelman et al., 2004).		
hes4	*hes5.1*	Positive in ectoderm	*hes5.1* overexpression: ↑*hes4* in animal caps (unknown stage, RT-PCR) (Cui, 2005)	
	hes6.1	Inhibits *hes4* post-transcriptionally and induces it indirectly (Koyano-Nakagawa et al., 2000). Hes6.1 binds Hes4 *in vitro* and in embryos, antagonizing its ability to repress *hes1* in a TLE/Groucho-independent way (Koyano-Nakagawa et al., 2000).	*hes6.1* overexpression: ↑*hes4* in whole embryos (neural plate stage). The induction of *hes4* is explained by the blockade of the *hes1/hes4* auto-regulatory negative feedback loop (Koyano-Nakagawa et al., 2000).	*hes6.1*DBM: ↑*hes4* in whole embryos (neural plate stage) like *hes6.1* overexpression. Therefore, *hes6.1* does not need to bind DNA to induce *hes4*. (Koyano-Nakagawa et al., 2000).
hes5.1	*hes4*	Negative in the ectoderm, ventral mesoderm, and anterior neural plate.	*hes4* overexpression: blocked *hes5.1* induction by *neurog2* in animal caps and by *NICD1* in VMZ explants (NF unknown, RT-PCR); ↓anterior neural *hes5.1* expression (ISH, neurula) (Cui, 2005).	*hes4-ΔWRPW:* ↑*hes5.1* in animal caps and VMZ explants; rescued the repression of *hes5.1* by *hes4* in VMZ explants (NF unknown, RT-PCR); weakly ↑*hes5.1* anterior neural domain (ISH, neurula) (Cui, 2005).
	hes6.1	Positive during PN.	*hes6.1* overexpression: ↑*hes5.1* PN domains [data not shown in (Koyano-Nakagawa et al., 2000)].	
	hes7.1	Represses *hes5.1* (probably directly) during the establishment of the presumptive MHB.	*hes7.1* overexpression: ↓*hes5.1* (neural plate stage) (Takada et al., 2005).	*GR-hes7.1-VP16* (Dex NF10.5–11) +/− CHX: ↑*hes5.1* domains at neural plate stage (Takada et al., 2005).

	hes7.3/esr5	Negative during mesoderm induction.		*hes7.1* MO: filled the MHB gap with *hes5.1* expression (PN medial stripes) and anteriorly expanded the *hes5.1* PN intermediate stripes (Takada et al., 2005). *hes7.3/esr5-ΔWRPW:* ↑*hes5.1* in animal caps mesodermalized by *nodal2* (NF10.5, sqRT-PCR) and in the MZ (NF10.5, ISH) (Kinoshita et al., 2011).
hes5.2	*hes6.1*	Positive during PN.	*hes6.1* overexpression: ↑*hes5.2* PN domains [data not shown in (Koyano-Nakagawa et al., 2000)].	
	hes7.1	Represses *hes5.2* (probably directly) during the establishment of the presumptive MHB.	*hes7.1* overexpression: ↓*hes5.2* (neural plate stage) (Takada et al., 2005).	*GR-hes7.1-VP16* (Dex NF10.5–11) +/– CHX: ↑*hes5.2* domains at neural plate stage (Tak[a]da et al., 2005).
hes5.4	*hes5.4*	Auto-repression		*hes5.4* MO: ectopic *hes5.4* in the neural plate (Riddiford and Schlosser, 2017).
hes5.7	*hes7.1*	Represses *hes5.7* (probably directly) during the establishment of the presumptive MHB.	*hes7.1* overexpression: ↓*hes5.7* (neural plate stage) (Takada et al., 2005).	*GR-hes7.1-VP16* (Dex NF10.5–11) +/– CHX: ↑*hes5.7* domains at neural plate stage (Takada et al., 2005). *hes7.1* MO: filled the MHB gap with *hes5.7* expression (medial stripes of PN) and anteriorly expanded the *hes5.7* PN intermediate stripes (Takada et al., 2005).
hes7.1	*hes5.1*	Negative during MHB specification	*hes5.1* overexpression: ↓*hes7.1* (MHB, ISH at neural plate stage (Takada et al., 2005)	
	hes7.1	Auto-repression (probably direct) during the establishment of the presumptive MHB.		*GR-hes7.1-VP16* (Dex NF10.5–11) +/– CHX: ↑*hes7.1* domains at neural plate stage (Takada et al., 2005).
hes7.2	*hes7.3/esr5*	Negative in somitogenesis. Negative feedback loop of Notch pathway	*hes7.3/esr5* overexpression: ↓*hes7.2* in the PSM (Jen et al., 1999)	*hes7.3/esr5-ΔWRPW*: derepressed *hes7.2* in the TZ in the PSM (Jen et al., 1999).
hes7.3/esr5	*notch1/* *hes5.1*	Negative during mesoderm induction	*NICD1:* unable to induce *hes7.3/esr5* in naive animal caps. *NICD1* or *hes5.1* overexpression: ↓*hes7.3/esr5* induced by the mesodermal inducer *nodal2* (animal caps, NF10.5, sqRT-PCR). *hes5.1* overexpression: ↓*hes7.3/esr5* in the IMZ (NF10.5, ISH) (Kinoshita et al., 2011).	*hes5.1-ΔWRPW:* ↑*hes7.3/esr5* (animal caps mesodermalized by *nodal2*, NF10.5, sqRT-PCR) (Kinoshita et al., 2011).

7.3.1. Establishing the Dorsal-Ventral Axis

The polarity of the initial dorsal-ventral (DV) axis is controlled by two antagonistic centers: the ventral center (VC) secretes morphogens (BMP4, Wnt8a) that induce ventral-posterior fates, and the dorsal center (DC) secretes antagonists and expresses repressors of ventral morphogens, protecting the dorsal region from being ventralized and posteriorized thus promoting dorsal-anterior fates. These centers have been characterized in amphibians (De Robertis, 2009) and fish (Thisse and Thisse, 2015). The DC is evident at the blastula stage and consists of: (1) the Nieuwkoop center (NC) in vegetal cells and (2) the Blastula Chordin- and Noggin-expressing (BCNE) center in marginal zone and animal cells. The BCNE gives rise to most of the brain and the organizer and secretes the neural inducers Noggin, Chordin, and Nodal3, which trigger brain induction shortly after mid-blastula transition (Wessely et al., 2001; Kuroda et al., 2004).

While the molecular establishment of the DC has been well documented (see Chapters 4 and 6), the early events leading to the establishment of the VC were largely unknown; our work found that Notch1 is involved (Acosta et al., 2011; Castro Colabianchi et al., 2018). The first clue was that NICD1 down-regulated *chordin* and *nodal3* in the BCNE. Strikingly, *RBPJDBM* did not affect their early expression, but *notch1* knock-down in ventral cells expanded their domains. This indicated that a ventral *notch1* activity restricts the BCNE to the dorsal side through an *RBPJ*-independent pathway (Acosta et al., 2011). Indeed, NICD1 destabilized a β-Catenin mutant lacking the GSK3 phosphorylation sites, whereas *notch1* knock-down increased its levels and ventrally expanded the domain in which β-Catenin was nuclear in the blastula, indicating that maternal *notch1* contributes to confining nuclear β-Catenin to the dorsal side. Moreover, when analyzed at tailbud or tadpole stages, *NICD1* mRNA injection resulted in a ventralized phenotype, whereas *notch1* knock-down, but not *RBPJDBM*, favored dorsal-anterior development. Notably, NICD1 blocked secondary axis induction by ventral injection of *ctnnb1* (*β-catenin*) mRNA, indicating that Notch1 has ventralizing properties because it interferes with the β-Catenin dorsalizing activity (Acosta et al., 2011).

Although our functional experiments revealed a ventral, non-canonical *notch1* activity, it was not clear if this were due to an asymmetric *notch1* mRNA distribution or regulation of Notch1 activity. We found that both *notch1* mRNA and protein are enriched in the ventral region of *Xenopus* embryos from fertilization to mid-blastula, with an opposite distribution of nuclear β-Catenin (Castro Colabianchi et al., 2018), consistent with the proposed role for Notch1 in destabilizing β-Catenin. This ventral enrichment of *notch1* mRNA and protein is the earliest localized sign of ventral development described so far in vertebrates, preceding the ventral localization of *wnt8a*, *bmp4*, and *ventx* mRNAs and dorsal localization of nuclear β-Catenin. Importantly, we noticed nuclear Notch1 in ventral cells during cleavage and mid-blastula stages, suggesting that besides the non-canonical role in destabilizing β-Catenin, Notch1 could be poised to trigger transcriptional activity. Through a gene reporter assay, we found that *RBPJ*-dependent transcriptional activity was higher on the ventral side at the onset of gastrulation. Functional experiments involving NICD1, full-length *notch1*, *RBPJDBM*, and *notch1* knock-down showed that *notch1* is necessary for the proper expression of VC genes such as *wnt8a*, *ventx*, and *bmp4*. Canonical, RBPJ-dependent Notch1 activities are mainly involved in controlling their expression, but non-canonical Notch1 activities might also contribute indirectly through β-Catenin destabilization and the known complex crosstalk between the DC and the VC (Castro Colabianchi et al., 2018). Interestingly, animal-dorsal expression of *foxi1*, which is necessary for ectoderm development, is independent of Wnt/β-Catenin signaling and is restricted to the dorsal region through a Notch1/RBPJ-dependent mechanism (Mir et al., 2008). Overall, this work supports the hypothesis that asymmetric Notch1 activity, including both canonical and non-canonical components, is involved in dorsal-ventral axis formation.

We proposed that Notch1 participates in forming the initial DV axis via a dual, ventralizing role (Figure 7.3A): (1) promoting the VC mainly through the canonical Notch/RBPJ pathway and (2) restricting the DC by destabilizing maternal β-Catenin independent of its phosphorylation by GSK3 and RBPJ. Through this non-canonical pathway, Notch1 ensures the elimination of β-Catenin from the ventral side that escapes from the GSK3-dependent degradation route. By inhibiting the early Wnt/β-Catenin pathway, Notch1 contributes to preventing hyperdorsalization and controls brain size by restricting the BCNE (Acosta et al., 2011; Castro Colabianchi et al., 2018). Interestingly, in mammalian embryonic stem cells, membrane-bound Notch1 associates with hypophosphorylated β-Catenin, decreasing its levels through the endocytic/lysosomal degradation pathway (Kwon et al., 2011). This finding reinforces the conclusions of our work, the first to study this non-canonical Notch pathway in embryonic axis formation in vertebrates.

7.3.2. Germ Layer Formation

In invertebrates, Notch signaling is a key pathway for the induction of the germ layers, whereas in vertebrates, germ layer induction and specification are controlled by several TFs and signaling pathways (Favarolo and López, 2018). In *Xenopus*, the presumptive array of germ layers can be roughly predicted along the animal-vegetal axis of the egg (Figure 7.3B). At cleavage stages, the animal cells approximate the ectoderm, the vegetal cells approximate the endoderm, and the intervening equator or marginal zone (MZ) mostly contributes to the mesoderm (Dale and Slack, 1987; Moody, 1987a; Moody, 1987b). Thus, the MZ, which is composed of an involuting (IMZ) and a non-involuting (NIMZ) region (Keller and Danilchik, 1988), constitutes a transition area between germ layers whose limits need to be defined during gastrulation (Figure 7.3B–D). Individual MZ cells of the early gastrula simultaneously express markers of two or three germ layers, and segregation is gradually refined as cells progressively and asynchronously commit to one germ layer (Wardle and Smith, 2004).

7.3.2.1. Refining Germ Layer Boundaries

In *Xenopus*, the boundaries between germ layers are refined by Notch signaling. In early gastrulae, *notch1* is expressed in both the IMZ and NIMZ (López et al., 2003; Miazga and McLaughlin, 2009), whereas *dll1* and *dlc* are only in the IMZ. The *dlc* domain forms a complete ring (Peres et al., 2006), whereas the *dll1* domain has a gap in the organizer region (López et al., 2005); subsequently, *dlc* also shows this gap (Peres et al., 2006). *dll1* is expressed in the pre-involuted IMZ but does not persist after involution (Wittenberger et al., 1999) (López et al., 2005) (Figure 7.1C) (Table 7.5). *rbpj.S-v2* transcripts are abundant just before gastrulation (Wettstein et al., 1997; Ito et al., 2007b), and its protein seems to regulate Notch function because it is required for *hes5.1* expression (Table 7.4) and is essential for gastrulation movements and mesoderm specification (Table 7.5) (Ito et al., 2007a). However, lineage tracing showed that perturbed Notch signaling did not transform one germ layer completely into another; only cells near the presumptive boundaries are competent to respond to Notch signaling (Contakos et al., 2005; Revinski et al., 2010). Therefore, the Notch pathway is not essential for the formation of germ layers in *Xenopus* but rather refines their segregation (Revinski et al., 2010).

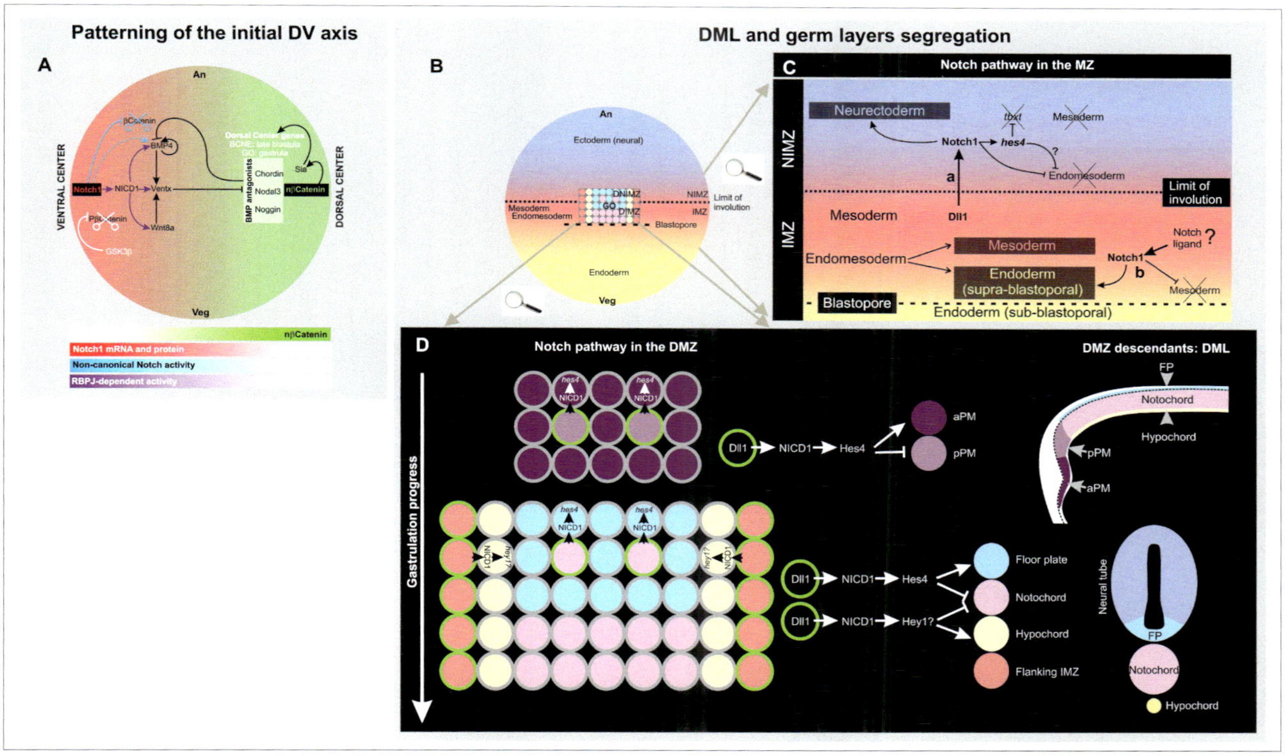

FIGURE 7.3 Notch plays early roles during patterning of the initial DV axis and during gastrulation in the germ layer and DML segregation. (A) Maternal *notch1* mRNA and Notch1 protein are enriched in the ventral region, exerting a ventralizing role by: (1) promoting ventral center development, mainly through the canonical Notch/RBPJ pathway and (2) preventing dorsal center development in the ventral side through a non-canonical pathway, independently of RBPJ, by destabilizing maternal βCatenin protein that escapes GSK3β-dependent phosphorylation (Pβcatenin) (modified from (Castro Colabianchi et al., 2018). (B–D) During gastrulation, Notch1 is required for the segregation of germ layers throughout the marginal zone (MZ), including dorsal midline (DML) components (adapted from (Favarolo and López, 2018) and aPM, pPM, and notochord arrangement in the DML based on (Yamaguti et al., 2005). (B) Diagram of a gastrulating embryo in dorsal view, showing the arrangement of presumptive germ layers (color-coded) along the An-Veg axis, the transition zone between them (marginal zone, MZ) (magnified in C), and the dorsal MZ containing the gastrula Organizer (GO) at the center, populated by the DML precursors (color-coded), magnified in D. IMZ, involuting marginal zone; NIMZ, non-involuting marginal zone; DIMZ, dorsal involuting marginal zone; DNIMZ, dorsal non-involuting marginal zone. (C) Dll1 from the IMZ activates the Notch1 pathway on the neighboring NIMZ cells, favoring neuroectoderm at the expense of mesodermal fates (a: type A decision), thus refining the limit of involution. In the IMZ, Notch1 promotes endomesoderm segregation, favoring endodermal at the expense of mesodermal fates (b, type B decision); the involved ligand is unknown. Inhibited markers and germ layers are crossed out. (D) In the DIMZ, Dll1 expressed in isolated cells activates Notch1/*hes4* in their neighbors, first favoring aPM at the expense of pPM and then favoring floor plate (FP) fates at the expense of the notochord, stopping involution. Dll1 from MZ cells flanking the DMZ activates Notch1 signaling, promoting hypochordal fate at the expense of the notochord. Because of its expression pattern, *hey1* is a good candidate for intervening in this choice.

TABLE 7.5
Core Components of the Notch Pathway in the Development of Germ Layers and the Dorsal Midline in *Xenopus*
See abbreviations in Table 7.3 legend.

Gene	Expression Role/Details	Gain-of-Function	Loss-of-Function (Dominant-Negative/Morpholino)
notch1	*notch1* is expressed in the IMZ and NIMZ during gastrulation (López et al., 2003; Miazga and McLaughlin, 2009). **Germ layers:** Balanced Notch1 signaling is required for delimiting the three germ layers and normal morphogenetic movements during gastrulation. Cell-fate choice upon activation of Notch signaling does not depend on proliferation during gastrulation. **DML:** Among the DML precursors, *notch1* favors FP and hypochord development at the expense of the notochord.	**Germ layers:** *notch1ΔE* +/− HUA at gastrula stage, analysis at tailbud stage: ↑epidermal, neural, and muscle tissues (Coffman et al., 1993). *notch1ΔE* (animal caps): ↑ectodermal response to neural induction; prolonged ectodermal competence for mesodermal induction beyond the onset of gastrulation (Coffman et al., 1993). Mouse NICD1, analysis at tailbud stage: ↓myosin (muscle differentiation marker) (Kopan et al., 1994). *GR-NICD1*, Dex NF2, but not Dex NF12: ↑neural plate (*sox2;* late neurula, NF18/19) and muscle tissues (12/101 marker; tailbud, NF25) (Glavic et al., 2004). GR-*NICD1* or GR-*RBPJ*-VP16, Dex NF10 or 12, analysis at late neurula/tailbud stages: ↑endodermal ↓mesodermal derivatives (Contakos et al., 2005). GR-*RBPJ*-VP16, Dex NF14, analysis at late neurula/tailbud stages: ↓endodermal ↑mesodermal derivatives (Contakos et al., 2005). *NICD1*, analysis at gastrula stages: DMZ-directed injection: delayed blastopore formation and closure; ↓mesoderm (*tbxt*), ↑neural ectoderm (*sox2*) ↑supra-blastoporal endoderm (*sox17*). Dorsal/animal-directed injection: *tbxt* domain expanded but more diffuse, probably due to ectopic prolonged competence for mesodermal induction in the ectoderm (Revinski et al., 2010). **DML:** *notch1ΔE* +/− HUA at gastrula stage: ↓notochord (tailbud stage) (Coffman et al., 1993). *NICD1*: Analysis at gastrula stage: ↓GO mesoderm (*chrd*), ↑presumptive FP (*hes4*) (López et al., 2003), (López et al., 2005). Analysis at neurula stage: ↓notochord (*chrd, tbxt*), ↑FP (*shh, foxa4*) (López et al., 2003). Analysis at tadpole stages: ↑hypochord (vegf,*spon1*, NF25,38) (Peyrot et al., 2011). *GR-NICD1*/time-controlled experiments with Dex: higher susceptibility to ↑FP and ↓notochord in the first than in the second half of gastrulation (López et al., 2003)	**Germ layers:** GR-*RBPJDBM*, Dex NF 10 or 12, analysis at late neurula/tailbud stages: reduced endodermal, expanded mesodermal derivatives (Contakos et al., 2005). GR-*RBPJDBM*, Dex NF14, analysis at late neurula/tailbud stages: expanded endodermal, reduced mesodermal derivatives (Contakos et al., 2005). *notch1* MO (dorsal injections), analysis at gastrula stage: delayed blastopore formation and closure, expanded the mesoderm (*tbxt*), animally shifted the neural-ectoderm (*sox2*), reduced the supra-blastoporal endoderm (*sox17*) (Revinski et al., 2010). **DML:** *notch1* MO: expanded the notochord (*chrd*+ cells, neurula stage) (López et al., 2003).

psen	**DML:** Ɣ-secretase is necessary to keep the morphogenetic movements at a normal pace during gastrulation.		DAPT (Ɣ-secretase inhibitor): delayed the convergent-extension movements typical of the notochordal cells during gastrulation (Revinski et al., 2010).
RBPJ	Both *RBPJ.S* variants (see Table 7.1) are expressed ubiquitously from the unfertilized egg to the tailbud stage, but *RBPJ.S-v2* transcripts are more abundant before gastrulation (Wettstein et al., 1997) (Ito et al., 2007b). **Germ layers:** *RBPJ*-dependent signaling favors the development of endoderm-derived cell types and disfavors the development of mesoderm-derived cell types during gastrulation but plays the opposite role since the neural plate stage. A critical time window that comprises gastrulation and ends at the onset of neurulation was described in particular for cardiac mesoderm specification, which is disfavored by Notch/RBPJ signaling during this period.	**Germ layers:** GR-*RBPJ-VP16*, analysis at late neurula and tailbud stages: Dex NF10 or 12: ↑markers of endodermal-derived cell types, ↓markers of mesodermal-derived cell types (paraxial, intermediate, and cardiac mesoderm). Dex NF14: ↓markers of endodermal-derived cell types, slightly ↑markers of intermediate mesoderm-derived cell types. Cardiac mesoderm unaffected (Contakos et al., 2005). GR-*RBPJ-VP16* (Dex NF10) +/- ER-*RBPJDBM* (E2 at NF10, 14, 15, 16); ISH of the cardiac field marker *nkx2.5* at tailbud stage: the loss of *nkx2.5* induced by activating Notch/RBPJ signaling at the onset of gastrulation was rescued by blocking Notch/RBPJ during gastrulation. This rescuing ability was gradually lost after the onset of neurulation (Miazga and McLaughlin, 2009). GR-*RBPJ*-VP16 (Dex NF10): ↓*gata4* (mid-gastrula) (Miazga and McLaughlin, 2009).	**Germ layers:** GR-*RBPJDBM*, analysis at late neurula and tailbud stages: Dex NF10 or 12: ↓markers of endodermal-derived cell types, ↑markers of mesodermal-derived cell types (paraxial, intermediate, and cardiac mesoderm). Dex NF14: ↑markers of endodermal-derived cell types, ↓markers of intermediate mesoderm-derived cell types. Cardiac mesoderm unaffected (Contakos et al., 2005) GR-*RBPJDBM* (Dex NF10), ISH analysis at NF11–11.5: precocious induction of heart field markers (*nkx2–5, gata4, tbx5)* (Miazga and McLaughlin, 2009). *RBPJDBM*, analysis at gastrula stage: mild expansion of *tbxt* (pan-mesodermal) and *myf5* (presumptive paraxial mesoderm) domains, variable ↓ and animal shift of *sox2*, variable results of *sox17* expression (endoderm) (Revinski et al., 2010). *RBPJDBM*, analysis at neural plate stage: mild tendency to ↑paraxial mesoderm markers *myf5* and *myoD* and ↓neural marker *sox2* (Revinski et al., 2010). *RBPJ.S v2* MO: abnormal gastrulation, neural fold disorganization, ↓mesodermal markers *tbxt, myod1, ventx1.2, chrd*, but not *gsc* (RT-sqPCR, NF 10.5). ↓*tbxt* in the IMZ (ISH). This effect was not rescued by *NICD1* or *DN-hes5.1* (Ito et al., 2007a). *RBPJ.S-v1* MO: no morphological defects, no changes in molecular markers (Ito et al., 2007a). Probably due to compensation by the other homeolog as neither MO used in this study is predicted to knock-down *RBPJ.L*. **DML:** *RBPJDBM:* ↑*chrd*+ in the GO and *chrd*+ notochord expression in neurula; ↓FP (*foxa4+/shh+* cells) (neurula)(López et al., 2003); ↓hypochord (*vegf, spon)* (NF28/35) (ISH) (Peyrot et al., 2011).

(Continued)

TABLE 7.5 (Continued)

Core Components of the Notch Pathway in the Development of Germ Layers and the Dorsal Midline in *Xenopus*

Gene	Expression Role/Details	Gain-of-Function	Loss-of-Function (Dominant-Negative/Morpholino)
dll1	Onset in the IMZ at early gastrula (NF10.5, ISH), with a gap of lower expression in the GO, with scattered *dll1*+ cells (López et al., 2005). Expression in the pre-involuted IMZ does not persist in involuted cells (Wittenberger et al., 1999; López et al., 2005). **Germ layers:** Dll1 signaling from pre-involuted IMZ activates Notch signaling in the NIMZ, promoting neural-ectodermal and inhibiting endomesodermal fates (Revinski et al., 2010). **DML:** Within the GO, Dll1 promotes FP development at the expense of notochordal fate.	**DML:** *dll1*, dorsal overexpression; analysis at gastrula stage: ↑FP precursors (*hes4*), ↓notochordal precursors (*chrd*) in the GO (López et al., 2005).	**Germ layers:** *dll1STU* (dorsal injection), analysis at gastrula stage: ↓*tbxt* in its normal domain, displaced by expanded supra-blastoporal endoderm (*sox17*) that took its place; ↓neural ectoderm (*sox2*) because of the animal shift and expansion of the presumptive mesoderm (*tbxt*) (Revinski et al., 2010). *dll1* MO (lateral injection), analysis at gastrula stage: impaired blastopore closure, ↓circumblastoporal *tbxt* expression (only a vegetal view was shown). Both effects were rescued by co-injection of *NICD1* mRNA (Kinoshita et al., 2011). **DML:** *dll1STU*, dorsal injection; analysis at gastrula stage: ↑notochordal precursors (*chrd*), ↓FP precursors (*hes4*) in the GO (López et al., 2005).
dlc	Onset in the IMZ at early gastrula (NF10.5, ISH) in a complete circumblastoporal ring; a lower expression gap appears at NF11 at the GO (Peres et al., 2006). Necessary for *hox* genes expression in the IMZ. There are restrictions, at least for *hox* genes, to respond to Dlc signaling outside the IMZ (see Table 7.12 for additional information).	*dlc* overexpression: did not affect *hoxc6* and *hoxd1* in the IMZ and could not expand them outside their normal region (ISH, gastrula stage) (Peres et al., 2006).	*dlc* MO: ↓*hoxb4, hoxc6, hoxb9*, and *bmp4* in the IMZ (ISH, gastrula stage); *tbxt* expression unaffected in the IMZ (ISH, late gastrula) (Peres et al., 2006).

Constitutively active Notch1 constructs and time-controlled Notch1/RBPJ activation or blockade resulted in a variety of changes in markers of ectodermal-, neural-, mesodermal-, and endodermal-derived cell types at neurula and tailbud stages (Tables 7.5, 7.9), indicating that the response to Notch1/RBPJ signaling changes over time. We perturbed the Notch pathway in several ways to address a possible role in controlling the boundaries between germ layers, analyzing the consequences during gastrulation when they segregate, and also in early neurulation (Revinski et al., 2010) (Table 7.5). Activation and blockade delayed gastrulation, indicating that Notch1 activity is tightly balanced to keep morphogenetic movements at a normal pace. Germ layers were specified, but they did not develop properly because their MZ boundaries were shifted. Consequently, cells at the boundaries allocated incorrectly and changed their specification to the incorrect germ layer. In *NICD1* mRNA-injected embryos, the presumptive neural ectoderm and supra-blastoporal endoderm were expanded at the expense of mesoderm, whereas *notch1* knock-down produced the opposite changes (Revinski et al., 2010). Both *dll1STU* (Revinski et al., 2010) and *dll1* knock-down (Kinoshita et al., 2011) inhibited the pan-mesodermal marker *tbxt* in its normal circumblastoporal domain. In embryos injected with *dll1STU*, this was accompanied by the animal displacement of the *tbxt* domain as expanded supra-blastoporal endoderm (*sox17*-positive) took its place. The neural ectoderm (*sox2* expressing) also was reduced in response to this animal-ward expansion of the presumptive mesoderm (Revinski et al., 2010).

We proposed that *notch1* is involved in the segregation between neural ectoderm, mesoderm, and endoderm by controlling their boundaries in the MZ (Revinski et al., 2010) (Figure 7.3B,C) in the following ways. (1) Refining the limit of involution between the IMZ and the NIMZ favored neural ectoderm at the expense of mesoderm (type A decisions). (b) In the IMZ, by refining mesoderm segregation from the supra-blastoporal endoderm, favoring endoderm over mesoderm (type B decisions). Strikingly, *dll1STU* shifted the limit of involution animal-wards, favoring endomesodermal development over neural ectoderm, but *notch1* knock-down expanded the mesoderm at the expense of both endoderm and neural ectoderm; perturbing Notch/RBPJ signaling during gastrulation had a similar effect (Contakos et al., 2005). This indicates that Dll1 is involved in type A but not in type B decisions. According to this model, pre-involuted IMZ cells present Dll1 to the neighboring cells on the other side of the limit of involution, thus preventing them from adopting the same fate (endomesoderm) by triggering the Notch pathway, which instead promotes neural ectoderm specification (Revinski et al., 2010). Interestingly, in animal caps assays, Notch1ΔE alone weakly induced neural ectoderm but strongly enhanced ectodermal competence for neural induction (Coffman et al., 1993). Therefore, Dll1 signaling from pre-involuted IMZ might enhance the competence of their neighbors above the limit of involution to respond to neural inducers and become neural instead of mesoderm, sharpening the boundary between both populations. Once the mesodermal cells involute, they no longer express Dll1, ending this activity. While it appears that Dll1 controls the limit of involution (type A decisions), it remains unknown which *notch1*-dependent mechanisms underlie mesodermal versus endodermal (type B) decisions. More work is needed to discern the possible role of the diverse Dll/Jag ligands and non-canonical Notch pathways in the segregation of germ layers.

7.3.2.2. Which Notch Targets Are Involved in Germ Layer Segregation?

The IMZ expresses several *hes* genes during gastrulation (Figure 7.2). Most of their patterns are similar to that of *dll1*, but the dorsal *hes5.1* and *hes5.3* boundaries are more distant from the organizer, and *hes7.2* is more abundant in the organizer. Only *hes4* is broadly expressed in the NIMZ, but *hes5.10* (at early gastrula) and *hes2* (at mid-gastrula) are expressed in scattered cells (Figure 7.2). Except for *hes5.8* and *hes5.9*, whose regulation by Notch signaling has not been studied, all the *hes5* genes expressed in the MZ, as well as *hes2*, *hes4*, and *hes7.2*, are positively regulated by Notch in several contexts, although a few of them were tested for Notch responsiveness in the MZ. In contrast, *hes6.1* and *hes7.3/esr5* are down-regulated in the neural plate and the IMZ, respectively (Tables 7.2, 7.3). Interestingly, *hes5.10* is later expressed throughout the non-neural ectoderm and responds positively to NICD1, although it is not clear whether RBPJ is involved (Deblandre et al., 1999). The early expression of *hes6.1, hes7.2,* and *hes7.3/esr5* in the IMZ might be related to their role during somitogenesis (see Section 3.6). Only a few of the *hes* genes expressed in the MZ have been experimentally tested for their role in the MZ (*hes4, hes5.1, hes5.6, hes6.1, hes7.3/esr5*) (Table 7.6). One clear candidate for positioning the limit of involution is *hes4*, which first is broadly expressed in the presumptive ectoderm of the blastula, then progressively confined to the boundary with the mesoderm during gastrulation, accumulating transcripts in the whole NIMZ with highest levels dorsally in a pattern complementary to *tbxt* (pan-mesoderm) (López et al., 2005; Aguirre et al., 2013). *hes4* might be one of the Notch targets involved in type A decisions because: (1) the *hes4* NIMZ domain was expanded by NICD1 (López et al., 2005) and down-regulated by blockade of Notch1/RBPJ signaling (unpublished results); (2) *hes4* overexpression blocked gastrulation movements, impeding MZ cell involution; (3) *hes4* overexpression repressed *tbxt* throughout the entire IMZ (López et al., 2005; Cui, 2005; Aguirre et al., 2013); and (4) *hes4* knock-down expanded the *tbxt* domain toward the animal pole, indicating that it is required for the correct placement of the ectoderm-mesoderm boundary (Aguirre et al., 2013) (Figure 7.3B,C) (Tables 7.5, 7.6).

Overexpression and dominant-negative experiments indicate that *hes7.3/esr5* promotes and *hes5.1* inhibits mesoderm specification and they repress each other (Kinoshita et al., 2011) (Tables 7.4, 7.6). In animal cap explants, NICD1 induced *hes5.1* but not *hes7.3/esr5* (Kinoshita et al., 2011) (Tables 7.2, 7.3), Nodal2 induced *hes7.3/esr5* but not *hes5.1*,

TABLE 7.6
hes/hey* Genes in the Development of Germ Layers and the Dorsal Midline in *Xenopus

See abbreviations in Table 7.3 legend.

Gene	Role/Details	Gain-of-Function	Loss-of-Function (Dominant-Negative/Morpholino)
hes1	Represses *myod1*, probably directly, downstream of mesodermal induction.	*hes1* overexpression, mesodermalized animal caps: ↓*tbxt* (pan-mesoderm) and *myoD* (paraxial mesoderm) but not GO mesodermal markers (*chrd*, *gsc*). Inhibition of *myoD* downstream of *tbxt* (Umbhauer et al., 2001). *hes1* overexpression, whole embryos: delayed gastrulation, trunk defects; ↓*myoD* but not *chrd* (gastrula) through the DNA-binding and repressor domains (Umbhauer et al., 2001).	*hes1-VP16* or *GR-hes1-VP16* (Dex 2 hs after NF11) +/– CHX, animal caps: ↑*myoD* but not *tbxt*, suggesting a direct regulation of *myod*.
hes4	**Germ layers:** *hes4* is normally expressed in the NIMZ since late blastula stage and restricts mesoderm specification to the IMZ during gastrulation. **DML:** As a mediator of Notch1 signaling in DML development, blocks the involution of *hes4*+ GO cells, favoring their incorporation into the dorsal NIMZ (notoplate/future FP) at the expense of the notochord. Both *hes4* homeologs are required for FP development (López et al., 2005; Aguirre et al., 2013). Required for aPM specification by restricting pPM and notochord specification (Yamaguti et al., 2005). *hes4* acts as a cell-autonomous repressor in the aPM, restricting contiguous cell fates, contributing to regionalize the axial mesoderm. It is also able to induce dorsal genes in a non-cell-autonomous way through the activity of the WRPW motif, to ensure an organizer environment (Murato et al., 2006).	**Germ layers:** *hes4.S* overexpression: ↓*tbxt* (pan-mesoderm) throughout the IMZ (López et al., 2005; Cui, 2005; Aguirre et al., 2013). **DML:** *hes4.S*, dorsal overexpression: ↓notochord specification (*chrd, tbxt)* in the GO; ↑FP precursors in gastrulae and neurulae (*foxa4*+ cells in GO and notoplate, *shh*+ cells in notoplate) (López et al., 2005). *hes4.L*, dorsal overexpression: ↓*chrd* (pPM, notochord), *not* (notochord), *dkk1*, and *hex* (anterior endoderm) in the GO; head defects (Yamaguti et al., 2005). *hes4.S*, ventral overexpression: ↓*tbxt*, induced ectopic GO markers (*chrd, foxa4)*, but these cells were unable to involute during gastrulation (López et al., 2005); induced a headless secondary axis (Aguirre et al., 2013). *hes4.L*, ventral overexpression: ↓ventral mesoderm specification (*ventx*); induced ectopic GO markers and a headless secondary axis (Yamaguti et al., 2005).	**Germ layers:** *hes4.L+S* MO: animal expansion of *tbxt* (pan-mesoderm) throughout the MZ. *hes4.S* might be more relevant in restricting *tbxt* expression to the IMZ (Aguirre et al., 2013). **DML:** *hes4.S* MO, dorsal injection: ↑notochordal precursors' population in the GO and notochord in neurulae (*chrd*, *tbxt*); ↓FP precursors population in gastrulae and neurulae (*foxa4*+ cells in GO and notoplate, *shh*+ cells in the notoplate). Reversed the effects of *NICD1* on DML markers (López et al., 2005). *hes4.L* MO or *hes4.S* MO: ↓FP *shh* (Murato et al., 2006). *hes4.L* MO: ↑aPM (*chrd*) and notochord (*chrd*, *not*) at the expense of the anterior pPM (*gsc*) (ISH NF14); did not affect anterior endodermal markers nor *ventx* (ventral mesoderm) (early gastrula stage) (Yamaguti et al., 2005).
hes5.1	Inhibits mesoderm specification.	*hes5.1* overexpression: ↓*tbxt* in mesodermalized animal caps (Ito et al., 2007a) and the IMZ at early gastrula (NF10.5, ISH) (Kinoshita et al., 2011).	*hes5.1 ΔWRPW:* ↑*tbxt* at early gastrula (NF10.5, ISH) (Kinoshita et al., 2011).

hes5.6	**Germ layers:** Possible regulation of the timing of cardiac field specification (Miazga and McLaughlin, 2009). **DML:** Possible role in DML development.	**Germ layers:** GR-*hes5.6*-VP16 (Dex NF10): ↓*gata4* (mid-gastrula) (Miazga and McLaughlin, 2009). **DML:** *hes5.6*, dorsal overexpression: ↑*chrd* (GO) (Taelman et al., 2004).	
hes6.1	Required for presumptive paraxial mesoderm specification, involving recruitment of TLE/Groucho co-repressors but not DNA binding.	*hes6.1* overexpression: ↑*tbxt* (pan-mesoderm), *myod1* (presumptive paraxial mesoderm), and *wnt8a* (lateral/ventral mesoderm) but did not affect the GO marker *chrd*. Expansion was restricted to the nearby MZ (ISH, gastrula). ↑*tbxt* and *wnt8a* but not *chrd* or endodermal markers (animal caps, RT-qPCR). Mesodermal induction by *hes6.1* mainly required FGF but also Nodal signaling (RT-qPCR, ISH) (Murai et al., 2007).	*hes6.1DBM:* same effects on mesoderm as *hes6.1* overexpression (gastrula). *hes6.1 ΔWRPW:* mesodermal markers unaffected (gastrula). *hes6.1* MO: ↓*myod1, myf5* (presumptive paraxial mesoderm) in the IMZ (without affecting *tbxt*, *wnt8a, chrd,* or the endodermal marker *sox17)*. This effect was rescued by *hes6.1* or *hes6.1DBM* but not by *hes6.1 ΔWRPW* (ISH, mid-gastrula) (Murai et al., 2007).
hes7.3/ esr5	Necessary for mesoderm specification.	*hes7.3/esr5* overexpression: ↑*tbxt* in the IMZ at early gastrula (NF10.5, ISH) (Kinoshita et al., 2011).	*hes7.3/esr5 ΔWRPW:* ↓*tbxt* at early gastrula (NF10.5, ISH) (Kinoshita et al., 2011).
hey1	Possible role in the segregation of DML precursors, repressing notochordal fates. Hey1 does not bind the co-repressor TLE/Groucho, as it lacks the typical WRPW motif of bHLH-O repressors (Pichon et al., 2004). For *chrd* downregulation, Hey1 acts as a DNA binding repressor, requiring the Orange domain and the C-terminal region for dimerization. Hey1 heterodimerizes *in vivo* with Hes1 and Hes4 and weakly with Hes2, but it does not bind Hes5.5 or Hes5.6 (Taelman et al., 2004).	*hey1* overexpression: stopped gastrulation, ↓*chrd* (GO), *tbxt* (pan-mesoderm; IMZ), and *not* in the notochord and GRP. It did not affect *not* in the notoplate but induced ectopic *not* in the limit of involution, outside the GO (Taelman et al., 2004). GR-*hey1* (Dex NF12): FP not affected at tailbud stage (Taelman et al., 2004). An earlier induction with Dex will be useful to address possible effects during DML segregation.	*hey1DBM:* did not repress *chrd* (Taelman et al., 2004).

and Hes5.1 reduced *hes7.3/esr5* induction by Nodal2. Therefore, it was proposed that a mutually antagonistic relationship between *hes5.1* and *hes7.3/esr5* controls the balance of mesoderm specification within the IMZ (Kinoshita et al., 2011). It will be interesting to determine whether *hes5.1* also mediates endodermal versus mesodermal choices.

hes5.5 and *hes5.6* are positively regulated by canonical Notch-RBPJ in the IMZ (Miazga and McLaughlin, 2009). While the function of *hes5.5* in this tissue has not been studied, Notch regulates the timing of heart field specification, possibly through *hes5.6* (Tables 7.5, 7.6) (Miazga and McLaughlin, 2009). While *hes6.1* is negatively regulated by Notch/RBPJ in the neural plate, it is not known whether this pathway controls *hes6.1* expression in the IMZ during gastrulation, although it is a direct Wnt/β-Catenin target and requires input from zygotic Wnt/β-Catenin signaling (Hufton et al., 2006; Kjolby and Harland, 2017). *hes6.1* favors paraxial mesoderm development (but not general mesoderm induction) by sequestering TLE/Groucho co-repressors, thus relieving *myod1* from repression in mesodermal precursors (Cossins et al., 2002; Murai et al., 2007) (Table 7.6).

7.3.3. Dorsal Midline Tissues

Cells that derive from the dorsal MZ/organizer region constitute the vertebrate dorsal midline (DML), an essential signaling center for development of the surrounding tissues. The DML gives rise to several derivatives: (1) the prechordal endomesoderm (PEM), a key signaling center for anterior neural development that emerges from the deep cells of the organizer to form the prechordal plate; (2) the notoplate (Figure 7.3B,D), which gradually converges and extends during gastrulation to form the floor plate (FP) of the neural tube; (3) the notochord; and (4) the dorsal midline of the endoderm, in *Xenopus* known as the gastrocoel roof plate (GRP), which functions as a left-right organizer. During neurulation, some GRP cells incorporate into the notochord and somites, while bilateral GRP rows gradually fuse into the hypochord, ventral to the notochord (Keller and Danilchik, 1988; Minsuk and Keller, 1997; Kiecker and Niehrs, 2001; Shook et al., 2004; López and Carrasco, 2006) (Figure 7.3D). Gene marker studies revealed that the precursors of these various derivatives are intermingled at the beginning of gastrulation but gradually segregate (Bouwmeester et al., 1996; Artinger et al., 1997; Yamaguti et al., 2005); this process is highly influenced by the Notch pathway.

Components of the Notch pathway are differentially expressed in the multipotent DML precursors that either involute (as the IMZ) or remain on the surface as the NIMZ, that is, notoplate (López et al., 2003; López et al., 2005). *hes4* is expressed in the dorsal NIMZ and then in the notoplate and FP. *dll1* is expressed in a compact domain throughout the IMZ, except for the organizer region, where only scattered cells express *dll1* and also *hes4* prior to involution. Once these dorsal IMZ cells involute, only *hes4* expression continues, restricted to the prechordal mesoderm but absent from the notochord. In fact, *hes4* is the only *hes1–7* gene expressed in the *Xenopus* DML during gastrulation and neurulation (Figure 7.2) (Tsuji et al., 2003; López et al., 2005; Yamaguti et al., 2005).

We perturbed the Notch pathway in several ways to address its role during DML development, including *hes4* overexpression and knock-down, constitutive NICD1 activation, time-controlled GR-NICD1 activation, blocking the whole *notch1* pathway by knock-down, the RBPJ-dependent pathway with *RBPJDBM*, Dll1 signaling with *dll1STU*, and Notch processing with *psen1* knock-down. Our results indicated that during gastrulation, *notch1/psen1/RBPJ/hes4* signaling favors notoplate over notochord fate (López et al., 2003; López et al., 2005) (Tables 7.5, 7.6). Other authors showed that Notch promotes hypochord over notochord by injecting *NICD1* and *RBPJDBM* (Peyrot et al., 2011) (Table 7.5). As the PEM mesodermal population segregates during gastrulation into two subdomains, *hes4* and *gsc* are expressed in the anterior prechordal mesoderm (aPM), whereas *chordin* is expressed in the posterior prechordal mesoderm (pPM) (Yamaguti et al., 2005). It was proposed that *hes4* initially ensures an organizer environment by inducing early organizer genes through a non-cell-autonomous activity that depends on the WRPW domain. Then, *hes4* is required for aPM specification, as it inhibits contiguous fates through a cell-autonomous repressive activity, restricting pPM and notochord (Yamaguti et al., 2005; Murato et al., 2006) (Table 7.5).

Based on these studies, we propose a model for how the Notch pathway allocates dorsal MZ descendants into the different DML tissues (Figure 7.3B,D). First, DML precursors choose between aPM or pPM fates. Dll1 from scattered cells in the organizer induces *hes4* in neighboring cells, which represses pPM fates, thus promoting aPM. As gastrulation proceeds and more posterior cells involute, multipotent precursors in the mid- and late organizer choose between FP, notochord, or hypochord fates. Dll1 from scattered cells in the boundary between the dorsal NIMZ and the pre-involuted IMZ interacts with the Notch1 receptor on the surrounding cells, activating *hes4* to repress the genes that promote notochord development and impede their involution so they gradually incorporate into the notoplate. By this mechanism, *dll1* executes a cell fate switch that favors notoplate development at the expense of notochord. Dll1 presented by the IMZ cells flanking the organizer activate Notch1 signaling in a pair of bilateral rows of dorsal IMZ cells, favoring hypochord over notochord; the down-stream mechanism is unknown, since *hes4* is not expressed by hypochord precursors. In addition, *notch1/hes4* expand the expression of *foxa4*, a positive notoplate/FP regulator (López et al., 2003; López et al., 2005), whereas *foxa4* knock-down suppress *hes4* in the FP (Murgan et al., 2014), suggesting they establish a positive feedback loop. The expression of *hey1*, which is positively regulated by Notch/RBPJ (Pichon et al., 2002; Rones et al., 2002) (Tables 7.2, 7.3), matches the time and spatial profile of hypochord development, with initial bilateral stripes in the GRP that later fuse at the midline

(Pichon et al., 2002; Shook et al., 2004) (Figure 7.2). Since *hey1* is also expressed in scattered cells in the involuted dorsal IMZ at early gastrula (see clone XL097h17 in Taverner et al., 2005), it will be interesting to study if it promotes hypochordal fates over notochord. *hey1* is also expressed in the FP during neurulation, but neither time-controlled *hey1* overexpression at late gastrula nor *hey1* knock-down affected FP development when analyzed at tailbud stages (Taelman et al., 2004). There is evidence that *hey1* might suppress neurogenesis in the FP by antagonizing proneural genes, thereby maintaining FP identity. Interestingly, *neurog2* overexpression revealed FP's potential to differentiate into neurons, and *hey1* overexpression suppressed primary neurogenesis in the neural plate. However, neither *hey1* knock-down alone nor combined with *hes4.L* knock-down induced ectopic neurogenesis in the FP, indicating that additional inhibitors might be required to inhibit neurogenesis (Taelman et al., 2004).

7.3.4. Primary Neurogenesis

Differentiation of multipotent neural progenitors into the diverse nervous system cell types is an orchestrated process ensuring that neurons and glia appear at the right time and place during development. Key players in this process are Notch pathway components and several bHLH proteins. Those encoded by "proneural" genes—members of the Neurogenin and Achaete-scute families—heterodimerize with the bHLH factor E47, bind the E box (CANNTG), and activate transcription, promoting competence for neuronal differentiation. Downstream, other bHLH transcriptional activators promote the determination of neurons (NeuroD family) or oligodendrocytes (Olig family). Notch-regulated Hes proteins typically repress proneural gene expression or activity, maintaining neural precursors in a proliferative and undifferentiated state, and allow astrocyte differentiation (Davis and Turner, 2001; Bertrand et al., 2002; Huang et al., 2014; Kageyama et al., 2007; Imayoshi and Kageyama, 2014).

Anamniotes develop through a larval period that requires a simple neuronal circuitry for swimming and escape reflexes to be functioning around hatching (Roberts, 1989). In *Xenopus*, a first wave of primary neurogenesis, which begins at late gastrula and peaks at neural plate stages, generates three bilateral pairs of longitudinal stripes of "primary neurons": motoneurons, interneurons, and sensory neurons that are responsible for these larval behaviors (Chitnis et al., 1995). Study of primary neurogenesis in *Xenopus* significantly contributed to the discovery of the molecular and cellular basis of vertebrate neurogenesis and provided an accessible paradigm to study the Notch pathway (Tables 7.7, 7.8). The neurogenesis gene regulatory network (GRN) built from this work is initially controlled by the balanced expression of "prepattern genes," such as those encoding Gli and Zic TFs (Lee et al., 1997; Marine et al., 1997; Brewster et al., 1998; Nakata et al., 1998). By refining proneural gene expression, the prepattern TFs roughly outline regions in the neural plate in which primary neuronal differentiation can or cannot occur; this confers neuronal differentiation competence to restricted domains (Zimmerman et al., 1993; Ma et al., 1996).

Proneural genes induce *notch1* and *dll1*, whose expression in the posterior neural plate begins around late gastrula in overlapping stripes that prefigure the placement of the primary neurons (Turner and Weintraub, 1994; Chitnis et al., 1995; Ma et al., 1996; Chitnis and Kintner, 1996) (Figure 7.1C). *notch1* is expressed by most cells in the proneural domain, whereas *dll1* is restricted to a subset of them (Chitnis et al., 1995; Ma et al., 1996; Chalmers et al., 2002). Primary neurogenesis is circumscribed by Notch-dependent lateral inhibition: the selected neuronal precursor expresses Dll1, which binds Notch1 in neighboring cells, activating the Notch/Psen/RBPJ/Maml pathway, resulting in the induction of Hes1–7 bHLH-O repressors that inhibit proneural genes and thereby repress neuronal fate in the neighbors (Chitnis et al., 1995; Ma et al., 1996; Bellefroid et al., 1996; Wettstein et al., 1997; Perron et al., 1999; Paganelli et al., 2001; Katada and Kinoshita, 2003; López et al., 2003; Nichane et al., 2008b; Revinski et al., 2010; Riddiford and Schlosser, 2017) (Tables 7.3, 7.4, 7.8). A negative feedback loop is established in the neuronal precursors that suppresses *dll1* expression in their neighbors through NICD1. The neuronal precursors continue to express proneural genes, which induce the bHLH-determination factor *neurod1*. Once *neurod1* is activated, the cells become refractory to lateral inhibition and undergo terminal differentiation into neurons (Chitnis et al., 1995; Chitnis and Kintner, 1996; Olson et al., 1998; Sjöqvist and Andersson, 2019). Another pathway for preventing lateral inhibition is through the upregulation of the zinc-finger TF myt1 by Neurog2 (Bellefroid et al., 1996). In other scenarios, when proneural TFs reach a certain threshold, they induce Ebf2 in selected progenitors, which stabilizes commitment to a neuronal fate by enhancing *dll1* expression and reinforcing *neurod1* expression (Dubois et al., 1998). *neurod1* appears to feed back to directly potentiate *dll1* expression, as it promotes ectopic *dll1* in whole embryos and induces *dll1* in animal caps in the absence of protein synthesis (Seo et al., 2007).

7.3.4.1. Notch Ligands

In the neural plate, *dll1* expression is stronger posteriorly, and *jag1* expression is stronger anteriorly (Table 7.7) (Figure 7.1C) (Kiyota et al., 2001). Jag1 normally restricts the differentiation of primary neurons, and combined *dll1* and *jag1* expression is indispensable for the normal primary neurogenesis pattern (Kiyota et al., 2001) (Table 7.7). Interestingly, Dll1 and Jag1 contain sequences encoding a putative nuclear localization signal. Moreover, GFP fusion proteins of both ligands naturally underwent proteolitic cleavage during gastrulation, releasing their intracellular domains (ICDs), which were detected in nuclei. However, only Jag1-ICD-GFP persisted in cell nuclei and repressed primary neurogenesis without activating *hes5.1* (Kiyota and Kinoshita, 2004). It was proposed that Jag1 inhibits primary

TABLE 7.7
Core Components of the Notch Pathway Involved in *Xenopus* Neurogenesis

See abbreviations in Table 7.3 legend. *ascl2*, *neurog1*, and *neurog2* are bHLH proneural genes; *neurod1* and *neurod4* encode bHLH neuronal determination TFs; *ebf2* encodes an HLH TF, positive regulator of neurogenesis; *pak3* acts downstream of proneural genes, withdrawing progenitors from the cell-cycle and promoting neuronal differentiation; *myt1* encodes a zinc-finger TF, positive regulator of neurogenesis; *tubb2b*, terminal neuronal differentiation marker; *sox2* and *sox3*, neural plate markers.

Gene	Expression Role/Details	Gain-of-Function	Loss-of-Function (Dominant-Negative/Morpholino/Antagonist)
notch1/psen1/RBPJ	*notch1* is expressed by most or all cells in the proneural domains since late gastrula (Chitnis et al., 1995). Inhibits PN in neural plate and placodes. RBPJ directly binds the RAM23 region of NICD1 (Wettstein et al., 1997), and this complex activates transcription of Notch target genes.	*notch1ΔE* or *NICD1*: ↓PN (neural plate and placodes); *NICD1*: ↓*dll1, neurog2, neurod4, myt1, pak3* (Chitnis et al., 1995; Ma et al., 1996; Bellefroid et al., 1996; Perron et al., 1999; Souopgui et al., 2002; Riddiford and Schlosser, 2017). *NICD1*: blocked ectopic neurogenesis induced by *neurod4, ascl2*, or *neurog2*; unable to block ectopic neurogenesis induced by *neurod1* or by co-injection of either proneural gene together with *myt1* (Chitnis and Kintner, 1996; Ma et al., 1996; Bellefroid et al., 1996; Perron et al., 1999). *notch1ΔE*: blocked terminal neuronal differentiation induced by *neurog2* but not by *neurod1* (Olson et al., 1998). *RBPJ* overexpression: weakly ↓PN (Wettstein et al., 1997). *psen1* overexpression:↑PN (Paganelli et al., 2001). See *hes5.1, hes5.4, hes6.1* in Table 7.3.	*notch1* MO: ↑PN (NF15, ISH) (López et al., 2003). *RBPJ*^{DBM}: ↑density of *neurog2*+, *dll1*+ cells, and primary neurons; reversed the inhibition of PN produced by *dll1* overexpression (Wettstein et al., 1997). DAPT (γ-secretase inhibitor): ↑PN (Revinski et al., 2010). See *hes5.1, hes6.1* in Table 8.3.
dll1	"Salt-and-pepper" expression begins at late gastrula, restricted to a subset of cells in the proneural domains, scattered within the deep layer of the neural ectoderm, preceding the onset of *neurod1* and *tubb2b*. Marks future neurons around the time they are committed to a neuronal fate (Chitnis et al., 1995; Ma et al., 1996; Chalmers et al., 2002). Within the proneural domains, Dll1 inhibits PN in neighboring cells by lateral inhibition, involving down-regulation of *dll1* (but not of *jag1*) in the receiving cells. Not involved through lateral inhibition in delaying or restricting anterior neurogenesis at neural plate stage.	*dll1* overexpression: ↓PN (Chitnis et al., 1995) without affecting *jag1* (Kiyota et al., 2001); ↓*myt1* (Bellefroid et al., 1996). See *hes5.1* in Table 7.3.	*dll1STU*: ↑PN even when proliferation was blocked with HUA; effect confined to proneural domains, reversed by *dll1* overexpression (Chitnis et al., 1995). ↑density of *neurog2*+, *myt1*+ cells in the proneural domains (Ma et al., 1996; Bellefroid et al., 1996). ↑density of supernumerary neurons induced by *ebf2* overexpression (Dubois et al., 1998). *dll1* MO: ↑PN (Nichane et al., 2008b). *dll1STU*: unable to promote neurogenesis in neuralized animal caps of anterior character overexpressing *ascl2* (NF16) (Papalopulu and Kintner, 1996). Unable to rescue the inhibition of neurogenesis in the trigeminal placode produced by *rax* overexpression (whole embryos) (Andreazzoli et al., 2003).

dlc	Expression in the pair of medial stripes of PN (Peres and Durston, 2006). In the neural plate, *dlc* is necessary for terminal neuronal differentiation in the medial stripe of PN.		*dlc MO:* ↓primary neurons (ISH *tubb2b*, NF21) without affecting *neurog2* in the medial stripe (ISH, NF21) (Peres and Durston, 2006).
jag1	Expressed in the ANB, in a pair of transversal, bilateral stripes corresponding to the future mesencephalon, trigeminal placodes, and two pairs of bilateral AP stripes in the posterior neural plate, one presumably corresponding to the developing intermediate neurons and the other in-between the medial and intermediate *dll1* stripes (Kiyota et al., 2001). *jag1* inhibits PN. This does not rely on *dll1* down-regulation.	*jag1* overexpression*:* ↓PN without affecting *dll1*; prevented the increase of PN induced by *dll1STU* (Kiyota et al., 2001). *jag1ICD:* translocated to the nucleus, ↓PN (Kiyota and Kinoshita, 2004). See *hes5.1* in Table 7.3.	*DN-jag1* (lacking the intracellular domain): ↑PN; this was rescued by *dll1* in a dose-dependent manner (Kiyota et al., 2001).

neurogenesis through Notch-canonical trans-signaling, involving *hes5.1* activation in receiving cells, and through Jag1-ICD cis-signaling, perhaps acting as a transcriptional regulator not involving *hes5.1* (Kiyota et al., 2001; Kiyota and Kinoshita, 2004).

In the posterior neural plate, another member of the *delta* gene family, *dlc*, which is only expressed by the medial stripes (Figure 7.1C), appears to be necessary for terminal differentiation of primary neurons (Peres and Durston, 2006) (Table 7.7).

7.3.4.2. Which *hes1–7* Genes Are Involved in Primary Neurogenesis?

There is a synexpression group transcribed in primary neurogenesis domains, including *hes6.1* and multiple *hes5* genes (Figure 7.2). Of these, *hes5.1*, *hes5.2*, and *hes5.4–5.7* are positively regulated by Notch signaling in some cases through paired RBPJ binding sites (Tables 7.2, 7.3). *hes5.5* needs direct additional input from proneural bHLH factors, whereas *hes5.1* is indirectly up-regulated by them. Therefore, Notch/RBPJ signaling is necessary for the expression of *hes5* genes in proneural domains *in vivo*, but they also require regulation by additional inputs. In contrast, *hes4*, which is expressed in other domains and is directly regulated by Notch through RBPJ binding sites, was not up-regulated by *neurog2* overexpression (Lamar and Kintner, 2005).

So far, gain- and loss-of-function experiments show that *hes2*, *hes4*, *hes5.1*, *hes5.4*, *hes1*, *hes5.5*, *hes5.6*, and *hey1* are able to suppress primary neurogenesis but in different domains of the neural plate (Table 7.8). *hes1*, *hes2*, and *hes4* are involved in the development of the neural border and/or its descendants (see subsequently) and, as well as *hey1*, they are up-regulated by Notch (Tables 7.2, 7.3). *hes4* and *hey1* are expressed in the midline of the neural plate (future floor plate), while *hes2* is expressed in the superficial layer of the intermediate and lateral primary neuron stripes (Figure 7.2), where neural precursors continue to proliferate (Sölter et al., 2006).

hes6.1 is expressed in scattered cells in the medial and lateral primary neuron domains (Figure 7.2). Interestingly, it is repressed by Notch/RBPJ signaling and is required for expression and activity of *neurog2* and the neuronal determination gene *neurod1*, thus relieving neuronal precursors from Notch-mediated lateral inhibition (Koyano-Nakagawa et al., 2000; Murai et al., 2011) (Tables 7.2, 7.3, 7.8). Moreover, *neurog2* and *neurod1* induced *hes6.1* in the absence of protein synthesis in animal caps (Seo et al., 2007), suggesting they directly activate *hes6.1*, establishing a positive feedback loop. It was proposed that Hes6.1 promotes primary neurogenesis through direct protein-protein antagonistic interactions with other Hes factors (e.g. Hes1, Hes4) that inhibit neuronal differentiation and by sequestering TLE/Groucho co-repressors that antagonize bHLH-O Hes proteins that directly repress proneural and neuronal determination genes (Murai et al., 2011).

7.3.4.3. Regulation of the Cell-Cycle

There is evidence that Notch1 might inhibit the withdrawal of neuroblasts from mitosis and prevent their differentiation through the negative regulation of *p21-activated kinase 3 (pak3)* (Souopgui et al., 2002). However, either blocking Dll1 or excessive Notch1/RBPJ signaling inhibited mitosis in the neural plate (Vernon et al., 2006). Experiments with mouse P19 cells and *Xenopus* embryos showed a differential sensitivity of the *dll1* and *neurod1* promoters to the Cdk-dependent phosphorylation status of Neurog2: while the *dll1* promoter can be activated by hypo-phosphorylated Neurog2 (in cells undergoing cycle lengthening) or by phospho-Neurog2 (in rapidly cycling progenitors), the *neurod1* promoter can only be activated by hypo-phosphorylated Neurog2. Hypo-phosphorylated Neurog2 was able and phospho-Neurog2 was unable to promote neuronal differentiation in the presence of NICD1, indicating that the Cdk-dependent Neurog2 phosphorylation status also determines its post-transcriptional sensitivity to Notch signaling (Hindley et al., 2012). Therefore, it was proposed that hypo-phosphorylated Neurog2 shifts the balance from progenitor maintenance to neuronal differentiation. Similarly, studies employing the proneural mouse Ascl1 in *Xenopus* mitotic and interphase egg extracts and *Xenopus* embryos undergoing primary neurogenesis indicate that the Cdk-dependent phosphorylation status of Ascl1 regulates its post-translational sensitivity to Notch signaling. Hypo-phosphorylated Ascl1 probably escapes Notch-mediated lateral inhibition through up-regulation of Myt1 (Ali et al., 2014). Hes1 is phosphorylated by CyclinB/Cdk1 and CyclinA/Cdk2 *in vitro*, suggesting it may be controlled by phosphorylation in the G2/M phase (Hardwick and Philpott, 2015). Phosphorylation by proline-directed kinases destabilizes Hes1 protein and decreases its inhibitory activity on PN *in vivo* (Hardwick and Philpott, 2019).

7.3.5. Neural Plate Border and Midbrain-Hindbrain Boundary

Neural induction subdivides the embryonic ectoderm into neural and non-neural regions, with an intervening transition zone known as the neural plate border (NB) zone. This zone gives rise to neural crest cells (NCCs) and cranial placodes and is positioned by intermediate BMP levels as well as local FGF and Wnt signaling that induce a number of NB specifier genes (Stuhlmiller and García-Castro, 2012; Pla and Monsoro-Burq, 2018; Grocott et al., 2012; Saint-Jeannet and Moody, 2014; Steventon et al., 2014). Members of the Notch pathway and *hes1–7* genes are expressed throughout the development of the NB and its derivatives (Figure 7.2) (Tables 7.9, 7.10). *dll1* is restricted to the NB by the counterbalanced activities of a positive regulator, Irx1, and a negative regulator, Snai1 (Glavic et al., 2004), and *hes4* expression in the NB laterally restricts the neural plate (Maharana and Schlosser, 2018). Interestingly, *hes3* can promote neural plate fate at the expense of NCC and cranial

TABLE 7.8
Role of *hes/hey* Genes in *Xenopus* Neurogenesis and Epidermal Differentiation

See abbreviations in Table 7.3 legend and marker details in Table 7.7 legend.

Gene	Role/Details	Gain-of-Function	Loss-of-Function (Dominant-Negative/Morpholino)
hes1	Inhibits neurogenesis by abolishing the activity of proneural/neuronal differentiation bHLH TFs and keeping neural progenitors in a proliferative state, independently of binding to TLE/Groucho co-repressors. Hes1 binds to N-box *in vitro* (Koyano-Nakagawa et al., 2000) and does not block the ability of Neurog2 to bind DNA or its coactivator E12 *in vitro* (Murai et al., 2011).	*hes1* overexpression: ↑*sox2* [not shown in (Nichane et al., 2008a)], induced ectopic *dll1* [not shown in (Nichane et al., 2008b)], ↓PN by abolishing *Neurog2* and *Neurod1* activity (Murai et al., 2011). mir-9 MO and *hes1* target protector MO: ↓neurogenesis (NF30), ↑proliferation (Bonev et al., 2011).	*hes1*-ΔWRPW: did not suppress PN (Murai et al., 2011).
hes2	Inhibits PN by repressing proneural gene transcription and other mechanisms such as binding to proneural bHLH proteins like NeuroD. In the retina, *hes2* promotes gliogenesis and inhibits neurogenesis through a DNA-binding mechanism. Immunoprecipitation in animal cap explants: XHes2 interacts with XNeuroD1, XNeuroD4, XHes1, and XHes6 but not with other bHLH proteins tested, including XNeurog2 (Sölter et al., 2006).	*hes2:* mRNA overexpression: ↓PN, ↓ *neurog2* (Sölter et al., 2006). Local overexpression by *in vivo* DNA lipofection in the developing retina: ↑glial population at the expense of neurons (Sölter et al., 2006).	*Hes2-ΔW-VP16:* mRNA: induced ectopic primary sensory neurons and *neurog2* (Sölter et al., 2006). *In vivo* lipofection of Hes2-ΔW-VP16 DNA: ↓glial population, ↑some neuronal types (Sölter et al., 2006). *hes2DBM:* did not affect retinal fates; DNA binding domain necessary for promoting gliogenesis in the retina (Sölter et al., 2006). *hes2 MO:* ↑*neurog2* in presumptive otic placode. *In vivo* lipofection of *hes2* MOs ↓glial population in the retina (Sölter et al., 2006).
hes4	Inhibits PN. Hes4 protein binds to N-box *in vitro* (electrophoretic mobility shift assay) (Koyano-Nakagawa et al., 2000).	*hes4* overexpression: ↓PN (*tubb2b*, neural plate stage) (Glavic et al., 2004) (Cui, 2005); ↓*neurog2* and *neurod1* in the anterior neural plate (Andreazzoli et al., 2003; Cui, 2005); ↑*dll1* independently of DNA binding (neural plate stage) (Nichane et al., 2008b); blocked the induction of *tubb2b* by *neurog2/noggin* in animal caps (NF unknown, RT-PCR) (Cui, 2005). *hGR-hes4* (Dex NF18): ↓PN (ISH *tubb2b*, neurula stage) (Taelman et al., 2006).	*hes4* MO: ↑ *neurog2, myt1*, and *dll1;* ↑PN in the trigeminal placode (early neurula) (Nagatomo and Hashimoto, 2007; Murato and Hashimoto, 2009); ↓proliferation, ↑apoptosis (late neurula) (Nagatomo and Hashimoto, 2007); ↓*dll1* (neural plate stage) (Nichane et al., 2008b).
hes5.1	Inhibits PN probably by repressing bHLH proneural genes through DNA binding and recruitment of the co-repressor TLE/Groucho. The DNA binding domain and the WRPW motif are necessary for inhibiting PN and preventing ectopic neurogenesis induced by the bHLH TF Atoh7 (Schneider et al., 2001).	*hes5.1* overexpression: ↓PN, even in the presence of the bHLH TF Atoh7 (ISH, neural plate stage) (Schneider et al., 2001).	*hes5.1 ΔWRPW:* ↑PN [data not shown in (Ito et al., 2007a)]. *hes5.1 ΔWRPW* or *hes5.1 ΔN* (basic domain deleted): did not prevent ectopic PN induced by the bHLH TF Atoh7 (ISH, neural plate stage) (Schneider et al., 2001).
hes5.4	During placodal development, *hes5.4* restricts the neurogenesis cascade upstream of proneural genes. Required for the expression of the neural progenitor gene *sox3* and neuronal differentiation downstream of proneural genes.	*hes5.4* overexpression: ↓*neurog1, neurog2, neurod1*, and *tubb2b* in the neural plate and placodes; ↓ neural marker *sox3* in placodes (ISH, neural plate) (Riddiford and Schlosser, 2017).	*hes5.4* MO: ↑*hes5.4, neurog1, neurog2, dll1*, and *pou4f1.2* placodal domains with occasional reductions of these markers; ↓placodal *sox3, neurod1*, and *tubb2b* (Riddiford and Schlosser, 2017).

(Continued)

TABLE 7.8 (Continued)
Role of *hes/hey* Genes in *Xenopus* Neurogenesis and Epidermal Differentiation

Gene	Role/Details	Gain-of-Function	Loss-of-Function (Dominant-Negative/Morpholino)
hes5.5	Inhibition of PN. Does not interact *in vitro* with Hey1 (Taelman et al., 2004).	GR-*hes5.5* (Dex NF12 or NF18): ↓PN (ISH neurula stage) (Taelman et al., 2004; Taelman et al., 2006).	
hes5.6	Inhibition of PN. Does not form homodimers *in vivo* nor heterodimerizes with other bHLH-O proteins (Hes1, Hes4, Hes5.6, Hey1). Heterodimerizes *in vivo* with bHLH proteins that promote neurogenesis (Neurog2, Neurod1, Neurod4) (IP assay from *Xenopus* embryo extracts) (Taelman et al., 2004).	GR-*hes5.6* (Dex NF12 or NF18): ↓PN (ISH neurula stage) (Taelman et al., 2004; Taelman et al., 2006).	
hes5.10	Inhibits differentiation of epidermal multiciliate cells in the inner layer of the non-neural ectoderm through a Dll1/Notch-dependent lateral inhibition mechanism. Possibly participates in the inhibition of neurogenesis in the outer layer of the ectoderm through a different mechanism.	*hes5.10* overexpression: ↓density of multiciliate cells in the epidermis (tailbud stage); ↓*dll1* in animal caps (mid-gastrula) (Deblandre et al., 1999); ↓*neurod1* and PN without affecting *neurog2* or neural specification; acts downstream of *neurod1*, because it prevented ectopic neurogenesis but not ectopic *neurod1* induced by *neurog2* (Chalmers et al., 2002).	*hes5.10DBM:* ↑density of multiciliate cells in the epidermis (tailbud stage), ↑*dll1* in animal caps (mid-gastrula) (Deblandre et al., 1999). *hes5.10* MO: unable to induce ectopic PN in the deep layer (Chalmers et al., 2002).
hes6.1	*hes6.1* promotes PN by antagonizing other Hes proteins that suppress neuronal differentiation. It does so in a post-transcriptional, TLE/Groucho-independent, and DNA binding-independent way (Koyano-Nakagawa et al., 2000; Murai et al., 2011). However, the full promotion of PN requires TLE/Groucho binding (Murai et al., 2011). See Table 7.4 for the regulation of *hes1* and *hes4* by *hes6.1*.	*hes6.1* overexpression: ↑*neurog2* domains (Koyano-Nakagawa et al., 2000) and primary neuron differentiation (Cossins et al., 2002) throughout the posterior neural plate; ↑*dll1*, *hes5.1*, and *hes5.7* domains, indicating that the increase in neuronal differentiation does not involve transcriptional repression of genes of the lateral inhibition program (Koyano-Nakagawa et al., 2000). Hes6.1 did not affect the binding of Neurog2 to its E12 coactivator during *in vitro* binding to E-box. Hes6.1 did not directly interact with Neurog2 (immunoprecipitation assay). Hes6.1 directly binds and impairs the ability of Hes1 to repress PN through a TLE/Groucho-independent mechanism (Murai et al., 2011).	*hes6.1* MO: ↓*tubb2b*, *neurog2*, and *neurod1* (neural plate stage); prevented the induction of ectopic neurons by *neurog2* or *neurod1* (Murai et al., 2011). *hes6.1* ΔWRPW: did not rescue the inhibition of PN produced by *hes6.1* MO (Murai et al., 2011). *hes6.1DBM:* promoted PN like *hes6.1* overexpression, indicating that DNA binding is not required for this activity (Cossins et al., 2002).
hey1	Possible role in suppressing neurogenesis in the FP by antagonizing proneural genes, contributing to promote or maintain FP identity. Additional inhibitors might be required to maintain neurogenesis inhibited in FP. Hey1 does not bind the co-repressor TLE/Groucho, as it lacks the typical WRPW motif of bHLH-O repressors (Pichon et al., 2004). For inhibition of PN, Hey1 acts as a DNA binding repressor, requiring the Orange domain and the C-terminal region for dimerization. Hey1 heterodimerizes *in vivo* with Hes1 and Hes4 and weakly with Hes2, but it does not bind Hes5.5 or Hes5.6. It weakly binds bHLH proteins that promote neuronal differentiation (Neurod1 and Neurod4) but does not	*hey1* or GR-*hey1* (Dex NF12 or NF18): ↓PN (Taelman et al., 2004; Taelman et al., 2006). *hey1* blocked *neurog2*'s ability to induce ectopic neurogenesis (Taelman et al., 2004).	*hey1DBM:* did not inhibit PN and could not block *neurog2*'s ability to induce ectopic neurogenesis (Taelman et al., 2004). *hey1* MO +/- *hes4.L* MO: did not induce ectopic neurogenesis in the FP. (Taelman et al., 2004). Note: MO sequences were not reported in this study. *hey1.L+S* MO: *tubb2b* unaffected (ISH in tailbuds) (Taelman et al., 2006).

placodes, suggesting it also participates in setting the neural plate/NB boundary, although this could not be confirmed by *hes3* knock-down, perhaps due to compensation by other *hes* genes (Hong and Saint-Jeannet, 2018).

7.3.5.1. The Role of *hes4* in NB and NCC Development

hes4 is expressed in prospective NCC territories, along with *notch1* (Glavic et al., 2004; Vega-López et al., 2015), and appears to act through partially counteracting pathways to promote and restrict *foxd3* expression to NCC (Maharana and Schlosser, 2018).

Neural induction subdivides the ectoderm into neural and non-neural regions. The transition zone is the neural border, from which neural crest cells, placodes, the bordering non-neural ectoderm, and dorsal neural tube segregate (Stuhlmiller and García-Castro, 2012; Pla and Monsoro-Burq, 2018). The pre-placodal ectoderm (PPE) forms a horseshoe-shaped domain surrounding the neural plate at its anterior end and later segregates into individual placodes (Stuhlmiller and García-Castro, 2012; Grocott et al., 2012; Saint-Jeannet and Moody, 2014; Steventon et al., 2014; Pla and Monsoro-Burq, 2018). NCC development begins during gastrulation, when the NB is induced and stabilized and progresses through sequential steps, with the NB-TFs controlling the onset of NCC induction within the NB at the end of gastrulation and during the neural plate stage. This is followed by NCC specification, which occurs through a multistep process during neurulation and involving the activation of a new set of genes encoding specific NCC transcription factors (NCC-TFs), which are not shared with contiguous populations. These NCC-TFs are activated in a stereotyped sequence, with an early step (activation of *sox9* since NF11; *snai2* and *foxd3* since NF12–12.5) and a maturation step, with the activation of late NCC-TFs (*sox10* from NF13–14, *twist1*). NCC migration begins at the end of neurulation (once the neural folds fuse at the midline, transforming the neural plate into the neural tube) and continues during organogenesis, during which post-migratory NCCs colonize target tissues and organs, where they differentiate into multiple cell types (Pegoraro and Monsoro-Burq, 2013; Pla and Monsoro-Burq, 2018).

There are conflicting interpretations concerning the role of *hes4* in NB/NCC development and its regulation by Notch signaling in these tissues (Tables 7.9, 7.10). Although *hes4* is already expressed in the NB at mid-gastrula (Tsuji et al., 2003) and is considered an NB specifier *in vivo*, this TF alone can not initiate NCC specification in animal cap explants (Milet et al., 2013). Analysis at advanced neurula stages indicated that Dll1 signals to presumptive NCC through Notch1/RBPJ inducing *hes4*, which represses *bmp4* to ensure optimal BMP signaling levels for NCC specification (Glavic et al., 2004). Others showed that during NCC specification, *hes4* restricts *dll1* in NCC for the survival and maintenance of precursors in a mitotic, undifferentiated state (Nagatomo and Hashimoto, 2007). Strikingly, another group showed that the NCC *hes4* domain was unaffected in neurulae following activation or blockade of the RBPJ-dependent pathway at mid-gastrula and noticed that *hes4* and *dll1* domains partially overlap in the anterior, lateral neural plate. They proposed that Dll1 favors NCC precursor proliferation rather than controlling the balance between primary neurons and NCC fates. Through a DNA-binding independent mechanism, *hes4* transiently and indirectly induces *dll1* expression, leading to NCC proliferation and differentiation. Through a cell-autonomous, DNA-binding-dependent mechanism, *hes4* up-regulates early NB genes and is required for NCC survival and maintenance in an undifferentiated state (Nichane et al., 2008a; Nichane et al., 2008b). More recently, other authors showed that in the pre-migratory NCC territory, *hes1* and *hes4* are positively regulated by Dll1/Notch/RBPJ whereas BMP down-regulates *hes4* (Nagatomo and Hashimoto, 2007; Vega-López et al., 2015). These authors propose that *hes1* and *hes4* are required for several processes during NCC development. First, both promote NCC specification at the expense of neural plate and epidermis independent of cell proliferation; then Hes4 acts as a transcriptional repressor during NCC specification and is later required for their survival during neurulation; finally, *hes4* is required cell-autonomously to initiate NCC migration and their differentiation into the cranial skeleton.

Wnt and FGF signaling are necessary for *hes4* expression, whereas BMP down-regulates it in the presumptive NCC at neural plate stages (Nichane et al., 2008a; Vega-López et al., 2015). Others identified three *hes4* expression phases during NB/NCC development. First, during NB induction (early gastrula), *hes4* expression is insensitive to BMP signaling but requires down-regulation of the Wnt pathway. Then, during NCC induction (mid-gastrula), it requires Wnt and down-regulation of BMP signaling. Finally, both pathways are required for *hes4* expression during NCC maintenance in early neurula (Steventon and Mayor, 2012). FGF signaling was proposed to regulate *hes4* and *dll1* in the NB through Stat3.1, which is phosphorylated by FGF/FGFR4. Whereas low Stat3.1 activity up-regulates *hes4*, high Stat3.1 activity promotes *dll1* expression and Dll1/Notch signaling (Nichane et al., 2010). Overall, it is clear that *hes4* is required for NCC development in *Xenopus*, but controversies still exist about the underlying mechanisms (for discussion, see Vega-López et al., 2015). Time-dependent opposite responses to the same experimental Notch perturbation, which were observed in other contexts in *Xenopus* (Contakos et al., 2005; Revinski et al., 2010), might underlie the conflicting results between different studies. Some reviews regard Notch signaling as an important source for NCC maintenance rather than as a key player in NB induction during gastrulation (Stuhlmiller and García-Castro, 2012; Pla and Monsoro-Burq, 2018). Curiously, most studies analyzed the effects of perturbing Notch signaling on *hes4* too late to address if this pathway plays an early role in establishing the NB *hes4* domain (Table 7.9). However, we observed a clear expansion of the NB *hes4* domain at mid-gastrula after constitutive Notch1 activation beginning at cleavage stages (López et al., 2005), suggesting that Notch participates in the establishment of

TABLE 7.9
Core Components of the Notch Pathway in *Xenopus* Neural Border and Its Descendants

krt12.4, non-neural ectoderm marker; *nkx1–2*, posterior NB MARKER (Kurata and Ueno, 2003); *sox2*, *sox3*, *soxd*, neural plate markers; *foxe3*, positive regulator of lens fate (Ogino et al., 2008); *pax3*, *msx1*, *zic1*, NB markers (Nichane et al., 2008b). For additional abbreviations, see main text and Table 7.3 legend.

Gene	Expression	Gain-of-Function	Loss-of-Function (Dominant-Negative/Morpholino)
notch1/RBPJ	*notch1* expression: NB, neural plate, and ectoderm (Coffman et al., 1993). Prospective NCC territories (neural plate stage) (Glavic et al., 2004).	**NB, NCC:** *notch1ΔE:* ↓*twist1, nkx1–2*, non-neural ectoderm (*krt12.4*) (neural plate stage, ISH) (Coffman et al., 1993; Kurata and Ueno, 2003); ↓branchial arches (tadpole) (Coffman et al., 1993). *NICD1:* ↑NB marker ***hes4*** (gastrula, ISH) (López et al., 2005), neural plate marker *sox2* (neural plate stage, ISH) (Revinski et al., 2010). *GR-NICD1* or *GR-RBPJ-Ank* (Dex NF12): ↑presumptive NCC (*snai2, foxd3,* ***hes4***); ↓*bmp4* (non-neural ectoderm) (late neurula) (Glavic et al., 2004). *GR-RBPJ-Ank* (Dex NF11): ***hes4***NCC domain unaffected (neural plate stage, ISH) (Nichane et al., 2008a). **PPE:** *NICD1:* ↑*hes5.4* in placodes and adjacent non-neural ectoderm, even in the absence of *six1/eya1* function (ISH, neural plate stage) (Riddiford and Schlosser, 2017). **Lens placode (*notch1 /notch2* ?):** GR-*RBPJ*-VP16 (Dex NF15): ectopic *foxe3* restricted to the anterior ectoderm (NF22–24) (Ogino et al., 2008). GR-*RBPJ*-VP16 or *NICD1* + GR-*otx2* (Dex NF15): massive ectopic *foxe3* (NF22–24) (Ogino et al., 2008).	**NB, NCC:** *RBPJ*^{DBM}*:* ↓neural plate (*sox2*) (neural plate stage, ISH) (Revinski et al., 2010). *notch1* MO: ↓presumptive neural plate (*sox2*) (late gastrula); ↑neural plate (*sox2*) (neural plate stage) (ISH) (Revinski et al., 2010). *GR-RBPJDBM* (Dex NF12): ↓presumptive NCC (*snai2, foxd3,* ***hes4***); ↑*bmp4* domain (late neurula) (Glavic et al., 2004). GR-RBPJDBM (Dex NF11): ***hes4***NCC domain unaffected (neural plate stage, ISH) (Nichane et al., 2008a) **PPE:** *RBPJDBM:* ↓*hes5.4* in neural plate and placodes (ISH, neural plate stage) (Riddiford and Schlosser, 2017). **Lens placode (*notch1 /notch2* ?):** GR-*RBPJ*DBM/Dex NF15: ↓*foxe3*, impaired lens development (NF22–24) (Ogino et al., 2008).
notch2	PPE (neural plate stage) (Murato and Hashimoto, 2009; Maharana and Schlosser, 2018).		
dll1	— Anterior neural border (ANB), surrounding the presumptive NCC at early neurula stage. (Glavic et al., 2004).	**NB, NCC:** *dll1* overexpression*:* ↑NCC (*snai2, sox10*) without affecting NB (*pax3, msx1*), neural plate (*soxd, sox3*), or placodal marker *six1* (Nichane et al., 2008b).	**NB, NCC:** *dll1STU:* ↓NCC (*snai2, foxd3,* ***hes4***); ↑*bmp4* (late neurula) (Glavic et al., 2004); ↓neural plate (*sox2*) (late gastrula, early neurula; ISH) (Revinski et al., 2010).. *dll1* MO: ↓NCC (*snai2, sox10*) (neural plate stage) (Nichane et al., 2008b). **Lens placode:** *dll1STU:* Head defects, ↓*foxe3* (Ogino et al., 2008)
dlc	Similar to *dll1* but at lower levels. Expressed in an arc corresponding to the ANB; later resolves into *dlc*+ cranial placodes (Glavic et al., 2004; Peres and Durston, 2006).		**NCC, placodes:** *dlc* MO*:* migration failure of NCC and placodal cells (Peres and Durston, 2006). *dlctr:* ↓*foxe3* (NF23) (Ogino et al., 2008).
jag1	ANB, surrounding the presumptive NCC at early neurula stage (Glavic et al., 2004).		

the NB through the positive regulation of the NB-specifier *hes4*, which receives multiple regulatory inputs from other pathways. Moreover, Notch1 can suppress *nkx1–2* in presumptive NCC; this gene encodes a transcriptional repressor thought to inhibit neural fate to allow NCC induction (Kurata and Ueno, 2003).

7.3.5.2. The Role of *hes* Genes in PPE Development

Notch signaling and genes of the *hes1–7* group are also involved in cranial placode development (Tables 7.9, 7.10). Six1 and its co-activator Eya1 are crucial regulators of placode development (Brugmann et al., 2004; Riddiford and Schlosser, 2016). *hes4* is required for establishing the preplacodal ectoderm; the expression of *notch2*, *six1*, and *eya1* in this tissue; and the development of the lens field (Murato and Hashimoto, 2009; Maharana and Schlosser, 2018) (Table 7.10), placing *hes4* upstream of the placodal program probably at the level of NB establishment. However, additional gene cascades converge in setting this program, since other PPE markers were not affected by *hes4* knock-down (Murato and Hashimoto, 2009). Recently, a NB gene regulatory network that cross-regulates with Six1/Eya1 was proposed for controlling PPE and NCC specification. This GRN includes Hes4 and other TFs expressed in the neural and non-neural ectoderm (Maharana and Schlosser, 2018). Downstream of this GRN are *hes2*, *hes5.4*, and *hes5.6*, which are expressed in the PPE. They are presumptive direct targets of Six1/Eya1, as they were up-regulated by them in the absence of protein synthesis, and placodal *hes5.4* and *hes5.6* expression require Six1/Eya1 function (Riddiford and Schlosser, 2016).

In neurogenic placodes, Six1/Eya1 control *dll1* in a dose-dependent manner. High Six1/Eya1 levels maintain proliferating placodal precursors, but as cells delaminate from the placodes, Six/Eya1 levels are reduced and the neurogenesis program is triggered, including the onset of *dll1* expression (Schlosser et al., 2008). Notch1/RBPJ is required for *hes5.4* expression during placodal development (Tables 7.2, 7.3) (Riddiford and Schlosser, 2016). The requirement of Notch signaling for *hes2* and *hes5.6* expression in this process is currently unknown, although *hes5.6* is induced by Notch1/RBPJ during gastrulation (Tables 7.2, 7.3). *hes2* was not induced by activation of the Notch pathway in naive or neuralized animal cap explants but was moderately induced in embryos, with ectopic expression restricted to the NB (Sölter et al., 2006) (Tables 7.2, 7.3).

PPE *hes5.4* expression requires two positive regulators: (1) high Six1/Eya1 levels activate *hes5.4* independently of Notch/RBPJ and (2) Notch/RBPJ activates *hes5.4* independently of Six1/Eya1, probably through lateral inhibition (Tables 7.2, 7.3, 7.9). Subsequently, *hes5.4* maintains placodal progenitors in an undifferentiated state, restricting primary neurogenesis upstream of proneural genes (Table 7.8). As Six1/Eya1 activity declines, *hes5.4* is required at low levels to promote neuronal differentiation. Intriguingly, *hes5.4* is also required for neuronal differentiation downstream of proneural genes, as terminal differentiation markers were frequently decreased after *hes5.4* knock-down (Table 7.8). The details of the mechanism underlying these opposing roles remain unresolved, but oscillation of *hes5.4* expression might be involved since *hes5.4* represses its own transcription (Tables 7.4, 7.8) (Riddiford and Schlosser, 2017).

7.3.5.3. Other Roles for Notch Pathway in Placode Development

Other evidence further supports Notch pathway involvement in cranial placode development (Table 7.9). Potentiated by Otx2, Notch can activate *dmrta1/2* in the anterior ectoderm. These genes encode TFs expressed in the presumptive olfactory placodes and are involved in olfactory neurogenesis (Parlier et al., 2013). An RBPJ binding site was found in the main enhancer of *foxe3*, a key TF required for lens placode development (Ogino et al., 2008; Kenyon et al., 1999). *dll1* and *dlc* are expressed in the adjacent presumptive retina, from where they presumably induce *foxe3* through Notch/RBPJ. While an antimorphic *dll1* produced severe head defects, making the results difficult to interpret, the antimorphic *dlc* produced a more restricted, lens-defective phenotype, indicating that *dlc* is involved in lens development (Ogino et al., 2008).

7.3.5.4. The Midbrain/Hindbrain Boundary

The midbrain/hindbrain boundary (MHB) is considered an organizing center because signals from this region induce and pattern the adjacent mesencephalon and hindbrain (Anderson and Stern, 2016). *hes7.1* is one of the first genes to demarcate the presumptive MHB at early gastrula stages. Notably, at neural plate stages, the *hes7.1* domain coincides with a *hes5.1*, *hes5.2*, and *hes5.7* expression gap (Shinga et al., 2001; Takada et al., 2005) (Figure 7.2) (Table 7.11). *hes7.1* is necessary for MHB establishment through repressing, probably directly, *hes5.1*, *hes5.2*, and *hes5.7* in this region, whereas these *hes5* genes (which are positively regulated by Dll/Notch signaling during primary neurogenesis) are thought to restrict *hes7.1* to the MHB (Shinga et al., 2001; Takada et al., 2005) (Tables 7.4, 7.11). Strikingly, NICD1 abolished and *RBPJDBM* did not affect MHB *hes7.1* expression (Takada et al., 2005). It would be interesting to address whether *notch1* normally down-regulates *hes7.1* by a non-canonical pathway.

7.3.6. Somitogenesis

The classic Clock and Wavefront hypothesis for vertebrate somitogenesis, which involves Notch signaling, was originally postulated by experimental work based on *Xenopus* (Cooke and Zeeman, 1976; Cooke, 1981). It explains the sequential formation of vertebrate somites from the posterior presomitic mesoderm that is due to an oscillation between permissive and non-permissive phases for segmentation, the so-called "segmentation clock" that is controlled by *hes* genes. Their proteins act as pacemakers that cell-autonomously cycle on and off through an autoregulatory

TABLE 7.10
***hes* Genes in the *Xenopus* Neural Border and Its Descendants**

For the role of *hes* genes in placodal neurogenesis, see *hes2*, *hes5.4* in Table 7.8. See abbreviations in Table 7.3 legend.

Gene	Expression/Role	Gain-of-Function	Loss-of-Function (Dominant-Negative/Morpholino)
hes1	At neural plate stage, *hes1* is expressed in thin longitudinal lines corresponding to the dorsal and ventral boundaries of NCC (Vega-López et al., 2015). Promotes NCC at the expense of contiguous fates (neural plate/epidermis), independently of cell proliferation during early NCC development (Vega-López et al., 2015)	GR-*hes1* (Dex s11): ↑NCC (*snai2*), ↓neural plate (*sox2*) and non-neural ectoderm (*krt12.4*) (ISH neurula); proliferation unaffected (pH3, neurula stage) (Vega-López et al., 2015). GR-*hes1* (Dex s11) +/– HUA: ↑NCC (*foxd3*), ↓neural plate (*sox2*) (ISH neurula) (Vega-López et al., 2015).	GR-*hes1* (Dex s11): ↓NCC (*snai2*), ↑neural plate (*sox2*) and non-neural ectoderm (*krt12.4*) (ISH neurula); proliferation unaffected (pH3, neurula stage) (Vega-López et al., 2015). GR-DN-*hes1* (Dex s11) +/– HUA: ↓NCC (*foxd3*), ↑neural plate (*sox2*) (ISH neurula) (Vega-López et al., 2015).
hes2	Scattered cells in the ectoderm/neural ectoderm (mid-gastrula) Posterior PPE, including the prospective otic and lateral line placodes (neural plate) (Sölter et al., 2006; Riddiford and Schlosser, 2016).		
hes3	Strongly expressed in a domain comprising the anterior and lateral edges of the neural plate, just adjacent to the NB (Hong and Saint-Jeannet, 2018). Sufficient to promote neural plate fates at the expense of NB fates (NCC and placodes) but normal requirement still unproven. Other genes of the Hes family might compensate for *hes3* knock-down.	GR-*hes3* (Dex NF10.5): Promoted neural plate at the expense of NB fates. Blocks NCC induction by *wnt8a* DNA or β-catenin DNA injection. In animal cap assays, blocked NCC and PPE gene induction by Wnt8 and Noggin and promoted neural plate fate (Hong and Saint-Jeannet, 2018).	*hes3* MO: two translation-blocking MOs targeting *hes3.L*, one translation-blocking MO targeting *hes3.S*, and one splice-blocking MO targeting *hes3.L* and *hes3.S*, tested alone or in different combinations, did not affect the expression of gene markers that were affected by the gain of function approach (Hong and Saint-Jeannet, 2018).
hes4	NB at mid-gastrula (Tsuji et al., 2003). One of the earliest PPE markers (mid-gastrula) (Murato and Hashimoto, 2009; Maharana and Schlosser, 2018). Prospective NCC territories (neural plate stage) (Vega-López et al., 2015) *hes4* is required during several steps of NCC development (specification, maintenance, onset of migration). Only the *hes4.L* homeolog is required in *X. laevis*. *hes4* is required for the establishment of the PPE and the development of the lens field.	*hes4* overexpression: ↓*bmp4*, ↑NCC (*snai2, msx1*) (neural plate stage) (Glavic et al., 2004) GR-*hes4* (Dex NF11.5–12): ↓NCC (*snai2, foxd3, sox9, sox10*), ↑neural plate (*soxd*), ↑NB (*msx1, pax3, zic1*), ↑*dll1*,↑proliferation (pH3), ↓apoptosis (neural plate stage); ↑ and ↓ different subsets of NCC cell-types (tailbud stage) (Nichane et al., 2008b; Nichane et al., 2008a). GR-*hes4.S*, GR-*hes4.L* (Dex NF11) +/– HUA treatments, pH3; GR-*hes4.S*-EnR, GR-*hes4.L*-EnR (Dex NF12.5): Hes4 acts as a transcriptional repressor, promoting NCC specification at the expense of contiguous fates (neural plate/epider	*hes4* MO: ↓NCC markers without affecting neural plate and epidermal markers (early neurula) (Nagatomo and Hashimoto, 2007) or with ↑ of neural plate and epidermal markers (Vega-López et al., 2015). Variable changes in the expression domains of the NCC specifier *foxd3*, with slight ↑*sox3* (neural plate) (Maharana and Schlosser, 2018). ↓*six1/eya1*, *notch2* (PPE) (neural plate stage); other PPE markers unaffected (*foxe3, dlx5*) (Maharana and Schlosser, 2018; Murato and Hashimoto, 2009); ↓lens field (*pax6, six3, pitx1*); malformed lens (tadpoles, NF42) (Murato and

		mis) independently of cell proliferation. It is later required for NCC survival during neurulation, and cell autonomously for their initial migration (Vega-López et al., 2015).	Hashimoto, 2009). In these studies, either a mixture of *hes4.L MO* + *hes4.S* MO (Nagatomo and Hashimoto, 2007; Nichane et al., 2008b; Maharana and Schlosser, 2018), *hes4.L* MO alone (Vega-López et al., 2015; Murato et al., 2007), or *hes4.S* MO alone (Murato et al., 2007) were used. *hes4.L* MO but not *hes4.S* MO affected NCC development (Murato et al., 2007). GR-DN-*hes4.S*, GR-DN-*hes4.L* (Dex NF11) +/− HUA, pH3; GR-*hes4.S*-E1A, *GR-hes4.L-E1A* (Dex NF12.5): Hes4 acts as a transcriptional repressor, promoting NCC specification at neural plate/epidermis fates' expense, independently of cell proliferation. It is later required for NCC survival during neurulation and cell-autonomously for their initial migration (Vega-López et al., 2015). GR-*hes4DBM* (Dex NF11.5–12): same results as GR-*hes4* (Dex NF11.5–12), except that it did not affect *soxd* and *msx1* (Nichane et al., 2008b).
hes5.4	Anterior PPE (neural plate stage); otic vesicle; olfactory, epibranchial, and lateral line placodes (tailbud stage) (Riddiford and Schlosser, 2016).	See Table 7.8.	See Table 7.8.
hes5.6	Anterior PPE (neural plate stage); otic vesicle; olfactory, epibranchial, and lateral line placodes (tailbud stage) (Riddiford and Schlosser, 2016).		

TABLE 7.11
hes Genes in the Establishment of the _Xenopus_ Midbrain/Hindbrain Boundary

See abbreviations in Table 7.3 legend.

Gene	Expression Role/Details	Gain-of-Function	Loss-of-Function (Dominant-Negative/Morpholino)
hes5.1	PN domains with a gap at the MHB. Restricts the MHB.	*hes5.1* overexpression: ↓MHB specification (*hes7.1* and *pax2*, ISH neural plate stage) (Takada et al., 2005)-	
hes5.2	PN domains with a gap at the MHB. Restricts the MHB.	*hes5.2* overexpression: ↓MHB specification (*pax2*, ISH at neural plate stage) (Takada et al., 2005).	
hes5.7	PN domains with a gap at the MHB. Restricts the MHB.	*hes5.7* overexpression: ↓MHB specification (*pax2*, ISH at neural plate stage) (Takada et al., 2005).	
hes7.1	Demarcates the presumptive MHB: restricted band of the inner layer in the center of the prospective neuroectoderm (onset: NF10.5) that is progressively resolved into a pair of bilateral stripes (Shinga et al., 2001). Establishes the presumptive MHB region as a prepattern gene, where it represses the putative direct target genes *hes5.1, hes5.2, hes5.7, dll1*, and *hes7.1*.	*hes7.1* overexpression: ↓MHB markers (*pax2, en2*) at neural plate stage, but ↑*en2* in some cases with lower doses, while they were unaffected at later stages (Shinga et al., 2001). ↓*hes5.1, hes5.2, hes5.7, dll1, neurog2* (neural plate stage) (Takada et al., 2005).	*hes7.1-VP16, mb-hes7.1* (basic region mutated), *hes7.1-ΔWRPW*: ↓MHB markers (neural plate stage); morphologically disrupted the MHB (tailbuds) (Shinga et al., 2001). *GR-hes7.1-VP16* (Dex NF10.5–11) +/– CHX: ↑*hes5.1, hes5.2, hes5.7, dll1*, and *hes7.1* domains (neural plate stage) (Takada et al., 2005). *hes7.1* MO: filled the MHB gap with *dll1, hes5.1*, and *hes5.7* (medial stripes of PN) and anteriorly expanded their intermediate stripes (PN domains) (Takada et al., 2005).

negative feedback loop. Cell oscillations slow down towards the anterior presomitic mesoderm, generating a kinematic wave of cycling gene expression. Opposite gradients of RA from the anterior presomitic mesoderm and FGF and Wnt signaling from the tailbud region define a so-called "determination wavefront," which sweeps through the presomitic mesoderm rostro-caudally. When cycling cells at the permissive phase are reached by the determination wavefront, they stop oscillating. This results in the striped activation of *mesp* genes. Consequently, the anterior presomitic mesoderm forms whorls of "somitomeres" whose gene expression prepattern delineates the future boundaries that lead to the formation of individual somites (Figure 7.1D). During vertebrate somitogenesis, intercellular Dll/Notch signaling acts first in the posterior presomitic mesoderm to synchronize the frequencies of neighboring cell-oscillators and then in the anterior presomitic mesoderm to position the future intersomitic boundary and define the anterior-posterior polarity of the somite (Cooke, 1981; Takahashi et al., 2000; Pourquié and Tam, 2001; Moreno and Kintner, 2004; Nagano et al., 2006; Sparrow, 2008; Gomez et al., 2008; Sasaki et al., 2011; Oates et al., 2012; Hubaud and Pourquié, 2014; Wahi et al., 2016; Janesick et al., 2017; Venzin and Oates, 2020; Naoki and Matsui, 2020).

7.3.6.1. Notch Ligands and Hes Genes in Somitogenesis

While the general segmentation mechanism is conserved across vertebrates, the individual *hes* and *delta* oscillating genes involved vary between species (Oates et al., 2012). In *Xenopus*, several Notch pathway genes are expressed in discrete stripes in the presomitic mesoderm. So far, only *dlc, hes5.3, hes5.5*, and *hes5.6* were reported as oscillatory in *Xenopus* (Table 7.12, Figure 7.1D). Since a large number of embryos must be analyzed to discern a changing pattern that results from oscillatory expression (Figure 7.1D, lower panel), some genes with cycling behavior might have been overlooked (Sparrow, 2008). Interestingly, *dlc* delineates somitomeres at late gastrula (Peres et al., 2006), significantly earlier than genes involved in somite segregation (Durston et al., 2018). Notably, *hes4, hes5.3, hes5.6*, and *dlc* show a left-right asynchrony in their somitomeric pattern (Davis et al., 2001; Li et al., 2003; Blewitt, 2009; Durston et al., 2018) (Figure 7.1D, lower panel). A careful examination led to the proposal that somitogenesis waves are propagated as counter-clockwise spirals, probably linked to the mechanisms imposing left-right asymmetries (Durston et al., 2018).

The consequences of the experimental perturbation of Dlc/Notch/RBPJ signaling in *Xenopus* are consistent with a crucial role in regulating somitogenesis (Jen et al., 1997; Sparrow et al., 1998; Peres et al., 2006) (Table 7.12). Although *dll1* is expressed in the tailbud in a poorly described segmental prepattern (Table 7.12), its possible role in somitogenesis has been overlooked. *notch1* shows continuous expression throughout the tailbud presomitic mesoderm but is restricted to one-half of mature somites, whereas *jag2* is expressed in the opposite pattern (Table 7.12), suggesting an interplay between *notch1* and *jag2* after somite segregation.

Among *hes* genes for which there is a precise description of their somitomeric expression pattern, *hes4* is restricted to posterior compartments, and others, including *dlc*, are restricted to anterior compartments (Figure 7.1D) (Jen et al., 1997). A paired RBPJ motif in the proximal promoter of *hes4*, including the intervening hexamer, is necessary for its somitomeric pattern together with its 3'UTR, which confers mRNA instability except in its striped domains in the presomitic mesoderm. Since the 3'UTR of *hes5.6* can impose this striped pattern on *hes4*, cyclic *hes5.6* expression might also be regulated by mRNA decay (Davis et al., 2001). The results summarized in Table 7.12 indicate that spatially controlled *dlc* expression is necessary for *Xenopus* somitogenesis and for setting the normal segmental prepattern of Notch targets related to the segmentation program (*hes7.2, hes7.3/esr5, hes4*). Notch/RBPJ represses *dlc* and *mespa*, whereas Dlc from the anterior half of somitomeres activates *hes4* in the posterior half through Notch/RBPJ.

hes7.3/esr5 is necessary for proper somitogenesis, including the refinement of *dlc, hes7.2*, and *mespa* expression into stripes in the so-called "transition zone" between the somitomeric and tailbud regions (Table 7.12) (Figure 7.1D) (Jen et al., 1999). It was proposed that the Notch pathway uniformly activates targets like *hes7.3/esr5* in the tailbud region. Then, a mechanism requiring *de novo* protein synthesis and HDAC represses *dlc, hes7.2*, and *hes7.3/esr5* in the transition zone, which introduces an expression gap that generates an on/off periodicity that is stably maintained in the somitomeres. *hes7.3/esr5* participates in a negative feedback loop, repressing *dlc* and *hes7.2* in posterior half-segments in the transition zone from where somitomeres arise. In contrast, rostral to the transition zone, *hes7.3/esr5* participates in a positive feedback loop, maintaining the segmental *dlc* prepattern in the somitomeric region (Jen et al., 1999).

hes6.1 shows a broad tailbud expression domain and a segmental prepattern in somitomeres (Table 7.12). Overexpression of *hes6.1* or a mutant DNA binding form severely disrupted somitogenesis and molecular markers (Cossins et al., 2002), suggesting that *hes6.1* must be spatially regulated in the presomitic mesoderm for proper segmentation, perhaps by protein-protein interactions rather than DNA binding. Interestingly, *hes6.1* is negatively regulated by Notch/RBPJ in the neural plate (Koyano-Nakagawa et al., 2000), so it will be interesting to study a possible interplay between *hes6.1* and the Notch pathway in somitogenesis.

7.3.6.2. Interplay between Notch and Other Genes and Pathways in Somitogenesis

Somite boundary formation is also regulated by the Notch pathway via repression of *protocadherin 8* (*pcdh8*) in the posterior half of somitomeres. Pcdh8, which is expressed in their anterior half, in turn regulates differential cell adhesion and prevents the intermingling of anterior and posterior cells between somitomeres, contributing to maintaining

TABLE 7.12
Notch Pathway Genes Expressed during *Xenopus* Somitogenesis

See abbreviations in Table 7.3 legend.

Gene	Expression Related to Somitogenesis and Mature Somites	Functional Evidence Related to Somitogenesis
notch1	Continuous in PSM and TBD. One-half of mature somites (Chitnis et al., 1995; McLaughlin et al., 2000; Rones et al., 2002; Xenbase, community submitted images: www.xenbase.org/, RRID:SCR_003280; Bowes et al., 2010).	**Gain-of-function:** *NICD1:* expanded *hes7.2* and *hes7.3/esr5* domains into the gaps between stripes in the somitomeric region (Jen et al., 1999; Peres et al., 2006); expanded *rnd1*, ↓*rnd3* (Goda et al., 2009). *notch1-ΔE:* ↓*ripply2.2* in the PSM (Kondow et al., 2006). GR-*NICD1;* Dex NF11 or 19: ↑*hey1* in somites, global disorganization of somite borders (Rones et al., 2002).
RBPJ		**Gain-of-function:** *RBPJ-Ank:* ↓*dlc* along the PSM (Sparrow et al., 1998); expanded the *hes7.2* and *hes7.3/esr5* domains into the gaps between stripes in the somitomeric region (Jen et al., 1999; Peres et al., 2006). GR-*RBPJ*-VP16 (Dex NF11–13, Dex NF15–19, ISH NF24–30): ↑*hey1* in somites, global disorganization of somite borders (Rones et al., 2002). **Loss-of-function:** *RBPJDBM*: expanded *dlc* domains into the gaps between somitomeric stripes without affecting *dlc* in the TBD (Jen et al., 1997; Sparrow et al., 1998); ↓*hes4* in somitomeres (Jen et al., 1997); ↓*hes7.2* along the PSM (Jen et al., 1999; Peres et al., 2006); ↓*hes7.3/esr5* in the somitomeric and TZ regions (Jen et al., 1999; Peres et al., 2006); produced a fusion of both *ripply2.2* somitomeric stripes (Kondow et al., 2006); ↓*ripply2.1* in the PSM (Chan et al., 2006); disrupted *rnd3* expression (Goda et al., 2009). GR-*RBPJDBM* (Dex NF11–13, Dex NF15–19, ISH NF24–30): ↓*hey1* in somites (Rones et al., 2002). ***RBPJ* regulation:** Celf1 controls somitogenesis through *RBPJ* mRNA decay (see main text for details) (Gautier-Courteille et al., 2004; Cibois et al., 2010; Cibois et al., 2013).
dll1	More restricted than *dlc*. Circumblastoporal collar/TBD. Two somitomeric stripes (from NF12): stronger in the posterior than in the anterior one (Chitnis et al., 1995; McLaughlin et al., 2000; Rones et al., 2002; Lamar and Kintner, 2005; Dingwell and Smith, 2006; Kondow et al., 2007; Wang et al., 2007).	***dll1* regulation:** *ripply2.1* overexpression: slight posterior shift of *dll1* in the PSM (Chan et al., 2006). *ripply2.1* MO: anterior shift of *dll1* in the PSM (Chan et al., 2006). *ripply2.2* MO: ↓*dll1* stripes in the PSM (Kondow et al., 2007).

dlc	Anterior part of prospective somites SO, S-I, S-II (somitomeres), and S-III (TZ), with spatial refinement: stripes progressively thinner from caudal to rostral PSM. Expression in somitomeres first appears at late gastrula (NF12). Broad domain in TBD. Oscillatory behavior in TZ and TBD. Left-right asynchrony in somitomeres, with the right side relatively more advanced (Jen et al., 1997; Sparrow et al., 1998; Jen et al., 1999; Moreno and Kintner, 2004; Peres et al., 2006; Sparrow, 2008; Durston et al., 2018).	**Gain-of-function:** *dlc* overexpression: segmentation defects (Jen et al., 1997; Peres et al., 2006); expanded *hes7.2* and *hes7.3/esr5* expression into the gaps between stripes in the somitomeric region (ISH NF20–22, late neurula/early tailbud stages) (Jen et al., 1999; Peres et al., 2006); did not affect *hoxc6* and *hoxd1* in the paraxial mesoderm (ISH, tailbud stage) (Peres et al., 2006) **Loss-of-function:** DN-*dlc*: segmentation defects (Jen et al., 1997); expanded the *dlc* domains into the gaps between somitomeric stripes, without affecting *dlc* in the TBD (Jen et al., 1997; Sparrow et al., 1998); ↓*hes4* in somitomeres (Jen et al., 1997); disrupted and ↓*rnd1* and *rnd3* expression domains (Goda et al., 2009). *dlc* MO: segmentation defects (Peres et al., 2006); ↓*hes7.2* in somitomeres (Jen et al., 1999; Peres et al., 2006) and *hoxb4* and *hoxc6* in the paraxial mesoderm during somitogenesis (ISH tailbud stage, NF21) (Peres et al., 2006). ***dlc* regulation:** *RBPJ-Ank:* ↓*dlc* along the PSM (Sparrow et al., 1998). *RBPJDBM*: expanded the *dlc* domains into the gaps between somitomeric stripes, without affecting *dlc* in the TBD (Jen et al., 1997; Sparrow et al., 1998). Knock-down of the whole *hox* paralogous group 1: disrupted somitogenesis and ↓*dlc* in the PSM (Peres et al., 2006). *mespa* overexpression: segmental pattern of *dlc* lost, ↓*dlc* in somitomeres; expression in TBD unaffected (Sparrow et al., 1998). *ripply2.1* overexpression: slight posterior shift of *dlc* in the PSM (Chan et al., 2006). *ripply2.1* MO: anterior shift of *dlc* in the PSM (Chan et al., 2006). *ripply2.2* overexpression: ↓*dlc* in the PSM through recruitment of the TLE4 co-repressor (Kondow et al., 2006). *ripply2.2* MO: ↑*dlc* expression in S0 and S-I and anteriorly shifted these domains (Kondow et al., 2007). *myod1* MO: ↓*dlc* in PSM stripes at early neurula (NF13); anteriorly shifted *dlc* stripes around the onset of somitogenesis (NF19) (Maguire et al., 2012). CHX treatment: continuous *dlc* expression in the PSM (Jen et al., 1999; Kim et al., 2000). RA or SU5402 (inhibitor of FGF signaling) treatments: caudal shift of the *dlc* TBD domain (Moreno and Kintner, 2004).
dll4	Intersomitic vessels (sprouting from the dorsal aorta) (NF39). Image from Kirmizitas et al. (2017), description from Xenbase (Bowes et al., 2010).	
jag2	One-half of mature somites. Xenbase, community submitted images (Bowes et al., 2010).	
hes1	Similar to *hes4* pattern, but with much lower expression (Davis et al., 2001).	***hes1* regulation:** GR-*NICD1*, GR-*RBPJ*-VP16 (Dex NF11–13, Dex NF15–19, ISH NF24–30): ↑*hey1* in somites, global disorganization of somite borders (Rones et al., 2002). GR-*RBPJDBM* (Dex NF11–13, Dex NF15–19, ISH NF24–30): ↓*hey1* in somites (Rones et al., 2002).

(Continued)

TABLE 7.12 (Continued)
Notch Pathway Genes Expressed during *Xenopus* Somitogenesis

Gene	Expression Related to Somitogenesis and Mature Somites	Functional Evidence Related to Somitogenesis
hes4	Posterior part of S-I and S-II (somitomeres), complementary to *dlc*. Left-right asynchrony in somitomeres observed from NF18–19, with the right side relatively more advanced. Very low expression in mature somites (Jen et al., 1997; Davis et al., 2001).	***hes4* regulation:** Segmental *hes4* prepattern requires a paired RBPJ motif (including the intervening hexamer) present in the proximal promoter of *hes4* and the 3' UTR, which confers general mRNA instability (Davis et al., 2001). *RBPJDBM* or DN-*dlc:* ↓*hes4* in somitomeres (Jen et al., 1997). *dlc* MO: ↓*hes4* (Peres et al., 2006). *ripply2.1* overexpression: slight posterior shift of *hes4* in the PSM. *ripply2.1* MO: anterior shift of *hes4* in the PSM (Chan et al., 2006). *ripply2.2* MO ↓*hes4* in the PSM (Kondow et al., 2007). CHX treatment: ↓*hes4* in the PSM (Kim et al., 2000).
hes5.3	1–2 somitomeric stripes. Broad domain in TZ+TBD. Oscillatory behavior in the PSM, with left-right asynchrony (Blewitt, 2009).	Microarray analysis showed that manipulating RA signaling significantly changed *hes5.3* levels (Janesick et al., 2017).
hes5.5	Anterior part of prospective somite S-II. Broad domain in TZ+TBD. Oscillatory behavior in TZ and TBD. (Li et al., 2003).	
hes5.6	Anterior part of prospective somite S-II. Broad domain in TZ+TBD. Oscillatory behavior in TZ and TBD. Left-right asynchrony in somitomeres, with the right side relatively more advanced (Li et al., 2003).	***hes5.6* regulation:** *hes5.6* 3' UTR can replace *hes4* 3'UTR to generate a striped pattern (Davis et al., 2001). CHX treatments (Li et al., 2003): 30 or 60 min (comprising the approximate time of formation of 1 somite): ↑*hes5.6*. *De novo* protein synthesis is not required for cyclic expression. CHX 120 min: *de novo* protein synthesis required for *hes5.6* repression to generate the typical striped pattern RA or SU5402 (inhibitor of FGF signaling) treatment: caudal shift of the *hes5.6* TBD domain (Moreno and Kintner, 2004).
hes6.1	Broad domain in TBD. 2–3 stripes in prospective somites (Koyano-Nakagawa et al., 2000; Cossins et al., 2002; Moreno and Kintner, 2004).	**Gain-of-function:** *hes6.1* overexpression: expanded the myotome (*myod1*+ cells) at neural plate stage, inhibited terminal myogenic differentiation (Ab 12/101), and severely disrupted somitogenesis (Cossins et al., 2002). **Loss-of-function:** *hes6.1DBM*: same effects as *hes6.1* overexpression, indicating that DNA binding is not required for these activities (Cossins et al., 2002). ***hes6.1* regulation:** RA or SU5402 (inhibitor of FGF signaling) treatment: caudal shift of the *hes6.1* TBD domain (Moreno and Kintner, 2004).

hes7.2	Anterior part of prospective somites S-II (somitomere) and S-III (TZ). Broad domain in TBD (Jen et al., 1999).	***hes7.2* regulation:** *dlc* overexpression, *RBPJ-Ank*, or *NICD1:* expanded the *hes7.2* domains into the gaps between stripes in the somitomeric region (Jen et al., 1999; Peres et al., 2006). *RBPJDBM:* ↓*hes7.2* throughout the PSM (Jen et al., 1999; Peres et al., 2006). CHX treatment: continuous *hes7.2* expression in the PSM (Jen et al., 1999).
hes7.3/esr5	Anterior part of prospective somites S-II (somitomere) and S-III (TZ). Broad domain in TBD. Segmented prepattern already detected at late gastrula (NF12) (Sparrow et al., 1998; Jen et al., 1999)	**Gain-of-function:** *hes7.3/esr5* overexpression: ↓*dlc* and *hes7.2* in the somitomeric region but activated *dlc* more anteriorly, with a net result of a rostral shift of the *dlc* domain (Jen et al., 1999). **Loss-of-function:** DN-*hes7.3/esr5:* impaired segmentation and expanded *dlc, hes7.2*, and *mespa* expression into the gaps between stripes in the TZ (Jen et al., 1999). ***hes7.3/esr5* regulation:** *dlc* overexpression, *RBPJ-Ank* or *NICD1:* expanded *hes7.3/esr5* domains into the gaps between stripes in the somitomeric region (Jen et al., 1999; Peres et al., 2006). *RBPJDBM:* ↓*hes7.3/esr5* in the somitomeric and TZ regions (Jen et al., 1999; Peres et al., 2006). CHX treatment: continuous *hes7.2* expression in the PSM (Jen et al., 1999). *mespa* overexpression: segmental *hes7.3/esr5* pattern lost. *hes7.3/esr5* in TBD unaffected (Sparrow et al., 1998). RA or SU5402 (inhibitor of FGF signaling) treatment: caudal shift of the *hes7.3/esr5* TBD domain (Moreno and Kintner, 2004). DN-RARα: rostral shift of the *hes7.3/esr5* TBD domain (Moreno and Kintner, 2004). RARβ2 MO: ↓somite number, ↑somite size, and rostrally expanded PSM markers, including *hes7.3/esr5* (Janesick et al., 2017). *ripply2.1* overexpression: slight posterior shift of *hes7.3/esr5* expression in the PSM (Chan et al., 2006). *ripply2.1* MO: anterior shift of *hes7.3/esr5* expression in the PSM (Chan et al., 2006).
hey1	Mature somites (apparently in the caudal half), from NF17 (Rones et al., 2002; Pichon et al., 2002).	

a segmental boundary (Kim et al., 2000). *ripply2.2*, which encodes a WRPW-containing protein, is also required for the formation of somite boundaries. While *ripply2.2* is required for *dll1* and *hes4* expression in their proper position in the presomitic mesoderm (Kondow et al., 2007) (Table 7.12), Notch/RBPJ signaling localizes *ripply2.2* in the anterior halves of somitomeres (Kondow et al., 2006) (Figure 7.1D), which in turn further restricts *dlc* to the anterior border of the most anterior somitomeres by recruiting the TLE4 co-repressor (Kondow et al., 2006). It was proposed that Tbx6 acts as a transcriptional activator of segmental target genes in posterior somitomeres but changes to a transcriptional repressor through binding to a Ripply2.2/TLE4 complex when Ripply2.2 accumulates above a threshold level in the most anterior somitomeres. In this way, Ripply2.2 and Tbx6 contribute to terminate the segmentation program in the anterior presomitic mesoderm (Kondow et al., 2007; Hitachi et al., 2008). Interestingly, the related *ripply2.1*, whose striped expression in the anterior presomitic mesoderm is also dependent on Notch/RBPJ signaling (Figure 7.1D) (Table 7.12), is not required for segmentation but for positioning the segmentation front via RA signaling (Chan et al., 2006). The Dlc/Notch/RBPJ pathway is also necessary for striped *rnd1* and *rnd3* expression in somitomeres (Figure 7.1D) (Table 7.12). They encode GTP binding proteins required for segmentation independently of *hes7.3/esr5*. *rnd1* is expressed in the anterior half of somitomeres, whereas *rnd3* is expressed in the boundary between the anterior and posterior halves, suggesting different roles in segmentation (Goda et al., 2009).

hox genes are necessary for somitogenesis and establish a reciprocal positive regulation with *dlc* in the paraxial mesoderm (Tables 7.15, 7.12). *dlc* might regulate the timer for temporal collinearity of *hox* gene expression, which in turn controls somite anterior-posterior identity (Peres et al., 2006; Durston et al., 2012).

The Notch pathway also is regulated during somitogenesis at other levels. For example, *celf1*, which encodes an RNA-binding protein that mediates sequence-specific mRNA deadenylation, binds the 3'UTR of *RBPJ* mRNA promoting its degradation; this is required to control the interplay between FGF and RA signaling that governs the determination front (Gautier-Courteille et al., 2004; Cibois et al., 2010; Cibois et al., 2013). RA treatments or FGF pathway blockade repressed *dlc, hes5.6, hes6.1*, and *hes7.3/esr5* in the tailbud, shifting their expression domain caudally (Moreno and Kintner, 2004). RARβ2 knock-down reduced somite number, increased somite size, and rostrally expanded presomitic mesoderm markers, including *hes7.3/esr5*. Microarray analysis showed that manipulating RA signaling significantly changed *hes5.3* levels (Janesick et al., 2017) (Table 7.12).

7.4. FUTURE DIRECTIONS

Work in *Xenopus* frequently has led the field in addressing the role of *notch1* and *dll1* in several developmental programs, as well as *dlc* in somitogenesis, but there is much less information about other Notch receptors and ligands during frog development. Expression patterns for some are available in limited types of tissues. For example, *notch2* is expressed in the PPE and during lens development, where it is positively controlled by *hes4* (Ogino et al., 2008; Murato and Hashimoto, 2009); *notch1*, *notch2*, *jag1*, and *jag2* are expressed in the liver during metamorphosis (Ueno et al., 2015), whereas *notch4* and *dll4* are implicated in the arterial endothelial program (Ciau-Uitz et al., 2010; Leung et al., 2013; Nimmo et al., 2013; Kirmizitas et al., 2017). RNAseq data (Session et al., 2016) (Figure 7.1C, right column) suggest that *notch3*, *jag1*, *jag2*, and *dll4* might have roles during embryogenesis, but we need to know their spatial distributions at different developmental timepoints. Finally, since the *notch4* gene model was not available at the time of the RNAseq study, a developmental expression profile is not yet available.

To dissect Notch signaling involvement in developmental processes more precisely, a combination of strategies will be necessary. For example, experiments employing *RBPJDBM*, *psen* MO, or γ-Secretase inhibitors impair the function of every *notch* paralogue; knock-down/knock-out strategies are needed to provide results specific for each paralogue. Strikingly, only *notch1* has been knocked down so far, both in studies by our group concerning germ layers, DML, and DV axis, and by others concerning ciliogenesis in the epidermis, GRP and left-right patterning (Sakano et al., 2010; Tözser et al., 2015; Tomankova et al., 2017). Since there is growing evidence of *RBPJ*-independent Ligand/Notch functions in several biological contexts (Hayward et al., 2005), including various aspects of *Xenopus* development (Revinski et al., 2010; Peres et al., 2006; Acosta et al., 2011), knock-down/knock-out approaches next need to address both canonical and non-canonical functions. As RBPJ has dual properties, activating or repressing Notch-targets depending on the ON/OFF status of Notch signaling, it might not always be straightforward to interpret the results of RBPJ blockade in complex contexts, including those in which multiple inputs from different ligands might take place. For example, in some studies, *RBPJDBM* produced milder or more variable effects in comparison to the blockade of one ligand or the receptor (Revinski et al., 2010) or did not affect the process under study (Takada et al., 2005; Peres et al., 2006; Nichane et al., 2008a). The effects of protecting *RBPJ* mRNA from Celf1-mediated degradation were compatible with a Notch/RBPJ gain-of-function in the posterior presomitic mesoderm and with a Notch/RBPJ loss-of-function in the anterior presomitic mesoderm (Cibois et al., 2013). In addition, the response to perturbing Notch signaling can change abruptly at certain developmental transitions (Contakos et al., 2005; Revinski et al., 2010), thus requiring a more detailed analysis of gene markers and phenotypes by time-controlled manipulations.

Work from different animal models and cell types show that the number of direct Notch/RBPJ targets outside the *hes/hey* families is constantly growing, including genes involved in proliferation, apoptosis, cell-fate choice, signaling pathways, metabolism, and cytoskeletal regulators (Bray and Bernard, 2010; Meier-Stiegen et al., 2010). The discovery of new targets and modulators in *Xenopus* will be essential to building Notch-regulated GRNs that control different developmental processes. microRNAs regulate Notch signaling during multiciliogenesis in the epidermis (Marcet et al., 2011), and RITA (RBPJ-interacting and tubulin-associated protein) negatively modulates Notch signaling through nuclear export of RBPJ during primary neurogenesis (Wacker et al., 2011). It will be exciting to extend the study of such modulations to the different contexts in which Notch signaling operates.

REFERENCES

Acosta, H., López, S.L., Revinski, D.R., Carrasco, A.E., 2011. Notch destabilises maternal beta-catenin and restricts dorsal-anterior development in Xenopus. *Development* 138, 2567–2579. https://doi.org/10.1242/dev.061143

Aguirre, C.E., Murgan, S., Carrasco, A.E., López, S.L., 2013. An intact brachyury function is necessary to prevent spurious axial development in *Xenopus laevis*. *PLoS One* 8, e54777. https://doi.org/10.1371/journal.pone.0054777

Ali, F.R., Cheng, K., Kirwan, P., Metcalfe, S., Livesey, F.J., Barker, R.A., Philpott, A., 2014. The phosphorylation status of Ascl1 is a key determinant of neuronal differentiation and maturation in vivo and in vitro. *Development* 141, 2216–2224. https://doi.org/10.1242/dev.106377

Anderson, C., Stern, C.D., 2016. Organizers in development. *Curr. Top. Dev. Biol.* 117, 435–454. https://doi.org/10.1016/BS.CTDB.2015.11.023

Andreazzoli, M., Gestri, G., Cremisi, F., Casarosa, S., Dawid, I.B., Barsacchi, G., 2003. Xrx1 controls proliferation and neurogenesis in Xenopus anterior neural plate. *Development* 130, 5143–5154. https://doi.org/10.1242/dev.00665

Artavanis-Tsakonas, S., Grimwade, B.G., Harrison, R.G., Markopoulou, K., Muskavitch, M.A.T., Schlesinger-Bryant, R., Wharton, K., Yedvobnick, B., 1983. The Notch locus of *Drosophila melanogaster*: A molecular analysis. *Dev. Genet.* 4, 233–254. https://doi.org/10.1002/dvg.1020040403

Artinger, M., Blitz, I., Inoue, K., Tran, U., Cho, K.W., 1997. Interaction of goosecoid and brachyury in Xenopus mesoderm patterning. *Mech. Dev*. 65, 187–196.

Beck, C., Slack, J.M., 1998. Analysis of the developing Xenopus tail bud reveals separate phases of gene expression during determination and outgrowth. *Mech. Dev.* 72, 41–52. https://doi.org/10.1016/S0925-4773(98)00015-X

Bellefroid, E.J., Bourguignon, C., Hollemann, T., Ma, Q., Anderson, D.J., Kintner, C., Pieler, T., 1996. X-MyT1, a Xenopus C2HC-type zinc finger protein with a regulatory function in neuronal differentiation. *Cell* 87, 1191–1202.

Bertrand, N., Castro, D.S., Guillemot, F., 2002. Proneural genes and the specification of neural cell types. *Nat. Rev. Neurosci.* 3, 517–530. https://doi.org/10.1038/nrn874

Blewitt, R., 2009. Enhancer of split-related-2 mRNA shows cyclic expression during somitogenesis in *Xenopus laevis*. *Biosci. Horizons* 2, 22–31. https://doi.org/10.1093/biohorizons/hzp006

Bonev, B., Pisco, A., Papalopulu, N., 2011. microRNA-9 reveals regional diversity of neural progenitors along the anterior-posterior axis. *Dev. Cell* 20, 19–32. https://doi.org/10.1016/j.devcel.2010.11.018

Bouwmeester, T., Kim, S., Sasai, Y., Lu, B., De Robertis, E.M., 1996. Cerberus is a head-inducing secreted factor expressed in the anterior endoderm of Spemann's organizer. *Nature* 382, 595–601. https://doi.org/10.1038/382595a0

Bowes, J.B., Snyder, K.A., Segerdell, E., Jarabek, C.J., Azam, K., Zorn, A.M., Vize, P.D., 2010. Xenbase: Gene expression and improved integration. *Nucleic Acids Res* 38, D607–D612. https://doi.org/10.1093/nar/gkp953

Bray, S.J., 2016. Notch signalling in context. *Nat. Rev. Mol. Cell Biol.* 17, 722–735. https://doi.org/10.1038/nrm.2016.94

Bray, S.J., Bernard, F., 2010. Notch targets and their regulation. *Curr. Top. Dev. Biol.* 92, 253–275. https://doi.org/10.1016/S0070-2153(10)92008-5

Brewster, R., Lee, J., Ruiz i Altaba, A., 1998. Gli/Zic factors pattern the neural plate by defining domains of cell differentiation. *Nature* 393, 579–583. https://doi.org/10.1038/31242

Bridges, C.B., Morgan, T.H., 1916. *Sex-Linked Inheritance in Drosophila*. Carnegie Institution of Washington, Washington. https://doi.org/10.5962/bhl.title.22854

Brugmann, S.A., Pandur, P.D., Kenyon, K.L., Pignoni, F., Moody, S.A., 2004. Six1 promotes a placodal fate within the lateral neurogenic ectoderm by functioning as both a transcriptional activator and repressor. *Development* 131, 5871–5881. https://doi.org/10.1242/dev.01516

Campos-Ortega, J.A., 1985. Genetics of early neurogenesis in *Drosophila melanogaster*. *Trends Neurosci* 8, 245–250. https://doi.org/10.1016/0166-2236(85)90097-9

Castro Colabianchi, A.M., Revinski, D.R., Encinas, P.I., Baez, M.V., Monti, R.J., Rodríguez Abinal, M., Kodjabachian, L., Franchini, L.F., López, S.L., 2018. Notch1 is asymmetrically distributed from the beginning of embryogenesis and controls the ventral center. *Development* 145, dev159368. https://doi.org/10.1242/dev.159368

Chalmers, A.D., Welchman, D., Papalopulu, N., 2002. Intrinsic differences between the superficial and deep layers of the Xenopus ectoderm control primary neuronal differentiation. *Dev. Cell* 2, 171–182.

Chan, T., Satow, R., Kitagawa, H., Kato, S., Asashima, M., 2006. Ledgerline, a novel *Xenopus laevis* gene, regulates differentiation of presomitic mesoderm during somitogenesis. *Zoolog. Sci.* 23, 689–697. https://doi.org/10.2108/zsj.23.689

Chen, J.-A., Voigt, J., Gilchrist, M., Papalopulu, N., Amaya, E., 2005. Identification of novel genes affecting mesoderm formation and morphogenesis through an enhanced large scale functional screen in Xenopus. *Mech. Dev.* 122, 307–331. https://doi.org/10.1016/J.MOD.2004.11.008

Chitnis, A., Henrique, D., Lewis, J., Ish-Horowicz, D., Kintner, C., 1995. Primary neurogenesis in Xenopus embryos regulated by a homologue of the Drosophila neurogenic gene Delta. *Nature* 375, 761–766. https://doi.org/10.1038/375761a0

Chitnis, A., Kintner, C., 1996. Sensitivity of proneural genes to lateral inhibition affects the pattern of primary neurons in Xenopus embryos. *Development* 122, 2295–2301.

Ciau-Uitz, A., Pinheiro, P., Gupta, R., Enver, T., Patient, R., 2010. Tel1/ETV6 specifies blood stem cells through the agency of VEGF signaling. *Dev. Cell* 18, 569–578. https://doi.org/10.1016/j.devcel.2010.02.009

Cibois, M., Gautier-Courteille, C., Kodjabachian, L., Paillard, L., 2013. A gene regulation network controlled by Celf1 protein-rbpj mRNA interaction in Xenopus somite segmentation. *Biol. Open* 2, 1078–1083. https://doi.org/10.1242/bio.20135629

Cibois, M., Gautier-Courteille, C., Vallée, A., Paillard, L., 2010. A strategy to analyze the phenotypic consequences of inhibiting the association of an RNA-binding protein with a specific RNA. *RNA* 16, 10–15. https://doi.org/10.1261/rna.1742610

Coffman, C.R., Harris, W., Kintner, C., 1990. Xotch, the Xenopus homolog of Drosophila notch. *Science* 249, 1438–1441.

Coffman, C.R., Skoglund, P., Harris, W.A., Kintner, C.R., 1993. Expression of an extracellular deletion of Xotch diverts cell fate in Xenopus embryos. *Cell* 73, 659–671.

Contakos, S.P., Gaydos, C.M., Pfeil, E.A., McLaughlin, K.A., 2005. Subdividing the embryo: A role for Notch signaling during germ layer patterning in *Xenopus laevis*. *Dev. Biol.* 288, 294–307. https://doi.org/10.1016/j.ydbio.2005.09.015

Cooke, J., 1981. The problem of periodic patterns in embryos. *Philos. Trans. R. Soc. Lond. B. Biol. Sci.* 295, 509–524. https://doi.org/10.1098/rstb.1981.0157

Cooke, J., Zeeman, E.C., 1976. A clock and wavefront model for control of the number of repeated structures during animal morphogenesis. *J. Theor. Biol.* 58, 455–476. https://doi.org/10.1016/S0022-5193(76)80131-2

Cossins, J., Vernon, A.E., Zhang, Y., Philpott, A., Jones, P.H., 2002. Hes6 regulates myogenic differentiation. *Development* 129, 2195–207.

Cui, Y., 2005. Hairy is a cell context signal controlling Notch activity. *Dev. Growth Differ.* 47, 609–625. https://doi.org/10.1111/j.1440-169X.2005.00823.x

Dale, L., Slack, J.M., 1987. Fate map for the 32-cell stage of *Xenopus laevis*. *Development* 99, 527–551.

Davis, R.L., Turner, D.L., 2001. Vertebrate hairy and enhancer of split related proteins: Transcriptional repressors regulating cellular differentiation and embryonic patterning. *Oncogene* 20, 8342–8357. https://doi.org/10.1038/sj.onc.1205094

Davis, R.L., Turner, D.L., Evans, L.M., Kirschner, M.W., 2001. Molecular targets of vertebrate segmentation: Two mechanisms control segmental expression of Xenopus hairy2 during somite formation. *Dev. Cell* 1, 553–565.

Deblandre, G.A., Wettstein, D.A., Koyano-Nakagawa, N., Kintner, C., 1999. A two-step mechanism generates the spacing pattern of the ciliated cells in the skin of Xenopus embryos. *Development* 126, 4715–4728.

De Robertis, E.M., 2009. Spemann's organizer and the self-regulation of embryonic fields. *Mech. Dev.* 126, 925–941. https://doi.org/10.1016/j.mod.2009.08.004

Dexter, J.S., 1914. The analysis of a case of continuous variation in Drosophila by a study of its linkage relations. *Am. Nat.* 48, 712–758.

Dingwell, K.S., Smith, J.C., 2006. Tes regulates neural crest migration and axial elongation in Xenopus. *Dev. Biol.* 293, 252–267. https://doi.org/10.1016/j.ydbio.2006.02.004

D'Souza, B., Meloty-Kapella, L., Weinmaster, G., 2010. Canonical and non-canonical Notch ligands. *Curr. Top. Dev. Biol.* 92, 73–129. https://doi.org/10.1016/S0070-2153(10)92003-6

Dubois, L., Bally-Cuif, L., Crozatier, M., Moreau, J., Paquereau, L., Vincent, A., 1998. XCoe2, a transcription factor of the Col/Olf-1/EBF family involved in the specification of primary neurons in Xenopus. *Curr. Biol.* 8, 199–209. https://doi.org/10.1016/S0960-9822(98)70084-3

Durston, A.J., Peres, J., Cohen, M.H., 2018. Spiral waves and vertebrate embryonic handedness. *J. Biosci.* 43, 375–390.

Durston, A.J., Wacker, S., Bardine, N., Jansen, H.J., 2012. Time space translation: A hox mechanism for vertebrate A-P patterning. *Curr. Genomics* 13, 300–307. https://doi.org/10.2174/138920212800793375

Favarolo, M.B., López, S.L., 2018. Notch signaling in the division of germ layers in bilaterian embryos. *Mech. Dev.* 154, 122–144. https://doi.org/10.1016/j.mod.2018.06.005

Fortini, M.E., 2009. Notch signaling: The core pathway and its posttranslational regulation. *Dev. Cell* 16, 633–647. https://doi.org/10.1016/j.devcel.2009.03.010

Gautier-Courteille, C., Le Clainche, C., Barreau, C., Audic, Y., Graindorge, A., Maniey, D., Osborne, H.B., Paillard, L., 2004. EDEN-BP-dependent post-transcriptional regulation of gene expression in Xenopus somitic segmentation. *Development* 131, 6107–6117. https://doi.org/10.1242/dev.01528

Gawantka, V., Pollet, N., Delius, H., Vingron, M., Pfister, R., Nitsch, R., Blumenstock, C., Niehrs, C., 1998. Gene expression screening in Xenopus identifies molecular pathways, predicts gene function and provides a global view of embryonic patterning. *Mech. Dev.* 77, 95–141. https://doi.org/10.1016/S0925-4773(98)00115-4

Gazave, E., Lapébie, P., Richards, G.S., Brunet, F., Ereskovsky, A. V., Degnan, B.M., Borchiellini, C., Vervoort, M., Renard, E., 2009. Origin and evolution of the Notch signalling pathway: An overview from eukaryotic genomes. *BMC Evol. Biol.* 9, 249. https://doi.org/10.1186/1471-2148-9-249

Glavic, A., Silva, F., Aybar, M.J., Bastidas, F., Mayor, R., 2004. Interplay between Notch signaling and the homeoprotein Xiro1 is required for neural crest induction in Xenopus embryos. *Development* 131, 347–359. https://doi.org/10.1242/dev.00945

Goda, T., Takagi, C., Ueno, N., 2009. Xenopus Rnd1 and Rnd3 GTP-binding proteins are expressed under the control of segmentation clock and required for somite formation. *Dev. Dyn.* 238, 2867–2876. https://doi.org/10.1002/dvdy.22099

Gomez, C., Özbudak, E.M., Wunderlich, J., Baumann, D., Lewis, J., Pourquié, O., 2008. Control of segment number in vertebrate embryos. *Nature* 454, 335–339. https://doi.org/10.1038/nature07020

Grocott, T., Tambalo, M., Streit, A., 2012. The peripheral sensory nervous system in the vertebrate head: A gene regulatory perspective. *Dev. Biol.* 370, 3–23. https://doi.org/10.1016/j.ydbio.2012.06.028

Groot, A.J., Vooijs, M.A., 2012. The role of Adams in notch signaling. *Adv. Exp. Med. Biol.* 727, 15–36. https://doi.org/10.1007/978-1-4614-0899-4_2

Hardwick, L.J.A., Philpott, A., 2015. Multi-site phosphorylation regulates NeuroD4 activity during primary neurogenesis: A conserved mechanism amongst proneural proteins. *Neural Dev.* 10, 15. https://doi.org/10.1186/s13064-015-0044-8

Hardwick, L.J.A., Philpott, A., 2019. N-terminal phosphorylation of xHes1 controls inhibition of primary neurogenesis in Xenopus. *Biochem. Biophys. Res. Commun.* 509, 557–563. https://doi.org/10.1016/J.BBRC.2018.12.135

Hayata, T., Blitz, I.L., Iwata, N., Cho, K.W.Y., 2009. Identification of embryonic pancreatic genes using Xenopus DNA microarrays. *Dev. Dyn.* 238, 1455–1466. https://doi.org/10.1002/dvdy.21868

Hayward, P., Brennan, K., Sanders, P., Balayo, T., DasGupta, R., Perrimon, N., Martinez Arias, A., 2005. Notch modulates Wnt signalling by associating with Armadillo/beta-catenin and regulating its transcriptional activity. *Development* 132, 1819–1830. https://doi.org/10.1242/dev.01724

Hayward, P., Kalmar, T., Martinez Arias, A., 2008. Wnt/Notch signalling and information processing during development. *Development* 135, 411–424. https://doi.org/10.1242/dev.000505

Heeg-Truesdell, E., LaBonne, C., 2006. Neural induction in Xenopus requires inhibition of Wnt-beta-catenin signaling. *Dev. Biol.* 298, 71–86. https://doi.org/10.1016/j.ydbio.2006.06.015

Hein, K., Mittler, G., Cizelsky, W., Kühl, M., Ferrante, F., Liefke, R., Berger, I.M., Just, S., Sträng, J.E., Kestler, H.A., Oswald, F., Borggrefe, T., 2015. Site-specific methylation of Notch1 controls the amplitude and duration of the Notch1 response. *Sci. Signal.* 8, ra30. https://doi.org/10.1126/scisignal.2005892

Heitzler, P., 2010. Biodiversity and noncanonical Notch signaling. *Curr. Top. Dev. Biol.* 92, 457–481. https://doi.org/10.1016/S0070-2153(10)92014-0

Hindley, C., Ali, F., McDowell, G., Cheng, K., Jones, A., Guillemot, F., Philpott, A., 2012. Post-translational modification of Ngn2 differentially affects transcription of distinct targets to regulate the balance between progenitor maintenance and differentiation. *Development* 139, 1718–1723. https://doi.org/10.1242/dev.077552

Hitachi, K., Danno, H., Tazumi, S., Aihara, Y., Uchiyama, H., Okabayashi, K., Kondow, A., Asashima, M., 2009. The Xenopus Bowline/Ripply family proteins negatively regulate the transcriptional activity of T-box transcription factors. *Int. J. Dev. Biol.* 53, 631–639. https://doi.org/10.1387/ijdb.082823kh

Hitachi, K., Kondow, A., Danno, H., Inui, M., Uchiyama, H., Asashima, M., 2008. Tbx6, Thylacine1, and E47 synergistically activate bowline expression in Xenopus somitogenesis. *Dev. Biol.* 313, 816–828. https://doi.org/10.1016/J.YDBIO.2007.10.015

Hong, C.-S., Saint-Jeannet, J.-P., 2018. The b-HLH transcription factor Hes3 participates in neural plate border formation by interfering with Wnt/β-catenin signaling. *Dev. Biol.* 442, 162–172. https://doi.org/10.1016/j.ydbio.2018.07.011

Howell, M., Inman, G.J., Hill, C.S., 2002. A novel Xenopus Smad-interacting forkhead transcription factor (XFast-3) cooperates with XFast-1 in regulating gastrulation movements. *Development* 129, 2823–2834.

Huang, C., Chan, J.A., Schuurmans, C., 2014. Proneural bHLH genes in development and disease. *Curr. Top. Dev. Biol.* 110, 75–127. https://doi.org/10.1016/B978-0-12-405943-6.00002-6

Hubaud, A., Pourquié, O., 2014. Signalling dynamics in vertebrate segmentation. *Nat. Rev. Mol. Cell Biol.* 15, 709–721. https://doi.org/10.1038/nrm3891

Hufton, A.L., Vinayagam, A., Suhai, S., Baker, J.C., 2006. Genomic analysis of Xenopus organizer function. *BMC Dev. Biol.* 6, 27. https://doi.org/10.1186/1471-213X-6-27

Imayoshi, I., Kageyama, R., 2014. Oscillatory control of bHLH factors in neural progenitors. *Trends Neurosci* 37, 531–538. https://doi.org/10.1016/j.tins.2014.07.006

Ito, M., Katada, T., Miyatani, S., Kinoshita, T., 2007a. XSu(H)2 is an essential factor for gene expression and morphogenesis of the Xenopus gastrula embryo. *Int. J. Dev. Biol.* 51, 27–36. https://doi.org/10.1387/ijdb.062211mi

Ito, M., Nishitani, E., Kinoshita, T., 2007b. Xenopus suppressor of Hairless 2 is involved in the cell fate decision during gastrulation through the transcriptional regulation of Xoct25/91. *Biochem. Biophys. Res. Commun.* 353, 644–649. https://doi.org/10.1016/j.bbrc.2006.12.087

Janesick, A., Tang, W., Nguyen, T.T.L., Blumberg, B., 2017. RARβ2 is required for vertebrate somitogenesis. *Development* 144, 1997–2008. https://doi.org/10.1242/dev.144345

Jen, W.C., Gawantka, V., Pollet, N., Niehrs, C., Kintner, C., 1999. Periodic repression of Notch pathway genes governs the segmentation of Xenopus embryos. *Genes Dev.* 13, 1486–1499. https://doi.org/10.1101/gad.13.11.1486

Jen, W.C., Wettstein, D., Turner, D., Chitnis, A., Kintner, C., 1997. The Notch ligand, X-Delta-2, mediates segmentation of the paraxial mesoderm in Xenopus embryos. *Development* 124, 1169–1178.

Jorissen, E., De Strooper, B., 2010. Gamma-secretase and the intramembrane proteolysis of Notch. *Curr. Top. Dev. Biol.* 92, 201–230. https://doi.org/10.1016/S0070-2153(10)92006-1

Kageyama, R., Ohtsuka, T., Kobayashi, T., 2007. The Hes gene family: Repressors and oscillators that orchestrate embryogenesis. *Development* 134, 1243–1251. https://doi.org/10.1242/dev.000786

Kao, H.Y., Ordentlich, P., Koyano-Nakagawa, N., Tang, Z., Downes, M., Kintner, C.R., Evans, R.M., Kadesch, T., 1998. A histone deacetylase corepressor complex regulates the Notch signal transduction pathway. *Genes Dev.* 12, 2269–2277. https://doi.org/10.1101/gad.12.15.2269

Karimi, K., Fortriede, J.D., Lotay, V.S., Burns, K.A., Wang, D.Z., Fisher, M.E., Pells, T.J., James-Zorn, C., Wang, Y., Ponferrada, V.G., Chu, S., Chaturvedi, P., Zorn, A.M., Vize, P.D., 2018. Xenbase: A genomic, epigenomic and transcriptomic model organism database. *Nucleic Acids Res* 46, D861–D868. https://doi.org/10.1093/nar/gkx936

Katada, T., Kinoshita, T., 2003. XMam1, the Xenopus homologue of mastermind, is essential to primary neurogenesis in *Xenopus laevis* embryos. *Int. J. Dev. Biol.* 47, 397–404. https://doi.org/10.1387/IJDB.14584777

Keller, R., Danilchik, M., 1988. Regional expression, pattern and timing of convergence and extension during gastrulation of *Xenopus laevis*. *Development* 103, 193–209.

Kenyon, K.L., Moody, S.A., Jamrich, M., 1999. A novel fork head gene mediates early steps during Xenopus lens formation. *Development* 126, 5107–5116.

Kiecker, C., Niehrs, C., 2001. The role of prechordal mesendoderm in neural patterning. *Curr. Opin. Neurobiol.* 11, 27–33.

Kim, S.H., Jen, W.C., De Robertis, E.M., Kintner, C., 2000. The protocadherin PAPC establishes segmental boundaries during somitogenesis in xenopus embryos. *Curr. Biol.* 10, 821–830.

Kinoshita, T., Haruta, Y., Sakamoto, C., Imaoka, S., 2011. Antagonistic role of XESR1 and XESR5 in mesoderm formation in *Xenopus laevis*. *Int. J. Dev. Biol.* 55, 25–31. https://doi.org/10.1387/ijdb.092990tk

Kirby, M.L., Lawson, A., Stadt, H.A., Kumiski, D.H., Wallis, K.T., McCraney, E., Waldo, K.L., Li, Y.-X., Schoenwolf, G.C., 2003. Hensen's node gives rise to the ventral midline of the foregut: Implications for organizing head and heart development. *Dev. Biol.* 253, 175–188. https://doi.org/10.1016/S0012-1606(02)00024-6

Kirmizitas, A., Meiklejohn, S., Ciau-Uitz, A., Stephenson, R., Patient, R., 2017. Dissecting BMP signaling input into the gene regulatory networks driving specification of the blood stem cell lineage. *Proc. Natl. Acad. Sci. U. S. A.* 114, 5814–5821. https://doi.org/10.1073/pnas.1610615114

Kiyota, T., Jono, H., Kuriyama, S., Hasegawa, K., Miyatani, S., Kinoshita, T., 2001. X-Serrate-1 is involved in primary neurogenesis in *Xenopus laevis* in a complementary manner with X-Delta-1. *Dev. Genes Evol.* 211, 367–376. https://doi.org/10.1007/s004270100165

Kiyota, T., Kinoshita, T., 2004. The intracellular domain of X-Serrate-1 is cleaved and suppresses primary neurogenesis in *Xenopus laevis*. *Mech. Dev.* 121, 573–585. https://doi.org/10.1016/j.mod.2004.03.034

Kjolby, R.A.S., Harland, R.M., 2017. Genome-wide identification of Wnt/β-catenin transcriptional targets during Xenopus gastrulation. *Dev. Biol.* 426, 165–175. https://doi.org/10.1016/j.ydbio.2016.03.021

Kondow, A., Hitachi, K., Ikegame, T., Asashima, M., 2006. Bowline, a novel protein localized to the presomitic mesoderm, interacts with Groucho/TLE in Xenopus. *Int. J. Dev. Biol.* 50, 473–479. https://doi.org/10.1387/ijdb.052138ak

Kondow, A., Hitachi, K., Okabayashi, K., Hayashi, N., Asashima, M., 2007. Bowline mediates association of the transcriptional corepressor XGrg-4 with Tbx6 during somitogenesis in Xenopus. *Biochem. Biophys. Res. Commun.* 359, 959–964. https://doi.org/10.1016/j.bbrc.2007.05.211

Kopan, R., Ilagan, M.X.G., 2009. The canonical Notch signaling pathway: Unfolding the activation mechanism. *Cell* 137, 216–233. https://doi.org/10.1016/j.cell.2009.03.045

Kopan, R., Nye, J.S., Weintraub, H., 1994. The intracellular domain of mouse Notch: A constitutively activated repressor of myogenesis directed at the basic helix-loop-helix region of MyoD. *Development* 120, 2385–2396.

Kovall, R.A., Blacklow, S.C., 2010. Mechanistic insights into Notch receptor signaling from structural and biochemical studies. *Curr. Top. Dev. Biol.* 92, 31–71. https://doi.org/10.1016/S0070-2153(10)92002-4

Koyano-Nakagawa, N., Kim, J., Anderson, D., Kintner, C., 2000. Hes6 acts in a positive feedback loop with the neurogenins to promote neuronal differentiation. *Development* 127, 4203–4216.

Kurata, T., Ueno, N., 2003. Xenopus Nbx, a novel NK-1 related gene essential for neural crest formation. *Dev. Biol.* 257, 30–40. https://doi.org/10.1016/s0012-1606(03)00060-5

Kuriyama, S., Lupo, G., Ohta, K., Ohnuma, S., Harris, W.A., Tanaka, H., 2006. Tsukushi controls ectodermal patterning and neural crest specification in Xenopus by direct regulation of BMP4 and X-delta-1 activity. *Development* 133, 75–88. https://doi.org/10.1242/dev.02178

Kuroda, H., Wessely, O., De Robertis, E.M., 2004. Neural induction in Xenopus: Requirement for ectodermal and endomesodermal signals via Chordin, Noggin, beta-Catenin, and Cerberus. *PLoS Biol.* 2, E92. https://doi.org/10.1371/journal.pbio.0020092

Kwon, C., Cheng, P., King, I.N., Andersen, P., Shenje, L., Nigam, V., Srivastava, D., 2011. Notch post-translationally regulates β-catenin protein in stem and progenitor cells. *Nat. Cell Biol.* 13, 1244–1251. https://doi.org/10.1038/ncb2313

Lahaye, K., Kricha, S., Bellefroid, E.J., 2002. XNAP, a conserved ankyrin repeat-containing protein with a role in the Notch pathway during Xenopus primary neurogenesis. *Mech. Dev.* 110, 113–124. https://doi.org/10.1016/S0925-4773(01)00570-6

Lai, E.C., 2004. Notch signaling: Control of cell communication and cell fate. *Development* 131, 965–973. https://doi.org/10.1242/dev.01074

Lamar, E., Deblandre, G., Wettstein, D., Gawantka, V., Pollet, N., Niehrs, C., Kintner, C., 2001. Nrarp is a novel intracellular component of the Notch signaling pathway. *Genes Dev.* 15, 1885–1899. https://doi.org/10.1101/gad.908101

Lamar, E., Kintner, C., 2005. The Notch targets Esr1 and Esr10 are differentially regulated in Xenopus neural precursors. *Development* 132, 3619–3630. https://doi.org/10.1242/dev.01937

Layden, M.J., Martindale, M.Q., 2014. Non-canonical Notch signaling represents an ancestral mechanism to regulate neural differentiation. *Evodevo* 5, 30. https://doi.org/10.1186/2041-9139-5-30

Lee, J., Platt, K.A., Censullo, P., Ruiz i Altaba, A., 1997. Gli1 is a target of Sonic hedgehog that induces ventral neural tube development. *Development* 124, 2537–2552.

Lehmann, R., Jiménez, F., Dietrich, U., Campos-Ortega, J.A., 1983. On the phenotype and development of mutants of early neurogenesis in *Drosophila melanogaster*: Wilhelm Roux's Arch. *Dev. Biol.* 192, 62–74. https://doi.org/10.1007/BF00848482

Leung, A., Ciau-Uitz, A., Pinheiro, P., Monteiro, R., Zuo, J., Vyas, P., Patient, R., Porcher, C., 2013. Uncoupling VEGFA functions in arteriogenesis and hematopoietic stem cell specification. *Dev. Cell* 24, 144–158. https://doi.org/10.1016/j.devcel.2012.12.004

Lewis, J., 1998. Notch signalling and the control of cell fate choices in vertebrates. *Semin. Cell Dev. Biol.* 9, 583–589. https://doi.org/10.1006/scdb.1998.0266

Li, Y., Fenger, U., Niehrs, C., Pollet, N., 2003. Cyclic expression of esr9 gene in Xenopus presomitic mesoderm. *Differentiation* 71, 83–89. https://doi.org/10.1046/j.1432-0436.2003.700608.x

López, S.L., Carrasco, A.E., 2006. Sonic hedgehog signalling in dorsal midline and neural development. In: Fisher, C.E., Howie, S.E.M. (Eds.), *Shh and Gli Signalling and Development*. Landes Biosciences and Springer New York, New York, NY, pp. 12–22.

López, S.L., Paganelli, A.R., Rosato Siri, M. V, Ocaña, O.H., Franco, P.G., Carrasco, A.E., 2003. Notch activates Sonic hedgehog and both are involved in the specification of dorsal midline cell-fates in Xenopus. *Development* 130, 2225–2238. https://doi.org/10.1242/dev.00443

López, S.L., Rosato-Siri, M. V, Franco, P.G., Paganelli, A.R., Carrasco, A.E., 2005. The Notch-target gene hairy2a impedes the involution of notochordal cells by promoting floor plate fates in Xenopus embryos. *Development* 132, 1035–1046. https://doi.org/10.1242/dev.01659

Ma, Q., Kintner, C., Anderson, D.J., 1996. Identification of neurogenin, a vertebrate neuronal determination gene. *Cell* 87, 43–52.

Maguire, R.J., Isaacs, H. V., Pownall, M.E., 2012. Early transcriptional targets of MyoD link myogenesis and somitogenesis. *Dev. Biol.* 371, 256–268. https://doi.org/10.1016/j.ydbio.2012.08.027

Maharana, S.K., Schlosser, G., 2018. A gene regulatory network underlying the formation of pre-placodal ectoderm in *Xenopus laevis*. *BMC Biol.* 16, 79. https://doi.org/10.1186/s12915-018-0540-5

Marcet, B., Chevalier, B., Luxardi, G., Coraux, C., Zaragosi, L.-E., Cibois, M., Robbe-Sermesant, K., Jolly, T., Cardinaud, B., Moreilhon, C., Giovannini-Chami, L., Nawrocki-Raby, B., Birembaut, P., Waldmann, R., Kodjabachian, L., Barbry, P., 2011. Control of vertebrate multiciliogenesis by miR-449 through direct repression of the Delta/Notch pathway. *Nat. Cell Biol.* 13, 694–701. https://doi.org/10.1038/ncb2241

Marine, J.-C., Bellefroid, J.E., Pendeville, H., Martial, J.A., Pieler, T., 1997. A role for Xenopus Gli-type zinc finger proteins in the early embryonic patterning of mesoderm and neuroectoderm. *Mech. Dev.* 63, 211–225. https://doi.org/10.1016/S0925-4773(97)00050-6

McLaughlin, K.A., Rones, M.S., Mercola, M., 2000. Notch regulates cell fate in the developing pronephros. *Dev. Biol.* 227, 567–580. https://doi.org/10.1006/dbio.2000.9913

Meier-Stiegen, F., Schwanbeck, R., Bernoth, K., Martini, S., Hieronymus, T., Ruau, D., Zenke, M., Just, U., 2010. Activated Notch1 target genes during embryonic cell differentiation depend on the cellular context and include lineage determinants and inhibitors. *PLoS One* 5, e11481. https://doi.org/10.1371/journal.pone.0011481

Miazga, C.M., McLaughlin, K.A., 2009. Coordinating the timing of cardiac precursor development during gastrulation: A new role for Notch signaling. *Dev. Biol.* 333, 285–296. https://doi.org/10.1016/j.ydbio.2009.06.040

Michiue, T., Yamamoto, T., Yasuoka, Y., Goto, T., Ikeda, T., Nagura, K., Nakayama, T., Taira, M., Kinoshita, T., 2017. High variability of expression profiles of homeologous genes for Wnt, Hh, Notch, and Hippo signaling pathways in *Xenopus laevis*. *Dev. Biol.* 426, 270–290. https://doi.org/10.1016/j.ydbio.2016.12.006

Milet, C., Maczkowiak, F., Roche, D.D., Monsoro-Burq, A.H., 2013. Pax3 and Zic1 drive induction and differentiation of multipotent, migratory, and functional neural crest in Xenopus embryos. *Proc. Natl. Acad. Sci. U. S. A.* 110, 5528–5533. https://doi.org/10.1073/pnas.1219124110

Minsuk, S.B., Keller, R.E., 1997. Surface mesoderm in Xenopus: A revision of the stage 10 fate map. *Dev. Genes Evol.* 207, 389–401. https://doi.org/10.1007/s004270050128

Mir, A., Kofron, M., Heasman, J., Mogle, M., Lang, S., Birsoy, B., Wylie, C., 2008. Long- and short-range signals control the dynamic expression of an animal hemisphere-specific gene in Xenopus. *Dev. Biol.* 315, 161–172. https://doi.org/10.1016/j.ydbio.2007.12.022

Mohr, O.L., 1919. Character changes caused by mutation of an entire region of a chromosome in Drosophila. *Genetics* 4, 275–282.

Moody, S.A., 1987a. Fates of the blastomeres of the 16-cell stage Xenopus embryo. *Dev. Biol.* 119, 560–578. https://doi.org/10.1016/0012-1606(87)90059-5

Moody, S.A., 1987b. Fates of the blastomeres of the 32-cell-stage Xenopus embryo. *Dev. Biol.* 122, 300–319. https://doi.org/10.1016/0012-1606(87)90296-x

Moreno, T.A., Kintner, C., 2004. Regulation of segmental patterning by retinoic acid signaling during Xenopus somitogenesis. *Dev. Cell* 6, 205–218. https://doi.org/10.1016/S1534-5807(04)00026-7

Morgan, T.H., 1917. The theory of the gene. *Am. Nat.* 51, 513–544. https://doi.org/10.1086/279629

Muñoz-Descalzo, S., Sanders, P.G.T., Montagne, C., Johnson, R.I., Balayo, T., Martinez Arias, A., 2010. Wingless modulates the ligand independent traffic of Notch through Dishevelled. *Fly (Austin).* 4, 182–193. https://doi.org/10.4161/fly.4.3.11998

Murai, K., Philpott, A., Jones, P.H., 2011. Hes6 is required for the neurogenic activity of neurogenin and NeuroD. *PLoS One* 6, e27880. https://doi.org/10.1371/journal.pone.0027880

Murai, K., Vernon, A.E., Philpott, A., Jones, P., 2007. Hes6 is required for MyoD induction during gastrulation. *Dev. Biol.* 312, 61–76. https://doi.org/10.1016/j.ydbio.2007.09.011

Murato, Y., Hashimoto, C., 2009. Xhairy2 functions in Xenopus lens development by regulating p27(xic1) expression. *Dev. Dyn.* 238, 2179–2192. https://doi.org/10.1002/dvdy.21883

Murato, Y., Nagatomo, K., Yamaguti, M., Hashimoto, C., 2007. Two alloalleles of *Xenopus laevis* hairy2 gene: Evolution of duplicated gene function from a developmental perspective. *Dev. Genes Evol.* 217, 665–673. https://doi.org/10.1007/s00427-007-0176-x

Murato, Y., Yamaguti, M., Katamura, M., Cho, K.W.Y., Hashimoto, C., 2006. Two modes of action by which Xenopus hairy2b establishes tissue demarcation in the Spemann-Mangold organizer. *Int. J. Dev. Biol.* 50, 463–471. https://doi.org/10.1387/ijdb.052106ym

Murgan, S., Castro Colabianchi, A.M., Monti, R.J., Boyadjián López, L.E., Aguirre, C.E., Stivala, E.G., Carrasco, A.E., López, S.L., 2014. FoxA4 favours notochord formation by inhibiting contiguous mesodermal fates and restricts anterior neural development in Xenopus embryos. *PLoS One* 9, e110559. https://doi.org/10.1371/journal.pone.0110559

Nagano, T., Takehara, S., Takahashi, M., Aizawa, S., Yamamoto, A., 2006. Shisa2 promotes the maturation of somitic precursors and transition to the segmental fate in Xenopus embryos. *Development* 133, 4643–4654. https://doi.org/10.1242/dev.02657

Nagatomo, K., Hashimoto, C., 2007. Xenopus hairy2 functions in neural crest formation by maintaining cells in a mitotic and undifferentiated state. *Dev. Dyn.* 236, 1475–1483. https://doi.org/10.1002/dvdy.21152

Nakata, K., Nagai, T., Aruga, J., Mikoshiba, K., 1998. Xenopus Zic family and its role in neural and neural crest development. *Mech. Dev.* 75, 43–51. https://doi.org/10.1016/S0925-4773(98)00073-2

Naoki, H., Matsui, T., 2020. Somite boundary determination in normal and clock-less vertebrate embryos. *Dev. Growth Differ.* 62, 177–187. https://doi.org/10.1111/dgd.12655

Nichane, M., de Crozé, N., Ren, X., Souopgui, J., Monsoro-Burq, A.H., Bellefroid, E.J., 2008a. Hairy2-Id3 interactions play an essential role in Xenopus neural crest progenitor specification. *Dev. Biol.* 322, 355–367. https://doi.org/10.1016/j.ydbio.2008.08.003

Nichane, M., Ren, X., Bellefroid, E.J., 2010. Self-regulation of Stat3 activity coordinates cell-cycle progression and neural crest specification. *EMBO J.* 29, 55–67. https://doi.org/10.1038/emboj.2009.313

Nichane, M., Ren, X., Souopgui, J., Bellefroid, E.J., 2008b. Hairy2 functions through both DNA-binding and non DNA-binding mechanisms at the neural plate border in Xenopus. *Dev. Biol.* 322, 368–380. https://doi.org/10.1016/j.ydbio.2008.07.026

Nieber, F., Hedderich, M., Jahn, O., Pieler, T., Henningfeld, K.A., 2013. NumbL is essential for Xenopus primary neurogenesis. *BMC Dev. Biol.* 13, 36. https://doi.org/10.1186/1471-213X-13-36

Nimmo, R., Ciau-Uitz, A., Ruiz-Herguido, C., Soneji, S., Bigas, A., Patient, R., Enver, T., 2013. MiR-142–3p controls the specification of definitive hemangioblasts during ontogeny. *Dev. Cell* 26, 237–249. https://doi.org/10.1016/j.devcel.2013.06.023

Oates, A.C., Morelli, L.G., Ares, S., 2012. Patterning embryos with oscillations: Structure, function and dynamics of the vertebrate segmentation clock. *Development* 139, 625–639. https://doi.org/10.1242/dev.063735

Ogino, H., Fisher, M., Grainger, R.M., 2008. Convergence of a head-field selector Otx2 and Notch signaling: A mechanism for lens specification. *Development* 135, 249–258. https://doi.org/10.1242/dev.009548

Olson, E.C., Schinder, A.F., Dantzker, J.L., Marcus, E.A., Spitzer, N.C., Harris, W.A., 1998. Properties of ectopic neurons induced by Xenopus neurogenin1 misexpression. *Mol. Cell. Neurosci.* 12, 281–299. https://doi.org/10.1006/mcne.1998.0712

Onai, T., Aramaki, T., Inomata, H., Hirai, T., Kuratani, S., 2015. On the origin of vertebrate somites. *Zool. Lett.* 1, 33. https://doi.org/10.1186/s40851-015-0033-0

Oswald, F., Rodriguez, P., Giaimo, B.D., Antonello, Z.A., Mira, L., Mittler, G., Thiel, V.N., Collins, K.J., Tabaja, N., Cizelsky, W., Rothe, M., Kühl, S.J., Kühl, M., Ferrante, F., Hein, K., Kovall, R.A., Dominguez, M., Borggrefe, T., 2016. A phospho-dependent mechanism involving NCoR and KMT2D controls a permissive chromatin state at Notch target genes. *Nucleic Acids Res.* 44, 4703–4720. https://doi.org/10.1093/nar/gkw105

Paganelli, A.R., Ocaña, O.H., Prat, M.I., Franco, P.G., López, S.L., Morelli, L., Adamo, A.M., Riccomagno, M.M., Matsubara, E., Shoji, M., Affranchino, J.L., Castaño, E.M., Carrasco, A.E., 2001. The Alzheimer-related gene presenilin-1 facilitates Sonic hedgehog expression in Xenopus primary neurogenesis. *Mech. Dev.* 107, 119–131. https://doi.org/10.1016/s0925-4773(01)00458-0

Papalopulu, N., Kintner, C., 1996. A posteriorising factor, retinoic acid, reveals that anteroposterior patterning controls the timing of neuronal differentiation in Xenopus neuroectoderm. *Development* 122, 3409–3418.

Parlier, D., Moers, V., Van Campenhout, C., Preillon, J., Leclère, L., Saulnier, A., Sirakov, M., Busengdal, H., Kricha, S., Marine, J.-C., Rentzsch, F., Bellefroid, E.J., 2013. The Xenopus doublesex-related gene Dmrt5 is required for olfactory placode neurogenesis. *Dev. Biol.* 373, 39–52. https://doi.org/10.1016/J.YDBIO.2012.10.003

Pegoraro, C., Monsoro-Burq, A.H., 2013. Signaling and transcriptional regulation in neural crest specification and migration: Lessons from Xenopus embryos. *WIREs Dev. Biol.* 2, 247–259. https://doi.org/10.1002/wdev.76

Peres, J.N., Durston, A.J., 2006. Role of X-Delta-2 in the early neural development of *Xenopus laevis*. *Dev. Dyn.* 235, 802–810. https://doi.org/10.1002/dvdy.20677

Peres, J.N., McNulty, C.L., Durston, A.J., 2006. Interaction between X-Delta-2 and Hox genes regulates segmentation and patterning of the anteroposterior axis. *Mech. Dev.* 123, 321–333. https://doi.org/10.1016/j.mod.2006.03.001

Perron, M., Opdecamp, K., Butler, K., Harris, W.A., Bellefroid, E.J., 1999. X-ngnr-1 and Xath3 promote ectopic expression of sensory neuron markers in the neurula ectoderm and have distinct inducing properties in the retina. *Proc. Natl. Acad. Sci. U. S. A.* 96, 14996–15001. https://doi.org/10.1073/PNAS.96.26.14996

Peyrot, S.M., Wallingford, J.B., Harland, R.M., 2011. A revised model of Xenopus dorsal midline development: Differential and separable requirements for Notch and Shh signaling. *Dev. Biol.* 352, 254–266. https://doi.org/10.1016/j.ydbio.2011.01.021

Pichon, B., Taelman, V., Bellefroid, E.J., Christophe, D., 2004. Transcriptional repression by the bHLH-Orange factor XHRT1 does not involve the C-terminal YRPW motif. *Biochim. Biophys. Acta-Gene Struct. Expr.* 1680, 46–52. https://doi.org/10.1016/j.bbaexp.2004.08.010

Pichon, B., Taelman, V., Kricha, S., Christophe, D., Bellefroid, E.J., 2002. XHRT-1, a hairy and enhancer of split related gene with expression in floor plate and hypochord during early Xenopus embryogenesis. *Dev. Genes Evol.* 212, 491–495. https://doi.org/10.1007/s00427-002-0270-z

Pla, P., Monsoro-Burq, A.H., 2018. The neural border: Induction, specification and maturation of the territory that generates neural crest cells. *Dev. Biol.* 444, S36–S46. https://doi.org/10.1016/j.ydbio.2018.05.018

Pollet, N., Muncke, N., Verbeek, B., Li, Y., Fenger, U., Delius, H., Niehrs, C., 2005. An atlas of differential gene expression during early Xenopus embryogenesis. *Mech. Dev.* 122, 365–439. https://doi.org/10.1016/j.mod.2004.11.009

Poulson, D.F., 1937. Chromosomal deficiencies and the embryonic development of *Drosophila melanogaster*. *Proc. Natl. Acad. Sci. U. S. A.* 23, 133–137.

Pourquié, O., Tam, P.P.L., 2001. A nomenclature for prospective somites and phases of cyclic gene expression in the presomitic mesoderm. *Dev. Cell* 1, 619–620. https://doi.org/10.1016/s1534-5807(01)00082-x

Revinski, D.R., Paganelli, A.R., Carrasco, A.E., López, S.L., 2010. Delta-Notch signaling is involved in the segregation of the three germ layers in *Xenopus laevis*. *Dev. Biol.* 339, 477–492. https://doi.org/10.1016/j.ydbio.2010.01.010

Riddiford, N., Schlosser, G., 2016. Dissecting the pre-placodal transcriptome to reveal presumptive direct targets of Six1 and Eya1 in cranial placodes. *Elife* 5. https://doi.org/10.7554/eLife.17666

Riddiford, N., Schlosser, G., 2017. Six1 and Eya1 both promote and arrest neuronal differentiation by activating multiple Notch pathway genes. *Dev. Biol.* 431, 152–167. https://doi.org/10.1016/J.YDBIO.2017.09.027

Roberts, A., 1989. The neurons that control axial movements in a frog embryo. *Am. Zool.* 29, 53–63. https://doi.org/10.1093/icb/29.1.53

Rones, M.S., Woda, J., Mercola, M., McLaughlin, K.A., 2002. Isolation and characterization of Xenopus Hey-1: A downstream mediator of notch signaling. *Dev. Dyn.* 225, 554–560. https://doi.org/10.1002/dvdy.10192

Saint-Jeannet, J.-P., Moody, S.A., 2014. Establishing the pre-placodal region and breaking it into placodes with distinct identities. *Dev. Biol.* 389, 13–27. https://doi.org/10.1016/j.ydbio.2014.02.011

Sakano, D., Kato, A., Parikh, N., McKnight, K., Terry, D., Stefanovic, B., Kato, Y., 2010. BCL6 canalizes Notch-dependent transcription, excluding Mastermind-like1 from selected target genes during left-right patterning. *Dev. Cell* 18, 450–462. https://doi.org/10.1016/j.devcel.2009.12.023

Sanalkumar, R., Dhanesh, S.B., James, J., 2010. Non-canonical activation of Notch signaling/target genes in vertebrates. *Cell. Mol. Life Sci.* 67, 2957–2968. https://doi.org/10.1007/s00018-010-0391-x

Sanders, P.G.T., Muñoz-Descalzo, S., Balayo, T., Wirtz-Peitz, F., Hayward, P., Martinez Arias, A., 2009. Ligand-independent traffic of Notch buffers activated Armadillo in Drosophila. *PLoS Biol.* 7, e1000169. https://doi.org/10.1371/journal.pbio.1000169

Sasaki, N., Kiso, M., Kitagawa, M., Saga, Y., 2011. The repression of Notch signaling occurs via the destabilization of mastermind-like 1 by Mesp2 and is essential for somitogenesis. *Development* 138, 55–64. https://doi.org/10.1242/dev.055533

Schlosser, G., Awtry, T., Brugmann, S.A., Jensen, E.D., Neilson, K., Ruan, G., Stammler, A., Voelker, D., Yan, B., Zhang, C., Klymkowsky, M.W., Moody, S.A., 2008. Eya1 and Six1 promote neurogenesis in the cranial placodes in a SoxB1-dependent fashion. *Dev. Biol.* 320, 199–214. https://doi.org/10.1016/J.YDBIO.2008.05.523

Schneider, M.L., Turner, D.L., Vetter, M.L., 2001. Notch signaling can inhibit Xath5 function in the neural plate and developing retina. *Mol. Cell. Neurosci.* 18, 458–472. https://doi.org/10.1006/mcne.2001.1040

Seo, S., Lim, J.-W., Yellajoshyula, D., Chang, L.-W., Kroll, K.L., 2007. Neurogenin and NeuroD direct transcriptional targets and their regulatory enhancers. *EMBO J.* 26, 5093–5108. https://doi.org/10.1038/sj.emboj.7601923

Session, A.M., Uno, Y., Kwon, T., Chapman, J.A., Toyoda, A., Takahashi, S., Fukui, A., Hikosaka, A., Suzuki, A., Kondo, M., van Heeringen, S.J., Quigley, I., Heinz, S., Ogino, H., Ochi, H., Hellsten, U., Lyons, J.B., Simakov, O., Putnam, N., Stites, J., Kuroki, Y., Tanaka, T., Michiue, T., Watanabe, M., Bogdanovic, O., Lister, R., Georgiou, G., Paranjpe, S.S., van Kruijsbergen, I., Shu, S., Carlson, J., Kinoshita, T., Ohta, Y., Mawaribuchi, S., Jenkins, J., Grimwood, J., Schmutz, J., Mitros, T., Mozaffari, S. V., Suzuki, Y., Haramoto, Y., Yamamoto, T.S., Takagi, C., Heald, R., Miller, K., Haudenschild, C., Kitzman, J., Nakayama, T., Izutsu, Y., Robert, J., Fortriede, J., Burns, K., Lotay, V., Karimi, K., Yasuoka, Y., Dichmann, D.S., Flajnik, M.F., Houston, D.W., Shendure, J., DuPasquier, L., Vize, P.D., Zorn, A.M., Ito, M., Marcotte, E.M., Wallingford, J.B., Ito, Y., Asashima, M., Ueno, N., Matsuda, Y., Veenstra, G.J.C., Fujiyama, A., Harland, R.M., Taira, M., Rokhsar, D.S., 2016.

Genome evolution in the allotetraploid frog *Xenopus laevis*. *Nature* 538, 336–343. https://doi.org/10.1038/nature19840

Shinga, J., Itoh, M., Shiokawa, K., Taira, S., Taira, M., 2001. Early patterning of the prospective midbrain: Hindbrain boundary by the HES-related gene XHR1 in Xenopus embryos. *Mech. Dev*. 109, 225–239. https://doi.org/10.1016/S0925-4773(01)00528-7

Shook, D.R., Majer, C., Keller, R., 2004. Pattern and morphogenesis of presumptive superficial mesoderm in two closely related species, *Xenopus laevis* and *Xenopus tropicalis*. *Dev. Biol*. 270, 163–185. https://doi.org/10.1016/j.ydbio.2004.02.021

Sjöqvist, M., Andersson, E.R., 2019. Do as I say, Not(ch) as I do: Lateral control of cell fate. *Dev. Biol*. 447, 58–70. https://doi.org/10.1016/j.ydbio.2017.09.032

Sölter, M., Locker, M., Boy, S., Taelman, V., Bellefroid, E.J., Perron, M., Pieler, T., 2006. Characterization and function of the bHLH-O protein XHes2: Insight into the mechanisms controlling retinal cell fate decision. *Development* 133, 4097–4108. https://doi.org/10.1242/dev.02567

Souopgui, J., Sölter, M., Pieler, T., 2002. XPak3 promotes cell cycle withdrawal during primary neurogenesis in *Xenopus laevis*. *EMBO J*. 21, 6429–6439.

Sparrow, D.B., 2008. Old wares and new: Five decades of investigation of somitogenesis in *Xenopus laevis*. *Adv. Exp. Med. Biol*. 638, 73–94. https://doi.org/10.1007/978-0-387-09606-3_4

Sparrow, D.B., Jen, W.C., Kotecha, S., Towers, N., Kintner, C., Mohun, T.J., 1998. Thylacine 1 is expressed segmentally within the paraxial mesoderm of the Xenopus embryo and interacts with the Notch pathway. *Development* 125, 2041–2051.

Steventon, B., Mayor, R., 2012. Early neural crest induction requires an initial inhibition of Wnt signals. *Dev Biol* 365, 196–207. https://doi.org/10.1016/j.ydbio.2012.02.029

Steventon, B., Mayor, R., Streit, A., 2014. Neural crest and placode interaction during the development of the cranial sensory system. *Dev. Biol*. 389, 28–38. https://doi.org/10.1016/j.ydbio.2014.01.021

Stuhlmiller, T.J., García-Castro, M.I., 2012. Current perspectives of the signaling pathways directing neural crest induction. *Cell. Mol. Life Sci*. 69, 3715–3737. https://doi.org/10.1007/s00018-012-0991-8

Taelman, V., Van Campenhout, C., Sölter, M., Pieler, T., Bellefroid, E.J., 2006. The Notch-effector HRT1 gene plays a role in glomerular development and patterning of the Xenopus pronephros anlagen. *Development* 133, 2961–2971. https://doi.org/10.1242/dev.02458

Taelman, V., Van Wayenbergh, R., Sölter, M., Pichon, B., Pieler, T., Christophe, D., Bellefroid, E.J., 2004. Sequences downstream of the bHLH domain of the Xenopus hairy-related transcription factor-1 act as an extended dimerization domain that contributes to the selection of the partners. *Dev. Biol*. 276, 47–63. https://doi.org/10.1016/j.ydbio.2004.08.019

Takada, H., Hattori, D., Kitayama, A., Ueno, N., Taira, M., 2005. Identification of target genes for the Xenopus Hes-related protein XHR1, a prepattern factor specifying the midbrain-hindbrain boundary. *Dev. Biol*. 283, 253–267. https://doi.org/10.1016/j.ydbio.2005.04.020

Takahashi, Y., Koizumi, K.I., Takagi, A., Kitajima, S., Inoue, T., Koseki, H., Saga, Y., 2000. Mesp2 initiates somite segmentation through the Notch signalling pathway. *Nat. Genet*. 25, 390–396. https://doi.org/10.1038/78062

Tanigaki, K., Honjo, T., 2010. Two opposing roles of RBP-J in Notch signaling. *Curr. Top. Dev. Biol*. 92, 231–252. https://doi.org/10.1016/S0070-2153(10)92007-3

Taverner, N. V., Kofron, M., Shin, Y., Kabitschke, C., Gilchrist, M.J., Wylie, C., Cho, K.W.Y., Heasman, J., Smith, J.C., 2005. Microarray-based identification of VegT targets in Xenopus. *Mech. Dev*. 122, 333–354. https://doi.org/10.1016/J.MOD.2004.10.010

Thisse, B., Thisse, C., 2015. Formation of the vertebrate embryo: Moving beyond the Spemann organizer. *Semin. Cell Dev. Biol*. 42, 1–9. https://doi.org/10.1016/j.semcdb.2015.05.007

Thurmond, J., Goodman, J.L., Strelets, V.B., Attrill, H., Gramates, L.S., Marygold, S.J., Matthews, B.B., Millburn, G., Antonazzo, G., Trovisco, V., Kaufman, T.C., Calvi, B.R., Perrimon, N., Gelbart, S.R., Agapite, J., Broll, K., Crosby, L., Dos Santos, G., Emmert, D., Falls, K., Jenkins, V., Sutherland, C., Tabone, C., Zhou, P., Zytkovicz, M., Brown, N., Garapati, P., Holmes, A., Larkin, A., Pilgrim, C., Urbano, P., Czoch, B., Cripps, R., Baker, P., 2019. FlyBase 2.0: The next generation. *Nucleic Acids Res*. 47, D759–D765. https://doi.org/10.1093/nar/gky1003

Tomankova, S., Abaffy, P., Sindelka, R., 2017. The role of nitric oxide during embryonic epidermis development of *Xenopus laevis*. *Biol. Open* 6, 862–871. https://doi.org/10.1242/bio.023739

Tözser, J., Earwood, R., Kato, A., Brown, J., Tanaka, K., Didier, R., Megraw, T.L., Blum, M., Kato, Y., 2015. TGF-β signaling regulates the differentiation of motile cilia. *Cell Rep*. 11, 1000–1007. https://doi.org/10.1016/j.celrep.2015.04.025

Tsuji, S., Cho, K.W.Y., Hashimoto, C., 2003. Expression pattern of a basic helix-loop-helix transcription factor Xhairy2b during *Xenopus laevis* development. *Dev. Genes Evol*. 213, 407–411. https://doi.org/10.1007/s00427-003-0338-4

Turner, D.L., Weintraub, H., 1994. Expression of achaete-scute homolog 3 in Xenopus embryos converts ectodermal cells to a neural fate. *Genes Dev*. 8, 1434–1447. https://doi.org/10.1101/gad.8.12.1434

Ueno, T., Ishihara, A., Yagi, S., Koike, T., Yamauchi, K., Shiojiri, N., 2015. Histochemical analyses of biliary development during metamorphosis of *Xenopus laevis* tadpoles. *Zool. Sci*. 32, 88–96. https://doi.org/10.2108/zs140104

Umbhauer, M., Boucaut, J.C., Shi, D.L., 2001. Repression of XMyoD expression and myogenesis by Xhairy-1 in Xenopus early embryo. *Mech. Dev*. 109, 61–68.

Vasiliu, D., Clamons, S., McDonough, M., Rabe, B., Saha, M., 2015. A regression-based differential expression detection algorithm for microarray studies with ultra-low sample size. *PLoS One* 10. https://doi.org/10.1371/journal.pone.0118198

Vega-López, G.A., Bonano, M., Tríbulo, C., Fernández, J.P., Agüero, T.H., Aybar, M.J., 2015. Functional analysis of Hairy genes in Xenopus neural crest initial specification and cell migration. *Dev. Dyn*. 244, 988–1013. https://doi.org/10.1002/dvdy.24295

Venzin, O.F., Oates, A.C., 2020. What are you synching about? Emerging complexity of Notch signaling in the segmentation clock. *Dev. Biol*. 460, 40–54. https://doi.org/10.1016/j.ydbio.2019.06.024

Vernon, A.E., Movassagh, M., Horan, I., Wise, H., Ohnuma, S., Philpott, A., 2006. Notch targets the Cdk inhibitor Xic1 to regulate differentiation but not the cell cycle in neurons. *EMBO Rep*. 7, 643–648. https://doi.org/10.1038/sj.embor.7400691

Wacker, S.A., Alvarado, C., von Wichert, G., Knippschild, U., Wiedenmann, J., Clauss, K., Nienhaus, G.U., Hameister, H., Baumann, B., Borggrefe, T., Knöchel, W., Oswald, F., 2011. RITA, a novel modulator of Notch signalling, acts via nuclear export of RBP-J. *EMBO J.* 30, 43–56. https://doi.org/10.1038/emboj.2010.289

Wahi, K., Bochter, M.S., Cole, S.E., 2016. The many roles of Notch signaling during vertebrate somitogenesis. *Semin. Cell Dev. Biol.* 49, 68–75. https://doi.org/10.1016/j.semcdb.2014.11.010

Wang, J., Li, S., Chen, Y., Ding, X., 2007. Wnt/β-catenin signaling controls Mespo expression to regulate segmentation during Xenopus somitogenesis. *Dev. Biol.* 304, 836–847. https://doi.org/10.1016/j.ydbio.2006.12.034

Wardle, F.C., Smith, J.C., 2004. Refinement of gene expression patterns in the early Xenopus embryo. *Development* 131, 4687–4696. https://doi.org/10.1242/dev.01340

Watanabe, M., Yasuoka, Y., Mawaribuchi, S., Kuretani, A., Ito, M., Kondo, M., Ochi, H., Ogino, H., Fukui, A., Taira, M., Kinoshita, T., 2017. Conservatism and variability of gene expression profiles among homeologous transcription factors in *Xenopus laevis*. *Dev. Biol.* 426, 301–324. https://doi.org/10.1016/j.ydbio.2016.09.017

Wessely, O., Agius, E., Oelgeschläger, M., Pera, E.M., De Robertis, E.M., 2001. Neural induction in the absence of mesoderm: Beta-catenin-dependent expression of secreted BMP antagonists at the blastula stage in Xenopus. *Dev. Biol.* 234, 161–173. https://doi.org/10.1006/dbio.2001.0258

Wettstein, D.A., Turner, D.L., Kintner, C., 1997. The Xenopus homolog of Drosophila suppressor of Hairless mediates Notch signaling during primary neurogenesis. *Development* 124, 693–702.

Wittenberger, T., Steinbach, O.C., Authaler, A., Kopan, R., Rupp, R.A., 1999. MyoD stimulates delta-1 transcription and triggers notch signaling in the Xenopus gastrula. *EMBO J.* 18, 1915–1922. https://doi.org/10.1093/emboj/18.7.1915

Yamaguti, M., Cho, K.W.Y., Hashimoto, C., 2005. Xenopus hairy2b specifies anterior prechordal mesoderm identity within Spemann's organizer. *Dev. Dyn.* 234, 102–113. https://doi.org/10.1002/dvdy.20523

Yan, B., Moody, S.A., 2007. The competence of Xenopus blastomeres to produce neural and retinal progeny is repressed by two endo-mesoderm promoting pathways. *Dev. Biol.* 305, 103–119. https://doi.org/10.1016/J.YDBIO.2007.01.040

Yan, B., Neilson, K.M., Moody, S.A., 2009. Notch signaling downstream of foxD5 promotes neural ectodermal transcription factors that inhibit neural differentiation. *Dev. Dyn.* 238, 1358–1365. https://doi.org/10.1002/dvdy.21885

Zhou, M., Yan, J., Ma, Z., Zhou, Y., Abbood, N.N., Liu, J., Su, L., Jia, H., Guo, A.-Y., 2012. Comparative and evolutionary analysis of the HES/HEY gene family reveal exon/intron loss and teleost specific duplication events. *PLoS One* 7, e40649. https://doi.org/10.1371/journal.pone.0040649

Zimmerman, K., Shih, J., Bars, J., Collazo, A., Anderson, D.J., 1993. XASH-3, a novel Xenopus achaete-scute homolog, provides an early marker of planar neural induction and position along the mediolateral axis of the neural plate. *Development* 119, 221–232.

8 The Development and Evolution of the Vertebrate Neural Crest

Insights from Xenopus

Joshua R. York and Carole LaBonne

CONTENTS

8.1. Introduction ... 125
8.2. Historical Background: The Importance of Amphibian Models in Early Neural Crest Research ... 126
8.3. Insights into Neural Crest Development from *Xenopus* in the Era of Molecular Biology and "-Omics" ... 127
8.3.1. Neural Crest Progenitors and Establishment of the Neural Plate Border ... 127
8.3.2. Neural Crest EMT and Migration ... 129
8.3.3. Neural Crest Lineage Diversification ... 131
8.4. The Origins of Neural Crest Potential ... 131
8.5. Future Directions and Outlook ... 133
References ... 134

8.1. INTRODUCTION

The first vertebrates appeared some 500 million years ago, and much of what fueled their origin and early diversification can be credited to a small population of stem cells that appear only transiently during embryonic development—the neural crest (Gans and Northcutt 1983; Hall 2018; York and McCauley 2020a, 2020b; Trainor 2013). Neural crest cells are centrally important to vertebrates because they are responsible for generating many of the morphological, physiological, and behavioral traits that define the vertebrate clade. Among these are cartilage and bone elements of the head, face, and neck; nearly all of the sensory neurons and glia of the peripheral nervous system; colorful patterns of pigmentation in skin, feathers, and scales; and even parts of the heart and teeth (Green et al. 2015; Jandzik et al. 2015; Square et al. 2016; Martik, Gandhi et al. 2019).

Neural crest cells are found in all vertebrate embryos, and their development follows a course of events that is roughly similar across even distantly related groups (Green et al. 2015; York and McCauley 2020a). The precursors of these cells arise at the lateral edges of the neural plate (presumptive central nervous system, or CNS), a region known as the neural plate border (Le Douarin and Kalcheim 1999; Hall 2008a). During neurulation, neural crest stem cells expressing a definitive set of gene-regulatory factors localize to the dorsal region ("crest") of the closing neural tube. From there, they undergo an epithelium-to-mesenchyme transition (EMT), delaminate from the neural epithelium, and migrate extensively throughout the embryo to sites where they will give rise to both ectomesenchymal (cartilage, bone, smooth muscle) and non-ectomesenchymal (neurons, glia, pigment) derivatives (Le Douarin and Kalcheim 1999; Hall 2008a, 2008b). By studying and comparing how these processes play out in different vertebrates, it is possible to build a comprehensive picture of a shared developmental program for neural crest development.

Much of our knowledge of neural crest developmental mechanisms has come from studies in a handful of vertebrate organisms, including mouse, chicken, zebrafish, and amphibians. These so-called model systems are the workhorses of modern embryology research, and for good reason. The embryos are relatively easy to obtain and rear in simple laboratory settings, they have well-annotated genomes and transcriptomes, and they are amenable to a wide range of molecular-genetic techniques. Among these, amphibian embryos—particularly those of the genus *Xenopus*—have historically been a "go-to" for scientists using perturbation experiments to gain mechanistic insights into developmental processes (Elsdale et al. 1960; Akira and Ide 1987; Sadaghiani and Thiébaud 1987). With the advent of powerful molecular biology and genomics approaches (Vize and Zorn 2017; Blum and Ott 2018; Kakebeen and Wills 2019), the contributions from this model system have only continued to grow. Indeed, studies of the neural crest in *Xenopus* and other amphibians have yielded important discoveries regarding the developmental and evolutionary origins of these stem cells, their role in sculpting the vertebrate body plan and producing evolutionary novelty, and their links to human congenital disorders and diseases (Piekarski et al. 2014; LaBonne and Zorn 2015; Greenberg et al. 2019).

DOI: 10.1201/9781003050230-9

8.2. HISTORICAL BACKGROUND: THE IMPORTANCE OF AMPHIBIAN MODELS IN EARLY NEURAL CREST RESEARCH

The neural crest was first described in chick embryos by Wilhelm His in 1868 (Figure 8.1A) (His 1868). He reported a band of cells—which he termed the *Zwischenstrang*—situated between presumptive epidermis (*Hornblatt*) and neural tube. In identifying what would become known as the neural crest (the term "neural crest" would not appear until 1879; Marshall 1879), His had also implicitly mapped out key domains of the early vertebrate ectoderm, which he later outlined in more detail (His 1879). It was around this time that other researchers such as Balfour, Sagemhel, Kastschenko, and Marshall began describing neural crest cells in other vertebrates (Balfour and Foster 1876; Marshall 1879; Kastschenko 1888).

Following the identification of neural crest in amphibians, these embryos became the focus of many of the classic studies of neural crest cells throughout the late 19th and early 20th centuries (Horstadius 1950). Studies in amphibians gained further prominence with the advent of experimental embryology (or *Entwicklungsmechanic*), led by Wilhelm Roux. This new crop of experimentalists sought to manipulate, rather than simply describe, embryonic development. Amphibians are ideal for such studies because the embryos are large, develop externally, and are tolerant of physical perturbations such as grafting, microsurgery, and lineage tracing. For example, Brachet (1907), and Baker and Graves (1939) showed in *Rana* and *Ambystoma*, respectively, that neural crest progenitors could be identified at open neural plate stages. Studies in amphibians provided important insights into the lineages and derivatives that neural crest cells give rise to. In a series of papers, DuShane (1935, 1938) and Niu (1947) used elegant transplantation experiments to show a neural crest contribution to melanocytes, whereas Harrison (1924), Detwiler and Kehoe (1939), and Raven (1935, 1937) cleverly used dyes and grafting experiments to reveal the neural crest origin of spinal ganglia (also called dorsal root ganglia) and the dental papillae of teeth. One of the most important and controversial discoveries was made by Julia Platt (Figure 8.1B), who showed that the neural crest—a cell population ostensibly of ectodermal origin—generated craniofacial features, such as cartilage and dentine, thought to be strictly derived from mesoderm (Platt 1894). This seminal finding, later to be confirmed by others such as Sven Hörstadius (Figure 8.1C) (Hörstadius and Sellman 1941), ran counter to the prevailing dogma of August Weissman's germ layer theory (Platt 1894; Hall 1998, 2008a, 2018). Raven would bolster the findings of Hörstadius and Platt by using grafting experiments to show that during gastrulation, the neural crest was, in his terms, "omnipotent" and could give rise to cell types associated with all three germ layers (Raven 1935).

These discoveries about the neural crest not only forced a reevaluation of one of the cornerstone theories of embryology, but they also revolutionized our understanding of vertebrate origins. The realization that many of the structures and cell types that define the vertebrate body plan are derived from neural crest made understanding the origins of these cells central to studies of vertebrate evolution. This was emphasized in the 1980s with the "New Head" hypothesis put forward by Gans and Northcutt (Gans and Northcutt 1983; Northcutt and Gans 1983; Gans 1989). In their model, the vertebrate head is a truly neomorphic structure, created in large part by neural crest (and placodes), and gave our early vertebrate ancestors a selective advantage by enabling the transition from a passive, filter-feeding lifestyle to one of active predation (Gans and Northcutt 1983; Northcutt and Gans 1983; Gans 1989). Collectively, this early work in amphibians would cement the importance of the neural crest to both evolutionary and developmental biology for decades to come.

FIGURE 8.1 Key figures in the early history of neural crest research. (A) Wilhelm His, (B) Julia Platt, and (C) Sven Hörstadius.

Source: Image from Marine Biological Laboratory Archives, http://creativecommons.org/licenses/by-nc-sa/3.0/

8.3. INSIGHTS INTO NEURAL CREST DEVELOPMENT FROM *XENOPUS* IN THE ERA OF MOLECULAR BIOLOGY AND "-OMICS"

Up until the 1970s and 1980s, nearly all of our understanding of the mechanisms of embryonic development in amphibians relied on either observation or relatively simple methods of experimentation and biochemical characterization (Nieuwkoop 1973; Gurdon and Hopwood 2003). While these approaches, led in large part by the pioneering work of Peter Nieuwkoop and David Newth, fundamentally advanced our understanding of vertebrate development, they could not provide a mechanistic, molecular-genetic understanding of these processes. The development of widely used molecular techniques such as *in situ* hybridization and immunohistochemistry (Clarke et al. 1984; Hemmati-Brivanlou et al. 1990) positioned *Xenopus* as a system that could rapidly provide key insights into the roles of molecules and genes during vertebrate development (Gurdon and Hopwood 2003). Within the last two decades, an expanded arsenal of powerful new technologies, including morpholinos and antisense RNA (Heasman et al. 2000), CRISPR/Cas9-mediated genome editing (Nakayama et al. 2013; Aslan et al. 2017; DeLay et al. 2018), unbiased genome-wide approaches (RNA-Seq, ChIP-Seq, ATAC-Seq) (Hellsten et al. 2010; Bright and Veenstra 2019; Gentsch and Smith 2019; Hontelez et al. 2019), transgenic lines (Alkobtawi et al. 2018; Li et al. 2019), proteomics (Wühr et al. 2014), and single-cell sequencing (Briggs et al. 2018) became available, greatly expanding the *Xenopus* developmental biologists' toolkit and providing unparalleled insights into embryo development.

Use of these tools in *Xenopus* has significantly enhanced our understanding of the complex, interconnected web of molecular-genetic interactions that drive neural crest development (Prasad et al. 2012). These gene-regulatory interactions progressively drive the development of the neural crest from an early stem cell state through lineage diversification and differentiation into derivatives such as cartilage, neurons, and pigment cells (Meulemans and Bronner-Fraser 2004; Prasad et al. 2012). Although the neural crest gene regulatory network (GRN) is spatially and temporally continuous and hence cannot be reduced into fully separable components, we can nonetheless identify distinct GRN "subcircuits" or "modules" that serve dedicated functions at certain times and places during development. For example, some of these subcircuits direct cell migration, whereas others govern pluripotency and lineage restriction. In the following, we describe our current understanding of these processes in *Xenopus*, focusing on the functional roles of evolutionarily conserved GRN subcircuits.

8.3.1. Neural Crest Progenitors and Establishment of the Neural Plate Border

Definitive neural crest cells can first be identified using molecular markers during gastrulation (~stage 11/12) in *Xenopus* (Mayor et al. 1995; O'Donnell et al. 2006). The importance of tissue interactions in establishing the neural crest was demonstrated in *Xenopus* as early as the 1940s using *in vivo* grafting studies that juxtaposed lateral archenteron roof (paraxial and lateral plate mesoderm precursors) to non-neural ectoderm (Raven and Kloos 1945). This was later confirmed using molecular markers: combining non-neural ectoderm and mesoderm explants was sufficient to induce expression of *snai2* (one of the first identified neural crest markers, formerly known as *slug*) and promote melanocyte formation (Bonstein et al. 1998). Such studies led to a search for the signals released by these tissues. As in other vertebrates, establishment of definitive neural crest involves the combined activity of several evolutionarily conserved signaling pathways (Mayor et al. 1995; LaBonne and Bronner-Fraser 1998; Klymkowsky et al. 2010; Prasad et al. 2012). This includes intermediate levels of BMP activity, in contrast to the high and low levels that characterize the epidermal and neural ectoderm, respectively (Wilson and Hemmati-Brivanlou 1995; Weinstein and Hemmati-Brivanlou 1997; LaBonne and Bronner-Fraser 1998; Marchant et al. 1998). Decreased levels of BMP signaling are achieved via the activity of secreted BMP inhibitors, such as noggin (Lamb et al. 1993; Zimmerman et al. 1996), chordin (Piccolo et al. 1996), follistatin (Fainsod et al. 1997; Iemura et al. 1998), and gremlin 1 (Hsu et al. 1998), as well as other genes and pathways that modulate BMP activity directly or indirectly (e.g. *snai2*, *tsku*, *dll1/notch1*, *traf4*, *snw1*) (Glavic et al. 2004; Kalkan et al. 2009; Wu and Hill 2009; Shi et al. 2011; Wu et al. 2011). Intermediate levels of BMP activity are not sufficient to establish a neural crest state, however (LaBonne and Bronner-Fraser 1998). There is evidence that both canonical and non-canonical Wnt signaling and Wnt-mediated FGF signaling are key players in this step (Mayor et al. 1997; Chang and Hemmati-Brivanlou 1998; Villanueva et al. 2002; Monsoro-Burq et al. 2003; Wu et al. 2005; Abu-Elmagd et al. 2006; Hong et al. 2008; Steventon et al. 2009; Borday et al. 2018). Inhibition of either signaling pathway abrogates expression of neural crest markers and results in loss of neural crest derivatives, whereas enhanced signaling can expand the pool of neural crest progenitors. Work in *Xenopus* has made major contributions to identifying the novel proteins and mechanisms that amplify or attenuate these signals, including Kctd15, Apoc1, Hes3, Dkk2, Adam33, Adam19, Szl, and Daam1 (Table 8.1). There is also evidence for roles for Dll1/Notch1 (Glavic et al. 2004; Kuriyama et al. 2006), and retinoic acid pathways (Li et al. 2018) in establishing neural crest, as well as from other genes typically studied at slightly later stages of neural crest ontogeny, such as *tfap2a* (Luo et al. 2003) and *adam33* (Wei et al. 2010). In addition to signals regulating nuclear gene expression, some studies have identified pathways and proteins linked to mechanical processes including control of the cytoskeleton and tight junction assembly as important for early neural crest formation. These include Rhov (Guémar et al. 2007) and Marveld3-mediated attenuation of Jnk/Mapk8 signaling (Vacca et al. 2018). Similar roles have been described for Kremen2, which mediates its effects through canonical Wnt signaling (Hassler et al. 2007). Taken together, the combined activity of these proteins and signaling pathways establishes an embryonic territory, the neural plate border, which constitutes the earliest recognizable domain of the neural crest in vertebrates.

The neural plate border region is characterized by the expression of a set of transcription factors implicated in intricate regulatory interactions that include positive and negative feedback loops, feed-forward loops, and mutual repression. Although there is substantial inter-species variation in spatiotemporal expression patterns and the exact types of regulatory interactions, all vertebrates nonetheless express a common suite of neural plate border genes. For *Xenopus*, these include members of the Pax (*pax3, pax7*) (Milet et al. 2013), Msx (*msx1, msx2*) (Khadka et al. 2003; Monsoro-Burq et al. 2005), and Zic (*zic1, zic2, zic3, zic5*) (Kuo et al. 1998; Nakata et al. 1998) families, as well as *dlx3* (Feledy et al. 1999; Luo et al. 2001), *gbx2.2* (Li et al. 2009), *irx1* (Glavic et al. 2004), *snai1* (Aybar et al. 2003), *tfap2a* (de Crozé et al. 2011), and *hes4/hairy2* (Vega-López et al. 2015). Recently, there has been evidence that miRNAs (*mir-301a, mir-338*) and miRNA-associated proteins are also expressed at the neural plate border and may be important for maintaining pluripotency (Gessert et al. 2010; Ward et al. 2018). Together, these factors progressively refine the spatial boundaries of the neural crest domain to distinguish it from adjacent tissues, activate downstream target genes responsible for maintaining the neural crest in a stem cell state, and eventually endow these cells with the ability to migrate (Bae et al. 2014).

Work in *Xenopus* has suggested that the core regulatory circuitry of the neural plate border includes *pax3*, *msx1*, *zic1*, and *tfap2a* (Hong and Saint-Jeannet 2007; Plouhinec et al. 2014; Pla and Monsoro-Burq 2018). There is evidence that these transcription factors are some of the earliest targets of signaling through Wnt, BMP, FGF, and retinoic acid pathways and pattern the neural plate border by mediating the effects of these signaling pathways on downstream targets (Tríbulo et al. 2003). Some neural plate border factors are capable of driving expression of downstream neural crest genes in neuralized (i.e. BMP-attenuated) ectodermal explants. For example, expression of either *tfap2a* or *msx1* in neuralized explants is sufficient to activate expression of neural crest genes *pax3*, *msx1*, *sox9*, *sox10*, *snai2* (Monsoro-Burq et al. 2005; Sato et al. 2005; Hong and Saint-Jeannet 2007; de Crozé et al. 2011). Gain- and loss-of-function experiments have revealed some of the epistatic relationships among neural plate border factors, as well as the signaling pathways through which they mediate regulatory control. For example, *msx1* is required for activation of *pax3*, whereas *tfap2a* can coordinate with canonical Wnt signaling to activate *pax3* (de Crozé et al. 2011). *pax3* then works with *msx1* to promote FGF signaling by activating early neural crest genes such as *snai2*. Unlike the Tfap2a and Msx1 factors, neither Pax3 nor Zic1 alone are capable of promoting formation of neural crest cells in neuralized explants. However their combined activity is sufficient to drive robust expression of *sox9*, *sox10*, *snai2*, and *meis3*, followed by *tubb2b* (*n-tubulin*) and *th* (*tyrosine hydroxylase*) during differentiation into neural and melanocyte derivatives, respectively (Maeda et al. 2001). Combined Pax3/Zic1 activity can also promote Cadherin switching, EMT, and production of melanocytes in ectodermal explants (Milet and Monsoro-Burq 2012) and can reprogram other tissues such as ventral ectoderm toward a neural crest progenitor state (Sato et al. 2005; Milet et al. 2013). Importantly, however, these effects on the neural crest are highly dose dependent. For example, although increased levels of Pax3 can promote ectopic neural crest, high *pax3* expression converts neural crest-fated ectoderm to an alternative fate, hatching gland (Hong and Saint-Jeannet 2007). Another important function of neural crest GRN components at these stages is to refine spatial boundaries between neural crest and other ectodermal populations. This can occur through transcriptional repression of *sox2* and *sox3* by Snai1 at the lateral edges of the neural plate (LaBonne and Bronner-Fraser 1998, 2000; Aybar et al. 2003; Langer et al. 2008) and by Nbx/Nkx1–2-mediated inhibition of *sox2* and *otx2* (Kurata and Ueno 2003). Conversely, Prdm12 is expressed in the preplacodal ectoderm and can inhibit expression of neural crest genes such as *foxd3*, *snai2*, and *sox8* (Matsukawa et al. 2015).

Beyond the core neural plate border subcircuit consisting of *pax3-msx1-zic1-tfap2a*, studies in *Xenopus* have uncovered additional factors essential for regulating the establishment of definitive neural crest. These include Pbx1 and Meis1 (Maeda et al. 2002), Pcdh7 (Rashid et al. 2018), Klf4/Neptune (Kurauchi et al. 2010), and Znf703, which act downstream of Pax3 (Hong and Saint-Jeannet 2017) and activate the neural crest genes *snai2* and *sox10*. Similarly, Hmga2 (Macri et al. 2009, 2016) and Fbxw7/Cdc4 (Almeida et al. 2010) act downstream of Pax3 and Msx1 and upstream of *myc*, *snai1*, and *snai2*. More recently, unbiased genome-wide analyses have described the global transcriptomic landscape at the neural plate border. These studies have suggested that, once activated, Pax3 creates an autoregulatory loop and, together with Zic1, activates definitive neural crest regulatory factors including *ednra, gbx2.2, sox8, sox9, sox10, twist1, tfap2b, snai2*, and *foxd3* (Bae et al. 2014; Plouhinec et al. 2014). Many of these interactions were found to promote Wnt and retinoic acid signaling, confirming observations from earlier work (Bae et al. 2014; Plouhinec et al. 2014). Moreover, some of these interactions are likely direct, as in the case with Pax3 and Zic1 directly regulating expression of *snai1/snai2* (Bae et al. 2014; Plouhinec et al. 2014).

The proteins Snai1 and Snai2 are widely recognized for their roles in regulating early neural crest stem cell development. They have been historically categorized as transcriptional repressors (Barrallo-Gimeno and Nieto 2005, 2009), and this activity is key to sharpening the medial-lateral boundaries of the early neural crest by repressing expression of pro-neural *soxb1* (*sox2, sox3*) factors (LaBonne and Bronner-Fraser 2000; Acloque et al. 2011). Another mechanism by which Snai1 and Snai2 factors control the spatial boundaries and balance of neural crest proliferation is by regulating apoptosis. In *Xenopus* Snai2 has been proposed to promote neural crest cell survival, which is antagonized by Msx2-mediated pro-apoptotic signals in the same domain (Tríbulo et al. 2004). Although numerous targets of Snail and Snai2 have been identified at various stages of neural crest ontogeny, much remains to be learned about how *snail* and *snai2* expression is regulated. There is evidence that BMP (Vallin et al. 2001), Wnt (Li, et al. 2019), and retinoic acid signaling (Tamanoue et al. 2010), as well as rho GTPases (Broders-Bondon et al. 2007), all activate expression of *snail* and *snai2*. By contrast, YY1 appears to repress *snai2* expression at the lateral edge of the neural plate border, thereby ensuring that the domain of Snai2 activity does not encroach on the presumptive epidermal ectoderm (Morgan et al. 2004).

The *SRY-related HMG-box* (*sox*) genes comprise a large superfamily of transcription factors that play diverse roles in metazoan embryogenesis. In vertebrates, the SoxE (Sox8, Sox9, Sox10) and SoxD (Sox5, Sox6) subfamilies play reiterated roles in neural crest development. In *Xenopus*, *sox8*, *sox9*, and *sox10* are expressed sequentially (Buitrago-Delgado et al. 2018) and direct neural crest stem cell formation (Spokony et al. 2002; Honoré et al. 2003; Lee et al. 2004; O'Donnell et al. 2006). Similarly, *sox5* is expressed in premigratory neural crest and morpholino-mediated knockdown leads to loss of neural crest gene expression (Nordin and LaBonne 2014).

Other factors involved in establishing neural crest stem cells include Tfap2 (alpha, beta and gamma, epsilon variants) (Gotoh et al. 2003; Zhang et al. 2006; de Crozé et al. 2011; Hong et al. 2014), Myc (Bellmeyer et al. 2003), Id3 (Light et al. 2005), and Hes4/Hairy2 (Nagatomo and Hashimoto 2007). These proteins are all required for neural crest formation and maintenance of pluripotency. *id3* is a key Myc transcriptional target in neural crest stem cells (Reynaud-Deonauth et al. 2002; Light et al. 2005) and controls lineage restriction in these cells by regulating the balance between proliferation and cell death (Kee and Bronner-Fraser 2005). In *Xenopus* neural crest, Id3 can be regulated in part by physical interactions with Hes4/Hairy2 (Nichane et al. 2008). Moreover, both Hes4/Hairy2 and Id3 regulate, and are regulated by, Stat3.1 signaling, with high levels of stat3.1 promoting pluripotent maintenance and low levels triggering differentiation (Nichane et al. 2010). The transcription factor Ets1 also drives neural crest specification and migration via interactions with Vgll3 (Simon et al. 2017) and promotes formation of cardiac neural crest in *Xenopus* (Nie and Bronner 2015).

More work is needed to identify the direct transcriptional targets of each of the key neural crest factors to fully understand their individual regulatory contributions, and *Xenopus* provides an ideal system for such studies. Importantly, a number of these factors can act as both transcriptional activators and repressors in a context-dependent manner, complicating efforts to build simple GRNs depicting their regulatory hierarchies. In addition, the activities of a number of neural crest regulatory factors have been shown to be regulated by post-translational modifications, including phosphorylation and SUMOylation (Taylor and LaBonne 2007). Studies in *Xenopus* have shown that although Sox9 and Sox10 can promote the neural crest stem cell state as well as glia and melanocyte formation, their activity can be modified by SUMOylation, which converts them to transcriptional repressors and promotes ear formation (Taylor and LaBonne 2005; Lee, Taylor-Jaffe et al. 2012). Similarly, the activity of Twist1 and Snail in the neural crest is regulated by ubiquitination (Vernon and LaBonne 2006; Lander et al. 2011, 2013). Understanding both post-translational regulation and transcriptional targets is thus essential to understanding how individual neural crest factors contribute to the control of developmental potential, the control of invasive and migratory behavior, and the capacity to give rise to specific lineage states.

8.3.2. Neural Crest EMT and Migration

The establishment of a definitive neural crest state requires gene regulatory circuits to maintain the broad developmental potential of these cells while endowing them with the capacity for migratory and invasive behavior and potentially biasing them in their capacity to transit to specific lineage fates. The set of identified GRN components functioning at these stages in *Xenopus* is quite large (Pegoraro and Monsoro-Burq 2013), and there is likewise substantial variation in gene regulatory activity across the vertebrate phylogeny. However, a common suite of proteins and signaling pathways for enabling EMT and migration are deployed by representatives of all major vertebrate lineages studied to date. These include Twist1, Snail/Snai2, Tfap2a/b, Foxd3, Ets1, Myc, Id3 and signaling pathways including Endothelin, BMP, Wnt, and FGF, as well as novel factors such as Anos1, Phb, Slc19a1/rfc, Zfand6/awp1, and Sp5 (Table 8.1).

The proteins Twist1 and Twist2 are bHLH transcription factors that play key roles in mesodermal development

TABLE 8.1
Novel/Non-Canonical Genes Identified as Being Essential for Regulating the Neural Crest GRN

Gene	Functional Roles	References
kctd15	Inhibits neural crest formation	Dutta and Dawid 2010
apoc1	Induction/Wnt signaling	Tamai et al. 2000
hes3	Induction and NPB establishment	Hong and Saint-Jeannet 2018
dkk2	NPB establishment/Wnt signaling	Devotta et al. 2018
adam33	Induction/Wnt signaling	Wei et al. 2010
adam19	Induction/Wnt signaling	Li et al. 2018
szl	Induction/Wnt signaling	Salic et al. 1997
anos1	Formation of cranial neural crest/placodes	Bae et al. 2018
zfand6/awp1	Early neural crest formation	Seo et al. 2014
phb	Early neural crest formation	Schneider et al. 2010; Deichmann et al. 2015
sp5	Early neural crest formation	Park et al. 2013
slc19a1/rfc	Epigenetic regulation	Li et al. 2011

across bilaterians. They are also instrumental in specifying a subset of neural crest cells. In *Xenopus*, morpholino-mediated depletion of Twist1 leads to defects in neural crest specification and migration and ultimately results in loss of neural crest-derived elements of the head skeleton (Lander et al. 2011). Forced expression of Twist1 in *Xenopus* also partially inhibits some aspects of neural crest development, in part by pushing them toward an ecto-mesenchymal fate at the expense of non-ectomesenchyme (Lander et al. 2011). Twist1, along with several other factors (Snai1, Sip1, Hif1a), regulates EMTs in diverse cell types, including neural crest in *Xenopus* (Linker et al. 2000; Lander et al. 2011; Barriga et al. 2013). Twist1 mediates this EMT program in part by physically interacting with and modulating the activity of Snai1/Snai2 proteins, which is regulated by Gsk3ß-mediated phosphorylation (Lander et al. 2013).

Snail family transcription factors are also important for regulating cell migration across metazoans (Nieto 2002). In most vertebrates, including *Xenopus*, both Snai1 and Snai2 factors are essential for regulating neural crest EMT (Carl, Dufton et al. 1999; LaBonne and Bronner-Fraser 2000; Aybar et al. 2003; Heeg-Truesdell and LaBonne 2004), although there is evidence that some of their functions are partially redundant and have been swapped during evolution (Locascio et al. 2002). The mechanisms by which Snai1 and Snai2 control neural crest EMT have been a topic of intense study. This typically occurs by direct transcriptional repression of genes that promote epithelial integrity (e.g. *cdh1*) (Cano et al. 2000; Langer et al. 2008), which is often enhanced by physical interaction with other EMT factors, including the Polycomb repressor complex (Tien et al. 2015), lmo4 (Ochoa et al. 2012; Ferronha et al. 2013), and Ajuba LIM-domain proteins (Langer et al. 2008). Moreover, the ability of Snai1 and Snai2 to exert their effects on neural crest formation and migration are highly dependent upon their stability, which is controlled by the F-box protein Ppa (Vernon and LaBonne 2006; Lander et al. 2011) and can be stabilized by Elp3 (Yang et al. 2016).

The forkhead box/winged helix transcription factor Foxd3 also plays critical roles in both controlling the neural crest stem cell state and in promoting EMT. Work in *Xenopus* has shown that knockdown of Foxd3 activity reduces expression of neural crest regulatory factors and causes defects in neural crest migration (Sasai et al. 2001). A similar result is obtained when *foxd3* is ectopically expressed, suggesting that precise levels of this transcription factor are required for normal neural crest development (Pohl and Knöchel 2001). Although homologs of *forkhead/fox* genes are found in invertebrates, the vertebrate *foxd3* paralog is uniquely suited for specification and EMT in the neural crest by having evolved a unique N-terminal sequence (Ono et al. 2014).

Because neural crest cells contribute derivatives throughout the body plan, they must migrate extensively to reach those sites (Theveneau and Mayor 2012b; Gouignard et al. 2018). *Xenopus* is the only tetrapod model in which this complex process can be studied both *in vitro* and *in vivo*, enabling fine-grained quantitative analysis of the migratory properties of neural crest cells (Borchers et al. 2000; Alfandari et al. 2010; Toro-Tapia et al. 2017; Barriga et al. 2019). The cranial neural crest in amphibians initiates migration in three discrete streams along the anterior-posterior axis, termed mandibular, hyoid, and branchial (Sadaghiani and Thiébaud 1987; Szabó et al. 2019). Trunk neural crest cells typically migrate individually or in small clusters (Krotoski et al. 1988; Theveneau and Mayor 2012b; Vega-Lopez et al. 2017; Shellard et al. 2018; Li, Vieceli et al. 2019).

In most vertebrates, neural crest migration requires a "Cadherin switch," in which pro-epithelial Cadherins, such as Cdh1, are degraded and transcriptionally repressed and replaced on the cell surface with Cadherins that enable migration (e.g. Cdh6, Cdh7, Cdh11) (Levi et al. 1991; Vallin et al. 1998; Nandadasa et al. 2009; Theveneau and Mayor 2012a; Langhe et al. 2016; Taneyhill and Schiffmacher 2017). Of these, Cdh11 is a target for cleavage by various proteases, and one of its soluble cleavage products interacts with Cdgfra, Cgfr1, and Crbb2—all of which promote migration (Mathavan et al. 2017). Interestingly, while *cdh11* is highly expressed in migratory neural crest in *Xenopus*, (Vallin et al. 1998; Borchers et al. 2001; Becker et al. 2013), *cdh1* is as well (Huang et al. 2016; Cousin 2017), raising questions about the role for classical Cadherin switching in neural crest migration in amphibians.

Neural crest migration requires active remodeling of the extracellular environment, as well as interplay between the extracellular environment, cell surface, nucleus, and cytoplasm. *Xenopus* is an advantageous system in which to dissect these complex interactions, particularly given the ability to carry out lateralized manipulations that use the contralateral side of the same embryo as a control for normal migration and to combine both *in vivo* and *ex vivo* experiments. A recent fundamental discovery that took advantage of these attributes showed the degree to which mechanical forces drive neural crest behavior. This elegant study showed that the mesoderm surrounding premigratory neural crest undergoes stiffening and that the mechanical forces of this stiffening are both necessary and sufficient to trigger migration (Barriga et al. 2018). Integrin, Vinculin, and Talin proteins sense these mechanical perturbations and then relay this information inside the cell to mediate gene regulatory changes necessary for migration. This work built on prior studies in *Xenopus* demonstrating the importance of interactions between cell surface proteins and the extracellular environment for controlling migration (Epperlein et al. 1988). Many of these interactions also require the activity of proteases, including Adam33, which has been shown to remodel fibronectin substrates and cadherins (Alfandari et al. 1997; Cai et al. 1998; McCusker et al. 2009; Cousin et al. 2012; Abbruzzese et al. 2016).

8.3.3. Neural Crest Lineage Diversification

One of the hallmarks of the neural crest is its capacity to generate remarkably diverse cell types and structures, including cartilage; bone; neurons; glia; melanocytes; and components of the endocrine system, heart, and teeth (Le Douarin and

Kalcheim 1999; Hall 2008a, 2018). Importantly, many of the same transcription factors and signaling pathways that play roles in controlling the neural crest stem cell state are redeployed to direct transit to specific lineage states, including SoxD/E families, Foxd3, Wnt, and BMP. However, in this new context, these transcription factors and pathways must control a new set of regulatory targets that will promote and characterize a differentiated state. As the list of neural crest-derived structures is vast, here we focus on two derivatives where work in *Xenopus* has made fundamental contributions that have informed work in other models: the craniofacial skeleton and melanocytes. The full set of neural crest derivatives have been comprehensively reviewed elsewhere (Le Douarin and Kalcheim 1999; Hall 2008a, 2018; Schock and LaBonne 2020).

The formation of the neural crest-derived head skeleton requires an early phase of patterning followed by differentiation via chondrogenesis and, eventually, osteogenesis (Gross and Hanken 2005). The three-dimensional structure of the head skeleton is delineated early in development by Wnt and BMP signaling (Jacox et al. 2016). These pathways create a Cartesian coordinate system of transcription factors and other signaling molecules that interact along the dorsal-ventral (DV) and anterior-posterior (AP) axes. The function of this system is to specify cell type and structural identity of specific skeletal elements. For the DV axis, this involves nested and combinatorial expression of *dlx*, *alx*, *msx*, *prrx*, *emx*, and *hand* transcription factors. The precise combination of these nested expression patterns is what confers skeletal identity. For example, *hand1*, *hand2*, *dlx1*, *dlx2*, *dlx5*, and *dlx6* are all expressed in the ventral pharynx and lower jaw precursors, whereas a *dlx1/dlx2*-positive and *hand1/hand2*-negative domain is expressed dorsally and generates the upper jaw skeleton (Square et al. 2015). Similarly, expression of *barx*, *nkx*, *gdf5*, and *zax* genes, along with signaling by Satb2 and Endothelins, prefigures the jaw joint in the intermediate pharynx (Newman et al. 1997; Reisoli et al. 2010; Square et al. 2015; Lukas and Olsson 2018; Lukas et al. 2020). Finally, recent work has shown that intercellular signaling by the Endothelin pathway—one of just a few signaling pathways that is truly unique to vertebrates—is crucial for establishing the DV patterning program, a feature conserved between *Xenopus* and lamprey, a jawless vertebrate (Bonano et al. 2006; Kawasaki-Nishihara et al. 2011; Square et al. 2016, 2020).

A similar patterning role for *hox* genes occurs along the AP axis. An anterior *hox*-negative expression domain generates oropharyngeal elements such as Meckel's cartilage, with increasingly nested *hox* expression domains more posteriorly prefiguring the gill-bearing ceratobranchial cartilages. The importance of this AP *hox* expression code in *Xenopus* was demonstrated by knockdown and overexpression experiments. When *hoxa2* was over-expressed in post-migratory neural crest, homeotic transformations were observed wherein the anterior jaw skeleton of the first arch was transformed into a hyoid skeleton, a second-arch derivative (Pasqualetti et al. 2000). Complete knockdown of the entire first *hox* paralogous group (*hoxa1, hoxb1*, and *hoxd1*) caused more severe defects, including an inability of cranial neural crest to colonize the pharyngeal arches (McNulty et al. 2005).

After patterning, differentiation of cranial neural crest cells requires expression of several transcription factors known to be important for chondrogenesis and osteogenesis across vertebrates (Square et al. 2016). These include members of the SoxE (Sox8, Sox9), SoxD (Sox5, Sox6), and Runx (Runx2) transcription factor families and FGF, BMP, and retinoic acid signaling pathways (Golub et al. 2000; Eimon and Harland 2001; Spokony et al. 2002; Lee et al. 2004; Kerney et al. 2007; Tahir et al. 2014). Interactions between neural crest cells and surrounding tissues (mesoderm and endoderm) are also required for proper differentiation, and this likely reflects the requirement for continued input from signaling pathways (Seufert and Hall 1990). Overall, these tissue interactions and transcription factor activities promote expression of proteins (e.g. Collagens, Aggrecans, Lectins) that mark the transition from progenitor regions to differentiated skeletal elements (Seufert et al. 1994; Evanson and Milos 1996; Kerney et al. 2010).

Melanocytes are responsible for creating the colorful patterns of pigmentation observed in the skin, feathers, and scales of phylogenetically diverse vertebrates, including amphibians (Dupin and Le Douarin 2003). In *Xenopus*, as in other vertebrates, the transcription factor Sox10 acts as a key regulator of melanocyte development by activating expression of *mitf* and is also required, along with Endothelin signaling, for proper melanocyte migration (Aoki et al. 2003; Kawasaki-Nishihara et al. 2011). Indeed, neural crest-derived melanocyte precursors in the head are readily identified in *Xenopus* embryos days before they actually differentiate by expression of the endothelin receptor, *ednrb2* (Square et al. 2016). The progression from melanocyte precursor to fully differentiated melanocyte in both *Xenopus* and *Rana* proceeds along the dorsal-ventral axis controlled by an antagonism between melanocyte-stimulating and melanocyte-inhibiting factors and their corresponding receptors, and defects in these features may underlie the albinism of natural *Xenopus* mutants (Fukuzawa and Ide 1987; Fukuzawa and Bagnara 1989). Signaling interactions between neural crest-derived melanocyte precursors and the surrounding tissues through which they migrate strongly influence both the migratory routes of these cells and their subsequent patterns of melanophore pigmentation (MacMillan 1976; Milos and Wilson 1986; Frunchak and Milos 1990).

8.4. THE ORIGINS OF NEURAL CREST POTENTIAL

One of the most remarkable features of the neural crest is its ability to produce such a large and diverse array of cell types and structures. This impressive developmental potential is most obvious when considering that neural crest cells produce derivatives associated with both ectodermal (neurons) and mesodermal (cartilage, bone) germ layers and also make contributions to endodermal organs (thymus, thyroid). This has led to their being considered by some a "fourth germ layer" (Hall 2008a; Le Douarin and Dupin 2014) (Figure 8.2A), and their multi-germ layer potential has at times fomented controversy and confusion in the field.

According to C.H. Waddington's classic model of developmental potential, stem cells transit through a series of decision points which progressively restrict their developmental potential until they ultimately commit to a specific lineage state (Waddington 1947, 1957; Slack 2002). Waddington (1947, 1957) depicted this process in his iconic landscape analogy. In Waddington's landscape, a pluripotent stem cell can be envisioned as a ball rolling down a hill, with each position at the bottom representing a unique differentiated state. As the ball descends, it will be influenced, in part by the landscape itself, to "choose" which of the bifurcating troughs and valleys it will travel down until it reaches a final position at the base. In Waddington's analogy, once a cell "descends" down particular pathways of identity, its future potential will always be less than what it was previously—the ball cannot roll back up the hill.

Viewed through the lens of germ layer theory, however, neural crest cells seem to be an exception to Waddington's model. Despite their origins in the ectoderm, neural crest cells exhibit developmental potential that is greater than that of ectoderm, giving rise to extensive mesodermal derivatives and contributing to tissues of otherwise endodermal origin (Le Douarin and Kalcheim 1999). Early work showed that a combination of inductive cues from paraxial mesoderm and non-neural ectoderm were able to trigger the formation of neural crest stem cells and derivatives from ectoderm (Raven and Kloos 1945; Bonstein et al. 1998; Marchant et al. 1998), suggesting that ectoderm could somehow "re-gain" developmental potential, in opposition to Waddington's model. More recent work has suggested a model that reconciles Waddington's landscape with the unique features of the vertebrate neural crest (Buitrago-Delgado et al. 2015).

Work in *Xenopus* has shown that many of the transcription factors and signaling pathways crucial for genesis of neural crest cells are first expressed in the pluripotent animal pole cells of the blastula. These include signals such as BMP and FGF and a large set of transcription factors including Myc, Id3, Snai1, Sox5, Tfap2a, Foxd3, Ets1, Pax3, and Zic1 that are co-expressed with canonical pluripotency factors such as Sox2, Sox3, Ventx2.2 (*Xenopus* equivalent of Nanog), and Oct25/Oct60 (*Xenopus* equivalents of Oct4) (Morrison and Brickman 2006; Nordin and LaBonne 2014; Buitrago-Delgado, Nordin et al. 2015; Buitrago-Delgado, Schock et al. 2018; Geary and

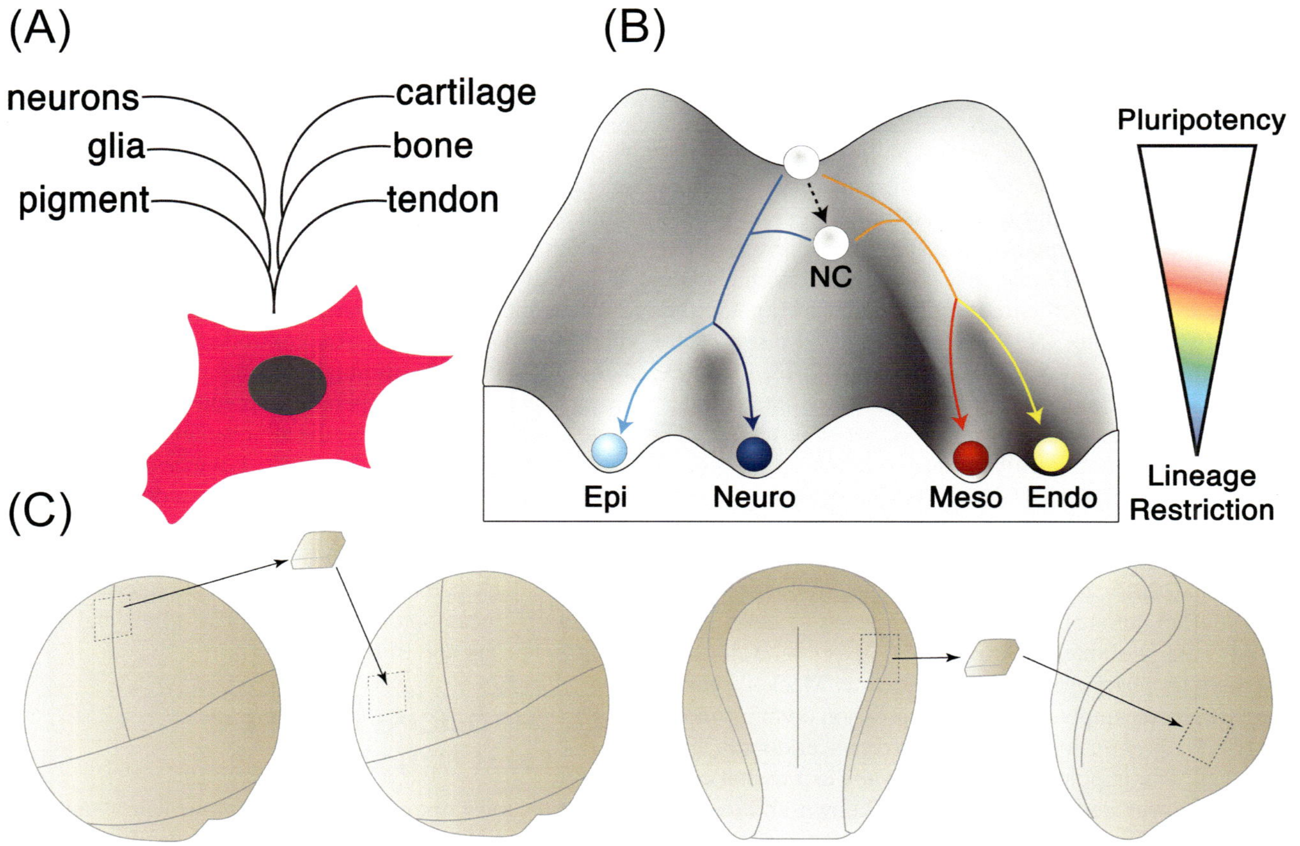

FIGURE 8.2 Insights into neural crest pluripotency from the past and present. (A) Examples of neural crest derivatives demonstrating pluripotency. (B) New model for how retention of pluripotency by neural crest cells can be reconciled with Waddington's landscape of cellular potential. (C) Diagram adapted from Hörstadius (Horstadius 1950) of Raven's transplantation experiments in amphibians demonstrating neural crest pluripotency *in vivo*.

LaBonne 2018). These findings indicate that neural crest cells and pluripotent blastula cells are similar at the gene regulatory level and provide novel insights into the broad multi-germ layer developmental potential that characterizes the neural crest. Indeed, work in *Xenopus* has demonstrated that neural crest cells can not only contribute to ectoderm and mesoderm but also endoderm (Buitrago-Delgado et al. 2015). The latter provides new context to the contributions of neural crest cells to endoderm-derived organs such as the adrenal gland, thyroid, and thymus (Pearse and Polak 1971; Le Douarin and Teillet 1974; Polak et al. 1974; Le Douarin and Jotereau 1975; Bockman and Kirby 1984; Le Douarin and Kalcheim 1999).

The similarities between neural crest and blastula stem cells extend beyond shared transcription factor circuitry. FGF/MAPK signaling is required for establishment of neural crest development and maintenance of pluripotency and proper lineage restriction of blastula stem cells. Both processes are accompanied by a decrease in MAPK signaling and increase in PI3K/Akt signaling (Geary and LaBonne 2018). This can be replicated *in vitro* as Pax3/Zic1-mediated reprogramming of animal cap cells to a neural crest state results in robust MAPK signaling with low levels of PI3K (Geary and LaBonne 2018).

Similar results are obtained with transcription factors. Both Snail and Sox5 regulate pluripotency factors and lineage restriction in the blastula and the formation of neural crest cells both *in vitro* and *in vivo*. Sox5 has been shown to partner with BMP R-Smads to regulate expression of target genes in the blastula (*ventx2.2*, *id3*), neural crest (*msx1*), and epidermis (*krt12.4/epk*) (Nordin and LaBonne 2014; Buitrago-Delgado et al. 2015). Similarly, inhibition of Snail function revealed that it regulates expression of most of the core pluripotency network (*sox2/3*, *oct25/60*, *ventx2.2*, *tfap2a*, *id3*) in the blastula *in vivo* and is essential to direct proper lineage restriction of pluripotent blastula cells toward an endomesodermal progenitor state (Buitrago-Delgado et al. 2015). Shared features between neural crest and blastula stem cells are also found at the epigenetic level. For example, Hdac1 activity is crucial for expression of the *sox-oct-myc-vent* pluripotency axis and proper lineage restriction of all three germ layers, as well as for establishment of the neural crest (Rao and LaBonne 2018). In addition, low levels of H3K9 and H3K27 acetylation are characteristic of both pluripotent blastula stem cells and neural crest cells (Rao and LaBonne 2018). Importantly, increased hdac1 activity was found to enhance reprogramming to a neural crest state, which has implications for regenerative medicine (Rao and LaBonne 2018).

While there are many similarities between neural crest and blastula stem cells, there are also key differences. One of these involves a change in the deployment of Sox transcription factors. SoxB1 (Sox2, Sox3), but not SoxE factors (Sox8, Sox9, Sox10), are active in pluripotent blastula cells (Buitrago-Delgado et al. 2018). Later, as expression of *soxB1* genes becomes restricted to the neural plate, *soxE* genes are turned on in neural crest cells (Buitrago-Delgado et al. 2018). This suggests a functional "hand-off" from SoxB1 to SoxE transcription factor activity during the emergence of neural crest progenitors from pluripotent blastula cells. Consistent with this idea, there are distinct functional requirements for these gene families in each population. Ectopic expression of *sox9* or *sox10* in the blastula disrupts pluripotency, whereas forced expression of *soxB1* genes results in loss of neural crest. By contrast, increased expression of *soxE* genes promote excess neural crest. Finally, there is evidence that this Sox factor "hand-off" is essential for the emergence of neural crest cells because SoxB1 factors cannot replace the activity of SoxE factors to rescue neural crest development (Buitrago-Delgado, Schock et al. 2018).

The extensive shared features between neural crest and blastula stem cells suggest a new model for neural crest origins in which neural crest cells do not "re-gain" developmental potential in response to inductive cues but rather retain aspects of the pluripotent regulatory state of the blastula stem cells from which they are derived. This new model helps explain why neural crest cells exhibit greater developmental potential than the ectoderm in which they originate. In this model, the neural crest should be thought of not as a population of ectoderm-derived cells but rather as a population derived from pluripotent blastula cells that retain their position at or near the top of Waddington's landscape (Figure 8.2B). Retention of this blastula-stage potential into later stages of development allows neural crest cells to add novelty to the basic chordate body plan by contributing cell types and features associated with all three germ layers. Interestingly, Raven recognized this capacity from his amphibian transplantation experiments over 80 years ago (Figure 8.2C), when he presciently observed that:

> The Neural Crest Material in the early stages was omnipotent [e.g. pluripotent] as to the "faculty of differentiation" but had no "tendencies of differentiation" of its own. At the transition from the stage of a yolk plug of intermediate size to one with a small yolk plug there is a change in the potency . . . the faculty of differentiation is restricted . . . as regards to tissues that characterize other germ layers (endomesoderm, notochord, intestine)
>
> **(Raven 1935; Horstadius 1950).**

8.5. FUTURE DIRECTIONS AND OUTLOOK

This chapter provides an overview of the many advances in our understanding of vertebrate neural crest cells derived from studies in amphibians, in particular *Xenopus*. This historical overview reveals just how much our knowledge of neural crest development owes to the over 150 years of research using this key model organism. That said, much remains to be learned about how the neural crest develops and evolves, and many outstanding questions remain. Many of these questions center on control of neural crest fate and potential, some of the very same issues faced by early experimentalists such as Raven and Hörstadius. How, for example, do neural crest cells ultimately decide what fate to adopt and when to do so? Are these choices binary and sequential, or are multiple paths open to all or some cells? What are the gene regulatory and epigenetic mechanisms that control these processes? How are decisions made at the single cell, rather than population, level? With an ever-expanding toolkit of powerful genomic technologies, answers to such questions are now within reach. A similar set of questions concerns the developmental potential of neural crest cells. To what extent is the global regulatory architecture of the neural crest shared with pluripotent blastula cells, and what are the shared versus unique regulatory features of each population? To what extent is the link between neural crest and blastula stem cells evolutionarily conserved across the vertebrate tree of life? Comparative embryology ("evo-devo") will be instrumental in addressing these and similar questions.

Overall, this is a very exciting time for studies focused on understanding the development and evolution of neural crest cells. The experimental strengths of *Xenopus* as a model system, combined with cutting-edge molecular and genomic tools, make it likely that this model will continue to be at the forefront of such studies and continue to produce exciting advances in our understanding of this fascinating cell type.

REFERENCES

Abbruzzese, G., S. F. Becker, J. Kashef and D. Alfandari (2016). "ADAM13 cleavage of cadherin-11 promotes CNC migration independently of the homophilic binding site." *Developmental Biology* **415**(2): 383–390.

Abu-Elmagd, M., C. Garcia-Morales and G. N. Wheeler (2006). "Frizzled7 mediates canonical Wnt signaling in neural crest induction." *Developmental Biology* **298**(1): 285–298.

Acloque, H., O. H. Ocaña, A. Matheu, K. Rizzoti, C. Wise, R. Lovell-Badge and M. A. Nieto (2011). "Reciprocal repression between Sox3 and snail transcription factors defines embryonic territories at gastrulation." *Developmental Cell* **21**(3): 546–558.

Akira, E. and H. Ide (1987). "Differentiation of neural crest cells of *Xenopus laevis* in clonal culture." *Pigment Cell Research* **1**(1): 28–36.

Alfandari, D., H. Cousin and M. Marsden (2010). "Mechanism of Xenopus cranial neural crest cell migration." *Cell Adhesion & Migration* **4**(4): 553–560.

Alfandari, D., T. G. Wolfsberg, J. M. White and D. W. DeSimone (1997). "ADAM 13: A novel ADAM expressed in somitic mesoderm and neural crest cells during *Xenopus laevis* development." *Developmental Biology* **182**(2): 314–330.

Alkobtawi, M., H. Ray, E. H. Barriga, M. Moreno, R. Kerney, A.-H. Monsoro-Burq, J.-P. Saint-Jeannet and R. Mayor (2018). "Characterization of Pax3 and Sox10 transgenic *Xenopus laevis* embryos as tools to study neural crest development." *Developmental Biology* **444**(Supplement): S202–S208.

Almeida, A. D., H. M. Wise, C. J. Hindley, M. K. Slevin, R. S. Hartley and A. Philpott (2010). "The F-box protein Cdc4/Fbxw7 is a novel regulator of neural crest development in *Xenopus laevis*." *Neural Development* **5**(1): 1.

Aoki, Y., N. Saint-Germain, M. Gyda, E. Magner-Fink, Y.-H. Lee, C. Credidio and J.-P. Saint-Jeannet (2003). "Sox10 regulates the development of neural crest-derived melanocytes in Xenopus." *Developmental Biology* **259**(1): 19–33.

Aslan, Y., E. Tadjuidje, A. M. Zorn and S.-W. Cha (2017). "High-efficiency non-mosaic CRISPR-mediated knock-in and indel mutation in F0 Xenopus." *Development* **144**(15): 2852–2858.

Aybar, M. J., M. A. Nieto and R. Mayor (2003). "Snail precedes slug in the genetic cascade required for the specification and migration of the Xenopus neural crest." *Development* **130**(3): 483–494.

Bae, C.-J., C.-S. Hong and J.-P. Saint-Jeannet (2018). "Anosmin-1 is essential for neural crest and cranial placodes formation in Xenopus." *Biochemical and Biophysical Research Communications* **495**(3): 2257–2263.

Bae, C.-J., B.-Y. Park, Y.-H. Lee, J. W. Tobias, C.-S. Hong and J.-P. Saint-Jeannet (2014). "Identification of Pax3 and Zic1 targets in the developing neural crest." *Developmental Biology* **386**(2): 473–483.

Baker, R. C. and G. O. Graves (1939). "The behavior of the neural crest in the forebrain region of Amblystoma." *Journal of Comparative Neurology* **71**(2): 389–415.

Balfour, F. M. and M. Foster (1876). "II. On the development of the spinal nerves in elasmobranch fishes." *Philosophical Transactions of the Royal Society of London* **166**: 175–195.

Barrallo-Gimeno, A. and M. A. Nieto (2005). "The Snail genes as inducers of cell movement and survival: Implications in development and cancer." *Development* **132**(14): 3151–3161.

Barrallo-Gimeno, A. and M. A. Nieto (2009). "Evolutionary history of the Snail/Scratch superfamily." *Trends in Genetics* **25**(6): 248–252.

Barriga, E. H., K. Franze, G. Charras and R. Mayor (2018). "Tissue stiffening coordinates morphogenesis by triggering collective cell migration in vivo." *Nature* **554**(7693): 523–527.

Barriga, E. H., P. H. Maxwell, A. E. Reyes and R. Mayor (2013). "The hypoxia factor Hif-1α controls neural crest chemotaxis and epithelial to mesenchymal transition." *Journal of Cell Biology* **201**(5): 759–776.

Barriga, E. H., A. Shellard and R. Mayor (2019). In vivo and in vitro quantitative analysis of neural crest cell migration. *Neural Crest Cells*, Springer: 135–152.

Becker, S., R. Mayor and J. Kashef (2013). "Cadherin-11 mediates contact inhibition of locomotion during Xenopus neural crest cell migration." *PLoS One* **8**(12): e85717.

Bellmeyer, A., J. Krase, J. Lindgren and C. LaBonne (2003). "The protooncogene c-myc is an essential regulator of neural crest formation in Xenopus." *Developmental Cell* **4**(6): 827–839.

Blum, M. and T. Ott (2018). "Xenopus: An undervalued model organism to study and model human genetic disease." *Cells Tissues Organs* **205**(5–6): 303–313.

Bockman, D. E. and M. L. Kirby (1984). "Dependence of thymus development on derivatives of the neural crest." *Science* **223**(4635): 498–500.

Bonano, M., C. Tribulo, T. Aguero, J. DeCalisto, R. Mayor, S. Sanchez and S. Aybar (2006). "Edn1/Ednra pathway in Xenopus neural crest development." *Developmental Biology* **295**(1): 430–430.

Bonstein, L., S. Elias and D. Frank (1998). "Paraxial-fated mesoderm is required for neural crest induction in Xenopus embryos." *Developmental Biology* **193**(2): 156–168.

Borchers, A., R. David and D. Wedlich (2001). "Xenopus cadherin-11 restrains cranial neural crest migration and influences neural crest specification." *Development* **128**(16): 3049–3060.

Borchers, A., H.-H. Epperlein and D. Wedlich (2000). "An assay system to study migratory behavior of cranial neural crest cells in Xenopus." *Development Genes and Evolution* **210**(4): 217–222.

Borday, C., K. Parain, H. Thi Tran, K. Vleminckx, M. Perron and A. H. Monsoro-Burq (2018). "An atlas of Wnt activity during embryogenesis in *Xenopus tropicalis*." *PLoS One* **13**(4): e0193606.

Brachet, A. (1907). "Recherches sur l'ontogénèse de la tête chez les Amphibiens." *Archives de biologie* **23**: 167–247.

Briggs, J. A., C. Weinreb, D. E. Wagner, S. Megason, L. Peshkin, M. W. Kirschner and A. M. Klein (2018). "The dynamics of gene expression in vertebrate embryogenesis at single-cell resolution." *Science* **360**(6392).

Bright, A. R. and G. J. C. Veenstra (2019). "Assay for Transposase-accessible chromatin-sequencing using Xenopus embryos." *Cold Spring Harbor Protocols* **2019**(1): pdb. prot098327.

Broders-Bondon, F., A. Chesneau, F. Romero-Oliva, A. Mazabraud, R. Mayor and J. P. Thiery (2007). "Regulation of XSnail2 expression by rho GTPases." *Developmental Dynamics* **236**(9): 2555–2566.

Buitrago-Delgado, E., K. Nordin, A. Rao, L. Geary and C. LaBonne (2015). "Shared regulatory programs suggest retention of blastula-stage potential in neural crest cells." *Science* **348**(6241): 1332–1335.

Buitrago-Delgado, E., E. N. Schock, K. Nordin and C. LaBonne (2018). "A transition from SoxB1 to SoxE transcription factors is essential for progression from pluripotent blastula cells to neural crest cells." *Developmental Biology* **444**(2): 50–61.

Cai, H., J. Krätzschmar, D. Alfandari, G. Hunnicutt and C. P. Blobel (1998). "Neural crest-specific and general expression of distinct metalloprotease: Disintegrins in early *Xenopus laevis* development." *Developmental Biology* **204**(2): 508–524.

Cano, A., M. A. Pérez-Moreno, I. Rodrigo, A. Locascio, M. J. Blanco, M. G. del Barrio, F. Portillo and M. A. Nieto (2000). "The transcription factor snail controls epithelial: Mesenchymal transitions by repressing E-cadherin expression." *Nature Cell Biology* **2**(2): 76–83.

Carl, T. F., C. Dufton, J. Hanken and M. W. Klymkowsky (1999). "Inhibition of neural crest migration in Xenopus using antisense slug RNA." *Developmental Biology* **213**(1): 101–115.

Chang, C. and A. Hemmati-Brivanlou (1998). "Neural crest induction by Xwnt7B in Xenopus." *Developmental Biology* **194**(1): 129–134.

Clarke, J., B. Hayes, S. Hunt and A. Roberts (1984). "Sensory physiology, anatomy and immunohistochemistry of Rohon-Beard neurones in embryos of *Xenopus laevis*." *The Journal of Physiology* **348**(1): 511–525.

Cousin, H. (2017). "Cadherins function during the collective cell migration of Xenopus cranial neural crest cells: Revisiting the role of E-cadherin." *Mechanisms of Development* **148**: 79–88.

Cousin, H., G. Abbruzzese, C. McCusker and D. Alfandari (2012). "ADAM13 function is required in the 3 dimensional context of the embryo during cranial neural crest cell migration in *Xenopus laevis*." *Developmental Biology* **368**(2): 335–344.

de Crozé, N., F. Maczkowiak and A. H. Monsoro-Burq (2011). "Reiterative AP2a activity controls sequential steps in the neural crest gene regulatory network." *Proceedings of the National Academy of Sciences* **108**(1): 155–160.

Deichmann, C., M. Link, M. Seyfang, V. Knotz, D. Gradl and D. Wedlich (2015). "Neural crest specification by Prohibitin1 depends on transcriptional regulation of prl3 and vangl1." *Genesis* **53**(10): 627–639.

DeLay, B. D., M. E. Corkins, H. L. Hanania, M. Salanga, J. M. Deng, N. Sudou, M. Taira, M. E. Horb and R. K. Miller (2018). "Tissue-specific gene inactivation in *Xenopus laevis*: Knockout of lhx1 in the kidney with CRISPR/Cas9." *Genetics* **208**(2): 673–686.

Detwiler, S. and K. Kehoe (1939). "Further observations on the origin of the sheath cells of Schwann." *Journal of Experimental Zoology* **81**(3): 415–435.

Devotta, A., C.-S. Hong and J.-P. Saint-Jeannet (2018). "Dkk2 promotes neural crest specification by activating Wnt/β-catenin signaling in a GSK3β independent manner." *Elife* **7**: e34404.

Dupin, E. and N. M. Le Douarin (2003). "Development of melanocyte precursors from the vertebrate neural crest." *Oncogene* **22**(20): 3016–3023.

DuShane, G. P. (1935). "An experimental study of the origin of pigment cells in Amphibia." *Journal of Experimental Zoology* **72**(1): 1–31.

Dushane, G. P. (1938). "Neural fold derivatives in the amphibia: Pigment cells, spinal ganglia and Rohon-Beard cells." *Journal of Experimental Zoology* **78**(4): 485–503.

Dutta, S. and I. B. Dawid (2010) "Kctd15 inhibits neural crest formation by attenuating Wnt/β-catenin signaling output." *Development* **137**(18): 3013–3018.

Eimon, P. M. and R. M. Harland (2001). "Xenopus Dan, a member of the Dan gene family of BMP antagonists, is expressed in derivatives of the cranial and trunk neural crest." *Mechanisms of Development* **107**(1–2): 187–189.

Elsdale, T., J. Gurdon and M. Fischberg (1960). "A description of the technique for nuclear transplantation in *Xenopus laevis*." *Development* **8**(4): 437–444.

Epperlein, H.-H., W. Halfter and R. P. Tucker (1988). "The distribution of fibronectin and tenascin along migratory pathways of the neural crest in the trunk of amphibian embryos." *Development* **103**(4): 743–756.

Evanson, J. E. and N. C. Milos (1996). "A monoclonal antibody against neural crest-stage *Xenopus laevis* lectin perturbs craniofacial development of Xenopus." *Journal of Craniofacial Genetics and Developmental Biology* **16**(2): 74–93.

Fainsod, A., K. Deißler, R. Yelin, K. Marom, M. Epstein, G. Pillemer, H. Steinbeisser and M. Blum (1997). "The dorsalizing and neural inducing gene follistatin is an antagonist of BMP-4." *Mechanisms of Development* **63**(1): 39–50.

Feledy, J. A., M. J. Beanan, J. J. Sandoval, J. S. Goodrich, J. H. Lim, M. Matsuo-Takasaki, S. M. Sato and T. D. Sargent (1999). "Inhibitory patterning of the anterior neural plate in Xenopus by homeodomain factors Dlx3 and Msx1." *Developmental Biology* **212**(2): 455–464.

Ferronha, T., M. A. Rabadán, E. Gil-Guiñon, G. Le Dréau, C. de Torres and E. Martí (2013). "LMO4 is an essential cofactor in the Snail2-mediated epithelial-to-mesenchymal transition of neuroblastoma and neural crest cells." *Journal of Neuroscience* **33**(7): 2773–2783.

Frunchak, Y. N. and N. C. Milos (1990). "Studies on cellular adhesion of *Xenopus laevis* melanophores: Pigment pattern formation and alteration in vivo by endogenous galactoside-binding lectin or its sugar hapten inhibitor." *Pigment Cell Research* **3**(2): 101–114.

Fukuzawa, T. and J. T. Bagnara (1989). "Control of melanoblast differentiation in amphibia by α-melanocyte stimulating hormone, a serum melanization factor, and a melanization inhibiting factor." *Pigment Cell Research* **2**(3): 171–181.

Fukuzawa, T. and H. Ide (1987). "Melanophore differentiation in the periodic albino mutant of *Xenopus laevis*." *Pigment Cell Research* **1**(3): 197–201.

Gans, C. (1989). "Stages in the origin of vertebrates: Analysis by means of scenarios." *Biological Reviews* **64**(3): 221–268.

Gans, C. and R. G. Northcutt (1983). "Neural crest and the origin of vertebrates: A new head." *Science* **220**(4594): 268–274.

Geary, L. and C. LaBonne (2018). "FGF mediated MAPK and PI3K/Akt signals make distinct contributions to pluripotency and the establishment of neural crest." *Elife* **7**: e33845.

Gentsch, G. E. and J. C. Smith (2019). "Mapping chromatin features of Xenopus embryos." *Cold Spring Harbor Protocols* **2019**(4): pdb. prot100263.

Gessert, S., V. Bugner, A. Tecza, M. Pinker and M. Kühl (2010). "FMR1/FXR1 and the miRNA pathway are required for eye and neural crest development." *Developmental Biology* **341**(1): 222–235.

Glavic, A., F. Silva, M. J. Aybar, F. Bastidas and R. Mayor (2004). "Interplay between Notch signaling and the homeoprotein Xiro1 is required for neural crest induction in Xenopus embryos." *Development* **131**(2): 347–359.

Golub, R., Z. Adelman, J. Clementi, R. Weiss, J. Bonasera and M. Servetnick (2000). "Evolutionarily conserved and divergent expression of members of the FGF receptor family among vertebrate embryos, as revealed by FGFR expression patterns in Xenopus." *Development Genes and Evolution* **210**(7): 345–357.

Gotoh, M., Y. Izutsu and M. Maéno (2003). "Complementary expression of AP-2 and AP-2rep in ectodermal derivatives of Xenopus embryos." *Development Genes and Evolution* **213**(7): 363–367.

Gouignard, N., C. Andrieu and E. Theveneau (2018). "Neural crest delamination and migration: Looking forward to the next 150 years." *Genesis* **56**(6–7): e23107.

Green, S. A., M. Simões-Costa and M. E. Bronner (2015). "Evolution of vertebrates as viewed from the crest." *Nature* **520**(7548): 474–482.

Greenberg, R. S., H. K. Long, T. Swigut and J. Wysocka (2019). "Single amino acid change underlies distinct roles of H2A: Z subtypes in human syndrome." *Cell* **178**(6): 1421–1436. e1424.

Gross, J. B. and J. Hanken (2005). "Cranial neural crest contributes to the bony skull vault in adult *Xenopus laevis*: Insights from cell labeling studies." *Journal of Experimental Zoology Part B: Molecular and Developmental Evolution* **304**(2): 169–176.

Guémar, L., P. de Santa Barbara, E. Vignal, B. Maurel, P. Fort and S. Faure (2007). "The small GTPase RhoV is an essential regulator of neural crest induction in Xenopus." *Developmental Biology* **310**(1): 113–128.

Gurdon, J. B. and N. Hopwood (2003). "The introduction of *Xenopus laevis* into developmental biology: Of empire, pregnancy testing and ribosomal genes." *International Journal of Developmental Biology* **44**(1): 43–50.

Hall, B. K. (1998). Germ layers and the germ-layer theory revisited." *Evolutionary Biology*, Springer: 121–186.

Hall, B. K. (2008a). *The neural crest and neural crest cells in vertebrate development and evolution*. New York, Springer.

Hall, B. K. (2008b). "The neural crest and neural crest cells: Discovery and significance for theories of embryonic organization." *Journal of Biosciences* **33**(5): 781–793.

Hall, B. K. (2018). "Germ layers, the neural crest and emergent organization in development and evolution." *Genesis* **56**(6–7): e23103.

Harrison, R. G. (1924). "Neuroblast versus sheath cell in the development of peripheral nerves." *The Journal of Comparative Neurology* **37**(1): 123–205.

Hassler, C., C.-M. Cruciat, Y.-L. Huang, S. Kuriyama, R. Mayor and C. Niehrs (2007). "Kremen is required for neural crest induction in Xenopus and promotes LRP6-mediated Wnt signaling." *Development* **134**(23): 4255–4263.

Heasman, J., M. Kofron and C. Wylie (2000). "βCatenin signaling activity dissected in the early Xenopus embryo: A novel antisense approach." *Developmental Biology* **222**(1): 124–134.

Heeg-truesdell, E. and C. Labonne (2004). "A slug, a fox, a pair of sox: Transcriptional responses to neural crest inducing signals." *Birth Defects Research Part C: Embryo Today: Reviews* **72**: 124–139.

Hellsten, U., R. M. Harland, M. J. Gilchrist, D. Hendrix, J. Jurka, V. Kapitonov, I. Ovcharenko, N. H. Putnam, S. Shu and L. Taher (2010). "The genome of the Western clawed frog *Xenopus tropicalis*." *Science* **328**(5978): 633–636.

Hemmati-Brivanlou, A., D. Frank, M. Bolce, B. Brown, H. Sive and R. Harland (1990). "Localization of specific mRNAs in Xenopus embryos by whole-mount in situ hybridization." *Development* **110**(2): 325–330.

His, W. (1868). *Untersuchungen über die erste Anlage des Wirbelthierleibes: die erste Entwickelung des Hühnchens im Ei*. Leipzig, FCW Vogel.

His, W. (1879). "Ueber die Anfänge des peripherischen Nervensystemes." *Archiv für Anatomie*: 455–482.

Hong, C. S., A. Devotta, Y. H. Lee, B. Y. Park and J. P. Saint-Jeannet (2014). "Transcription factor AP2 epsilon (Tfap2e) regulates neural crest specification in Xenopus." *Developmental Neurobiology* **74**(9): 894–906.

Hong, C.-S., B.-Y. Park and J.-P. Saint-Jeannet (2008). "Fgf8a induces neural crest indirectly through the activation of Wnt8 in the paraxial mesoderm." *Development* **135**(23): 3903–3910.

Hong, C.-S. and J.-P. Saint-Jeannet (2007). "The activity of Pax3 and Zic1 regulates three distinct cell fates at the neural plate border." *Molecular Biology of the Cell* **18**(6): 2192–2202.

Hong, C. S. and J. P. Saint-Jeannet (2017). "Znf703, a novel target of Pax3 and Zic1, regulates hindbrain and neural crest development in Xenopus." *Genesis* **55**(12): e23082.

Hong, C.-S. and J.-P. Saint-Jeannet (2018). "The b-HLH transcription factor Hes3 participates in neural plate border formation by interfering with Wnt/β-catenin signaling." *Developmental Biology* **442**(1): 162–172.

Honoré, S. M., M. J. Aybar and R. Mayor (2003). "Sox10 is required for the early development of the prospective neural crest in Xenopus embryos." *Developmental Biology* **260**(1): 79–96.

Hontelez, S., I. van Kruijsbergen and G. J. C. Veenstra (2019). "ChIP-sequencing in Xenopus embryos." *Cold Spring Harbor Protocols* **2019**(1): pdb. prot097907.

Horstadius, S. O. (1950). *The neural crest, its properties and derivatives in the light of experimental research*. Oxford, Oxford University Press.

Hörstadius, S. O. and S. Sellman (1941). "Experimental studies on the determination of the chondrocranium in *Amblystoma mexicanum*." *Arkiv för Zoologi* **33**: 1–8.

Hsu, D. R., A. N. Economides, X. Wang, P. M. Eimon and R. M. Harland (1998). "The Xenopus dorsalizing factor Gremlin identifies a novel family of secreted proteins that antagonize BMP activities." *Molecular Cell* **1**(5): 673–683.

Huang, C., M. C. Kratzer, D. Wedlich and J. Kashef (2016). "E-cadherin is required for cranial neural crest migration in *Xenopus laevis*." *Developmental Biology* **411**(2): 159–171.

Iemura, S.-I., T. S. Yamamoto, C. Takagi, H. Uchiyama, T. Natsume, S. Shimasaki, H. Sugino and N. Ueno (1998). "Direct binding of follistatin to a complex of bone-morphogenetic protein and its receptor inhibits ventral and epidermal cell fates in early Xenopus embryo." *Proceedings of the National Academy of Sciences* **95**(16): 9337–9342.

Jacox, L., J. Chen, A. Rothman, H. Lathrop-Marshall and H. Sive (2016). "Formation of a 'pre-mouth array' from the extreme anterior domain is directed by neural crest and Wnt/PCP signaling." *Cell Reports* **16**(5): 1445–1455.

Jandzik, D., A. T. Garnett, T. A. Square, M. V. Cattell, J. K. Yu and D. M. Medeiros (2015). "Evolution of the new vertebrate head by co-option of an ancient chordate skeletal tissue." *Nature* **518**(7540): 534–537.

Kakebeen, A. and A. Wills (2019). "Advancing genetic and genomic technologies deepen the pool for discovery in *Xenopus tropicalis*." *Developmental Dynamics* **248**(8): 620–625.

Kalkan, T., Y. Iwasaki, C. Y. Park and G. H. Thomsen (2009). "TRAF4 is a positive regulator of TGFβ signaling that affects neural crest formation." *Molecular Biology of the Cell* **20**: 3436–3450.

Kastschenko, N. (1888). "Zur Entwicklungsgeschichte des selachierembryos." *Anatomischer Anzeiger* **3**: 445–467.

Kawasaki-Nishihara, A., D. Nishihara, H. Nakamura and H. Yamamoto (2011). "ET3/Ednrb2 signaling is critically involved in regulating melanophore migration in Xenopus." *Developmental Dynamics* **240**(6): 1454–1466.

Kee, Y. and M. Bronner-Fraser (2005). "To proliferate or to die: Role of Id3 in cell cycle progression and survival of neural crest progenitors." *Genes & Development* **19**(6): 744–755.

Kerney, R., J. B. Gross and J. Hanken (2007). "Runx2 is essential for larval hyobranchial cartilage formation in *Xenopus laevis*." *Developmental Dynamics* **236**(6): 1650–1662.

Kerney, R., B. K. Hall and J. Hanken (2010). "Regulatory elements of Xenopus col2a1 drive cartilaginous gene expression in transgenic frogs." *International Journal of Developmental Biology* **54**(1): 141.

Khadka, D., T. Luo and T. D. Sargent (2003). "Msx1 and Msx2 have shared essential functions in neural crest but may be dispensable in epidermis and axis formation in Xenopus." *International Journal of Developmental Biology* **50**(5): 499–502.

Klymkowsky, M., C. Cortez Rossi and K. B. Artinger (2010). "Mechanisms driving neural crest induction and migration in the zebrafish and *Xenopus laevis*." *Cell Adhesion & Migration* **4**(4): 595–608.

Krotoski, D. M., S. E. Fraser and M. Bronner-Fraser (1988). "Mapping of neural crest pathways in *Xenopus laevis* using inter-and intra-specific cell markers." *Developmental Biology* **127**(1): 119–132.

Kuo, J. S., M. Patel, J. Gamse, C. Merzdorf, X. Liu, V. Apekin and H. Sive (1998). "Opl: A zinc finger protein that regulates neural determination and patterning in Xenopus." *Development* **125**(15): 2867–2882.

Kurata, T. and N. Ueno (2003). "Xenopus Nbx, a novel NK-1 related gene essential for neural crest formation." *Developmental Biology* **257**(1): 30–40.

Kurauchi, T., Y. Izutsu and M. Maéno (2010). "Involvement of Neptune in induction of the hatching gland and neural crest in the Xenopus embryo." *Differentiation* **79**(4–5): 251–259.

Kuriyama, S., G. Lupo, K. Ohta, S.-I. Ohnuma, W. A. Harris and H. Tanaka (2006). "Tsukushi controls ectodermal patterning and neural crest specification in Xenopus by direct regulation of BMP4 and X-delta-1 activity." *Development* **133**(1): 75–88.

LaBonne, C. and M. Bronner-Fraser (1998). "Neural crest induction in *Xenopus*: Evidence for a two-signal model." *Development* **125**(13): 2403–2414.

LaBonne, C. and M. Bronner-Fraser (2000). "Snail-related transcriptional repressors are required in Xenopus for both the induction of the neural crest and its subsequent migration." *Developmental Biology* **221**(1): 195–205.

LaBonne, C. and A. M. Zorn (2015). "Modeling human development and disease in Xenopus." *Developmental Biology* **408**(2): 179.

Lamb, T. M., A. K. Knecht, W. C. Smith, S. E. Stachel, A. N. Economides, N. Stahl, G. D. Yancopolous and R. M. Harland (1993). "Neural induction by the secreted polypeptide noggin." *Science* **262**(5134): 713–718.

Lander, R., T. Nasr, S. D. Ochoa, K. Nordin, M. S. Prasad and C. LaBonne (2013). "Interactions between Twist and other core epithelial: Mesenchymal transition factors are controlled by GSK3-mediated phosphorylation." *Nature Communications* **4**(1): 1–11.

Lander, R., K. Nordin and C. LaBonne (2011). "The F-box protein Ppa is a common regulator of core EMT factors Twist, Snail, Slug, and Sip1." *Journal of Cell Biology* **194**(1): 17–25.

Langer, E. M., Y. Feng, H. Zhaoyuan, F. J. Rauscher III, K. L. Kroll and G. D. Longmore (2008). "Ajuba LIM proteins are snail/slug corepressors required for neural crest development in Xenopus." *Developmental Cell* **14**(3): 424–436.

Langhe, R. P., T. Gudzenko, M. Bachmann, S. F. Becker, C. Gonnermann, C. Winter, G. Abbruzzese, D. Alfandari, M.-C. Kratzer and C. M. Franz (2016). "Cadherin-11 localizes to focal adhesions and promotes cell: Substrate adhesion." *Nature Communications* **7**(1): 1–10.

Le Douarin, N. M. and E. Dupin (2014). The neural crest, a fourth germ layer of the vertebrate embryo: Significance in chordate evolution. *Neural Crest Cells*, Elsevier: 3–26.

Le Douarin, N. M. and F. V. Jotereau (1975). "Tracing of cells of the avian thymus through embryonic life in interspecific chimeras." *The Journal of Experimental Medicine* **142**(1): 17–40.

Le Douarin, N. M. and C. Kalcheim (1999). *The neural crest*. New York, Cambridge University Press.

Le Douarin, N. M. and M.-A. M. Teillet (1974). "Experimental analysis of the migration and differentiation of neuroblasts of the autonomic nervous system and of neurectodermal mesenchymal derivatives, using a biological cell marking technique." *Developmental Biology* **41**(1): 162–184.

Lee, P.-C., K. M. Taylor-Jaffe, K. M. Nordin, M. S. Prasad, R. M. Lander and C. LaBonne (2012). "SUMOylated SoxE factors recruit Grg4 and function as transcriptional repressors in the neural crest." *Journal of Cell Biology* **198**(5): 799–813.

Lee, Y.-H., Y. Aoki, C.-S. Hong, N. Saint-Germain, C. Credidio and J.-P. Saint-Jeannet (2004). "Early requirement of the transcriptional activator Sox9 for neural crest specification in Xenopus." *Developmental Biology* **275**(1): 93–103.

Levi, G., B. M. Gumbiner and J. P. Thiery (1991). "The distribution of E-cadherin during *Xenopus laevis* development." *Development* **111**(1): 159–169.

Li, B., S. Kuriyama, M. Moreno and R. Mayor (2009). "The posteriorizing gene Gbx2 is a direct target of Wnt signalling and the earliest factor in neural crest induction." *Development* **136**(19): 3267–3278.

Li, C., R. Hu, N. Hou, Y. Wang, Z. Wang, T. Yang, Y. Gu, M. He, Y. Shi and J. Chen (2018). "Alteration of the retinoid acid-CBP signaling pathway in neural crest induction contributes to enteric nervous system disorder." *Frontiers in Pediatrics* **6**: 382.

Li, J., M. Perfetto, C. Materna, R. Li, H. T. Tran, K. Vleminckx, M. K. Duncan and S. Wei (2019). "A new transgenic reporter line reveals Wnt-dependent Snai2 re-expression and cranial neural crest differentiation in Xenopus." *Scientific Reports* **9**(1): 1–14.

Li, J., Y. Shi, J. Sun, Y. Zhang and B. Mao (2011). "Xenopus reduced folate carrier regulates neural crest development epigenetically." *PLoS One* **6**(11): e27198.

Li, Y., F. M. Vieceli, W. G. Gonzalez, A. Li, W. Tang, C. Lois and M. E. Bronner (2019). "In vivo quantitative imaging provides insights into trunk neural crest migration." *Cell Reports* **26**(6): 1489–1500. e1483.

Light, W., A. E. Vernon, A. Lasorella, A. Iavarone and C. LaBonne (2005). "Xenopus Id3 is required downstream of Myc for the formation of multipotent neural crest progenitor cells." *Development* **132**(8): 1831–1841.

Linker, C., M. Bronner-Fraser and R. Mayor (2000). "Relationship between gene expression domains of Xsnail, Xslug, and Xtwist and cell movement in the prospective neural crest of Xenopus." *Developmental Biology* **224**(2): 215–225.

Locascio, A., M. Manzanares, M. J. Blanco and M. A. Nieto (2002). "Modularity and reshuffling of Snail and Slug expression during vertebrate evolution." *Proceedings of the National Academy of Sciences of the United States of America* **99**(26): 16841–16846.

Lukas, P. and L. Olsson (2018). "Bapx1 is required for jaw joint development in amphibians." *Evolution & Development* **20**(6): 192–206.

Lukas, P., J. Schmidt and L. Olsson (2020). "Knockdown of zax in *Xenopus laevis* leads to craniofacial malformations and the absence of the intramandibular joint." *Vertebrate Zoology* **70**: 9–22.

Luo, T., Y.-H. Lee, J.-P. Saint-Jeannet and T. D. Sargent (2003). "Induction of neural crest in Xenopus by transcription factor AP2α." *Proceedings of the National Academy of Sciences* **100**(2): 532–537.

Luo, T., M. Matsuo-Takasaki and T. D. Sargent (2001). "Distinct roles for Distal-less genes Dlx3 and Dlx5 in regulating ectodermal development in Xenopus." *Molecular Reproduction and Development: Incorporating Gamete Research* **60**(3): 331–337.

MacMillan, G. J. (1976). "Melanoblast-tissue interactions and the development of pigment pattern in Xenopus larvae." *Development* **35**(3): 463–484.

Macri, S., M. Onorati, G. Manfioletti and R. Vignali (2009). "Xenopus HMGA2 is required for neural crest cell survival and migration." *Mechanisms of Development* **126**: S106.

Macri, S., L. Simula, I. Pellarin, S. Pegoraro, M. Onorati, R. Sgarra, G. Manfioletti and R. Vignali (2016). "Hmga2 is required for neural crest cell specification in *Xenopus laevis*." *Developmental Biology* **411**(1): 25–37.

Maeda, R., A. Ishimura, K. Mood, E. K. Park, A. M. Buchberg and I. O. Daar (2002). "Xpbx1b and Xmeis1b play a collaborative role in hindbrain and neural crest gene expression in Xenopus embryos." *Proceedings of the National Academy of Sciences* **99**(8): 5448–5453.

Maeda, R., K. Mood, T. L. Jones, J. Aruga, A. M. Buchberg and I. O. Daar (2001). "Xmeis1, a protooncogene involved in specifying neural crest cell fate in Xenopus embryos." *Oncogene* **20**(11): 1329–1342.

Marchant, L., C. Linker, P. Ruiz, N. Guerrero and R. Mayor (1998). "The inductive properties of mesoderm suggest that the neural crest cells are specified by a BMP gradient." *Developmental Biology* **198**(2): 319–329.

Marshall, A. M. (1879). "I. Note on the development of the olfactory nerve and olfactory organ of vertebrates." *Proceedings of the Royal Society of London* **28**(190–195): 324–329.

Martik, M. L., S. Gandhi, B. R. Uy, J. A. Gillis, S. A. Green, M. Simoes-Costa and M. E. Bronner (2019). "Evolution of the new head by gradual acquisition of neural crest regulatory circuits." *Nature* **574**(7780): 675–678.

Mathavan, K., V. Khedgikar, V. Bartolo and D. Alfandari (2017). "The ectodomain of cadherin-11 binds to erbB2 and stimulates Akt phosphorylation to promote cranial neural crest cell migration." *PLoS One* **12**(11): e0188963.

Matsukawa, S., K. Miwata, M. Asashima and T. Michiue (2015). "The requirement of histone modification by PRDM12 and Kdm4a for the development of pre-placodal ectoderm and neural crest in Xenopus." *Developmental Biology* **399**(1): 164–176.

Mayor, R., N. Guerrero and C. Martinez (1997). "Role of FGF and noggin in neural crest induction." *Developmental Biology* **189**(1): 1–12.

Mayor, R., R. Morgan and M. G. Sargent (1995). "Induction of the prospective neural crest of Xenopus." *Development* **121**(3): 767–777.

McCusker, C., H. Cousin, R. Neuner and D. Alfandari (2009). "Extracellular cleavage of cadherin-11 by ADAM metalloproteases is essential for Xenopus cranial neural crest cell migration." *Molecular Biology of the Cell* **20**(1): 78–89.

McNulty, C. L., J. N. Peres, N. Bardine, W. M. van den Akker and A. J. Durston (2005). "Knockdown of the complete Hox paralogous group 1 leads to dramatic hindbrain and neural crest defects." *Development* **132**(12): 2861–2871.

Meulemans, D. and M. Bronner-Fraser (2004). "Gene-regulatory interactions in neural crest evolution and development." *Developmental Cell* **7**(3): 291–299.

Milet, C., F. Maczkowiak, D. D. Roche and A. H. Monsoro-Burq (2013). "Pax3 and Zic1 drive induction and differentiation of multipotent, migratory, and functional neural crest in Xenopus embryos." *Proceedings of the National Academy of Sciences* **110**(14): 5528–5533.

Milet, C. and A. H. Monsoro-Burq (2012). "Neural crest induction at the neural plate border in vertebrates." *Developmental Biology* **366**(1): 22–33.

Milos, N. C. and H. C. Wilson (1986). "Cell surface carbohydrate involvement in controlling the adhesion and morphology of neural crest cells and melanophores of *Xenopus laevis*." *Journal of Experimental Zoology* **238**(2): 211–224.

Monsoro-Burq, A.-H., R. B. Fletcher and R. M. Harland (2003). "Neural crest induction by paraxial mesoderm in Xenopus embryos requires FGF signals." *Development* **130**(14): 3111–3124.

Monsoro-Burq, A.-H., E. Wang and R. Harland (2005). "Msx1 and Pax3 cooperate to mediate FGF8 and WNT signals during Xenopus neural crest induction." *Developmental Cell* **8**(2): 167–178.

Morgan, M. J., J. M. Woltering, P. M. I. der Rieden, A. J. Durston and J. P. Thiery (2004). "YY1 regulates the neural crest-associated slug gene in *Xenopus laevis*." *Journal of Biological Chemistry* **279**(45): 46826–46834.

Morrison, G. M. and J. M. Brickman (2006). "Conserved roles for Oct4 homologues in maintaining multipotency during early vertebrate development." *Development* **133**(10): 2011–2022.

Nagatomo, K. I. and C. Hashimoto (2007). "Xenopus hairy2 functions in neural crest formation by maintaining cells in a mitotic and undifferentiated state." *Developmental Dynamics* **236**(6): 1475–1483.

Nakata, K., T. Nagai, J. Aruga and K. Mikoshiba (1998). "Xenopus Zic family and its role in neural and neural crest development." *Mechanisms of Development* **75**(1–2): 43–51.

Nakayama, T., M. B. Fish, M. Fisher, J. Oomen-Hajagos, G. H. Thomsen and R. M. Grainger (2013). "Simple and efficient CRISPR/Cas9-mediated targeted mutagenesis in *Xenopus tropicalis*." *Genesis* **51**(12): 835–843.

Nandadasa, S., Q. Tao, N. R. Menon, J. Heasman and C. Wylie (2009). "N- and E-cadherins in Xenopus are specifically required in the neural and non-neural ectoderm, respectively, for F-actin assembly and morphogenetic movements." *Development* **136**(8): 1327.

Newman, C. S., M. W. Grow, O. Cleaver, F. Chia and P. Krieg (1997). "Xbap, a vertebrate gene related to bagpipe, is expressed in developing craniofacial structures and in anterior gut muscle." *Developmental Biology* **181**(2): 223–233.

Nichane, M., N. de Crozé, X. Ren, J. Souopgui, A. H. Monsoro-Burq and E. J. Bellefroid (2008). "Hairy2-Id3 interactions play an essential role in Xenopus neural crest progenitor specification." *Developmental Biology* **322**(2): 355–367.

Nichane, M., X. Ren and E. J. Bellefroid (2010). "Self-regulation of Stat3 activity coordinates cell-cycle progression and neural crest specification." *The EMBO Journal* **29**(1): 55–67.

Nie, S. and M. E. Bronner (2015). "Dual developmental role of transcriptional regulator Ets1 in Xenopus cardiac neural crest vs. heart mesoderm." *Cardiovascular Research* **106**(1): 67–75.

Nieto, M. A. (2002). "The snail superfamily of zinc-finger transcription factors." *Nature Reviews Molecular Cell Biology* **3**(3): 155–166.

Nieuwkoop, P. (1973). "The 'organization center' of the amphibian embryo: Its origin, spatial organization, and morphogenetic action." *Advances in Morphogenesis*, Elsevier. **10**: 1–39.

Niu, M. C. (1947). "The axial organization of the neural crest, studied with particular reference to its pigmentary component." *Journal of Experimental Zoology* **105**(1): 79–113.

Nordin, K. and C. LaBonne (2014). "Sox5 is a DNA-binding cofactor for BMP R-Smads that directs target specificity during patterning of the early ectoderm." *Developmental Cell* **31**(3): 374–382.

Northcutt, R. G. and C. Gans (1983). "The genesis of neural crest and epidermal placodes: A reinterpretation of vertebrate origins." *The Quarterly Review of Biology* **58**(1): 1–28.

Ochoa, S. D., S. Salvador and C. Labonne (2012). "The LIM adaptor protein LMO4 is an essential regulator of neural crest development." *Developmental Biology* **361**(2): 313–325.

O'Donnell, M., C.-S. Hong, X. Huang, R. J. Delnicki and J.-P. Saint-Jeannet (2006). "Functional analysis of Sox8 during neural crest development in Xenopus." *Development* **133**(19): 3817–3826.

Ono, H., Z. Kozmik, J. K. Yu and H. Wada (2014). "A novel N-terminal motif is responsible for the evolution of neural crest-specific gene-regulatory activity in vertebrate FoxD3." *Developmental Biology* **385**(2): 396–404.

Park, D. S., J. H. Seo, M. Hong, W. Bang, J. K. Han and S. C. Choi (2013). "Role of Sp5 as an essential early regulator of neural crest specification in xenopus." *Developmental Dynamics* **242**(12): 1382–1394.

Pasqualetti, M., M. Ori, I. Nardi and F. M. Rijli (2000). "Ectopic Hoxa2 induction after neural crest migration results in homeosis of jaw elements in Xenopus." *Development* **127**(24): 5367.

Pearse, A. and J. M. Polak (1971). "Neural crest origin of the endocrine polypeptide (APUD) cells of the gastrointestinal tract and pancreas." *Gut* **12**(10): 783–788.

Pegoraro, C. and A. H. Monsoro-Burq (2013). "Signaling and transcriptional regulation in neural crest specification and migration: Lessons from xenopus embryos." *Wiley Interdisciplinary Reviews: Developmental Biology* **2**(2): 247–259.

Piccolo, S., Y. Sasai, B. Lu and E. M. De Robertis (1996). "Dorsoventral patterning in Xenopus: Inhibition of ventral signals by direct binding of chordin to BMP-4." *Cell* **86**(4): 589–598.

Piekarski, N., J. B. Gross and J. Hanken (2014). "Evolutionary innovation and conservation in the embryonic derivation of the vertebrate skull." *Nature Communications* **5**(1): 1–9.

Pla, P. and A. H. Monsoro-Burq (2018). "The neural border: Induction, specification and maturation of the territory that generates neural crest cells." *Developmental Biology* **444**: S36–S46.

Platt, J. B. (1894). "Ectodermic origin of the cartilages of the head." *Anatomischer Anzeiger* **1893**: 506–509.

Plouhinec, J. L., D. D. Roche, C. Pegoraro, A. L. Figueiredo, F. Maczkowiak, L. J. Brunet, C. Milet, J. P. Vert, N. Pollet, R. M. Harland and A. H. Monsoro-Burq (2014). "Pax3 and Zic1 trigger the early neural crest gene regulatory network by the direct activation of multiple key neural crest specifiers." *Developmental Biology* **386**(2): 461–472.

Pohl, B. S. and W. Knöchel (2001). "Overexpression of the transcriptional repressor FoxD3 prevents neural crest formation in Xenopus embryos." *Mechanisms of Development* **103**(1–2): 93–106.

Polak, J. M., A. Pearse, C. Le Lievre, J. Fontaine and N. Le Douarin (1974). "Immunocytochemical confirmation of the neural crest origin of avian calcitonin-producing cells." *Histochemistry* **40**(3): 209–214.

Prasad, M. S., T. Sauka-Spengler and C. LaBonne (2012). "Induction of the neural crest state: Control of stem cell attributes by gene regulatory, post-transcriptional and epigenetic interactions." *Developmental Biology* **366**(1): 10–21.

Rao, A. and C. LaBonne (2018). "Histone deacetylase activity has an essential role in establishing and maintaining the vertebrate neural crest." *Development* **145**(15): dev163386.

Rashid, D., P. Puettmann, E. Roy and R. S. Bradley (2018). "Neural crest development in Xenopus requires Protocadherin 7 at the lateral neural crest border." *Mechanisms of Development* **149**: 41.

Raven, C. P. (1935). "Zur Entwicklung der Ganglienleiste. IV." *Wilhelm Roux'Archiv für Entwicklungsmechanik der Organismen* **132**(4): 509–575.

Raven, C. P. (1937). "Experiments on the origin of the sheath cells and sympathetic neuroblasts in amphibia." *Journal of Comparative Neurology* **67**(2): 221–240.

Raven, C. P. and J. Kloos (1945). "Induction by medial and lateral pieces of the archenteron roof with special reference to the determination of the neural crest." *Acta Morphologica Neerlando-Scandinavica* **5**: 348–362.

Reisoli, E., S. De Lucchini, I. Nardi and M. Ori (2010). "Serotonin 2B receptor signaling is required for craniofacial morphogenesis and jaw joint formation in Xenopus." *Development* **137**(17): 2927–2937.

Reynaud-Deonauth, S., H. Zhang, A. Afouda, S. Taillefert, P. Beatus, M. Kloc, L. D. Etkin, J. Fischer-Lougheed and G. Spohr (2002). "Notch signaling is involved in the regulation of Id3 gene transcription during Xenopus embryogenesis." *Differentiation* **69**(4–5): 198–208.

Sadaghiani, B. and C. H. Thiébaud (1987). "Neural crest development in the *Xenopus laevis* embryo, studied by interspecific transplantation and scanning electron microscopy." *Developmental Biology* **124**(1): 91–110.

Salic, A. N., K. L. Kroll, L. M. Evans and M. W. Kirschner (1997). "Sizzled: A secreted Xwnt8 antagonist expressed in the ventral marginal zone of Xenopus embryos." *Development* **124**(23): 4739–4748.

Sasai, N., K. Mizuseki and Y. Sasai (2001). "Requirement of FoxD3-class signaling for neural crest determination in Xenopus." *Development* **128**(13): 2525–2536.

Sato, T., N. Sasai and Y. Sasai (2005). "Neural crest determination by co-activation of Pax3 and Zic1 genes in Xenopus ectoderm." *Development* **132**(10): 2355–2363.

Schneider, M., A. Schambony and D. Wedlich (2010). "Prohibitin1 acts as a neural crest specifier in Xenopus development by repressing the transcription factor E2F1." *Development* **137**(23): 4073–4081.

Schock, E. N. and C. LaBonne (2020). "Sorting sox: Diverse roles for sox transcription factors during neural crest and craniofacial development." *Frontiers in Physiology* **11**: 1564.

Seo, J.-H., D.-S. Park, M. Hong, E.-J. Chang and S.-C. Choi (2014). "Essential role of AWP1 in neural crest specification in Xenopus." *International Journal of Developmental Biology* **57**(11–12): 829–836.

Seufert, D. W. and B. K. Hall (1990). "Tissue interactions involving cranial neural crest in cartilage formation in *Xenopus laevis* (Daudin)." *Cell Differentiation and Development* **32**(2): 153–165.

Seufert, D. W., J. Hanken and M. W. Klymkowsky (1994). "Type II collagen distribution during cranial development in *Xenopus laevis*." *Anatomy and Embryology* **189**(1): 81–89.

Shellard, A., A. Szabó, X. Trepat and R. Mayor (2018). "Supracellular contraction at the rear of neural crest cell groups drives collective chemotaxis." *Science* **362**(6412): 339–343.

Shi, J., C. Severson, J. Yang, D. Wedlich and M. W. Klymkowsky (2011). "Snail2 controls mesodermal BMP/Wnt induction of neural crest." *Development* **138**(15): 3135–3145.

Simon, E., N. Thézé, S. Fédou, P. Thiébaud and C. Faucheux (2017). "Vestigial-like 3 is a novel Ets1 interacting partner and regulates trigeminal nerve formation and cranial neural crest migration." *Biology Open* **6**(10): 1528–1540.

Slack, J. M. (2002). "Conrad Hal Waddington: The last renaissance biologist?" *Nature Reviews Genetics* **3**(11): 889–895.

Spokony, R. F., Y. Aoki, N. Saint-Germain, E. Magner-Fink and J.-P. Saint-Jeannet (2002). "The transcription factor Sox9 is required for cranial neural crest development in Xenopus." *Development* **129**(2): 421–432.

Square, T., D. Jandzik, M. Cattell, A. Coe, J. Doherty and D. M. Medeiros (2015). "A gene expression map of the larval *Xenopus laevis* head reveals developmental changes underlying the evolution of new skeletal elements." *Developmental Biology* **397**(2): 293–304.

Square, T., D. Jandzik, M. Cattell, A. Hansen and D. M. Medeiros (2016). "Embryonic expression of endothelins and their receptors in lamprey and frog reveals stem vertebrate origins of complex Endothelin signaling." *Scientific Reports* **6**: 34282.

Square, T. A., D. Jandzik, J. L. Massey, M. Romášek, H. P. Stein, A. W. Hansen, A. Purkayastha, M. V. Cattell and D. M. Medeiros (2020). "Evolution of the endothelin pathway drove neural crest cell diversification." *Nature* **585**(7826): 563–568.

Square, T., D. Jandzik, M. Romášek, R. Cerny and D. Medeiros (2016). "The origin and diversification of the developmental mechanisms that pattern the vertebrate head skeleton." *Developmental Biology* **427**(2): 219–229.

Steventon, B., C. Araya, C. Linker, S. Kuriyama and R. Mayor (2009). "Differential requirements of BMP and Wnt signalling during gastrulation and neurulation define two steps in neural crest induction." *Development* **136**(5): 771–779.

Szabó, A., E. Theveneau, M. Turan and R. Mayor (2019). "Neural crest streaming as an emergent property of tissue interactions during morphogenesis." *PLoS Computational Biology* **15**(4): e1007002.

Tahir, R., A. Kennedy, S. H. Elsea and A. J. Dickinson (2014). "Retinoic acid induced-1 (Rai1) regulates craniofacial and brain development in Xenopus." *Mechanisms of Development* **133**: 91–104.

Tamai, K., M. Semenov, Y. Kato, R. Spokony, C. Liu, Y. Katsuyama, F. Hess, J.-P. Saint-Jeannet and X. He (2000). "LDL-receptor-related proteins in Wnt signal transduction." *Nature* **407**(6803): 530–535.

Tamanoue, Y., M. Yamagishi, I. Hongo and H. Okamoto (2010). "Polypyrimidine tract-binding protein is required for the repression of gene expression by all-trans retinoic acid." *Development, Growth & Differentiation* **52**(5): 469–479.

Taneyhill, L. A. and A. T. Schiffmacher (2017). "Should I stay or should I go? Cadherin function and regulation in the neural crest." *Genesis* **55**(6): e23028.

Taylor, K. M. and C. LaBonne (2005). "SoxE factors function equivalently during neural crest and inner ear development and their activity is regulated by SUMOylation." *Developmental Cell* **9**(5): 593–603.

Taylor, K.M. and LaBonne. C. (2007). Modulating the activity of neural crest regulatory factors. *Current Opinion in Genetics and Development* **17**(4): 326–331.

Theveneau, E. and R. Mayor (2012a). "Cadherins in collective cell migration of mesenchymal cells." *Current Opinion in Cell Biology* **24**(5): 677–684.

Theveneau, E. and R. Mayor (2012b). "Neural crest delamination and migration: From epithelium-to-mesenchyme transition to collective cell migration." *Developmental Biology* **366**(1): 34–54.

Tien, C.-L., A. Jones, H. Wang, M. Gerigk, S. Nozell and C. Chang (2015). "Snail2/Slug cooperates with Polycomb repressive complex 2 (PRC2) to regulate neural crest development." *Development* **142**(4): 722–731.

Toro-Tapia, G., S. Villaseca, J. Leal, A. Beyer, J. Fuentealba and M. Torrejón (2017). "Xenopus as a model organism to study heterotrimeric G-protein pathway during collective cell migration of neural crest." *Genesis* **55**(1–2): e23008.

Trainor, P. A. (2013). *Neural crest cells: Evolution, development and disease*. Cambridge, Academic Press.

Tríbulo, C., M. J. Aybar, V. H. Nguyen, M. C. Mullins and R. Mayor (2003). "Regulation of Msx genes by a Bmp gradient is essential for neural crest specification." *Development* **130**(26): 6441–6452.

Tríbulo, C., M. J. Aybar, S. S. Sánchez and R. Mayor (2004). "A balance between the anti-apoptotic activity of Slug and the apoptotic activity of msx1 is required for the proper development of the neural crest." *Developmental Biology* **275**(2): 325–342.

Vacca, B., E. Sanchez-Heras, E. Steed, S. L. Busson, M. S. Balda, S.-I. Ohnuma, N. Sasai, R. Mayor and K. Matter (2018). "Control of neural crest induction by MarvelD3-mediated attenuation of JNK signalling." *Scientific Reports* **8**(1): 1–15.

Vallin, J., J.-M. Girault, J. P. Thiery and F. Broders (1998). "Xenopus cadherin-11 is expressed in different populations of migrating neural crest cells." *Mechanisms of Development* **75**(1–2): 171–174.

Vallin, J., R. Thuret, E. Giacomello, M. M. Faraldo, J. P. Thiery and F. Broders (2001). "Cloning and characterization of three Xenopus slug promoters reveal direct regulation by lef/β-catenin signaling." *Journal of Biological Chemistry* **276**(32): 30350–30358.

Vega-López, G. A., M. Bonano, C. Tríbulo, J. P. Fernández, T. H. Agüero and M. J. Aybar (2015). "Functional analysis of Hairy genes in Xenopus neural crest initial specification and cell migration." *Developmental Dynamics* **244**(8): 988–1013.

Vega-Lopez, G. A., S. Cerrizuela and M. J. Aybar (2017). "Trunk neural crest cells: Formation, migration and beyond." *International Journal of Developmental Biology* **61**(1–2): 5–15.

Vernon, A. E. and C. LaBonne (2006). "Slug stability is dynamically regulated during neural crest development by the F-box protein Ppa." *Development* **133**(17): 3359–3370.

Villanueva, S., A. Glavic, P. Ruiz and R. Mayor (2002). "Posteriorization by FGF, Wnt, and retinoic acid is required for neural crest induction." *Developmental Biology* **241**(2): 289–301.

Vize, P. D. and A. M. Zorn (2017). "Xenopus genomic data and browser resources." *Developmental Biology* **426**(2): 194–199.

Waddington, C. H. (1947). *Organisers and genes*. Cambridge, Cambridge University Press.

Waddington, C. H. (1957). *The strategy of the genes*. Crows Nest, Australia, G. Allen and Unwin.

Ward, N. J., D. Green, J. Higgins, T. Dalmay, A. Münsterberg, S. Moxon and G. N. Wheeler (2018). "microRNAs associated with early neural crest development in *Xenopus laevis*." *BMC Genomics* **19**(1): 1–13.

Wei, S., G. Xu, L. C. Bridges, P. Williams, J. M. White and D. W. DeSimone (2010). "ADAM13 induces cranial neural crest by cleaving class B Ephrins and regulating Wnt signaling." *Developmental Cell* **19**(2): 345–352.

Weinstein, D. C. and A. Hemmati-Brivanlou (1997). "Neural induction in *Xenopus laevis*: Evidence for the default model." *Current Opinion in Neurobiology* **7**(1): 7–12.

Wilson, P. A. and A. Hemmati-Brivanlou (1995). "Induction of epidermis and inhibition of neural fate by Bmp-4." *Nature* **376**(6538): 331–333.

Wu, J., J. Yang and P. S. Klein (2005). "Neural crest induction by the canonical Wnt pathway can be dissociated from anterior: Posterior neural patterning in Xenopus." *Developmental Biology* **279**(1): 220–232.

Wu, M. Y. and C. S. Hill (2009). "13-P055 A factor needed for proper BMP gradient formation identified in an overexpression screen for neural crest regulators in *Xenopus laevis* embryos." *Mechanisms of Development* (126): S211.

Wu, M. Y., M.-C. Ramel, M. Howell and C. S. Hill (2011). "SNW1 is a critical regulator of spatial BMP activity, neural plate border formation, and neural crest specification in vertebrate embryos." *PLoS Biology* **9**(2): e1000593.

Wühr, M., R. M. Freeman Jr, M. Presler, M. E. Horb, L. Peshkin, S. P. Gygi and M. W. Kirschner (2014). "Deep proteomics of the *Xenopus laevis* egg using an mRNA-derived reference database." *Current Biology* **24**(13): 1467–1475.

Yang, X., J. Li, W. Zeng, C. Li and B. Mao (2016). "Elongator protein 3 (Elp3) stabilizes Snail1 and regulates neural crest migration in Xenopus." *Scientific Reports* **6**: 26238.

York, J. R. and D. W. McCauley (2020a). "Functional genetic analysis in a jawless vertebrate, the sea lamprey: Insights into the developmental evolution of early vertebrates." *Journal of Experimental Biology* **223**(Supplement): jeb206433.

York, J. R. and D. W. McCauley (2020b). "The origin and evolution of neural crest cells." *Open Biology* **10**(1): 190285.

Zhang, Y., T. Luo and T. D. Sargent (2006). "Expression of TFAP2β and TFAP2γ genes in *Xenopus laevis*." *Gene Expression Patterns* **6**(6): 589–595.

Zimmerman, L. B., J. M. De Jesús-Escobar and R. M. Harland (1996). "The Spemann organizer signal noggin binds and inactivates bone morphogenetic protein 4." *Cell* **86**(4): 599–606.

9 The Use of *Xenopus* Oocytes to Study the Biophysics and Pharmacological Properties of Receptors and Channels

Ataúlfo Martínez-Torres and Elizabeth Pereida-Jaramillo

CONTENTS

9.1. Historical Background 143
9.2. Past Observations 143
9.3. Present Status 144
9.4. Insight of Native Ion Currents of the Follicle 147
9.5. Future Directions 147
Acknowledgments 148
References 148

9.1. HISTORICAL BACKGROUND

Oocytes of the frog *Xenopus laevis* have been widely used as an experimental system for expressing ion channels and receptors cloned from diverse sources. Relatively simple electrophysiological techniques, such as the two-electrode voltage-clamp configuration, allow the study of biophysical and pharmacological characteristics of voltage- and neurotransmitter-gated ion channels. Development of the oocyte expression system began by serendipity in the mid-70s as Ricardo Miledi and co-workers explored the early development of acetylcholine responses at the frog neuromuscular junction. At some point, the biological preparation was so difficult to perform due to the small size of the animals that impaling electrodes in tadpoles became impractical to determine at what age the responses to the neurotransmitter emerged; that is when the researchers decided to explore if frog oocytes responded to acetylcholine. To their surprise, many oocytes elicited an electric response when exposed to acetylcholine (Kusano et al. 1977), and many other endogenous ion currents were subsequently discovered (Arellano et al. 1995, 1996; Miledi et al. 1989; Parker and Miledi, 1987, 1988). Other studies disclosed the presence of receptors in follicle-enclosed oocytes, such as noradrenaline, angiotensin, vasointestinal peptide, and follicle stimulating hormone (Woodward and Miledi 1987; Miledi and Woodward 1989b; Hershey et al. 1991; Woodward and Miledi 1991). These studies paved the way to "transplant" ion channels by injecting mRNA isolated from the brain and later from cloned cDNAs isolated from diverse tissues and species (Miledi et al. 1982, 1983; Barnard et al. 1982). Another application, also developed by Miledi, showed that plasma membranes isolated from brain tissue can be "transplanted" into oocytes, retaining fully functional receptor and channel activity that can be studied more than ten years after the samples are obtained (Eusebi et al. 2009; Miledi et al. 2002). This has allowed researchers to determine the pharmacological and functional changes of receptors in neurological disorders such as epilepsy, Alzheimer's disease, and autism and opens the possibility for applying a combination of proteomics and electrophysiology to establish a relationship between the pathological state and molecular imbalances.

There are nearly 10,000 citations in PubMed using the keywords *Xenopus* oocyte and ion channel. This gives an idea of the impact that this methodology has on the field and the difficulty to offer a thorough review on the subject. Nonetheless, here we provide an account of past observations and the present status of the field, offering new perspectives to be explored.

9.2. PAST OBSERVATIONS

The earliest use of the *Xenopus* oocyte as an expression system was demonstrated by Sir John Gurdon (Gurdon et al. 1971), who published a series of papers that firmly established that oocytes can synthesize a protein from a foreign mRNA. (Lane et al. 1971) conclude that

> when injected into a living cell, the 9 s RNA is fairly stable and has the properties of a hemoglobin messenger. The messenger requires no reticulocyte-specific factors for translation, and the translational machinery of the oocyte will accept the messenger RNA from a totally different cell type, from another species.

DOI: 10.1201/9781003050230-10

Using this information while working at University College London, Miledi and co-workers tried to express ion channels and receptors involved in synaptic transmission; after using diverse strategies, they realized that the critical step was the preparation of high-quality mRNA. In the early days of molecular cloning, that methodology was a tedious procedure involving ultracentrifugation of Cesium chloride gradients and other biochemical tricks that made the effort a heroic act. Selection of mRNA by Oligo dT chromatography to enrich the poly-A$^+$ mRNA fraction was crucial to prove that acetylcholine receptors from denervated muscle injected in oocytes indeed have the same electrophysiological and pharmacological characteristics as the receptors embedded in the muscle membrane (Miledi et al. 1982). After those initial steps, many brain ion channels and receptors were expressed after mRNA injection (Gundersen et al. 1983, 1984; Sumikawa et al. 1984). Ultracentrifugation of sucrose density gradients of mRNA allowed the identification of specific fractions containing mRNAs with different coding potential for functional receptors; this strategy was used to separate different functional fractions and became an expression cloning approach to synthesize cDNA clones for proteins essential for synaptic transmission (e.g. GABA-A and nicotinic receptors, as well as K^+, Na^+, and Cl-channels). Today, these studies are rarely found in the recent literature because of the current availability of gene clones, synthetic genes, and other resources that permit a straightforward experimental approach to assess protein function. But they led to major advances in the field.

For many years, studies aiming to understand the function and pharmacology of ion channels and receptors were potentiated by using *Xenopus* oocytes. One of the most effective approaches to studying the inner workings of these proteins consisted of a combination of site-directed mutagenesis and functional analysis in the oocyte. This approach revealed important structural characteristics, including voltage sensors, neurotransmitters, and pharmacologically relevant binding sites, and allowed researchers to evaluate functional effects such as gating, desensitization, and inactivation.

9.3. PRESENT STATUS

From the mid-80s, researchers in the field primarily focused on the *Xenopus* oocyte to understand how ion channels and neurotransmitter receptors work. Key biophysical and structural traits of these molecular entities that have been approached experimentally are voltage sensing, gating, neurotransmitter binding sites, and permeability (see subsequently). When oocytes began to be used as an experimental tool for expressing ion channels, they were ready to be exploited during the molecular cloning era.

Voltage-gated ion channels constitute a diverse class of proteins that are crucial for neuronal excitability and plasticity. They gate an ion channel upon sensing changes in the plasma membrane voltage. "Gating" is an intrinsic electrophysiological characteristic of ion channels that refers to the opening (by activation) or closing (by deactivation or inactivation) of the protein complex that forms the conducting pore of the ions. Gating involves changes in conformation in response to changes in membrane voltage or to an agonist. This biophysical characteristic of ion channels has been widely studied using *Xenopus* oocytes, from the first reports of their molecular cloning (Mackinnon et al. 1988; Takumi et al. 1988; Murai et al. 1989) to recent approaches that combine site-directed mutagenesis and structural models of crystallized proteins (Carvalho-de-Souza and Bezanilla 2019; Hou et al. 2019; Rinné et al. 2019).

Another fundamental structural component of both sodium- and potassium-selective channels is the so-called "voltage sensor." Studies in frog oocytes were critical for deciphering the molecular structure and function of voltage sensors that were found to be highly conserved among channel families and from different species (Noda et al. 1986; Stühmer et al. 1988; Tempel et al. 1988; Timpe et al. 1988; Stühmer et al. 1989, 1989), which currently remains an intensive field of study (Catterall et al. 2017; Ori et al. 2020).

Mapping the binding sites of the agonist to ligand-gated ion channels was essential to understand their molecular structure. *Xenopus* oocytes played a central role in determining fine pharmacology and specific interaction sites for nicotinic, GABA, serotonin, and glycine receptors (Blair et al. 1988; Bertrand et al. 1990; Grenningloh et al. 1990; Kuhse et al. 1990; Maricq et al. 1991; Amin et al. 1994), as well as for the family of glutamate receptors AMPA/kainite and NMDA (Bettler et al. 1990; Boulter et al. 1990; Egebjerg et al. 1991). High-resolution structures of these receptor complexes have confirmed many of the observations first determined by electrophysiology in oocytes (Kesters et al. 2013; Hassaine et al. 2014; Miller and Aricescu 2014).

Detailed electrophysiological characterization of *Xenopus* follicle-enclosed oocytes was prompted when it was realized that they could be very useful to study ion channels and receptors. Many studies have shown that electrical responses to neurotransmitters and hormones generated by ovarian follicular cells require maintenance of the electrical communication between the oocyte and its surrounding cells. Therefore, the responses originate in the membrane of the follicular cells and these cells express membrane receptors and ion channels. Several "native" ion currents and receptors that are fundamental for neural transmission and other physiological processes are endogenously expressed by *Xenopus* oocytes. Here we provide some examples of ion channels in the oocyte membrane whose currents were identified years ago. The presence of these currents is supported by new proteomic (Table 9.1) or other analyses and is associated with human diseases (Session et al. 2016; Peshkin et al. 2019).

One case is the calcium-dependent chloride channel TMEM16A (Anoctamin 1), originally discovered in the oocyte (Miledi 1982; Miledi and Parker 1984) but known to play a crucial role in regulating anxiety-related behaviors, for it is expressed in cholinergic neurons of the medial habenula

TABLE 9.1
Endogenous Ion Channels in the *Xenopus* Oocyte

Ion Current or Receptor	Pharmacology	Associated Gene Found in Oocytes	Characteristics of the Channel	Associated Diseases	References
Voltage-activated sodium-current	Amiloride sensitive	*scnn1b, scnn1g*	Epithelial sodium channel (ENaC)	Liddle syndrome	(Peshkin et al. 2019; Yang et al. 2014; Parker and Miledi 1987) Xenbase
Voltage-activated potassium-current	Ba^{2+} sensitive, hyperpolarization activated inward current	Kcnj5	*Activation of protein kinase C, ATP sensitive*	Familial hyperaldosteronism type III and type of long QT syndrome 1 and 13	(Parker and Miledi 1988; Peshkin et al. 2019) Xenbase
	Ba^{2+} and chlofilium sensitive	Kcnn2 Kcnq1	Small K^+ channel IsK, minK	Long QT syndrome 1 Jervell and Lange-Nielsen syndrome	(Parker and Miledi 1988; Peshkin et al. 2019; Dixit, et al. 2020; Sun et al. 2015) Xenbase
	———	Kcnk7	twik channel	———	(Peshkin et al. 2019) Xenbase
Voltage-activated calcium-current	Blocked by dihydropyridines, benzothiazepines	Cacna1s Cacnb1	L-type calcium channel	Periodic paralysis hypokalemic 1 Malignant hyperthermia 5 Thyrotoxic periodic paralysis1 Malignant hyperthermia	(Peshkin et al. 2019) Xenbase
	Blocked by Cd^{2+}, inactivation by injection of AMPc	Cacna1h Cacna1i (Crata) Cachd1	T type calcium channel	Childhood absence epilepsy Idiopathic generalized epilepsy	(Peshkin et al. 2019) Xenbase
	Blocked by bioallethrin	Cacna2d1, Cacna2d2	P type calcium channel	Familial short QT syndrome Short QT syndrome	(Peshkin et al. 2019; Antzelevitch et al. 2007; Pippucci et al. 2013) Xenbase
Mechano-sensitive current	———	Trpc4ap	Short transient receptor	———	(Peshkin et al. 2019) Xenbase
	Blocked by SET2	Trpv2	Activated by temperature and chemical, mechanic stimuli	———	(Peshkin et al. 2019) Xenbase
	Blocked by amiloride, gentamicin, Gd^{3+}, ruthenium red	Trpv4	Activated by temperature and chemical mechanic stimuli	BrachyolmialCharcot-Marie tooth disease axonal type 2C Metatropic dysplasia Parastremmatic dwarfism Spondyloepiphyseal dysplasia Maroteaux type spondylometaphysea ldysplasia Kozlowski type	(Peshkin et al. 2019; Rosenbaum et al. 2020; McCray et al. 2014) Xenbase
	Activated by intracellular Ca^{2+}	Trpm4	Melastatin 4, ATP-binding	———	(Wang, et al. 2018; Gao and Liao 2019; Peshkin et al. 2019) Xenbase
Nonselective ion-current	Current reduced by Ca^{2+} and Mg^{2+} and negative voltage membrane	Gj1a, Gja4, Gj1b	gj1a(Cxc43), gja4(Cxc37), gj1b(Cxc32)	Oculodentodigital dysplasia Heart malformations Cancers Erythorokeratodermia variabilis et progressive 1 Myocardial infarction Charcot Marie tooth neuropathy	(Oshima 2014; Kim et al. 2019; Kelly et al. 2016; Macari et al. 2000; Peshkin et al. 2019) Xenbase
Chloride ion channels (Cl^-)	Activated by intracellular Ca^{2+}, blocked by niflumic acid, DIDS	Tmem16A (Ano1)	Ca^{2+} activated (CaCC)	Frontal sinus squamous cell carcinoma Frontal sinus cancer	(Miledi 1982; Miledi and Parker 1984; Peshkin et al. 2019) Xenbase
	———	Clic1 Clic3	Intracellular receptor	Gallbladder cancer Pnicilliosis Gallbladder cancer	(Peshkin et al. 2019) Xenbase

(*Continued*)

TABLE 9.1 (Continued)

Ion Current or Receptor	Pharmacology	Associated Gene Found in Oocytes	Characteristics of the Channel	Associated Diseases	References
Organic transporters	Activated by pH changes	Clcn3 Clcn7	H^+/Cl^- exchange transporter, outwardly rectifying	Dent disease1 Cystic fibrosis Autosomal dominant Osteopetrosis 2	(Peshkin et al. 2019; Reyes et al. 2004; Ochoa-de la Paz et al. 2013) Xenbase
	Sensitive to La^{3+}	Tmem184b, Tmem184c	Organic solute transporter	Amyloidosis Finnish type Mandibulofacial dysostosis with alopecia Acrocallosal syndrome	(Peshkin et al. 2019) Xenbase
Water (aquaporins)	Sensitive to pCMBS and phloretin	Aqp1 Aqp9	Water-permeable channels	Blood group Colton system Obstructive hydrocephalous Hydrarthrosis polyhydramnios	(Castañeyra-Ruiz et al. 2019; Peshkin et al. 2019; Kalani, et al. 2012; Nagahara et al. 2010; Preston et al. 1992) Xenbase
Glycine receptor	Activated by glycine, blocked by picrotoxin	Glra3	Ionotropic receptor	Susceptibility to epilepsy idiopathic generalized 13	(Zeilhofer et al. 2018; Peshkin et al. 2019) Xenbase

(Cho et al. 2020). There is also an outwardly rectifying chloride current in *X. laevis* and *X. tropicalis* oocytes identified by increasing the concentration of anions in the extracellular medium (Reyes et al. 2004; Ochoa-de la Paz et al. 2013). This current corresponds to the chloride/H^+ exchanger, ClC-5, located in the plasma membrane and endosomes, and is mutated in Dent's disease, where endocytosis is defective in renal proximal tubes (Günther et al. 1998; Piwon et al. 2000). Other studies have overexpressed ClC-5 and mutant versions of the channel overriding the native current, thus allowing the study of the mutant channel's characteristics (Scheel et al. 2005; Chang et al. 2019).

The *cacna1c* gene encodes for several versions of the voltage-activated calcium channel CaV1.2 subindex. Calcium ions are important for many cellular functions, including regulating the electrical activity of cells, cell-to-cell communication, muscle contraction, and gene regulation. In the frog oocyte, CACNA1C is present and may well provide the entry pathway for calcium necessary for the activation of TMEM16A, which elicits the transient outward chloride current (T*out* = subindex), but this has not been proven (Miledi 1982; Miledi and Parker 1984). This gene is expressed in heart, muscle, and brain and is mutated in several conditions such as QT syndrome and Timothy's syndrome. Other members of the *cacna* gene family are expressed in the oocyte, including the P-type calcium channel (Cacna2d1 and Cacna2d2) related to short QT syndrome and epilepsy (Antzelevitch et al. 2007; Pippucci et al. 2013).

Connexins are the main components of gap junctions. They provide direct links between cells and play a central role in many cell functions in all tissues; they even participate in transcriptional regulation (Oshima 2014; Ribeiro-Rodrigues et al. 2017; Kim et al. 2019). In the frog oocyte, Cx43, Cx37, and Cx32 are expressed and modulated by calcium, magnesium, and negative membrane voltage. Cx43 is involved in oculodentodigital dysplasia and heart malformations, and its activity is exacerbated in many types of cancer, whereas Cx37 is altered in erythrokeratodermia (Macari et al. 2000; Kelly et al. 2016).

Aquaporins 1 and 9 have been identified in *Xenopus* oocytes. These water-transporting channels are sensitive to pCMBSS and phloretin and are associated with obstructive hydrocephaly (Aquaporin 1) and hydrarthrosis polyhydramnios (Aquaporin 9) (Nagahara et al. 2010; Kalani et al. 2012; Castañeyra-Ruiz et al. 2019). Aquaporins form a family of water-permeable channels that were originally cloned from red blood cells by Peter Agre, who showed that upon overexpression in oocytes, these cells increase the osmotic water permeability (Preston et al. 1992). Despite the endogenous expression of aquaporins in frog oocytes, Agre performed a series of classic experiments to show unequivocally the functional characteristics of aquaporins.

At least two sodium channels were identified in the *X. tropicalis* oocyte proteome: the *scnn1b* gene, which encodes for the β subunit of the epithelial sodium channel ENaC, and the *scnn1g* gene, which encodes for the γ subunit. ENaC is assembled as a heterotrimer composed of homologous subunits α, β, and γ or δ, β, and γ. ENaC is constitutively active and is not voltage dependent like the classic neuronal sodium channels. Mutations associated with the ENaC genes give rise to Liddle syndrome, a rare genetic disorder that is characterized by high blood pressure (hypertension) which is resistant to pharmacological treatment (Yang et al. 2014). The ENaC currents may well correspond to those previously described (Parker and Miledi 1987) that are sporadically present in the oocyte and are resistant to tetrodotoxin. On the other hand, the transient receptor potential melastatin (TRPM) family belongs to the superfamily of TRP cation channels. The TRPM subfamily is composed of eight members that are involved in diverse biological functions such as temperature sensing, inflammation, insulin secretion, and redox sensing (Gao and Liao 2019; Huang et al. 2020). The TRPM4 channel is expressed in the oocyte. This channel is widely expressed and regulates calcium oscillations after T cell activation and prevents cardiac conduction and smooth muscle contraction (Wang et al. 2018). TRPM4 is activated by intracellular calcium and voltage; thus, the activation mechanisms converge with those of TMEM16A (or *Tout* in the oocyte; Miledi, 1982). Finally, TRPV4 has also been found in the proteomic analysis of *Xenopus* oocytes. This calcium-permeable non-selective channel performs multiple physiological roles in diverse organs; the channel is also gated by osmotic pressure, mechanical pressure, and biochemical signaling. Mutations in the gene encoding TRPV4 induce skeletal dysplasia, osteoarthritis, and neurological motor disorders (McCray et al. 2014; Rosenbaum et al. 2020); however, the study of the endogenous TRPV4 has escaped analysis in oocytes.

Follicle-enclosed oocytes present several ion currents, including a barium-sensitive potassium current that is activated upon depolarization of the membrane. This current is removed by collagenase treatment, indicating the need for the intercommunication between the oocyte and surrounding cells (Parker and Miledi 1988). In the oocyte proteome, two potassium channels have been identified: Kcnq1, which has multiple functions such as the regulation of gastric acid secretion, thyroid hormone synthesis, salt and glucose homeostasis, and cell volume (Dixit et al. 2020), and Kcnn2, a member of the calcium-activated potassium channel family. The latter channel is targeted by the ubiquitin-protein ligase E3A, whose function is reduced in Angelman's syndrome; thus, in the disease, Kcnn2 function increases, leading to changes in synaptic function (Sun et al. 2015, 2020).

Oocytes rarely express ligand-gated ion channels of the Cys-loop family. One case is the α3 subunit of the glycine receptor that is capable of forming homomeric receptors with high affinity to glycine but generates small chloride currents upon exposure to glycine (Kuhse et al. 1990; Nikolic et al. 1998). Thus, considering the limited number of selective agonists and antagonists of the glycine receptors (Zeilhofer et al. 2018), the oocyte may be valuable to screen for new active molecules.

9.4. INSIGHT OF NATIVE ION CURRENTS OF THE FOLLICLE

As mentioned, a poorly explored advantage of the oocyte as an experimental model for understanding the role of receptors and ion channels is the follicle-enclosed oocyte. In the ovary, cells that surround the oocyte include epithelial and follicular cells that maintain a close physical interaction and exchange biochemical signals mediated by calcium and cAMP, for example (Miledi and Woodward 1989a). Follicle cells produce important modulators such as serotonin, dopamine and noradrenaline, gonadotropins (follicle-stimulating hormone, luteinizing hormone and growth hormone), prostaglandins, and neuropeptides (oxytocin, atrial natriuretic peptide, corticotropin releasing factor, gonadotropin-releasing hormone, calcitonin gene-related peptide) and express zinc-gated ion channels (Miledi et al. 1989). Follicles also express acetylcholine and angiotensin receptors that couple to inositol-3-phosphate, which in turn releases calcium and gates TMEM16A (Miledi and Parker 1984; Woodward and Miledi 1987; Parker and Miledi 1988; Miledi and Woodward 1989a; Arellano et al. 1995; Arellano et al. 1996). Studies on these endogenous receptors and the ion-currents elicited upon activation are very important for understanding follicular physiology, inter- and intracellular communication, and biochemical signaling and place the follicular oocyte as an important model for studying diverse processes which will help us understand the physiology of diverse cellular systems.

9.5. FUTURE DIRECTIONS

Miledi and co-workers developed a novel, powerful assay that relies upon the ability of cell membranes to spontaneously form vesicles and the ease with which these vesicles fuse with the plasma membrane when they are injected in the oocyte (Eusebi et al. 2009). This assay offers the possibility of studying the characteristics of ion channels and neurotransmitter receptors embedded in their original lipid environment from samples that were frozen years before or from freshly resected brain tissue (Palma et al. 2005, 2006, 2007). This assay has shed some light on the characteristics of important components of the synaptic function and how they are altered in human disorders such as autism, epilepsy, and Alzheimer's disease (Miledi et al. 2004; Limon et al. 2008; Roseti et al. 2008).

A combination of proteomic, transcriptomic, and functional assays of glutamate receptors "transplanted" from postmortem samples of schizophrenic brain showed the electrophysiological and pharmacological characteristics of AMPA receptors from the dorsolateral prefrontal cortex and its impairment in the disease (Zeppillo et al. 2020). Another example of the power of this assay is the microtransplantation of brain samples from Rett syndrome patients. This disease is caused by mutations within the methylcytosine-binding protein 2 (*MECP2*) gene, although evidence has shown linkage to the *DKL5* and *FOXG1* genes. Rett syndrome is associated with the X chromosome, and patients develop language and communication problems and learning and coordination deficits (Brunetti and Lumsden 2020; Sandweiss et al. 2020). In a study by Ruffolo et al. (2020), membranes from the prefrontal cortex of Rett syndrome patients showed an imbalance in the excitatory/inhibitory ratio given by AMPA and GABA responses, a modification of GABA currents towards a more depolarizing value, and differences in the AMPA/GABA ratio. Interestingly, a transgenic mouse model of Rett syndrome exhibited similar functional impairments (Ruffolo et al. 2020).

Because the human neurotransmitter receptors are "microtransplanted" in their native cell lipid environment, this method extends the applications of *Xenopus* oocytes as an expression system for approaching many channelopathies.

ACKNOWLEDGMENTS

Work supported by grants from CONACYT (A1S7659) and PAPIIT-DGAPA-UNAM IN204520 to AMT. EPJ is supported by a fellowship from CONACYT (446622) and is a student of Programa de Doctorado en Ciencias Biomédicas-UNAM. Jessica González Norris kindly edited the English version of the manuscript.

REFERENCES

Amin, J., I. M. Dickerson, and D. S. Weiss. 1994. The agonist binding site of the gamma-aminobutyric acid type A channel is not formed by the extracellular cysteine loop. *Molecular Pharmacology* 45:317–323.

Antzelevitch, C., A. Wilde, L. Eckardt, M. Hiraoka, and D. Corrado. 2007. Diagnostic and genetic aspects of the Brugada and other inherited arrhythmias syndromes. *Journal Electrocardiology* 40:S11–S14. https://doi.org/10.1016/j.jelectrocard.2006.10.019.

Arellano, R. O., R. M. Woodward, and R. Miledi. 1995. A monovalent cationic conductance that is blocked by extracellular divalent cations in *Xenopus* oocytes. *Journal of Physiology* 484:593–604. https://doi.org/10.1113/jphysiol.1995.sp020689.

Arellano, R. O., R. M. Woodward, and R. Miledi. 1996. Ion channels and membrane receptors in follicle-enclosed *Xenopus* oocytes. *Ion Channels*, ed. Toshio Narahashi, 203–259. Boston, MA: Springer US. https://doi.org/10.1007/978-1-4899-1775-1_6.

Barnard, E. A., R. Miledi, and K. Sumikawa. 1982. Translation of exogenous messenger RNA coding for nicotinic acetylcholine receptors produces functional receptors in *Xenopus* oocytes. *Proceedings of the Royal Society of London: Series B, Biological Sciences* 215:241–246. https://doi.org/10.1098/rspb.1982.0040.

Bertrand, D., M. Ballivet, and D. Rungger. 1990. Activation and blocking of neuronal nicotinic acetylcholine receptor reconstituted in *Xenopus* oocytes. *Proceedings of the National Academy of Sciences* 87:1993–1997. https://doi.org/10.1073/pnas.87.5.1993.

Bettler, B., J. Boulter, I. Hermans-Borgmeyer, A. O'Shea-Greenfield, E. S. Daneris, C. Moll, U. Borgmeyer, M. Hollmann, and S. Heinemann. 1990. Cloning of a novel glutamate receptor subunit, GluR5: Expression in the nervous system during development. *Neuron* 5:583–595. https://doi.org/10.1016/0896-6273(90)90213-y.

Blair, L. A., E. S. Levian, J. Marshall, V. E. Dionne, and E. A. Barnard. 1988. Single subunits of the GABAA receptor form ion channels with properties of the native receptor. *Science* 242:577–579. https://doi.org/10.1126/science.2845583.

Boulter, J., M. Hollmann, A. O'Shea-Greenfield, M. Hartley, E. Daneris, C. Maron, and S. Heinemann. 1990. Molecular cloning and functional expression of glutamate receptor subunit genes. *Science* 249:1033–1037. https://doi.org/10.1126/science.2168579.

Brunetti, S., and D. E. Lumsden. 2020. Rett syndrome as a movement and motor disorder. *European Journal Pediatric of Neurology* 28:29–37. https://doi.org/10.1016/j.ejpn.2020.06.020.

Carvalho-de-Souza, J. L., and F. Bezanilla. 2019. Noncanonical mechanism of voltage sensor coupling to pore revealed by tandem dimers of Shaker. *Nature Communications* 10:3584. https://doi.org/10.1038/s41467-019-11545-7.

Castañeyra-Ruiz, L., L. G. Hernández-Abad, E. M. Carmona-Calero, A. Castañeyra-Perdomo, and I. González-Marrero. 2019. AQP1 overexpression in the CSF of obstructive hydrocephalus and inversion of its polarity in the choroid plexus of a chiari malformation type II case. *Journal Neuropathology Experimental Neurology* 78:641–647. https://doi.org/10.1093/jnen/nlz033.

Catterall, W. A., G. Wisedchaisri, and N. Zheng. 2017. The chemical basis for electrical signaling. *Nature Chemical Biology* 13:455–463. https://doi.org/10.1038/nchembio.2353.

Chang, M. H., M. R. Brown, Y. Liu, V. G. Gainullin, P. C. Harris, M. F. Romero, and J. C. Lieske. 2019. Cl– and H+ coupling properties and subcellular localizations of wildtype and disease-associated variants of the voltage-gated Cl-/H+ exchanger ClC-5. *Journal of Biological Chemistry* 295:1464–1473. https://doi.org/10.1074/jbc.RA119.011366.

Cho, C. H., S. Lee, A. Kim, O. Yarishkin, K. Ryoo, Y. S. Lee, H. G. Jung, E. Yang, D. Y. Lee, B. Lee, H. Kim, U. Oh, H. I. Im, E. M. Hwang, and J. Y. Park. 2020. TMEM16A expression in cholinergic neurons of the medial habenula mediates anxiety-related behaviors. *EMBO Reports* 21:e48097. https://doi.org/10.15252/embr.201948097.

Dixit, G., C. Dabney-Smith, and G. A. Lorigan. 2020. The membrane protein KCNQ1 potassium ion channel: Functional diversity and current structural insights. *Biochimical et Biophysical Acta (BBA)—Biomembranes* 1862:183148. https://doi.org/10.1016/j.bbamem.2019.183148.

Egebjerg, J., B. Bettler, I. Hermans-Borgmeyer, and S. Heinemann. 1991. Cloning of a cDNA for a glutamate receptor subunit activated by kainate but not AMPA. *Nature* 351:745–748. https://doi.org/10.1038/351745a0.

Eusebi, F., E. Palma, M. Amici, and R. Miledi. 2009. Microtransplantation of ligand-gated receptor-channels from fresh or frozen nervous tissue into *Xenopus* oocytes: A potent tool for expanding functional information. *Progress in Neurobiology* 88:32–40. https://doi.org/10.1016/j.pneurobio.2009.01.008.

Gao, Y., and P. Liao. 2019. TRPM4 channel and cancer. *Cancer Letters* 454:66–69. https://doi.org/10.1016/j.canlet.2019.04.012.

Grenningloh, G., I. Pribilla, P. Prior, G. Multhaup, K. Beyreuther, O. Taleb, and H. Betz. 1990. Cloning and expression of the 58 kd beta subunit of the inhibitory glycine receptor. *Neuron* 4:963–970. https://doi.org/10.1016/0896-6273(90)90149-a.

Gundersen, C. B., R. Miledi, and I. Parker. 1983. Serotonin receptors induced by exogenous messenger RNA in *Xenopus* oocytes. *Proceedings of the Royal Society of London: Series B, Biological Sciences* 219:103–109. https://doi.org/10.1098/rspb.1983.0062.

Gundersen, C. B., R. Miledi, and I. Parker. 1984. Messenger RNA from human brain induces drug- and voltage-operated channels in *Xenopus* oocytes. *Nature* 308:421–424. https://doi.org/10.1038/308421a0.

Günther, W., A. Lüchow, F. Cluzeaud, A. Vandewalle, T. J. Jentsch. 1998. ClC-5, the chloride channel mutated in Dent's disease, colocalizes with the proton pump in endocytotically active kidney cells. *Proceedings of the National Academy of Sciences of the United States of America* 95:8075–8080. https://doi.org/10.1073/pnas.95.14.8075.

Gurdon, J. B., C. D. Lane, H. R. Woodland, and G. Marbaix. 1971. Use of frog eggs and oocytes for the study of messenger RNA and its translation in living cells. *Nature* 233:177–182. https://doi.org/10.1038/233177a0.

Hassaine, G., C. Deluz, L. Grasso, R. Wyss, M. B. Tol, R. Hovius, A. Graff, H. Stahlberg, T. Tomizaki, A. Desmyter, C. Moreau, X. D. Li, F. Poitevin, H. Vogel, H. Nury. 2014. X-ray structure of the mouse serotonin 5-HT3 receptor. *Nature* 512:276–281. https://doi.org/10.1038/nature13552.

Hershey, A. D., L. Polenzani, R. M. Woodward, R. Miledi, and J. E. Krause. 1991. Molecular and genetic characterization, functional expression, and mRNA expression patterns of a rat substance P receptor. *Annals of the New York Academy of Sciences* 632:63–73. https://doi.org/10.1111/j.1749-6632.1991.tb33095.x.

Hou, P., J. Shi, K. M. White, Y. Gao, and J. Cui. 2019. ML277 specifically enhances the fully activated open state of KCNQ1 by modulating VSD-pore coupling. *Elife* 8:e48576. https://doi.org/10.7554/eLife.48576.

Huang, Y., R. Fliegert, A. H. Guse, W. Lü, and J. Du. 2020. A structural overview of the ion channels of the TRPM family. *Cell Calcium* 85:102111. https://doi.org/10.1016/j.ceca.2019.102111.

Kalani, M. Y., A. S. Filippidis, and H. L. Rekate. 2012. Hydrocephalus and aquaporins: The role of aquaporin-1. *Acta Neurochirurgica Supplement* 113:51–54. https://doi.org/10.1007/978-3-7091-0923-6_11.

Kelly, J. J., J. L. Esseltine, Q. Shao, E. W. Jabs, J. Sampson, M. Auranen, D. Bai, and D. W. Laird. 2016. Specific functional pathologies of Cx43 mutations associated with oculodentodigital dysplasia. *Molecular of Biology Cell* 27:2172–2185. https://doi.org/10.1091/mbc.E16-01-0062.

Kesters, D., A. J. Thompson, M. Brams, R. van Elk, R. Spurny, M. Geitmann, J. M. Villalgordo, A. Guskov, U. H. Danielson, S. C. Lummis, A. B. Smit, and C. Ulens. 2013. Structural basis of ligand recognition in 5-HT3 receptors. *The EMBO Journal* 14:49–56. https://doi.org/10.1038/embor.2012.189.

Kim, J. C., M. Pérez-Hernández, F. J. Alvarado, S. R. Maurya, J. Montnach, Y. Yin, M. Zhang, X. Lin, C. Vasquez, A. Heguy, F. X. Liang, S. H. Woo, G. E. Morley, E. Rothenberg, A. Lundby, H. H. Valdivia, M. Cerrone, and M. Delmar. 2019. Disruption of Ca2+i homeostasis and connexin 43 hemichannel function in the right ventricle precedes overt arrhythmogenic cardiomyopathy in plakophilin-2-deficient mice. *Circulation* 140:1015–1030. https://doi.org/10.1161/CIRCULATIONAHA.119.039710.

Kuhse, J., V. Schmieden, and H. Betz. 1990. Identification and functional expression of a novel ligand binding subunit of the inhibitory glycine receptor. *Journal of Biological Chemistry* 265:22317–22320.

Kusano, K, R. Miledi, and J. Stinnakre. 1977. Acetylcholine receptors in the oocyte membrane. *Nature* 270:22–29.

Lane, C. D., G. Marbaix, and J. B. Gurdon. 1971. Rabbit haemoglobin synthesis in frog cells: The translation of reticulocyte 9 s RNA in frog oocytes. *Journal of Molecular and Biology* 61:73–91. https://doi.org/10.1016/0022-2836(71)90207-5.

Limon, A., J. M. Reyes-Ruiz, and R. Miledi. 2008. Microtransplantation of neurotransmitter receptors from postmortem autistic brains to *Xenopus* oocytes. *Proceedings of the National Academy of Sciences* 105:10973–10977. https://doi.org/10.1073/pnas.0804386105.

Macari, F., M. Landau, P. Cousin, B. Mevorah, S. Brenner, R. Panizzon, D. F. Schorderet, D. Hohl, and M. Huber. 2000. Mutation in the gene for connexin 30.3 in a family with erythrokeratodermia variabilis. *American Journal Human Genetics* 67:1296–1301. https://doi.org/10.1016/S0002-9297(07)62957-7.

Mackinnon, R., P. H. Reinhart, and M. M. White. 1988. Charybdotoxin block of Shaker K+ channels suggests that different types of K+ channels share common structural features. *Neuron* 1:997–1001. https://doi.org/10.1016/0896-6273(88)90156-0.

Maricq, A. V., A. S. Peterson, A. J. Brake, R. M. Myers, and D. Julius. 1991. Primary structure and functional expression of the 5HT3 receptor, a serotonin-gated ion channel. *Science* 254:432–4337. https://doi.org/10.1126/science.1718042.

McCray, B. A., A. Schindler, J. E. Hoover-Fong, and C. J. Sumner. 2014. Autosomal dominant TRPV4 disorders. *GeneReviews*. www.ncbi.nlm.nih.gov/books/NBK201366/.

Miledi, R. 1982. A calcium-dependent transient outward current in *Xenopus laevis* oocytes. *Proceedings of the Royal Society of London: Series B, Biological Sciences* 215:491–497. https://doi.org/10.1098/rspb.1982.0056.

Miledi, R., Z. Dueñas, A. Martínez-Torres, C. H. Kawas, and F. Eusebi. 2004. Microtransplantation of functional receptors and channels from the Alzheimer's brain to frog oocytes. *Proceedings of the National Academy of Sciences* 101:1760–1763. https://doi.org/10.1073/pnas.0308224100.

Miledi, R., F. Eusebi, A. Martínez-Torres, E. Palma, and F. Trettel. 2002. Expression of functional neurotransmitter receptors in *Xenopus* oocytes after injection of human brain membranes. *Proceedings of the National Academy of Sciences* 99:13238–13942. https://doi.org/10.1073/pnas.192445299.

Miledi, R., and I. Parker. 1984. Chloride current induced by injection of calcium into *Xenopus* oocytes. *Journal of Physiology* 357:173–183. https://doi.org/10.1113/jphysiol.1984.sp015495.

Miledi, R., I. Parker, and K. Sumikawa. 1982. Properties of acetylcholine receptors translated by cat muscle mRNA in *Xenopus* oocytes. *The EMBO Journal* 1:1307–1312.

Miledi, R., I. Parker, and K. Sumikawa. 1983. Recording of single gamma-aminobutyrate-and acetylcholine-activated receptor channels translated by exogenous mRNA in *Xenopus* oocytes. *Proceedings of the Royal Society of London: Series B, Biological Sciences* 218:481–484. https://doi.org/10.1098/rspb.1983.0053.

Miledi, R., I, Parker, and R. M. Woodward. 1989. Membrane currents elicited by divalent cations in *Xenopus* oocytes. *Journal of Physiology* 417:173–195. https://doi.org/10.1113/jphysiol.1989.sp017796.

Miledi, R., and R. M. Woodward. 1989a. Membrane currents elicited by prostaglandins, atrial natriuretic factor and oxytocin in follicle-enclosed *Xenopus* oocytes. *Journal of Physiology* 416:623–643. https://doi.org/10.1113/jphysiol.1989.sp017781.

Miledi, R., and R. M. Woodward. 1989b. Effects of defolliculation on membrane current responses of *Xenopus* oocytes. *Journal of Physiology* 416:601–621.

Miller, P. S., and A. S. Aricescu. 2014. Crystal structure of a human GABAA receptor. *Nature* 512:270–275. https://doi.org/10.1038/nature13293.

Murai, T., A. Kakizuka, T. Takumi, H. Ohkubo, and S. Nakanishi. 1989. Molecular cloning and sequence analysis of human genomic DNA encoding a novel membrane protein which exhibits a slowly activating potassium channel activity. *Biochemical and Biophysical Research Communications* 161:176–181. https://doi.org/10.1016/0006-291x(89)91577-5.

Nagahara, M., Y. Waguri-Nagaya, T. Yamagami, M. Aoyama, T. Tada, K. Inoue, K. Asai, and T. Otsuka. 2010. TNF-alpha-induced aquaporin 9 in synoviocytes from patients with OA and RA. *Rheumatology* 49:898–906. https://doi.org/10.1093/rheumatology/keq028.

Nikolic, Z., B. Laube, R. G. Weber, P. Lichter, P. Kioschis, A. Poustka, C. Mülhardt, and C. M. Becker. 1998. The human glycine receptor subunit alpha3: Glra3 gene structure, chromosomal localization, and functional characterization of alternative transcripts. *Journal of Biological*

Chemistry 273:19708–19714. https://doi.org/10.1074/jbc.273.31.19708.

Noda, M., T. Ikeda, T. Kayano, H. Suzuki, H. Takeshima, M. Kurasaki, H. Takahashi, and S. Numa. 1986. Existence of distinct sodium channel messenger RNAs in rat brain. *Nature* 320:188–192. https://doi.org/10.1038/320188a0.

Ochoa-de la Paz, L. D., A. E. Espino-Saldaña, R. Arellano-Ostoa, J. P. Reyes, R. Miledi, and A. Martinez-Torres. 2013. Characterization of an outward rectifying chloride current of *Xenopus tropicalis* oocytes. *Biochimica et Biophysica Acta (BBA)—Biomembranes* 1828:1743–1753. https://doi.org/10.1016/j.bbamem.2013.03.013.

Ori, H., H. Hazan, E. Marder, and S. Marom. 2020. Dynamic clamp constructed phase diagram for the Hodgkin and Huxley model of excitability. *Proceedings of the National Academy of Sciences* 117:33575–33582. https://doi.org/10.1073/pnas.1916514117.

Oshima, A. 2014. Structure and closure of connexin gap junction channels. *FEBS Letters* 588:1230–1237. https://doi.org/10.1016/j.febslet.2014.01.042.

Palma, E., M. Amici, F. Sobrero, G. Spinelli, S. Di Angelantonio, D. Ragozzino, A. Mascia, C. Scoppetta, V. Esposito, R. Miledi, and F. Eusebi. 2006. Anomalous levels of Cl-transporters in the hippocampal subiculum from temporal lobe epilepsy patients make GABA excitatory. *Proceedings of the National Academy of Sciences* 103:8465–8468. https://doi.org/10.1073/pnas.0602979103.

Palma, E., D. Ragozzino, S. Di Angelantonio, A. Mascia, F. Maiolino, M. Manfredi, G. Cantore, V. Esposito, G. Di Genaro, P. Quarato, R. Miledi, and F. Eusebi. 2007. The antiepileptic drug levetiracetam stabilizes the human epileptic GABAA receptors upon repetitive activation. *Epilepsia* 48:1842–1849. https://doi.org/10.1111/j.1528-1167.2007.01131.x.

Palma, E., G. Spinelli, G. Torchia, A. Martínez-Torres, D. Ragozzino, R. Miledi, and F. Eusebi. 2005. Abnormal GABAA receptors from the human epileptic hippocampal subiculum microtransplanted to Xenopus oocytes. *Proceedings of the National Academy of Sciences* 102:2514–2518. https://doi.org/10.1073/pnas.0409687102.

Parker, I., and R. Miledi. 1987. Tetrodotoxin-sensitive sodium current in native *Xenopus* oocytes. *Proceedings of the Royal Society of London: Series B, Biological Sciences* 232:289–296. https://doi.org/10.1098/rspb.1987.0075.

Parker, I., and R. Miledi. 1988. Transient potassium current in native *Xenopus* oocytes. *Proceedings of the Royal Society of London: Series B, Biological Sciences* 234:45–53. www.jstor.org/stable/36289.

Peshkin, L., A. Lukyanov, M. Kalocsay, R. M. Gage, D. Z. Wang, T. J. Pells, K. Karimi, P. D. Vize, M. Wühr, and M. W. Kirschner. 2019. The protein repertoire in early vertebrate embryogenesis. *Developmental Biology*. https://doi.org/10.1101/571174 (accessed March 11, 2019) Xenbase.

Pippucci, T., A. Parmeggiani, F. Palombo, A. Maresca, A. Angius, L. Crisponi, F. Cucca, R. Liguori, M. L. Valentino, M. Seri, and V. A. Carelli. 2013. Novel null homozygous mutation confirms CACNA2D2 as a gene mutated in epileptic encephalopathy. *PLoS One* 8:e82154. https://doi.org/10.1371/journal.pone.0082154.

Piwon, N., W. Günther, M. Schwake, M. R. Bösl, and T. J. Jentsch. 2000. ClC-5 Cl-channel disruption impairs endocytosis in a mouse model for Dent's disease. *Nature* 408:369–373. https://doi.org/10.1038/35042597.

Preston, G. M., T. P. Carroll, W. B. Guggino, and P. Agre. 1992. Appearance of water channels in *Xenopus* oocytes expressing red cell CHIP28 protein. *Science* 256:385–387. https://doi.org/10.1126/science.256.5055.385.

Reyes, J. P., C. Y. Hernandez-Carballo, P. Pérez-Cornejo, U. Meza, R. Espinosa-Tanguma, and J. Arreola. 2004. Novel outwardly rectifying anion conductance in *Xenopus* oocytes. *European Journal of Physiology* 449:271–277. https://doi.org/10.1007/s00424-004-1324-z.

Ribeiro-Rodrigues, T. M., T. Martins-Marques, S. Morel, B. R. Kwak, and H. Girão. 2017. Role of connexin 43 in different forms of intercellular communication: Gap junctions, extracellular vesicles and tunnelling nanotubes. *Journal of Cellular Science* 130:3619–3630. https://doi.org/10.1242/jcs.200667.

Rinné, S., A. K. Kiper, K. S. Vowinkel, D. Ramírez, M. Schewe, M. Bedoya, D. Aser, I. Gensler, M. F. Netter, P. J. Stansfeld, T. Baukrowitz, W. Gonzalez, and N. Decher. 2019. The molecular basis for an allosteric inhibition of K+-flux gating in K2P channels. *Elife* 8:e39476. https://doi.org/10.7554/eLife.39476.

Rosenbaum, T., M. Benítez-Angeles, R. Sánchez-Hernández, S. L. Morales-Lázaro, M. Hiriart, L. E. Morales-Buenrostro, and F. Torres-Quiroz. 2020. TRPV4: A physio and pathophysiologically significant ion channel. *International Journal Molecular Science* 21:3837. https://doi.org/10.3390/ijms21113837.

Roseti, C., K. Martinello, S. Fucile, V. Piccari, A. Mascia, G. Di Gennaro, P. P. Querato, M. Manfredi, V. Esposito, G. Cantore, A. Arcella, M. Simonato, B. B. Fredholm, C. Limatola, R. Miledi, and F. Eusebi. 2008. Adenosine receptor antagonists alter the stability of human epileptic GABAA receptors. *Proceedings of the National Academy of Sciences* 105:15118–15123. https://doi.org/10.1073/pnas.0807277105.

Ruffolo, G., P. Cifelli, C. Miranda-Lourenço, E. De Felice, C. Limatola, A. M. Sebastião, M. J. Diógenes, E. Aronica, and E. Palma. 2020. Rare diseases of neurodevelopment: Maintain the mystery or use a dazzling tool for investigation? The case of Rett syndrome. *Neuroscience* 439:146–152. https://doi.org/10.1016/j.neuroscience.2019.06.015.

Sandweiss, A. J., V. L. Brandt, and H. Y. Zoghbi. 2020. Advances in understanding of Rett syndrome and MECP2 duplication syndrome: Prospects for future therapies. *Lancet Neurology* 19:689–698. https://doi.org/10.1016/S1474-4422(20)30217-9.

Scheel, O., A. A. Zdebik, S. Lourdel, and T. J. Jentsch. 2005. Voltage-dependent electrogenic chloride/proton exchange by endosomal CLC proteins. *Nature* 436:424–427. https://doi.org/10.1038/nature03860.

Session, A. M., U. Yoshinobu, K. Taejoon, J. A. Chapman, A. Toyoda, S. Takahashi, A. Fukui, A. Hikosaka, A. Suzuki, M. Kondo, S. J. van Heeringen, I. Quigley, S. Heinz, H. Ogino, H. Ochi, U. Hellsten, J. B. Lyons, O. Simakov, N. Putnam, J. Stites, Y. Kuroki, T. Tanaka, M. Watanabe, O. Bogdanovic, R. Lister, G. Georgiou, S. S. Paranjpe, I. van Krujisbergen, S. Shu, J. Carlson, T. Kinoshita, Y. Ohta, S. Mawaribuchi, J. Jenkins, J. Grimwood, J. Schmutz, T. Mitros, S. V. Mozaffari, Y. Suzuki, Y. Haramoto, T. S. Yamamoto, C. Takaqi, R. Heald, K. Miller, C. Haudenschild, J. Kitzman, T. Nakayama, Y. Izutsu, J. Robert, J. Fortriede, K. Burns, V. Lotay, K. Karimi, Y. Yasuoka, D. S. Dichmann, M. F. Flajnik, D. W. Houston, J. Shendure, L. DuPasquier, P. D. Vize, A. M. Zorn, M. Ito, E. M. Marcotte, J. B. Wallingford, Y. Ito, M. Asashima, N. Ueno, Y. Matsuda, G. J. C. Veenstra, A. Fujiyama, R. M. Harland, M. Taira, D. S. Rokhsar. 2016.

Genome evolution in the allotetraploid frog *Xenopus laevis*. *Nature* 538:336–343. https://doi.org/10.1038/nature19840. Xenbase.

Stühmer, W., J. P. Ruppersberg, K. H. Schröter, B. Sakmann, M. Stocker, K. P. Giese, A. Perschke, A. Baumann, and O. Pongs. 1989. Molecular basis of functional diversity of voltage-gated potassium channels in mammalian brain. *The EMBO Journal* 8:3235–3234.

Stühmer, W., M. Stocker, B. Sakmann, P. Seeburg, A. Baumann, A. Grupe, and O. Pongs. 1988. Potassium channels expressed from rat brain cDNA have delayed rectifier properties. *FEBS Letters* 242:199–206. https://doi.org/10.1016/0014-5793(88)81015-9.

Sumikawa, K., I. Parker, and R. Miledi. 1984. Messenger RNA from rat brain induces noradrenaline and dopamine receptors in *Xenopus* oocytes. *Proceedings of the Royal Society of London: Series B, Biological Sciences* 223:255–260. https://doi.org/10.1098/rspb.1984.0093.

Sun, J., Y. Liu, M. Baudry, and X. Bi. 2020. SK2 channel regulation of neuronal excitability, synaptic transmission, and brain rhythmic activity in health and diseases. *Biochemical and Biophysical Acta Molecular Cellular Response* 1867:118834. https://doi.org/10.1016/j.bbamcr.2020.118834.

Sun, J., G. Zhu, Y. Liu, S. Standley, A. Ji, R. Tunuguntla, Y. Wang, C. Claus, Y. Luo, M. Baudry, and X. Bi. 2015. UBE3A regulates synaptic plasticity and learning and memory by controlling SK2 channel endocytosis. *Cell Reports* 12:449–461. https://doi.org/10.1016/j.celrep.2015.06.023.

Takumi, T., H. Ohkubo, and S. Nakanishi. 1988. Cloning of a membrane protein that induces a slow voltage-gated potassium current. *Science* 242:1042–1045. https://doi.org/10.1126/science.3194754.

Tempel, B. L., Y. N. Jan, and L. Jan. 1988. Cloning of a probable potassium channel gene from mouse brain. *Nature* 332:837–839. https://doi.org/10.1038/332837a0.

Timpe, L. C., Y. N. Jan, and L. Y. Jan. 1988. Four cDNA clones from the Shaker locus of Drosophila induce kinetically distinct A-type potassium currents in *Xenopus* oocytes. *Neuron* 1:659–667. https://doi.org/10.1016/0896-6273(88)90165-1.

Wang, C., K. Naruse, and K. Takahashi. 2018. Role of the TRPM4 channel in cardiovascular physiology and pathophysiology. *Cells* 7:62. https://doi.org/10.3390/cells7060062.

Woodward, R. M., and R. Miledi. 1987. Membrane currents elicited by porcine vasoactive intestinal peptide (VIP) in follicle-enclosed *Xenopus* oocytes. *Proceedings of the Royal Society of London: Series B, Biological Sciences* 231:489–497. https://doi.org/10.1098/rspb.1987.0057.

Woodward, R. M., and R. Miledi. 1991. Angiotensin II receptors in *Xenopus* oocytes. *Proceedings of the Royal Society of London: Series B. Biological Sciences* 244:11–19. https://doi.org/10.1098/rspb.1991.0044.

Yang, K. Q., Y. Xiao, T. Tian, L. G. Gao, and X. L. Zhou. 2014. Molecular genetics of Liddle's syndrome. *Clinica Chimica Acta*. Sep 25;436:202–206. doi: 10.1016/j.cca.2014.05.015. Epub 2014 May 29.

Yang, X., Niu, N., Liang, C., Wu, M. M., Tang, L. L., Wang, Q. S., Lou, J., Song B. L., Zheng W. W., Ma, H. P., and Zhang, Z. R. 2020. Simulation of epithelial sodium channels in endothelial cells by bone morphogenetic protein-4 contributes to salt-sensitive hypertension in rats. *Oxid Med Cell Longev* July 16;2020:1–13. doi: 10.1155/2020/3921897. Published online 2020 Oct 28.

Zeilhofer, H. U., M. A. Acuña, J. Gingras, and G. E. Yévenes. 2018. Glycine receptors and glycine transporters: Targets for novel analgesics? *Cellular and Molecular Life Sciences* 75:447–465. https://doi.org/10.1007/s00018-017-2622-x.

Zeppillo, T., A. Shulmann, F. Macciardi, B. E. Hjelm, M. Föcking, P. A. Sequeira, I. Guella, D. Cotter, and W. E. Bunney. 2020. Functional impairment of cortical AMPA receptors in schizophrenia. *Schizophrenia Resourses* S0920–9964:30147-X. https://doi.org/10.1016/j.schres.2020.03.037.

Section II

Systems Biology and the Genomic Era

10 The Continuing Evolution of the *Xenopus* Genome

Mariko Kondo and Masanori Taira

CONTENTS

10.1. Genome History of the Genus *Xenopus* ... 155
- 10.1.1. Heterosis ... 156
- 10.1.2. Recognition of Allotetraploidy ... 156

10.2. Genome Sequencing of *Xenopus*—The Highlights ... 157
- 10.2.1. Genome Sequencing of *X. tropicalis* ... 157
- 10.2.2. Sequencing and Assembly of the *X. laevis* Genome ... 157
- 10.2.3. Comparison between *X. tropicalis* and *X. laevis* Genomes and Chromosomes ... 158
- 10.2.4. The Real Story of the Identification of L and S Subgenomes and Chromosomes ... 158
- 10.2.5. Asymmetrical Evolution of the L and S Subgenomes ... 159

10.3. Studies Utilizing the Genome Sequences of *Xenopus* ... 161
- 10.3.1. W- and Z-Specific Regions in the Sex Chromosomes ... 161
- 10.3.2. Analyses of Hox Genes ... 162
- 10.3.3. Detailed Gene Organization in the *Xenopus laevis* Subgenomes ... 163

10.4. Conclusions ... 165

10.5. Future Directions ... 166
- 10.5.1. Comparative Genome Analysis ... 166
- 10.5.2. Artificial Hybrid Analysis ... 166
- 10.5.3. Artificial Allopolyploid Analysis ... 166

Acknowledgments ... 167

References ... 167

Clawed frogs of the genus *Xenopus* inhabit a large area of Africa, where 29 extant species are currently recognized (as of May 2021, https://amphibiaweb.org). Of them, the African clawed frog *Xenopus laevis* was originally imported into Europe from its native South Africa and has been used for research in laboratories since the 1930s, initially mainly for endocrinology studies and subsequently for developmental biology research (reviewed in Gurdon and Hopwood, 2000), and used as a model system to study cell cycle, oogenesis, early development, and so on, by biochemical and molecular biological approaches. *X. laevis* has various advantages as a model animal, such as easy rearing and breeding, a wide range of survival temperatures, resistance to infectious disease, and many others, which might be due to its allotetraploidy. However, from the late 1990s, when developmental genetics started to dominate the field of developmental biology using knockout mice and mutagenesis screening of zebrafish, the allotetraploidy of *X. laevis* became a disadvantage. Therefore, scientists tried to replace *X. laevis* with a diploid *Xenopus* species, *X. tropicalis*, which has a simpler genome and shorter generation time. *X. tropicalis* was previously called *Silurana tropicalis*, but later *Silurana* was treated as a subgenus; thus, this species was renamed *Xenopus Silurana tropicalis*, commonly called *Xenopus tropicalis*. The genus *Xenopus* now consists of two subgenera, *Xenopus* and *Silurana*. Although *X. tropicalis* is a suitable diploid model system in amphibians, *X. laevis* still remains useful, and hence both *X. tropicalis* and *X. laevis* have continued being used for research.

10.1. GENOME HISTORY OF THE GENUS *XENOPUS*

The evolutionary history of *Xenopus* frogs leading to the current phylogenetic relationships consist of bifurcating speciation and allopolyploidization by interspecies hybridization followed by polyploidization (Figure 10.1A) (reviewed in Kobel and Du Pasquier, 1986; Evans et al., 2004; Evans, 2008). Polyploidization is caused by duplication of the entire genome, otherwise called whole genome duplication (WGD). Polyploidy originating from duplication of a genome from a single ancestral species is called autopolyploidy, whereas allopolyploidy refers to genome duplication of hybrid genomes via interspecific crossing, which is the case for *Xenopus* species. Allotetraploid genomes, therefore, contain two different genomes derived from diploid ancestor species,

DOI: 10.1201/9781003050230-12

each called a subgenome. According to whole genome analysis of *X. laevis* (Session et al., 2016), in the subgenus *Xenopus* lineage, bifurcating speciation of a diploid species occurred 34 million years ago (Mya) to generate two species, the so-called "L" and "S" species (see subsequently for the explanation of "L" and "S"). Then, allotetraploidization occurred 17~18 Mya between the two species to generate a new allotetraploid species. This is the common ancestor of the subgenus *Xenopus* and had subgenomes "L" and "S." The "same" genes in the original L and S species are called orthologs, but after allotetraploidization, those genes came to reside in a single species, and now the former orthologs became "homeologs" in subgenomes L and S. After allotetraploidization, repeated specification occurred to generate several *Xenopus* species, as shown in Figure 10.1A. Some of them underwent additional rounds of allopolyploidization to generate octoploid and dodecaploid species.

WGD is one of the driving forces behind evolution. This is postulated to have occurred twice in the common ancestor of vertebrates 500~600 Mya (Meyer and Van de Peer, 2005) and a third time in the ancestor species of bony fishes about 306 Mya (Inoue et al., 2015) and is considered to have contributed greatly to diversity. However, over the years since WGD, chromosomes have reorganized and become diploid, referred to as "diploidization," and it is now impossible to distinguish whether the ancestral vertebrate species was autopolyploid or allopolyploid. Compared to those species, allotetraploidization in the *Xenopus* lineage occurred relatively very recently, and thereby subgenomes L and S were clearly identified in *X. laevis* by whole genome analysis. Thus, though allotetraploidy of *X. laevis* has long been considered a disadvantage for genetics, the identification of subgenomes made *X. laevis* a very useful model for studying subgenome evolution after allopolyploidization.

10.1.1. Heterosis

Heterosis, or hybrid vigor, is observed in hybrids that have superior phenotypes such as more vigorous growth or higher progeny yields compared to their parents. This phenomenon is also observed in allopolyploid plants, for example, wheat (Chen, 2013). In animals, heterosis is commonly known in mules whose sire is a donkey and dam is a horse, and they are particularly useful as working animals for their physical strength. However, interspecies hybrids are generally infertile, mostly due to the impairment of meiosis. If WGD occurs in F1 hybrids (i.e. allopolyploidization), they could be fertile and possibly lead to the generation of a new species (Chen, 2013). Therefore, allopolyploid animals are expected to exhibit heterosis, but this has not been actually demonstrated. But still, this can be inferred from two lines of circumstantial evidence for *Xenopus*. First, the parent species L and S of the subgenus *Xenopus* are extinct, possibly due to domination of their descendant allotetraploids over the parental species; second, the habitat of diploid (not polyploid) *X. tropicalis* is limited to the region including Nigeria and Ivory Coast, whereas species of subgenus *Xenopus* inhabit a much wider region (Tinsley et al., 1996; Evans et al., 2004). Therefore, it may have been an advantage for *Xenopus* species to have become allotetraploid (or higher).

10.1.2. Recognition of Allotetraploidy

What led to the current understanding that *X. tropicalis* is diploid and *X. laevis* is allotetraploid? Analyses of various species from the *Xenopus* genus and other genera in Pipidae showed that their number of chromosomes and the DNA content per cell were in proportions of approximately 1:2:4:6 and that the basal level is those of *X. tropicalis*, which has 20 chromosomes (n = 10) and 3.55 pg DNA/cell, indicating the existence of polyploid species (Kobel, 1981). The finding that the numbers of *X. laevis* was about twice of *X. tropicalis* suggested tetraploidy. Since around 1990, *X. laevis* has become widely recognized as being allotetraploid, after gene cloning became popular in the late 1980s and the presence of two different sequences for many genes were revealed. For example, for the *hoxb7* gene (formerly known as *XlHbox2*), the clone called p52 identified by Wright et al. (1987) was a different version of the gene identified by Müller et al. (1984), called MM3. These two sequences showed high identity to each other, raising the possibility that this reflected polymorphism, but analyses in *X. laevis*/*X. borealis* interspecific hybrids showed that these two genes do not segregate from each other, which led to the conclusion that these are not alleles of the same gene (Fritz et al., 1989). At that time, the first identified gene was given a postfix "a," and the second gene was given "b"; thus, they were usually called the a- and b-genes. However, analysis of one of the two genes was (or appeared to be) practical for developmental investigations; because of the high nucleotide identities of coding sequences as well as translated amino acid identities between the two versions, cross-hybridization was inevitable for standard Northern blots and whole-mount *in situ* hybridization, thus making distinguishing the two difficult. The difference in the functions of the two versions was probably not always studied, but in the case of GATA-1, precise analyses showed that GATA-1a and GATA-1b share the same function in stimulating erythropoiesis, but only the latter inhibits neurogenesis when overexpressed (Xu et al., 1997). As another example, two *gbx2* genes, *gbx2.2.L* (*Xgbx2a*) and *gbx2.1.S* (*Xgbx2b*), when overexpressed, caused malformation of the head and notochord but showed differences in their temporal and spatial expression patterns (Tour et al., 2001).

Now, after the completion of the whole genome sequencing of *X. laevis*, the two versions a and b are acknowledged as homeologs that arise from the different subgenomes and are called the L and S genes. Back to the two versions of *XlHbox2/hoxb7*, p52 and MM3 are now identified as *hoxb7.L* and *hoxb7.S*, respectively (see Kondo et al., 2017). *GATA-1a* and *GATA-1b* are *gata1.L* and *gata1.S*, respectively (Watanabe et al., 2017). Note that a- and b-genes do not necessarily correspond to L and S genes, respectively, because of the difference between their definitions: the chronological order of identification for a and b versus the position on the L or S chromosomes.

10.2. GENOME SEQUENCING OF *XENOPUS*—THE HIGHLIGHTS

Whole genome sequencing (WGS) is a powerful tool for analyzing the evolution of the genome. It can also reveal the full genetic make-up of an organism, allowing for comprehensive analysis of the relationships among many genes and their regulatory sequences. WGS of vertebrate organisms has been done in humans and other major model organisms such as mice, zebrafish, and medaka that are useful for genetic analysis (Table 10.1).

As mentioned earlier, *X. laevis* has been a very useful experimental model animal in embryology and cell biology from the 1950s to the present. However, its long generation time (more than a year) and allotetraploidy made it unsuitable for genetic analysis, whereas the closely related and diploid *X. tropicalis* came into use as a model organism around the year 2000. Mutagenesis screens using *X. tropicalis* identified new or uncharacterized genes (Goda et al., 2006; Chung et al., 2014; Nakayama et al., 2017). The linkage map of *X. tropicalis* was constructed using simple sequence length polymorphism (SSLP) markers (Wells et al., 2011) and became the first amphibian to have its entire genome decoded (Hellsten et al., 2010). This left the African clawed frog *X. laevis* as the only one of the major model organisms whose genome had not been decoded (Table 10.1).

TABLE 10.1
Publications of Whole Genome Sequences of Human, Model, and Polyploid Organisms

Year	Organism	Reference	Comments
1994 to 1997	yeast	Dujon et al., 1994; Dietrich et al., 1997 and others	*Saccharomyces cerevisiae*; each chromosome sequence was published one by one
1998	*C. elegans*	*C. elegans* Sequencing Consortium, 1998	*Caenorhabditis elegans*, nematode
2000	*Drosophila melanogaster*	Myers et al., 2000	Fruit fly
2001	human	International Human Genome Sequencing Consortium, 2001	
2002	mouse	Mouse Genome Sequencing Consortium, 2002	*Mus musculus*
2004	chick	International Chicken Genome Sequencing Consortium, 2004	*gallus*
2005	zebrafish	Woods et al., 2005	*Danio rerio* 309
2007	medaka	Kasahara et al., 2007	*Oryzias latipes* 309
2010	*Xenopus tropicalis*	Hellsten et al., 2010	Western clawed frog, diploid
2014	*Arabidopsis thaliana*	Poczai et al., 2014	Flowering plant
2014	rainbow trout	Berthelot et al., 2014	Tetraploidization 100 Mya
2014	carp	Xu et al., 2014	*Cyprinus carpio*, allotetraploidization 8.2 Mya
2016	Atlantic salmon	Lien et al., 2016	*Salmo salar*, tetraploidization 80 Mya
2016	*Xenopus laevis*	Session et al., 2016	African clawed frog, allotetraploidization 17–18 Mya, subgenomes L and S
2020	goldfish	Chen et al., 2020	*Carassius auratus*, allotetraploid, subgenomes A and B

10.2.1. Genome Sequencing of *X. tropicalis*

The genome of *X. tropicalis* was estimated to be about 1.7 Gbp, and the whole genome sequencing was carried out using a frog of the inbred Nigerian strain and published in 2010 (Hellsten et al., 2010) (see details for *X. tropicalis* strains in Igawa et al., 2015). To produce the draft assembly (Xentr4.1), plasmids, fosmids, and bacterial artificial chromosomes (BACs) containing genomic DNA were subjected for sequencing using the Sanger method. Together with EST and cDNA data from many resources, the number of protein-coding genes was estimated to be 20,000 to 21,000. *X. tropicalis* has 20 chromosomes (n = 10), and at the time of the publication of the whole genome sequence, a linkage map with 10 linkage groups had been constructed (Wells et al., 2011). Scaffolds and linkage groups were mapped onto these chromosomes (Hellsten et al., 2010). As this was the first amphibian species to be fully sequenced, the genome served as an interesting example for comparison of chromosomal structures and sequences with two tetrapods: human and chicken. The analyses showed that synteny was conserved over long stretches of chromosomes and elucidated that fusions and fissions occurred lineage specifically. Thus, WGS not only provided information on the species itself but also added insight into chromosomal and genome evolution. As a resource, the genome sequence of *X. tropicalis* is very useful for gene cloning, designing antisense morpholinos, enhancer and promoter analyses, and genetic screens for mutants as well as ChIP-sequencing and RNA-sequencing.

10.2.2. Sequencing and Assembly of the *X. laevis* Genome

Even though *X. tropicalis* appeared to be a suitable amphibian model for biological studies, *X. laevis* still has its advantages, as mentioned previously, including ease of maintenance and husbandry under laboratory conditions, ease of obtaining embryos, and ease of performing biological and biochemical experiments. Therefore, in 2010, genome projects of *X. laevis* began independently in Japan and the United States. At the 14th International *Xenopus* Conference in France,

2012, these two teams reached an agreement to collaborate; as the project progressed, other researchers joined to form an international consortium.

The genome sequencing of *X. laevis* was challenging not only because of its estimated large genome size (3 Gbp, nearly twice as much of *X. tropicalis* and comparable to human, with a total of 36 chromosomes), but the biggest obstacle was to distinguish the highly conserved pairs of homeologous genes and to correctly assemble the whole genome. To overcome the problem with heterozygosity, in which allelic variations potentially disturb homeolog-specific assembly, the highly inbred "J-strain" was chosen as the source of DNA for sequencing. The J-strain was made by inbreeding frogs by C. Katagiri, S. Tochinai, and their colleagues in Hokkaido University, Japan (see Extended Data Figure 1a in Session et al., 2016, for the history of the J-strain). Originally, this strain was made as a model animal for immunological analyses and has been used in several studies (Izutsu and Yoshizato, 1993; Robert and Ohta, 2009). At the time the whole genome sequencing project started, inbreeding had been conducted for more than 30 generations, and even before that, long-term immunological rejection did not occur in skin transplantation assays (Izutsu and Yoshizato, 1993). Therefore, polymorphic sequences could most likely be attributed to the difference between homeologs.

As *X. laevis* has a female heterozygous (ZW) sex determination system (Chang and Witschi, 1956), DNA from females of the J-strain was shotgun and mate pair sequenced with Illumina. Contigs and scaffolds were constructed incorporating end-sequencing of BAC and fosmid clones. These sequences were further assembled to chromosome-scale using several methods: FISH analyses with BACs and *in vivo* and *in vitro* chromatin conformation capture methods (HiC and the Chicago method, respectively) (Lieberman-Aiden et al., 2009; Putnam et al., 2016). FISH analyses identified several chimeric scaffolds, which were then separated by estimating the fusion sites and reassembled. Chimeric sequences were also identified by randomly and arbitrarily selecting 987 genes and testing the integrity of their gene structures. In the case of tandemly repeated homologous gene clusters, such as Hox clusters and *mix/bix* gene clusters and regions whose sequence could not be fully determined by automatic assembly of shotgun sequences, they were filled by manually selecting the corresponding BAC and fosmid clones and determining their entire sequences. Finally, the ver. 9.1 assembly contained chromosome-level sequences for each of the 18 chromosomes. Through meticulous verification of the assembly, feedback, and re-assembly, the genome sequence of *X. laevis* became probably one of the most reliable genomes of its size. The current version (as of April 2021) is ver. 10.1, which was reassembled using PacBio long read sequences (Genbank GCA_017654675.1).

10.2.3. Comparison between *X. tropicalis* and *X. laevis* Genomes and Chromosomes

Before determining the whole genome sequence of *X. laevis*, the correspondence between the chromosomes of *X. tropicalis* (XTR) and *X. laevis* (XLA) was identified by FISH analysis using cDNAs of *X. laevis* (Uno et al., 2013), as well as by the detailed chromosome map of *X. laevis* constructed with the assignment of BAC clones (Matsuda et al., 2015; Session et al., 2016). FISH analyses with cDNA probes determined the pairs of homeologous chromosomes of *X. laevis*, since homeologous chromosomes harbored basically the same genes.

As mentioned before, *X. tropicalis* is diploid and has 20 chromosomes ($2n = 2x = 20$); the chromosome pairs are numbered as XTR1 to XTR10 (Khokha et al., 2009). The allotetraploid *X. laevis* has 36 chromosomes ($2n = 4x = 36$), consisting of nine pairs of homoeologous chromosomes. Of them, eight pairs each corresponded to one *X. tropicalis* chromosome (XTR1–XTR8), whereas the remaining XLA chromosomes corresponded to a fusion between XTR9 and XTR10 (Uno et al., 2013). Based on the correspondence with *X. tropicalis* chromosomes, the nine homoeologous chromosome sets of *X. laevis* were renumbered as XLA1, XLA2, XLA3, and so on. As one of the homeologous chromosomes is longer than the other, according to relative lengths measured with the karyotypes (Matsuda et al., 2015), the longer chromosomes were suffixed by adding the letter L (for "long") and the shorter chromosomes with S (for "short"), making XLA1L, XLA1S, XLA2L, and so on (see Figure 10.1B for the nomenclature and relationships of the chromosomes). The chromosomes homologous to XTR9 and XTR10 were named XLA9_10L and XLA9_10S to reflect their fused status, or simply XLA9L and XLA9S. As stated in Matsuda et al. (2015), cytogeneticists may prefer the simple nomenclature, but XLA9_10L and XLA9_10S are more convenient to perform direct genome and chromosome comparisons between *X. laevis* and *X. tropicalis*.

Chromosome fusion sites in XLA9_10 were identified by a comparison of the synteny of genes of XLA9_10L and XLA9_10S and those at the ends of XTR9 and XTR10 (Session et al., 2016). Since the fusion regions in XLA9_10L and XLA9_10S were identical and the basic number of chromosomes in the family Pipidae (including the genus *Xenopus*) was $n = 10$, the chromosome fusion was suggested to have occurred after the divergence from *X. tropicalis* and before the speciation of the two diploid ancestor species of *X. laevis*. Analyses showed that two chromosomes corresponding to XTR9 and XTR10 in an ancestral frog fused tandemly (without any obvious gene loss) and the centromere was repositioned during karyotype evolution (see Extended Data Figure 2 in Session et al., 2016).

10.2.4. The Real Story of the Identification of L and S Subgenomes and Chromosomes

X. laevis homeologous chromosomes were designated with postfixes L and S just according to the difference in lengths of chromosomes, as mentioned previously. The description of chromosomes and subgenomes in Session et al. (2016) may give the impression that the subgenomes coincidentally corresponded to the sets of the L (longer) and S (shorter) chromosomes of each homeologous pair, but in fact, we had identified the subgenomes before naming the chromosomes. Here is the real story of how L and S were adopted.

As mentioned before, several hundreds of long scaffolds had been assigned to 18 chromosomes by FISH analysis using BACs located in the scaffolds as probes. The homeologous relationships of the scaffolds could be determined by their chromosomal locations as well as the presence of homeologous genes that the scaffolds contained. However, to distinguish the subgenomes consisting of nine chromosomes each, derived from the two diploid ancestor species, we needed to find some "marks" specific to either of the parental chromosome sets. For this purpose, we searched for "fossil" transposons which were horizontally transferred into either of the parental species and expanded but became inactivated before allotetraploidization (Figure 10.1C, D). After identifying and analyzing a large number of fossil transposon sequences in the scaffolds, we found that two PIF/harbinger-type and one Tc1/mariner-type DNA transposons were distributed unevenly in either of each homeologous pair of scaffolds, suggesting that they were subgenome-specific fossil transposons. The biased distribution of fossil transposons on either of actual homeologous chromosomes was confirmed by FISH using the Tc1/mariner-type transposon as a probe (Figure 10.1E) (Session et al., 2016). This result indicated that the set of nine chromosomes belonging to each of the extinct ancestor species remained largely intact as a subgenome in *X. laevis* without massive reorganization.

Once we successfully identified the subgenomes, we needed to name them. Naming them "A" and "B" would probably have been the first choice in general, but the suffixes "a" and "b" were already used randomly for homeologs, so we had to find another pair of suffixes. In parallel with this, we were analyzing gene syntenies between homeologous scaffolds and finding cases that either of the homeologs was absent, being a single-copy gene or "singleton." Furthermore, we realized that singleton genes and tandemly repeated genes (gene expansions; see subsequently) were biased toward the Tc1/mariner-*negative* subgenome. Meanwhile, we noticed from karyotypes that the Tc1/mariner-*positive* chromosomes all appeared shorter than their counterparts (Figure 10.1E). Therefore, we precisely measured the relative lengths of all chromosomes, and found that this observation was correct (see Table 1 in Matsuda et al., 2015). The biased deletion of genes from the Tc1/mariner-positive chromosomes matches with those chromosomes being shorter.

If the chromosomes tend to have fewer gene deletions/more gene expansions, they will become relatively longer than those with more gene deletions/less gene expansions. Based on this, we proposed to name the homeologous chromosomes "L" for "longer" and "S" for "shorter," as well as the corresponding subgenomes "L" and "S." The difference in length between homeologous chromosomes is consistent with the characteristics of the subgenomes.

10.2.5. Asymmetrical Evolution of the L and S Subgenomes

By distinguishing the L and S subgenomes, we compared the divergence of protein-coding gene sequences and estimated that the two ancestor species arose about 34 Mya (Figure 10.1A) (Session et al., 2016). A draft genome sequence of *X. borealis*, which is related to *X. laevis* and shares a common allotetraploid ancestor, allowed us to estimate their speciation about 17 Mya. Together with estimation of expansion and cessation periods of the L- and S-specific transposons, the allotetraploidization event occurred about 17–18 Mya (Figure 10.1A) (Session et al., 2016).

Based on the assembly sequence ver. 9.1 and gene annotation ver. 1.8, the number of protein-coding genes in the genome of *X. laevis* was estimated to be 45,099 in total, about twice that of the diploid *X. tropicalis* (Session et al., 2016). According to the allotetraploid nature of *X. laevis*, a single gene in *X. tropicalis* is supposed to correspond to a pair of its orthologous homeologs in *X. laevis*: for example, a gene on XTR1 is present as copies on both XLA1L and XLA1S. If a pair of homeologs became a singleton in *X. laevis*, the relationship would be one to one. Therefore, genes of *X. tropicalis* corresponding to those of *X. laevis* in either a one-to-two or one-to-one relationship were counted. As a result, of the 15,613 protein-coding genes of *X. tropicalis*, 8806 corresponded to homeolog pairs of *X. laevis*, whereas the remaining 6807 genes corresponded to singleton genes. The retention rate of homeolog pairs is calculated to be 56%.

Comparison of homeologous gene loci between L and S shows that the L chromosomes are more similar to *X. tropicalis*, while the S chromosomes have several large inversions (Figure 10.1B), suggesting that the S chromosomes are more prone to accumulating changes (Session et al., 2016). Furthermore, focusing on the genes with one-to-one correspondence between *X. tropicalis* and *X. laevis* (singletons), we found that more genes were lost from the S subgenome (8.3% of the genes in L and 31.5% in S were lost). Differences in expression among homeologs were examined by RNA-seq analysis of 14 developmental stages and 14 adult tissues. The results showed that zygotic expression levels of L genes were, on average, approximately 25% higher than those of the corresponding S genes. In addition, genes with lower expression levels tended to accumulate more mutations, suggesting that they are pseudogenizing or perhaps subfunctionalizing.

Another typical example for asymmetry of the L and S subgenomes is the rDNA (45S pre-ribosomal RNA gene) cluster. The number of rDNA genes in the cluster was originally estimated to be about 450 (Brown and Dawid, 1968) and, more recently, about 750 (Michalak et al., 2015). The nucleolus is formed at the rDNA cluster, and therefore the rDNA region is also called the nucleolar organizing region (NOR). The NOR in *X. laevis* was known to be on XLA3L (the same as the previously called chromosome 12) (Schmid and Steinlein, 1991). FISH analysis using rDNA as a probe confirmed that the rDNA gene cluster was detected only on XLA3L, not on XLA3S (see Extended Data Figure 5a in Session et al., 2016). In general, of the two NORs inherited from the parental species of a hybrid, only one is actively transcribed, the phenomenon called nucleolar dominance (reviewed in McStay, 2006; Preuss and Pikaard, 2007). This suggests that the parental L species of *X. laevis* had nucleolar dominance, leading to asymmetric evolution of the rDNA cluster.

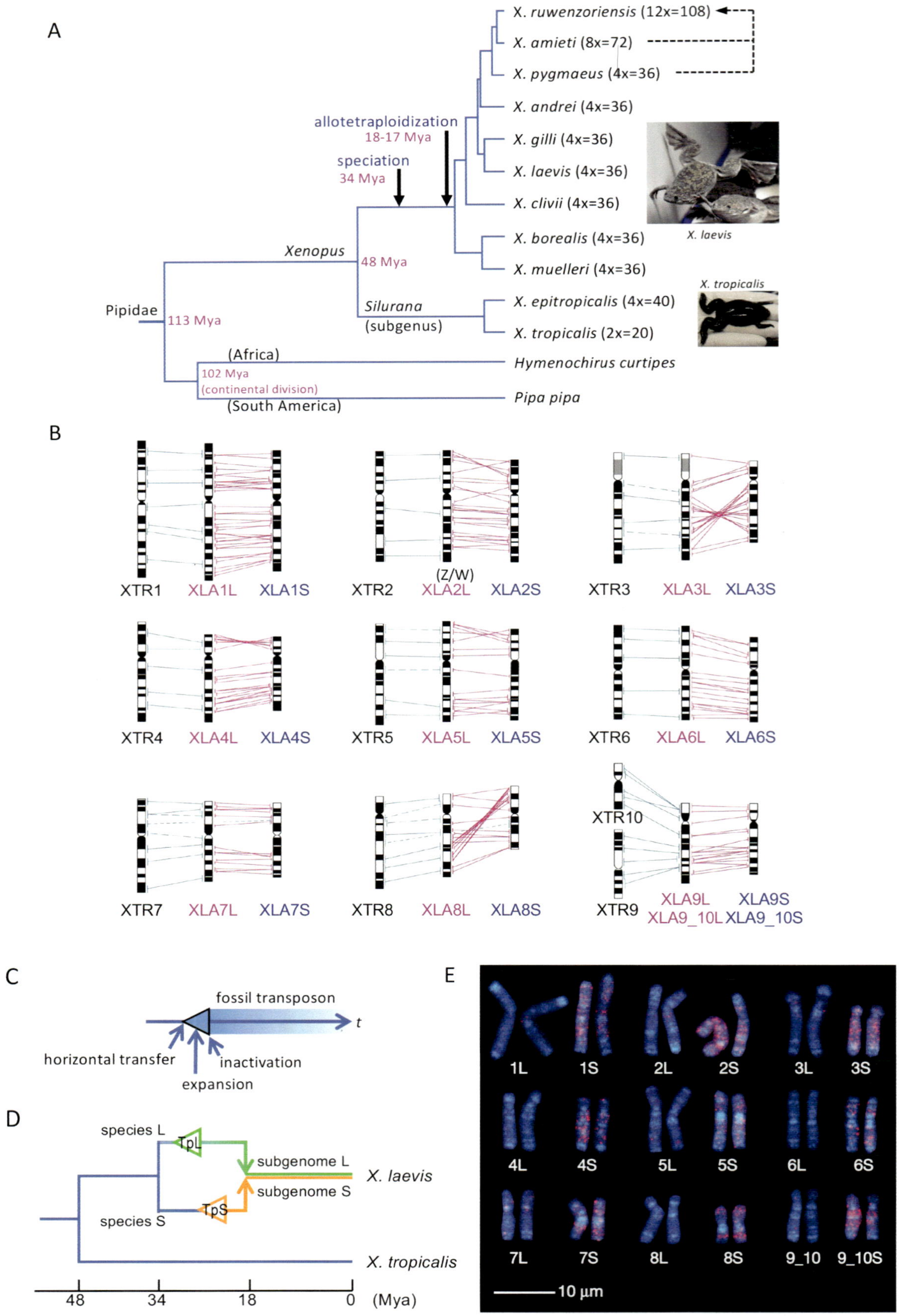

FIGURE 10.1 Phylogenetic relationships of *Xenopus* species, chromosomal relationships between *X. tropicalis* and *X. laevis*, and the identification of subgenomes L and S using fossil transposons. (A) Family Pipidae and the genus *Xenopus*. Divergence times adopted

FIGURE 10.1 (Continued)
from Session et al. (2016) are indicated. Phylogenetic relationships of *Xenopus* species are adopted from Evans et al. (2004). Species of *Xenopus* other than those of subgenus *Silurana* (including *X. epitropicalis* and *X. tropicalis*) belong to the subgenus *Xenopus*. *X. ruwenzoriensis* speciated by hybridization between *X. amieti* and *X. pygmaeus* followed by polyploidization (dotted lines with arrow). (B) Orthologous and homeologous relationships between *X. tropicalis* chromosomes (XTR) and *X. laevis* (XLA) L (long) and S (short) chromosomes based on chromosome-scale sequence assembly and BAC FISH data. XLA9L and XLA9S are also called XLA9_10L and XLA9_10S, which correspond to XTR9 and XTR10 due to chromosome fusion. XLA2L corresponds to both Z and W chromosomes. (C) Schematic representation of the fate of transposons. Transposons can be horizontally transferred into a genome, then increase their number to invade the entire genome (expansion). Meanwhile, transposons become inactivated by a host immune system and the inactivated transposons, called fossil transposons, are gradually mutated over time (t). (D) Allotetraploidization and subgenome specific fossil transposons in the *X. laevis* lineage. L- and S-specific fossil transposons (TpL and TpS, respectively) were identified in *X. laevis*. (E) FISH analysis with the TpS probe. Either homeologous chromosome pair is labeled (magenta). Notably, TpS-positive chromosomes are found to be the shorter chromosome in each homologous pair, leading to the idea that each of longer (L) and shorter (S) chromosome sets corresponds to either of the parental chromosome sets.

Source: Panels B and E were copied from Session et al. (2016).

Overall, the two subgenomes evolved differently, with the L subgenome conserving sequences of the ancestral species and the S subgenome having a higher percentage of broken genes due to deletions and rearrangements. Thus, we showed for the first time that the subgenomes in an allotetraploid evolve asymmetrically. Diploidization is the process in which polyploids become diploid, and the loss of homeologous genes that is observed in *X. laevis* is probably an early event in diploidization. However, it still remains unclear what the main evolutionary force driving the differences between the L and S subgenomes was.

10.3. STUDIES UTILIZING THE GENOME SEQUENCES OF *XENOPUS*

Further precise studying of the whole genome led to a series of new findings to understand the changes in the *X. laevis* genome after allotetraploidization. Individual studies were performed by different groups, focusing on the sex determining Z and W loci, Hox clusters, tandemly repeated genes, gene families, and so on in the genomes of *X. laevis* and *X. tropicalis*.

10.3.1. W- and Z-Specific Regions in the Sex Chromosomes

The sex of *X. laevis* is determined by a ZZ-ZW chromosomal system, in which females are heterogametic, as identified by genetic analysis (Chang and Witschi, 1956). The sex-determining gene, *DM-W* (the same as *dmw*), which is female specific and thus is on the W-chromosome, was identified in 2008 (Yoshimoto et al., 2008). This gene is one of the members of the *dmrt* family, whose members are known to be involved in sexual development, such as *Drosophila melanogaster doublesex* (*dsx*), *Caenorhabditis elegans mab-3*, and *Mus musculus dmrt1*. The sex chromosomes of *X. laevis* were not identified for a long time, but from FISH analysis using *dmw* as a probe, a single W-chromosome was distinguished and *dmw* was located at the q-subtelomeric region (Yoshimoto et al., 2008). The sex chromosome is XLA2L, but there are two versions: the one containing *dmw* is the W-chromosome, and the other lacking *dmw* is the Z-chromosome.

To identify possible W or Z region-containing clones, a fosmid library was screened by colony hybridization with *dmw*-flanking sequences as probes. These clones were analyzed for the presence of *dmw*. A number of these clones with or without *dmw* and BACs in this region were chosen for full sequencing (Session et al., 2016; Mawaribuchi et al., 2017). Using the full sequences of these clones helped to correct the mistakes in the computational ver. 9.1 assembly of the *X. laevis* genome and revealed the structure of the W- or Z-specific regions. The W-specific sequence containing *dmw* is 278 kb in length, whereas the Z-specific sequence is 83 kb (Figure 10.2A) (Mawaribuchi et al., 2017). In total, three W-specific genes including *dmw* and one Z-specific gene were found (Figure 10.2A). Comparisons of these genes with their autosomal homologous genes/sequences in *X. laevis* and *X. tropicalis* were performed to clarify their evolutionary histories. The result for *dmw* was consistent with previous reports that *dmw* may have emerged as a duplicate of the *dmrt1* gene (Yoshimoto et al., 2008), to be specific, from *dmrt1.S* (reported as *dmrt1*β) (Bewick et al., 2010) on XLA1S, and integrated into the XLA2L chromosome. The other two W-specific and one Z-specific genes were suggested to be duplicates from homologous counterparts on the L chromosomes, but from separate chromosomes. Therefore, it appears that all four W- or Z-specific genes were integrated independently. Interestingly, *dmw* is not found in *X. tropicalis* (Yoshimoto et al., 2008) nor in *X. borealis* (Bewick et al., 2010). This suggests that *dmw* emerged after the divergence of *X. laevis* and *X. borealis* or was lost in the *X. borealis* lineage and that sex-determination mechanisms of related species are different.

A similar situation is found in the fish *Oryzias latipes*, whose male sex-determining gene is *dmy/dmrt1bY* (Matsuda et al., 2002; Nanda et al., 2002), which arose evolutionarily as a duplicated version of the autosomal *dmrt1* gene. This gene is, however, not the universal sex-determining gene in the genus *Oryzias* (Kondo et al., 2003; Kondo et al., 2004). These examples demonstrate the variability of sex-determining systems. It would be quite interesting to identify the sex-determining genes of *X. tropicalis* and *X. borealis* as well as other *Xenopus* species to learn about the evolution of sex determination in *Xenopus*.

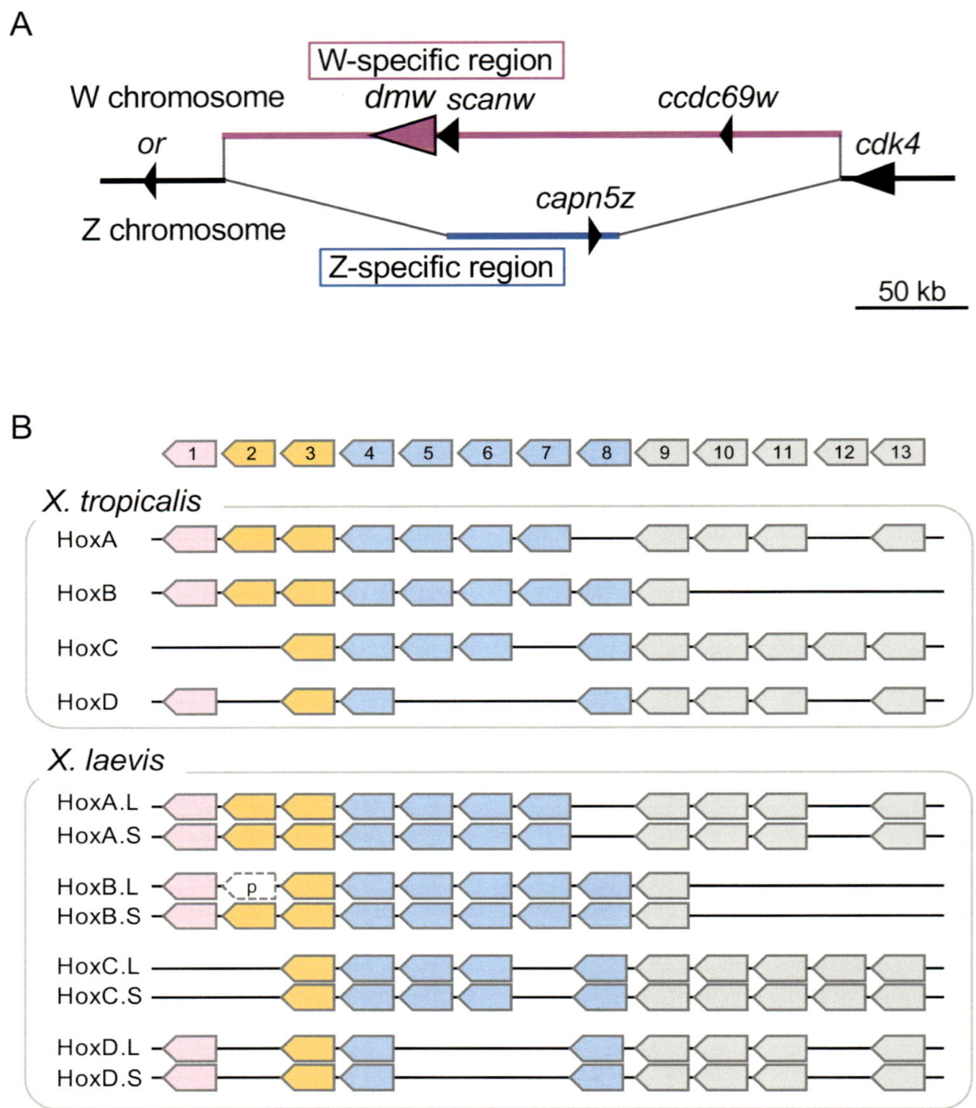

FIGURE 10.2 Structures of the W- and Z-specific regions and the Hox clusters. (A) W- and Z-specific regions at XLA2Lq32–33 are indicated by magenta and blue lines and correspond to the W and Z chromosomes, respectively. The flanking regions on the left and right sides (black lines) are shared between the W and Z chromosomes and harbor olfactory receptor (*or*) or *cdk4* genes. Modified from (Mawaribuchi et al., 2017). (B) Comparison of Hox clusters between *X. tropicalis* and *X. laevis*. *X. tropicalis* has four clusters (Hox A, HoxB, HoxC, and HoxD), while *X. laevis* has four each in the L and S subgenomes (indicated with ".L" and ".S"). *hox* genes of 13 paralogous groups are present. All *hox* genes are retained in *X. laevis* after allotetraploidization, but for *hoxb2.L*, which is pseudogenized (indicated as "p").

Source: Modified from (Kondo et al., 2017).

10.3.2. Analyses of Hox Genes

The *hox* genes and the Hox clusters are often identified and characterized for genome analysis or evolutionary studies. *hox* genes encode transcription factors, and in the genome of most animals, these genes are clustered in a specific region called the Hox cluster. The genes and the structure of the clusters are well conserved in bilaterians. While the chordate ancestor had a single Hox cluster as in the cephalochordate *Branchiostoma floridae* (Garcia-Fernández and Holland, 1994), the number of Hox clusters has increased to four through two rounds of WGD in the vertebrate lineage (reviewed in Holland, 2013). Thus, cartilaginous fishes, bony fishes, and tetrapods have four Hox clusters in their genome. An additional third round of WGD occurred in the teleost lineage; this was initially proposed based on the existence of short duplicated segments of gene synteny in the teleost genomes (Aparicio et al., 2002; Meyer and Van de Peer, 2005). The strongest evidence for the third round of WGD is that teleosts have seven to eight Hox clusters. Most teleosts examined have seven clusters by losing one of the two HoxC clusters in the lineage including medaka fish and fugu or by losing one of the two HoxD clusters in the lineage including zebrafish (reviewed in Kuraku and Meyer, 2009; Pascual-Anaya et al., 2013).

The third round of WGD in the teleost lineage happened about 306 Mya (Inoue et al., 2015), which is a much more ancient event than the WGD in *X. laevis* about 18 Mya. As expected, the *X. laevis* genome has eight Hox clusters consisting of pairs of HoxA, B, C, and D clusters on L and S chromosomes, exactly doubling the four Hox clusters present in *X. tropicalis* (Figure 10.2B) (Session et al., 2016; Kondo et al., 2017). The 76 *hox* genes in *X. laevis* included *hoxb2p.L*, which appeared to be pseudogenized, since its complete coding sequence could not be identified and indels were found when compared to its homolog *hoxb2.S*. Although the Hox clusters are highly conserved, comprehensive gene expression analyses of *hox* genes revealed differences in expression patterns between L and S *hox* genes during development or in adult tissues, which suggest subfunctionalization. In addition, the *hoxb.L* genes in the HoxB.L cluster appear to be rapidly diverging compared to *hoxb.S* genes (Kondo et al., 2017). This trend is the opposite of other homeologous gene sets, in which generally S genes tend to be subjected to mutations, pseudogenization, or partial or total deletions (Session et al., 2016).

Collinearity is the idea that the order of genes in the Hox cluster is correlated to spatial and temporal sequential gene expression during development. Spatial collinearity has been demonstrated in various organisms in which the anterior *hox* genes (located to the 3' end of the cluster) are expressed anterior to posterior genes (located to the 5' end). However, the definition of "temporal collinearity" is ambiguous, that is, it is not clear whether "gene expression" refers to the accumulation of mRNA or activation of gene transcription. Moreover, due to this ambiguity, the evidence supporting temporal collinearity has been weak; as a result, the "collinearity" of a complete set of genes within a cluster has never been proven. We analyzed the developmental expression patterns of all *hox* genes of *X. laevis* by RNA-seq and grouped them according to their profiles by clustering analysis, that is, when mRNA accumulation starts and reaches the maximum amount. We realized that temporal collinearity hypothesis could not be supported (Kondo et al., 2017).

To further examine the temporal collinearity hypothesis, since genomic and transcriptome data of *X. tropicalis* are available and there are no homeologs to distinguish in this species, we used *X. tropicalis* to investigate two aspects of "gene expression" to test the temporal collinearity hypothesis. First, with the help of high-resolution transcriptome analysis in *X. tropicalis* (Owens et al., 2016), we examined the order of genes whose transcript level reaches a certain threshold. Second, the timing of the start of *de novo* transcription was determined using RT-qPCR to detect pre-spliced transcripts for all genes in the HoxA cluster and some from other clusters (Kondo et al., 2019). As a result, we were able to demonstrate that the temporal collinearity theory is not experimentally supported. These analyses were possible because of the recent enrichment of genome and transcriptome information of the two species. Methodologically, the detection of *de novo* transcripts was based on designing qPCR primers at introns or exon/intron junctions using genome sequence information.

10.3.3. Detailed Gene Organization in the *Xenopus laevis* Subgenomes

From the late 1980s to the 1990s, many interesting genes were identified in *X. laevis*, but some of them were not found in mice or humans as orthologs, and some other genes had multiple paralogs, probably by local gene duplication or expansion. To clarify orthologous, paralogous, and homeologous relationships, elucidation of chromosomal localization and syntenic gene organizations of these genes was inevitable. Our gene annotation group in the international consortium of the *X. laevis* genome project examined such well-studied genes encoding transcription factors (Watanabe et al., 2017; Haramoto et al., 2017), peptide growth factors (Michiue et al., 2017; Suzuki et al., 2017a), signal transduction components (Suzuki et al., 2017b), cell cycle regulators (Tanaka et al., 2017), and others. In addition, genomic organization was further clarified for large gene clusters, such as the olfactory receptor gene clusters (see Extended Data Figure 5a in Session et al., 2016), and the type I and II keratin gene clusters (Suzuki et al., 2017). In the following, we introduce some remarkable findings from these analyses.

10.3.3.1. *nodal5, nodal3,* and *vg1* Clusters

Extreme asymmetry between subgenomes L and S was found in the *nodal5* and *nodal3* clusters (Session et al., 2016; Suzuki et al., 2017a), as well as the *vg1* cluster (Suzuki et al., 2017a) (Figure 10.3A, B, C). These are examples of clusters in which all functional genes were lost from the S subgenome. *nodal5* and *nodal3* are TGFß family members, and the numbers of the genes are expanded (Figure 10.3A, B). Probably more than five copies of *nodal5* genes exist on XLA3L (personal communications from Dr. Shuji Takahashi and Dr. Yoshikazu Haramoto). Other TGFß family members, *vg1* and *derriere*, are duplicates of a common ancestor gene and became subfunctionalized paralogs; *vg1* is a maternal factor in the egg, whereas *derriere* is zygotically expressed. *vg1* is present as a single copy gene in *X. tropicalis*, whereas in *X. laevis*, the *vg1.L* gene has expanded, forming a gene cluster, but *vg1.S* was pseudogenized (Figure 10.3C). Notably, there are two types of the expanded *vg1* genes, a functional Ser20 type (S20) and a non-functional Pro20 type (P20) as paralogs (Suzuki et al., 2017a), in which *vg1* (P20) was first identified as a maternal mRNA (Rebagliati et al., 1985). These three clusters have common features: (1) the numbers of expanded genes, including pseudogenes and mutated genes, in the subgenome L of *X. laevis* are more or less the same or more than those in *X. tropicalis*, whereas the corresponding genes are missing or pseudogenized in subgenome S and (2) the expanded genes appeared to be functionally equivalent (not subfunctionalized), implying that these gene expansions

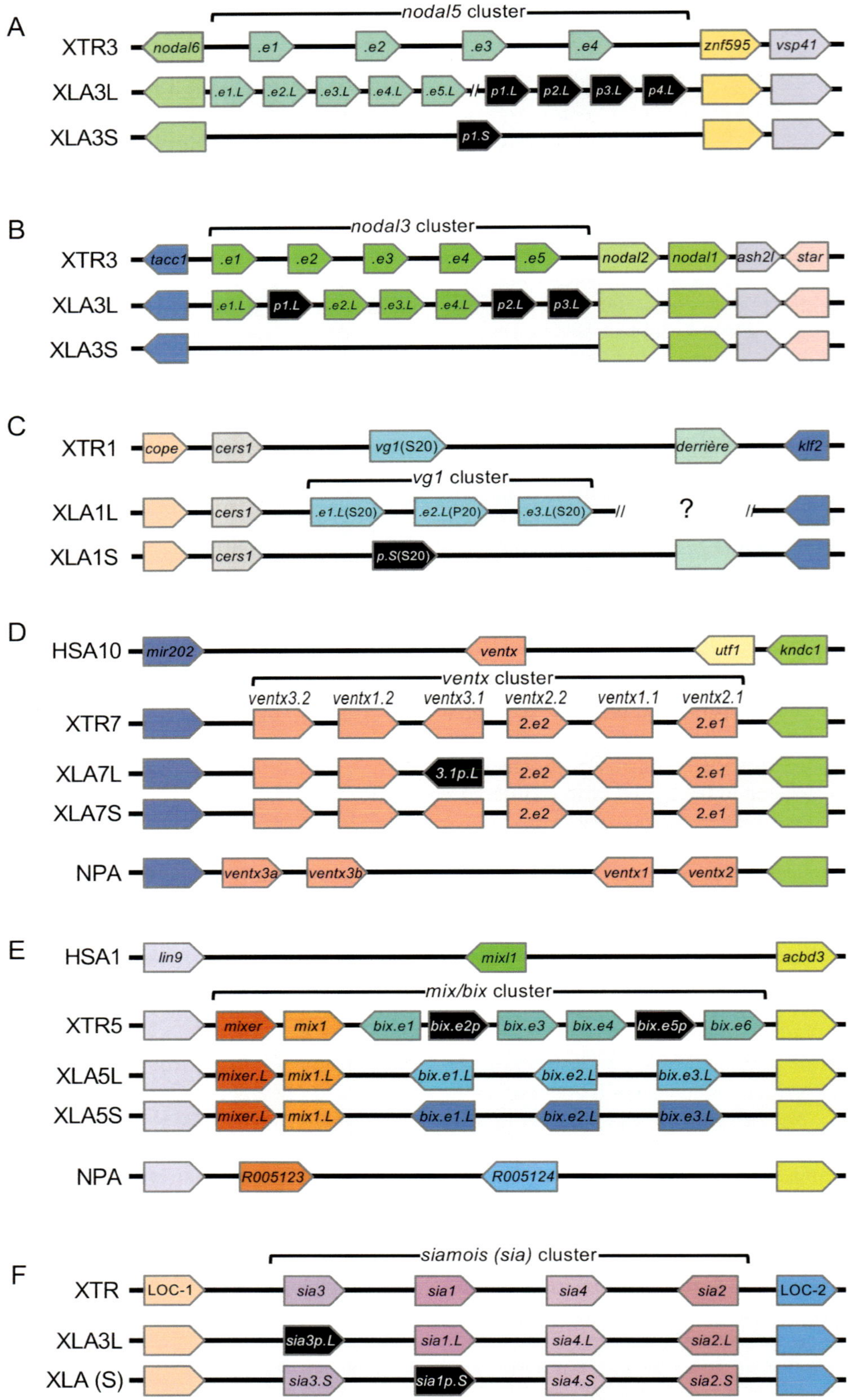

FIGURE 10.3 Gene clusters with cluster-specific gene expansions. Comparison of the genomic regions of gene clusters and their flanking genes on the chromosomes of *X. tropicalis* (XTR), *X. laevis* (XLA), *Homo sapiens* (human; HSA), and *Nanorana parkeri* (Tibetan frog; NPA). Pentagon arrows show the genes and their transcriptional directions. In the gene names, "*e*" indicates independent gene expansion/homogenization in each cluster, indicating that there is no one-to-one orthologous relationships between the genes of those clusters. "*p*" indicates pseudogenes and depicted as black pentagon arrows. (A) *nodal5* cluster. The sequence of the *nodal5* cluster on

FIGURE 10.3 (Continued)
XLA3L is not complete as indicated by double slashes. (B) *nodal3* cluster. (C) *vg1* cluster. The *derrière* gene is unidentified on XLA1L due to the incomplete genomic sequence (double slashes with a question mark). (D) *ventx* cluster. (E) *mix/bix* cluster. The same region in teleosts and chicken are similar to HSA without gene expansions (see Watanabe et al., 2017). R005123 and R005124 of *N. parkeri* are similar to *mix1/mixer* and *bix* genes, respectively. (F) *siamois* (*sia*) cluster. *sia1* and *sia2* are also known by the names *siamois* and *twin*, respectively. LOC-1 and LOC-2 stand for *LOC100490626* and *LOC100490281*, respectively.

Source: Panels are modified from: A (Session et al., 2016); B and C (Suzuki et al., 2017a); D and E (Watanabe et al., 2017); and F (Haramoto et al., 2017).

brought about an increase in gene dose but not functional diversity. These features are reminiscent of the rDNA cluster (see previously), which maintains high gene doses by homogenization through gene conversion (Hori et al., 2021) and is located only in XLA3L, not S. Because tandemly repeated genes tend to increase or decrease in number due to uneven crossovers during meiosis, it may be easier to maintain an appropriate gene dose at a single locus.

10.3.3.2. *Ventx, mix/bix,* and *siamois* Clusters

Gene clusters were not always deleted from subgenome S, as already shown for the Hox clusters, which consist of different types of paralogous genes, not as in the *nodal* clusters. Therefore, it is possible to assume that subfunctionalization of the expanded genes in the cluster may be the reason for retaining full sets (or close to full sets) on both L and S chromosomes in *X. laevis*, as shown in Figure 10.3D, E and F. The *ventx* cluster in *X. tropicalis* contains six genes, whereas in *X. laevis*, the six genes were doubled and maintained with orthologous relationships between subgenomes L and S except for *ventx2.1/2.2* (see the following for details) and pseudogenized *ventx3.1p.L* (Figure 10.3D) (Watanabe et al., 2017). Because another frog, *Nanorana parkeri* (Tibetan frog) (Sun et al., 2015), has three orthologous genes, *ventx1*, *ventx2*, and *ventx3a/3b*, the *ventx* genes were expanded and subfunctionalized possibly in the frog lineage or earlier (Watanabe et al., 2017). In the common ancestor of *Xenopus*, *ventx1*, *ventx2*, and *ventx3* were tandemly duplicated and rearranged. Curiously, after duplication, *ventx2.1* and *ventx2.2* were presumably homogenized by gene conversion in each cluster of *X. tropicalis* and *X. laevis* (see Supplementary Figure 18 in Watanabe et al., 2017) and could be designated as *ventx2.e1* and *vent2.e2* (Figure 10.3D). Two of the *ventx* genes in *X. laevis*, *Xvent-1* and *Xvent-2* (now identified as *ventx1.2.S* and *ventx2.1.L*, respectively), have been described to have significant differences in their responses to BMP4 (Dosch et al., 1997), suggesting that *ventx1* and *ventx2* are subfunctionalized. *X. laevis Xvex-1* (ventx3.2.S), which is again involved in BMP4 signaling (Shapira et al., 1999), also seems to be subfunctionalized. This is because the similarity of the homeodomain and entire region of Xvex-1 to either those of Xvent-1 or Xvent-2 is much lower than between these latter two (Shapira et al., 1999; Watanabe et al., 2017).

The *mix/bix* cluster is similar to the *ventx* cluster (Figure 10.3E). Previously, the mesodermal/endodermal genes *mix1*, *mixer*, and *bix* were individually identified and analyzed in different studies (see references in Watanabe et al., 2017), but those genes were found to form a single gene cluster each in subgenomes L and S (see Figure 3d and Extended Data Figure 7a in Session et al., 2016; Watanabe et al., 2017). That there is only one *mix/bix* ortholog in humans and teleosts suggests *Xenopus*- or frog-specific gene expansions (Figure 10.3E) (Watanabe et al., 2017). In the common ancestor of *Xenopus*, *mix1*, *mixer*, and *bix* were subfunctionalized, and *bix* was tandemly expanded. Similar to *ventx2.1* and *ventx2.2*, *bix* genes were likely homogenized by gene conversion in each cluster (see Extended Figure 7a in Session et al., 2016).

In the *siamois* gene cluster, *sia1*, *sia2*, *sia3*, and *sia4* were expanded (Figure 10.3F) (Haramoto et al., 2017) and subfunctionalized (Laurent et al., 1997) in the common ancestor of *Xenopus*. In *X. laevis*, after allotetraploidization, the homeologous relationships between the four paralogs in the clusters are maintained between subgenomes L and S, though one gene in each cluster was pseudogenized (Haramoto et al., 2017).

These data of *ventx*, *mix/bix*, and *siamois* gene clusters, along with the Hox clusters, suggest that the retainment of multiple tandemly repeated genes in both the L and S subgenomes is likely due to the genes having different roles. The data also suggest that these gene clusters consisting of tandemly repeated subfunctionalized paralogs are much more conserved between the L and S subgenomes compared to those consisting of functionally equivalent genes only in the L subgenome, such as *rDNA*, *nodal5*, *nodal3*, and *vg1*.

Thus, there are two patterns of evolutional changes of gene clusters between the subgenomes after allotetraploidization in *X. laevis*. First, tandemly repeated equivalent genes that increase gene dose may have to be kept at an optimal level in a single locus. Second, the gene clusters that consist of subfunctionalized paralogs appear to maintain them between L and S. This may be because there is some coordination of gene expression between gene clusters, requiring the whole region containing the cluster to be conserved. These hypotheses remain to be tested.

10.4. CONCLUSIONS

WGD is considered one of the driving forces of evolution. Especially, allopolyploidization occurs through hybridization coupled with WGD, which could generate a new species with heterosis by combining two different genomes from parental species. During evolution, multiple genome duplications have occurred in different lineages of *Xenopus*, leading to speciation (see Figure 10.1A). In general, evolution is considered to be brought about by the accumulation of mutations (base substitutions, deletions and insertions) in the genome sequence, which is a continuous and gradual change. On the other hand, allopolyploidization causes the sudden combination of genetic diversities from two different species. This means that allopolyploidization, given that the resulting heterosis is in favor of natural selection, can lead to "discontinuous evolution" whereby a new species is suddenly born that is superior to the parent species. One can imagine that the two rounds of WGD in the common ancestor of vertebrates were possibly allopolyploidization to make two consecutive giant leaps forward in evolution.

Allotetraploidy of the *Xenopus laevis* genome is a disadvantage for genomic analysis but became an advantage to analyze genome evolution of allopolyploidized species, since this complex genome was successfully decoded, and moreover, subgenomes L and S were clearly identified. Luckily, there are no massive reciprocal translocations between chromosomes, thereby revealing asymmetric evolution between the subgenomes, in which subgenome L is more conserved than subgenome S. Typical asymmetries are gene clusters of *rDNA*, *nodal3*, *nodal5*, and *vg1(S20)*, all of which were retained only in subgenome L. The need for appropriate gene expression dosage may be a reason for maintaining these asymmetric clusters. By contrast, in a gene cluster of subfunctionalized paralogs, such as the Hox clusters and the *ventx* cluster, almost all homeologous genes are retained in the subgenomes. Using updated versions of the *X. laevis* and *X. tropicalis* WGS will give us the chance to discover other types of asymmetric evolution of subgenomes. The WGS of *X. tropicalis* and *X. laevis* has also provided many insights. We now know (almost) all of the genomic composition of these two species, and we are able to extract the information that is necessary for detailed analysis.

10.5. FUTURE DIRECTIONS

What we learned from the genomes of *X. tropicalis* and *X. laevis* goes beyond the analyses of these two species and has pointed out that the other members of this genus will be interesting subjects for evolutionary studies and much more. We will discuss here one of those, how allopolyploidization impacts initial and early subgenome evolution. To address this issue, there are two alternative but complementary ways: one is comparative genome analysis between *X. laevis* and *X. borealis*, and the other is an experimental approach using hybrid formation and polyploidization.

10.5.1. Comparative Genome Analysis

Currently, the completion of the WGS of *X. borealis* (SRA Accession No. SRX1606064) is underway. A full analysis of this genome and comparison with the *X. laevis* genome would definitely reveal how allopolyploid genomes evolve. That is, because *X. laevis* and *X. borealis* speciated just ~1 million years after allopolyploidization (Session et al., 2016), early events (mutations and rearrangements) which were common between *X. laevis* and *X. borealis* can be separated from later events which are not shared. Furthermore, genome sequences and transcriptome analyses of other *Xenopus* species with higher ploidy levels (see Figure 10.1A) would give information on what kinds of subgenomes are present, how genomes cope with many subgenomes, how gene expression is regulated among the homeologs, and so on.

10.5.2. Artificial Hybrid Analysis

To experimentally analyze allopolyploidization, the *Xenopus* genus provides us a good system, because it is possible to make hybrids and allopolyploids between them as reported (see subsequently). When divergent genomes are merged by hybridization or allopolyploidization, a "genomic shock" occurs in which the subgenomes come into conflict with each other (McClintock, 1984; Bird et al., 2018). Studies of plant genomes have shown gene loss, genomic rearrangements, and reactivation of previously silenced transposable elements (reviewed in Bird et al., 2018), as well as nucleolar dominance (reviewed in Preuss and Pikaard, 2007). These early alterations probably lead to "subgenome dominance" and thereby to asymmetric evolution of subgenomes.

In the *Xenopus* genus, naturally occurring interspecific hybrids have been reported (for example, Picker et al., 1996; Fischer et al., 2000; Yager, 1996), exemplifying the pre-events of allopolyploidization recurring during the course of evolution. Experimentally, hybrids between *Xenopus* species have been generated for biochemical analysis (for example, Honjo and Reeder, 1973; Brown et al., 1977; De Robertis and Black, 1979; Kobel et al., 1981; Bürki, 1985). More recently, *Xenopus* interspecific hybrids have been examined at molecular and gene levels for nucleolar dominance (Michalak et al., 2015) or changes in the epigenome (Elurbe et al., 2017) to identify early effects of merging of two genomes. Of note, Elrube et al. (2017) found derepression of young DNA transposons, which is analogous to the reaction of plant genomes at genomic shock. Further analyses of these interspecific hybrids of *Xenopus* are expected to reveal the mechanism of (sub)genome dominance after genomic shock, which may trigger asymmetric evolution of subgenomes if hybridization will be followed by polyploidization.

10.5.3. Artificial Allopolyploid Analysis

While hybrids are in general infertile, females of experimentally generated *Xenopus* interspecies hybrids can be partially fertile. This is because the diploid hybrid females produce two types of eggs, aneuploid and diploid (2x) (Müller, 1977; Kobel and Du Pasquier, 1986). Diploid eggs are produced from endoreduplicated (DNA is replicated in the nucleus without division) oocytes (4x) and are larger in size. When these diploid eggs are fertilized with haploid sperm from a non-hybrid male, triploid individuals develop (Kobel and Du Pasquier, 1975). The triploid females similarly produce triploid eggs (3x) that can be fertilized again with haploid sperm to yield tetraploids (Kobel, 1996). On the other hand, males of the *Xenopus* interspecies hybrids tested are sterile, and the number of sperm is remarkably lower than non-hybrids (Kobel, 1996; Malone et al., 2007). Nevertheless, they may produce some viable sperm with the ability to fertilize (Kobel, 1996). The ploidy of these sperm is unknown, but they are estimated to be aneuploid, since endoreduplication has not been demonstrated in male *Xenopus* hybrids (Kobel and Du Pasquier, 1986). In contrast, it has been reported that the sperm of a hybrid between *X. laevis* and *X. muelleri* appeared to be much larger than its parental species (Malone et al., 2007), and they could be diploid. If hybrids are able to produce diploid eggs as well as diploid sperm, and if they fertilize, an allopolyploid (allotetraploid) individual may be produced simply in the next generation.

Amphibians have a long history of being experimental animals, and numerous research techniques have been developed over the years. Besides *Xenopus*, it was reported that amphidiploids (allotetraploids) developed from heat-shock treatment of *Rana brevipoda* eggs inseminated with sperm of *Rana nigromaculata* (Kawamura and Nishioka, 1960). Another approach has also been reported in which autotetraploid individuals of *R. nigromaculata* and amphidiploids (allotetraploids) of *R. nigromaculata* and *R. brevipoda* were generated (Kawamura and Nishioka, 1983). These were made possible by utilizing "traditional" methods for amphibian embryo manipulation, namely applying heat shock to inhibit first cleavage after insemination to double the whole genome (2x to 4x, for example), or early cold, heat, or high-hydrostatic pressure treatment to suppress extrusion of the second polar body to double the maternal genome (see references in Kawahara, 1978). To produce allotetraploids, the first step was to produce autotetraploid *R. nigromaculata* by heat-shock treatment of normally fertilized eggs. The sperm (2x) from this male autotetraploid *R. nigromaculata* were then used for fertilization of eggs from diploid female *R. brevipoda*, followed by early cold treatment to double the maternal genome. This enables the fertilized egg to possess two full sets of paternal as well as maternal chromosomes. These methods would also be applicable to generate allopolyploids of *Xenopus* species. Furthermore, application of hydrostatic pressure or late cold shock inhibits first cleavage (Reinschmidt et al., 1979; Geach et al., 2012), and by the former treatment, autooctoploid *X. laevis* individuals (designated as tetraploid in the literature) have been produced (Reinschmidt et al., 1979). Normal fertilized hybrid embryos subjected to these treatments would presumably grow into allopolyploid individuals. Therefore, theoretically, application of any of these methods could generate allooctoploids of *X. laevis* and *X. borealis* as well as allohexaploids using either of them and *X. tropicalis*, three species whose whole genome sequence is/will be available. It would be interesting to use the hybrids or allopolyploids to detect the immediate effects of genomic shock and subgenome dominance, making use of the genetic information at hand.

All things considered, *X. laevis* and its related species could be a model for evolutionary studies of the genome and for polyploidy. With the increase in molecular information, experimental hybrid formation, or allopolyploidization between two different *Xenopus* species, they are excellent tools for analyzing genomic shock, such as asymmetric changes of epigenetic marks and gene expression patterns. These kinds of analyses will shed light on how asymmetric evolution of subgenomes initially took place and may also provide some clues to imagine what happened in the two rounds of WGD in the common ancestor of vertebrates, as well as in the third round of WGD in the common ancestor of teleosts.

ACKNOWLEDGMENTS

The authors are grateful to Dr. Shuji Takahashi (Hiroshima University) and Dr. Yoshikazu Haramoto (National Institute of Advanced Industrial Science and Technology) for providing unpublished data. This work was supported in part by Japan Society for the Promotion of Science KAKENHI grant numbers: (15K14521 to M.K.) and (25251026, 18H02447, and 21K06126 to M.T.).

REFERENCES

Aparicio, S., Chapman, J., Stupka, E., Putnam, N., Chia, J.M., Dehal, P., Christoffels, A., Rash, S., Hoon, S., Smit, A., Sollewijn Gelpke, M.D., Roach, J., Oh, T., Ho, I.Y., Wong, M., Detter, C., Verhoef, F., Predki, P., Tay, A., Lucas, S., Richardson, P., Smith, S.F., Clark, M.S., Edwards, Y.J.K., Doggett, N., Zharkikh, A., Tavtigian, S. V., Pruss, D., Barnstead, M., Evans, C., Baden, H., Powell, J., Glusman, G., Rowen, L., Hood, L., Tan, Y.H., Elgar, G., Hawkins, T., Venkatesh, B., Rokhsar, D., Brenner, S., 2002. Whole-genome shotgun assembly and analysis of the genome of *Fugu rubripes*. *Science* 297, 1301–1310. https://doi.org/10.1126/science.1072104

Berthelot, C., Brunet, F., Chalopin, D., Juanchich, A., Bernard, M., Noël, B., Bento, P., Da Silva, C., Labadie, K., Alberti, A., Aury, J.M., Louis, A., Dehais, P., Bardou, P., Montfort, J., Klopp, C., Cabau, C., Gaspin, C., Thorgaard, G.H., Boussaha, M., Quillet, E., Guyomard, R., Galiana, D., Bobe, J., Volff, J.N., Genêt, C., Wincker, P., Jaillon, O., Crollius, H.R., Guiguen, Y., 2014. The rainbow trout genome provides novel insights into evolution after whole-genome duplication in vertebrates. *Nat. Commun.* 5, 3657. https://doi.org/10.1038/ncomms4657

Bewick, A.J., Anderson, D.W., Evans, B.J., 2010. Evolution of the closely related, sex-related genes DM-W and DMRT1 in African clawed frogs (Xenopus). *Evolution (N. Y).* 65, 698–712. https://doi.org/10.1111/j.1558-5646.2010.01163.x

Bird, K.A., VanBuren, R., Puzey, J.R., Edger, P.P., 2018. The causes and consequences of subgenome dominance in hybrids and recent polyploids. *New Phytol.* 220, 87–93. https://doi.org/10.1111/nph.15256

Brown, D.D., Dawid, I.B., 1968. Specific gene amplification in oocytes. *Science* 160, 272–280. https://doi.org/10.1126/science.160.3825.272

Brown, D.D., Dawid, I.B., Reeder, R.H., 1977. *Xenopus borealis* misidentified as *Xenopus mulleri*. *Dev. Biol.* 59, 266–267. https://doi.org/10.1016/0012-1606(77)90263-9

Bürki, E., 1985. The expression of creatine kinase isozymes in *Xenopus tropicalis*, *Xenopus laevis laevis*, and their viable hybrid. *Biochem. Genet.* 23, 73–88. https://doi.org/10.1007/BF00499114

C. elegans Sequencing Consortium, 1998. Genome sequence of the nematode *C. elegans*: A platform for investigating biology. *Science* 282, 2012–2018. https://doi.org/10.1126/science.282.5396.2012

Chang, C.Y., Witschi, E., 1956. Genic control and hormonal reversal of sex differentiation in Xenopus. *Proc. Soc. Exp. Biol. Med.* 93, 140–144. https://doi.org/10.3181/00379727-93-22688

Chen, D., Zhang, Q., Tang, W., Huang, Z., Wang, G., Wang, Y., Shia, J., Xu, H., Lin, L., Li, Z., Chi, W., Huang, L., Xia, J., Zhang, X., Guo, L., Wang, Y., Ma, P., Tang, J., Zhou, G., Liu, M., Liu, F., Hua, X., Wang, B., Shen, Q., Jiang, Q., Lin, J., Chen, X., Wang, H., Dou, M., Liu, L., Pan, H., Qi, Y., Wu, B., Fang, J., Zhou, Y., Cen, W., He, W., Zhang, Q., Xue, T., Lin, G., Zhang, W., Liu, Z., Qu, L., Wang, A., Ye, Q., Chen, J., Zhang, Y., Ming, R., Van Montagu, M., Tang, H., Van de Peer, Y., Chen, Y., Zhang, J., 2020. The evolutionary origin and domestication history of goldfish (*Carassius auratus*). *Proc. Natl. Acad. Sci. U.S.A.* 117, 29775–29785. https://doi.org/10.1073/pnas.2005545117/-/DCSupplemental

Chen, Z.J., 2013. Genomic and epigenetic insights into the molecular bases of heterosis. *Nat. Rev. Genet.* 14, 471–482. https://doi.org/10.1038/nrg3503

Chung, H.A., Medina-Ruiz, S., Harland, R.M., 2014. Sp8 regulates inner ear development. *Proc. Natl. Acad. Sci. U. S. A.* 111, 6329–6334. https://doi.org/10.1073/pnas.1319301111

De Robertis, E.M., Black, P., 1979. Hybrids of *Xenopus laevis* and Xenopus borealis express proteins from both parents. *Dev. Biol.* 68, 334–339.

Dietrich, F.S., Mulligan, J., Hennessy, K., Yelton, M.A., Allen, E., Araujo, R., Aviles, E., Berno, A., Brennan, T., Carpenter, J., Chen, E., Cherry, J.M., Chung, E., Duncan, M., Guzman, E., Hartzell, G., Hunicke-Smith, S., Hyman, R.W., Kayser, A., Komp, C., Lashkari, D., Lew, H., Lin, D., Mosedale, D., Davis, R.W., 1997. The nucleotide sequence of *Saccharomyces cerevisiae* chromosome V. *Nature* 387, 78–81.

Dosch, R., Gawantka, V., Delius, H., Blumenstock, C., Niehrs, C., 1997. Bmp-4 acts as a morphogen in dorsoventral mesoderm patterning in Xenopus. *Development* 124, 2325–2334. https://doi.org/10.1242/dev.124.12.2325

Dujon, B., Alexandraki, D., André, B., Ansorge, W., Baladron, V., Ballesta, J.P.G., Banrevi, A., Bolle, P.A., Bolotin-Fukuhara, M., Bossier, P., Bou, G., Boyer, J., Buitrago, M.J., Cherét, G., Colleaux, L., Dalgnan-Fornier, B., del Rey, F., Dion, C., Domdey, H., Düsterhöft, A., Düsterhus, S., Entian, K.-D., Erfle, H., Esteban, P.F., Feldmann, H., Fernandes, L., Fobo, G.M., Fritz, C., Fukuhara, H., Gabel, C., Gaillon, L., Carcia-Cantalejo, J.M., Garcia-Ramirez, J.J., Gent, M.E., Ghazvini, M., Goffeau, A., Gonzaléz, A., Grothues, D., Guerreiro, P., Hegemann, J., Hewitt, N., Hilger, F., Hollenberg, C.P., Horaitis, O., Indge, K.J., Jacquier, A., James, C.M., Jauniaux, J.C., Jimenez, A., Keuchel, H., Kirchrath, L., Kleine, K., Kötter, P., Legrain, P., Liebl, S., Louis, E.J., Maia e Silva, A., Marck, C., Monnier, A.-L., Möstl, D., Müller, S., Obermaier, B., Oliver, S.G., Pallier, C., Pascolo, S., Pfeiffer, F., Philippsen, P., Planta, R.J., Pohl, F.M., Pohl, T.M., Pöhlmann, R., Portetelle, D., Purnelle, B., Puzos, V., Rad, M.R., Rasmussen, S.W., Remacha, M., Revuelta, J.L., Richard, G.-F., Rieger, M., Rodrigues-Pousada, C., Rose, M., Rupp, T., Santos, M.A., Schwager, C., Sensen, C., Skala, J., Soares, H., Sor, F., Stegemann, J., Tettelin, H., Thierry, A., Tzermia, M., Urrestarazu, L.A., van Dyck, L., van Vliet-Reedijk, J.C., Valens, M., Vandenbo, M., Vilela, C., Vissers, S., von Wettstein, D., Voss, H., Wiemann, S., Xu, G., Zimmermann, J., Haasemann, M., Becker, I., Mewes, H.W., 1994. Complete DNA sequence of yeast chromosome XI. *Nature* 369, 371–378. https://doi.org/10.1038/369371a0

Elurbe, D.M., Paranjpe, S.S., Georgiou, G., van Kruijsbergen, I., Bogdanovic, O., Gibeaux, R., Heald, R., Lister, R., Huynen, M.A., van Heeringen, S.J., Veenstra, G.J.C., 2017. Regulatory remodeling in the allo-tetraploid frog *Xenopus laevis*. *Genome Biol.* 18, 198. https://doi.org/10.1186/s13059-017-1335-7

Evans, B.J., 2008. Genome evolution and speciation genetics of clawed frogs (Xenopus and Silurana). *Front. Biosci.* 13, 4687–4706. https://doi.org/10.2741/3033

Evans, B.J., Kelley, D.B., Tinsley, R.C., Melnick, D.J., Cannatella, D.C., 2004. A mitochondrial DNA phylogeny of African clawed frogs: Phylogeography and implications for polyploid evolution. *Mol. Phylogenet. Evol.* 33, 197–213. https://doi.org/10.1016/j.ympev.2004.04.018

Fischer, W.J., Koch, W.A., Elepfandt, A., 2000. Sympatry and hybridization between the clawed frogs *Xenopus laevis laevis* and *Xenopus muelleri* (Pipidae). *J. Zool.* 252, 99–107. https://doi.org/10.1111/j.1469-7998.2000.tb00824.x

Fritz, A.F., Cho, K.W.Y., Wright, C.V.E., Jegalian, B.G., de Robertis, E.M., 1989. Duplicated homeobox genes in Xenopus. *Dev. Biol.* 131, 584–588. https://doi.org/https://doi.org/10.1016/S0012-1606(89)80029-6

Garcia-Fernàndez, J., Holland, P.W., 1994. Archetypal organization of the amphioxus Hox gene cluster. *Nature* 370, 563–566. https://doi.org/10.1038/370563a0

Geach, T.J., Stemple, D.L., Zimmerman, L.B., 2012. Genetic analysis of *Xenopus tropicalis*. *Methods Mol. Biol.* 917, 69–110. https://doi.org/10.1007/978-1-61779-992-1_5

Goda, T., Abu-Daya, A., Carruthers, S., Clark, M.D., Stemple, D.L., Zimmerman, L.B., 2006. Genetic screens for mutations affecting development of *Xenopus tropicalis*. *PLoS Genet.* 2, 811–825. https://doi.org/10.1371/journal.pgen.0020091

Gurdon, J.B., Hopwood, N., 2000. The introduction of *Xenopus laevis* into developmental biology: Of empire, pregnancy testing and ribosomal genes. *Int. J. Dev. Biol.* 44, 43–50. https://doi.org/10.1387/ijdb.10761846

Haramoto, Y., Saijyo, T., Tanaka, T., Furuno, N., Suzuki, A., Ito, Y., Kondo, M., Taira, M., Takahashi, S., 2017. Identification and comparative analyses of Siamois cluster genes in *Xenopus laevis* and *tropicalis*. *Dev. Biol.* 426, 374–383. https://doi.org/10.1016/j.ydbio.2016.07.015

Hellsten, U., Harland, R.M., Gilchrist, M.J., Hendrix, D., Jurka, J., Kapitonov, V., Ovcharenko, I., Putnam, N.H., Shu, S., Taher, L., Blitz, I.L., Blumberg, B., Dichmann, D.S., Dubchak, L., Amaya, E., Detter, J.C., Fletcher, R., Gerhard, D.S., Goodstein, D., Graves, T., Grigoriev, I. V., Grimwood, J., Kawashima, T., Lindquist, E., Lucas, S.M., Mead, P.E., Mitros, T., Ogino, H., Ohta, Y., Poliakov, A. V., Pollet, N., Robert, J., Salamov, A., Sater, A.K., Schmutz, J., Terry, A., Vize, P.D., Warren, W.C., Wells, D., Wills, A., Wilson, R.K., Zimmerman, L.B., Zorn, A.M., Grainger, R., Grammer, T., Khokha, M.K., Richardson, P.M., Rokhsar, D.S., 2010. The genome of the western clawed frog *Xenopus tropicalis*. *Science* 328, 633–636. https://doi.org/10.1126/science.1183670

Holland, P.W.H., 2013. Evolution of homeobox genes: Wiley interdiscip. *Rev. Dev. Biol.* 2, 31–45. https://doi.org/10.1002/wdev.78

Honjo, T., Reeder, R.H., 1973. Preferential transcription of *Xenopus laevis* ribosomal RNA in interspecies hybrids between *Xenopus laevis* and *Xenopus mulleri*. *J. Mol. Biol.* 80, 217–228. https://doi.org/10.1016/0022-2836(73)90168-X

Hori, Y., Shimamoto, A., Kobayashi, T., 2021. The human ribosomal DNA array is composed of highly homogenized tandem clusters. *Genome Res* August, gr.275838.121. https://doi.org/10.1101/gr.275838.121

Igawa, T., Watanabe, A., Suzuki, A., Kashiwagi, A., Kashiwagi, K., Noble, A., Guille, M., Simpson, D.E., Horb, M.E., Fujii, T., Sumida, M., 2015. Inbreeding ratio and genetic relationships among strains of the western clawed frog, *Xenopus tropicalis*. *PLoS One* 10, e0133963. https://doi.org/10.1371/journal.pone.0133963

Inoue, J., Sato, Y., Sinclair, R., Tsukamoto, K., Nishida, M., 2015. Rapid genome reshaping by multiple-gene loss after whole-genome duplication in teleost fish suggested by mathematical modeling. *Proc. Natl. Acad. Sci.* 112, 14918–14923. https://doi.org/10.1073/pnas.1507669112

International Chicken Genome Sequencing Consortium, 2004. Sequence and comparative analysis of the chicken genome provide unique perspectives on vertebrate evolution. *Nature* 432, 695–716. https://doi.org/10.1038/nature03154

International Human Genome Sequencing Consortium, 2001. Initial sequencing and analysis of the human genome. *Nature* 409, 860–921.

Izutsu, Y., Yoshizato, K., 1993. Metamorphosis-dependent recognition of larval skin as non-self by inbred adult frogs (*Xenopus laevis*). *J. Exp. Zool.* 266, 163–167. https://doi.org/10.1002/jez.1402660211

Kasahara, M., Naruse, K., Sasaki, S., Nakatani, Y., Qu, W., Ahsan, B., Yamada, T., Nagayasu, Y., Doi, K., Kasai, Y., Jindo, T., Kobayashi, D., Shimada, A., Toyoda, A., Kuroki, Y., Fujiyama, A., Sasaki, T., Shimizu, A., Asakawa, S., Shimizu, N., Hashimoto, S.I., Yang, J., Lee, Y., Matsushima, K., Sugano, S., Sakaizumi, M., Narita, T., Ohishi, K., Haga, S., Ohta, F., Nomoto, H., Nogata, K., Morishita, T., Endo, T., Shin-I, T., Takeda, H., Morishita, S., Kohara, Y., 2007. The medaka draft genome and insights into vertebrate genome evolution. *Nature* 447, 714–719. https://doi.org/10.1038/nature05846

Kawahara, H., 1978. Production of triploid and gynogenetic diploid Xenopus by cold treatment. *Dev. Growth Differ.* 20, 227–236. https://doi.org/10.1111/j.1440-169X.1978.00227.x

Kawamura, T., Nishioka, M., 1983. Reproductive capacity of male autotetraploid *Rana nigromaculata* and male and female amphidiploids produced from them by mating with female diploid Rana brevipoda. *Sci. Rep. Lab. Amphib. Biol. Hiroshima Univ.* 6, 1–45.

Kawamura, T., Nishioka, M., 1960. Amphidiploid frogs produced by artificial means. *J. Sci. Hiroshima Univ., Ser. B. Div. I* 18, 195–220.

Khokha, M.K., Krylov, V., Reilly, M.J., Gall, J.G., Bhattacharya, D., Cheung, C.Y.J., Kaufman, S., Dang, K.L., Macha, J., Ngo, C., Prakash, N., Schmidt, P., Tlapakova, T., Trivedi, T., Tumova, L., Abu-Daya, A., Geach, T., Vendrell, E., Ironfield, H., Sinzelle, L., Sater, A.K., Wells, D.E., Harland, R.M., Zimmerman, L.B., 2009. Rapid gynogenetic mapping of *Xenopus tropicalis* mutations to chromosomes. *Dev. Dyn.* 238, 1398–1406. https://doi.org/10.1002/dvdy.21965

Kobel, H.R., 1996. Allopolyploid speciation. In: Tinsley, R.C., Kobel, H.R. (Eds.), *The Biology of Xenopus*. Clarendon Press, Oxford, pp. 391–401.

Kobel, H.R., 1981. Evolutionary trends in Xenopus (Anura Pipidae). *Monit. Zool. Ital. Suppl.* 15, 119–131. https://doi.org/10.1080/03749444.1981.10736632

Kobel, H.R., Du Pasquier, L., 1986. Genetics of polyploid Xenopus. *Trends Genet.* 2, 310–315. https://doi.org/10.1016/0168-9525(86)90286-6

Kobel, H.R., Du Pasquier, L., 1975. Production of large clones of histocompatible, fully identical clawed toads (Xenopus). *Immunogenetics* 2, 87–91.

Kobel, H.R., Du Pasquier, L., Tinsley, R.C., 1981. Natural hybridization and gene introgression between *Xenopus gilli* and *Xenopus laevis laevis* (Anura: Pipidae). *J. Zool.* 194, 317–322. https://doi.org/10.1111/j.1469-7998.1981.tb04584.x

Kondo, M., Matsuo, M., Igarashi, K., Haramoto, Y., Yamamoto, T., Yasuoka, Y., Taira, M., 2019. De novo transcription of multiple Hox cluster genes takes place simultaneously in early *Xenopus tropicalis* embryos. *Biol. Open* 8, bio038422. https://doi.org/10.1242/bio.038422

Kondo, M., Nanda, I., Hornung, U., Asakawa, S., Shimizu, N., Mitani, H., Schmid, M., Shima, A., Schartl, M., 2003. Absence of the candidate male sex-determining gene dmrt-1b(Y) of medaka from other fish species. *Curr. Biol.* 13, 416–420. https://doi.org/10.1016/s0960-9822(03)00125-8

Kondo, M., Nanda, I., Hornung, U., Schmid, M., Schartl, M., 2004. Evolutionary origin of the medaka Y chromosome. *Curr. Biol.* 14, 1664–1669. https://doi.org/10.1016/j

Kondo, M., Yamamoto, T., Takahashi, S., Taira, M., 2017. Comprehensive analyses of hox gene expression in *Xenopus laevis* embryos and adult tissues. *Dev. Growth Differ.* 59, 526–539. https://doi.org/10.1111/dgd.12382

Kuraku, S., Meyer, A., 2009. The evolution and maintenance of Hox gene clusters in vertebrates and the teleost-specific genome duplication. *Int. J. Dev. Biol.* 53, 765–773. https://doi.org/10.1387/ijdb.072533km

Laurent, M.N., Blitz, I.L., Hashimoto, C., Rothbächer, U., Cho, K.W.Y., 1997. The Xenopus homeobox gene Twin mediates Wnt induction of Goosecoid in establishment of Spemann's organizer. *Development* 124, 4905–4916.

Lieberman-Aiden, E., Van Berkum, N.L., Williams, L., Imakaev, M., Ragoczy, T., Telling, A., Amit, I., Lajoie, B.R., Sabo, P.J., Dorschner, M.O., Sandstrom, R., Bernstein, B., Bender, M.A., Groudine, M., Gnirke, A., Stamatoyannopoulos, J., Mirny, L.A., Lander, E.S., Dekker, J., 2009. Comprehensive mapping of long-range interactions reveals folding principles of the human genome. *Science* 326, 289–293. https://doi.org/10.1126/science.1181369

Lien, S., Koop, B.F., Sandve, S.R., Miller, J.R., Kent, M.P., Nome, T., Hvidsten, T.R., Leong, J.S., Minkley, D.R., Zimin, A., Grammes, F., Grove, H., Gjuvsland, A., Walenz, B., Hermansen, R.A., Von Schalburg, K., Rondeau, E.B., Di Genova, A., Samy, J.K.A., Olav Vik, J., Vigeland, M.D., Caler, L., Grimholt, U., Jentoft, S., Inge Våge, D., De Jong, P., Moen, T., Baranski, M., Palti, Y., Smith, D.R., Yorke, J.A., Nederbragt, A.J., Tooming-Klunderud, A., Jakobsen, K.S., Jiang, X., Fan, D., Hu, Y., Liberles, D.A., Vidal, R., Iturra, P., Jones, S.J.M., Jonassen, I., Maass, A., Omholt, S.W., Davidson, W.S., 2016. The Atlantic salmon genome provides insights into rediploidization. *Nature* 533, 200–205. https://doi.org/10.1038/nature17164

Malone, J.H., Chrzanowski, T.H., Michalak, P., 2007. Sterility and gene expression in hybrid males of *Xenopus laevis* and *X. muelleri*. *PLoS One* 2, e781. https://doi.org/10.1371/journal.pone.0000781

Matsuda, M., Nagahama, Y., Shinomiya, A., Sato, T., Matsuda, C., Kobayashi, T., Morrey, C.E., Shibata, N., Asakawa, S., Shimizu, N., Hori, H., Hamaguchi, S., Sakaizumi, M., 2002. DMY is a Y-specific DM-domain gene required for male development in the medaka fish. *Nature* 417, 559–563. https://doi.org/10.1038/nature751

Matsuda, Y., Uno, Y., Kondo, M., Gilchrist, M.J., Zorn, A.M., Rokhsar, D.S., Schmid, M., Taira, M., 2015. A new nomenclature of *Xenopus laevis* chromosomes based on the phylogenetic relationship to silurana/*Xenopus tropicalis*. *Cytogenet. Genome Res*. 145, 187–191. https://doi.org/10.1159/000381292

Mawaribuchi, S., Takahashi, S., Wada, M., Uno, Y., Matsuda, Y., Kondo, M., Fukui, A., Takamatsu, N., Taira, M., Ito, M., 2017. Sex chromosome differentiation and the W- and Z-specific loci in *Xenopus laevis*. *Dev. Biol*. 426, 393–400. https://doi.org/10.1016/j.ydbio.2016.06.015

McClintock, B., 1984. The significance of responses of the genome to challenge. *Science* 226, 792–801. https://doi.org/10.1126/science.15739260

McStay, B., 2006. Nucleolar dominance: A model for rRNA gene silencing. *Genes Dev*. 20, 1207–1214. https://doi.org/10.1101/gad.1436906

Meyer, A., Van de Peer, Y., 2005. From 2R to 3R: Evidence for a fish-specific genome duplication (FSGD). BioEssays 27, 937–945. https://doi.org/10.1002/bies.20293

Michalak, K., Maciak, S., Kim, Y.B., Santopietro, G., Oh, J.H., Kang, L., Garner, H.R., Michalak, P., 2015. Nucleolar dominance and maternal control of 45S rDNA expression. *Proc. R. Soc. B Biol. Sci*. 282, 20152201. https://doi.org/10.1098/rspb.2015.2201

Michiue, T., Yamamoto, T., Yasuoka, Y., Goto, T., Ikeda, T., Nagura, K., Nakayama, T., Taira, M., Kinoshita, T., 2017. High variability of expression profiles of homeologous genes for Wnt, Hh, Notch, and Hippo signaling pathways in *Xenopus laevis*. *Dev. Biol*. 426, 270–290. https://doi.org/10.1016/j.ydbio.2016.12.006

Mouse Genome Sequencing Consortium, 2002. Initial sequencing and comparative analysis of the mouse genome. *Nature* 420, 520–562.

Müller, M.M., Carrasco, A.E., De Robertis, E.M., 1984. A homeo-box-containing gene expressed during oogenesis in Xenopus. *Cell* 39, 157–162. https://doi.org/10.1016/0092-8674(84)90201-0

Müller, W.P., 1977. Diplotene chromosomes of Xenopus hybrid oocytes. *Chromosom* 59, 273–282.

Myers, E.W., Sutton, G.G., Delcher, A.L., Dew, I.M., Fasulo, D.P., Flanigan, M.J., Kravitz, S.A., Mobarry, C.M., Reinert, K.H.J., Remington, K.A., Anson, E.L., Bolanos, R.A., Chou, H.H., Jordan, C.M., Halpern, A.L., Lonardi, S., Beasley, E.M., Brandon, R.C., Chen, L., Dunn, P.J., Lai, Z., Liang, Y., Nusskern, D.R., Zhan, M., Zhang, Q., Zheng, X., Rubin, G.M., Adams, M.D., Venter, J.C., 2000. A whole-genome assembly of Drosophila. *Science* 287, 2196–2204. https://doi.org/10.1126/science.287.5461.2196

Nakayama, T., Nakajima, K., Cox, A., Fisher, M., Howell, M., Fish, M.B., Yaoita, Y., Grainger, R.M., 2017. No privacy, a *Xenopus tropicalis* mutant, is a model of human Hermansky-Pudlak syndrome and allows visualization of internal organogenesis during tadpole development. *Dev. Biol*. 426, 472–486. https://doi.org/10.1016/j.ydbio.2016.08.020

Nanda, I., Kondo, M., Hornung, U., Asakawa, S., Winkler, C., Shimizu, A., Shan, Z., Haaf, T., Shimizu, N., Shima, A., Schmid, M., Schartl, M., 2002. A duplicated copy of DMRT1 in the sex-determining region of the Y chromosome of the medaka, Oryzias latipes. *Proc. Natl. Acad. Sci. U. S. A*. 99, 11778–11783. https://doi.org/10.1073/pnas.182314699

Owens, N.D.L., Blitz, I.L., Maura, A., Gilchrist, M.J., Ken, W.Y., Khokha, M.K., Owens, N.D.L., Blitz, I.L., Lane, M.A., Patrushev, I., Overton, J.D., Gilchrist, M.J., 2016. Measuring absolute RNA copy numbers at high temporal resolution reveals transcriptome kinetics in development. *Cell Rep*. 14, 632–647. https://doi.org/10.1016/j.celrep.2015.12.050

Pascual-Anaya, J., D'Aniello, S., Kuratani, S., Garcia-Fernàndez, J., 2013. Evolution of hox gene clusters in deuterostomes. *BMC Dev. Biol*. 13, 26. https://doi.org/10.1186/1471-213X-13-26

Picker, M.D., Harrison, J.A., Wallace, D., 1996. Natural hybridization between *Xenopus laevis laevis* and *X. gilli* in the south-western Cape Province, South Africa. In: Tinsley, R.C., Kobel, H.R. (Eds.), *The Biology of Xenopus*. Clarendon Press, Oxford, pp. 61–71.

Poczai, P., Cernák, I., Varga, I., Hyvönen, J., 2014. Analysis of the genome sequence of the flowering plant *Arabidopsis thaliana*. *Genet. Resour. Crop Evol*. 61, 796–815. https://doi.org/10.1134/S1022795411020074

Preuss, S., Pikaard, C.S., 2007. rRNA gene silencing and nucleolar dominance: Insights into a chromosome-scale epigenetic on/off switch. *Biochim. Biophys. Acta: Gene Struct. Expr*. 1769, 383–392. https://doi.org/10.1016/j.bbaexp.2007.02.005

Putnam, N.H., O'Connell, B.L., Stites, J.C., Rice, B.J., Blanchette, M., Calef, R., Troll, C.J., Fields, A., Hartley, P.D., Sugnet, C.W., Haussler, D., Rokhsar, D.S., Green, R.E., 2016. Chromosome-scale shotgun assembly using an in vitro method for long-range linkage. *Genome Res* 26, 342–350. https://doi.org/10.1101/gr.193474.115

Rebagliati, M.R., Weeks, D.L., Harvey, R.P., Melton, D.A., 1985. Identification and cloning of localized maternal RNAs from Xenopus eggs. *Cell* 42, 769–777. https://doi.org/https://doi.org/10.1016/0092-8674(85)90273-9

Reinschmidt, D.C., Simon, S.J., Volpe, E.P., Tompkins, R., 1979. Production of tetraploid and homozygous diploid amphibians by suppression of first cleavage. *J. Exp. Zool*. 210, 137–143. https://doi.org/10.1002/jez.1402100115

Robert, J., Ohta, Y., 2009. Comparative and developmental study of the immune system in Xenopus. *Dev. Dyn*. 238, 1249–1270. https://doi.org/10.1002/dvdy.21891

Schmid, M., Steinlein, C., 1991. Chromosome banding in Amphibia-XVI: High-resolution replication banding patterns in *Xenopus laevis*. *Chromosoma* 101, 123–132. https://doi.org/10.1007/BF00357062

Session, A.M., Uno, Y., Kwon, T., Chapman, J.A., Toyoda, A., Takahashi, S., Fukui, A., Hikosaka, A., Suzuki, A., Kondo, M., Van Heeringen, S.J., Quigley, I., Heinz, S., Ogino, H., Ochi, H., Hellsten, U., Lyons, J.B., Simakov, O., Putnam, N., Stites, J., Kuroki, Y., Tanaka, T., Michiue, T., Watanabe, M., Bogdanovic, O., Lister, R., Georgiou, G., Paranjpe, S.S., Van Kruijsbergen, I., Shu, S., Carlson, J., Kinoshita, T., Ohta, Y., Mawaribuchi, S., Jenkins, J., Grimwood, J., Schmutz, J., Mitros, T., Mozaffari, S.V., Suzuki, Y., Haramoto, Y., Yamamoto, T.S., Takagi, C., Heald, R., Miller, K., Haudenschild, C., Kitzman, J., Nakayama, T., Izutsu, Y., Robert, J., Fortriede, J., Burns, K., Lotay, V., Karimi, K., Yasuoka, Y., Dichmann, D.S., Flajnik, M.F., Houston, D.W., Shendure, J., Dupasquier, L., Vize, P.D., Zorn, A.M., Ito, M., Marcotte, E.M., Wallingford, J.B., Ito, Y., Asashima, M., Ueno, N., Matsuda, Y., Veenstra, G.J.C., Fujiyama, A., Harland, R.M., Taira, M., Rokhsar, D.S., 2016. Genome evolution in the allotetraploid frog *Xenopus laevis*. *Nature* 538, 336–343. https://doi.org/10.1038/nature19840

Shapira, E., Marom, K., Yelin, R., Levy, A., Fainsod, A., 1999. A role for the homeobox gene Xvex-1 as part of the BMP-4 ventral signaling pathway. *Mech. Dev.* 86, 99–111. https://doi.org/10.1016/S0925-4773(99)00120-3

Sun, Y.B., Xiong, Z.J., Xiang, X.Y., Liu, S.P., Zhou, W.W., Tu, X.L., Zhong, L., Wang, L., Wu, D.D., Zhang, B.L., Zhu, C.L., Yang, M.M., Chen, H.M., Li, F., Zhou, L., Feng, S.H., Huang, C., Zhang, G.J., Irwin, D., Hillis, D.M., Murphy, R.W., Yang, H.M., Che, J., Wang, J., Zhang, Y.P., 2015. Whole-genome sequence of the Tibetan frog *Nanorana parkeri* and the comparative evolution of tetrapod genomes. *Proc. Natl. Acad. Sci. U. S. A.* 112, E1257–E1262. https://doi.org/10.1073/pnas.1501764112

Suzuki, A., Uno, Y., Takahashi, S., Grimwood, J., Schmutz, J., Mawaribuchi, S., Yoshida, H., Takebayashi-Suzuki, K., Ito, M., Matsuda, Y., Rokhsar, D., Taira, M., 2017a. Genome organization of the vg1 and nodal3 gene clusters in the allotetraploid frog *Xenopus laevis*. *Dev. Biol.* 426, 236–244. https://doi.org/10.1016/j.ydbio.2016.04.014

Suzuki, A., Yoshida, H., van Heeringen, S.J., Takebayashi-Suzuki, K., Veenstra, G.J.C., Taira, M., 2017b. Genomic organization and modulation of gene expression of the TGF-β and FGF pathways in the allotetraploid frog *Xenopus laevis*. *Dev. Biol.* 426, 336–359. https://doi.org/10.1016/j.ydbio.2016.09.016

Suzuki, K.-I.T., Suzuki, M., Shigeta, M., Fortriede, J.D., Takahashi, S., Mawaribuchi, S., Yamamoto, T., Taira, M., Fukui, A., 2017. Clustered Xenopus keratin genes: A genomic, transcriptomic, and proteomic analysis. *Dev. Biol.* 426, 384–392. https://doi.org/10.1016/j.ydbio.2016.10.018

Tanaka, T., Ochi, H., Takahashi, S., Ueno, N., Taira, M., 2017. Genes coding for cyclin-dependent kinase inhibitors are fragile in Xenopus. *Dev. Biol.* 426, 291–300. https://doi.org/10.1016/j.ydbio.2016.06.019

Tinsley, R.C., Loumont, C., Kobel, H.R., 1996. Geographical distribution and ecology. In: Tinsley, R.C., Kobel, H.R. (Eds.), *The Biology of Xenopus*. Clarendon Press, Oxford, pp. 35–59.

Tour, E., Pillemer, G., Gruenbaum, Y., Fainsod, A., 2001. The two Xenopus Gbx2 genes exhibit similar, but not identical expression patterns and can affect head formation. *FEBS Lett.* 507, 205–209. https://doi.org/10.1016/S0014-5793(01)02963-5

Uno, Y., Nishida, C., Takagi, C., Ueno, N., Matsuda, Y., 2013. Homoeologous chromosomes of *Xenopus laevis* are highly conserved after whole-genome duplication. *Heredity (Edinb).* 111, 430–436. https://doi.org/10.1038/hdy.2013.65

Watanabe, M., Yasuoka, Y., Mawaribuchi, S., Kuretani, A., Ito, M., Kondo, M., Ochi, H., Ogino, H., Fukui, A., Taira, M., Kinoshita, T., 2017. Conservatism and variability of gene expression profiles among homeologous transcription factors in *Xenopus laevis*. *Dev. Biol.* 426, 301–324. https://doi.org/10.1016/j.ydbio.2016.09.017

Wells, D.E., Gutierrez, L., Xu, Z., Krylov, V., Macha, J., Blankenburg, K.P., Hitchens, M., Bellot, L.J., Spivey, M., Stemple, D.L., Kowis, A., Ye, Y., Pasternak, S., Owen, J., Tran, T., Slavikova, R., Tumova, L., Tlapakova, T., Seifertova, E., Scherer, S.E., Sater, A.K., 2011. A genetic map of *Xenopus tropicalis*. *Dev. Biol.* 354, 1–8. https://doi.org/10.1016/j.ydbio.2011.03.022

Woods, I.G., Wilson, C., Friedlander, B., Chang, P., Reyes, D.K., Nix, R., Kelly, P.D., Chu, F., Postlethwait, J.H., Talbot, W.S., 2005. The zebrafish gene map defines ancestral vertebrate chromosomes. *Genome Res.* 15, 1307–1314. https://doi.org/10.1101/gr.4134305

Wright, C.V., Cho, K.W., Fritz, A., Bürglin, T.R., De Robertis, E.M., 1987. A *Xenopus laevis* gene encodes both homeobox-containing and homeobox-less transcripts. *EMBO J.* 6, 4083–4094.

Xu, P., Zhang, X., Wang, X., Li, J., Liu, G., Kuang, Y., Xu, J., Zheng, X., Ren, L., Wang, G., Zhang, Y., Huo, L., Zhao, Z., Cao, D., Lu, C., Li, C., Zhou, Y., Liu, Z., Fan, Z., Shan, G., Li, X., Wu, S., Song, Lipu, Hou, G., Jiang, Y., Jeney, Z., Yu, D., Wang, L., Shao, C., Song, Lai, Sun, J., Ji, P., Wang, Jian, Li, Q., Xu, L., Sun, F., Feng, J., Wang, C., Wang, S., Wang, B., Li, Y., Zhu, Y., Xue, W., Zhao, L., Wang, Jintu, Gu, Y., Lv, W., Wu, K., Xiao, J., Wu, J., Zhang, Z., Yu, J., Sun, X., 2014. Genome sequence and genetic diversity of the common carp, *Cyprinus carpio*. *Nat. Genet.* 46, 1212–1219. https://doi.org/10.1038/ng.3098

Xu, R.-H., Kim, J., Taira, M., Lin, J.-J., Zhang, C.-H., Sredni, D., Evans, T., Kung, H.-F., 1997. Differential regulation of neurogenesis by the two Xenopus GATA-1 genes. *Mol. Cell. Biol.* 17, 436–443.

Yager, D.D., 1996. Sound production and acoustic communication in *Xenopus borealis*. In: Tinsley, R.C., Kobel, H.R. (Eds.), *The Biology of Xenopus*. Clarendon Press, Oxford, pp. 121–141.

Yoshimoto, S., Okada, E., Umemoto, H., Tamura, K., Uno, Y., Nishida-Umehara, C., Matsuda, Y., Takamatsu, N., Shiba, T., Ito, M., 2008. A W-linked DM-domain gene, DM-W, participates in primary ovary development in *Xenopus laevis*. *Proc. Natl. Acad. Sci. U.S.A.* 105, 2469–2474.

11 Dynamics of Chromatin Remodeling during Embryonic Development

Gert Jan C. Veenstra

CONTENTS

11.1. Chromatin State: from Packaging DNA to Regulatory Substrate ... 173
11.2. Overview of Epigenome Maps: What Is Where, When? ... 174
11.3. Chromatin Accessibility: Opening the Door for Gene Expression ... 176
11.3.1. Histone Variants and Linker Histones Modulate Nucleosome Dynamics and Accessibility ... 177
11.3.2. Chromatin Assembly and the Onset of Zygotic Genome Activation ... 177
11.3.3. Chromatin Accessibility and Transcription Factor Binding ... 177
11.4. Histone Modifications: Acquisition and Dynamics during Early Development ... 178
11.4.1. Oocyte and Egg Storage Histones and Replication-Coupled Chromatin Assembly Dynamics ... 178
11.4.2. Acquisition of the Antagonistic H3K4me3 and H3K27me3 Marks in the Embryo ... 178
11.4.3. Heterochromatin Marks at Repetitive Elements and Near Genes ... 180
11.5. DNA Methylation: Repression versus Modulation of Histone Modifications ... 180
11.5.1. DNA Methylation Dynamics ... 180
11.5.2. DNA Methylation, Repression and Cross-Talk with Histone Modifications ... 181
11.6. Future Directions ... 181
References ... 181

11.1. CHROMATIN STATE: FROM PACKAGING DNA TO REGULATORY SUBSTRATE

Chromatin is the macromolecular complex of genomic DNA with proteins and RNA, as it is found in the cell nucleus. The histone proteins account for the majority of the protein content of chromatin, contributing roughly the same mass as chromosomal DNA. The basic structural unit of chromatin is the nucleosome, which consists of an octamer of four histone proteins (histones H2A, H2B, H3, and H4), each present twice, around which 145–147 base pairs of DNA are wrapped in a left-handed coil (Zhou et al. 2019). Both the nucleosomes and the higher-order folding of nucleosomal DNA contribute to packaging, condensing, and storing chromosomal DNA in an orderly fashion. In addition, rather than mere packaging material, chromatin also represents the *in vivo* substrate of all processes involving DNA, whether it be transcription, DNA replication, DNA repair, recombination, mitosis, or meiosis. The specific way chromatin is organized at a particular locus therefore holds great regulatory potential at the molecular level (Perino and Veenstra 2016). First, nucleosome positions matter; gene-regulatory regions tend to have a nucleosome-free region or exhibit a reduced nucleosome stability. This is associated with increased accessibility of the DNA for other molecules, for example, transcription factors (Section 3). Second, the histones are subject to extensive post-translational modifications; the modifications on the N-terminal tails of histone H3 have been investigated extensively (Bogdanovic et al. 2012; Perino and Veenstra 2016). Histone modifications are reversible, involving enzymes with "writer" and "eraser" activities. Moreover, many of the modifications are bound by specific proteins ("readers"), which bring about the molecular functions of the histone marks. In Section 4, we will discuss the dynamics of these modifications. Third, the DNA itself can be reversibly modified both within and between nucleosomes; this occurs predominantly by methylation of cytosine within CpG dinucleotides. Methylated DNA is bound by methyl-CpG binding proteins. In addition, methylation status has ramifications for the modifications of the histone tails (Sections 4 and 5). Last, chromatin is organized in loops and topologically associating domains (TADs), which constitute interaction neighborhoods. Within these neighborhoods, regulatory sequences such as promoters and enhancers show relatively high interaction frequencies. Many chromatin-associated proteins and protein complexes play a part in how chromatin and the regulatory sequences and genes it contains are regulated, involving architectural proteins, chromatin remodeling enzymes, DNA- and chromatin-modifying enzymes, histone chaperones, and DNA-binding proteins and their co-factors.

This chapter will focus on the roles of chromatin in gene regulation during early embryonic development. We will highlight insights obtained in *Xenopus* but will also discuss the similarities and differences between species where relevant. We will start with an overview of aspects of chromatin that have been studied in *Xenopus tropicalis* and *Xenopus laevis*, including epigenome maps that represent a valuable community resource (Section 2). This will be followed by

DOI: 10.1201/9781003050230-13

sections discussing chromatin dynamics in the context of embryonic development (Sections 3–5). We will conclude with future perspectives (Section 6).

11.2. OVERVIEW OF EPIGENOME MAPS: WHAT IS WHERE, WHEN?

Chromatin state, or epigenome, refers to the way chromatin is organized across the genome in a particular cell type or stage of development. Each of the organization levels of chromatin can be characterized using various techniques in combination with next generation sequencing (Bright and Veenstra 2019; Hontelez et al. 2019; Quigley and Heinz 2019; Gilchrist et al. 2020). The genome-wide regulatory landscapes are referred to as epigenomic maps or chromatin state maps (Tables 11.1 and 11.2; Figure 11.1). Conceptually, chromatin state maps can be used in two different ways: (1) inspection of particular loci in gene-centric analyses, for example, identifying relevant regulatory elements of a gene at a particular stage of development, and (2) genome-wide analyses of global transitions and the impact of perturbations. The first chromatin state maps of vertebrate embryos were obtained using *Xenopus* (Akkers

TABLE 11.1

Histone Modification Maps of *Xenopus* Embryos, Obtained by ChIP-seq, Chromatin Immunoprecipitation Followed by Sequencing. The modifications are denoted by histone (H3, H4), modified residue (K, lysine; number, amino acid position), modification (me1, monomethylated; me2, dimethylated; me3, trimethylated, ac, acetylated). Other abbreviations: X.tro, *Xenopus tropicalis*; X.lae, *Xenopus laevis*; NF, Nieuwkoop-Faber stage of development; AC, animal caps.

Epigenomic Feature	Genomic Locations	Xenopus Species	Stage (NF)	Reference
H3K4me1	Active or poised regulatory elements (promoters and enhancers)	X.tro	8, 9, 10½, 12½, 16, 30	(Gupta et al. 2014; Hontelez et al. 2015)
		X.lae	10½	(Elurbe et al. 2017)
H3K4me3	Active or poised promoters (broad domains at some loci)	X.tro	8, 9, 10½, 12½, 16, 30	(Akkers et al. 2009; van Heeringen et al. 2014; Hontelez et al. 2015)
		X.lae	10½, 12, AC	(Elurbe et al. 2017; Quigley and Kintner 2017; Kuznetsov et al. 2019)
H3K9ac	Active promoters	X.tro	8, 9, 10½, 12½, 16, 30	(Hontelez et al. 2015)
H3K9me2	Repetitive elements	X.tro	9, 10½, 12½, 16, 30	(Hontelez et al. 2015; van Kruijsbergen et al. 2017)
H3K9me3	Repetitive elements	X.tro	9, 10½, 12½, 16, 30	(Hontelez et al. 2015; van Kruijsbergen et al. 2017)
H3K27ac	Active regulatory elements, enhancers and promoter-distal elements	X.tro	8, 9, 10½	(Gupta et al. 2014)
		X.lae	12, AC	(Quigley and Kintner 2017; Kuznetsov et al. 2019)
H3K27me3	Repressed or poised regulatory elements (broad domains at some loci)	X.tro	8, 9, 10½, 12½, 16, 30	(Akkers et al. 2009; van Heeringen et al. 2014; Hontelez et al. 2015)
		X.lae	12	(Kuznetsov et al. 2019)
H3K36me3	Gene bodies of transcribed genes	X.tro	9, 10½, 12½, 16, 30	(Hontelez et al. 2015)
		X.lae	10½	(Elurbe et al. 2017)
H4K20me3	Repressive, repetitive elements	X.tro	9, 10½, 12½, 16, 30	(Hontelez et al. 2015; van Kruijsbergen et al. 2017)

TABLE 11.2

Non-Histone Chromatin State Maps Abbreviations: WGBS, Whole genome bisulfite sequencing; BioCAP, capture of unmethylated CpGs followed by sequencing; ATAC-seq, assay for transposase-accessible chromatin, followed by sequencing; HiC, chromosome conformation capture assay for detecting all-to-all interactions; X.tro, *Xenopus tropicalis*; X.lae, *Xenopus laevis*; NF, Nieuwkoop-Faber stage of development; AC, animal caps.

Epigenomic Feature	Genomic Locations	Technique	Xenopus Species	Stage (NF)	Reference
DNA methylation	Throughout the genome, except CpG islands (H3K4me3 and H3K27me3-regulated loci)	BioCAP	X.tro	11–12	(Long et al. 2013)
		WGBS		9, 10½, 12½, 30, 43	(Hontelez et al. 2015; Bogdanovic et al. 2016)
		WGBS	X.lae	10½	(Session et al. 2016; Elurbe et al. 2017)
Chromatin accessibility	Active regulatory regions	ATAC-seq	X.tro	9, 10½, 12½, 16, AC, DMZ	(Bright et al. 2021)
			X.lae	AC	(Esmaeili et al. 2020)
Chromosome topology	Topologically associating domains (TADs), interaction neighborhoods of chromosomes	HiC	X.tro	8, 9, 10, 11, 12, 13, 15, 17, 23	(Niu et al. 2021)
			X.lae	10½, AC	(Quigley and Kintner 2017)

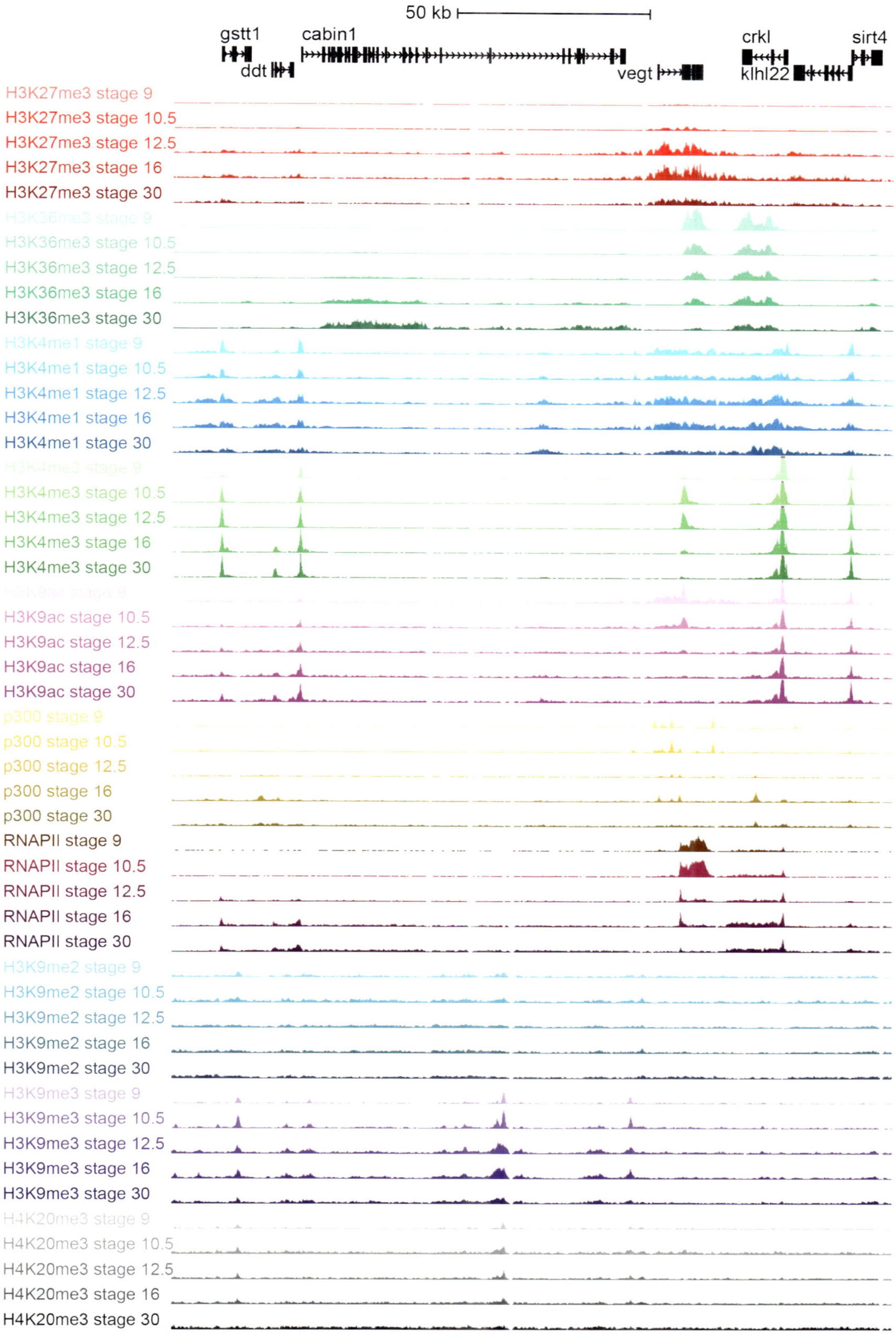

FIGURE 11.1 Genome browser view of chromatin state maps of early *X. tropicalis* development. Depicted is the *vegt* locus with neighboring genes and H3K4me1 (active and poised regulatory elements, including promoters and enhancers); H3K4me3 (active and poised promoters); H3K9ac (mainly active promoters); H3K27me3 (Polycomb repression); H3K36me3 (transcribed transcription units); RNA polymerase II; Ep300 (active putative enhancers); and H3K9me2, H3K9me3, and H4K20me3 (heterochromatin, repetitive elements).

et al. 2009). Chromatin state maps of zebrafish, mouse, and human embryos would follow suit (Vastenhouw et al. 2010; Lindeman et al. 2011; Liu et al. 2016; Xia et al. 2019). The functional genomic elements associated with particular chromatin states in *Xenopus* are very similar to those in other vertebrate species, despite substantial differences in GC content and CpG density between genomes. However, the differences in overall nucleotide composition do affect the relative frequencies of promoter elements. For example, there is a more frequent use of the TATA box relative to the GC-rich Sp1 motif in *Xenopus* compared to human promoters (van Heeringen et al. 2011). Nonetheless, similar to human promoters, frog promoters frequently feature a relatively high density of unmethylated CpG dinucleotides (Bogdanovic et al. 2011; Long et al. 2013; Hontelez et al. 2015). Also similar to the promoters in human and other genomes, active promoters in frogs feature accessible chromatin and are enriched for methylated histone H3 lysine 4 (H3K4me1, H3K4me3) and acetylated H3K9 (H3K9ac; Tables 11.1, 11.2) (Akkers et al. 2009; van Heeringen et al. 2014; Hontelez et al. 2015; Bright et al. 2021). Compared with promoters, enhancers typically show an intermediate level of chromatin accessibility when active and are marked by H3K4me1 (but not H3K4me3) and H3K27ac. The H3K27 acetylation mark is deposited by the p300 (Ep300) coactivator. H3K4 methylation is considered permissive; active regulatory elements usually exhibit H3K4 methylation, but some inactive or "poised" elements have these modifications as well. The H3K27me3 mark (mutually exclusive with H3K27ac) is involved in repression of transcription by the Polycomb Repression Complex 2 (PRC2). It is often found in broad domains of unmethylated CpG islands at developmentally and spatially regulated genes. Trimethylation of H3K36 is a proxy for elongating RNA polymerase II, whereas methylated H3K9 and H4K20 (H3K9me2, H3K9me3, H4K20me3) are modifications associated with heterochromatin and repetitive elements (Hontelez et al. 2015; van Kruijsbergen et al. 2017). The epigenome maps of both *X. tropicalis* and *X. laevis* allow for studies of genome evolution (Elurbe et al. 2017) and provide a rich resource for both whole genome and gene-centric analyses of embryonic development.

11.3. CHROMATIN ACCESSIBILITY: OPENING THE DOOR FOR GENE EXPRESSION

11.3.1. Histone Variants and Linker Histones Modulate Nucleosome Dynamics and Accessibility

With the exception of sperm, histones are abundant nuclear proteins in all cells. In *Xenopus* oocytes and early embryos, these proteins are even more abundant. Normally cells contain a mass of core histones that is approximately equal to the mass of their genomic DNA, but oocytes and early embryos contain excess histone protein, stored in the cytoplasm. The abundance of stored histone proteins and histone-encoding mRNAs in *Xenopus* embryos made them attractive targets for early studies of the histones in relation to their incorporation in embryonic chromatin and the mid-blastula transition (Woodland and Adamson 1977; Woodland 1980; Koster et al. 1988). In agreement with these early studies, quantitative proteomics has established that oocytes and early embryos contain over 1012 molecules of core histone proteins (Smits et al. 2014). This is more than enough to package newly replicated DNA into chromatin until well after the onset of embryonic transcription during blastula stages.

Most of the histone protein is incorporated in chromatin by DNA replication-coupled chromatin assembly mechanisms. In somatic cells, this happens through S phase-specific expression of canonical histone genes, tightly coordinated with the need for chromatin assembly during DNA replication. However, additional non-canonical histone genes exist. They encode variant histones that serve distinct functions and that can be incorporated in chromatin independent of replication-coupled chromatin assembly (Martire and Banaszynski 2020). Variant histones are subject to dynamic exchange (deposition and eviction), which is mediated by histone chaperones. For example, in the histone H3 family of proteins, histones H3.1/3.2 are incorporated during S phase by the CAF-1 complex. By contrast, H3.3, which differs at just five and four amino acids with H3.1 and H3.2, respectively, is selectively deposited at gene-regulatory regions by the HIRA histone chaperone complex in a DNA synthesis-independent manner. This causes nucleosomes to be more dynamic at regulatory regions. Depletion of histone H3.3 by morpholinos leads to problems with blastopore closure in *Xenopus* embryos (Szenker et al. 2012; Sitbon et al. 2020). One H3.3-specific residue, S31, is critical for rescue of this phenotype. This serine is phosphorylated, and a phospho-mimetic S31D mutation not only rescues H3.3 depletion but also increases H3.3 K27 acetylation. This modification is strongly associated with gene activation. This suggests that the gastrulation phenotype is caused by a requirement for efficient H3K27 acetylation that depends on histone exchange and S31 phosphorylation (Sitbon et al. 2020).

H2A.Z (H2az1/H2az2) is one of many variants of histone H2A that is predominantly found at gene-regulatory regions, where it facilitates access of transcription factors and other chromatin-associated proteins to DNA (Martire and Banaszynski 2020). Mutations in Srcap, a chromatin remodeler mediating the incorporation of H2A.Z into chromatin, cause Floating-Harbor syndrome (FHS). There are two H2A.Z variants, which differ by just three amino acids. Knockdown of H2A.Z.2 (H2az2) mimics the FHS craniofacial phenotype in *Xenopus* (Greenberg et al. 2019). Also, for this molecule, a single amino acid difference with H2A.Z.1 is critical for H2A.Z.2 to rescue the phenotype. In contrast to the amino acid substitutions in H3.3, this amino acid difference in H2A.Z.2 may affect the stability of nucleosomes. The altered H2A.Z-containing nucleosomes appear critical for a specific category of AT-rich enhancers that regulate neural crest gene expression (Greenberg et al. 2019). H2A.Z specifically recruits the Pwwp2a protein to DNA (Pünzeler et al. 2017). Knockdown of this protein in *Xenopus* also results in craniofacial defects and aberrant neural crest migration, suggesting that this protein mediates the function of H2A.Z in neural crest gene expression.

Linker histones bind to the dyad axis of nucleosomes (where DNA enters and exits the nucleosome) and generally have a role in chromatin compaction (Prendergast and Reinberg 2021). However, this does not mean that all linker histones repress transcription. Studies in *Xenopus* have made important early contributions in the emerging roles of variant linker histones in embryonic development (Smith et al. 1988; Dworkin-Rastl et al. 1994; Steinbach et al. 1997). Canonical histone H1 is virtually absent in oocytes and early embryos. Instead, they contain a variant linker histone encoded by the *h1–8* gene (also referred to as linker histone B4, H1M, or *h1foo*). This early embryonic linker histone is gradually diluted and replaced by canonical (somatic) H1 during development. In animal cap explants, somatic H1 promotes the loss of mesodermal competence, suggesting that somatic H1 but not maternal H1–8/B4 represses genes involved in mesoderm induction (Steinbach et al. 1997). H1–8/B4 has a relatively low affinity for chromatin compared to somatic H1 (Ura et al. 1996), which may explain the transcriptionally permissive nature of chromatin associated with developmental competence in early embryos. These data illustrate how variant core and linker histone are critically important for chromatin accessibility and gene regulation during development.

11.3.2. Chromatin Assembly and the Onset of Zygotic Genome Activation

The nucleosomes that are formed by the very abundant maternal stores of histone proteins generally provide a barrier towards transcription; this is especially the case in early development, in which chromatin assembly is very repressive towards transcription. Before ovulation, oocytes are arrested in prophase of meiosis I and transcriptionally active. Meiotic maturation under the influence of progesterone leads to completion of meiosis I and an arrest at metaphase of meiosis II. These eggs are transcriptionally quiescent. After fertilization, the maternal gene products of the egg need to sustain the embryo until well into blastula stages, when the embryo starts to transcribe its own genes (zygotic genome activation, ZGA). Injected promoter-reporter templates recapitulate this pattern of endogenous transcription: they are active in oocytes but repressed in embryos until the mid-blastula stage, when they become active again. This was first demonstrated with an RNA polymerase III-dependent tRNA promoter and subsequently with a number of RNA polymerase II-dependent promoters, such as the *c-Myc* promoter, one of the histone *H2B* promoters, and the *CMV* promoter (Newport and Kirschner 1982; Prioleau et al. 1994; Veenstra et al. 1999). Excess competitor DNA that is co-injected with the promoter construct interferes effectively with chromatin assembly, as judged from supercoiling assays, and relieves the repression of transcription before the mid-blastula stage. Injection of the general transcription factor TBP stimulates this precocious transcription by one to two orders of magnitude (Prioleau et al. 1994; Veenstra et al. 1999). This is because TBP (Tbp) is only minimally present before the mid-blastula stage. TBP is also virtually absent in transcriptionally active oocytes, but oocytes contain TBP2 (Tbpl2), a TBP replacement factor in oocytes that is largely degraded during oocyte maturation (Jallow et al. 2004; Akhtar and Veenstra 2009). In cleavage-stage embryos, neither TBP nor TBP2 is abundant, but TBP accumulates during blastula stages due to translation of maternal *tbp* mRNA. Interestingly, in cleavage-stage embryos with exogenous TBP, injected promoter constructs become transiently active, to be repressed coincident with chromatin assembly (Prioleau et al. 1994; Veenstra et al. 1999). Chromatin assembly is much more efficient in egg extract than in oocyte extract (Wang and Shechter 2016). Moreover, depletion of free histones leads to an earlier ZGA (Amodeo et al. 2015). This suggests that repression of transcription before ZGA occurs at multiple levels: chromatin assembly is repressive towards transcription, and, in addition, the abundance of TBP and TBP2 determines the capacity of the transcription machinery in early development.

11.3.3. Chromatin Accessibility and Transcription Factor Binding

Chromatin accessibility, as interrogated using ATAC-sequencing, is observed from late blastula stages onward (Bright et al. 2021), suggesting an important role for maternal factors in chromatin opening. Indeed, maternal factors may contribute to chromatin opening, pluripotency, and germ layer specification (Paraiso et al. 2020). Many regulatory elements that are active in early embryos contain binding sites for the pluripotency factors Oct4 (Pou5f3.1, 3.2, and 3.3) and Sox2/3. Of these proteins, Pou5f3.2, Pou5f3.3, and Sox3 are maternally expressed and are required for establishing chromatin accessibility at approximately 40% of putative regulatory elements (Gentsch et al. 2019). Another maternal factor, Foxh1, is bound to regulatory elements as early as the 32-cell stage (Charney et al. 2017a). In early blastula embryos, Foxh1 recruits Tle, a co-repressor that can associate with the histone deacetylase Hdac1. In prospective endoderm, early Foxh1-Tle binding to regulatory elements of endoderm-expressed genes precedes the recruitment of the zygotically expressed Foxa activator and the co-activator Ep300. This suggests a molecular "hand-off" between maternal and zygotic forkhead transcription factors that coincides with a switch from repression to activation of transcription (Charney et al. 2017a). Maternal factors Vegt and Otx1 also bind to regulatory elements in cleavage stage embryos and orchestrate endoderm formation together with Foxh1 (Paraiso et al. 2019). The mediator of canonical Wnt signaling, β-Catenin (Ctnnb1), is also associated with DNA before the mid-blastula stage (Blythe et al. 2010; Afouda et al. 2020). It recruits Prmt2, a histone methyl transferase, methylating histone H3 on arginine 8 and priming target genes such as *sia1*, *nodal5*, and *nodal6* for expression in the dorsal region of blastula-stage embryos (Blythe et al. 2010). In addition, β-Catenin works together with Foxh1 and Nodal/Tgfβ signaling in the activation of Wnt target genes, as shown by knockdown and pharmacological inhibition experiments (Afouda et al. 2020). The endoderm-specific transcription factor Sox17 and β-Catenin also co-occupy many regulatory elements, which modifies the response to Wnt/β-Catenin signaling in endoderm (Mukherjee et al. 2020).

These data illustrate how some transcription factors gain early access to generally inaccessible chromatin to orchestrate embryonic gene regulation. Chromatin accessibility at promoter and enhancer regions generally increases during blastula and gastrula stages (Bright et al. 2021). At the early gastrula stage, extensive chromatin accessibility is observed in different regions of the embryo, supporting the extensive lineage potential of the cells in these regions. Within this relatively open chromatin, transcription factors operate in a combinatorial fashion (Bright et al. 2021), allowing the cells to respond to inductive signaling. This involves so-called feed-forward circuitry in which (maternal) factors activate genes both directly and indirectly (Gazdag et al. 2016; Charney et al. 2017b; Afouda et al. 2020). Collectively, these studies document how establishing chromatin accessibility is a key step in establishing developmental competence and the ability to respond to inductive signals.

11.4. HISTONE MODIFICATIONS: ACQUISITION AND DYNAMICS DURING EARLY DEVELOPMENT

11.4.1. Oocyte and Egg Storage Histones and Replication-Coupled Chromatin Assembly Dynamics

The histone proteins are extensively modified in cell-type and locus-specific ways. The modifications provide a scaffold for binding proteins that are recruited to sequences wrapped in nucleosomes with these modifications. These "reader" proteins can be effector molecules, involved in activation or repression of transcription (reviewed in Smith and Shilatifard 2010; Soshnev et al. 2016). As discussed above (3.1), oocytes and early embryos contain storage forms of histone protein that are not associated with genomic DNA. Early experiments indicated that pre-deposition (i.e. maternally stored) histones H3 and H4 are acetylated (Woodland 1979). These marks are linked to gene activity when present in chromatin, but H4, and to some extent H3, is deacetylated upon deposition. This corresponds to removal of some of the storage modifications. Mass spectrometry and western blotting experiments showed that these are not the only modifications on storage histones, some of which are not removed during deposition. H2A and H3 arginine methyl marks, for example, are no different before and after deposition in chromatin. Conversely, some modifications are not present in storage histones. Examples of these are H3K4 methylation (permissive mark for transcription), H3K27 trimethylation (facultative heterochromatin), and H3K9 trimethylation (constitutive heterochromatin), which are not found on pre-deposition histones (Nicklay et al. 2009; Shechter et al. 2009).

DNA replication-coupled chromatin assembly exerts a major effect on histone modifications in chromatin, as it determines to what extent the existing marks are maintained. During DNA replication in *Drosophila* embryos, chromatin of nascent DNA does not contain H3K4me3 and H3K27me3 until one hour after DNA replication, when they are re-established (Petruk et al. 2012). In mammalian cells, some di- and tri-methylation modifications in "old" nucleosomes may be recycled, as they are reduced two-fold upon DNA replication, to be gradually restored during the ensuing cell cycle (Alabert et al. 2015). In both cases, this means that there is a major delay in restoring histone modifications after DNA replication. Contributing to such a delay, in both *Drosophila* embryos and differentiating mouse ES cells, is the Utx histone H3K27demethylase, which prevents re-establishing of H3K27me3 on newly replicated DNA (Petruk et al. 2013; Petruk et al. 2017). These mechanisms have not been studied in *Xenopus*, but such delays are likely to have a dramatic impact on histone modification dynamics during the cleavage stages, when the cell cycles are extremely short.

11.4.2. Acquisition of the Antagonistic H3K4me3 and H3K27me3 Marks in the Embryo

H3K4me3 and H3K27me3 are the marks associated with respectively the Trithorax and Polycomb group genes, originally identified in *Drosophila*. Using mass spectrometry, the permissive H3K4me3 promoter mark appears to be relatively abundant in late blastula embryos. In later development, its share in embryonic chromatin decreases (Schneider et al. 2011). H3K27me3, the Polycomb repressive mark, on the other hand, is quite low in blastula embryos but increases during subsequent development. This suggests that chromatin becomes less permissive and more repressive as development proceeds. As a general notion, this is in line with chromatin accessibility data that link open chromatin to multi-lineage potential in blastula and gastrula embryos (Bright et al. 2021). Chromatin immunoprecipitation (ChIP) experiments have provided independent results on these dynamics. H3K4 methylation accumulates rapidly in early blastula embryos, whereas H3K27me3 follows a bit later: it accumulates between mid-blastula and late gastrula stages (Akkers et al. 2009; van Heeringen et al. 2014; Hontelez et al. 2015). We have referred to this as a hierarchy of activation and repression (Figure 11.2), as opposed to co-occurring H3K4me3 and H3K27me3, referred to as bivalency. Bivalent chromatin does occur in *Xenopus* embryos, as determined by ChIP-re-ChIP experiments, but it is not a predominant feature, and it tends to be quickly resolved spatially within the embryo (Akkers et al. 2009). These observations initially caused some debate because bivalency had attracted some attention as a special type of chromatin state.

As it turns out, the characteristics observed in *Xenopus* are conserved among vertebrates. In zebrafish, the histone modification dynamics appear quite similar to those in *Xenopus* (Lindeman et al. 2011). Even in mouse and human preimplantation embryos, which are never completely devoid of histone modifications, the permissive H3K4me3 mark at ZGA precedes a wave of H3K27 methylation (Liu et al. 2016; Xia et al. 2019). Bivalency is also much less frequent in mouse embryos compared to mouse embryonic stem cells (Liu et al. 2016).

In cleavage-stage *Xenopus* embryos, H3K4me3 and H3K27me3 appear to be absent. It should be noted that any constitutive chromatin-associated epitope will double with every cleavage division of the embryo. Therefore, it is important to address by how much H3K4me3 increases on a per-cell or per-unit-of-chromatin basis. This is not necessarily resolved in ChIP-sequencing experiments. Even if a modification is hardly present, it is still possible that what little is present is highly localized in particular genomic regions, producing highly enriched ChIP-sequencing peaks. ChIP with quantitative PCR allows quantifying the percentage recovery relative to input DNA for a particular sequence, which then can be compared across stages. The median recovery of H3K4me3-associated DNA relative to genomic DNA is about 200 lower at stage 7 compared to stage 9 (Figure 11.2A; Hontelez et al. 2015). This is a very conservative estimate, since it is based on the median of a collection of genes that are not all expressed very abundantly. Because the highest recovery (observed at stage 9) cannot exceed 100% of histones at a promoter, and as the average chromatin fragment size in the ChIP experiments accommodates maximally two nucleosomes (four H3 tails per allele), this suggests an upper limit of H3K4me3 abundance of one modified histone tail molecule per promoter (both alleles) per 25 cells at stage 7. This is the upper limit of what the abundance of H3K4me3 could be given the data at hand. From such a minimal amount of the H3K4me3 modification, the embryo goes on to increase H3K4 methylation preceding and coincident with an increase of transcription. This increase thus necessarily involves *de novo* methylation of H3K4 where H3K4me3 did not exist. Based on experiments involving α-amanitin, a drug that inhibits RNA polymerase II, H3K4 trimethylation largely depends on maternal factors operating on DNA methylation-free CpG islands, whereas non-CpG island promoters tend to require new transcription for the acquisition of H3K4me3 (Hontelez et al. 2015).

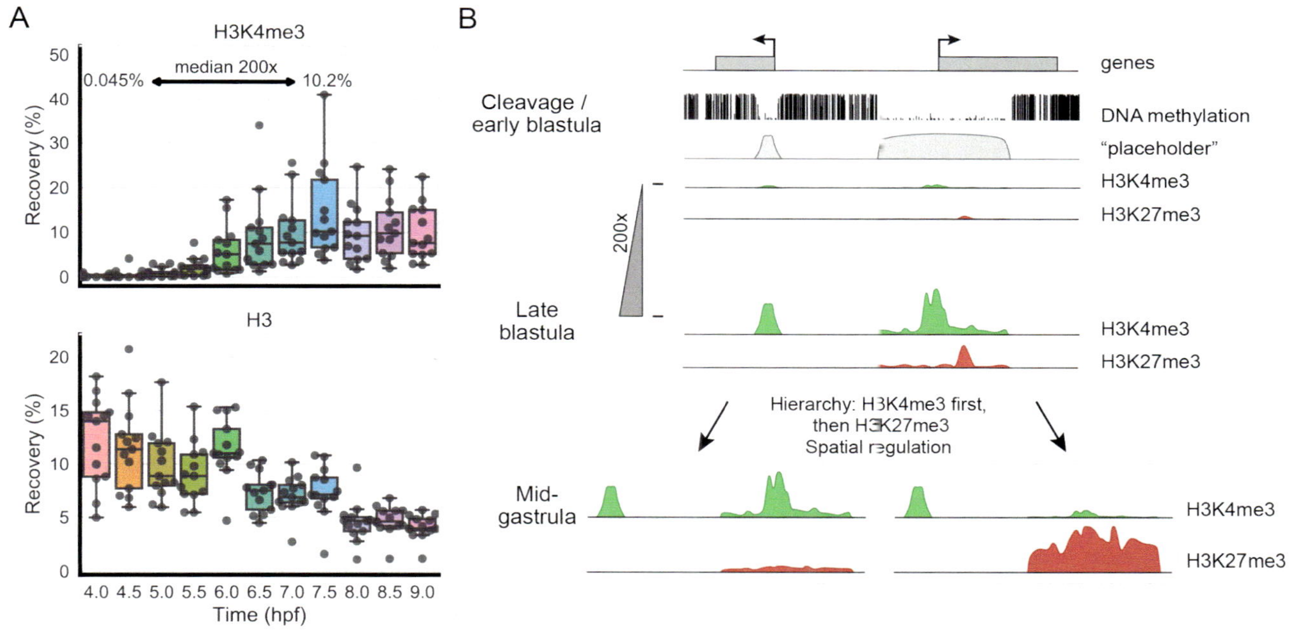

FIGURE 11.2 (A) Recovery of DNA (% of input DNA) in H3K4me3 (upper panel) and histone H3 (lower panel) ChIP-qPCR at promoters of selected loci *(rnf146, torla, zic1, cdc14b, eomes, odc1, xrcc1, drosha, gdf3, tbxt, tbx2, fastkd3, eef1a1o)* in *X. tropicalis* embryos. The plot shows re-analyzed data published by Hontelez et al. (2015). The median recovery of H3K4me3 increases over 200-fold between four hours post-fertilization (4 hpf, stage 7) and 7.5 hpf (stage 9). The overall recovery of histone H3 decreases, possibly because of increasing amounts of chromatin per ChIP or increasing levels of modifications not recognized by the pan-H3 antibody. (B) Model of DNA methylation and hierarchical acquisition of H3K4me3 (first) and H3K27me3 (subsequent), with spatial resolution of temporary bivalent modifications (co-occurring H3K4me3 and H3K27me3). These modifications accumulate preferentially in hypo-methylated regions under the influence of maternal factors (that is, not affected by α-amanitin). This may be regulated by "placeholder" nucleosomes in cleavage stage embryos, similar to zebrafish, which do not contain H3K4me3 or H3K27me3 but mono-methylated H3K4 (H3K4me1) and the H2A.Z(FV) histone variant. The two boxes in the bottom row represent two epigenomic states present in different cells, referred to as spatial regulation.

At face value, this major increase in H3K4me3 during blastula stages appears to be at odds with nuclear transfer experiments that suggest that histone modifications can be transmitted through early development. Ever since the Nobel prize-winning work of Sir John Gurdon, *Xenopus* has been an exceedingly powerful system to study the mechanisms underlying the reprogramming of somatic nuclei by injection into eggs or oocytes. The major implication of the reprogramming achievements is that the cellular changes of differentiation are not genetic and irreversible but epigenetic and reversible in nature (Gurdon 2013). Nuclear reprogramming, however, is not very efficient due to epigenetic barriers that stabilize cellular identity. These barriers involve epigenetic memory of the donor cell type that manifests itself in donor cell type-specific gene expression in nuclear transfer embryos. This is referred to as ON-memory, whereas OFF-memory reflects a failure to activate genes in nuclear transfer embryos that were repressed in donor cells. Reducing H3K4 methylation (mono-, di-, and tri-methylation) by overexpression of the Kdm5b demethylase in donor cells (endoderm) reduced ON-memory (endoderm-specific gene expression in ectoderm) in nuclear transfer embryos and improved their development (Hörmanseder et al. 2017). This raises the question of how ON-memory by H3K4me3 is transmitted through the stages of short cell cycles before the blastula stage. Sperm nuclei have both H3K4me3 and H3K27me3 at loci where they also accumulate in blastula embryos (Oikawa et al. 2020). The presence of H3K4me3 at the same loci in sperm and blastula chromatin does not necessarily mean that the histone modifications are transmitted through the cleavage stages, because H3K4me3 preferentially accumulates in DNA methylation-free regions (Bogdanovic et al. 2011; van Heeringen et al. 2014; Hontelez et al. 2015). Conversely, H3K4me3 can influence DNA methylation. For example, RNAi of H3K4 methyl-transferase in human embryo carcinoma cells caused CpG island hyper-methylation on a subset of loci in addition to a loss of H3K4me3 (Putiri et al. 2014). If such cross-talk is operational in donor nuclei, it is possible that H3K4me3 indirectly mediates ON-memory in nuclear transfer embryos by stabilizing DNA methylation-free CpG islands, which subsequently stage preferential H3K4 re-methylation at the blastula stage. Interestingly, in zebrafish embryos, "placeholder" nucleosomes with a histone H2A variant and H3K4me1 (but not H3K4me3) have been found to occupy regions lacking DNA methylation in cleavage-stage embryos (Murphy et al. 2018). Upon ZGA, many of the placeholder-occupied regions become active and acquire H3K4me3 (Murphy et al. 2018). Loss of these placeholder nucleosomes, however, results in the accumulation of DNA methylation on CpG islands. The pre-ZGA placeholder nucleosomes have not been studied in *Xenopus* but are consistent with acquisition of H3K4me3 and H3K27me3 in pre-existing hypomethylated regions under the influence of maternal factors during blastula and gastrula stages in *Xenopus* (Bogdanovic et al. 2011; van Heeringen et al. 2014; Hontelez et al. 2015). This is an attractive model (Figure 11.2B) that can potentially explain H3K4 methylation-dependent ON-memory in nuclear transfer embryos. Other possibilities include the selective retention of H3K4me3 on a small subset of genes during the cleavage stages (exceptions to the dynamics shown in Figure 11.2A) and differences between the *Xenopus* species used in these experiments.

11.4.3. Heterochromatin Marks at Repetitive Elements and Near Genes

H3K9 methylation (me2, me3) and H4K20me3 decorate the epigenome in a punctate manner. They co-occupy many of the transposon sequences that are interspersed in the genome and are enriched at putative centromeric and telomeric regions (van Kruijsbergen et al. 2017). The modifications mark a subset of all transposons. Some retrotransposon families gain H3K9me3 and H4K20me3 between blastula and gastrula stages. By contrast, specific DNA transposon families are enriched for H4K20me3 and the Polycomb mark H3K27me3 early in development, only to lose these marks later in development. Interestingly, these DNA transposons acquire H3K27me3 earlier than gene promoters do. Taking intra-family diversity as a proxy for age, young transposable elements (high similarity within the sub-family) were preferentially associated with H4K20me3 and H3K9me3, in line with a role in repressing these genomic parasites (van Kruijsbergen et al. 2017). Although repetitive elements may be the primary drivers of heterochromatinization, the interspersed localization of transposable elements may cause them to influence gene regulation. The two major methyl transferases responsible for H4K20me3, Kmt5b/c (Suv4–20h1/h2), repress one of the three Oct4 paralogs (*pou5f3.2*, also referred to as *oct25*) in *X. laevis* (Nicetto et al. 2013). Knockdown of these enzymes caused defects in neural differentiation. Simultaneous morpholino targeting of *pou5f3.2* rescued these defects in neural ectoderm gene expression. In mouse embryonic stem cells, a similar role for Kmt5b/c was identified in downregulation of Oct4 during the exit of pluripotency (Nicetto et al. 2013). These data show that the role of H4K20me3 is not restricted to repression in the context of constitutive heterochromatin and that it contributes to dynamic gene regulation and lineage commitment.

11.5. DNA METHYLATION: REPRESSION VERSUS MODULATION OF HISTONE MODIFICATIONS

11.5.1. DNA Methylation Dynamics

DNA methylation predominantly occurs at the majority of CG (CpG) dinucleotides in vertebrate genomes. Deamination of 5-methyl cytosine produces thymine, which explains why CpG dinucleotides are relatively rare in the genome except for sequences where CpGs are kept in an unmethylated state, so-called CpG islands. Unmethylated CpG islands are found at many promoters, and they are associated with large H3K27me3-enriched domains (Bogdanovic et al. 2011; Long et al. 2013).

There is not much detailed information available on DNA methylation dynamics during the earliest stages of *Xenopus* development. In zebrafish, for example, the oocyte and sperm patterns of DNA methylation differ. In cleavage-stage fish embryos, the maternal patterns are remodeled to mimic those found in sperm and later stages of development (Jiang et al. 2013; Potok et al. 2013). Different from pre-implantation development of mammals, neither zebrafish nor *Xenopus* exhibit a global genome-wide DNA hypomethylation around ZGA. Based on Southern blotting in combination with methylation-sensitive restriction enzymes, it was found that *X. laevis* genomic DNA is globally hypermethylated from early blastula (stage 6) onwards, similar to sperm (Veenstra and Wolffe 2001). In mouse, Tet1, a dioxygenase enzyme involved in DNA demethylation, demethylates the genome after fertilization. *Xenopus*, however, do not have an ortholog of mammalian *Tet1*. *Xenopus* species do have *Tet2* and *Tet3*, which are abundantly expressed from neurula stages onwards. Based on morpholino injections, *Tet3* is important for eye and neural development (Xu et al. 2012). Whole genome bisulfite sequencing (WGBS) has been performed from late blastula stages (stage 9) onward, confirming the overall stability of DNA methylation over developmental time (Hontelez et al. 2015; Bogdanovic et al. 2016). Around 600 enhancer elements, however, are actively demethylated around stage 30, at the so-called phylotypic stage. A similar phenomenon is observed in zebrafish and mouse embryos (Bogdanovic et al. 2016).

11.5.2. DNA Methylation, Repression and Cross-Talk with Histone Modifications

DNA methylation is generally thought to be repressive towards transcription. Using morpholinos in *X. laevis* embryos to knock down the DNA methyl transferase Dnmt1, the protein was found to contribute to repression of transcription before the mid-blastula transition. A catalytically dead form of the protein, however, could rescue this phenotype, showing that this function is unrelated to DNA methylation (Dunican et al. 2008). Although DNA methylation of CpG-dense promoters is strongly repressive in oocytes (Jones et al. 1998), this is not necessarily true in blastula- and gastrula-stage embryos. In a comparison of methylated versus unmethylated transgenes, methylated reporters were strongly repressed in oocytes and tailbud-stage embryos but not in blastula and gastrula embryos where they were expressed despite being methylated (Bogdanovic et al. 2011). This does not mean that DNA methylation is inconsequential in early development, as the H3K4me3 and H3K27me3 histone modifications preferentially accumulate in hypomethylated regions from blastula stages onwards (Hontelez et al. 2015).

11.6. FUTURE DIRECTIONS

Research in *Xenopus* has been at the frontier of chromatin research. As discussed previously, this included early discoveries of the histone proteins as the packaging material of genomic DNA, key insights in cloning and nuclear transfer experiments, and the first epigenome maps of vertebrate embryos. There are promising indications that *Xenopus* will continue to play this role. Recently the chromosome topology of chromosomal DNA during development has been described (Quigley and Kintner 2017; Niu et al. 2021). Moreover, single cell transcriptomics and single-cell chromatin accessibility assays allow resolving gene regulatory landscape at the level of individual cells, providing the resolution to disentangle developmental trajectories and gene regulation in embryos at the level of individual cell types (Briggs et al. 2018; Kakebeen et al. 2020; Bright et al. 2021). New insights can also be gleaned from comparisons between *X. laevis* and *X. tropicalis*. These two species are generally considered similar in their developmental programs. Comparisons between the two species are interesting in the context of genome evolution. *X. laevis* as a species originated from inter-specific hybridization 17–18 million years ago, resulting in a duplicate, allo-tetraploid genome (Session et al. 2016). Kilobase-sized deletions are prevalent in this genome, especially in chromosomes derived from one of the parental species. In addition, reactivation of transposons may have played a major role in reshaping the epigenomic landscape (Elurbe et al. 2017). This indicates that epigenomic comparisons of *Xenopus* species may contribute new insights in genome evolution and chromatin dynamics over ultra-long time scales.

All these exciting developments have been shaped in some way or another by new technologies. Computational analyses have become essential to extract insight from ever-increasing amounts of data. All these developments and compelling recent findings on chromatin dynamics indicate that *Xenopus* species have been and will continue to be powerful models for development and disease.

REFERENCES

Afouda BA, Nakamura Y, Shaw S, Charney RM, Paraiso KD, Blitz IL, Cho KWY, Hoppler S. 2020. Foxh1/nodal defines context-specific direct maternal Wnt/β-catenin target gene regulation in early development. *iScience* **23**: 101314.

Akhtar W, Veenstra GJ. 2009. TBP2 is a substitute for TBP in Xenopus oocyte transcription. *BMC Biology* **7**: 45.

Akkers RC, van Heeringen SJ, Jacobi UG, Janssen-Megens EM, Francoijs KJ, Stunnenberg HG, Veenstra GJ. 2009. A hierarchy of H3K4me3 and H3K27me3 acquisition in spatial gene regulation in Xenopus embryos. *Developmental Cell* **17**: 425–434.

Alabert C, Barth TK, Reverón-Gómez N, Sidoli S, Schmidt A, Jensen ON, Imhof A, Groth A. 2015. Two distinct modes for propagation of histone PTMs across the cell cycle. *Genes & Development* **29**: 585–590.

Amodeo AA, Jukam D, Straight AF, Skotheim JM. 2015, Mar 10. Histone titration against the genome sets the DNA-to-cytoplasm threshold for the Xenopus midblastula transition. *Proceedings of the National Academy of Sciences of the United States of America* **112**(10): E1086–E1095. doi: 10.1073/pnas.1413990112. Epub 2015 Feb 23. PMID: 25713373; PMCID: PMC4364222.

Blythe SA, Cha SW, Tadjuidje E, Heasman J, Klein PS. 2010. Beta-catenin primes organizer gene expression by recruiting a histone H3 arginine 8 methyltransferase, Prmt2. *Developmental Cell* **19**: 220–231.

Bogdanovic O, Long SW, van Heeringen SJ, Brinkman AB, Gomez-Skarmeta JL, Stunnenberg HG, Jones PL, Veenstra GJ. 2011. Temporal uncoupling of the DNA methylome and transcriptional repression during embryogenesis. *Genome Research* **21**: 1313–1327.

Bogdanovic O, Smits AH, de la Calle Mustienes E, Tena JJ, Ford E, Williams R, Senanayake U, Schultz MD, Hontelez S, van Kruijsbergen I et al. 2016. Active DNA demethylation at enhancers during the vertebrate phylotypic period. *Nature Genetics* **48**: 417–426.

Bogdanovic O, van Heeringen SJ, Veenstra GJ. 2012. The epigenome in early vertebrate development. *Genesis (New York, NY: 2000)* **50**: 192–206.

Briggs JA, Weinreb C, Wagner DE, Megason S, Peshkin L, Kirschner MW, Klein AM. 2018. The dynamics of gene expression in vertebrate embryogenesis at single-cell resolution. *Science (New York, NY)* **360**.

Bright AR, van Genesen S, Li Q, Grasso A, Frölich S, van der Sande M, van Heeringen SJ, Veenstra GJC. 2021. Combinatorial transcription factor activities on open chromatin induce embryonic heterogeneity in vertebrates. *The EMBO Journal*: e104913.

Bright AR, Veenstra GJC. 2019. Assay for transposase-accessible chromatin-sequencing using Xenopus embryos. *Cold Spring Harbor Protocols* **2019**.

Charney RM, Forouzmand E, Cho JS, Cheung J, Paraiso KD, Yasuoka Y, Takahashi S, Taira M, Blitz IL, Xie X et al. 2017a. Foxh1 occupies cis-regulatory modules prior to dynamic transcription factor interactions controlling the mesendoderm gene program. *Developmental Cell* **40**: 595–607.e594.

Charney RM, Paraiso KD, Blitz IL, Cho KWY. 2017b. A gene regulatory program controlling early Xenopus mesendoderm formation: Network conservation and motifs. *Seminars in Cell & Developmental Biology* **66**: 12–24.

Dunican DS, Ruzov A, Hackett JA, Meehan RR. 2008. xDnmt1 regulates transcriptional silencing in pre-MBT Xenopus embryos independently of its catalytic function. *Development* **135**: 1295–1302.

Dworkin-Rastl E, Kandolf H, Smith RC. 1994. The maternal histone H1 variant, H1M (B4 protein), is the predominant H1 histone in Xenopus pregastrula embryos. *Developmental Biology* **161**: 425–439.

Elurbe DM, Paranjpe SS, Georgiou G, van Kruijsbergen I, Bogdanovic O, Gibeaux R, Heald R, Lister R, Huynen MA, van Heeringen SJ et al. 2017. Regulatory remodeling in the allo-tetraploid frog *Xenopus laevis*. *Genome Biology* **18**: 198.

Esmaeili M, Blythe SA, Tobias JW, Zhang K, Yang J, Klein PS. 2020. Chromatin accessibility and histone acetylation in the regulation of competence in early development. *Developmental Biology* **462**: 20–35.

Gazdag E, Jacobi UG, van Kruijsbergen I, Weeks DL, Veenstra GJ. 2016. Activation of a T-box-Otx2-Gsc gene network independent of TBP and TBP-related factors. *Development* **143**: 1340–1350.

Gentsch GE, Spruce T, Owens NDL, Smith JC. 2019. Maternal pluripotency factors initiate extensive chromatin remodelling to predefine first response to inductive signals. *Nature Communications* **10**: 4269.

Gilchrist MJ, Cho KWY, Veenstra GJC. 2020. Genomics methods for Xenopus embryos and tissues. *Cold Spring Harbor Protocols* **2020**: 097915.

Greenberg RS, Long HK, Swigut T, Wysocka J. 2019. Single amino acid change underlies distinct roles of H2A.Z subtypes in human syndrome. *Cell* **178**: 1421–1436.e1424.

Gupta R, Wills A, Ucar D, Baker J. 2014. Developmental enhancers are marked independently of zygotic Nodal signals in Xenopus. *Developmental Biology* **395**: 38–49.

Gurdon JB. 2013. The egg and the nucleus: A battle for supremacy. *Development* **140**: 2449–2456.

Hontelez S, van Kruijsbergen I, Georgiou G, van Heeringen SJ, Bogdanovic O, Lister R, Veenstra GJ. 2015. Embryonic transcription is controlled by maternally defined chromatin state. *Nature Communications* **6**: 10148.

Hontelez S, van Kruijsbergen I, Veenstra GJC. 2019. ChIP-sequencing in Xenopus embryos. *Cold Spring Harbor Protocols* **2019**.

Hörmanseder E, Simeone A, Allen GE, Bradshaw CR, Figlmüller M, Gurdon J, Jullien J. 2017. H3K4 methylation-dependent memory of somatic cell identity inhibits reprogramming and development of nuclear transfer embryos. *Cell Stem Cell* **21**: 135–143.e136.

Jallow Z, Jacobi UG, Weeks DL, Dawid IB, Veenstra GJ. 2004. Specialized and redundant roles of TBP and a vertebrate-specific TBP paralog in embryonic gene regulation in Xenopus. *Proceedings of the National Academy of Sciences of the United States of America* **101**: 13525–13530.

Jiang L, Zhang J, Wang JJ, Wang L, Zhang L, Li G, Yang X, Ma X, Sun X, Cai J et al. 2013. Sperm, but not oocyte, DNA methylome is inherited by zebrafish early embryos. *Cell* **153**: 773–784.

Jones PL, Veenstra GJ, Wade PA, Vermaak D, Kass SU, Landsberger N, Strouboulis J, Wolffe AP. 1998. Methylated DNA and MeCP2 recruit histone deacetylase to repress transcription. *Nature Genetics* **19**: 187–191.

Kakebeen AD, Chitsazan AD, Williams MC, Saunders LM, Wills AE. 2020. Chromatin accessibility dynamics and single cell RNA-Seq reveal new regulators of regeneration in neural progenitors. *eLife* **9**.

Koster JG, Destrée OH, Westerhoff HV. 1988. Kinetics of histone gene expression during early development of *Xenopus laevis*. *Journal of Theoretical Biology* **135**: 139–167.

Kuznetsov JN, Aguero TH, Owens DA, Kurtenbach S, Field MG, Durante MA, Rodriguez DA, King ML, Harbour JW. 2019. BAP1 regulates epigenetic switch from pluripotency to differentiation in developmental lineages giving rise to BAP1-mutant cancers. *Science Advances* **5**: eaax1738.

Lindeman LC, Andersen IS, Reiner AH, Li N, Aanes H, Ostrup O, Winata C, Mathavan S, Muller F, Alestrom P et al. 2011. Prepatterning of developmental gene expression by modified histones before zygotic genome activation. *Developmental Cell* **21**: 993–1004.

Liu X, Wang C, Liu W, Li J, Li C, Kou X, Chen J, Zhao Y, Gao H, Wang H et al. 2016. Distinct features of H3K4me3 and H3K27me3 chromatin domains in pre-implantation embryos. *Nature* **537**: 558–562.

Long HK, Sims D, Heger A, Blackledge NP, Kutter C, Wright ML, Grutzner F, Odom DT, Patient R, Ponting CP et al. 2013. Epigenetic conservation at gene regulatory elements revealed by non-methylated DNA profiling in seven vertebrates. *eLife* **2**: e00348.

Martire S, Banaszynski LA. 2020. The roles of histone variants in fine-tuning chromatin organization and function. *Nature Reviews Molecular Cell Biology* **21**: 522–541.

Mukherjee S, Chaturvedi P, Rankin SA, Fish MB, Wlizla M, Paraiso KD, MacDonald M, Chen X, Weirauch MT, Blitz IL et al. 2020. Sox17 and β-catenin co-occupy Wnt-responsive enhancers to govern the endoderm gene regulatory network. *eLife* **9**.

Murphy PJ, Wu SF, James CR, Wike CL, Cairns BR. 2018. Placeholder nucleosomes underlie germline-to-embryo DNA methylation reprogramming. *Cell* **172**: 993–1006.e1013.

Newport J, Kirschner M. 1982. A major developmental transition in early Xenopus embryos. II: Control of the onset of transcription. *Cell* **30**: 687–696.

Nicetto D, Hahn M, Jung J, Schneider TD, Straub T, David R, Schotta G, Rupp RA. 2013. Suv4–20h histone methyltransferases promote neuroectodermal differentiation by silencing the pluripotency-associated Oct-25 gene. *PLoS Genetics* **9**: e1003188.

Nicklay JJ, Shechter D, Chitta RK, Garcia BA, Shabanowitz J, Allis CD, Hunt DF. 2009. Analysis of histones in *Xenopus laevis*. II: Mass spectrometry reveals an index of cell type-specific modifications on H3 and H4. *The Journal of Biological Chemistry* **284**: 1075–1085.

Niu L, Shen W, Shi Z, Tan Y, He N, Wan J, Sun J, Zhang Y, Huang Y, Wang W, Fang C, Li J, Zheng P, Cheung E, Chen Y, Li L, Hou C. 2021 Jul. Three-dimensional folding dynamics of the Xenopus tropicalis genome. *Nature Genetics* **53**(7): 1075–1087. doi: 10.1038/s41588-021-00878-z. Epub 2021 Jun 7. PMID: 34099928; PMCID: PMC8270788

Oikawa M, Simeone A, Hormanseder E, Teperek M, Gaggioli V, O'Doherty A, Falk E, Sporniak M, D'Santos C, Franklin VNR et al. 2020. Epigenetic homogeneity in histone methylation underlies sperm programming for embryonic transcription. *Nature Communications* **11**: 3491.

Paraiso KD, Blitz IL, Coley M, Cheung J, Sudou N, Taira M, Cho KWY. 2019. Endodermal maternal transcription factors establish super-enhancers during zygotic genome activation. *Cell Reports* **27**: 2962–2977.e2965.

Paraiso KD, Cho JS, Yong J, Cho KWY. 2020. Early Xenopus gene regulatory programs, chromatin states, and the role of maternal transcription factors. *Current Topics in Developmental Biology* **139**: 35–60.

Perino M, Veenstra GJ. 2016. Chromatin control of developmental dynamics and plasticity. *Developmental Cell* **38**: 610–620.

Petruk S, Black KL, Kovermann SK, Brock HW, Mazo A. 2013. Stepwise histone modifications are mediated by multiple enzymes that rapidly associate with nascent DNA during replication. *Nature Communications* **4**: 2841.

Petruk S, Cai J, Sussman R, Sun G, Kovermann SK, Mariani SA, Calabretta B, McMahon SB, Brock HW, Iacovitti L et al. 2017. Delayed accumulation of H3K27me3 on nascent DNA is essential for recruitment of transcription factors at early stages of stem cell differentiation. *Molecular Cell* **66**: 247–257.e245.

Petruk S, Sedkov Y, Johnston DM, Hodgson JW, Black KL, Kovermann SK, Beck S, Canaani E, Brock HW, Mazo A. 2012. TrxG and PcG proteins but not methylated histones remain associated with DNA through replication. *Cell* **150**: 922–933.

Potok ME, Nix DA, Parnell TJ, Cairns BR. 2013. Reprogramming the maternal zebrafish genome after fertilization to match the paternal methylation pattern. *Cell* **153**: 759–772.

Prendergast L, Reinberg D. 2021. The missing linker: Emerging trends for H1 variant-specific functions. *Genes & Development* **35**: 40–58.

Prioleau MN, Huet J, Sentenac A, Méchali M. 1994. Competition between chromatin and transcription complex assembly regulates gene expression during early development. *Cell* **77**: 439–449.

Pünzeler S, Link S, Wagner G, Keilhauer EC, Kronbeck N, Spitzer RM, Leidescher S, Markaki Y, Mentele E, Regnard C et al. 2017. Multivalent binding of PWWP2A to H2A.Z regulates mitosis and neural crest differentiation. *The EMBO Journal* **36**: 2263–2279.

Putiri EL, Tiedemann RL, Liu C, Choi JH, Robertson KD. 2014. Impact of human MLL/COMPASS and polycomb complexes on the DNA methylome. *Oncotarget* **5**: 6338–6352.

Quigley IK, Heinz S. 2019. Generating a three-dimensional genome from Xenopus with Hi-C. *Cold Spring Harbor Protocols* **2019**.

Quigley IK, Kintner C. 2017. Rfx2 stabilizes Foxj1 binding at chromatin loops to enable multiciliated cell gene expression. *PLoS Genetics* **13**: e1006538.

Schneider TD, Arteaga-Salas JM, Mentele E, David R, Nicetto D, Imhof A, Rupp RA. 2011. Stage-specific histone modification profiles reveal global transitions in the Xenopus embryonic epigenome. *PLoS One* **6**: e22548.

Session AM, Uno Y, Kwon T, Chapman JA, Toyoda A, Takahashi S, Fukui A, Hikosaka A, Suzuki A, Kondo M et al. 2016. Genome evolution in the allotetraploid frog *Xenopus laevis*. *Nature* **538**: 336–343.

Shechter D, Nicklay JJ, Chitta RK, Shabanowitz J, Hunt DF, Allis CD. 2009. Analysis of histones in *Xenopus laevis*. I: A distinct index of enriched variants and modifications exists in each cell type and is remodeled during developmental transitions. *The Journal of Biological Chemistry* **284**: 1064–1074.

Sitbon D, Boyarchuk E, Dingli F, Loew D, Almouzni G. 2020. Histone variant H3.3 residue S31 is essential for Xenopus gastrulation regardless of the deposition pathway. *Nature Communications* **11**: 1256.

Smith E, Shilatifard A. 2010. The chromatin signaling pathway: Diverse mechanisms of recruitment of histone-modifying enzymes and varied biological outcomes. *Molecular Cell* **40**: 689–701.

Smith RC, Dworkin-Rastl E, Dworkin MB. 1988. Expression of a histone H1-like protein is restricted to early Xenopus development. *Genes & Development* **2**: 1284–1295.

Smits AH, Lindeboom RG, Perino M, van Heeringen SJ, Veenstra GJ, Vermeulen M. 2014. Global absolute quantification reveals tight regulation of protein expression in single Xenopus eggs. *Nucleic Acids Research* **42**: 9880–9891.

Soshnev AA, Josefowicz SZ, Allis CD. 2016. Greater than the sum of parts: Complexity of the dynamic epigenome. *Molecular Cell* **62**: 681–694.

Steinbach OC, Wolffe AP, Rupp RA. 1997. Somatic linker histones cause loss of mesodermal competence in Xenopus. *Nature* **389**: 395–399.

Szenker E, Lacoste N, Almouzni G. 2012. A developmental requirement for HIRA-dependent H3.3 deposition revealed at gastrulation in Xenopus. *Cell Reports* **1**: 730–740.

Ura K, Nightingale K, Wolffe AP. 1996. Differential association of HMG1 and linker histones B4 and H1 with dinucleosomal DNA: Structural transitions and transcriptional repression. *The EMBO Journal* **15**: 4959–4969.

van Heeringen SJ, Akhtar W, Jacobi UG, Akkers RC, Suzuki Y, Veenstra GJ. 2011. Nucleotide composition-linked divergence of vertebrate core promoter architecture. *Genome Research* **21**: 410–421.

van Heeringen SJ, Akkers RC, van Kruijsbergen I, Arif MA, Hanssen LL, Sharifi N, Veenstra GJ. 2014. Principles of nucleation of H3K27 methylation during embryonic development. *Genome Research* **24**: 401–410.

van Kruijsbergen I, Hontelez S, Elurbe DM, van Heeringen SJ, Huynen MA, Veenstra GJC. 2017. Heterochromatic histone modifications at transposons in *Xenopus tropicalis* embryos. *Developmental Biology* **426**: 460–471.

Vastenhouw NL, Zhang Y, Woods IG, Imam F, Regev A, Liu XS, Rinn J, Schier AF. 2010. Chromatin signature of embryonic pluripotency is established during genome activation. *Nature* **464**: 922–926.

Veenstra GJ, Destree OH, Wolffe AP. 1999. Translation of maternal TATA-binding protein mRNA potentiates basal but not activated transcription in Xenopus embryos at the mid-blastula transition. *Molecular and Cellular Biology* **19**: 7972–7982.

Veenstra GJ, Wolffe AP. 2001. Constitutive genomic methylation during embryonic development of Xenopus. *Biochimica et biophysica acta* **1521**: 39–44.

Wang WL, Shechter D. 2016. Chromatin assembly and transcriptional cross-talk in *Xenopus laevis* oocyte and egg extracts. *The International Journal of Developmental Biology* **60**: 315–320.

Woodland HR. 1979. The modification of stored histones H3 and H4 during the oogenesis and early development of *Xenopus laevis*. *Developmental Biology* **68**: 360–370.

Woodland HR. 1980. Histone synthesis during the development of Xenopus. *FEBS Letters* **121**: 1–10.

Woodland HR, Adamson ED. 1977. The synthesis and storage of histones during the oogenesis of *Xenopus laevis*. *Developmental Biology* **57**: 118–135.

Xia W, Xu J, Yu G, Yao G, Xu K, Ma X, Zhang N, Liu B, Li T, Lin Z et al. 2019. Resetting histone modifications during human parental-to-zygotic transition. *Science (New York, NY)* **365**: 353–360.

Xu Y, Xu C, Kato A, Tempel W, Abreu JG, Bian C, Hu Y, Hu D, Zhao B, Cerovina T et al. 2012. Tet3 CXXC domain and dioxygenase activity cooperatively regulate key genes for Xenopus eye and neural development. *Cell* **151**: 1200–1213.

Zhou K, Gaullier G, Luger K. 2019. Nucleosome structure and dynamics are coming of age. *Nature Structural & Molecular Biology* **26**: 3–13.

12 Gene Regulatory Networks Controlling *Xenopus* Embryogenesis

Ken W.Y. Cho and Ira L. Blitz

CONTENTS

12.1. Historical Background 185
12.2. Past Observations 186
12.2.1. Brief Overview of *Xenopus* Development 186
12.2.2. GRNs during Germ Layer Formation 186
12.2.3. GRNs during Gastrulation 188
12.2.4. GRN Subcircuits in *Xenopus* Embryos 188
12.3. Present Status of the Field 190
12.4. Our Contribution to This Field—Insights 192
12.5. Future Directions and Important Questions 192
References 193

12.1. HISTORICAL BACKGROUND

The orchestration of cohorts of gene expression at specific times and places during animal development needs to be precisely repeated in every cell in every generation to create and sustain the life of the organism. Regulatory programs controlling gene expression in time and space, known as a gene regulatory network (GRN), are critical to understand the mechanisms governing developmental processes. A GRN displays interactions between regulatory effectors, such as transcription factors (TFs) and cis-regulatory modules (CRMs), and explains how target gene expression is precisely triggered to be induced or repressed (Peter and Davidson, 2015). CRMs are typically 100–1000 base pairs in length and dock a number of TFs, acting as a functional unit, to regulate expression of a nearby gene (Davidson, 2006). In general, a gene produces a specific pattern of transcription in space and time by integrating the enhancers, insulators and silencers, and multiple CRMs can combine to produce complex patterns of gene expression (Gray et al., 1994). All of this information is hardwired into the animal genome, and GRN study provides a mechanistic understanding of these biological processes, often displaying TF and gene interactions using a wiring diagram.

Genetic modification of individual genes has served as a powerful tool to uncover critical functions in both phenotype and the developmental regulation to achieve it. The uncovering of the hierarchical regulatory interactions of the segmentation gene network in *Drosophila* has led to a powerful illustration of how TFs organize networks of subordinate genes to guide the behavior of embryonic cells during development (Schroeder et al., 2004). These discoveries also inform us that developmental genes function in networks, and it is their function within these networks that ultimately generates biological outcomes. Thus, identifying the structure of a network and its functions and activities is a necessary step toward comprehending the cause of various embryonic and cellular behaviors at a system-wide level, as well as predicting phenotypes in disease.

The first theoretical model describing the mechanisms controlling gene regulation in higher eukaryotes was postulated by Britten and Davidson (1969). They hypothesized that individual TFs regulate the expression of diverse batteries of genes to provide specific phenotypic outcomes. The authors presented the structure of the regulatory network as a wiring diagram to describe how genes interact in a network. Because of the unidirectional information flow (input => output) within genomic control systems, a GRN diagram is distinct from other types of networks, such as protein-protein interaction networks or metabolic networks, which typically lack clear directionality. Subsequently, Davidson and coworkers expanded their approach and performed more systematic investigation of a GRN, including all the genes operating in a single network controlling the specification of endomesoderm cells in early sea urchin development (Peter and Davidson, 2010). Their endomesoderm GRN diagram depicted relationships between nearly 50 TFs and illustrated how changes at a single node (one interaction point between a gene and its regulator) in the network can impact the transcription of multiple downstream genes (Nam and Davidson, 2012). By linking the direct input and output information of each interaction (based on the spatiotemporal expression patterns of the genes, DNA binding information, and perturbation analysis), it becomes feasible to predict the potential outcome of a given regulatory interaction (Faure et al., 2012).

Establishment of similar regulatory network diagrams for vertebrate embryonic development is a necessary next

DOI: 10.1201/9781003050230-14

step to understand the control of cell fate specification at the transcriptional level. The diploid frog *Xenopus tropicalis* is an ideal system to build and study developmental GRN. *Xenopus* has long been used as a model for vertebrate early development and has contributed greatly to the elucidation of gene regulation. *Xenopus* embryos can be experimentally manipulated at the earliest stages of development to allow gain-of-function and loss-of-function analyses with relative ease. The readily available genomic tools make the *Xenopus* system ideally suited for the systematic identification of genes affected by such manipulations. These advantages point to the utility of the *Xenopus* system for the collection of the large number of data sets to be used for GRN study.

Identifying the molecular parts encoded in the genome is only a first step towards understanding how context-specific biological processes are controlled. As we shall see, the knowledge of how these parts are combined and function together is an equally important question. A little over a decade ago, efforts were made to compile the available molecular data into GRNs describing *Xenopus* mesendoderm and Spemann organizer development (Loose and Patient, 2004; Koide et al., 2005). In 2017, additional findings were applied to update the mesendoderm GRN from fertilization to the beginning of gastrulation, linking TFs and critical signaling pathways with transcriptional targets (Charney et al., 2017a). Close attention was paid to focus on "direct" interactions between TFs and target genes to build the GRN regulating mesendoderm.

An important consideration in creating GRNs is to identify direct links between a TF and its direct target genes to understand the underlying network substructures regulating gene expression. We recommend the following three criteria to be used to create an interpretable GRN (Koide et al., 2005). First, an obvious and critically important criterion is that a putative direct target gene must be expressed temporally and spatially in a manner consistent with the expression of the TF proposed to control it. For example, if the TF is an activator, the target gene should be coexpressed in the same or overlapping region where the TF is expressed. Conversely, if the TF is a repressor, target gene expression should be excluded (or reduced) from the region where the TF is expressed. Second, a strong correlation must exist between perturbation of a regulatory TF and the expression changes of the suspected target genes. Regulation can be measured following gain- and/or loss-of-function experiments (e.g., injection of mRNA encoding a TF or a translation blocking antisense morpholino oligonucleotide) by analyzing changes in target gene RNA expression (e.g., RNA-seq, RT-PCR, northern blotting, *in situ* hybridization). Third, a direct physical interaction between the TF and a CRM controlling the proposed target needs to be validated experimentally because perturbation experiments alone are insufficient to distinguish between direct and indirect effects. Methods for assessing direct interaction include chromatin immunoprecipitation (ChIP), gel electrophoretic mobility shift assay (EMSA), DNase footprinting, or reporter gene assays (evaluating appropriate binding site mutations). This linkage of direct TF binding to regulated genes is often lacking in the published literature, making it difficult to state with confidence whether the induction of a proposed target gene is direct or indirect. Only connections that satisfy all three criteria should be defined as direct. However, we recognize that DNA binding is only suggestive of functional regulation and that the "gold standard" evidence is to mutate the TF binding site and examine the effect on gene expression *in vivo*. This is particularly important as ChIP-seq analysis finds abundant neutral binding sites (in addition to those involved in regulation) that don't affect gene expression (Bardet et al., 2013). These criteria should be carefully considered whenever building GRNs.

12.2. PAST OBSERVATIONS

12.2.1. Brief Overview of *Xenopus* Development

After fertilization, the *Xenopus* zygote undergoes multiple rounds of cell division to give rise to an embryo with smaller cells (blastomeres) but without increasing the overall volume of the embryo. During the early stages of this process, some maternally deposited mRNAs and proteins are asymmetrically inherited by individual blastomeres through cytokinesis (see Chapter 3). These differentially localized materials specify the germ layer cell identities by initiating their respective germ layer-specific GRN programs along the animal-vegetal axis. Other maternal components are asymmetrically distributed along the future dorsal-ventral axis, which is initiated by the site of sperm entry. During gastrulation, the germ layers undergo patterning and are subdivided into smaller more specified cell fate territories that eventually define organ/tissue primordia. The broadly defined mesendodermal territory (vegetal and equatorial regions of the embryo) forms distinct endodermal and mesodermal lineages. Initially, the mesoderm is broadly defined into dorsal mesoderm (Spemann's organizer) and ventral mesoderm territories. Demarcation of the boundary is not distinct. Further interactions subdivide mesoderm into more distinct domains that give rise to head mesoderm, notochord, somite, lateral plate, and ventral mesenchyme. The ectoderm also subdivides into neural and epidermal ectoderm territories via neuralizing signals emanating from Spemann's organizer, which secretes Bmp, Wnt, and Nodal antagonists to promote naïve ectoderm (presumptive epidermal tissue) into neural tissue (see Chapter 4). As embryogenesis progresses, neural subdivisions are formed. Neural crest and pre-placodal ectoderm also arise from the neural plate border region that is formed between the neural plate and the epidermis. At the same time, endodermal regions subdivide, providing distinct anteroposterior characteristics to give rise to endodermal organ precursors. GRN structural features underlying germ layer specification, Spemann's organizer formation, and mesendoderm patterning in *Xenopus* are discussed in the following section.

12.2.2. GRNs during Germ Layer Formation

Germ layer specification (the delineation of ectoderm, mesoderm, endoderm) is one of the first cell lineage commitment events in animal development. This occurs shortly after the activation of the zygotic genome roughly around the mid-blastula transition stage (Blitz and Cho, 2021). With the exception of amniotes, germ layer specification depends on unequally distributed maternal determinants present in the egg before fertilization (reviewed in Paraiso et al., 2020), which are the first inputs that specify the differentiation of germ layer cell types. In *Xenopus*, maternal RNAs and proteins are specifically enriched animally (future ectoderm) or vegetally (future endoderm) in the egg, which are subsequently asymmetrically inherited by different blastomeres. Ectoderm germ layer specification requires the maternally expressed forkhead domain TF Foxi2, which is highly enriched as mRNA in the animal region of the *Xenopus* embryo (Cha et al., 2012) and is required for the zygotic expression of ectodermal genes such as *lhx5* and *cdh1* (*e-cadherin*). Additionally, maternal Foxi2 has been shown to directly activate the zygotic expression of *foxi1* (a gene encoding a closely related Foxi TF) by binding to the *foxi1* promoter. Like Foxi2, Foxi1 is an important regulator of the ectodermal gene expression program (Suri et al., 2005; Mir et al., 2007). The mechanism by which these related Fox TFs execute the ectodermal specification GRN program is currently unknown. A signaling pathway important for ectodermal cell patterning is Bmp signaling, which is mediated by signaling mediators Smad1/5/9 and Smad4 (hereafter called Smad1/4 complex) (review Harland, 2000; Hawley et al., 1995). In the future, ChIP-seq analysis of Foxi2 and Smad1/4, as well as the knockdown of these TFs, are likely to uncover the beginnings of the ectodermal GRN structure.

The popular model of mesoderm and endoderm layer specification in *Xenopus* places maternal T-box transcription factor Vegt at the top of the hierarchy of mesoderm/endoderm gene regulatory cascade (Zhang et al., 1998; Kofron et al., 1999; Xanthos et al., 2001) (Figure 12.1A). *Vegt* mRNA, which is maternally transcribed and anchored, is released into the vegetal half equivalent to the future endoderm and mesoderm of blastula stage embryos after fertilization. Vegt controls the zygotic transcription of the *Xenopus* Nodal ligands, accumulation of which initiates Nodal signaling. Since high concentrations of Nodal specify endoderm and lower concentrations induce only mesoderm (Green and Smith, 1990), one popular model proposes that graded concentrations of Vegt regulate graded amounts of Nodal ligands along the animal-vegetal axis to specify distinct mesoderm and endoderm cell fates (Kimelman and Griffin, 1998). While this model is attractive, it is too simplistic (Kofron et al., 2004). Recent work uncovered the

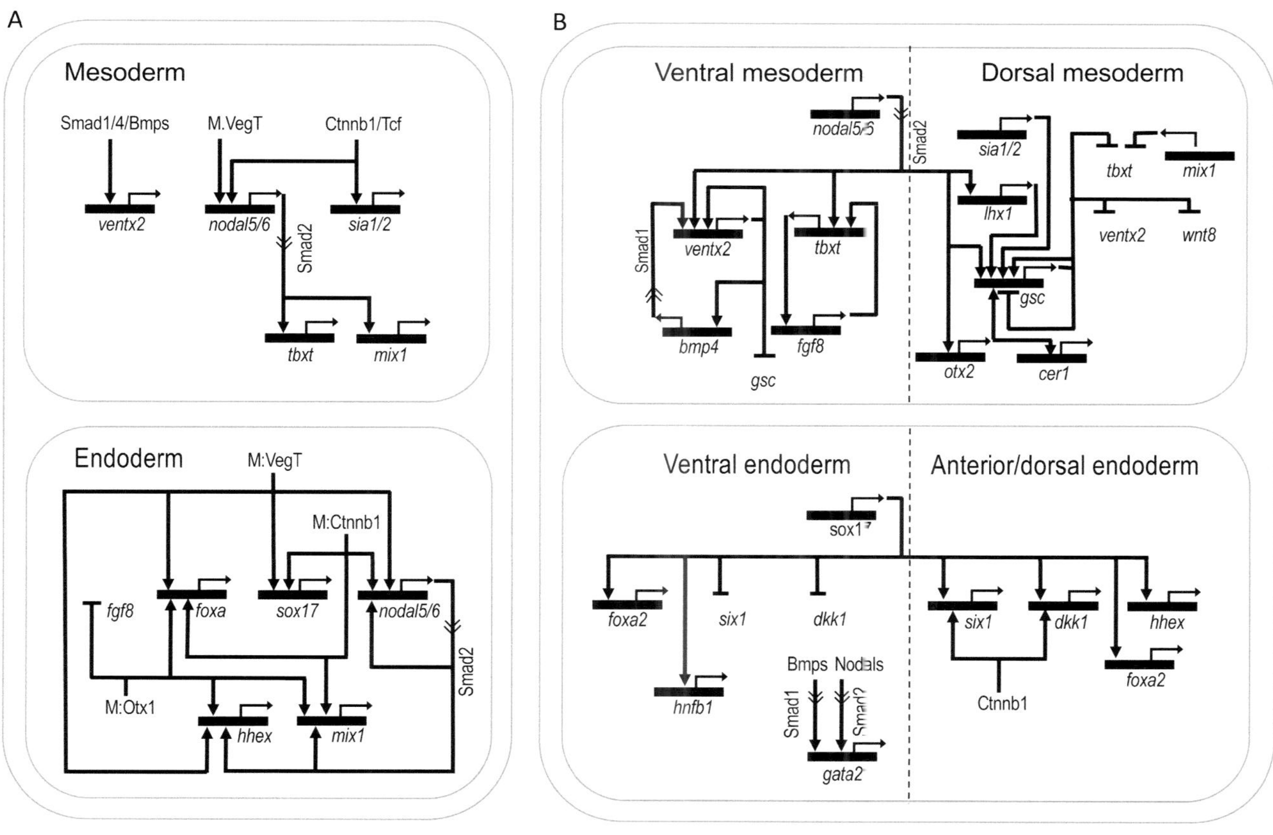

FIGURE 12.1 Gene regulatory networks for initial mesoderm and endoderm specification in *Xenopus*. (A) Early blastula mesoderm and endoderm. Maternal TFs are critical for setting up initial mesoderm and endoderm GRNs. Solid lines indicate linkages confirmed at the cis-regulatory level. (B) Early gastrula mesoderm and endoderm. Note that mesoderm is subdivided into two dorsal and ventral mesoderm regions, whereas endoderm is subdivided into anterior/dorsal and ventral endoderm.

importance of the maternal TF Otx1 in endoderm formation (Paraiso et al., 2019). Like *vegt*, *otx1* mRNA is localized to the vegetal region, and Otx1 works with other maternal TFs (e.g., Vegt and Foxh1) to promote the transcription of endodermally expressed genes. Additionally, Otx1 is a dual-function TF, as it also has a repressive role in endoderm by suppressing *fgf20* and *fgf8* expression. This is important because Fgf signaling promotes mesoderm development.

12.2.3. GRNs during Gastrulation

While germ layer specification occurs along the animal-vegetal axis, cortical rotation of the cytoplasm shortly after fertilization transports Dishevelled, a component of Wnt signaling, to the dorsal side. This results in activation of a Wnt signaling cascade dorsally as revealed by enrichment of Ctnnb1 (β-Catenin) to the nucleus, which causes transcriptional activation of homeobox genes *sia1/2* at the blastula stage in the mesendodermal region (Laurent et al., 1997; Brannon et al., 1997). More specifically, Ctnnb1 forms a complex with Tcf/Lef members of the HMG box TF family on the regulatory regions (CRMs) of the *sia1/2* genes and the *nodal5/6* genes prior to zygotic genome activation (Figure 12.1A) (Laurent et al., 1997; Nishita et al., 2000; Rex et al., 2002). This activation of *sia1/2* and *nodal5/6* provides an essential spatial activating input that is needed to induce the expression of the Spemann organizer gene *goosecoid* (*gsc*) (Watabe et al., 1995; Laurent et al., 1997), which marks this signaling center required for the formation of primary body axis (Spemann and Mangold, 1924). In terms of gene regulation, CRMs of *gsc* integrate multiple different inputs and synergistically respond to these inputs (Figure 12.1B). Sia1/2 directly bind to a CRM called the proximal element located upstream of the *gsc* promoter at pregastrula stages and activate its transcription (Laurent et al., 1997). In addition to the Sia1/2 input, a TF complex of Smad2/4-Foxh1 serves as an input for signaling pathway activators, including Gdf1/3 and a number of Nodal ligands, via another CRM called the distal element. The combination of these different inputs synergistically activates *gsc* transcription during late blastula stages.

At the gastrula stage, shortly after *gsc* induction, additional transcription of genes encoding TFs including Mix1, Otx2, and Lhx1 occurs in the organizer region (Mochizuki et al., 2000). All of these zygotically expressed factors later bind to the same *gsc* CRMs and maintain the expression of *gsc* through a feedforward loop until the neurula stage (Charney et al., 2017a; Paraiso et al., 2020). At about the same time when *gsc* expression begins in the dorsal marginal zone of late blastula-stage embryos, *tbxt (brachyury)* expression encompasses the entire marginal zone (Smith et al., 1991). Thus, *gsc* and *tbxt* are briefly coexpressed within a population of the Spemann's organizer cells. However, the *gsc* and the *tbxt* expression domains quickly segregate to generate two non-overlapping expression domains, due to Gsc's ability to bind to a CRM of *tbxt* to suppress its expression in the organizer (Artinger et al., 1997; Figure 12.1B). This results in generating spatially distinct gene expression domains. Gsc-expressing cells will become the prechordal plate mesoderm, and Tbxt-expressing cells will become the chordamesoderm and the remaining ventral mesoderm.

Regarding *tbxt* regulation, its initial expression is induced by intermediate levels of Nodal (Green and Smith, 1990), but its sustained expression is regulated by a positive feedback regulation via a CRM of *tbxt* that responds to Foxh1/Smad2/4 input and later by Fgf signaling, which helps sustain *tbxt* expression after gastrulation (Latinkić et al., 1997). Gsc additionally represses the expression of *wnt8a* and *ventx2* genes in the organizer to promote head formation by inhibiting the posteriorizing function of these factors (Yao and Kessler, 2001; Latinkić et al., 1997; Yasuoka et al., 2014). These repressive actions of Gsc illustrate how broadly overlapping expression of TFs can segregate into two spatially distinct domains (spatial exclusion) with clear expression boundaries to ensure specification of new cell states.

In *Xenopus* endoderm, the maternal TF Ctnnb1 (β-catenin) is required for high expression levels of several nodal genes at the onset of zygotic transcription. Both inputs of Ctnnb1 and Foxh1/Smad2 promote the expression of endodermal Sox17, Foxa, Gata, and Mix TF families, (Afouda et al., 2020; Figure 12.1). A recent genomic study shows that Sox17 and Ctnnb1 synergistically activate a subset of endodermal enhancers, and this activation is context dependent, sometimes requiring Tcf/Lef, and sometimes functioning independently of Tcf/Lef (Mukherjee et al., 2020). In addition to promoting expression of mesendoderm genes like *gsc* and *hhex*, Sox17 also represses ectodermal and mesodermal gene transcription in endoderm and promotes an endodermal GRN program by activating endodermal target gene expression.

12.2.4. GRN Subcircuits in *Xenopus* Embryos

Developmental genes, which control expression of large gene batteries, are rapidly induced or repressed to control the expression levels and duration of specific gene expression in both space and time. Much of the complexity of the transcriptional regulation involves the interplay of TFs on CRMs that reside close to target genes. It is through these mechanisms that feedback and feedforward loops are created. Here we examine common topological network subcircuits that are found during early *Xenopus* mesendoderm development.

12.2.4.1. Autoregulation Subcircuits

The intrinsic ability of critical genes to self-regulate their expression in development is important in maintaining cell fate and differentiation. Autoregulatory mechanisms come in a variety of forms. They can be positive or negative, and they may be direct or indirect. Positive autoregulation occurs when a TF enhances its own rate of production directly or indirectly. A common consequence of this type of regulation is to sustain the expression of target genes for a prolonged period to provide cell lineage memory. Alternatively, the regulation may allow a surge of gene transcription due to the production of more TF molecules, which cause amplification of the TF-regulated output response. Positive autoregulation can also lead to a stochastic difference in expression to specify cell fates due to amplification of noise or cell-cell variations (Alon, 2007; Peter and Davidson, 2015). The simplest positive feedback system requires only a single gene (node) (Figure 12.2A), in which the product binds to a CRM on its own gene and positively auto-stimulates its transcription. Examples include the positive autoregulation of Nodal signaling in the endoderm and dorsal mesoderm (Chiu et al., 2014) and the positive autoregulation of *ventx2* on the ventral side of the embryo (Henningfeld et al., 2002). Multi-gene positive feedback loops (Figure 12.2B) were found in the Spemann organizer region, where Otx2 induces the expression of *gsc*, which promotes *otx2* expression to maintain the activity of the prechordal pate mesoderm during gastrulation (Yasuoka et al., 2014). In ventral mesoderm, where sustained expression of *fgf2* and *bmp4* is needed, Ventx2 stimulates the transcription of *bmp4*, which in turn produces more Bmp4 ligand that enhances Bmp signal to further stimulate production of *ventx2* (Schuler-Metz et al., 2000; von Bubnoff et al., 2005). Similarly, Tbxt directly binds to CRMs of the *fgf8/20* genes and stimulates production of these Fgf ligands, which in turn actives more *tbxt* expression (Schulte-Merker and Smith, 1995; Latinkić et al., 1997). These examples show how the expression of both *tbxt* and *ventx2* are sustained from blastula onward during mesoderm patterning using two different feedback loops, that is, self-activation and multi-gene feedback loop regulation. It also reveals how different network subcircuits are tightly woven into the network system and multiple inputs regulate the expression of individual genes.

Negative autoregulation subcircuits may occur when a TF down-regulates the transcription of its own gene. Upon a rapid initial rise in concentration, the TF's level reaches a threshold for regulation of its own promoter and represses the transcriptional rate of its own gene. Thus, the concentration of the TF protein encoded by the gene locks into a steady-state level that is close to its repression threshold (Alon, 2007; Peter and Davidson, 2015). Regulation of the *gsc* gene is subject to a negative autoregulation (Yasuoka et al., 2014). When the network circuits of early mesendoderm GRN were examined, negative autoregulation was found to be notably rarer compared to positive autoregulation (Charney et al. 2017a).

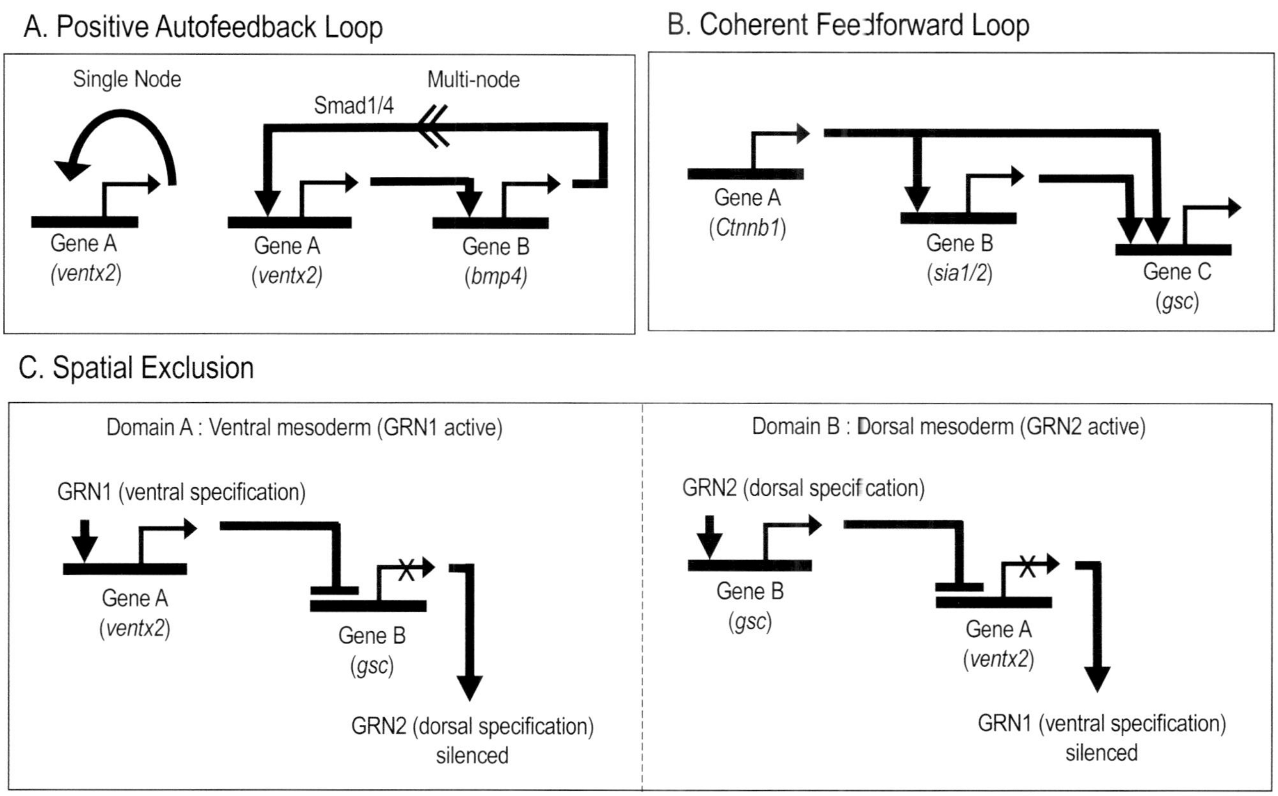

FIGURE 12.2 Subcircuit architectures in developmental gene regulatory networks. (A) Positive feedback subcircuits. (B) Coherent feedforward loop. (C) Spatial exclusion. Domain A will become ventral mesoderm when GRN1 is active, which controls the expression of *ventx2*. Ventx2 represses the expression of *gsc*, which regulates GRN2. Domain B will become dorsal mesoderm when Gsc is active, which represses the expression of *ventx2*.

12.2.4.2. Feedforward Loop Subcircuits

To understand the function of feedforward loops (FFLs), it is necessary to understand how the actions of gene A and B are integrated to regulate the activity of gene C. Two common input functions are an AND gate, in which the presence of both products A and B is required, and an OR gate, in which binding of either A or B is sufficient to activate gene C (Peter and Davidson, 2015). An AND gate can be beneficial in the tight control of factor C expression, as factor C is only activated when both factor A and B are expressed. On the other hand, an OR gate enables the sustained expression of factor C despite the loss of the initial factor A. Much of the essential behavior of FFLs appears to use either an AND or an OR gate system.

We investigated the types of FFLs that are utilized by the *Xenopus* mesendodermal GRN. Among 89 three-gene FFLs, 63 were type I coherent FFLs (Charney et al., 2017a). The coherent FFL network structure assumes product A activates product B, and the activation of product C requires positive inputs from both products A and B. In general, the coherent FFL allows a cell to respond rapidly to stimuli in both positive (ON) or negative (OFF) directions and/or regulate the temporal onset of gene expression (Mangan et al., 2003). In the majority of cases examined during early mesendoderm development (Charney et al., 2017a), the initial activators are frequently maternal TFs such as Ctnnb1, Foxh1, Smad2/3, or Vegt (gene A). These maternal factors positively stimulate the expression of early- and mid-blastula zygotic genes such as *wnt8a*, *sia1*, *sia2*, *mix1*, *gsc*, and the *nodal* genes (gene B), which, in turn, activate the expression a larger number of later expressed mesendodermal genes (gene C) together with sustained inputs from Ctnnb1, Foxh1, Smad2/3, or Vegt (gene A). Thus, in this case, the direct, primary, activated zygotic targets of maternal TFs can function to maintain the expression of later, secondary, activated genes. Coherent FFLs offers an effective way to robustly regulate the temporal onset of genes so that sequential activation of genes can be attained. The prominence of the coherent FFL during mesendoderm formation suggests that temporal regulation of the cascade may be extremely important during early embryogenesis, where events are changing rapidly. Additionally, this subcircuit can provide lineage memory to the cell: Only when cells first express gene A, they attain a specific cell fate upon expression of gene B.

12.2.4.3. Spatial Exclusion Subcircuits

Repression of alternative cell fates is an important and general feature of development (Figure 12.2C). As cells differentiate, it is common for gene expression domains to initially have broad overlap, but then their expression patterns are gradually segregated into distinct spatial domains, representing different cell specification states. This is accomplished by silencing the expression of alternative GRN subcircuits. This subcircuit requires the expression of a gene encoding a repressor that specifically targets a key molecule in an alternative GRN (Peter and Davidson, 2015). This can also be accomplished by two genes encoding transcription factors that mutually and directly repress one another. Typically, exclusion subcircuits are revealed experimentally when expression of the repressor is disrupted, leading to ectopic activation of the alternative regulatory state. Examples of spatial exclusion subcircuits include the interaction between Gsc-Hhex in specification of prechordal plate mesoderm and endoderm formation (Brickman et al., 2000), and Gsc-Ventx2 (Yasuoka et al., 2014) and Gsc-Tbx2 (Artinger et al., 1997), some of which are mutual exclusion interactions in dorsal and ventral mesoderm formation.

12.3. PRESENT STATUS OF THE FIELD

Despite past advances, all GRNs are still far from complete. This is in part because the past network connections provide only limited previews of the selected interactions that were chosen with *a priori* knowledge—large-scale genomic data were not fully integrated into the network analysis. Even the most intensely studied vertebrate mesendoderm GRN will inevitably miss the involvement of TFs and many important interactions. However, with the accumulation of large "omics" data sets, generation of a GRN based on a combination of computational and genomic methods is feasible. Mechanistic GRN building requires two key pieces of information—native TF binding data to CRMs and the expression output of the gene as a consequence of TF binding to the CRM. By integrating high-throughput omics data detecting these two events, it is feasible to build a genome-scale GRN. Specifically, RNA-seq transcriptome profiling studies can reveal the timing and scale of gene activation, and expression changes in gain- and loss-of-function experiments provide a wealth of potential regulatory connections between TFs and CRMs of these target genes. ChIP-seq identifies physical sites of TF binding, and ATAC-seq and DNase-seq datasets can provide information about the accessibility of CRMs, thus providing evidence for direct physical interactions.

Recently, a new approach was reported capable of building a *Xenopus tropicalis* mesendoderm GRN after integrating over 150 transcriptomic RNA-seq and ChIP/ATAC-seq data sets (Jansen et al., 2022). The integration of large genomic datasets derived from different data types has generally been difficult in the past. The first challenge is to obtain the data themselves, which is labor intensive. The second is a significant difficulty of integrating different data types—the RNA-seq datasets, which report expression levels of transcripts/genes, and the DNA datasets which report genomic regions outside of the transcription units that are bound by TFs or are contained in open chromatin. The third is to find ways to interrogate the integrated data in ways that permit building GRNs. In order to integrate different types of large data sets, a type of machine learning, called a self-organizing map (SOM), was applied, which is a type of unsupervised neural network that also permits the visualization of high-dimensional data (Kohonen, 2001). A SOM assesses complex relationships between many data items and groups them together into clusters, displayed as hexagons in the figure, to build a two-dimensional map (Figure 12.3).

For this work, a *Xenopus* RNA-based SOM was generated after incorporating 95 transcriptomic (RNA-seq) data sets to identify distinct sets of related gene expression profiles that co-vary across different experimental conditions (Figure 12.3). Each of the RNA SOM's hexagons contains a cluster representing multiple genes (transcripts) that show similar mRNA expression behavior across diverse experimental conditions including perturbation of TF expression, time course data, and differential spatial expression. A *Xenopus* DNA SOM was also generated after training 63 ChIP-seq and ATAC-seq data sets to obtain chromatin regulatory information by capturing DNA segments from across the genome that show similar TF binding, epigenetic histone marks, and chromatin regions with accessible DNA. In the DNA SOM, DNA regulatory regions are therefore clustered into different hexagons, sorted according to similarities in TF binding behavior and/or epigenetic signatures. These DNA and RNA SOMs consisting of individual hexagonal clusters are subject to further clustering into "higher-order" groups of hexagonal clusters (metaclusters) that share similar behaviors between individual hexagonals. This continuity-constrained metaclustering makes the statistical analysis more powerful by providing a more consistent clustering (Kiang and Kumar, 2001).

In order to build GRNs, a linked self-organizing map (linked SOM) method (Jansen et al., 2019) was applied that integrates the clustering of multiple SOMs by associating the individual partitioned genomic regions within the metaclusters of the DNA SOM to transcription units contained within the RNA SOM metaclusters. The linked metaclusters between DNA SOM and RNA SOM were subjected to TF binding motif enrichment searches to predict the involvement of candidate TFs, which then were used to generate genome-wide network connections. The resulting TF-CRM connections were then weighed based on statistical (DNA-RNA multicluster enrichment) and other criteria (presence of Ep300 binding indicative of active enhancers) to generate a *Xenopus* mesendodermal GRN. An *in vivo* validation experiment using a limited number of reporter genes shows that a high percentage (>90%) of the linkages identified were functional. This method not only identified newly predicted connections involved mesendoderm regulation, but also identified novel and in some cases unanticipated combinatorial interactions of TFs in mediating gene expression within the GRN.

An alternative approach to build GRNs is to use single cell (sc)RNA-seq data and base the network on cell lineage gene co-expression profiles. This approach assumes that genes that participate in similar biological processes (i.e. cell fate) will share regulatory programs, and consequently, these genes are co-expressed and regulate each other's activity (Ruprecht et al., 2017). Therefore, such a network would be built based on correlations between gene expression behaviors rather than direct mechanistic regulation between TFs and target genes. Because GRNs based on these interaction data are non-mechanistic, as they lack solid evidence for direct physical interactions, it is more difficult to make inferences of causality. However, this approach can provide a general idea of which genes are participating in

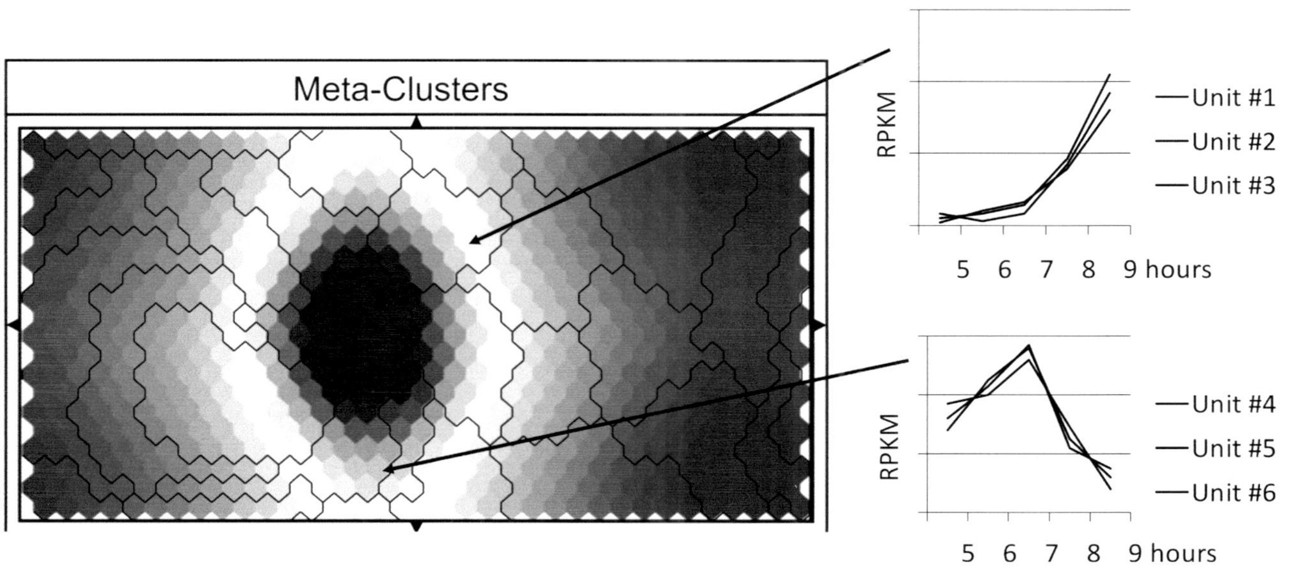

FIGURE 12.3 Metaclusters in a self-organizing map (SOM). An RNA-based SOM was generated after incorporating multiple RNA-seq data sets. Each RNA SOM's unit, displayed as hexagons, contains a cluster representing multiple genes that show similar mRNA expression behavior across diverse experimental conditions. Groups of hexagons are combined into metaclusters (presented in black lines) that share distinct sets of related gene expression profiles that co-vary across different experimental conditions. Right panels illustrate groups of genes that share similar temporal expression profiles in each hexagon. Unit numbers represent distinct transcripts.

a specific biological process and provide a framework with constraints for building GRNs. *Xenopus tropicalis* scRNA-seq data were recently reported describing the emergence of cell states across early embryogenesis (Briggs et al., 2018). While these scRNA-seq data are useful in identifying the lineages and the number of cell states appearing during early *Xenopus* development, the current data are not of high enough resolution to build GRNs due to the low sequencing depth of scRNA-seq and frequent dropouts of transcripts. Despite these challenges, the scRNA-seq approach will become an integral part of GRN building as the integration of these data with other high-throughput data (e.g., bulk RNA-seq, DNA-seq) becomes easier.

12.4. OUR CONTRIBUTION TO THIS FIELD—INSIGHTS

Hans Spemann and Hilde Mangold discovered the organizer in 1924, a tissue above the dorsal blastopore lip in the amphibian early gastrula that induced development of a secondary body axis containing a central nervous system. In 1991, we cloned the first gene expressed specifically in the organizer, which encodes Goosecoid (Gsc), a homeodomain TF, and began investigating the molecular mechanism regulating *gsc* expression using a reporter gene microinjection assay (Watabe et al., 1995; Laurent et al., 1997). The importance of synergistic inputs directly into *gsc* CRMs, mediated by TFs that are regulated by Tgf-β and Wnt growth factors, was demonstrated for Spemann organizer formation. Our first attempt to build a GRN controlling mesendoderm development was published in 2005 (Koide et al., 2005), and in 2017, we updated the network with new information (Charney et al., 2017a). While we had established the most comprehensive vertebrate mesendodermal GRN at the time, it provided only a limited view of the process due to the inherent limitations of the traditional one-gene-at-a-time approach used to address biological questions. To initiate systems-level analyses of GRN function, we adopted computational modeling and deep sequencing to infer a more comprehensive architecture of the mesendoderm GRN (Chiu et al., 2014; Charney et al., 2017b; Paraiso et al., 2019; Afouda et al., 2020; Mukherjee et al., 2020; Jansen et al., 2022).

Over the past several years, we have spent considerable energy to develop genomic resources to better understand GRNs in early development. We have determined the absolute amount of all known transcripts and the kinetics of their accumulation in *X. tropicalis* embryos from 1-cell to tadpole (organogenesis) stages (Owens et al., 2016). We mapped lncRNAs that are expressed during *Xenopus* development (Forouzmand et al., 2017) and created a manually curated catalog of all 1240 TFs encoded by the *X. tropicalis* genome (Blitz et al., 2017). Through such analyses, we have investigated the roles of maternal TFs in mesendoderm gene activation during zygotic genome activation. Upon examining the DNA binding behaviors of several critical maternal TFs (Fox, Sox, T-box, homeodomain, Lef/Tcf families) important for germ layer specification, we and others discovered that these maternal TFs bind to the genome at select sites before any sign of gene transcription (Charney et al., 2017b; Paraiso et al., 2019; Gentsch et al., 2019). Furthermore, many of these TFs (Foxh1, Otx1, Vegt, Tcf, Sox3/7) co-bind and form enhanceosomal complexes on endodermal CRMs and are responsible for the activity of super-enhancers (SEs) (Paraiso et al., 2019; Gentsch et al., 2019). These data support the notion that maternal TFs act as pioneer factors, binding to select CRMs to pre-mark the genome in advance of activation of the subsequent germ layer-specific GRN.

12.5. FUTURE DIRECTIONS AND IMPORTANT QUESTIONS

The insights into regulatory functions of GRNs will uncover new network connections that will generate specific hypotheses on the molecular components that are identified. GRN building based on genomic data will provide an enormous number of new links between TFs and CRMs, and it is necessary to test the validity of the connections experimentally using high-throughput approaches. Development of both single- and multi-locus gene perturbation assays through the use of morpholinos, genome editing techniques such as CRISPR/Cas9 (Blitz et al., 2013; Nakayama et al., 2013; Guo et al., 2014) and RNA cleaving Cas13 derivatives (Kushawah et al., 2020) will be extremely valuable in accelerating the rate of validation analysis to build GRNs based on genomic data.

A comprehensive understanding of the logic of GRNs should one day inform synthetic biology approaches, by which we can create or manipulate regulatory genetic circuits to modify cell functions. For example, by manipulating key nodes in a GRN, we could prevent or reverse a diseased cell state. Changing the identity of terminally differentiated cells or progenitor cells toward other cell types for regenerative medicine may be achieved by redesigning network circuits regulating the specification of cell fates. This realization can be also assisted by the use of CRISPR/Cas systems. Based on the current state of the GRN field, we do not have sufficient information to understand how GRNs control the diverse biological functions in various organisms, nor are we in the position to predict the changes in GRN function underlying human disease. However, provisional *Xenopus* GRNs regulating neural crest, ectodermal placodes, blood differentiation, eye, and podocyte development have also been reported (Seal et al., 2020, Maharana and Schlosser, 2018; Ciau-Uitz and Patient, 2019; Lee et al., 2014; Zuber et al., 2003; White et al., 2010). In the future, our current early mesendodermal GRN network should be expanded, refined, and linked to later GRNs to expand our current knowledge on the architecture of GRNs that control biology.

Genome-wide association studies (GWAS) have been a useful approach to link specific genetic variations such as single-nucleotide polymorphisms (SNPs) with particular diseases. Many GWAS publications have identified disease- and trait-associated SNPs in the human genome, and SNPs mapped within transcriptional regulatory regions were shown to affect the expression of linked genes (Pomerantz et al., 2009; Musunuru et al., 2010). This raises the possibility that SNPs within CRMs are causes of many human genetic diseases and phenotypic traits. Recent genome-wide studies examining DNase I hypersensitive sites harboring GWAS SNPs showed enrichment of SNPs in regulatory regions, supporting the involvement of cis regulatory DNA variations in human diseases (Maurano et al., 2012). This study also revealed that common variants associated with specific diseases (e.g., inflammation, cancer, diabetes) are often enriched in defined sets of TF motifs, indicating that cohorts of TFs form shared GRN architectures and that alterations in TF binding to these sites cause these diseases. Thus, seemingly unconnected SNPs may be associated with related diseases, perhaps by altering TF activity that affects common GRNs. In sum, we anticipate that GRN study will contribute broadly to understanding the etiology of human health and disease.

REFERENCES

Afouda, B.A., Y. Nakamura, S. Shaw, R.M. Charney, K.D. Paraiso, I.L. Blitz, K.W.Y. Cho, and Hoppler, S. 2020. Foxh1/nodal defines context-specific direct maternal Wnt/β-Catenin target gene regulation in early development. *iScience* 23:101314.

Alon, U. 2007. Network motifs: Theory and experimental approaches. *Nat Rev Genet* 8:450–461.

Artinger, M., I. Blitz, K. Inoue, U. Tran, and K.W. Cho. 1997. Interaction of goosecoid and brachyury in Xenopus mesoderm patterning. *Mech Dev* 65:187–196.

Bardet, A.F., J. Steinmann, S. Bafna, J.A. Knoblich, J. Zeitlinger, and A. Stark. 2013. Identification of transcription factor binding sites from ChIP-seq data at high resolution. *Bioinformatics* 29:2705–2713.

Blitz, I.L., J. Biesinger, X. Xie, and K.W.Y. Cho. 2013. Biallelic genome modification in F0 *Xenopus tropicalis* embryos using the CRISPR/Cas system. *Genesis* 51:827–834.

Blitz, I.L., and K.W.Y. Cho. 2021. Control of zygotic genome activation in Xenopus. *Curr Top Dev Biol* 145, in press.

Blitz, I.L., K.D. Paraiso, I. Patrushev, W.T. Chiu, K.W.Y. Cho, and M.J. Gilchrist. 2017. A catalog of *Xenopus tropicalis* transcription factors and their regional expression in the early gastrula stage embryo. *Dev Biol* 426:409–417.

Brannon, M., M. Gomperts, L. Sumoy, R.T, Moon, and D. Kimelman. 1997. A beta-catenin/XTcf-3 complex binds to the siamois promoter to regulate dorsal axis specification in Xenopus. *Genes Dev* 11:2359–2370.

Brickman, J.M., C.M. Jones, M. Clements, J.C. Smith., and R.S.P. Beddington. 2000. Hex is a transcriptional repressor that contributes to anterior identity and suppresses Spemann organiser function. *Development* 127:2303–2315.

Briggs, J.A., C. Weinreb, D.E. Wagner, S. Megason, L. Peshkin, M.W. Kirschner, and A.M. Klein. 2018. The dynamics of gene expression in vertebrate embryogenesis at single-cell resolution. *Science* 360:6392.

Britten, R.J., and E.H. Davidson. 1969. Gene regulation for higher cells: A theory. *Science* 165:349–357.

Cha, S.W., M. McAdams, J. Kormish, C.C. Wylie, and M. Kofron. 2012. Foxi2 is an animally localized maternal mRNA in Xenopus, and an activator of the zygotic ectoderm activator Foxi1e. *PLoS One* 7:e41782.

Charney, R.M., E. Forouzmand, J.S. Cho, J. Cheung, K.D. Paraiso, Y. Yasuoka, S. Takahashi, M. Taira, I.L. Blitz, X. Xie, and K.W.Y. Cho. 2017b. Foxh1 occupies cis-regulatory modules prior to dynamic transcription factor interactions controlling the mesendoderm gene program. *Dev Cell* 40:595–607.

Charney, R.M., K.D. Paraiso, I.L. Blitz, and K.W.Y. Cho. 2017a. A gene regulatory program controlling early Xenopus mesendoderm formation: Network conservation and motifs. *Semin Cell Dev Biol* 66:12–24.

Chiu, W.T., R. Charney Le, I.L. Blitz, M.B. Fish, Y. Li, J. Biesinger, X. Xie, and K.W.Y. Cho. 2014. Genome-wide view of TGF/Foxh1 regulation of the early mesendoderm program. *Development* 141:4537–4547.

Ciau-Uitz, A., and R. Patient. 2019. Gene regulatory networks governing the generation and regeneration of blood. *Journal of Computational Biology* 26:719–725.

Davidson, E.H. 2006 *The Regulatory Genome: Gene Regulatory Networks in Development and Evolution*. Academic Press, San Diego, CA.

Faure, E., I.S. Peter, and E.H. Davidson. 2012. A new software package for predictive gene regulatory network modeling and redesign. *J Comput Biol* 20:419–423.

Forouzmand, E., N.D.L. Owens, I.L. Blitz, K.D. Paraiso, M.K. Khokha, M.J. Gilchrist, X. Xie, and K.W.Y. Cho. 2017. Developmentally regulated long non-coding RNAs in *Xenopus tropicalis*. *Dev Biol* 426:401–408.

Gentsch, G.E., T. Spruce, N.D.L. Owens, and J.C. Smith. 2019. Maternal pluripotency factors initiate extensive chromatin remodelling to predefine first response to inductive signals. *Nat Commun* 10:4269.

Gray, S., P. Szymanski, and M. Levine. 1994. Short-range repression permits multiple enhancers to function autonomously within a complex promoter. *Genes Dev* 8:1829–1838.

Green, J., and J.C. Smith. 1990. Graded changes in dose of a Xenopus activin a homologue elicit stepwise transitions in embryonic cell fate. *Nature* 347:391–394.

Guo, X., T. Zhang, Z. Hu, Y. Zhang, Z. Shi, Q. Wang, Y. Cui, F. Wang, H. Zhao, Y. Chen. 2014. Efficient RNA/Cas9-mediated genome editing in Xenopus tropicalis. *Development* 141:707–714.

Harland, R. 2000. Neural induction. *Curr Opin Genet Dev* 10: 357–362.

Hawley, S.H., K Wünnenberg-Stapleton, C. Hashimoto, M.N. Laurent, T. Watabe, B.W. Blumberg, and K.W. Cho. 1995. Disruption of BMP signals in embryonic Xenopus ectoderm leads to direct neural induction. *Genes Dev* 9:2923–2935.

Henningfeld, K.A., H. Friedle, S. Rastegar, and W. Knöchel. 2002. Autoregulation of Xvent-2B; direct interaction and functional cooperation of Xvent-2 and Smad1. *J Biol Chem* 277:2097–2103.

Jansen, C., R.N. Ramirez, N.C. El-Ali, D. Gomez-Cabrero, J. Tegner, M. Merkenschlager, A. Conesa, and A. Mortazavi. 2019. Building gene regulatory networks from scATAC-seq and scRNA-seq using linked self organizing maps. *PLoS Comput Biol* 15:e1006555.

Jansen, C., K.D. Paraiso, J.J. Zhou, I.L., Blitz, M.B. Fish, R.M. Charney, J.S. Cho, Y. Yasuoka, N. Sudou, S.R. Bright, M. Wlizla, G.J.C Veenstra, M. Taira, A.M. Zorn, A. Mortazavi, and K.W.Y. Cho. 2022. Uncovering the mesendoderm gene regulatory network through multi-omic data integration. Cell Rep. 38:110364. *BioRxiv*. https://doi: 10.1016/j.celrep.2022.110364.

Kiang, M.Y., and A. Kumar. 2001. An evaluation of self-organizing map networks as a robust alternative to factor analysis in data mining applications. *Information Systems Research* 12:177–194.

Kimelman D., and K.J. Griffin. 1998. Mesoderm induction: A post-modern view. *Cell* 94:419–421.

Kofron, M., T. Demel, J. Xanthos, J. Lohr, B. Sun, H. Sive, S. Osada, C. Wright, C. Wylie, and J. Heasman. 1999. Mesoderm induction in Xenopus is a zygotic event regulated by maternal VegT via TGFbeta growth factors. *Development* 126:5759–5770.

Kofron, M., J. Xanthos, and J. Heasman. 2004. Maternal VegT and ß-catenin: Patterning the Xenopus blastula. In Grunz, Horst (Ed.), *The Vertebrate Organizer*. Springer, Berlin, New York.

Kohonen, T. 2001. *Self-Organizing Maps*. Springer, Berlin, Heidelberg.

Koide, T., T. Hayata, and K.W. Cho. 2005. Xenopus as a model system to study transcriptional regulatory networks. *Proc Natl Acad Sci USA* 102:4943–4948.

Kushawah G., L. Hernandez-Huertas, D.P.J. Abugattas-Nuñez, J.R. Martinez-Morales, M.L. DeVore, H. Hassan, I. Moreno-Sanchez et al. 2020. CRISPR-Cas13d induces efficient mRNA knockdown in animal embryos. *Dev Cell* 54:805–817.

Latinkić, B.V., M. Umbhauer, K.A. Neal, W. Lerchner, J.C. Smith, and V. Cunliffe. 1997. The Xenopus brachyury promoter is activated by FGF and low concentrations of activin and suppressed by high concentrations of activin and by paired-type homeodomain proteins. *Genes Dev* 11:3265–3276.

Laurent, M.N., I.L. Blitz, C. Hashimoto, U. Rothbächer, and K.W.Y. Cho. 1997. The Xenopus homeobox gene twin mediates Wnt induction of goosecoid in establishment of Spemann's organizer. *Development* 124:4905–4916.

Lee, H.K., H.S. Lee, and S.A. Moody. 2014. Neural transcription factors: From embryos to neural stem cells. *Mol Cells* 37:705–712.

Loose, M., and R.A. Patient. 2004. A genetic regulatory network for Xenopus mesendoderm formation. *Dev Biol* 271:467–478.

Maharana, S.K., and G. Schlosser. 2018. A gene regulatory network underlying the formation of pre-placodal ectoderm in *Xenopus laevis*. *BMC Biol* 16:79.

Mangan, S., A. Zaslaver, and U. Alon. 2003. The coherent feedforward loop serves as a sign-sensitive delay element in transcription networks. *J Mol Biol* 334:197–204.

Maurano, M.T., R. Humbert, E. Rynes, R.E. Thurman, E. Haugen, H. Wang, A.P. Reynolds, R. Sandstrom, H. Qu, J. Brody, A. Shafer, F. Neri, K. Lee, T. Kutyavin, S. Stehling-Sun, A.K. Johnson, T.K. Canfield, E. Giste, M. Diegel, D. Bates, R.S. Hansen, S. Neph, P.J. Sabo, S. Heimfeld, A. Raubitschek, S. Ziegler, C. Cotsapas, N. Sotoodehnia, I. Glass, S.R. Sunyaev, R. Kaul, J.A. Stamatoyannopoulos. 2012. Systematic localization of common disease-associated variation in regulatory DNA. *Science* 337:1190–1195.

Mir, A., M. Kofron, A.M. Zorn, M. Bajzer, M Haque, J. Heasman, and C.C. Wylie. 2007. Foxi1e activates ectoderm formation and controls cell position in the Xenopus blastula. *Development* 134:779–788.

Mochizuki, T., A.A. Karavanov, P.E. Curtiss, K.T. Ault, N. Sugimoto, T. Watabe, K. Shiokawa, M. Jamrich, K.W. Cho, I.B. Dawid, and M. Taira. 2000. Xlim-1 and LIM domain binding protein 1 cooperate with various transcription factors in the regulation of the goosecoid promoter. *Dev Biol* 224:470–485.

Mukherjee, S., P. Chaturvedi, S.A. Rankin, M.B. Fish, M. Wlizla, K.D. Paraiso, M. MacDonald, X. Chen, M.T. Weirauch, I.L. Blitz, K.W. Cho, and A.M. Zorn. 2020. Sox17 and β-catenin co-occupy Wnt-responsive enhancers to govern the endoderm gene regulatory network. *Elife* 9:e58029.

Musunuru, K., et al. 2010. From noncoding variant to phenotype via SORT1 at the 1p13 cholesterol locus. *Nature* 466:714–719.

Nakayama, T., M.B. Fish, M. Fisher, J. Oomen-Hajagos, G.H. Thomsen, R.M. Grainger. 2013. Simple and efficient CRISPR/Cas9-mediated targeted mutagenesis in Xenopus tropicalis. *Genesis* 51:835–843.

Nam, J., and E.H. Davidson. 2012. Barcoded DNA-tag reporters for multiplex cis-regulatory analysis. *PLoS One* 7:e35934.

Nishita, M., M.K. Hashimoto, S. Ogata, M.N. Laurent, N. Ueno, H. Shibuya, and K.W.Y. Cho. 2000. Interaction between Wnt and TGF-beta signalling pathways during formation of Spemann's organizer. *Nature* 403:781–785.

Owens, N.D.L., I.L. Blitz, M.A. Lane, I. Patrushev, J.D. Overton, M.J. Gilchrist, K.W.Y. Cho, and M.K. Khokha. 2016. Measuring absolute RNA copy numbers at high temporal resolution reveals transcriptome kinetics in development. *Cell Reports* 14:632–647.

Paraiso, K.D., I.L. Blitz, M. Coley, J. Cheung, N. Sudou, M. Taira, and K.W.Y. Cho. 2019. Endodermal maternal transcription factors establish super-enhancers during zygotic genome activation. *Cell Rep* 27:2962–2977.

Paraiso, K.D., J.S. Cho, J. Yong, and K.W.Y. Cho. 2020. Early Xenopus gene regulatory programs, chromatin states, and the role of maternal transcription factors. *Curr Top Dev Biol* 139:35–60.

Peter, I.S., and E.H. Davidson. 2010. The endoderm gene regulatory network in sea urchin embryos up to mid-blastula stage. *Dev Biol* 340:188–199.

Peter, I.S., and E.H. Davidson. 2015. *Genomic Control Process: Development and Evolution*. Academic Press, Amsterdam.

Pomerantz, M.M., N. Ahmadiyeh, L. Jia, P. Herman, M.P. Verzi, H. Doddapaneni, C.A. Beckwith, J.A. Chan, A. Hills, M. Davis, K. Yao, S.M. Kehoe, H.J. Lenz, C.A. Haiman, C. Yan, B.E. Henderson, B. Frenkel, J. Barretina, A. Bass, J. Tabernero, J. Baselga, M.M. Regan, J.R. Manak, R. Shivdasani, G.A. Coetzee, M.L. Freedman, 2009. The 8q24 cancer risk variant rs6983267 shows long-range interaction with MYC in colorectal cancer. *Nat Genet* 41:882–884.

Rex, M., E. Hilton, and R. Old. 2002. Multiple interactions between maternally-activated signalling pathways control Xenopus nodal-related genes. *Int J Dev Biol* 46:217–226.

Ruprecht C, S. Proost, M. Hernandez-Coronado, C. Ortiz-Ramirez, D. Lang, S.A. Rensing, J.D. Becker, K. Vandepoele, and M. Mutwil. 2017. Phylogenomic analysis of gene co-expression networks reveals the evolution of functional modules. *Plant J* 90:447–465.

Schroeder, M.D., M. Pearce, J. Fak, H. Fan, U. Unnerstall, E. Emberly, N. Rajewsky, E.D. Siggia, and U. Gaul. 2004. Transcriptional control in the segmentation gene network of Drosophila. *PLoS Biology* 2:e271.

Schuler-Metz, A., S. Knöchel, E. Kaufmann, and W. Knöchel. 2000. The homeodomain transcription factor Xvent-2 mediates autocatalytic regulation of BMP-4 expression in Xenopus embryos. *J. Biol Chem* 275:34365–34374.

Schulte-Merker, S., and J.C. Smith. 1995. Mesoderm formation in response to brachyury requires FGF signalling. *Curr Biol* 5:62–67.

Seal, S., and A.H. Monsoro-Burq. 2020. Insights into the early gene regulatory network controlling neural crest and placode fate choices at the neural border. *Front Physiol* 11:608812.

Smith, J.C., B.M. Price, J.B. Green, D. Weigel, and B.G. Herrmann. 1991. Expression of a Xenopus homolog of brachyury (T) is an immediate-early response to mesoderm induction. *Cell* 67:79–87.

Spemann, H. and H. Mangold. 1924. Uber Induktion von Embryonalanlagen durch Implantation artfremder Organisatoren. *Wilhelm's Roux's Arch. Dev. Biol* 100:599–638.

Suri, C., T. Haremaki, and D.C. Weinstein. 2005. Xema, a foxi-class gene expressed in the gastrula stage Xenopus ectoderm, is required for the suppression of mesendoderm. *Development* 132:2733–2742.

von Bubnoff, A., D.A. Peiffer, I.L. Blitz, T. Hayata, S. Ogata, Q. Zeng, M. Trunnell, and K.W. Cho. 2005. Phylogenetic footprinting and genome scanning identify vertebrate BMP response elements and new target genes. *Dev Biol* 281: 210–226.

Watabe, T., S. Kim, A. Candia, U. Rothbächer, C. Hashimoto, K. Inoue, and K.W.Y. Cho. 1995. Molecular mechanisms of Spemann's organizer formation: Conserved growth factor synergy between Xenopus and mouse. *Genes Dev* 9:3038–3050.

White, J.T., B. Zhang, M.D. Cerqueira, U. Tran, and O. Wessely. 2010. Notch signaling, wt1 and foxc2 are key regulators of the podocyte gene regulatory network in Xenopus. *Development* 137:1863–1873.

Xanthos, J.B., M. Kofron, C. Wylie, and J. Heasman. 2001. Maternal VegT is the initiator of a molecular network specifying endoderm in *Xenopus laevis*. *Development* 128:167–180.

Yao, J., and D.S. Kessler. 2001. Goosecoid promotes head organizer activity by direct repression of Xwnt8 in Spemann's organizer. *Development* 128:2975–2987.

Yasuoka, Y., Y. Suzuki, S. Takahashi, H. Someya, N. Sudou, Y. Haramoto, K.W. Cho, M. Asashima, S. Sugano, and M. Taira. 2014. Occupancy of tissue-specific cis-regulatory modules by Otx2 and TLE/Groucho for embryonic head specification. *Nat Commun* 5:4322.

Zhang, J., D.W. Houston, M.L. King, C. Payne, C. Wylie, and J. Heasman. 1998. The role of maternal VegT in establishing the primary germ layers in Xenopus embryos. *Cell* 94: 515–524.

Zuber, M.E., G. Gestri, A.S. Viczian, G. Barsacchi, and W.A. Harris. 2003. Specification of the vertebrate eye by a network of eye field transcription factors. *Development* 130:5155–5167.

13 The Development of High-Resolution Proteomic Analyses in *Xenopus*

Elizabeth Van Itallie and Leonid Peshkin

CONTENTS

13.1. Introduction 197
13.2. The Database 198
13.3. Nomenclature and Gene Symbols 209
13.4. Proteomics and Cell Biology 200
13.5. *X. laevis* versus *X. tropicalis* 200
13.6. Developmental Atlas of Protein Expression 201
13.7. Post-Translational Modifications 201
13.8. Single-Cell Proteomics 202
13.9. Discussion and Future Directions 203
Acknowledgments 204
Reference 204

13.1. INTRODUCTION

In many instances, proteins are the final product of gene expression and are responsible for the vast majority of biological processes. *Xenopus* oocytes and embryos are excellent model systems for research questions in the fields of biochemistry, cell biology, and developmental biology. In all of these areas, the specific functions and biochemical properties of hundreds of proteins have been characterized in detail. However, our knowledge of complex cellular processes such as the mitotic spindle, nuclear localization, and gastrulation, among so many others, are now at the stage where to make progress, we need to understand endogenous and perturbed phenotypes and phenomena at the level of thousands of proteins simultaneously to parse the intricacies of a biological system. The amino acid sequence and cellular expression of a specific protein do not solely determine its function: a protein's function is often regulated by differential post-translational modification, by proteolytic cleavage, and/or by chemical modification (e.g. phosphorylation, acetylation, or glycosylation) and its subcellular and temporal localization within the cell. Despite the widely acknowledged importance of these contextual properties of proteins for cell function and embryogenesis, the methodologies to study thousands of proteins and their modifications simultaneously currently suffer from experimental accessibility and depth of coverage compared to transcriptomics approaches. As a result, the biological knowledge resulting from the application of proteomics to *Xenopus* biology is limited, and the systemic vision of how omics-scale measurement of protein can expand biological knowledge is largely missing. In addition to highlighting the advances that have been made in this field, this chapter also seeks to (1) educate the reader about proteomics, (2) emphasize the strengths and weaknesses of the *Xenopus* system with regard to proteomics, and (3) provide ideas for the future. Throughout this chapter, "proteomics" is taken to mean mass spectrometry (MS)-based proteomics unless stated otherwise.

In this chapter we will focus on bottom-up tandem mass spectrometry-based proteomics in *Xenopus* research. Briefly, proteins are digested into peptides, and these peptides are ionized and sprayed into a mass spectrometer. The peptides have specific properties based on their mass and ionic charge. Peptides are measured first in their intact state (MS1) and then again after they are fragmented (MS/MS). The amino acid sequences of the peptide are then determined by comparing the mass spectrometer measurements and a reference sequence database of theoretical peptides. These peptide measurements are then summarized into protein measurements. These steps only result in identification. How are differences in protein concentrations/modifications under experimental and control conditions quantified? Analogous to "barcodes" in sequencing, digested peptides from different samples can be covalently labeled with various chemical groups with identical masses that vary in terms of distribution of heavy isotopes around their structure, that is, isobaric reporters. All of these reporters have the same mass before the peptides are fragmented, but after fragmentation, the masses are different. In this way, it can be determined what fraction of the signal is attributable to different conditions for a given peptide. The most common isobaric labeling reagents are iTRAQ and TMT MS-3. More recently, promising TMT-C+ technology has been developed (Sonnett et al., 2018a).

DOI: 10.1201/9781003050230-15

What is so challenging about this technology? The first issue is that large amounts of protein material are required. This is because there is currently no way to amplify proteins (analogous to reverse transcription PCR followed by deep sequencing for RNA quantification, i.e., RNAseq) and because the sample preparation, and especially ionization, results in material loss. In this case, the *Xenopus* model system, and in particular *X. laevis*, is an ideal model since a single egg or embryo contains ~25 μg of non-yolk protein (Gurdon et al., 1986). A single egg is sufficient for shallow protein measurement, and hundreds more sibling embryos can easily be collected from a single clutch, which is more than sufficient for deeper measurement. Starting material amount is even more of a concern when measuring posttranslational modifications. This is because an enrichment step has to be done to study posttranslational modifications since even the most abundant modification, phosphorylation, is present only in about 1% of peptides (Peuchen et al., 2016; Peuchen et al., 2017; Presler et al., 2017). The second challenge is depth of measurement. Unlike RNAseq, tandem mass spectrometry is not a parallelized measurement; the measurement of each peptide comes at the expense of an unmeasured one. Usually, the most abundant peptides are measured. Fortunately, egg/embryo lysis conditions that effectively remove abundant yolk proteins have been developed (Gupta et al., 2018; Wühr et al., 2014), which has made mass spectrometry-based proteomics possible in *Xenopus*. Measuring more deeply requires more instrument time, which means higher cost per experiment. A third challenge is the bioinformatics. In nucleotide sequencing, *a priori* knowledge of the genome or transcriptome is not needed to obtain nucleotide reads. In contrast, protein mass spectrometry measures the mass and charge of peptides and peptide fragments. The measurements that result from intact and fragmented peptides are matched to a set of theoretical digested peptides from a database of protein references. This means that having an accurate reference database is critical to identify and then measure a specific peptide. To make matters worse, mere completeness of the reference database is not sufficient; inclusion of artifact sequences (that could never be observed *in vivo*) actually hampers identification of other peptides.

There are other non-technical reasons successful application of mass spectrometry to perform quantitative proteomics in *Xenopus* has materialized only recently. Proteomics by mass spectrometry is lagging behind transcriptomics, which has been made routine mostly by the availability of Illumina sequencing machines and standardized protocols and standard operating procedures throughout industrial and academic sequencing facilities. The best MS instruments produced by Thermo-Fisher are expensive and are impossible for a non-specialist to use. When modern MS instruments are available via a "fee-for-service" model at facility cores, both the pre- and post-instrument processing pipelines are tailored to human and mouse samples. Perhaps most importantly, the bioinformatics support for proteomics in non-model organisms is lacking. Given this situation, it is not surprising that successful applications of mass spectrometry proteomics in *Xenopus* so far have resulted from close collaborations between *Xenopus* experts and MS specialists. A few notable examples of such collaborations are the labs of Moody with Nemes (Lombard-Banek et al., 2016; Onjiko et al., 2015), Huber with Dovichi (Peuchen et al., 2017; Sun et al., 2016), Kirschner with Gygi (Wühr et al., 2014; Peshkin et al., 2015), Veenstra with Vermeulen (Lindeboom et al., 2019; Smits et al., 2014), and Klein with Garcia (Saha-Shah et al., 2019). Very few laboratories have the expertise in both advanced mass spectrometry and *Xenopus* methods, with the one notable exception being Martin Wühr's lab at Princeton (Gupta et al., 2018; Sonnett et al., 2018a, Sonnett et al., 2018b).

One key milestone on the way towards *Xenopus* mass spectrometry proteomics was sequencing the genomes of both *Xenopus* species actively used in research—*X. laevis* and *X. tropicalis* (Hellsten et al., 2010; Session et al., 2016)—which naturally benefits systems-level analysis at the genome, RNA, and protein levels. Being able to attribute spectra to peptides relies on the availability of a complete and accurate list of protein sequences, for which a complete genome with a high-quality set of gene models is the key. At the point of being released, the genome quality is judged by various metrics of the DNA sequence itself, not so much the quality of gene models. Notably, proteomics studies feed back to the gene model quality by providing an accurate catalogue of detected peptides and respective proteins that can be used to validate and adjust the intron-exon-UTR annotation of gene structures. Both the *X. laevis* and *X. tropicalis* genomes have undergone many iterations of genome assembly and even more of refining the gene models. *X. laevis* is an allotetraploid species, whereas *X. tropicalis* is a true diploid. Unsurprisingly, given their respective complexity, the annotation of the *X. tropicalis* genome (~24K gene models) is currently in a much better state than that of *X. laevis* (~44K gene models), which presents a challenge when comparing the number of distinct proteins characterized across different studies.

13.2. THE DATABASE

For the methods described in this perspective, one can only detect a protein of a pre-defined sequence, and thus the spectra must be matched against a set of potential peptide sequences. Thus, having a sound and complete reference set of sequences (sometimes referred to as "the database" in the field) is particularly important. The reason for soundness deserves a special explanation, since one might expect that mixing in arbitrary sequences (e.g. sea urchin, since sea urchin peptide material is never observed in *Xenopus* samples) with *Xenopus* ones would be benign. However, such nonsense sequences affect the peptide-to-spectra matching algorithms by producing spuriously matched peptides, and at a given false discovery rate (FDR), fewer real proteins will remain. Having identical or similar sequences (from allo-alleles or highly conserved protein families such as histones) is also detrimental to the search outcomes. Currently, there are multiple disjoint sources of protein sequences in *Xenopus*, which we list in Table 13.1. These databases are largely but not entirely mutually redundant. One of the outstanding challenges for the field is to arrive at a single gold standard set of sequences to be used across studies. Selecting such a set has not been accomplished even in more common model species so far, and for a good reason. The genome-based sequences are likely to provide "phantom" sequences due to imperfect gene models and missing sequences due to unidentified splice variants. The federated collections such as UniProt will have curated a high-quality set of sequences, but it is bound to be incomplete. Presence of allo-alleles in *X. laevis* presents a separate challenge since allele-specific expression might be of interest. As an alternative to all these, we took a genome-free approach in which we compiled a set of all transcripts detected in *X. laevis* and used coding frame-aware translation to predict expressed proteins (Wühr et al., 2014).

There are several groups of sequences that warrant special consideration when selecting and/or compiling a set of reference sequences, provided in the following in a list that is by no means complete:

- Genes which are simply missing in the genomic sequence because the DNA sequencing pipeline is imperfect. Until genome assembly improves, this issue will remain problematic.
- Genes which are in the current genome assembly scaffolds but are missed by gene modeling and thus are missing from the set of predicted mRNAs and proteins. This issue might be rectified in the next release of the genome assembly. If you are aware of such a gene/protein, just add its sequence to the database before you search against it.
- Proteins that undergo atypical post-translational sequence editing (not covalent addition of a certain functional group like phosphorylation), such as amino acid sequence cleavage on the way to a mature protein, for example, signaling peptides or proteasomally processed immuno-peptides. Translating cDNA in these cases does not predict the correct protein sequence. Currently, in these cases, the correct protein or peptide sequence needs to be added to the database.
- Selenoproteins—a few dozen proteins (e.g. selenoproteins S, K, N, O, and I) use a non-canonical amino acid—"selenocysteine"—coded for by one of three stop codons, and this peculiarity of context-dependent genetic code is seldom taken into account when gene models are built. Usually, this imperfection results in a premature termination of a reference sequence, which is thus missing a portion of the actual sequence.
- There are 13 abundant and functionally important proteins encoded by mitochondrial DNA which are not part of nuclear genome-derived gene models and should always be appended to a database.

13.3. NOMENCLATURE AND GENE SYMBOLS

Accurate detection of proteins is the necessary first step, but biological interpretation of the observations requires biological knowledge of the proteins. For high-dimensional data, often the first analysis is to look for enrichment of different pathways, cell locations, or protein functions. In order to do this, gene symbol annotation is necessary. Systems biology rationalizes the organization of biological function at the level of gene sets such as molecular pathways or protein complexes. Such sets are mainly defined and studied using human cells and human gene nomenclature. Sequences in other species are traditionally named using homology to already named sequences with the assumption that sequence homology often implies functional homology. When gene models are released with a genome, the assigned gene symbols take into account species- and field-specific context. For example, a gene identified as Xelaev18018806m at the release of *X. laevis* genome v1.3 is assigned the symbol *nodal5.3.L*, which reflects the *Xenopus*-specific *nodal* gene family expansion. However, what matters for gene set analysis is that this gene is best matched by the human sp|Q96S42|NODAL_HUMAN

TABLE 13.1
A List of Available Resources to Obtain a Reference Set of Protein Sequences for Peptide-Spectra Matching of *X. laevis* Data

Source	Description	Source URL	Size	Remarks and Criticism
Genome-derived	Based on gene models	http://ftp.xenbase.org/pub/Genomics/JGI/	45K seqs 21 Mb	Only as complete and accurate as gene models are
"Phrog"	RNA-seq derived (Wühr et al., 2014)	https://scholar.princeton.edu/wuehr/sample_prep	80K seqs 30.5 Mb	Redundant, wildcards in sequences
UniProt	Mostly uncurated, assembled from misc.	www.uniprot.org/uniprot/?query=organism%3A%22XENLA	61K seqs 23 Mb	Severely incomplete and redundant, only 5K reviewed
Marcotte Lab	Compiled from miscellaneous sources, then curated	https://github.com/marcottelab/pivo	25 K seqs 18 Mb	Contains allo-alleles fused into a single sequence from non-redundant peptides

and thus maps into the symbol "nodal." When the pipeline for such reciprocal BLAST-based gene symbol assignment was developed and applied to *X. laevis* (Savova et al., 2017), 17,000 unique human gene symbols matched. Not surprisingly, in *X. tropicalis*, which has a smaller genome size but no allo-alleles (allo-alleles are essentially redundant in terms of genome complexity as assayed by homology to human), the set of gene models match a similar number of human gene symbols as for *X. laevis*.

Even though the *Xenopus* genome assemblies are in relatively good shape, much remains to be done in terms of gene annotation and cataloging of expressed protein forms. A hybrid approach to generating a complete non-redundant set of reference protein sequences that combines genome-based and genome-free (mRNA-based) methods is an unresolved problem. Methods such as ribosome profiling (RIBO-seq) or ribosome nascent chain sequencing (RNC-seq) could be used to identify the set of mRNAs that are actually associated with ribosomes and thereby being translated (Savova et al., 2017; Zhao et al., 2019) and to identify current unannotated protein coding regions (Verbruggen et al., 2019).

13.4. PROTEOMICS AND CELL BIOLOGY

Xenopus laevis is an important model for studies of cell biology and, as reported in other chapters in this book, has been used extensively to understand the cellular regulation of mitosis, DNA replication, transcription and translation, RNA and protein localization, and embryogenesis, among many other processes. Recent advances in proteomics have enriched our understanding of several cellular processes, and a few are summarized here.

One elegant study took advantage of the fact that not only is the *Xenopus* stage VI oocyte a very large cell, but the nucleus (germinal vesicle) is also very large. Wühr et al. removed the germinal vesicle from the oocyte with forceps and measured the distribution of the proteome between the nucleus and the cytoplasm (2015). We found that protein localization can be mostly explained by empirical measurement of native protein size; the majority of large proteins are found exclusively in either the nucleus or the cytoplasm, and smaller proteins are found equally distributed between the two. This measurement of native protein size was only possible because of the ability to make undiluted cell lysates from crushed oocytes. Molecular weight determined by polypeptide length did not predict localization well, which is evidence that many proteins natively exist in complexes. In addition to providing insights into the properties of proteins in undiluted cytoplasm, this work produced an important resource for nuclear versus cytoplasmic localization of proteins.

Presler et al. used egg activation as a model of the first 20 minutes following release from meiotic metaphase arrest (2017). We detected very few previously unknown degradation events and no examples of protein synthesis. Instead, we found that expulsion of proteins from the eggs, assumed to occur via fusion of cortical granules with the cell membrane, constitutes the largest change to protein level during this period. We were able to compare these changes in absolute concentration amounts because of our previous work determining the concentration of proteins in the egg (Wühr et al., 2014). This is an example of using proteomics to reexamine bulk biochemical measurements of the egg and early embryo with the genomics-era ability to identify and classify the proteins that constitute these changes.

Affinity purification is another proteomics strategy that has been used by the *Xenopus* community. Lee et al. used dissected animal caps to identify novel binding partners of inner and outer dynein arm subunits in liquid-like organelles (Dynein Axonemal Particles or DynAPs) before their assembly into cilia (2020). Identification of novel protein localization in DynAPs explained why certain proteins that are not found in motile cilia themselves can still give rise to ciliopathies when they are mutated. Drew et al. used the same animal cap system, and a methodology termed DIF-FRAC, to identify RNA-binding proteins in epidermal tissues at the time point when motile cilia have formed (2020). They found evidence that there is RNA associated with DynAPs which is consistent with reports of RNA in other cytoplasmic liquid-like organelles. These two papers exemplify taking advantage of the strengths of the *Xenopus* simple explant system to enrich for the cellular complexes or tissue of interest. Animal cap explants can be differentiated into many different tissue types, and this is a way to overcome the challenges of tissue type heterogeneity of the embryo while also using a vertebrate animal system (Chang, 2016).

13.5. *X. LAEVIS* VERSUS *X. TROPICALIS*

Currently, MS-based proteomics in *Xenopus* is almost entirely done in *X. laevis* and not *X. tropicalis*, likely due to the preponderance of laboratories that use *X. laevis* for their other experimental approaches. Although *X. tropicalis* animals are smaller (oocytes are only half size in diameter), they have many advantages for proteomics: their development is faster, so reaching more advanced developmental points is easier; there are many transgenic strains available; and CRISPR-based genome editing works well (Nakayama et al., 2014; Naert et al., 2020; Horb et al., 2021). The simplicity of the genome makes bioinformatics easier and should result in the measurement of more functionally distinct proteins and post-translational modifications. This is because not all present peptides can be measured, and the presence of allo-alleles increases the fraction of redundant peptides; thus, spectra matching is more complex in *X. laevis* for the same level of sample biological complexity. Additionally, owing to being a true diploid, the complexity of the protein and peptide mixture is simpler in *X. tropicalis* than in *X. laevis*, which improves the efficiency in both the sample workflow (labeling, fractionation) and the bioinformatics. Being diploid also means that the genome assembly and annotation is at a more advanced state in *X. tropicalis*, so interpretation of the results is also more robust. In our experiments (unpublished), we see 40% of collected *X. tropicalis* spectra successfully matched to a sequence, in contrast with *X. laevis*, in which spectral matching rates are typically under 25%. Finally, the first edition of the Single Cell Atlas covering transcriptomic profiles of 200+ differentiated states over ten developmental stages has been done in *X. tropicalis* (Briggs et al., 2018) and could be cross-referenced more easily with proteomics data from *X. tropicalis* for interpretation of proteomics outcomes.

13.6. DEVELOPMENTAL ATLAS OF PROTEIN EXPRESSION

Over the past several years, we have completed several projects in *Xenopus* proteomics which resulted in resources for the *Xenopus* community made available through Xenbase (Karimi et al., 2018). Figure 13.1 illustrates the workflow and resulting data: whole embryos of *X. laevis* were collected at multiple developmental stages in order to characterize the proteins used in normal development. In the most recent project (Peshkin et al., 2019a), we profiled stages spanning early development from mature oocyte and unfertilized egg (NF-0) through blastula (NF-9), gastrula (NF-12), neurula (NF 17–24), and tailbud (NF-30) (Nieuwkoop et al., 1994). The last time point (NF-42) is taken long after the heartbeat has started and the tadpole has hatched and most of the cardiovascular and digestive (liver, pancreas) system has been established. Our processing pipeline for quantitatively measuring levels of protein is as previously described (Peshkin et al., 2015; Gupta et al., 2018). Proteins were digested into peptides, and the change of abundance was measured by isobaric labeling, followed by MultiNotch MS3 analysis (Wühr et al., 2015); absolute protein abundance was estimated via MS1 ion-current (Wühr et al., 2014). Protein abundance levels were measured at ten key stages (stage VI oocyte, egg = NF 0, 9, 12, 17, 22, 24, 26, 30, 42). Our primary dataset consists of 14,940 protein profiles. We collected and profiled the data in three independent biological replicates and, for the purposes of presenting the data at the gene-centric pages of Xenbase, combined all information using our "BACIQ" pipeline (Peshkin et al., 2019b), which produces the most likely patterns of relative protein abundance and 90% confidence intervals for these, as shown.

The confidence intervals narrow as more peptides are measured as long as the respective peptide changes are concordant. Additionally, peptides measured with a higher signal provide more confidence compared to low-signal peptides. It is possible, as illustrated in Figure 13.1B, to have higher confidence in one part of the interval and lower in another because peptide measurements have better agreement in one part of the trajectory. Having confidence intervals in addition to the average dynamics pattern provides for a more accurate interpretation in the context of embryonic development.

An interesting question is: How do protein and mRNA dynamics relate to one another? A master equation for a protein's abundance would imply its accumulation to be proportional to the respective mRNA concentration minus the protein loss, which is proportional to the protein abundance but independent of mRNA concentrations (Peshkin et al., 2015; Peshkin et al., 2019a). Many proteins that we measured in the embryo follow this functional relation, though the first order rate constants for synthesis, and the zero order rate constants of course differ for each protein (Peshkin et al., 2015). Figure 13.1C represents three discordant cases in which the levels of protein do not follow this simple pattern. This figure shows the accumulation of two homeologues of disulfide isomerases: *anterior gradient 2* (*agr2*) and *arginase 1* (*agr1*). Agr2 is particularly important for mucin secretion in the *Xenopus* cement gland, where it was first discovered. *agr2* mRNA and its protein are rapidly synthesized after stage 17 at the time of the appearance of the cement gland and stop accumulating as the cement gland is fully formed. Sometime after stage 17, the mRNA levels decrease steadily as the protein increases slightly, a result counter to our expectation that protein dynamics can be explained by mRNA dynamics alone. Then, the mRNA levels drop to 10% of their maximal level, whereas protein levels continue to increase, a second discrepancy. The protein data are of high confidence and are virtually the same

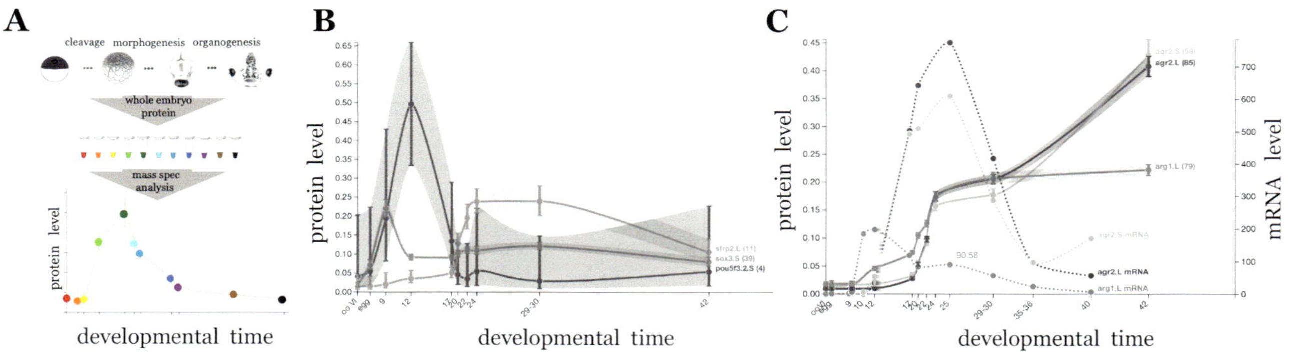

FIGURE 13.1 (A) schematic of the workflow—developmental stages of normally developing *X. laevis* embryos are compared to one another via a quantitative proteomics pipeline to reveal relative protein changes. (B) Protein expression encoded by three developmentally important genes (*sfrp2*, *sox3*, and *pou5f3*) with respective confidence intervals. The number in parentheses after the gene symbol indicates the number of peptides. (C) Examples of protein dynamics not readily explained by respective changes in mRNA expression, which is superimposed via dotted lines.

for the two isoforms, and the pattern is also very similar for Arg1. Arginase catalyzes the hydrolysis of arginine to ornithine and urea. This kind of discrepancy was anticipated by the many early studies postulating translational control, in which stored mRNA is regulated post-transcriptionally. Now this phenomenon can be studied on a case-by-case basis with this genome-wide dataset. What is perhaps most striking about mRNA-protein discordance is that it is relatively rare even in this comprehensive data set.

13.7. POST-TRANSLATIONAL MODIFICATIONS

Since many processes in biology are controlled by post-translational protein modification, quantitative measurement of post-translational modifications is perhaps the most important contribution that mass spectrometry-based proteomics can make. Widely used RNA-sequencing methods do not inform at all on the post-translational status of proteins. The strengths of *Xenopus* as a system are particularly well suited to studies of post-translational modifications. As mentioned earlier, deep quantitative measurement of any post-translational modification requires an experimental enrichment for peptides that are post-translationally modified. This means that the requirement for starting material is at least ten times higher for studies of post-translational modifications than for studies of the underlying protein levels. The ability to collect hundreds or even thousands of embryos from the same clutch, to experimentally manipulate hundreds of embryos by injection, and the size of *Xenopus* eggs and embryos mean this requirement for material is achievable.

The first study of post-translational modifications with mass spectrometry in *Xenopus* was a non-quantitative study cataloging phosphorylation sites on proteins in four early developmental stages up to the stage 10.5 gastrula (McGivern et al., 2009), where both conserved and novel phosphorylation sites were catalogued on hundreds of proteins. The first quantitative time series measurement of phosphorylation and acetylation reported on dynamics of only a few dozen changes (Peshkin et al., 2015) across early embryogenesis, gastrulation, and neurulation. The number of profiled peptides was quite small because they were measured without enrichment and reflected modifications of such abundant proteins as glycolytic enzymes and histones. Two groups measured phosphorylation with replicates and phosphopeptide enrichment during very early development: one measured ~3500 phosphorylated sites across the 20 minutes following fertilization and estimated absolute occupancy—the fraction of a protein with a specific modified residue—for ~500 phosphorylations (Presler et al., 2017); another measured ~9000 relative phosphorylated sites across the stage VI oocyte through the first cleavage (Peuchen et al., 2016; Peuchen et al., 2017). Both groups obtained evidence for decreasing proline directed phosphorylation following fertilization. Presler et al. additionally reported ubiquitinated peptides, and Qu et al. measured the de-N-glycoproteome of the egg and the stage 41 tadpole (Qu et al., 2020).

Presler et al. also reported a novel method for estimating the occupancy of post-translational modifications with confidence intervals (2017). The method combined previous approaches that only (1) took advantage of the inherent reciprocal relationship between the dynamics of the non-phosphorylated version of the phospho-peptide and the phosphorylated version in cases where phosphorylation levels change between conditions/timepoints (Olsen et al., 2010) or (2) used phosphatase treatment to artificially induce changes in the phosphorylated version of a phospho-peptide (Lim et al., 2017) into one method that was applicable for multiplexed experiments. It also took advantage of a system of equations that describes the relationships between phosphorylated and non-phosphorylated dynamics to report confidence intervals on the estimation of occupancy. Knowing the absolute amount of change in a modification for a protein either across development or downstream of a perturbation is much more useful for interpretation than having relative changes only. This is because a five-fold relative change could actually be an increase from 0.2% to 1% occupancy, whereas a two-fold relative change could actually be an increase from 45% to 90%. Importantly, since estimating occupancy for a significant number of phosphorylation events with confidence requires artificially inducing changes with phosphatase treatment, occupancy estimation needs to be considered in the initial experimental design.

There is extensive unrealized potential for the intersection of the *Xenopus* system and post-translational modifications measured by mass spectrometry. The published studies have utilized the strengths of natural cell cycle arrests of the stage 6 oocyte and egg for measuring the endogenous dynamics, but no study to date has included cell cycle perturbations. The stage 6 oocyte with its low amount of mitotic phosphorylation could be used to study low-occupancy phosphorylation events that are often unmeasured at the expense of high-occupancy cell cycle phosphorylations. Phosphorylation, acetylation, methylation, and ubiquitination have never been quantitatively measured after the first division. Studies in cell culture models have measured extensive phosphorylation of proteins involved in regulation of gene expression (Rigbolt et al., 2011). The importance of regulated gene expression in development is well understood, but the importance of phosphorylation for the function and regulation of protein expression is less well studied. Many proteins that do not change in level during development could be regulated by phosphorylation, which makes this an important next area of investigation in *Xenopus*.

13.8. SINGLE-CELL PROTEOMICS

Several years ago, our group developed a high-throughput droplet-microfluidic approach for barcoding the RNA from individual cells for subsequent analysis by next-generation sequencing: an inDrop platform that encapsulates cells into droplets with a lysis buffer, reverse transcription (RT) reagents, and barcoded oligonucleotide primers (Klein et al.,

2015). This platform has been widely adopted, enabling fundamental discoveries and new insight across many disciplines in biomedicine. We later applied this platform to profiling ten early developmental stages that provided a new view of development and differentiation of over 200 cell types in early *Xenopus* embryo (Briggs et al., 2018)—work which was recognized along with zebrafish counterparts as the Science Magazine Breakthrough of 2018 (Harland, 2018; Pennisi, 2018). Yet, as revealing as mRNA expression profiles might be, they do not fully capture and sometimes are strongly discordant with the protein expression, as we (Peshkin et al., 2015) and others (Smits et al., 2014) have shown. It is particularly relevant in the early embryo in which a maternal protein dowry has been synthesized in the mother and endocytosed by oocytes; as a result, many proteins might have no trace of the respective mRNA expression in cells resulting from progressive oocyte cleavages (Peshkin et al., 2015; Sonnett et al., 2018b). Therefore, it would be very attractive to profile protein expression at the single-cell level. A critical barrier on this path is the absence of *reverse translation* biochemistry. While a molecule of mRNA can be reverse-transcribed into cDNA and then amplified exponentially for detection and quantitation, there has so far been no discovery of a counterpart reaction for proteins, despite some intriguing proposed methods (Cook et al., 1977; Martin, 2006; Nashimoto, 2001). Attempts are being made to work directly with miniscule material quantities available from a single cell by reducing the loss of protein during sample preparation steps, but these show very few reproducibly measurable proteins and a limited dynamic range (Cheung et al., 2020). *Xenopus* offers a way out of this dilemma by providing large single cells. Technically speaking, single-cell proteomics in *Xenopus* begins with an egg (Smits et al., 2014; Lindeboom et al., 2019), since a 1.43-mm-diameter oocyte of *X. laevis* (Leibovich et al., 2020) is in fact a single cell. After a few initial divisions, blastomeres are still large—recently an effort has been made to examine individual blastomeres in early embryos and characterize lineage-specific protein deposit. The Nemes and Moody labs profiled single blastomeres with mass spectrometry and were able to quantify ~400 proteins of individual cells in the 16-cell stage (Onjiko et al., 2015; Lombard-Banek et al., 2016). First, they were able to minimize derivatization steps to enhance analytical sensitivity and use label-free quantification (Lombard-Banek et al., 2016) to obtain a low limit of detection and quantification for proteins in complex cell digests. Separately, both the Nemes and Moody (Onjiko et al., 2015) and Klein and Garcia (Saha-Shah et al., 2019) groups used pulled glass pipettes to capture material from specific blastomere cells, and animal-vegetal differences have been reproducibly measured. The Dovichi lab has also compared the protein expression among different blastomeres and observed that the blastomere-to-blastomere heterogeneity in 8-, 16-, 32-, and 50-cell embryos increases with development stage (Sun et al., 2016), reflecting progressive cellular differentiation.

13.9. DISCUSSION AND FUTURE DIRECTIONS

In this chapter, we discussed the successes of mass spectrometry-based proteomics in *Xenopus*, including methodology—sample preparation and bioinformatics analysis—and biological findings. It has been recognized for decades that *Xenopus* complements the more commonly used mouse and zebrafish animal models in the study of human biology and disease. *Xenopus* is particularly attractive for proteomics studies due to the ease of obtaining large amounts of protein without artificial perturbations needed, for example, synchronization of cells (with regard to cell cycle) or embryos (with regard to fertilization). While even a single egg or an early embryo in *X. laevis* contains sufficient amounts of non-yolk protein for some measurements of relative and absolute (Smits et al., 2014) protein abundance, ten embryos provide a protein amount sufficient for most protein measurements. Over 1000 sibling embryos can be obtained from a single clutch, and an *in-vitro* fertilization can be done with near-perfect (within one minute) synchrony.

This brief chapter could not exhaustively cover every aspect of *Xenopus* proteomics, and we encourage the reader to follow through other recent method reviews (Gupta et al., 2018; Sonnett et al., 2018b; Gilchrist et al., 2020). The field of *Xenopus* proteomics is in its infancy, and we have many reasons to expect great advances in developmental and cellular biology from bringing together the strengths of *Xenopus* as a system with advances in mass spectrometric instrumentation, biochemistry, machine learning, and analytics. The key component that will enable these advances is bioinformatics. This is true both for making existing methods more accessible to scientists without advanced computational skills and for creating new approaches to analysis of the data. Hosting a proteomic processing pipeline at Xenbase could go a long way to achieving this by maintaining the most complete and current set of reference proteins, using the optimal set of parameters for the spectra searches, and facilitating the archival and interactive access to the resulting datasets; it would also be beneficial for ongoing refinement of gene models.

We can anticipate some of the developments in the field. The push for single-cell proteomics will inevitably come full circle—having studied the repertoire of cell types and distilled abundant and specific cell type markers, we will develop ways to enrich single cells of a given type from dissociated embryos to extract enough material for cell-type-focused deep proteomics. In addition, we can expect the single-cell mRNA sequencing data for both *Xenopus* species to improve, which will allow for a more thorough classification of spatial expression across development that can then be used to deconvolve bulk protein data. It will be useful to not only increase our catalogue of cell type specific mRNAs but also have more certainty about which mRNAs—and potentially proteins—are expressed in all embryonic cell types.

As CRISPR knock-ins become commonplace in *Xenopus*, constructs for lineage tracing and controllable gene regulatory networks will create more opportunities for proteomic studies. We can imagine that in the same way that RNA-sequencing studies of phenotypes resulting from perturbations are now commonplace (Kjolby et al., 2017), soon so will be multiplexed proteomics studies of control and perturbed embryos, perhaps at multiple developmental stages. Not much has been accomplished to date in the way of proteomic measurements of the adult frogs, but embryological studies would greatly benefit from the tissue protein expression atlas in *Xenopus*, which is sure to both confirm widespread gene functional homology but also bring surprises with regard to the tissue expression differences across species. To date, *Xenopus* has not been considered a model organism for aging research, even though some research into reproductive system aging in *Xenopus* has been done (Brocas et al., 1961; Kara, 1994). It is not clear what the lifespan is or whether *Xenopus* undergoes considerable senescence, and thus a proteomic study of aging tissues presents an attractive direction.

The 2012 Nobel Prize in Physiology or Medicine was awarded to Sir John Gurdon for his work on nuclear transfer and reprogramming in *Xenopus*, which he chose for the ease of injection of somatic nuclei into oocytes. This well-established system presents another unfulfilled opportunity for deep quantitative proteomic characterization.

ACKNOWLEDGMENTS

LP and EVI were supported by NIH grant R01-HD091846.

REFERENCES

Briggs JA, Weinreb C, Wagner DE, Megason S, Peshkin L, Kirschner MW, Klein AM. The dynamics of gene expression in vertebrate embryogenesis at single-cell resolution. *Science*. 2018 Jun 1;360(6392):eaar5780. doi: 10.1126/science.aar5780. PMID: 29700227; PMCID: PMC6038144.

Brocas J, Verzar F. The aging of *Xenopus laevis*, a South African frog. *Gerontologia*. 1961;5:228–240. doi: 10.1159/000211062. PMID: 13873222.

Chang C. Animal cap assay for TGF-β signaling. *Methods Mol Biol*. 2016;1344:261–274. doi: 10.1007/978-1-4939-2966-5_16. PMID: 26520130.

Cheung TK, Lee CY, Bayer FP, McCoy A, Kuster B, Rose CM. Defining the carrier proteome limit for single-cell proteomics. *Nat Methods*. 2020 Dec 7. doi: 10.1038/s41592-020-01002-5. PMID: 33288958.

Cook ND. The case for reverse translation. *J Theor Biol*. 1977 Jan 7;64(1):113–135. doi: 10.1016/0022-5193(77)90116-3. PMID: 319301.

Drew K, Lee C, Cox RM, Dang V, Devitt CC, McWhite CD, Papoulas O, Huizar RL, Marcotte EM, Wallingford JB. A systematic, label-free method for identifying RNA-associated proteins in vivo provides insights into vertebrate ciliary beating machinery. *Dev Biol*. 2020 Nov 1;467(1–2):108–117. doi: 10.1016/j.ydbio.2020.08.008. PMID: 32898505; PMCID: PMC 7668317.

Gilchrist MJ, Veenstra GJC, Cho KWY. Transcriptomics and proteomics methods for Xenopus embryos and tissues. *Cold Spring Harb Protoc*. 2020 Feb 3;2020(2):098350. doi: 10.1101/pdb.top098350. PMID: 31772075; PMCID: PMC7362671.

Gupta M, Sonnett M, Ryazanova L, Presler M, Wühr M. Quantitative proteomics of Xenopus embryos I, sample preparation. *Methods Mol Biol*. 2018;1865:175–194. doi: 10.1007/978-1-4939-8784-9_13. PMID: 30151767; PMCID: PMC6564683.

Gurdon JB, Wakefield L. Microinjection of amphibian oocytes and eggs for the analysis of transcription. *Microinjection and Organelle Transplantation Techniques, eds Celis JE, Graessmann A, Loyter A (Academic, London)*. 1986: 269–299.

Harland RM. A new view of embryo development and regeneration. *Science*. 2018 Jun 1;360(6392):967–968. doi: 10.1126/science.aat8413. PMID: 29853675.

Hellsten U, Harland RM, Gilchrist MJ, Hendrix D, Jurka J, Kapitonov V, Ovcharenko I, Putnam NH, Shu S, Taher L, Blitz IL, Blumberg B, Dichmann DS, Dubchak I, Amaya E, Detter JC, Fletcher R, Gerhard DS, Goodstein D, Graves T, Grigoriev IV, Grimwood J, Kawashima T, Lindquist E, Lucas SM, Mead PE, Mitros T, Ogino H, Ohta Y, Poliakov AV, Pollet N, Robert J, Salamov A, Sater AK, Schmutz J, Terry A, Vize PD, Warren WC, Wells D, Wills A, Wilson RK, Zimmerman LB, Zorn AM, Grainger R, Grammer T, Khokha MK, Richardson PM, Rokhsar DS. The genome of the Western clawed frog *Xenopus tropicalis*. *Science*. 2010 Apr 30;328(5978):633–636. doi: 10.1126/science.1183670. PMID: 20431018; PMCID: PMC2994648.

Horb M, Abu-Daya, A, Wlizla, M, Noble, A, Guille M. This Volume. Chapter 14. *Advances in Genome Editing Tools*. 2021.

Kara TC. Ageing in amphibians. *Gerontology*. 1994;40(2–4): 161–173. doi: 10.1159/000213585. PMID: 7926854.

Karimi K, Fortriede JD, Lotay VS, Burns KA, Wang DZ, Fisher ME, Pells TJ, James-Zorn C, Wang Y, Ponferrada VG, Chu S, Chaturvedi P, Zorn AM, Vize PD. Xenbase: A genomic, epigenomic and transcriptomic model organism database. *Nucleic Acids Res*. 2018 Jan 4;46(D1):D861–D868. doi: 10.1093/nar/gkx936. PMID: 29059324; PMCID: PMC5753396.

Kjolby RAS, Harland RM. Genome-wide identification of Wnt/β-catenin transcriptional targets during *Xenopus* gastrulation. *Dev Biol*. 2017 Jun 15;426(2):165–175. doi: 10.1016/j.ydbio.2016.03.021. PMID: 27091726; PMCID: PMC6288011.

Klein AM, Mazutis L, Akartuna I, Tallapragada N, Veres A, Li V, Peshkin L, Weitz DA, Kirschner MW. Droplet barcoding for single-cell transcriptomics applied to embryonic stem cells. *Cell*. 2015 May 21;161(5):1187–1201. doi: 10.1016/j.cell.2015.04.044. PMID: 26000487; PMCID: PMC4441768.

Lee C, Cox RM, Papoulas O, Horani A, Drew K, Devitt CC, Brody SL, Marcotte EM, Wallingford JB. Functional partitioning of a liquid-like organelle during assembly of axonemal dyneins. *Elife*. 2020 Dec 2;9:e58662. doi: 10.7554/eLife.58662. PMID: 33263282; PMCID: PMC7785291.

Leibovich A, Edri T, Klein SL, Moody SA, Fainsod A. Natural size variation among embryos leads to the corresponding scaling in gene expression. *Dev Biol*. 2020 Jun 15;462(2):165–179. doi: 10.1016/j.ydbio.2020.03.014. PMID: 32259520; PMCID: PMC8073595.

Lim MY, O'Brien J, Paulo JA, Gygi SP. Improved method for determining absolute phosphorylation stoichiometry using Bayesian statistics and isobaric labeling. *J Proteome Res*. 2017 Nov 3;16(11):4217–4226. doi: 10.1021/acs.jproteome.7b00571. PMID: 28985074; PMCID: PMC6301010.

Lindeboom RGH, Smits AH, Perino M, Veenstra GJC, Vermeulen M. Mass spectrometry-based absolute quantification of single *Xenopus* embryo proteomes. *Cold Spring Harb Protoc*.

2019 Jun 3;2019(6). doi: 10.1101/pdb.prot098376. PMID: 30104410.

Lombard-Banek C, Reddy S, Moody SA, Nemes P. Label-free quantification of proteins in single embryonic cells with neural fate in the cleavage-stage frog (*Xenopus laevis*) embryo using capillary electrophoresis electrospray ionization high-resolution mass spectrometry (CE-ESI-HRMS). *Mol Cell Proteomics*. 2016 Aug;15(8):2756–2768. doi: 10.1074/mcp.M115.057760. PMID: 27317400; PMCID: PMC4974349.

Martin MT. *Methods and Compositions for Reverse Translation*. 2006. patents.google.com/patent/US7169894

McGivern JV, Swaney DL, Coon JJ, Sheets MD. Toward defining the phosphoproteome of *Xenopus laevis* embryos. *Dev Dyn*. 2009 Jun;238(6):1433–1443. doi: 10.1002/dvdy.21941. PMID: 19384857; PMCID: PMC2865133.

Naert T, Tulkens D, Edwards NA, Carron M, Shaidani NI, Wizla M, Boel A, Demuynck S, Horb ME, Coucke P, Willaert A, Zorn AM, Vleminckx K. Maximizing CRISPR/Cas9 phenotype penetrance applying predictive modeling of editing outcomes in *Xenopus* and zebrafish embryos. *Sci Rep*. 2020 Sep 4;10(1):14662. doi: 10.1038/s41598-020-71412-0. PMID: 32887910; PMCID: PMC7473854.

Nakayama T, Blitz IL, Fish MB, Odeleye AO, Manohar S, Cho KW, Grainger RM. Cas9-based genome editing in *Xenopus tropicalis*. *Methods Enzymol*. 2014;546:355–375. doi: 10.1016/B978-0-12-801185-0.00017-9. PMID: 25398349; PMCID: PMC4284096.

Nashimoto M. The RNA/protein symmetry hypothesis: Experimental support for reverse translation of primitive proteins. *J Theor Biol*. 2001 Mar 21;209(2):181–187. doi: 10.1006/jtbi.2000.2253. PMID: 11401460.

Nieuwkoop PD, Faber J. *Normal Table of Xenopus laevis (Daudin)*. 1994. Garland Publishing Inc, New York ISBN 0-8153-1896-0.

Olsen JV, Vermeulen M, Santamaria A, Kumar C, Miller ML, Jensen LJ, Gnad F, Cox J, Jensen TS, Nigg EA, Brunak S, Mann M. Quantitative phosphoproteomics reveals widespread full phosphorylation site occupancy during mitosis. *Sci Signal*. 2010 Jan 12;3(104):ra3. doi: 10.1126/scisignal.2000475. PMID: 20068231.

Onjiko RM, Moody SA, Nemes P. Single-cell mass spectrometry reveals small molecules that affect cell fates in the 16-cell embryo. *Proc Natl Acad Sci USA*. 2015 May 26;112(21):6545–6550. doi: 10.1073/pnas.1423682112. PMID: 25941375; PMC ID: PMC4450407.

Pennisi E. Development cell by cell. *Science*. 2018 Dec 21; 362(6421):1344–1345. doi: 10.1126/science.362.6421.1344. PMID: 30573610.

Peshkin L, Gupta M, Ryazanova L, Wühr M. Bayesian confidence intervals for multiplexed proteomics integrate ion-statistics with peptide quantification concordance. *Mol Cell Proteomics*. 2019a Oct;18(10):2108–2120. doi: 10.1074/mcp.TIR119.001317. PMID: 31311848; PMCID: PMC6773559.

Peshkin L, Lukyanov A, Kalocsay M, Gage RM, Wang DZ, Pells TJ, Karimi K, Vize PD, Wühr M, Kirschner MW. The protein repertoire in early vertebrate embryogenesis. *BioRxiv*. 2019b:571174. doi.org/10.1101/571174.

Peshkin L, Wühr M, Pearl E, Haas W, Freeman RM Jr, Gerhart JC, Klein AM, Horb M, Gygi SP, Kirschner MW. On the relationship of protein and mRNA dynamics in vertebrate embryonic development. *Dev Cell*. 2015 Nov 9;35(3):383–394. doi: 10.1016/j.devcel.2015.10.010. PMID: 26555057; PMCID: PMC4776761.

Peuchen EH, Cox OF, Sun L, Hebert AS, Coon JJ, Champion MM, Dovichi NJ, Huber PW. Phosphorylation dynamics dominate the regulated proteome during early *Xenopus* development. *Sci Rep*. 2017 Nov 15;7(1):15647. doi: 10.1038/s41598-017-15936-y. PMID: 29142207; PMCID: PMC5688136.

Peuchen EH, Sun L, Dovichi NJ. Optimization and comparison of bottom-up proteomic sample preparation for early-stage *Xenopus laevis* embryos. *Anal Bioanal Chem*. 2016 Jul;408(17):4743–4749. doi: 10.1007/s00216-016-9564-2. PMID: 27137514; PMCID: PMC4926613.

Presler M, Van Itallie E, Klein AM, Kunz R, Coughlin ML, Peshkin L, Gygi SP, Wühr M, Kirschner MW. Proteomics of phosphorylation and protein dynamics during fertilization and meiotic exit in the *Xenopus* egg. *Proc Natl Acad Sci USA*. 2017 Dec 12;114(50):E10838–E10847. doi: 10.1073/pnas.1709207114. PMID: 29183978; PMCID: PMC5740657.

Qu Y, Dubiak KM, Peuchen EH, Champion MM, Zhang Z, Hebert AS, Wright S, Coon JJ, Huber PW, Dovichi NJ. Quantitative capillary zone electrophoresis-mass spectrometry reveals the N-glycome developmental plan during vertebrate embryogenesis. *Mol Omics*. 2020 Jun 15;16(3):210–220. doi: 10.1039/d0mo00005a. PMID: 32149324; PMCID: PMC7299754.

Rigbolt KT, Prokhorova TA, Akimov V, Henningsen J, Johansen PT, Kratchmarova I, Kassem M, Mann M, Olsen JV, Blagoev B. System-wide temporal characterization of the proteome and phosphoproteome of human embryonic stem cell differentiation. *Sci Signal*. 2011 Mar 15;4(164):rs3. doi: 10.1126/scisignal.2001570. PMID: 21406692.

Saha-Shah A, Esmaeili M, Sidoli S, Hwang H, Yang J, Klein PS, Garcia BA. Single cell proteomics by data-independent acquisition to study embryonic asymmetry in *Xenopus laevis*. *Anal Chem*. 2019 Jul 16;91(14):8891–8899. doi: 10.1021/acs.analchem.9b00327. PMID: 31194517; PMCID: PMC6688503.

Savova V, Pearl EJ, Boke E, Nag A, Adzhubei I, Horb ME, Peshkin L. Transcriptomic insights into genetic diversity of protein-coding genes in *X. laevis*. *Dev Biol*. 2017 Apr 15;424(2):181–188. doi: 10.1016/j.ydbio.2017.02.019. PMID: 28283406; PMCID: PMC5405699.

Session AM, Uno Y, Kwon T, Chapman JA, Toyoda A, Takahashi S, Fukui A, Hikosaka A, Suzuki A, Kondo M, van Heeringen SJ, Quigley I, Heinz S, Ogino H, Ochi H, Hellsten U, Lyons JB, Simakov O, Putnam N, Stites J, Kuroki Y, Tanaka T, Michiue T, Watanabe M, Bogdanovic O, Lister R, Georgiou G, Paranjpe SS, van Kruijsbergen I, Shu S, Carlson J, Kinoshita T, Ohta Y, Mawaribuchi S, Jenkins J, Grimwood J, Schmutz J, Mitros T, Mozaffari SV, Suzuki Y, Haramoto Y, Yamamoto TS, Takagi C, Heald R, Miller K, Houston DW, Shendure J, DuPasquier L, Vize PD, Zorn AM, Ito M, Marcotte EM, Wallingford JB, Ito Y, Asashima M, Ueno N, Matsuda Y, Veenstra GJ, Fujiyama A, Harland RM, Taira M, Rokhsar DS. Genome evolution in the allotetraploid frog *Xenopus laevis*. *Nature*. 2016 Oct 20;538(7625):336–343. doi: 10.1038/nature19840. PMID: 27762356; PMCID: PMC5313049.

Sheets MD, Fritz B, Hartley RS, Zhang Y. Polyribosome analysis for investigating mRNA translation in *Xenopus* oocytes, eggs and embryos. *Methods*. 2010 May;51(1):152–156. doi: 10.1016/j.ymeth.2010.01.023. PMID: 20096782; PMCID: PMC2868116.

Smits AH, Lindeboom RG, Perino M, van Heeringen SJ, Veenstra GJ, Vermeulen M. Global absolute quantification reveals tight regulation of protein expression in single *Xenopus* eggs.

Nucleic Acids Res. 2014 Sep;42(15):9880–9891. doi: 10.1093/nar/gku661. PMID: 25056316; PMCID: PMC4150773.

Sonnett M, Gupta M, Nguyen T, Wühr M. Quantitative proteomics for *Xenopus* embryos II, data analysis. *Methods Mol Biol*. 2018b;1865:195–215. doi: 10.1007/978-1-4939-8784-9_14. PMID: 30151768; PMCID: PMC6534117.

Sonnett M, Yeung E, Wühr M. Accurate, sensitive, and precise multiplexed proteomics using the complement reporter ion cluster. *Anal Chem*. 2018a Apr 17;90(8):5032–5039. doi: 10.1021/acs.analchem.7b04713. PMID: 29522331; PMCID: PMC6220677.

Sun L, Dubiak KM, Peuchen EH, Zhang Z, Zhu G, Huber PW, Dovichi NJ. Single cell proteomics using frog (*Xenopus laevis*) blastomeres isolated from early stage embryos, which form a geometric progression in protein content. *Anal Chem*. 2016 Jul 5;88(13):6653–6657. doi: 10.1021/acs.analchem.6b01921. PMID: 27314579; PMCID: PMC4940028.

Verbruggen S, Ndah E, Van Criekinge W, Gessulat S, Kuster B, Wilhelm M, Van Damme P, Menschaert G. PROTEOFORMER 2.0: Further developments in the ribosome profiling-assisted proteogenomic hunt for new proteoforms. *Mol Cell Proteomics*. 2019 Aug 9;18(8 suppl 1):S126–S140. doi: 10.1074/mcp.RA118.001218. PMID: 31040227; PMCID: PMC6692777.

Wühr M, Freeman RM Jr, Presler M, Horb ME, Peshkin L, Gygi S, Kirschner MW. Deep proteomics of the *Xenopus laevis* egg using an mRNA-derived reference database. *Curr Biol*. 2014 Jul 7;24(13):1467–1475. doi: 10.1016/j.cub.2014.05.044. PMID: 24954049; PMCID: PMC4090281.

Wühr M, Güttler T, Peshkin L, McAlister GC, Sonnett M, Ishihara K, Groen AC, Presler M, Erickson BK, Mitchison TJ, Kirschner MW, Gygi SP. The nuclear proteome of a vertebrate. *Curr Biol*. 2015 Oct 19;25(20):2663–2671. doi: 10.1016/j.cub.2015.08.047. PMID: 26441354; PMCID: PMC4618192.

Zhao J, Qin B, Nikolay R, Spahn CMT, Zhang G. Translatomics: The global view of translation. *Int J Mol Sci*. 2019 Jan 8;20(1):212. doi: 10.3390/ijms20010212. PMID: 30626072; PMCID: PMC6337585.

14 Advances in Genome Editing Tools

Marko E. Horb, Anita Abu-Daya, Marcin Wlizla, Anna Noble, and Matt Guille

CONTENTS

14.1. Generation of Transgenic *Xenopus* 207
14.1.1. Beginnings 207
14.1.2. Technology Development 207
14.1.3. pTransgenesis: Streamlining Transgene Construction 208
14.2. Uses of Transgenic *Xenopus* 209
14.3. Genetics and Gene Editing in *Xenopus* 210
14.3.1. Beginnings 210
14.3.2. *X. tropicalis* and Forward Genetic Screens 210
14.3.3. TILLING 211
14.3.4. Insertional Mutagenesis 211
14.3.5. Reverse Genetics Using dsBreaks 212
14.4. Future Directions 214
14.5. Conclusions 215
Acknowledgments 216
References 216

14.1. GENERATION OF TRANSGENIC *XENOPUS*

14.1.1. Beginnings

Transgenic animals carry exogenous DNA integrated into their genome and can be designed to fulfill diverse experimental objectives, from using fluorescent proteins to label specific cells/tissues or to observe activity of signaling pathways to using dominant-negative and constitutively active mutants to disrupt gene activity. Prior to the development of stable transgenic lines, experiments to track or disrupt gene function were carried out via microinjection of RNA, DNA, morpholino oligonucleotides, or antibodies and through treatments with small molecules, but these all have related limitations. The mosaic distribution and progressive dilution with every cell division of microinjected reagents typically limits their use to the study of genes involved in early development (Amaya, 2005). Although small molecule treatments can be applied at any stage of development, care must be taken that the molecules used are specific to the genes being disrupted and that the phenotypes observed are not due to a disruption of a broader range of targets than intended (Vogt et al., 2011). These experimental limitations can be circumvented using transgenics. The current transgenic toolset available to *Xenopus* researchers permits efficient generation of stable, non-mosaic animal lines and inclusion of elements for temporal control of transgene activity. These allow their use through all stages of development and make possible the design of transgenic lines able to disrupt gene activity in a highly specific manner.

It has been over three decades since the initial reports that exogenous plasmid DNA was integrated into the *Xenopus* genome following its microinjection into fertilized eggs with successful transmission through the male germline (Etkin and Pearman, 1987; Etkin and Roberts, 1983; Rusconi and Schaffner, 1981). However, microinjection of linearized plasmid DNA is not a practical approach for generation of stable transgenic lines; it is highly inefficient, with only 1% of injected F0 animals showing mosaic integration of the foreign DNA into their germlines (Yergeau et al., 2010). Innovation has followed three distinct ways to improve the transgenic methodologies in *Xenopus*: first, development of alternative approaches to make genomic integration more efficient; second, by streamlining the generation of transgenic plasmids carrying the exogenous DNA; and third, by incorporating elements for temporal regulation of transgene activity, thus expanding on the versatility of this technology. This chapter outlines technological advances in *Xenopus* transgenesis, followed by an overview of how transgenics have impacted research using this popular model organism.

14.1.2. Technology Development

Restriction enzyme-mediate integration (REMI) was the first method that made *Xenopus* transgenesis practical (Kroll and Amaya, 1996). In REMI, transgenic plasmid DNA is incubated together with isolated sperm nuclei, a restriction enzyme, and *Xenopus* interphase egg extract and then injected into mature, unfertilized eggs. One to 16% of the injected eggs survive past feeding tadpole stages. However, this is not really a complication, as it is possible to inject

DOI: 10.1201/9781003050230-16

thousands of eggs at a time, and of those normal survivors, as many as 36% show stable, non-mosaic transgene expression. Furthermore, F0 animals that show strong non-mosaic expression almost invariably transmit the transgene through their germline (Marsh-Armstrong et al., 1999). This was a major improvement in transgenic efficiency over the results obtained by injection of linearized plasmid DNA into fertilized embryos and also works efficiently in *Xenopus tropicalis* (Offield et al., 2000). The major limitation of REMI is that it is highly mutagenic, capable of generating up to four different transgene integration sites with multiple integrations in the injected individual, complicating interpretation of phenotypic analyses (Bronchain et al., 1999; Marsh-Armstrong et al., 1999). Furthermore, studies analyzing late developmental stages are not trivial in the F0 generation due to the small number of survivors (Chesneau et al., 2008). In 2000, REMI was simplified by eliminating the need for restriction enzyme and egg extract while retaining similar efficiency (Sparrow et al., 2000).

Following REMI, the successful innovations in transgenesis techniques focused on increasing rates of survival and normal development, mainly through injection of transgenic reagents into embryos rather than eggs, resulting in less technically demanding procedures and greater normal development. First, a relatively simple method relying on co-injection of the commercially available rare-cutting meganuclease, I-SceI, and transgenic DNA carrying the 18 base-pair long I-SceI recognition site was efficient in both *Xenopus laevis* and *Xenopus tropicalis* (Pan et al., 2006). The observed survival rate past metamorphosis was greater than 55%, with 10% to 12% of the survivors showing non-mosaic transgene expression (Pan et al., 2006). Furthermore, only one to eight copies of the transgene integrated at up to two distinct integration sites.

A second similarly simple technique involves microinjection of mRNA encoding the φC31 bacteriophage integrase with a transgenic plasmid DNA containing a 34 base-pair long bacterial attachment site, *attB* (Allen and Weeks, 2005; Li et al., 2012). The integrase mediates recombination between the *attB* site and a 39 base-pair long phage-dependent attachment site *attP*; recombination, however, can also occur at pseudo-*attP* sites with similarity as low as 24% to the phage *attP* sequence, though at much lower efficiencies (Groth et al., 2004; Thyagarajan et al., 2001). The main advantage of this approach is that it is thought to result in integration of a single transgene into the *Xenopus* genome, although reports of successful germline transmission are lacking (Allen and Weeks, 2005; Li et al., 2012). The European *Xenopus* Resource Centre (EXRC) has an engineered transgenic line that contains an *attP* docking site within a functional cyan fluorescent protein coding sequence. The use of this line in conjunction with φC31 integrase may provide a highly efficient way to generate novel transgenics, screened by loss of cyan fluorescence (Horb et al., 2019).

Finally, two approaches based on the use of transposase-driven transgene integration, using *Sleeping Beauty* or *Tol2*, have been effective in *Xenopus*. Both involve co-injection of the transposase, either as mRNA or protein, with the transposon or DNA encoding the transgenic package flanked by transposase target sequences (Hamlet et al., 2006; Shibano et al., 2007; Sinzelle et al., 2006; Yergeau et al., 2009). Uniquely, this method can also be used in a variant of gene and enhancer trap, in which forward genetic experiments are achieved by secondarily remobilizing the transgene following reintroduction of the transposase (Lane et al., 2013; Yergeau et al., 2011a, 2012).

14.1.3. pTransgenesis: Streamlining Transgene Construction

Besides novel ways of promoting transgene integration into the *Xenopus* genome, a major innovation came through use of modular Gateway cloning to streamline the generation of transgenic plasmids. The pTransgenesis system uses multisite Gateway technology to allow for rapid generation of transgenic plasmids via recombination of a destination vector containing DNA sequences required for genome integration and three entry clones, a fluorescent transgenesis reporter, a promoter, and a coding sequence (Love et al., 2011b). The destination vectors allow a choice among transgenic methods, including I-SceI meganuclease, Tol2 transposase, and φC31 integrase. Novel entry clones can be easily generated through simple recombination of a PCR product and donor vector, and existing entry clones can be mixed and matched to generate a diverse range of transgenic plasmids, making this a very powerful and flexible system.

There are, however, a number of issues with this system. First, the particular donor vectors used to generate pTransgenesis entry clones stopped being commercially available shortly after publication. Nonetheless, they can be procured from the Zebrafish International Resource Center (ZIRC) as part of the Tol2kit used for transgenic plasmid creation in zebrafish (Kwan et al., 2007). Second, only three of the four destination vectors contain chicken beta-globin HS4 insulator sequences that have been shown to reduce integration site effects on transgene expression (Allen and Weeks, 2005). Third, one of the vectors includes two I-SceI target sequences flanking the transgene, which may increase the efficiency of transgenesis but risks integration of the vector backbone independent of the transgene sequence. Fourth, the pTransgenesis system, as constructed, is not compatible with the *Xenopus* ORFeome (Grant et al., 2015). The recombination sites used in the ORFeome constructs are the same ones as those used in the pTransgenesis promoter entry plasmids and combining both systems misplaces the ORF within the transgenic construct. Instead, a newer two-plasmid system can be used in conjunction with the ORFeome to rapidly generate transgenic plasmids (Sterner et al., 2019). Like pTransgenesis, this system is versatile and permits selection of the transgenic technique: I-SceI, Tol2, and φC31. One of the destination vectors, pDXTR, allows for rapid gateway recombination with the ORFeome plasmids and includes the Tet-On system for inducible transgene expression. However, it requires the use of restriction enzyme-based or Gibson Assembly cloning to introduce the promoters of interest (Das and Brown, 2004a; Kerney et al., 2012; Rankin et al., 2011). Alternatively, another destination vector, pDXTP, can be used in conjunction with pDXTR. pDXTP is compatible with the promoter entry vectors included in pTransgenesis, allowing rapid recombination of the promoter upstream of the doxycycline inducible transcription factor rtTA, thus maintaining the ability to use the Tet-On system to control transgene expression. One shortcoming of the pDXTR and pDXTP plasmids is that they are relatively large, and both contain several insulator sequences. This makes pDXTR particularly prone to self-recombination following transformation into *E. coli*, which can be prevented by growing the transformed bacteria at a lower temperature.

The ability to regulate transgene activity in both time and space expands the breadth of experimental questions that can be investigated using transgenics in *Xenopus*. A number of cell- and tissue-specific promoters have been characterized and used for spatial regulation of transgene activity in *Xenopus* (Horb et al., 2019). In addition to the Tet-On system, temporal control of transgene expression in *Xenopus* has been achieved using the temperature-inducible *hsp70* promoter or the modified dual-component GAL4-UAS system with the GAL4 fused to the ligand binding domain of the progesterone receptor (PR) (Beck et al., 2006; Horb et al., 2019). Finally, transgenic lines expressing Cre have been used as drivers to induce switches in fluorescence following a cross to lines with *loxP* sites (Roose et al., 2009; Waldner et al., 2006). These diverse methods of transgenesis, transgenic vector construction, and transgene activity regulation provide a solid framework within which *Xenopus* researchers can design experiments to study biological processes through genome modification.

14.2. USES OF TRANSGENIC *XENOPUS*

The way transgenesis would expand the experimental toolkit in *Xenopus* was made evident with the first transgenesis paper by Kroll and Amaya, in which they examined the temporal requirement of FGF signaling in early *Xenopus* development. Previous experiments suggested that FGF signaling was necessary for primary mesoderm induction as well for later processes, including maintenance of mesoderm fate and neural induction and patterning. Mesoderm induction occurs quite early in embryogenesis and can be studied using mRNA injections, but later events require FGF signaling to be perturbed after mesoderm induction. Transgenesis was used to express a dominant negative FGF receptor after mesoderm induction and showed that while FGF was required for maintenance of mesoderm fate, it was not required for neural induction and patterning (Kroll and Amaya, 1996). Its lack of involvement in neural induction contradicted previous data, revealing how transgenics could improve data quality in the frog. In addition to dominant negative protein expression, other loss-of-function approaches, such as shRNA, have also proven effective in *Xenopus* transgenics (Edholm and Robert, 2018).

Using inducible or tissue-specific promoters allows research on developmental processes that occur several days to months after fertilization, including regeneration and metamorphosis. Regeneration is thought to occur through the reactivation of the same program involved in normal development, but this cannot be studied in traditional knockout experiments, since the target tissue/organ may be perturbed by loss of function early in development. In *Xenopus*, tadpole tails regenerate upon amputation from three days after fertilization until metamorphosis, except during a short refractory period at four to five days. Using a heat shock-inducible *hsp70* promoter, Beck et al. (2003) showed that reactivation of the BMP and Notch signaling pathways during this refractory period promoted regeneration, whereas their inhibition at other stages blocked regeneration. In F0 transgenics, they found variability in this ability, possibly due to integration site and transgene expression levels, but in F1 transgenics, the phenotype was more consistent. These results suggested that generating stable transgenic lines produces more robust results.

Metamorphosis in *Xenopus* is a model for human perinatal endocrinology when multiple hormones regulate many aspects of tissue growth, development, remodeling, and maturation (Buchholz, 2015). This late-stage event occurs 30–60 days after fertilization, and transgenesis in *Xenopus* was essential for *in vivo* functional studies (Marsh-Armstrong et al., 2004; Mukhi et al., 2008, 2009; Schreiber et al., 2001). In particular, binary-inducible transgenic systems have proven useful for temporal and tissue-specific control of transgenes during metamorphosis (Buchholz, 2012; Das and Brown, 2004b). The tetracycline (Tet)-inducible system allows for tight control of transgene expression by simple addition of doxycycline (Dox) to the water. This system requires two different transgenes: one promoter (tissue-specific or ubiquitous) to control expression of rtTA (a Dox-dependent transcription factor) and a second tetracycline-inducible (TRE) promoter to control expression of the gene of interest. This system elucidated several aspects of metamorphosis, including gene switching, transdifferentiation of pancreatic acinar cells to ductal cells, and limb development (Brown et al., 2005; Cai et al., 2007; Mukhi and Brown, 2011; Mukhi et al., 2010).

Another benefit of *Xenopus* transgenics is the ability to use promoters from other species, including rat, mouse, and zebrafish, to drive expression in a tissue-specific manner (Beck and Slack, 1999; Love et al., 2011a). This can be used to study the ability of factors to convert one tissue to another. Combining such expression with a secondary reporter to label the tissue generated allows monitoring of transdifferentiation events in real time. For example, using a murine transthyretin promoter to drive expression of pancreatic transcription factors in the liver, combined with the rat elastase promoter driving GFP, it was found that only two pancreatic transcription factors, Ptf1a and Pdx1, were able to convert liver to pancreas (Horb et al., 2003; Jarikji et al.,

2007). First, both *Xenopus* and murine *pdx1* genes acted similarly in converting liver to pancreas in the *Xenopus* tadpole, but both proteins required an extra VP16 activation domain. The ability of other pancreatic transcription factors was tested similarly, and only one other transcription factor, Ptf1a, had similar activity. These results reveal the power of using multiple transgenes to test the functional ability of different factors *in vivo* in a tissue-specific manner allowing for analysis of cell fate conversions and commitment.

One of the major embryological benefits of *Xenopus* is transplantation that creates chimeric embryos, and transgenic *Xenopus* embryos have proven very useful for these experiments. Creating chimeric embryos between wild type and transgenic embryos has been used to study pancreas development, regeneration, and the origin of muscle satellite cells (Daughters et al., 2011; Gargioli and Slack, 2004; Jarikji et al., 2009). The developing pancreas is derived from separate dorsal and ventral buds arising from the roof and floor of the archenteron. These buds can be selectively labeled in chimeric embryos of wild type and pElas:GFP transgenics. Using such an approach, Jarikji et al. (2009) found that the ventral pancreatic cells migrate into the dorsal pancreas after fusion, whereas the dorsal pancreatic cells do not. In another example, grafting specific regions from a transgenic CMV:GFP neurula stage embryo onto a wild type host enabled Gargioli and Slack (2004) to follow the fate of individual tissues (neural, notochord, or somites) during tail regeneration. They found that notochord and spinal cord regenerate from the same tissue, whereas muscle cells regenerate from a small population of satellite cells. Other lines that are beneficial for such experiments include ROSA26:GFP and Brainbow lines for long-term fate mapping (Gross et al., 2006).

In addition to following transplanted cells, transgenes can be used to isolate specific embryonic cells or nuclei, facilitating analysis of transcriptomes or proteomes. One method uses two transgenes, one labeling nuclei of target cells with a biotin ligase receptor and the second expressing the BirA biotin ligase to biotinylate the target nuclei for isolation. Using cell-type-specific DNA elements, this approach generated proteomic profiles of *Xenopus* cardiac nuclei (Amin et al., 2014). Transgenic *Xenopus* in which particular cell types are labeled were used to purify these cells for downstream analysis; this has relied on the speed and simplicity of disaggregating cells in *Xenopus* prior to FACS sorting, which was combined with RNAseq to identify key regulators of tail regeneration (Kakebeen et al., 2020).

Although these transgenic lines allow the biochemistry of development to be studied, there are others that focus more on cell biology. Some lines label various subcellular structures, while others allow real-time spatiotemporal analysis of signaling pathway outputs, including Wnt, calcium, epigenetic changes, and oxidative stress (Love et al., 2013; Offner et al., 2020; Suzuki et al., 2016; Takagi et al., 2013; Tran and Vleminckx, 2014).

Successful transgenic strategies depend on having identified the appropriate DNA control elements to drive expression of transgenes in the required temporal and spatial manner. This has proven the most challenging element of transgene design despite all of the information now available regarding epigenetic modification and conservation of genomes (Kim et al., 2019). While BAC and fosmid transgenesis have been used successfully to address this challenge (Fish et al., 2011; Ochi et al., 2012), gene editing, which is discussed in the following, has produced an alternative approach that avoids the challenges of handling large, fragile constructs and the possibility of extremely distant control elements: inserting a transgene into the endogenous locus so that its expression is controlled by all of the DNA elements and epigenetic mechanisms that regulate the gene normally.

14.3. GENETICS AND GENE EDITING IN *XENOPUS*

Traditionally, *Xenopus* were used mainly for embryological experiments, with limited use as a genetic model. This was largely due to the allotetraploidy of *X. laevis*, which was the species used until recently; its long time to sexual maturity; and the resulting amount of effort required to breed successive generations. The publication of the *X. tropicalis* genome in 2009 and the *X. laevis* genome in 2016 together with new genome editing technologies that have become available in the last seven years have led to the generation of many new *Xenopus* mutants. In this section, we outline a brief history of *Xenopus* mutants and give an overview of gene editing successes in *Xenopus* and future directions.

14.3.1. Beginnings

Prior to the modern era of gene editing, naturally occurring *Xenopus* mutants were identified in the laboratory through successive inbreeding or gynogenetic screens. Several naturally occurring mutations were identified and studied, including *anucleate*, which lacked nucleoli and was instrumental in cloning ribosomal RNA genes (Elsdale et al., 1958; Wallace, 1960). The second spontaneous *X. laevis* mutant to be identified was the *periodic albinism* mutant, which produces white/yellow embryos that are excellent for gene expression analysis and commonly used by the *Xenopus* community (Hoperskaya, 1975). Recently, this mutation was mapped to a 1.9kb deletion in the *hps4* (Hermansky Pudlak syndrome type 4) gene (Fukuzawa, 2021). Other developmental mutants were identified in offspring derived from nuclear transfer animals at the Geneva *Xenopus* Centre (Droin, 1992). These naturally occurring mutants set the foundation for future genetic studies in *Xenopus*.

14.3.2. *X. tropicalis* and Forward Genetic Screens

Interest in *Xenopus* genetics was reignited after the introduction to the laboratory of a closely related species, *Xenopus tropicalis* (Abu-Daya et al., 2012). *X. tropicalis* is the only diploid species in the genus, with one of the smallest known haploid genomes, 1.5×10^9 bp in 10 chromosomes ($2n = 20$).

Genetic screens in zebrafish have yielded many insights into vertebrate developmental biology, but the realization that the teleost genome had undergone a duplication event generated interest in developing a genetic model organism with a more canonically organized genome that was evolutionarily closer to mammals (Glasauer and Neuhauss, 2014). In 1999 at the NIH Non-Mammalian Model Meeting, *Xenopus* was one of the model organisms discussed, and subsequently consensus was reached on ten priority areas for large-scale genomic and genetic resource development for *Xenopus*. These included establishing the viability of *X. tropicalis* as a genetic model organism through pilot genetic screens, including chemical mutagenesis, large-scale radiation-induced deletions, and insertional mutagenesis, and developing resources such as a genetic map and sequencing the *X. tropicalis* genome (Klein et al., 2002). As a result, a number of laboratories undertook forward screens for mutations affecting *X. tropicalis* development (Goda et al., 2006; Noramly et al., 2005).

Since amphibians do not have imprinting and fertilization is external, it is possible to use gynogenesis to obtain diploid embryos without paternal genetic contributions (Tompkins, 1978). Gynogenesis allows F1 female progeny of F0 mutagenized animals to be screened for recessive mutations without the need to obtain adult F2 animals for sib crosses. The technique is simple: irradiated, macerated wild type testes are used for *in vitro* fertilization; the sperm initiate egg cleavage, but their genetic material is destroyed, creating haploid embryos. Diploidy is restored by cold shock or pressure early after fertilization, which inhibits polar body formation and prevents the loss of maternal chromosomes duplicated during meiosis II (Geach et al., 2012). Disadvantages of gynogenesis are that the frequency of mutation depends on the distance from centromeres, ranging from 50% near centromeres to around 10% near telomeres, not the classical Mendelian ratio of 25%. Also, background abnormal gastrulation is higher than in normal fertilization. Gynogenetic screens thus focus on later development, uncovering mutations that affected organogenesis rather than early patterning. In the pilot gynogenetic screens, over 100 potential mutant phenotypes were observed, including defects in heartbeat, motility, pigmentation, otolith formation, haematopoiesis, gut coiling, axis formation, and left-right asymmetry (Goda et al., 2006; Noramly et al., 2005). Once a candidate mutation was identified in a female, an F2 generation was raised to adulthood to confirm the heritability of the phenotype in classic sib crosses.

The first mutation cloned affected cardiac function (Abu-Daya et al., 2009). Homozygous *muzak* tadpoles had no heartbeat, caused by a nonsense mutation in *myh6*. The resulting premature stop codon caused nonsense mediated decay of *myh6* mRNA, and the lack of myosin heavy chain prevented sarcomere formation in *muzak* cardiomyocytes. Another mutation that affected sarcomere assembly was *dicky ticker;* homozygous embryos were completely paralyzed and had no heartbeat. The genetic lesion was a missense mutation in the muscle-specific chaperone *unc45b*, required for the correct folding of the head domain of heavy chain myosins (Geach and Zimmerman, 2010). A model for human disease was *no privacy*, a recessive, non-lethal pigmentation mutant. The phenotype is characterized by significantly reduced pigmentation; the genetic mutation was identified as a 10 base pair deletion in the *hps6* homologue of the *Hermansky-Pudlak Syndrome 6* gene (Nakayama et al., 2017).

Improved genomic resources sped up positional cloning of mutations. In the five years after the identification of *muzak*, six more chemically induced mutants were mapped. These included *kaleidoscope*, characterized by variegated retinal epithelium and head cartilage defects caused by a splicing mutation in the ATPase copper transporting alpha (*atp7a*) gene, which is implicated in Menkes disease; *white heart*, characterized by haematopoiesis defects caused by a mutation in the *smad4.1* gene; *cyd vicious* characterized by a severe eye phenotype and very poor melanocyte migration from the neural tube, mapped to the *DSIF elongation factor subunit (supt5h)* gene; the otolith formation mutants *komimi*, a splicing mutation in the *otoconin90 (oc90)* gene; and *seasick*, a nonsense mutation in the vesicle transport adaptor protein *ap3d1* (Abu-Daya et al., 2012).

14.3.3. TILLING

The screens described previously were all "forward genetic" screens; at the same time, a "reverse genetics" project searched for mutations in specific genes by TILLING (targeting induced local lesions in genomes) using males produced by ENU mutagenesis (Stemple, 2004). Capillary sequencing was initially used to search for mutations in specific genes requested by the *Xenopus* community (Goda et al., 2006). This approach did not produce many mutants, since the F1 males tested were mosaic for mutations due to treating mature sperm, not spermatogonia, with ENU and PCR amplification introduced allele bias. However, a TILLING screen on F1 animals produced by spermatogonial ENU mutagenesis produced a nonsense mutation in the retinal anterior homeobox (*rax*) gene, which resulted in eyeless tadpoles (Fish et al., 2014). With the increasing availability of next-generation sequencing, it became possible to sequence the whole exome of mutants. This new approach uncovered mutations in more than 300 genes, although these were not necessarily in specific genes requested by *Xenopus* researchers.

14.3.4. Insertional Mutagenesis

Another strategy to produce *Xenopus* mutants was insertional mutagenesis. If a transgene integrates into a coding sequence or an important regulatory region, it will disrupt the function of that gene. Insertional mutagenesis is attractive because cloning the site of integration is simpler and faster than positional cloning. The basic approach using Sleeping Beauty is described previously, and a pilot study showed that in these "hopper frogs," the transposon was indeed excised and reintegrated. In approximately 80% of cases, the re-integration was within 3MB of the donor locus (Yergeau et al., 2011b).

Despite these promising initial results, no insertional mutations were produced in these screens. However, a functionally disruptive mutation was discovered accidently when an attempt was made to breed a *Xenopus tropicalis* line, carrying a nkx2.5-GFP insert, to homozygosity (Abu-Daya et al., 2011). Approximately 25% of metamorphic froglets completely lacked forelimbs, including the scapula and clavicle. *In situ* hybridization showed that mutant tadpoles did not express *tbx5* in the prospective forelimb region, although cardiac expression was unaffected. The integration site was cloned by ligation-mediated PCR in the first intron of the *nephronectin* gene, a secreted ligand of Alpha8Beta1, which was necessary for metanephros formation in mouse but had not been implicated in limb generation prior to this point. Early experiments thus showed the potential of using the frog model for genetics experiments; however, before this potential could be transformed into a major screening program as in other models, the landscape was changed by the availability and low cost of large-scale sequencing together with gene editing methods.

14.3.5. Reverse Genetics Using dsBreaks

Reverse genetics really became feasible in *Xenopus* with the introduction of new gene editing techniques that allowed targeted induction of double strand (ds) breaks in the genome (Lei et al., 2013; Tandon et al., 2017). The first study used Zinc Finger Nucleases to demonstrate the possibility of targeted mutations in the *Xenopus* genome (Young et al., 2011). The first published studies with TALENS were in 2012, and there have now been many different TALEN mutants created in both *Xenopus* species (2 in *X. laevis* and 14 in *X. tropicalis*) (Ishibashi et al., 2012; Lei et al., 2012). Initially, most of these studies analyzed F0 mutants with mosaic mutations, but phenotypes were successfully observed. The first germline TALEN mutants generated were F1 compound heterozygotes; these *X. tropicalis pax6* mutants displayed phenotypes similar to human aniridia patients, showing how *Xenopus* can be used to model human disease (Nakayama et al., 2015). The first *X. laevis* germline mutant was made using oocyte host transfer to improve mutation rates in both *tyr.L* and *tyr.S* genes (Ratzan et al., 2017). This paper showed that one can generate mutants effectively in the allotetraploid *X. laevis*. The first homozygous null TALEN mutants were generated in the *X. tropicalis* protein arginine histone methyltransferase 1 gene (*prmt1*) (Shibata et al., 2019). Prmt1 knockout mice die shortly after implantation, preventing functional analysis. Since *Xenopus* embryos develop externally, $prmt1^{-/-}$ embryos survive, allowing for functional analysis, but show delayed growth after five days and eventually die nine days later. Multiple genes can also be mutated using TALENs. Knockout of *rnf43* and *znf3* together (but not individually) led to limb deformities in F0 *X. tropicalis* (Szenker-Ravi et al., 2018). TALEN use peaked in terms of *Xenopus* publications in 2016, but already it had been overtaken by a different method of making ds breaks.

Shortly after TALENs, the CRISPR-Cas technology became available. The ease with which F0 *Xenopus* crispants can be made, the high level of indels caused, and the strong penetrance of resulting phenotypes led this method to become widely used in *Xenopus*. The first *Xenopus* CRISPR mutants were published in 2013, and in the last five years, over 50 different mutants have been published (Blitz et al., 2013; Nakayama et al., 2013). Due to its diploidy and growth at higher temperatures (where Cas is more effective), most CRISPR mutants have been F0 mosaic mutants generated in *X. tropicalis*; they affect a wide range of processes, including immunology, cancer, kidney, neurogenesis, limb, metamorphosis, regeneration, and eye development (Table 14.1). Examples of using *Xenopus* and CRISPR-Cas to underpin the understanding of human genetic disease include a series of studies analyzing the effects of gene variants causing congenital heart defects by the Khoka lab (Bhattacharya et al., 2015; Deniz et al., 2018). Reversade and co-workers investigated tetra-amelia syndrome, which causes lung aplasia and a lack of limbs. The deletion of *rspo2* by CRISPR-Cas caused amelia, validating the link between the gene and disease. Moreover, deletion of two transmembrane ligases associated with the disease caused formation of ectopic limbs, also revealing a master regulator of limb number (Szenker-Ravi et al., 2018). This example demonstrates how both clinical and discovery research is enhanced by analysis of patient-directed gene knockouts in *Xenopus*. For the RhoGEF TRIO, *Xenopus* crispants targeting one of its two human mutation hotspots was used to link that hotspot with

TABLE 14.1

Genetically Altered *Xenopus* Made Using Targeted Nucleases

Gene(s)	Year	Species	F0/F1	Reference/Notes
ZFNs				
egfp, nog	2011	tropicalis	F0	(Young et al., 2011)
tyr	2012	tropicalis	F0/F1	(Nakajima et al., 2012)
nog	2017	tropicalis	F1	(Young et al., 2017) homozygous null
TALENs				
tyr	2012	tropicalis	F0	(Ishibashi et al., 2012)
nog, ptf1a, ets1	2012	tropicalis	F0	(Lei et al., 2012)
egfp	2013	laevis	F0	(Sakuma et al., 2013)
tyr, pax6	2013	laevis	F0	(Suzuki et al., 2013)
ndrg1a	2013	tropicalis	F0	(Zhang et al., 2013)
tyr, nog, mmp-9.2	2013	tropicalis	F0	(Nakajima et al., 2013)
tyr, egfp	2014	laevis	F0	(Sakane et al., 2014)
sp8	2014	tropicalis	F0	(Chung et al., 2014) agrees with MO data
tyr	2015	tropicalis	F1	(Nakajima and Yaoita, 2015) deadsouth/germ cell; F0 crossed with ZFN albino
thra	2015	tropicalis	F0/F1	(Choi et al., 2015)

(Continued)

TABLE 14.1 (Continued)

Gene(s)	Year	Species	F0/F1	Reference/Notes
tyr, pax6	2015	laevis	F0	(Miyamoto et al., 2015) oocyte host transfer
cygb	2015	laevis	F0	(Nakade et al., 2015)
pax6	2015	tropicalis	F1	(Nakayama et al., 2015) compound het
apc	2015	tropicalis	F0	(Nieuwenhuysen et al., 2015)
thra	2015	tropicalis	F0	(Wen and Shi, 2015)
dot1l	2015	tropicalis	F0	(Wen et al., 2015)
ouro1, ouro2, foxn1	2016	tropicalis	F0	(Nakai et al., 2016)
thra	2017	tropicalis	F2	(Choi et al., 2017)
hps6	2017	tropicalis	F1	(Nakayama et al., 2017) agrees with MO data, ENU or natural mutation
p2ry4	2017	laevis	F0	(Harata et al., 2019)
mad1	2017	tropicalis	F2	(Okada et al., 2017)
tyr	2017	laevis	F1	(Ratzan et al., 2017) L and S mutants from oocyte host transfer. Compound hets
thra	2017	tropicalis	F2	(Wen et al., 2017)
tbxt/tbxt.2	2018	tropicalis	F1/F4	(Gentsch et al., 2018) F1 mutants for tbxt; tbxt/tbxt2 double knockouts in F4
thra/thrb	2018	tropicalis	F2/F1	(Nakajima et al., 2018)
mecom	2018	tropicalis	F2	(Okada and Shi, 2018)
pomc	2020	tropicalis	F1	(Shewade et al., 2020) compound het
prmt1	2020	tropicalis	F2	(Shibata et al., 2019)
thra	2021	tropicalis	F2	(Tanizaki et al., 2021b)
CRISPR				
tyr	2013	tropicalis	F0/F1	(Blitz et al., 2013)
tyr	2013	tropicalis	F0/F1	(Nakayama et al., 2013)
cela1.2, ets1, ets2, hspa5, hhex, pgat, pdx1, ptf1a, tm4sf4, tyr	2014	tropicalis	F0/F1	(Guo et al., 2014)
foxj1, pax8, dnah9, galnt1, ctnnb1	2015	tropicalis	F0	(Bhattacharya et al., 2015) agrees with MO data
tubb2b, tyr	2015	tropicalis	F0/F1	(Shi et al., 2015) knockin GFP, ela-GFP
ptf1a, tyr	2015	laevis	F0	(Wang et al., 2015) agrees with MO data
gsc, tyr	2016	tropicalis	F0/F1	(Blitz et al., 2016) leapfrogging, F0 and F1 phenotypes
cfap299	2016	laevis	F0	(Jaffe et al., 2016)
pax3, snai1, ctnnb1, sox9, tfap2a, pax3, zic1, snai2,	2016	tropicalis	F0	(Liu et al., 2016)
rb1, rbl1	2016	tropicalis	F0	(Naert et al., 2016) dual knockout
ctnnb1	2017	both	F0/F1	(Aslan et al., 2017) knock in by oocyte host transfer
xnc genes	2017	laevis	F0	(Banach et al., 2017)
rho	2017	laevis	F0/F1	(Feehan et al., 2017)

(Continued)

TABLE 14.1 (Continued)

Gene(s)	Year	Species	F0/F1	Reference/Notes
tyr	2017	tropicalis	F0	(Park et al., 2017) base editing
npffr1.1	2017	laevis	F0	(Waqas et al., 2017)
lhx1, slc45a2	2018	laevis	F0	(DeLay et al., 2018)
thrb	2018	tropicalis	F0	(Sakane et al., 2018)
fgfr4	2018	tropicalis	F0	(Sempou et al., 2018)
tbx5	2018	tropicalis	F0	(Steimle et al., 2018)
rnf43, znrf3, rspo2	2018	tropicalis	F0	(Szenker-Ravi et al., 2018) supernumery limbs, linked to clinic
katnal2	2018	tropicalis	F0	(Willsey et al., 2018) agrees with MO data
xnc10	2019	laevis		(Banach et al., 2019) cell line knockout
dsp	2019	laevis	F0	(Bharathan and Dickinson, 2019) agrees with MO data
l1cam, crb2	2019	tropicalis	F0	(Date et al., 2019)
dnmbp	2019	laevis	F0	(DeLay et al., 2019) MO as well
tbx4	2019	tropicalis	F0	(Kariminejad et al., 2019) link to clinic, limb defect
neurod2	2019	tropicalis	F0	(Sega et al., 2019) link to clinic; RNA overexpression
six1	2019	tropicalis	F0	(Sullivan et al., 2019)
rpe65, gnat1	2019	laevis	F0	(Wen et al., 2019)
ctnnd1	2020	tropicalis	F0	(Alharatani et al., 2020)
trio	2020	tropicalis	F0	(Barbosa et al., 2020) link to clinic, domain specific effect
pax9	2020	tropicalis	F0	(Farley-Barnes et al., 2020)
tnnc1	2020	tropicalis	F0	(Landim-Vieira et al., 2020)
dlg5	2020	tropicalis	F0	(Marquez et al., 2021) link to clinic
rb1, rbl1, tp53	2020	tropicalis	F0	(Naert et al., 2020a) cancer
multiple	2020	both		(Naert et al., 2020b) designing efficient guides
junB	2020	tropicalis	F0/F1	(Nakamura et al., 2020) compound hets
cfap43, foxj1	2020	laevis	F0	(Rachev et al., 2020) agrees with MO data
daam2	2020	tropicalis	F0	(Schneider et al., 2020) link to clinic
thrb	2020	tropicalis	F2	(Shibata et al., 2020a)
thra/thrb	2020	tropicalis	F2	(Shibata et al., 2020b) double knockout
adprhl1	2020	laevis	F0	(Smith et al., 2020) agrees with MO data
ednra, ednrb2, edn1, edn3	2020	laevis	F0	(Square et al., 2020)
nr3c1	2020	tropicalis	F2	(Sterner et al., 2020)
ncoa3	2020	tropicalis	F1	(Tanizaki et al., 2021a) compound hets
gja8, dnase2b	2020	laevis	F0	(Viet et al., 2020)
Msmb.3	2020	tropicalis	F0	(Wang et al., 2020) agrees with MO data
mtnr1a	2020	tropicalis	F0/F1	(Wiechmann et al., 2020)

(Continued)

TABLE 14.1 (Continued)

Gene(s)	Year	Species	F0/F1	Reference/Notes
dyrk1a	2020	tropicalis	F0	(Willsey et al., 2020) agrees with MO data
chd1	2020	laevis	F0	(Wyatt et al., 2020)
cep70	2021	laevis	F0	(Kim et al., 2021) agrees with MO data
HDR				
krt12.2, npm3	2014	laevis	F0	(Nakade et al., 2014) CRISPR/TALEN, TAL-PITCh and CRIS-PITCh
cfap299	2016	laevis	F0	(Jaffe et al., 2016) plasmid
ctnnb1	2017	both	F0	(Aslan et al., 2017) CRISPR, oocyte host transfer, oligo
myh6, gapdh	2018	tropicalis	F0	(Mao et al., 2018) CRISPR, donor plasmid
xnc10	2019	laevis	F0	(Banach et al., 2019) CRISPR, donor plasmid
slc45a2	2020	both	F0	(Nakayama et al., 2020) CRISPR, ssDNA
Base editing				
tyr, tp53	2017	laevis	F0	(Park et al., 2017) Cas9-linked cytidine deaminase BE3
tyr, tbx5, apc, cyp1b1, kcnj2, tbx22, gdf5, hhex, sftpb, ptf1a	2019	tropicalis	F0	(Shi et al., 2019) Cas9-linked cytidine deaminase BE3

Legend: Examples of published, genetically altered *Xenopus* show how the number of such animals is growing. Uncommon or novel aspects of particular studies are highlighted in the "Notes" column.

a specific phenotype in TRIO patients, showing that such experiments can extend beyond simple loss of function (Barbosa et al., 2020). Other examples include *tbx4*, *rho*, *neurod2*, *daam2*, and *dlg5* (Feehan et al., 2017; Kariminejad et al., 2019; Marquez et al., 2021; Schneider et al., 2020; Sega et al., 2019).

The first germline CRISPR mutants were compound heterozygotes produced using the leapfrogging technique, which bypasses embryonic lethality by transplanting the endoderm (the germ cells' location) from an F0 mutant into wild type; crossing two F0 *gsc* mutants produced this way resulted in the expected phenotypic mutants (Blitz et al., 2016). Compound mutations in *mtrn1a* in F1 *X. tropicalis* caused rod photoreceptor degeneration, while *junB* F1 mutants showed defective tail regeneration (Nakamura et al., 2020; Wiechmann et al., 2020). The first F2 homozygous null CRISPR mutants were made in genes affecting metamorphosis, including *thra*, *thrb*, and *nr3c1*, and more such homozygous null mutants are in the pipeline (Shibata et al., 2020b, 2020a; Sterner et al., 2020). The biggest drawbacks to making germline mutants in *Xenopus* are the effort, time, and cost needed to raise them through metamorphosis. To increase the generation of germline mutants, the National *Xenopus* Resource (NXR) has embarked on a project to generate over 200 mutants. To date, they have produced over 120 mutants, and they also host visiting researchers to come and work on these mutants or create new mutants. These mutants are cataloged on Xenbase and available to all researchers interested in working with them.

14.4. FUTURE DIRECTIONS

CRISPR-Cas gene editing revolutionized site-specific mutation in *Xenopus*, but the use of this new technology to generate site-specific integration is more challenging. Random integration of exogenous DNA through transgenesis works efficiently in *Xenopus*, but site-specific integration of large insertions initially proved difficult to achieve. Initial reports showed such an approach is feasible in *Xenopus*, but the methods were inefficient and unreliable, producing imprecise, mosaic mutations (Aslan et al., 2017; Jaffe et al., 2016; Mao et al., 2018; Nakade et al., 2014; Nakayama et al., 2020). Three of these reports utilized plasmid DNA containing either one or two target sites for sgRNA, allowing cleavage of circular vector in the embryo. In these studies, insertions occurred in both orientations, revealing imprecise integration. In the fourth study, a ssDNA oligonucleotide was injected with Cas9 into oocytes, which were then matured and implanted into another female, and the laid eggs were then fertilized *in vitro*; this is known as the oocyte host transfer (OHT) approach (Aslan et al., 2017). This approach was the most successful, with around 10% precision integration, but only a short ssDNA donor was tested, preventing insertion of fluorescent tags. The problem is that OHT is difficult, and few embryos survive to adulthood. A recent study showed that homology directed repair (HDR) is possible in embryo injections when using long single-stranded DNA (lssDNA) (Nakayama et al., 2020). Though few adult animals were tested for germline transmission, they showed successful insertion of point mutations and fluorescent proteins. One study showed that it is possible, in *X. laevis*, to selectively knock-in constructs into either the L or S homeologue of a gene (Jaffe et al., 2016). The future of HDR insertion in *Xenopus* is to generate novel, tagged proteins to allow for real-time visualization as well to generate precise point mutations to model human disease.

If lssDNA co-injection with CRISPR-Cas9 proves as successful at other loci as in this first report, it may come to dominate making specific, targeted changes to the genome. There are, however, alternatives that avoid making ds breaks but can make small alterations. Base editing involves targeting a Cas9 nickase fused to either a cytidine or adenine deaminase to a specific site in the genome, there converting T-A to C-G or G-C to A-T, respectively. This method has been used successfully in *Xenopus*, producing editing rates of up to 20.5%, but not extensively taken up by the community, most likely due to the limited changes available and the lack of single base-pair accuracy (Park et al., 2017).

Prime-editing is a related method that allows slightly larger changes to be made using Cas9 nickase fused to reverse transcriptase and a complex multifunctional prime-editing sgRNA (Anzalone et al., 2020). At the time of writing, this technique is to our knowledge yet to be used in *Xenopus* with success.

14.5. CONCLUSIONS

The range of methods currently available to *Xenopus* researchers for efficient genome modification is extensive and contains robust techniques for both transgenesis and mutagenesis. When combined and used together with other experimental manipulations commonly used in *Xenopus*, such as tissue-targeted microinjection, for example, they permit the pursuit of experimental questions not easily explored in other model systems. Mutations in human *PKD1* are associated with autosomal-dominant polycystic kidney disease (ADPKD), and homozygous mutations or deletions of *pkd1* in mice are embryonic lethal (Blackburn and Miller, 2019). Similarly, CRISPR-Cas9 induced mutations in *X. laevis pkd1* result in F0 edema and eventual lethality (Figure 14.1A). To bypass the embryonic lethality and more specifically target the germline for generation of mutant lines, *pkd1* sgRNAs can be microinjected into the vegetal pole at the 16-cell stage (Figure 14.1B). The effects of *pkd1* mutations on early kidney morphogenesis can, however, be easily observed in *Xenopus* if induced in the *pax8:GFP* transgenic background in which the pronephros is labeled with GFP. This is even more powerful if combined with unilateral mutagenesis where the *pkd1* sgRNA injected side shows disrupted pronephric morphology, and the other side injected with a control sgRNA provides a morphologically normal internal control (Figure 14.1C,D). *Xenopus*, which has always been a powerful system in many respects, has bravely leapt into the heart of the genetic age.

ACKNOWLEDGMENTS

We wish to thank Ira Blitz for providing helpful comments during the writing of this chapter. We apologize to those individuals whose publications we missed in compiling this

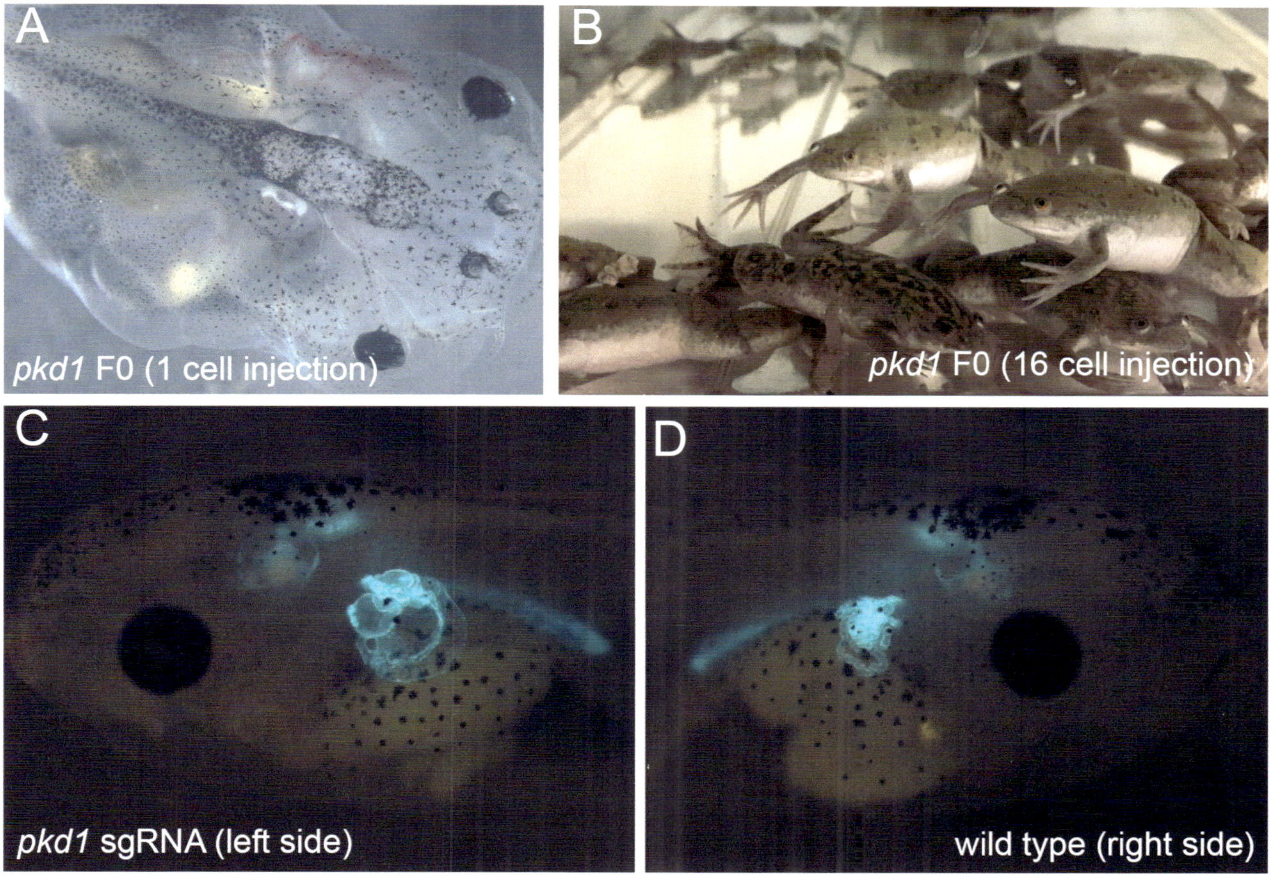

FIGURE 14.1 CRISPR knockout of *pkd1* in *X. laevis*. (A) Ten-day-old F0 *pkd1* knockout tadpole with severe edema; these tadpoles die before 14 days. (B) Six-month-old F0 *pkd1* frogs generated by injecting the sgRNA into vegetal blastomeres at the 16-cell stage. These F0 tadpoles do not generate edema and survive to adulthood. (C) Left side of *Xla.Tg(pax8:GFP)*^{Ogino} transgenic tadpole injected with *pkd1* sgRNA. Pronephric tubules are dilated. (D) Uninjected right side of tadpole showing normal pronephric tubules.

information. The NXR is funded by grants from the National Institutes of Health (P40OD010997, R24OD030008, and R01HD084409). The EXRC is supported by grants from the Wellcome Trust (212942/Z/18/Z) and BBSRC (BB/R014841/1).

REFERENCES

Abu-Daya, A., Khokha, M.K., and Zimmerman, L.B. (2012). The hitchhiker's guide to Xenopus genetics. *Genesis* (New York, NY: 2000) *50*, 164–175.

Abu-Daya, A., Nishimoto, S., Fairclough, L., Mohun, T.J., Logan, M.P.O., and Zimmerman, L.B. (2011). The secreted integrin ligand nephronectin is necessary for forelimb formation in *Xenopus tropicalis*. *Developmental Biology 349*, 204–212.

Abu-Daya, A., Sater, A.K., Wells, D.E., Mohun, T.J., and Zimmerman, L.B. (2009). Absence of heartbeat in the *Xenopus tropicalis* mutation muzak is caused by a nonsense mutation in cardiac myosin myh6. *Dev Biol 336*, 20–29.

Alharatani, R., Ververi, A., Beleza-Meireles, A., Ji, W., Mis, E., Patterson, Q.T., Griffin, J.N., Bhujel, N., Chang, C.A., Dixit, A., et al. (2020). Novel truncating mutations in CTNND1 cause a dominant craniofacial and cardiac syndrome. *Hum Mol Genet 29*, 1900–1921.

Allen, B.G., and Weeks, D.L. (2005). Transgenic *Xenopus laevis* embryos can be generated using phiC31 integrase. *Nature Methods 2*, 975–979.

Amaya, E. (2005). Xenomics. *Genome Res 15*, 1683–1691.

Amin, N.M., Greco, T.M., Kuchenbrod, L.M., Rigney, M.M., Chung, M.-I., Wallingford, J.B., Cristea, I.M., and Conlon, F.L. (2014). Proteomic profiling of cardiac tissue by isolation of nuclei tagged in specific cell types (INTACT). *Development 141*, 962–973.

Anzalone, A.V., Koblan, L.W., and Liu, D.R. (2020). Genome editing with CRISPR: Cas nucleases, base editors, transposases and prime editors. *Nature Biotechnology 38*, 824–844.

Aslan, Y., Tadjuidje, E., Zorn, A.M., and Cha, S.-W. (2017). High-efficiency non-mosaic CRISPR-mediated knock-in and indel mutation in F0 Xenopus. *Development* (Cambridge, England) *144*, 2852–2858.

Banach, M., Edholm, E.-S., Gonzalez, X., Benraiss, A., and Robert, J. (2019). Impacts of the MHC class I-like XNC10 and innate-like T cells on tumor tolerance and rejection in the amphibian Xenopus. *Carcinogenesis 40*, 924–935.

Banach, M., Edholm, E.-S., and Robert, J. (2017). Exploring the functions of nonclassical MHC class Ib genes in *Xenopus laevis* by the CRISPR/Cas9 system. *Dev Biol 426*, 261–269.

Barbosa, S., Greville-Heygate, S., Bonnet, M., Godwin, A., Fagotto-Kaufmann, C., Kajava, A.V., Laouteouet, D., Mawby, R., Wai, H.A., Dingemans, A.J.M., et al. (2020). Opposite modulation of RAC1 by mutations in TRIO is associated with distinct, domain-specific neurodevelopmental disorders. *Am J Hum Genetics 106*, 338–355.

Beck, C.W., Christen, B., Barker, D., and Slack, J.M.W. (2006). Temporal requirement for bone morphogenetic proteins in regeneration of the tail and limb of Xenopus tadpoles. *Mechanisms of Development 123*, 674–688.

Beck, C.W., Christen, B., and Slack, J.M.W. (2003). Molecular pathways needed for regeneration of spinal cord and muscle in a vertebrate. *Dev Cell 5*, 429–439.

Beck, C.W., and Slack, J.M. (1999). Gut specific expression using mammalian promoters in transgenic *Xenopus laevis*. *Mechanisms of Development 88*, 221–227.

Bharathan, N.K., and Dickinson, A.J.G. (2019). Desmoplakin is required for epidermal integrity and morphogenesis in the *Xenopus laevis* embryo. *Developmental Biology 450*, 115–131.

Bhattacharya, D., Marfo, C.A., Li, D., Lane, M., and Khokha, M.K. (2015). CRISPR/Cas9: An inexpensive, efficient loss of function tool to screen human disease genes in Xenopus. *Developmental Biology 408*, 196–204.

Blackburn, A.T.M., and Miller, R.K. (2019). Modeling congenital kidney diseases in *Xenopus laevis*. *Disease Models & Mechanisms 12*, dmm038604-12.

Blitz, I.L., Biesinger, J., Xie, X., and Cho, K.W.Y. (2013). Biallelic genome modification in F(0) *Xenopus tropicalis* embryos using the CRISPR/Cas system. *Genesis* (New York, NY: 2000) *51*, 827–834.

Blitz, I.L., Fish, M.B., and Cho, K.W.Y. (2016). Leapfrogging: Primordial germ cell transplantation permits recovery of CRISPR/Cas9-induced mutations in essential genes. *Development* (Cambridge, England) *143*, 2868–2875.

Bronchain, O.J., Hartley, K.O., and Amaya, E. (1999). A gene trap approach in Xenopus. *Curr Biol 9*, 1195, S1.

Brown, D.D., Cai, L., Das, B., Marsh-Armstrong, N., Schreiber, A.M., and Juste, R. (2005). Thyroid hormone controls multiple independent programs required for limb development in *Xenopus laevis* metamorphosis. *Proceedings of the National Academy of Sciences of the United States of America 102*, 12455–12458.

Buchholz, D.R. (2012). Tet-on binary systems for tissue-specific and inducible transgene expression. In *Xenopus Protocols* (Humana Press). *Methods Mol Biol 917*, 265–275.

Buchholz, D.R. (2015). More similar than you think: Frog metamorphosis as a model of human perinatal endocrinology. *Developmental Biology 408*, 188–195.

Cai, L., Das, B., and Brown, D.D. (2007). Changing a limb muscle growth program into a resorption program. *Developmental Biology 304*, 260–271.

Chesneau, A., Sachs, L.M., Chai, N., Chen, Y., Pasquier, L., Loeber, J., Pollet, N., Reilly, M., Weeks, D.L., and Bronchain, O.J. (2008). Transgenesis procedures in Xenopus. *Biol Cell 100*, 503–529.

Choi, J., Ishizuya-Oka, A., and Buchholz, D.R. (2017). Growth, development, and intestinal remodeling occurs in the absence of thyroid hormone receptor α in tadpoles of *Xenopus tropicalis*. *Endocrinology 158*, 1623–1633.

Choi, J., Suzuki, K.-I.T., Sakuma, T., Shewade, L., Yamamoto, T., and Buchholz, D.R. (2015). Unliganded thyroid hormone receptor α regulates developmental timing via gene repression in *Xenopus tropicalis*. *Endocrinology 156*, 735–744.

Chung, H.A., Medina-Ruiz, S., and Harland, R.M. (2014). Sp8 regulates inner ear development. *Proc National Acad Sci 111*, 6329–6334.

Das, B., and Brown, D.D. (2004a). Controlling transgene expression to study *Xenopus laevis* metamorphosis. *P Natl Acad Sci Usa 101*, 4839–4842.

Das, B., and Brown, D.D. (2004b). Controlling transgene expression to study *Xenopus laevis* metamorphosis. *Proceedings of the National Academy of Sciences of the United States of America 101*, 4839–4842.

Date, P., Ackermann, P., Furey, C., Fink, I.B., Jonas, S., Khokha, M.K., Kahle, K.T., and Deniz, E. (2019). Visualizing flow in an intact CSF network using optical coherence tomography: Implications for human congenital hydrocephalus. *Sci Rep-Uk 9*, 6196.

Daughters, R.S., Chen, Y., and Slack, J.M.W. (2011). Origin of muscle satellite cells in the Xenopus embryo. *Development* (Cambridge, England) *138*, 821–830.

DeLay, B.D., Baldwin, T.A., and Miller, R.K. (2019). Dynamin binding protein is required for *Xenopus laevis* kidney development. *Front Physiol 10*, 143.

DeLay, B.D., Corkins, M.E., Hanania, H.L., Salanga, M., Deng, J.M., Sudou, N., Taira, M., Horb, M.E., and Miller, R.K.

(2018). Tissue-specific gene inactivation in *Xenopus laevis*: Knockout of lhx1 in the kidney with CRISPR/Cas9. *Genetics 208*, 673–686.

Deniz, E., Mis, E.K., Lane, M., and Khokha, M.K. (2018). CRISPR/Cas9 F0 screening of congenital heart disease genes in *Xenopus tropicalis*. In *Xenopus*, (Humana Press), *Methods Mol Biol* 1865, 163–174.

Droin, A. (1992). The developmental mutants of *Xenopus*. *Int J Dev Bio 36*, 455–464.

Edholm, E.S., and Robert, J. (2018). RNAi-mediated loss of function of Xenopus immune genes by transgenesis. *Cold Spring Harb Protocols* pdb.prot101519.

Elsdale, T.R., Fischberg, M., and Smith, S. (1958). A mutation that reduces nucleolar number in *Xenopus laevis*. *Exp Cell Res 14*, 642–643.

Etkin, L.D., and Pearman, B. (1987). Distribution, expression and germ line transmission of exogenous DNA sequences following microinjection into *Xenopus laevis* eggs. *Development 99*, 15–23.

Etkin, L., and Roberts, M. (1983). Transmission of integrated sea urchin histone genes by nuclear transplantation in *Xenopus laevis*. *Science 221*, 67–69.

Farley-Barnes, K.I., Deniz, E., Overton, M.M., Khokha, M.K., and Baserga, S.J. (2020). Paired box 9 (PAX9), the RNA polymerase II transcription factor, regulates human ribosome biogenesis and craniofacial development. *PLoS Genet 16*, e1008967.

Feehan, J.M., Chiu, C.N., Stanar, P., Tam, B.M., Ahmed, S.N., and Moritz, O.L. (2017). Modeling dominant and recessive forms of retinitis pigmentosa by editing three rhodopsin-encoding genes in *Xenopus laevis* using CRISPR/Cas9. *Sci Rep-Uk 7*, 6920.

Fish, M.B., Nakayama, T., Fisher, M., Hirsch, N., Cox, A., Reeder, R., Carruthers, S., Hall, A., Stemple, D.L., and Grainger, R.M. (2014). Xenopus mutant reveals necessity of rax for specifying the eye field which otherwise forms tissue with telencephalic and diencephalic character. *Developmental Biology 395*, 317–330.

Fish, M.B., Nakayama, T., and Grainger, R.M. (2011). Simple, fast, tissue-specific bacterial artificial chromosome transgenesis in Xenopus. *Genesis* (New York, NY: 2000) *50*, 307–315.

Fukuzawa, T. (2021). Periodic albinism of a widely used albino mutant of *Xenopus laevis* caused by deletion of two exons in the Hermansky-Pudlak syndrome type 4 gene. *Genes Cells 26*, 31–39.

Gargioli, C., and Slack, J.M.W. (2004). Cell lineage tracing during Xenopus tail regeneration. *Development* (Cambridge, England) *131*, 2669–2679.

Geach, T.J., Stemple, D.L., and Zimmerman, L.B. (2012). Genetic analysis of *Xenopus tropicalis*. *Methods in Molecular Biology* (Clifton, NJ) *917*, 69–110.

Geach, T.J., and Zimmerman, L.B. (2010). Paralysis and delayed Z-disc formation in the *Xenopus tropicalis* unc45b mutant *dicky ticker*. *BMC Developmental Biology 10*, 75.

Gentsch, G.E., Spruce, T., Monteiro, R.S., Owens, N.D.L., Martin, S.R., and Smith, J.C. (2018). Innate immune response and off-target mis-splicing are common morpholino-induced side effects in Xenopus. *Dev Cell 44*, 597–610, e10.

Glasauer, S.M.K., and Neuhauss, S.C.F. (2014). Whole-genome duplication in teleost fishes and its evolutionary consequences. *Mol Genet Genomics 289*, 1045–1060.

Goda, T., Abu-Daya, A., Carruthers, S., Clark, M.D., Stemple, D.L., and Zimmerman, L.B. (2006). Genetic screens for mutations affecting development of *Xenopus tropicalis*. *PLoS Genetics 2*, e91.

Grant, I.M., Balcha, D., Hao, T., Shen, Y., Trivedi, P., Patrushev, I., Fortriede, J.D., Karpinka, J.B., Liu, L., Zorn, A.M., et al. (2015). The Xenopus ORFeome: A resource that enables functional genomics. *Dev Biol 408*, 345–357.

Gross, J.B., Hanken, J., Oglesby, E., and Marsh-Armstrong, N. (2006). Use of a ROSA26: GFP transgenic line for long-term Xenopus fate-mapping studies. *Journal of Anatomy 209*, 401–413.

Groth, A.C., Fish, M., Nusse, R., and Calos, M.P. (2004). Construction of transgenic Drosophila by using the site-specific integrase from phage φC31. *Genetics 166*, 1775–1782.

Guo, X., Zhang, T., Hu, Z., Zhang, Y., Shi, Z., Wang, Q., Cui, Y., Wang, F., Zhao, H., and Chen, Y. (2014). Efficient RNA/Cas9-mediated genome editing in *Xenopus tropicalis*. *Development* (Cambridge, England) *141*, 707–714.

Hamlet, M.R.J., Yergeau, D.A., Kuliyev, E., Takeda, M., Taira, M., Kawakami, K., and Mead, P.E. (2006). Tol2 transposon-mediated transgenesis in *Xenopus tropicalis*. *Genesis 44*, 438–445.

Harata, A., Hirakawa, M., Sakuma, T., Yamamoto, T., and Hashimoto, C. (2019). Nucleotide receptor P2RY4 is required for head formation via induction and maintenance of head organizer in *Xenopus laevis*. *Dev Growth Differ 61*, 186–197.

Hoperskaya, O. (1975). The development of animals homozygous for a mutation causing periodic albinism (ap) in *Xenopus laevis*. *Journal of Embryology and Experimental Morphology 34*, 253–264.

Horb, M.E., Shen, C.N., Tosh, D., and Slack, J.M.W. (2003). Experimental conversion of liver to pancreas. *Current Biology: CB 13*, 105–115.

Horb, M.E., Wlizla, M., Abu-Daya, A., McNamara, S., Gajdasik, D., Igawa, T., Suzuki, A., Ogino, H., Noble, A., France, C. de R.B.X. team in, et al. (2019). Xenopus resources: Transgenic, inbred and mutant animals, training opportunities, and web-based support. *Frontiers in Physiology 10*, 387.

Ishibashi, S., Cliffe, R., and Amaya, E. (2012). Highly efficient bi-allelic mutation rates using TALENs in *Xenopus tropicalis*. *Biology Open 1*, 1273–1276.

Jaffe, K.M., Grimes, D.T., Schottenfeld-Roames, J., Werner, M.E., Ku, T.-S.J., Kim, S.K., Pelliccia, J.L., Morante, N.F.C., Mitchell, B.J., and Burdine, R.D. (2016). c21orf59/kurly controls both cilia motility and polarization. *Cell Reports 14*, 1841–1849.

Jarikji, Z., Horb, L.D., Shariff, F., Mandato, C.A., Cho, K.W.Y., and Horb, M.E. (2009). The tetraspanin Tm4sf3 is localized to the ventral pancreas and regulates fusion of the dorsal and ventral pancreatic buds. *Development* (Cambridge, England) *136*, 1791–1800.

Jarikji, Z.H., Vanamala, S., Beck, C.W., Wright, C.V.E., Leach, S.D., and Horb, M.E. (2007). Differential ability of Ptf1a and Ptf1a-VP16 to convert stomach, duodenum and liver to pancreas. *Developmental Biology 304*, 786–799.

Kakebeen, A.D., Chitsazan, A.D., Williams, M.C., Saunders, L.M., and Wills, A.E. (2020). Chromatin accessibility dynamics and single cell RNA-Seq reveal new regulators of regeneration in neural progenitors. *Elife 9*, e52648.

Kariminejad, A., Szenker-Ravi, E., Lekszas, C., Tajsharghi, H., Moslemi, A.-R., Naert, T., Tran, H.T., Ahangari, F., Rajaei, M., Nasseri, M., et al. (2019). Homozygous null TBX4 mutations lead to posterior amelia with pelvic and pulmonary hypoplasia. *Am J Hum Genetics 105*, 1294–1301.

Kerney, R.R., Brittain, A.L., Hall, B.K., and Buchholz, D.R. (2012). Cartilage on the move: Cartilage lineage tracing during tadpole metamorphosis. *Development, Growth & Differentiation 54*, 739–752.

Kim, D., An, H., Shearer, R.S., Sharif, M., Fan, C., Choi, J., Ryu, S., and Park, Y. (2019). A principled strategy for mapping enhancers to genes. *Sci Rep-Uk 9*, 11043.

Kim, S.K., Brotslaw, E., Thome, V., Mitchell, J., Ventrella, R., Collins, C., and Mitchell, B. (2021). A role for Cep70 in centriole amplification in multiciliated cells. *Dev Biol 471*, 10–17.

Klein, S.L., Strausberg, R.L., Wagner, L., Pontius, J., Clifton, S.W., and Richardson, P. (2002). Genetic and genomic tools for Xenopus research: The NIH Xenopus initiative. *Null 225*, 384–391.

Kroll, K.L., and Amaya, E. (1996). Transgenic Xenopus embryos from sperm nuclear transplantations reveal FGF signaling requirements during gastrulation. *Development* (Cambridge, England) *122*, 3173–3183.

Kwan, K.M., Fujimoto, E., Grabher, C., Mangum, B.D., Hardy, M.E., Campbell, D.S., Parant, J.M., Yost, H.J., Kanki, J.P., and Chien, C. (2007). The Tol2kit: A multisite gateway-based construction kit for Tol2 transposon transgenesis constructs. *Dev Dynam 236*, 3088–3099.

Landim-Vieira, M., Johnston, J.R., Ji, W., Mis, E.K., Tijerino, J., Spencer-Manzon, M., Jeffries, L., Hall, E.K., Panisello-Manterola, D., Khokha, M.K., et al. (2020). Familial dilated cardiomyopathy associated with a novel combination of compound heterozygous TNNC1 variants. *Front Physiol 10*, 1612.

Lane, M.A., Kimber, M., and Khokha, M.K. (2013). Breeding based remobilization of Tol2 transposon in *Xenopus tropicalis*. *PLoS One 8*, e76807.

Lei, Y., Guo, X., Deng, Y., Chen, Y., and Zhao, H. (2013). Generation of gene disruptions by transcription activator-like effector nucleases (TALENs) in *Xenopus tropicalis* embryos. *Cell & Bioscience 3*, 21.

Lei, Y., Guo, X., Liu, Y., Cao, Y., Deng, Y., Chen, X., Cheng, C.H.K., Dawid, I.B., Chen, Y., and Zhao, H. (2012). Efficient targeted gene disruption in Xenopus embryos using engineered transcription activator-like effector nucleases (TALENs). *Proceedings of the National Academy of Sciences of the United States of America 109*, 17484–17489.

Li, Y.E., Allen, B.G., and Weeks, D.L. (2012). Xenopus protocols, post-genomic approaches. *Methods Mol Biology 917*, 219–230.

Liu, Z., Cheng, T.T.K., Shi, Z., Liu, Z., Lei, Y., Wang, C., Shi, W., Chen, X., Qi, X., Cai, D., et al. (2016). Efficient genome editing of genes involved in neural crest development using the CRISPR/Cas9 system in Xenopus embryos. *Cell & Bioscience 6*, 22.

Love, N.R., Chen, Y., Ishibashi, S., Kritsiligkou, P., Lea, R., Koh, Y., Gallop, J.L., Dorey, K., and Amaya, E. (2013). Amputation-induced reactive oxygen species are required for successful Xenopus tadpole tail regeneration. *Nature Publishing Group 15*, 222–228.

Love, N., Thuret, R., Chen, Y., Ishibashi, S., Sabherwal, N., Paredes, R., Alves-Silva, J., Dorey, K., Noble, A.M., Guille, M.J., et al. (2011a). pTransgenesis: A cross-species, modular transgenesis resource. *Development* (Cambridge, England) *138*, 5451–5458.

Love, N.R., Thuret, R., Chen, Y., Ishibashi, S., Sabherwal, N., Paredes, R., Alves-Silva, J., Dorey, K., Noble, A.M., Guille, M.J., et al. (2011b). pTransgenesis: A cross-species, modular transgenesis resource. *Development 138*, 5451–5458.

Mao, C.-Z., Zheng, L., Zhou, Y.-M., Wu, H.-Y., Xia, J.-B., Liang, C.-Q., Guo, X.-F., Peng, W.-T., Zhao, H., Cai, W.-B., et al. (2018). CRISPR/Cas9-mediated efficient and precise targeted integration of donor DNA harboring double cleavage sites in *Xenopus tropicalis*. *The FASEB Journal 32*, 6495–6509.

Marquez, J., Mann, N., Arana, K., Deniz, E., Ji, W., Konstantino, M., Mis, E.K., Deshpande, C., Jeffries, L., McGlynn, J., et al. (2021). DLG5 variants are associated with multiple congenital anomalies including ciliopathy phenotypes. *J Med Genet 58*, 453–464.

Marsh-Armstrong, N., Cai, L., and Brown, D.D. (2004). Thyroid hormone controls the development of connections between the spinal cord and limbs during *Xenopus laevis* metamorphosis. *Proc National Acad Sci 101*, 165–170.

Marsh-Armstrong, N., Huang, H., Berry, D.L., and Brown, D.D. (1999). Germ-line transmission of transgenes in *Xenopus laevis*. *Proc National Acad Sci 96*, 14389–14393.

Miyamoto, K., Suzuki, K.T., Suzuki, M., Sakane, Y., Sakuma, T., Herberg, S., Simeone, A., Simpson, D., Jullien, J., Yamamoto, T., et al. (2015). The expression of TALEN before fertilization provides a rapid knock-out phenotype in *Xenopus laevis* founder embryos. *PLoS One 10*, e0142946.

Mukhi, S., and Brown, D.D. (2011). Transdifferentiation of tadpole pancreatic acinar cells to duct cells mediated by Notch and stromelysin-3. *Developmental Biology 351*, 311–317.

Mukhi, S., Cai, L., and Brown, D.D. (2010). Gene switching at *Xenopus laevis* metamorphosis. *Developmental Biology 338*, 117–126.

Mukhi, S., Horb, M.E., and Brown, D.D. (2009). Remodeling of insulin producing beta-cells during *Xenopus laevis* metamorphosis. *Developmental Biology 328*, 384–391.

Mukhi, S., Mao, J., and Brown, D.D. (2008). Remodeling the exocrine pancreas at metamorphosis in *Xenopus laevis*. *Proceedings of the National Academy of Sciences of the United States of America 105*, 8962–8967.

Naert, T., Colpaert, R., Nieuwenhuysen, T.V., Dimitrakopoulou, D., Leoen, J., Haustraete, J., Boel, A., Steyaert, W., Lepez, T., Deforce, D., et al. (2016). CRISPR/Cas9 mediated knockout of rb1 and rbl1 leads to rapid and penetrant retinoblastoma development in *Xenopus tropicalis*. *Scientific Reports 6*, 35264.

Naert, T., Dimitrakopoulou, D., Tulkens, D., Demuynck, S., Carron, M., Noelanders, R., Eeckhout, L., Isterdael, G.V., Deforce, D., Vanhove, C., et al. (2020a). RBL1 (p107) functions as tumor suppressor in glioblastoma and small-cell pancreatic neuroendocrine carcinoma in *Xenopus tropicalis*. *Oncogene 39*, 2692–2706.

Naert, T., Tulkens, D., Edwards, N.A., Carron, M., Shaidani, N.-I., Wlizla, M., Boel, A., Demuynck, S., Horb, M.E., Coucke, P., et al. (2020b). Maximizing CRISPR/Cas9 phenotype penetrance applying predictive modeling of editing outcomes in Xenopus and zebrafish embryos. *Sci Rep-Uk 10*, 14662.

Nakade, S., Sakuma, T., Sakane, Y., Hara, Y., Kurabayashi, A., Kashiwagi, K., Kashiwagi, A., Yamamoto, T., and Obara, M. (2015). Homeolog-specific targeted mutagenesis in *Xenopus laevis* using TALENs. *In Vitro Cellular & Developmental Biology: Animal 51*, 879–884.

Nakade, S., Tsubota, T., Sakane, Y., Kume, S., Sakamoto, N., Obara, M., Daimon, T., Sezutsu, H., Yamamoto, T., Sakuma, T., et al. (2014). Microhomology-mediated end-joining-dependent integration of donor DNA in cells and animals using TALENs and CRISPR/Cas9. *Nature Communications 5*, 5560.

Nakai, Y., Nakajima, K., Robert, J., and Yaoita, Y. (2016). Ouro proteins are not essential to tail regression during *Xenopus tropicalis* metamorphosis. *Genes Cells 21*, 275–286.

Nakajima, K., Nakai, Y., Okada, M., and Yaoita, Y. (2013). Targeted gene disruption in the *Xenopus tropicalis* genome using designed TALE nucleases. *Zool Sci 30*, 455–460.

Nakajima, K., Nakajima, T., Takase, M., and Yaoita, Y. (2012). Generation of albino *Xenopus tropicalis* using zinc-finger nucleases. *Development, Growth & Differentiation 54*, 777–784.

Nakajima, K., Tazawa, I., and Yaoita, Y. (2018). Thyroid hormone receptor α-and β-knockout *Xenopus tropicalis* tadpoles reveal subtype-specific roles during development. *Endocrinology 159*, 733–743.

Nakajima, K., and Yaoita, Y. (2015). Development of a new approach for targeted gene editing in primordial germ cells using TALENs in Xenopus. *Biology Open 4*, 259–266.

Nakamura, M., Yoshida, H., Takahashi, E., Wlizla, M., Takebayashi-Suzuki, K., Horb, M.E., and Suzuki, A. (2020). The AP-1 transcription factor junB functions in Xenopus tail regeneration by positively regulating cell proliferation. *Biochemical and Biophysical Research Communications 522*, 990–995.

Nakayama, T., Fish, M.B., Fisher, M., Oomen-Hajagos, J., Thomsen, G.H., and Grainger, R.M. (2013). Simple and efficient CRISPR/Cas9-mediated targeted mutagenesis in *Xenopus tropicalis*. *Genesis* (New York, NY: 2000) *51*, 835–843.

Nakayama, T., Fisher, M., Nakajima, K., Odeleye, A.O., Zimmerman, K.B., Fish, M.B., Yaoita, Y., Chojnowski, J.L., Lauderdale, J.D., Netland, P.A., et al. (2015). Xenopus pax6 mutants affect eye development and other organ systems, and have phenotypic similarities to human aniridia patients. *Developmental Biology 408*, 328–344.

Nakayama, T., Grainger, R.M., and Cha, S. (2020). Simple embryo injection of long single-stranded donor templates with the CRISPR/Cas9 system leads to homology-directed repair in *Xenopus tropicalis* and *Xenopus laevis*. *Genesis 58*, e23366.

Nakayama, T., Nakajima, K., Cox, A., Fisher, M., Howell, M., Fish, M.B., Yaoita, Y., and Grainger, R.M. (2017). No privacy, a *Xenopus tropicalis* mutant, is a model of human Hermansky-Pudlak syndrome and allows visualization of internal organogenesis during tadpole development. *Dev Biol 426*, 472–486.

Nieuwenhuysen, T.V., Naert, T., Tran, H.T., Imschoot, G.V., Geurs, S., Sanders, E., Creytens, D., Roy, F.V., and Vleminckx, K. (2015). TALEN-mediated apc mutation in *Xenopus tropicalis* phenocopies familial adenomatous polyposis. *Oncoscience 2*, 555–566.

Noramly, S., Zimmerman, L., Cox, A., Aloise, R., Fisher, M., and Grainger, R.M. (2005). A gynogenetic screen to isolate naturally occurring recessive mutations in *Xenopus tropicalis*. *Mechanisms of Development 122*, 273–287.

Ochi, H., Tamai, T., Nagano, H., Kawaguchi, A., Sudou, N., and Ogino, H. (2012). Evolution of a tissue-specific silencer underlies divergence in the expression of pax2 and pax8 paralogues. *Nature Communications 3*, 848–847.

Offield, M.F., Hirsch, N., and Grainger, R.M. (2000). The development of *Xenopus tropicalis* transgenic lines and their use in studying lens developmental timing in living embryos. *Dev Camb Engl 127*, 1789–1797.

Offner, T., Daume, D., Weiss, L., Hassenklöver, T., and Manzini, I. (2020). Whole-brain calcium imaging in larval Xenopus. *Cold Spring Harb Protoc*, pdb.prot106815.

Okada, M., Miller, T.C., Wen, L., and Shi, Y.-B. (2017). A balance of Mad and Myc expression dictates larval cell apoptosis and adult stem cell development during Xenopus intestinal metamorphosis. *Cell Death Dis 8*, e2787–e2787.

Okada, M., and Shi, Y. (2018). EVI and MDS/EVI are required for adult intestinal stem cell formation during postembryonic vertebrate development. *Faseb J 32*, 431–439.

Pan, F.C., Chen, Y., Loeber, J., Henningfeld, K., and Pieler, T. (2006). I-SceI meganuclease-mediated transgenesis in Xenopus. *Developmental Dynamics: An Official Publication of the American Association of Anatomists 235*, 247–252.

Park, D.-S., Yoon, M., Kweon, J., Jang, A.-H., Kim, Y., and Choi, S.-C. (2017). Targeted base editing via RNA-guided cytidine deaminases in *Xenopus laevis* embryos. *Mol Cells 40*, 823–827.

Rachev, E., Schuster-Gossler, K., Fuhl, F., Ott, T., Tveriakhina, L., Beckers, A., Hegermann, J., Boldt, K., Mai, M., Kremmer, E., et al. (2020). CFAP43 modulates ciliary beating in mouse and Xenopus. *Dev Biol 459*, 109–125.

Rankin, S.A., Zorn, A.M., and Buchholz, D.R. (2011). New doxycycline-inducible transgenic lines in Xenopus. *Developmental Dynamics: An Official Publication of the American Association of Anatomists 240*, 1467–1474.

Ratzan, W., Falco, R., Salanga, C., Salanga, M., and Horb, M.E. (2017). Generation of a *Xenopus laevis* F1 albino J strain by genome editing and oocyte host-transfer. *Developmental Biology 426*, 188–193.

Roose, M., Sauert, K., Turan, G., Solomentsew, N., Werdien, D., Pramanik, K., Senkel, S., Ryffel, G.U., and Waldner, C. (2009). Heat-shock inducible Cre strains to study organogenesis in transgenic *Xenopus laevis*. *Transgenic Research 18*, 595–605.

Rusconi, S., and Schaffner, W. (1981). Transformation of frog embryos with a rabbit beta-globin gene. *Proc National Acad Sci 78*, 5051–5055.

Sakane, Y., Iida, M., Hasebe, T., Fujii, S., Buchholz, D.R., Ishizuya-Oka, A., Yamamoto, T., and Suzuki, K.-I.T. (2018). Functional analysis of thyroid hormone receptor beta in *Xenopus tropicalis* founders using CRISPR-Cas. *Biology Open 7*, bio030338.

Sakane, Y., Sakuma, T., Kashiwagi, K., Kashiwagi, A., Yamamoto, T., and Suzuki, K.T. (2014). Targeted mutagenesis of multiple and paralogous genes in *Xenopus laevis* using two pairs of transcription activator-like effector nucleases. *Dev Growth Differ 56*, 108–114.

Sakuma, T., Hosoi, S., Woltjen, K., Suzuki, K., Kashiwagi, K., Wada, H., Ochiai, H., Miyamoto, T., Kawai, N., Sasakura, Y., et al. (2013). Efficient TALEN construction and evaluation methods for human cell and animal applications. *Genes to Cells: Devoted to Molecular & Cellular Mechanisms 18*, 315–326.

Schneider, R., Deutsch, K., Hoeprich, G.J., Marquez, J., Hermle, T., Braun, D.A., Seltzsam, S., Kitzler, T.M., Mao, Y., Buerger, F., et al. (2020). DAAM2 variants cause nephrotic syndrome via actin dysregulation. *Am J Hum Genetics 107*, 1113–1128.

Schreiber, A.M., Das, B., Huang, H., Marsh-Armstrong, N., and Brown, D.D. (2001). Diverse developmental programs of *Xenopus laevis* metamorphosis are inhibited by a dominant negative thyroid hormone receptor. *Proceedings of the National Academy of Sciences of the United States of America 98*, 10739–10744.

Sega, A.G., Mis, E.K., Lindstrom, K., Mercimek-Andrews, S., Ji, W., Cho, M.T., Juusola, J., Konstantino, M., Jeffries, L., Khokha, M.K., et al. (2019). De novo pathogenic variants in neuronal differentiation factor 2 (NEUROD2) cause a form of early infantile epileptic encephalopathy. *J Med Genet 56*, 113–122.

Sempou, E., Lakhani, O.A., Amalraj, S., and Khokha, M.K. (2018). Candidate heterotaxy gene FGFR4 is essential for patterning of the left-right organizer in Xenopus. *Front Physiol 9*, 1705.

Shewade, L.H., Schoephoerster, J.A., Patmann, M.D., Kulkarni, S.S., and Buchholz, D.R. (2020). Corticosterone is essential for survival through frog metamorphosis. *Endocrinology 161*, bqaa193.

Shi, Z., Wang, F., Cui, Y., Liu, Z., Guo, X., Zhang, Y., Deng, Y., Zhao, H., and Chen, Y. (2015). Heritable CRISPR/Cas9-

mediated targeted integration in *Xenopus tropicalis*. *The FASEB Journal 29*, 4914–4923.

Shi, Z., Xin, H., Tian, D., Lian, J., Wang, J., Liu, G., Ran, R., Shi, S., Zhang, Z., Shi, Y., et al. (2019). Modeling human point mutation diseases in *Xenopus tropicalis* with a modified CRISPR/Cas9 system. *FASEB Journal: Official Publication of the Federation of American Societies for Experimental Biology 33*, 6962–6968.

Shibano, T., Takeda, M., Suetake, I., Kawakami, K., Asashima, M., Tajima, S., and Taira, M. (2007). Recombinant Tol2 transposase with activity in Xenopus embryos. *Febs Lett 581*, 4333–4336.

Shibata, Y., Okada, M., Miller, T.C., and Shi, Y.-B. (2019). Knocking out histone methyltransferase PRMT1 leads to stalled tadpole development and lethality in *Xenopus tropicalis*. *Biochimica Et Biophysica Acta Bba: Gen Subj 1864*, 129482.

Shibata, Y., Tanizaki, Y., and Shi, Y.-B. (2020a). Thyroid hormone receptor beta is critical for intestinal remodeling during *Xenopus tropicalis* metamorphosis. *Cell Biosci 10*, 46.

Shibata, Y., Wen, L., Okada, M., and Shi, Y.-B. (2020b). Organ-specific requirements for thyroid hormone receptor ensure temporal coordination of tissue-specific transformations and completion of Xenopus metamorphosis. *Thyroid 30*, 300–313.

Sinzelle, L., Vallin, J., Coen, L., Chesneau, A., Pasquier, D.D., Pollet, N., Demeneix, B., and Mazabraud, A. (2006). Generation of trangenic *Xenopus laevis* using the Sleeping Beauty transposon system. *Transgenic Res 15*, 751–760.

Smith, S.J., Towers, N., Demetriou, K., and Mohun, T.J. (2020). Defective heart chamber growth and myofibrillogenesis after knockout of adprhl1 gene function by targeted disruption of the ancestral catalytic active site. *PLoS One 15*, e0235433.

Sparrow, D.B., Latinkic, B., and Mohun, T.J. (2000). A simplified method of generating transgenic Xenopus. *Nucleic Acids Res 28*, 12e–112.

Square, T.A., Jandzik, D., Massey, J.L., Romášek, M., Stein, H.P., Hansen, A.W., Purkayastha, A., Cattell, M.V., and Medeiros, D.M. (2020). Evolution of the endothelin pathway drove neural crest cell diversification. *Nature 585*, 563–568.

Steimle, J.D., Rankin, S.A., Slagle, C.E., Bekeny, J., Rydeen, A.B., Chan, S.S.-K., Kweon, J., Yang, X.H., Ikegami, K., Nadadur, R.D., et al. (2018). Evolutionarily conserved Tbx5-Wnt2/2b pathway orchestrates cardiopulmonary development. *Proc National Acad Sci 115*, E10615–E10624.

Stemple, D.L. (2004). TILLING: A high-throughput harvest for functional genomics. *Nat Rev Genet 5*, 145–150.

Sterner, Z.R., Rankin, S.A., Wlizla, M., Choi, J.A., Luedeke, D.M., Zorn, A.M., and Buchholz, D.R. (2019). Novel vectors for functional interrogation of Xenopus ORFeome coding sequences. *Genesis 57*, e23329.

Sterner, Z.R., Shewade, L.H., Mertz, K.M., Sturgeon, S.M., and Buchholz, D.R. (2020). Glucocorticoid receptor is required to survive through metamorphosis in the frog *Xenopus tropicalis*. *Gen Comp Endocr 291*, 113419.

Sullivan, C.H., Majumdar, H.D., Neilson, K.M., and Moody, S.A. (2019). Six1 and Irx1 have reciprocal interactions during cranial placode and otic vesicle formation. *Developmental Biology 446*, 68–79.

Suzuki, K.-I.T., Isoyama, Y., Kashiwagi, K., Sakuma, T., Ochiai, H., Sakamoto, N., Furuno, N., Kashiwagi, A., and Yamamoto, T. (2013). High efficiency TALENs enable F0 functional analysis by targeted gene disruption in *Xenopus laevis* embryos. *Biology Open 2*, 448–452.

Suzuki, M., Takagi, C., Miura, S., Sakane, Y., Suzuki, M., Sakuma, T., Sakamoto, N., Endo, T., Kamei, Y., Sato, Y., et al. (2016). In vivo tracking of histone H3 lysine 9 acetylation in *Xenopus laevis* during tail regeneration. *Genes Cells 21*, 358–369.

Szenker-Ravi, E., Altunoglu, U., Leushacke, M., Bosso-Lefèvre, C., Khatoo, M., Tran, H.T., Naert, T., Noelanders, R., Hajamohideen, A., Beneteau, C., et al. (2018). RSPO2 inhibition of RNF43 and ZNRF3 governs limb development independently of LGR4/5/6. *Nature 557*, 564–569.

Takagi, C., Sakamaki, K., Morita, H., Hara, Y., Suzuki, M., Kinoshita, N., and Ueno, N. (2013). Transgenic *Xenopus laevis* for live imaging in cell and developmental biology. *Development, Growth & Differentiation 55*, 422–433.

Tandon, P., Conlon, F., Furlow, J.D., and Horb, M.E. (2017). Expanding the genetic toolkit in Xenopus: Approaches and opportunities for human disease modeling. *Developmental Biology 426*, 325–335.

Tanizaki, Y., Bao, L., Shi, B., and Shi, Y.-B. (2021a). A role of endogenous histone acetyltransferase steroid hormone receptor coactivator 3 in thyroid hormone signaling during Xenopus intestinal metamorphosis. *Thyroid 31*, 692–702.

Tanizaki, Y., Shibata, Y., Zhang, H., and Shi, Y.-B. (2021b). Analysis of thyroid hormone receptor α-knockout tadpoles reveals that the activation of cell cycle program is involved in thyroid hormone-induced larval epithelial cell death and adult intestinal stem cell development during *Xenopus tropicalis* metamorphosis. *Thyroid 31*, 128–142.

Thyagarajan, B., Olivares, E.C., Hollis, R.P., Ginsburg, D.S., and Calos, M.P. (2001). Site-specific genomic integration in mammalian cells mediated by phage φC31 integrase. *Mol Cell Biol 21*, 3926–3934.

Tompkins, R. (1978). Triploid and gynogenetic diploid *Xenopus laevis*. *J Exp Zool 203*, 251–255.

Tran, H.T., and Vleminckx, K. (2014). Design and use of transgenic reporter strains for detecting activity of signaling pathways in Xenopus. *Methods* (San Diego, Calif) *66*, 422–432.

Viet, J., Reboutier, D., Hardy, S., Lachke, S.A., Paillard, L., and Gautier-Courteille, C. (2020). Modeling ocular lens disease in Xenopus. *Dev Dynam 249*, 610–621.

Vogt, J., Traynor, R., and Sapkota, G.P. (2011). The specificities of small molecule inhibitors of the TGFß and BMP pathways. *Cell Signal 23*, 1831–1842.

Waldner, C., Sakamaki, K., Ueno, N., Turan, G., and Ryffel, G.U. (2006). Transgenic *Xenopus laevis* strain expressing cre recombinase in muscle cells. *Developmental Dynamics: An Official Publication of the American Association of Anatomists 235*, 2220–2228.

Wallace, H. (1960). The development of anucleolate embryos of *Xenopus laevis*. *J Embryol Exp Morph 8*, 405–413.

Wang, C., Qi, X., Zhou, X., Sun, J., Cai, D., Lu, G., Chen, X., Jiang, Z., Yao, Y., Chan, W.Y., et al. (2020). RNA-Seq analysis on ets1 mutant embryos of *Xenopus tropicalis* identifies microseminoprotein beta gene 3 as an essential regulator of neural crest migration. *Faseb J 34*, 12726–12738.

Wang, F., Shi, Z., Cui, Y., Guo, X., Shi, Y.-B., and Chen, Y. (2015). Targeted gene disruption in *Xenopus laevis* using CRISPR/Cas9. *Cell & Bioscience 5*, 15.

Waqas, S.F.H., Noble, A., Hoang, A.C., Ampem, G., Popp, M., Strauß, S., Guille, M., and Röszer, T. (2017). Adipose tissue macrophages develop from bone marrow: Independent progenitors in *Xenopus laevis* and mouse. *J Leukocyte Biol 102*, 845–855.

Wen, L., Fu, L., Guo, X., Chen, Y., and Shi, Y.-B. (2015). Histone methyltransferase Dot1L plays a role in postembryonic development in *Xenopus tropicalis*. *The FASEB Journal 29*, 385–393.

Wen, L., and Shi, Y.-B. (2015). Unliganded thyroid hormone receptor α controls developmental timing in *Xenopus tropicalis*. *Endocrinology 156*, 721–734.

Wen, L., Shibata, Y., Su, D., Fu, L., Luu, N., and Shi, Y.-B. (2017). Thyroid hormone receptor α controls developmental timing and regulates the rate and coordination of tissue-specific metamorphosis in *Xenopus tropicalis*. *Endocrinology 158*, 1985–1998.

Wen, R.H., Stanar, P., Tam, B., and Moritz, O.L. (2019). Autophagy in *Xenopus laevis* rod photoreceptors is independently regulated by phototransduction and misfolded RHOP23H. *Autophagy 15*, 1970–1989.

Wiechmann, A.F., Martin, T.A., and Horb, M.E. (2020). CRISPR/Cas9 mediated mutation of the mtnr1a melatonin receptor gene causes rod photoreceptor degeneration in developing *Xenopus tropicalis*. *Scientific Reports 10*, 13757.

Willsey, H.R., Walentek, P., Exner, C.R.T., Xu, Y., Lane, A.B., Harland, R.M., Heald, R., and Santama, N. (2018). Katanin-like protein Katnal2 is required for ciliogenesis and brain development in Xenopus embryos. *Dev Biol 442*, 276–287.

Willsey, H.R., Xu, Y., Everitt, A., Dea, J., Exner, C.R.T., Willsey, A.J., State, M.W., and Harland, R.M. (2020). The neurodevelopmental disorder risk gene DYRK1A is required for ciliogenesis and control of brain size in Xenopus embryos. *Development 147*, dev189290.

Wyatt, B.H., Raymond, T.O., Lansdon, L.A., Darbro, B.W., Murray, J.C., Manak, J.R., and Dickinson, A.J.G. (2020). Using an aquatic model, *Xenopus laevis*, to uncover the role of chromodomain 1 in craniofacial disorders. *Genesis*, e23394.

Yergeau, D.A., Hamlet, M.R.J., Kuliyev, E., Zhu, H., Doherty, J.R., Archer, T.D., Subhawong, A.P., Valentine, M.B., Kelley, C.M., and Mead, P.E. (2009). Transgenesis in Xenopus using the Sleeping Beauty transposon system. *Developmental Dynamics: An Official Publication of the American Association of Anatomists 238*, 1727–1743.

Yergeau, D.A., Kelley, C.M., Kuliyev, E., Zhu, H., Hamlet, M.R.J., Sater, A.K., Wells, D.E., and Mead, P.E. (2011a). Remobilization of Sleeping Beauty transposons in the germline of *Xenopus tropicalis*. *Mobile Dna-Uk 2*, 15.

Yergeau, D.A., Kelley, C.M., Kuliyev, E., Zhu, H., Hamlet, M.R.J., Sater, A.K., Wells, D.E., and Mead, P.E. (2011b). Remobilization of Sleeping Beauty transposons in the germline of *Xenopus tropicalis*. *Mobile DNA 2*, 15.

Yergeau, D.A., Kelley, C.M., Zhu, H., Kuliyev, E., and Mead, P.E. (2010). Transposon transgenesis in Xenopus. *Methods* (San Diego, Calif.) *51*, 92–100.

Yergeau, D.A., Kelley, C.M., Zhu, H., Kuliyev, E., and Mead, P.E. (2012). Xenopus protocols, post-genomic approaches. *Methods Mol Biology 917*, 111–127.

Young, J.J., Cherone, J.M., Doyon, Y., Ankoudinova, I., Faraji, F.M., Lee, A.H., Ngo, C., Guschin, D.Y., Paschon, D.E., Miller, J.C., et al. (2011). Efficient targeted gene disruption in the soma and germ line of the frog *Xenopus tropicalis* using engineered zinc-finger nucleases. *Proceedings of the National Academy of Sciences of the United States of America 108*, 7052–7057.

Young, J.J., Kjolby, R.A.S., Wu, G., Wong, D., Hsu, S., and Harland, R.M. (2017). Noggin is required for first pharyngeal arch differentiation in the frog *Xenopus tropicalis*. *Dev Biol 426*, 245–254.

Zhang, T., Guo, X., and Chen, Y. (2013). Retinoic acid-activated Ndrg1a represses Wnt/β-catenin signaling to allow Xenopus pancreas, oesophagus, stomach, and duodenum specification. *PLoS ONE 8*, e65058.

Section III

From Basic Biological Insights to Human Disease

15 Formation of the Left-Right Axis; Insights from the *Xenopus* Model

Axel Schweickert and Tim Ott

CONTENTS

15.1. Historical Background 225
15.2. Present State of the Field 225
15.2.1. The Nodal Cascade Dictates Laterality 225
15.2.2. The Left-Right Organizer 226
15.2.3. The Leftward Flow Target: Dand5 226
15.2.4. Post-Transcriptional Regulation of the *dand5* mRNA 228
15.2.5. Evolution of Dand5 as Flow Target 229
15.3. Future Directions and Important Questions 229
References 230

15.1. HISTORICAL BACKGROUND

Organ asymmetries are found throughout the animal kingdom, referred to as asymmetric positioning, asymmetric morphology, or both, as exemplified by the vertebrate heart (Blum et al., 2014). The evolutionary origin of organ asymmetries may have arisen with the necessity for a longer-than-body length gut that allows efficient retrieval of nutrients and the need to stow this long gut in the body cavity in an orderly manner (Blum et al., 2014; Blum and Ott, 2018a). Vertebrate organ asymmetries (*situs solitus*) are quite sophisticated. In humans, the apex of the asymmetrically built heart, with two atria and ventricles each that connect to lung and body circulation, points to the left. The lung in turn, due to space restrictions, has fewer lobes on the left than on the right side (in humans two and three, respectively). Stomach and spleen are found on the left and the liver on the right. The small and large intestines coil in a chiral manner. In very rare cases (1:10.000), the organ situs is inverted (*situs inversus*). Heterotaxia describes another rare situation (about 1:10.000), in which subsets of organs show normal or aberrant positioning and/or morphology, which inevitably are associated with severe disease syndromes (Duncan and Khokha, 2016; Grimes and Burdine, 2017; Hamada et al., 2002)

The knowledge of human organ asymmetry date back to the antiquity, as mentioned in Aristotle's history of animals. Interestingly, scientific approaches to understand left-right (LR) development were performed in the 19th century using chick embryos (Blum et al., 1999). The first systematic experimental analysis of the LR axis was conducted in the early 1920s by Hans Spemann and co-workers before their focus shifted towards the gastrula organizer phenomenon. They used two main approaches: (1) Regional ablations and transplantations at gastrula and neurula stages showed that the left side contains a specific information which is required for proper LR development. (2) Experimental induction of double axes, twinned embryos, to address the pathological outcomes in human conjoined twins, in which specifically the right, but not the left, twin exhibits heterotaxia in 50% of cases (see Blum et al., 2009, 1999; Tisler et al., 2017a, for details and original references). Based on the work of many laboratories, including ours, we now understand the molecular basis of both observations by Spemann and co-workers. Basically, two spatially and temporally distinct processes were interfered with at the time: propagation of Nodal signaling in the left lateral plate mesoderm (LPM) and ciliary-based symmetry breakage at the left-right organizer (LRO), respectively (see the following).

15.2. PRESENT STATE OF THE FIELD

15.2.1. THE NODAL CASCADE DICTATES LATERALITY

Modern research on the molecular basis of LR asymmetry principally started with the identification of *nodal*, which encodes a Tgfß growth factor transiently expressed exclusively in the left LPM. *nodal* transcription is regulated by a positive feedback loop. In addition, Nodal induces the expression of the Tgfß feedback inhibitor *lefty*. Lefty in turn terminates Nodal signaling by binding to Nodal and to its cognitive receptor complex (Shiratori and Hamada, 2014). An additional Nodal target, the homeobox transcription factor *pitx2*, which we accidentally identified in a PCR screen for *goosecoid*-related homeobox genes in mouse (Campione et al., 1999), became an important tool to study the LR axis. Unlike the transient *nodal* and *lefty* asymmetry, left-sided *pitx2* expression persists to later developmental stages and mediates left positional identity throughout organogenesis (Blum et al., 1999; Campione et al., 1999). Because transcriptional regulation is direct in all cases, expression of

DOI: 10.1201/9781003050230-18

nodal and its targets spreads rapidly in the left LPM and is therefore called the Nodal cascade.

15.2.2. The Left-Right Organizer

Left-sided Nodal cascade induction implies that left positional information was already generated by an earlier mechanism, which is termed symmetry breakage. In most vertebrates (fish, frog, and mammals) but also in sea urchins, cilia functions are ultimately linked to symmetry breakage (Blum et al., 2009; Blum and Ott, 2018b; Little and Norris, 2020). This notion is supported by the occurrence of heterotaxia/*situs* inversion in patients suffering from ciliopathies. Later, it was postulated by Afzelius in the 1980s that during development, motile cilia are required for LR axis determination (Afzelius, 1981). An extensive clinical picture of LR defects can be observed when specifically cilia motility is affected, summarized as primary ciliary dyskinesia (PCD; Wallmeier et al., 2020). Mutations in the axonemal *dynein 11*, for example, are responsible for Kartagener syndrome, which commonly includes the development of sinusitis, bronchiectasis, and laterality defects (Afzelius, 1976). Based on a substantial body of work using various model organisms and human genetics, a conserved mechanism has been suggested for symmetry breakage. At neurula stages, a small epithelium in the posterior midline is characterized by the presence of motile mono-cilia and was termed left-right organizer (LRO). LRO cilia rotate in a clockwise fashion, which is unusual because most other motile cilia show a stroke-type motion, for example, multiciliated cells of the respiratory tract. The molecular basis of cilia rotation and determination of its direction is not understood and remains to be elucidated. But, importantly, a leftward fluid flow of extracellular fluids is generated, which serves as positional information to specify the LR axis in mammals, fish, and amphibians but, interestingly, not in reptiles and birds (Blum and Ott, 2018a; Kajikawa et al., 2020). In *Xenopus*, the gastrocoel roof plate (GRP; Shook et al., 2004) constitutes the ciliated LRO, which develops inside of neurula embryos in a dorso-posterior position of the embryonic gut. Probably because of its hidden presence, our identification of the frog LRO occurred almost 10 years later than in mouse (Schweickert et al., 2007). The GRP cells have a mesodermal fate and only transiently line the gut before integrating into the notochord and somites (Shook et al., 2004). Interestingly, this fate mirrors two distinct cell populations which also differ in position and, importantly, in function. Notochordal cells are found along the midline, projecting motile and polarized cilia, whereas somitic cells localize bilaterally and have non-polarized and non-motile cilia (Figure 15.1A; Boskovski et al., 2013; Schweickert et al., 2007; Shook et al., 2004). This scenario is also found in mouse (Little and Norris, 2020) and is interpreted as central flow-generating and bilaterally sensory LRO cells (cLRO and sLRO, respectively; Figure 15.1A).

The mechanism by which the directionality of leftward flow is initially detected is still under debate. Two scenarios have been proposed, which either favor a morphogen or a mechano-sensory model. The first concept implicates one or more secreted factors being extracellularly transported to the left, where they activate receptor-based signaling (Hirokawa et al., 2012; Tanaka et al., 2005). The second model suggests that physical forces exerted by the flow result in a left-sided bending of non-motile, mechanosensory cilia, which transduces a laterality defining signal into the sensory cells (McGrath et al., 2003; Shinohara and Hamada, 2017; Tabin and Vogan, 2003). Although the picture of these upstream processes is incomplete, the cellular and molecular target of flow-induced signaling is well described.

15.2.3. The Leftward Flow Target: Dand5

We identified Dand5, a Cerberus-related, extracellular Wnt, Bmp, and Nodal inhibitor, as the molecular component by which laterality is fixed (Schweickert et al., 2010). Both *dand5* and *nodal* are co-expressed in left and right lateral sLRO cells (Vonica and Brivanlou, 2007). In post-flow stages, *dand5* mRNA is specifically downregulated in left sLRO cells. Reduction of *dand5* mRNA is flow dependent because embryos with blocked cilia motility or increased viscosity in the archenteron lose *dand5* asymmetry (Schweickert et al., 2010). Hence, a model was proposed in which Dand5 blocks the morphogen Nodal on both sides in pre-flow stages. Due to flow-induced asymmetry, Dand5 concentrations decrease in left sLRO cells, and as a consequence, Nodal is released from suppression (Figure 15.1B; Schweickert et al., 2010) and diffuses to the left LPM, resulting in Nodal cascade induction.

The frog system is particularly suited to analyze this complex mechanism, as well as many other LR processes (Blum et al., 2009). Besides many more general advantages outlined in this book, one experimental feature unique to the *Xenopus* embryo is the ability to specifically target left or right cell lineages via simple microinjections at early blastomere stages (2–32 cells). Although this possibility is of experimental value to study many developmental questions, its unique strength is particularly emphasized during the analysis of the LR body axis (Blum et al., 2009; Blum and Ott, 2019). Providing loss- and gain-of-function experiments in a one-sided manner allows one to decipher the signaling pathways and mechanisms required to establish LR asymmetry. In addition, targeting along the dorso-ventral axis allows one to manipulate tissue-specific processes in space and time. Figure 15.1C depicts a generalized injection setup for targeting distinct LR-relevant tissues like the flow-generating central cLRO cells (C1 lineage), the flow-sensing lateral sLRO cells of the C2 lineage, and the LPM cells (C3 lineage).

The relevance of such experimental options is highlighted by our analysis of *dand5* regulation. In this context, an antisense morpholino-oligo (MO) specific against endogenous *dand5* mRNA (*dand5* MO) turned out to be extremely valuable for analyzing flow-dependent processes. In the normal scenario, *dand5* is repressed in a flow-dependent mechanism

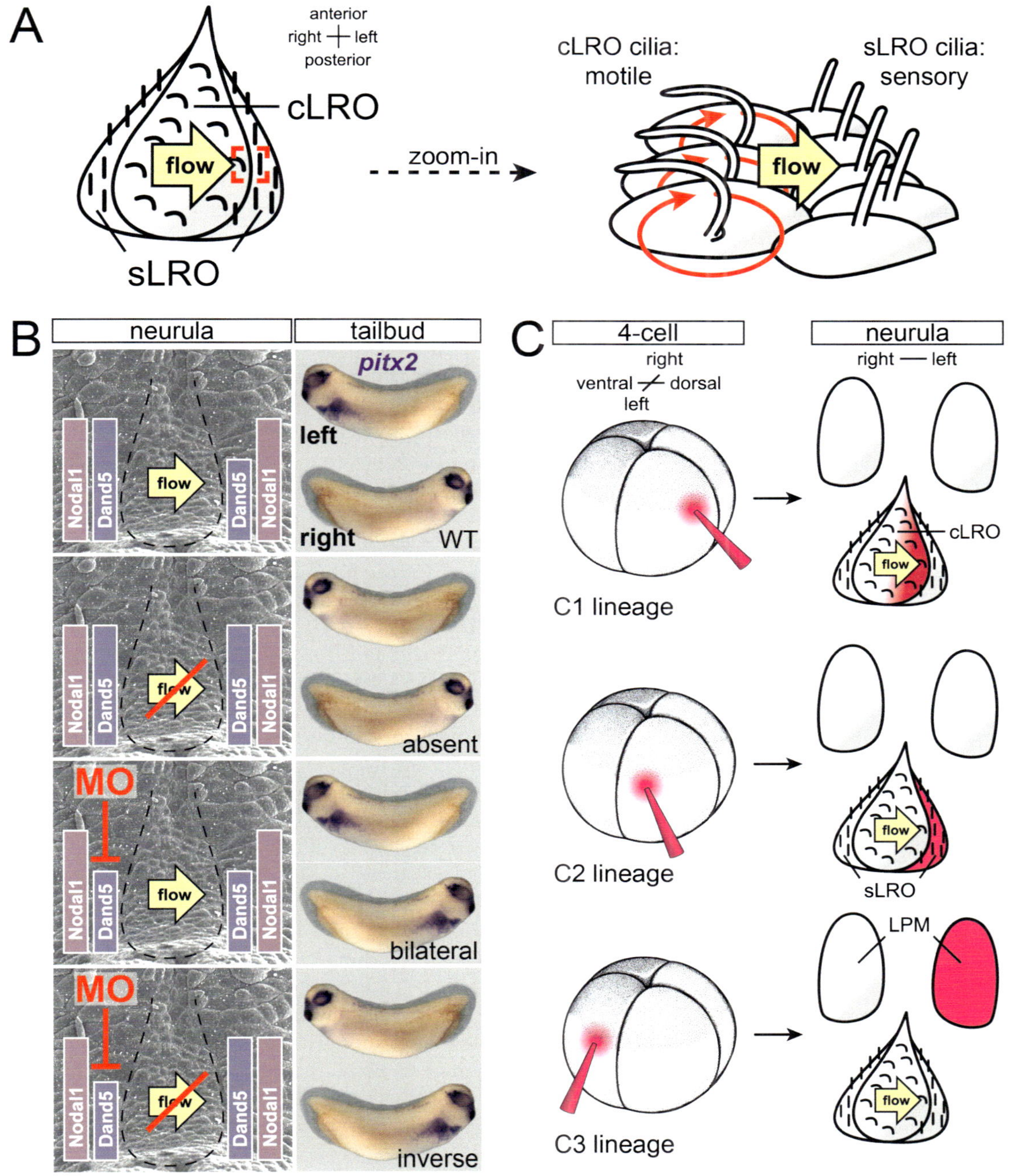

FIGURE 15.1 Two distinct LRO cells and the role of the flow target Dand5. (A) The central part of the vertebrate left right organizer (cLRO) is characterized by flow-generating cells which project motile cilia. Flow is sensed by lateral, sensory LRO cells with non-motile cilia (sLRO). (B) At pre-flow stages or in right sLRO cells, Nodal and Dand5 proteins interact in a stoichiometric equilibrium, preventing signaling or diffusion of the morphogen Nodal. In left sLRO cells, the amount of Dand5 is diminished by flow-induced signaling, freeing Nodal, which then triggers the expression of the Nodal cascade (here visualized by *pitx2)* in the left LPM. When flow is experimentally switched off, left-sided Nodal inhibition by Dand5 is maintained, and therefore leftness is not determined. This effect can be rescued by left-sided *dand5* loss of function (not shown). Injecting a *dand5* morpholino-oligo (MO) in the right lineage results in bilateral Nodal release and consequently bilateral *pitx2* expression. Intriguingly, *situs* inversions are easily and very efficiently generated by experimentally turning off leftward flow and right-sided *dand5* MO injections. A REM picture of the frog LRO is shown. (C) Specific targeting of distinct processes required for laterality development. Note that only left-sided injection schemes into the prospective marginal zone of four-cell-stage embryos are shown. Nomenclature of injection side (C1, C2, C3) corresponds to and is adapted from fate maps of a 16- and 32-cell-stage embryo (Moody, 1987). cLRO cells are targeted by injecting a dorsal blastomere near the cell junction (C1), reflecting the future dorsal midline. Choosing a more laterally located injection site of the dorsal blastomere (C2) will hit sLRO cells. (C) Lateral plate mesoderm (LPM), the signaling center for asymmetric organ morphogenesis, is most efficiently targeted by injecting ventral blastomeres in a lateral position.

in the left sLRO cells (Schweickert et al., 2010). Then, further diminishing the Dand5 activity by injecting *dand5* MO into the left lineage does not impact LR asymmetry. However, right-sided *dand5* knockdown induces an ectopic Nodal-cascade in the right LPM (Figure 15.1B). Both observations are in line with the key role of *dand5* in assigning the left positional information by its flow-induced repression. Importantly, basically all experimental manipulations of upstream processes, that is, impairing flow or flow-induced signaling, are rescued by left-sided *dand5* MO administrations (Figure 15.1B; Schweickert et al., 2010). Laterality defects induced by preventing leftward flow and thus *dand5* repression by increasing the viscosity of extracellular milieu or by interfering with cilia motility are rescued to wild-type levels by left-sided *dand5* knockdown. When injected into the right lineage, organ situs was inverted in almost 100% of cases (Schweickert et al., 2010). Similarly, we could demonstrate that laterality defects in conjoined twins are based on mispositioning of LRO tissue and thus misregulation of *dand5* (Tisler et al., 2017b). Therefore, using the *dand5* MO in combination with other treatments allows one to demonstrate the specificity of the experiments, and, even more importantly, it allows one to place a particular gene function upstream or downstream of flow-induced *dand5* inhibition. Any factor whose function is required for Dand5-mediated repression of Nodal should be rescuable by Dand5 knockdown. So far, such flow effectors are missing. Unfortunately, the best candidate available to date, *pkd2*, which encodes a critical calcium channel of the ciliary sensor complex (Yoshiba et al., 2012; Yoshiba and Hamada, 2014), is already required for the induction of the LRO anlage at blastula/gastrula stages (Vick et al., 2009), preventing its analysis during flow stages. Together with yet-to-be-identified factors of the flow-induced signaling pathway, a more detailed picture of symmetry breakage is still missing.

15.2.4. Post-Transcriptional Regulation of the *dand5* mRNA

As in *Xenopus*, flow-dependent Dand5 repression in mouse and fish embryos was experimentally demonstrated by a left-sided reduction of *dand5* mRNA (Hojo et al., 2007; Nakamura et al., 2012). In mouse, it was shown that *dand5* mRNA reduction was conveyed via its 3'UTR in a flow dependent manner (Nakamura et al., 2012), indicating a post-transcriptional mechanism. However, timing and penetrance of *dand5* asymmetry suggested that mRNA decay might not be the sole regulatory mechanism that controls Dand5 protein levels. This reasoning is backed by observations in frog embryos. At late neurula stages (st. 19–21) *dand5* asymmetry is most prevalently detected, although the Nodal cascade was already active in the left LPM (Schweickert et al., 2010). Thus, *dand5* mRNA decay was too late to be functionally relevant. Also, the frequency of left-sided *dand5* mRNA reduction is not complete and can be detected in 80% of wild type specimens at most, whereas leftness at the level of asymmetric gene expression and organogenesis was observed in 95% of cases. Obviously, these frequencies do not match, supporting the idea that flow-dependent and robust *dand5* repression requires an additional process beyond detectable left-sided *dand5* mRNA degradation. These observations prompted our hypothesis that flow sensation triggers a *dand5* mRNA-specific post-transcriptional regulatory program, which acts through translational repression followed by mRNA decay (Maerker et al., 2020).

Because the mechanisms by which the flow is sensed or by which the signaling pathway is initiated are unclear, we decided to tackle the problem from another perspective and addressed the question of how *dand5* is regulated in a flow-dependent manner. As outlined previously, post-transcriptional inhibition of *dand5* mRNA translation was a very likely possibility, and we searched for factors that could provide this activity. We identified the RNA binding protein Bicaudal C (Bicc1) as the crucial post-transcriptional regulator of *dand5* (Maerker et al., 2020). To date, a variety of Bicc1 functions have been reported, ranging from repression to protection of mRNAs (Rothé et al., 2015; Tran et al., 2010; Zhang et al., 2014, 2013). Interestingly, Bicc1 was already implicated in LR asymmetry in mouse and frog because of its requirement to polarize flow-generating LRO cilia via planar cell polarity (Maisonneuve et al., 2009). However, *bicc1* is co-expressed with *dand5* and *nodal* in sLRO cells (Maisonneuve et al., 2009) and inhibits the translation of a reporter mRNA containing the *dand5* 3'UTR (Zhang et al., 2013). We recently showed that 139 nucleotides of the most proximal *dand5* 3'UTR were sufficient to mediate Bicc1 regulation; we termed the sequence the Bicc1 responsive element (Bicc1RE; Maerker et al., 2020). *In vivo*, a target protector MO (tpMO), which blocks accessibility to the Bicc1RE, prevented left-sided Nodal cascade induction, suggesting that Bicc1 mediates *dand5* repression. Remarkably, flow-dependent decay of *dand5* mRNA was not affected by the tpMO, strongly suggesting that translation inhibition is central to Bicc1 function. We postulate that flow-induced signaling triggers a yet-unknown posttranslational modification of Bicc1 that switches its activity towards translational repression of *dand5* (Figure 15.2; Maerker et al., 2020). Interestingly, we demonstrated that Dicer, a ribonuclease that is critical for miR processing, was expressed in sLRO cells and is required for flow-induced *dand5* repression. In agreement, *dicer* morphants lack asymmetric Nodal cascade induction. We further showed that Bicc1 and Dicer acted in a synergistic manner, suggesting that both post-transcriptional mechanisms co-operate during symmetry breakage (Maerker et al., 2020). Taken together, our *Xenopus* work provides a new framework for how symmetry breakage is implemented from cilia-generated leftward fluid flow to the asymmetric release of the Nodal morphogen. Surprisingly, post-transcriptional regulation is the initial target of flow, which opens a new route for sophisticated analysis in the future.

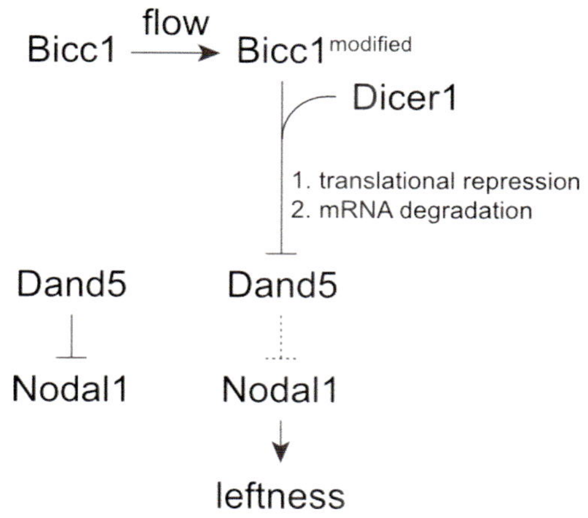

FIGURE 15.2 Flow-induced post-transcriptional repression of *dand5* determines laterality. The RNA binding protein Bicc1 is co-expressed with *nodal1* and *dand5* in sLRO cells. Flow-induced signaling modifies left-sided Bicc1 activity. Modified Bicc1, in concert with Dicer, post-transcriptionally suppress Dand5 protein expression by interfering with mRNA translation and decay. Therefore, Nodal repression by Dand5 is lifted, which consequently determines the left body axis.

15.2.5. Evolution of Dand5 as Flow Target

Symmetry breakage using cilia motility very likely reflects the basic mechanism during deuterostomian evolution, as it is already present in sea urchins or tunicates (Blum et al., 2009; Blum and Ott, 2018b). A flow-responsive Dand5-Nodal module, however, seems to be restricted to higher chordates (Kajikawa et al., 2020). Although Nodal function during LR patterning is conserved throughout the deuterostomian tree of life and even beyond (Blum et al., 2014), a role for Dand5 has only been described in the cephalochordate *Branchostoma* and most vertebrate model organisms (Blum et al., 2014; Zhu et al., 2019). However, potentially any Nodal inhibitor could theoretically be recruited for symmetry breakage, as long as its flow-dependent repression can be ensured. We compared protein sequences of the LR-relevant Nodal signaling module components, consisting of ligand, inhibitor and receptor, from species with cilia-driven symmetry breakage versus a Wnt signaling pathway reference. Intriguingly, Nodal homologues expressed at the LRO and Dand5 differ quite substantially between vertebrate species, which is obviously counterintuitive given the importance and conservation of the process (Table 15.1). Of note, *dand5* 3'UTRs sequences are even more diverse, although flow-induced regulation should be conveyed by this sequence. In species that express multiple Nodal genes, kin relations do not match with LR functions as well (Kajikawa et al., 2020). In contrast, the Nodal receptor Actvr2 or the Wnt3a ligand depict a high homology throughout the vertebrates, underscoring their conserved functions (Table 15.1). Multi-pathway inhibitors of the Cerberus family seem to generally show a high divergence, which is also observed for Dkk (Table 15.1). One might speculate that this could be the underlying reason for the rapid co-evolution of the Nodal/Dand5 module. To our knowledge, no systematic analysis of protein properties in a species-specific experimental setup has been undertaken so far. The *Xenopus* system would be particularly suited to characterize the sequence differences during LR development using a gain-of function approach.

A percent identity matrix of the core LR signaling module, consisting of ligand (Nodal), secreted inhibitor (Dand5), and receptor (Acvr2b), with protein sequences from human (*Homo sapiens*, Has), mouse (*Mus musculus*, Mus), zebrafish (*Danio rerio*, Dre), frog (*Xenopus laevis*, Xla), and *Branchiostoma* (*Branchiostoma floridae*, Bfl), is compared with a reference module of the Wnt-signaling pathway. Surprisingly low sequence conservation of the LR-relevant morphogenes stands out. Sequences used as indicated by the accession numbers behind the species identifier.

15.3. FUTURE DIRECTIONS AND IMPORTANT QUESTIONS

Because of the increasingly efficient abilities to identify potential human disease genes, including genome sequencing and computing power, vertebrate model organisms are central to functionally validate candidates. Here we describe the first asymmetric molecular events following cilia-based flow reception, leading to asymmetric signaling and finally resulting in post-transcriptional *dand5* suppression. At any step, human mutations could occur, impacting LR asymmetry and potentially human health. This notion is underscored by the rare appearance of laterality defects in patients with *pkd2* mutations, which is thought to be a central part of the ciliary sensor complex (Bataille et al., 2011). The same is true for patients with deleterious *pkd1l1* variants (Vetrini et al., 2016), which are believed to complex with Pkd2 on the protein level (Field et al., 2011; Kamura et al., 2011). In

TABLE 15.1
High Sequence Diversity of the Nodal-Dand5 Module

	LR Signaling Module	Hsa	Mmu	Xla	Dre	Bfl	Reference Module	Hsa	Mmu	Xla	Dre	Bfl
	Nodal						Wnt3a					
Ligand	Hsa (BC112025)	100.00					Hsa (AB060284)	100.00				
	Mmu (NM_013611)	80.12	100.00				Mmu (NM_009522)	96.02	100.00			
	Xla (NM_001085796)	34.35	37.01	100.00			Xla (NM_001085874)	84.09	84.66	100.00		
	Dre (AY240027)	35.20	36.70	43.19	100.00		Dre (AY613787)	85.23	85.51	87.78	100.00	
	Bfl (XM_035814606)	30.63	30.88	28.17	31.20	100.00	Bfl (AF361013)	63.92	64.77	66.19	63.01	100.00
	Dand5						Dkk					
Inhibitor	Hsa (NM_152654)	100.00					Hsa (NM_012242)	100.00				
	Mmu (NM_201227)	63.39	100.00				Mmu (JN966751)	82.71	100.00			
	Xla (NM_001098726)	30.12	30.91	100.00			Xla (NM_001085592)	56.57	54.12	100.00		
	Dre (NM_212969)	31.39	29.85	26.11	100.00		Dre (AB023488)	50.85	51.69	50.64	100.00	
	Bfl (EU67025	31.74	29.17	27.96	23.79	100.00	Bfl (HM590023)	35.65	34.76	34.63	34.22	100.00
	Acvr2b						Frd1					
Receptor	Hsa (NM_001106)	100.00					Hsa (NM_003505)	100.00				
	Mmu (BC106189)	99.41	100.00				Mmu (NM_021457)	94.55	100.00			
	Xla (NM_001090580)	82.32	82.28	100.00			Xla (NM_001085738)	85.79	85.97	100.00		
	Dre (NM_131210)	79.37	79.13	79.72	100.00		Dre (NM_001130614)	81.78	81.41	82.21	100.00	
	Bfl (XM_035801604)	63.10	63.22	62.82	61.88	100.00	Bfl (XM_035805275)	66.18	65.82	68.53	70.57	100.00

addition, human *dand5* mutations have been identified associated with congenital heart and laterality defects, including a point mutation in the *dand5* coding region (Cristo et al., 2017). However, based on our data, sequence mutations in the *dand5* 3'UTR could be medically relevant as well. We believe that *Xenopus* provides an excellent system in which to analyze nearly every stage of LR development with very good spatial and temporal resolution and unique experimental approaches (Blum and Ott, 2019). The *Xenopus* embryo is therefore highly suitable for functional validation of human LR disease genes.

REFERENCES

Afzelius, B., 1976. A human syndrome caused by immotile cilia. *Science* 193, 317–319. https://doi.org/10.1126/science.1084576

Afzelius, B., 1981. Genetical and ultrastructural aspects of the immotile-cilia syndrome. *American Journal of Human Genetics* 33(6), 852–864.

Bataille, S., Demoulin, N., Devuyst, O., Audrézet, M.-P., Dahan, K., Godin, M., Fontès, M., Pirson, Y., Burtey, S., 2011. Association of PKD2 (polycystin 2) mutations with left-right laterality defects. *Am J Kidney Dis* 58, 456–460. https://doi.org/10.1053/j.ajkd.2011.05.015

Blum, M., Beyer, T., Weber, T., Vick, P., Andre, P., Bitzer, E., Schweickert, A., 2009. Xenopus, an ideal model system to study vertebrate left-right asymmetry. *Developmental Dynamics : An Official Publication of the American Association of Anatomists* 238, 1215–1225. https://doi.org/10.1002/dvdy.21855

Blum, M., Feistel, K., Thumberger, T., Schweickert, A., 2014. The evolution and conservation of left-right patterning mechanisms. *Development (Cambridge, England)* 141, 1603–1613. https://doi.org/10.1242/dev.100560

Blum, M., Ott, T., 2018a. Animal left-right asymmetry. *Curr Biol* 28, R301–R304. https://doi.org/10.1016/j.cub.2018.02.073

Blum, M., Ott, T., 2018b. The power of strain: Organizing left-right cilia. *Developmental Cell* 45, 277–279. https://doi.org/10.1016/j.devcel.2018.04.015

Blum, M., Ott, T., 2019. Xenopus: An undervalued model organism to study and model human genetic disease. *Cells Tissues Organs* 205, 303–313. https://doi.org/10.1159/000490898

Blum, M., Steinbeisser, H., Campione, M., Schweickert, A., 1999. Vertebrate left-right asymmetry: old studies and new insights. *Cell Mol Biology Noisy-le-grand France* 45, 505–516.

Boskovski, M.T., Yuan, S., Pedersen, N.B., Goth, C.K., Makova, S., Clausen, H., Brueckner, M., Khokha, M.K., 2013. The heterotaxy gene GALNT11 glycosylates Notch to orchestrate cilia type and laterality. *Nature* 504, 456–459. https://doi.org/10.1038/nature12723

Campione, M., Steinbeisser, H., Schweickert, A., Deissler, K., Bebber, F. van, Lowe, L.A., Nowotschin, S., Viebahn, C., Haffter, P., Kuehn, M.R., Blum, M., 1999. The homeobox gene Pitx2: Mediator of asymmetric left-right signaling in vertebrate heart and gut looping. *Dev Camb Engl* 126, 1225–1234.

Cristo, F., Inácio, J.M., Almeida, S. de, Mendes, P., Martins, D.S., Maio, J., Anjos, R., Belo, J.A., 2017. Functional study of DAND5 variant in patients with congenital heart disease and laterality defects. *BMC medical genetics* 18, 77. https://doi.org/10.1186/s12881-017-0444-1

Duncan, A.R., Khokha, M.K., 2016. Xenopus as a model organism for birth defects—Congenital heart disease and heterotaxy. *Seminars in Cell and Developmental Biology*. https://doi.org/10.1016/j.semcdb.2016.02.022

Field, S., Riley, K.-L., Grimes, D.T., Hilton, H., Simon, M., Powles-Glover, N., Siggers, P., Bogani, D., Greenfield, A., Norris, D.P., 2011. Pkd1l1 establishes left-right asymmetry and physically interacts with Pkd2. *Development (Cambridge, England)* 138, 1131–1142. https://doi.org/10.1242/dev.058149

Grimes, D.T., Burdine, R.D., 2017. Left-right patterning: breaking symmetry to asymmetric morphogenesis. *Trends in genetics : TIG*. https://doi.org/10.1016/j.tig.2017.06.004

Hamada, H., Meno, C., Watanabe, D., Saijoh, Y., 2002. Establishment of vertebrate left-right asymmetry. *Nature Reviews Genetics* 3, 103–113. https://doi.org/10.1038/nrg732

Hirokawa, N., Tanaka, Y., Okada, Y., 2012. Cilia, KIF3 molecular motor and nodal flow. *Current Opinion in Cell Biology* 24, 31–39. https://doi.org/10.1016/j.ceb.2012.01.002

Hojo, M., Takashima, S., Kobayashi, D., Sumeragi, A., Shimada, A., Tsukahara, T., Yokoi, H., Narita, T., Jindo, T., Kage, T., Kitagawa, T., Kimura, T., Sekimizu, K., Miyake, A., Setiamarga, D., Murakami, R., Tsuda, S., Ooki, S., Kakihara, K., Naruse, K., Takeda, H., 2007. Right-elevated expression of charon is regulated by fluid flow in medaka Kupffer's vesicle. *Development Growth and Differentiation* 49, 395–405. https://doi.org/10.1111/j.1440-169x.2007.00937.x

Kajikawa, E., Horo, U., Ide, T., Mizuno, K., Minegishi, K., Hara, Y., Ikawa, Y., Nishimura, H., Uchikawa, M., Kiyonari, H., Kuraku, S., Hamada, H., 2020. Nodal paralogues underlie distinct mechanisms for visceral left-right asymmetry in reptiles and mammals. *Nature Ecology & Evolution* 4, 261–269. https://doi.org/10.1038/s41559-019-1072-2

Kamura, K., Kobayashi, D., Uehara, Y., Koshida, S., Iijima, N., Kudo, A., Yokoyama, T., Takeda, H., 2011. Pkd1l1 complexes with Pkd2 on motile cilia and functions to establish the left-right axis. *Development (Cambridge, England)* 138, 1121–1129. https://doi.org/10.1242/dev.058271

Little, R.B., Norris, D.P., 2020. Right, left and cilia: How asymmetry is established. *Semin Cell Dev Biol*. https://doi.org/10.1016/j.semcdb.2020.06.003

Maerker, M., Getwan, M., Dowdle, M.E., Pelliccia, J.L., McSheene, J.C., Yartseva, V., Minegishi, K., Vick, P., Giraldez, A.J., Hamada, H., Burdine, R.D., Sheets, M.D., Schweickert, A., Blum, M., 2020. Bicc1 and dicer regulate left-right patterning through post-transcriptional control of the Nodal-inhibitor dand5. *Biorxiv* 1(29), 924456. https://doi.org/10.1101/2020.01.29.924456

Maisonneuve, C., Guilleret, I., Vick, P., Weber, T., Andre, P., Beyer, T., Blum, M., Constam, D.B., 2009. Bicaudal C, a novel regulator of Dvl signaling abutting RNA-processing bodies, controls cilia orientation and leftward flow. *Development (Cambridge, England)* 136, 3019–3030. https://doi.org/10.1242/dev.038174

McGrath, J., Somlo, S., Makova, S., Tian, X., Brueckner, M., 2003. Two populations of node monocilia initiate left-right asymmetry in the mouse. *Cell* 114, 61–73. https://doi.org/10.1016/s0092-8674(03)00511-7

Moody, S.A., 1987. Fates of the blastomeres of the 32-cell-stage Xenopus embryo. *Dev Biol* 122, 300–319. https://doi.org/10.1016/0012-1606(87)90296-x

Nakamura, T., Saito, D., Kawasumi, A., Shinohara, K., Asai, Y., Takaoka, K., Dong, F., Takamatsu, A., Belo, J.-A., Mochizuki, A., Hamada, H., 2012. Fluid flow and interlinked feedback loops establish left-right asymmetric decay of Cerl2 mRNA. *Nature communications* 3, 1322. https://doi.org/10.1038/ncomms2319

Rothé, B., Leal-Esteban, L., Bernet, F., Urfer, S., Doerr, N., Weimbs, T., Iwaszkiewicz, J., Constam, D.B., 2015. Bicc1 polymerization regulates the localization and silencing of bound mRNA. *Mol Cell Biol* 35, 3339–3353. https://doi.org/10.1128/mcb.00341-15

Schweickert, A., Vick, P., Getwan, M., Weber, T., Schneider, I., Eberhardt, M., Beyer, T., Pachur, A., Blum, M., 2010. The nodal inhibitor Coco is a critical target of leftward flow in Xenopus. *Current Biology: CB* 20, 738–743. https://doi.org/10.1016/j.cub.2010.02.061

Schweickert, A., Weber, T., Beyer, T., Vick, P., Bogusch, S., Feistel, K., Blum, M., 2007. Cilia-driven leftward flow determines laterality in Xenopus. *Current Biology: CB* 17, 60–66. https://doi.org/10.1016/j.cub.2006.10.067

Shinohara, K., Hamada, H., 2017. Cilia in left-right symmetry breaking. *Cold Spring Harbor Perspectives in Biology*. https://doi.org/10.1101/cshperspect.a028282

Shiratori, H., Hamada, H., 2014. TGFβ signaling in establishing left-right asymmetry. *Semin Cell Dev Biol* 32, 80–84. https://doi.org/10.1016/j.semcdb.2014.03.029

Shook, D.R., Majer, C., Keller, R., 2004. Pattern and morphogenesis of presumptive superficial mesoderm in two closely related species, *Xenopus laevis* and *Xenopus tropicalis*. *Developmental Biology* 270, 163–185. https://doi.org/10.1016/j.ydbio.2004.02.021

Tabin, C.J., Vogan, K.J., 2003. A two-cilia model for vertebrate left-right axis specification. *Genes and Development* 17, 1–6. https://doi.org/10.1101/gad.1053803

Tanaka, Y., Okada, Y., Hirokawa, N., 2005. FGF-induced vesicular release of Sonic hedgehog and retinoic acid in leftward nodal flow is critical for left-right determination. *Nature* 435, 172–177. https://doi.org/10.1038/nature03494

Tisler, M., Schweickert, A., Blum, M., 2017a. Xenopus, an ideal model organism to study laterality in conjoined twins. *Genesis (New York, NY : 2000)* 55. https://doi.org/10.1002/dvg.22993

Tisler, M., Thumberger, T., Schneider, I., Schweickert, A., Blum, M., 2017b. Leftward flow determines laterality in conjoined twins. *Current Biology: CB* 27, 543–548. https://doi.org/10.1016/j.cub.2016.12.049

Tran, U., Zakin, L., Schweickert, A., Agrawal, R., Döger, R., Blum, M., Robertis, E.M.D., Wessely, O., 2010. The RNA-binding protein bicaudal C regulates polycystin 2 in the kidney by antagonizing miR-17 activity. *Development (Cambridge, England)* 137, 1107–1116. https://doi.org/10.1242/dev.046045

Vetrini, F., D'Alessandro, L.C.A., Akdemir, Z.C., Braxton, A., Azamian, M.S., Eldomery, M.K., Miller, K., Kois, C., Sack, V., Shur, N., Rijhsinghani, A., Chandarana, J., Ding, Y., Holtzman, J., Jhangiani, S.N., Muzny, D.M., Gibbs, R.A., Eng, C.M., Hanchard, N.A., Harel, T., Rosenfeld, J.A., Belmont, J.W., Lupski, J.R., Yang, Y., 2016. Bi-allelic mutations in PKD1L1 are associated with laterality defects in humans. *Am J Hum Genetics* 99, 886–893. https://doi.org/10.1016/j.ajhg.2016.07.011

Vick, P., Schweickert, A., Weber, T., Eberhardt, M., Mencl, S., Shcherbakov, D., Beyer, T., Blum, M., 2009. Flow on the right side of the gastrocoel roof plate is dispensable for symmetry breakage in the frog *Xenopus laevis*. *Dev Biol* 331, 281–291. https://doi.org/10.1016/j.ydbio.2009.05.547

Vonica, A., Brivanlou, A.H., 2007. The left-right axis is regulated by the interplay of Coco, Xnr1 and derrière in Xenopus embryos. *Developmental Biology* 303, 281–294. https://doi.org/10.1016/j.ydbio.2006.09.039

Wallmeier, J., Nielsen, K.G., Kuehni, C.E., Lucas, J.S., Leigh, M.W., Zariwala, M.A., Omran, H., 2020. Motile ciliopathies. *Nature Reviews: Disease Primers* 6, 77–29. https://doi.org/10.1038/s41572-020-0209-6

Yoshiba, S., Hamada, H. 2014. Roles of cilia, fluid flow, and Ca2+ signaling in breaking of left-right symmetry. *Trends Genet* 30, 10–17. https://doi.org/10.1016/j.tig.2013.09.001

Yoshiba, S., Shiratori, H., Kuo, I.Y., Kawasumi, A., Shinohara, K., Nonaka, S., Asai, Y., Sasaki, G., Belo, J.-A., Sasaki, H., Nakai, J., Dworniczak, B., Ehrlich, B.E., Pennekamp, P., Hamada, H., 2012. Cilia at the node of mouse embryos sense fluid flow for left-right determination via Pkd2. *Science*. https://doi.org/10.1126/science.1222538

Zhang, Y., Cooke, A., Park, S., Dewey, C.N., Wickens, M., Sheets, M.D., 2013. Bicaudal-C spatially controls translation of vertebrate maternal mRNAs. *RNA* 19, 1575–1582. https://doi.org/10.1261/rna.041665.113

Zhang, Y., Park, S., Blaser, S., Sheets, M.D., 2014. Determinants of RNA binding and translational repression by the Bicaudal-C regulatory protein. *Journal of Biological Chemistry* 289, 7497–7504. https://doi.org/10.1074/jbc.m113.526426

Zhu, X., Shi, C., Zhong, Y., Liu, X., Yan, Q., Wu, X., Wang, Y., Li, G., 2019. Cilia-driven asymmetric Hedgehog signalling determines the amphioxus left-right axis by controlling Cerberus/Dand5 expression. *Development (Cambridge, England)* dev.182469. https://doi.org/10.1242/dev.182469

16 Discovering the Function of Congenital Heart Disease Genes

Delfina Pearledith González and Mustafa K. Khokha

CONTENTS

16.1. Historical and Medical Background of CHD ... 233
16.1.1. The Roots of CHD ... 233
16.1.2. Surgical Interventions and Clinical Therapies in CHD ... 234
16.1.3. Difficulties in Diagnosing and Studying CHD ... 234
16.2. Past Molecular and Genetic Studies Contributing to Understanding CHD/HTX ... 234
16.2.1. Previous Studies of Vertebrate Cardiogenesis in *Xenopus* ... 235
16.2.2. Identifying Additional Candidate CHD Genes ... 235
16.3. Approaches to Studying CHD/HTX in *Xenopus* ... 236
16.3.1. Morphological and Developmental Benefits of Studying CHD/HTX in *Xenopus* ... 236
16.3.2. Using CRISPR/Cas9 Genome Editing as a Screening Tool in *Xenopus* ... 236
16.3.3. Left-Right Patterning: A Beautiful Pathway to Guide CHD/HTX Disease Mechanism Analysis ... 236
16.4. Patient-Derived CHD Disease Mechanisms Uncovered in *Xenopus* ... 237
16.4.1. Connecting Known Pathways in a New Context: Glycosylation in NOTCH Signaling and Cilia Cell Fate Determination ... 237
16.4.2. Uncovering Novel Roles for Well-Studied Genes: Nucleoporins in Cilia and Centrosomal Biology ... 237
16.4.3. Exploring Old Mysteries in Developmental Cell Signaling ... 238
16.4.4. Shared Molecular Pathways for Multiple Diseases: Neurodevelopmental Disorders, Craniofacial Abnormalities, and CHD ... 239
16.5. Where to Go from Here? The Future of Studying CHD in *Xenopus* ... 240
References ... 240

Congenital heart disease (CHD) is a class of developmental anomalies affecting the structure and function of the heart. CHD includes (1) septal defects (holes between chambers of the heart), (2) hypoplasia or hypertrophy (reduction or overgrowth of cardiac tissues), (3) abnormal connection of the major vessels, (4) misorientation of the heart along the left-right axis, or (5) any combination of 1–4. Heterotaxy (HTX), a misorientation of internal organs across the left-right (LR) axis, particularly affects the heart, and many HTX patients suffer from a particularly severe form of CHD. It is critical to understand the pathogenic mechanisms that contribute to both CHD and HTX to improve patient care. Advances in sequencing technologies has expedited identification of CHD/HTX candidate genes, but we lack functional studies. Developmental and molecular studies of these candidate genes in *Xenopus* have uncovered surprising connections between cell and developmental pathways that contribute to the LR pathways that underlie CHD and HTX. These and future studies pave the way for improvements in developmental biology and CHD/HTX patient care.

16.1. HISTORICAL AND MEDICAL BACKGROUND OF CHD

CHD and other congenital birth defects have been documented for centuries (Afzelius 1977; Gelb 2015). Modern medicine brought on a new age of surgical therapies. Advances have improved patient diagnosis, survival, and outcomes. Deeper understanding of the mechanisms leading to CHD can mitigate difficulties diagnosing the diseases on a purely phenotypic or genetic basis.

16.1.1. The Roots of CHD

Scientists and physicians have documented CHD cases as early as the times of Leonardo DaVinci (Castañeda 2005). It wasn't until the mid-1800s and early 1900s that physician Thomas Bevill Peacock and pathologist Maude E. Abbot categorically organized a wide range of CHD case reports into distinct categories (Peacock 1858; Abbot 1915). Pediatric cardiologist Helen Brooke Taussig published the first CHD textbook,

DOI: 10.1201/9781003050230-19

Congenital Malformations of the Heart, in 1947. Her observations of naturally occurring CHD in animals led to an effort to study cardiac development in other species to better understand human CHD pathology (Gelb 2015). All these early CHD scientists recognized the potential role of inheritance and linkage between CHD, arrested development, and non-cardiac developmental disorders, including neurodevelopmental disorders (NDDs). Dr. Taussig and cardiologist John Maurice Hardman Campbell also observed that neither the pattern of genetic inheritance of CHD nor environmental causes of CHD could be simply explained, leaving them to suggest that genetic complexity could be at play (Gelb 2015; Taussig 1947).

While some doctors and scientists worked to understand the underlying causes of CHD for future diagnosis and therapies, others were working on clinical strategies to ease more immediate patient suffering.

16.1.2. Surgical Interventions and Clinical Therapies in CHD

Dr. Daniel Hale Williams performed the first open-heart surgery in 1893 to repair a stab wound (Fenderson and Miller 1971). However, extracardiac corrective surgeries for CHD did not successfully take place until 1938 (Gross and Hubbard 1939), with further improvements by Dr. Taussig and her contemporaries in 1945 (Crafoord and Nylin 1945; Blalock and Taussig 1945). In order to correct the wider spectrum of CHD structural defects, surgeons needed access to the inside of the heart. To do so, they needed to stop the heart from beating, stop blood flow to the heart, and finally restart the heart to recover the patient. Eventually, surgeons developed methods for cardiopulmonary bypass using a pump and an external oxygenator, enabling them to stop the heart and, with cooling, also slow the patient's metabolic requirements (Castaneda et al. 1984). With these advancements, surgeons were able to do open-heart surgery on patients for corrective rather than palliative procedures for internal cardiac defects.

Further combination of palliative and corrective procedures drastically reduced hospital mortality rates due to CHD between the late 1950s and late 1990s (Castañeda 2005). Not only was this better for patient survival, but the now-single-week-long hospitalization reduced psychological, logistical, and financial impacts on patients and their families.

16.1.3. Difficulties in Diagnosing and Studying CHD

Although doctors achieved great innovations in surgical interventions, there is still much to improve. One vexing problem is the variability in outcomes for a given anatomical diagnosis. Patients with CHD are cohorted based on cardiac anatomy. Based on surgical and peri-operative care, most patients with the same anatomy have similar clinical outcomes. However, some patients do more poorly than expected based on their anatomical diagnosis alone. We propose that these outliers may relate through genotype rather than phenotype.

As sequencing technology has advanced, researchers identify more candidate genes and unique alleles. Remarkably, many of these candidate genes have no known role in cardiac development, no role in embryonic development, or no identified role in biology. A greater number of unique alleles also complicates genotype/phenotype correlations and reduces genetic cohort sizes. This is due to the complexity of cardiac development, in which multiple genes are critical. Genetic studies of CHD patient cohorts have identified high rates of locus heterogeneity, or mutation contributions from multiple genetic loci (Fakhro et al. 2011; Zaidi et al. 2013; Homsy et al. 2015; Jin et al. 2017). This, in combination with the low rate of second alleles, makes it difficult to demonstrate disease causality (Fakhro et al. 2011), creating a problem for identifying gene function. Therefore, there is a pressing need to discover gene function.

The cohort size necessary to establish disease causality on a purely genetic basis is often too large to be practical and delays discoveries. We propose that the optimal solution is to initiate studies of candidate genes in high-throughput model systems. Importantly, a better understanding of the molecular mechanisms of CHD candidate genes will allow doctors and researchers to exploit genotype to inform care, prognosis, and risk of recurrence.

16.2. PAST MOLECULAR AND GENETIC STUDIES CONTRIBUTING TO UNDERSTANDING CHD/HTX

The prevalence and impact of congenital heart disease on infant health makes it important to understand in more depth. As of 2006, about 8 million, or ~ 6%, of total worldwide births were affected by a genetically or partly genetically derived serious birth defect (Christianson, Howson, and Modell 2006). CHD is the most common birth defect, impacting about 20/10,000 live births in the United States (Mai et al. 2019). The CHD spectrum can range from severe cardiac structural defects at birth to minor abnormalities not identified until adulthood (Marino et al. 2012; Mussatto et al. 2014). Surgical correction of severe congenital cardiac anomalies is necessary to improve patient survival in about 25% of CHD cases identified at birth (Virani et al. 2020). However, the patients that survive to adulthood may experience additional health impacts seemingly unrelated to their CHD or surgery, such as infertility, pulmonary disease, neurodevelopmental disorders, and other non-cardiac congenital defects (Homsy et al. 2015; Mussatto et al. 2014; Marino et al. 2012).

HTX is a class of related birth defects that affects proper establishment of LR asymmetry of the internal organs. HTX affects 1/10,000 live births (Lin et al. 2014) and 3% of CHD patients, (Sutherland and Ware 2009) often leading to more severe forms of CHD. As the asymmetric orientation and formations of the vessels and chambers of the heart are important for its normal function, about 90% of HTX patients also have CHD (Lin et al. 2014). This relation highlights the importance of understanding CHD in the context of earlier developmental pathways, such as LR patterning, in addition to cardiogenesis.

16.2.1. Previous Studies of Vertebrate Cardiogenesis in *Xenopus*

Disruptions at various stages in early embryonic development can cause cardiac structural anomalies. Much work has gone into characterizing cardiac tissue specification and morphogenesis across many organisms. Several studies on cardiac morphogenesis performed in *Xenopus* models have connected CHD patient variants directly with cardiac development.

Vertebrate cardiac development follows a conserved developmental pathway. Briefly, the mesoderm is specified at gastrulation (Sater and Jacobson 1989). As the embryo elongates, mesodermal cardiac precursors migrate to the midline and specify into the cardiac progenitors of the first and second heart fields. The first heart field eventually forms the two atria and the trabeculated ventricle. The second heart field becomes the outflow tract (Buckingham, Meilhac, and Zaffran 2005; Gessert and Kühl 2009).

Several known cardiogenesis genes originally discovered through genetic screens in model organisms and CHD patient genetic analyses have been studied in *Xenopus*. *nkx2–5* and its co-factor *gata4* have been associated with atrio-septal defects (ASDs) (Cleaver, Patterson, and Krieg 1996; Durocher, Schwartz, and Nemer 1997; Schott et al. 1998; Benson et al. 1999; Bartlett et al. 2007). *msh-2 (nkx2–5)* was originally identified in a screen for novel homeobox genes in *Drosophila melanogaster* (Bodmer, Jan, and Jan 1990). The gene was renamed *tinman* (*tin*) in *Drosophila* due to *tin* mutants lacking cardiac primordia (Bodmer 1993). *Xenopus nkx-2.5*, like *tinman*, is expressed in heart and gut tissues (Tonissen et al. 1994). The *nkx2–5* co-factor *gata4* was originally identified in a cDNA screen for novel GATA family members in *Xenopus* expressed in the developing heart (Kelley et al. 1993). *X. laevis* studies showed that both *nkx2–5* and *gata4* overexpression resulted in embryos with enlarged hearts. Authors proposed that *nkx2–5* and *gata4* cause this myocardial hyperplasia through pathways involved in myocardial progenitor proliferation or recruitment (Cleaver, Patterson, and Krieg 1996).

In contrast, the T-box family gene *tbx5* was first associated with cardiac development via patient studies. *Tbx5* was initially identified in mice and was found to be expressed in early cardiac and forelimb tissues, but researchers focused on its role in limb formation (Gibson-Brown et al. 1996; Chapman et al. 1996). Patients with Holt-Oram syndrome (HOS) have defects in limb formation and cardiac septation. A study looking for the HOS gene mapped *TBX5* within a deleted region of chromosome 12q2, the locus responsible for HOS (Basson et al. 1997; Mcdermott et al. 2005). Studies in *Xenopus* showed that both overexpression and knockdown of *tbx5* was associated with heart tube formation abnormalities and heart-looping anomalies, loss of cardiac mass, and edema (Brown et al. 2005; Horb and Thomsen 1999).

Another T-box gene, *tbx20*, was identified in a screen for novel family members involved in FGF patterning of the mesoderm in developing zebrafish (Kevin J.P. Griffin et al. 1998) and was found to be expressed in the same tissues as *nkx2–5*, *gata4*, and *tbx5* (Griffin et al. 2000; Brown et al. 2005). Families with *TBX20* mutations present with cardiomyopathy, ASD, and mitral valve malformations (Kirk et al. 2007). *hrT (tbx20)* knockdown in zebrafish causes abnormal cardiac chamber formation (Szeto, Griffin, and Kimelman 2002). *tbx20* knockdown in *Xenopus* is associated with heart looping abnormalities, edema, and cardiac chamber differentiation, though specification of cardiac tissue is normal (Brown et al. 2005).

Similar to *tbx5*, *ets1* was not directly associated with cardiac structuring until patients with a CHD-related syndrome underwent genetic analysis. *ETS1* was initially identified as a potential oncogene (Leprince et al. 1983). It was subsequently associated with cranial neural crest migration, vasculogenesis, immune cell differentiation, and endothelial differentiation (Meyer et al. 1997; Tahtakran and Selleck 2003; Sumanas and Lin 2006; Barton et al. 1998; Wang et al. 2005). Later studies mapped *ETS1* within the region of chromosome 11 that is affected in patients with Jacobsen syndrome, a syndrome that includes CHD (Penny et al. 1995; Grossfeld et al. 2004; Ye et al. 2009). When investigated under in the context of Jacobsen syndrome, researchers found that *ets1* is necessary for cardiac mesoderm and neural crest specification (Ye et al. 2009; Nie and Bronner 2015). Studies in *Xenopus* found *ets1* to be expressed in several tissues, including neural crest and the developing heart (Meyer et al. 1997). Depleting *ets1* in the mesoderm, cardiac mesoderm, or cardiac neural crest leads to loss of endocardial specification, poor cardiac morphogenesis (single-chamber heart, no trabeculation, and no aortic septation) due to delayed heart tube formation, and small malformed outflow tracts, respectively (Nie and Bronner 2015). These CHD genes and the roles they play in cardiac morphogenesis account for only a small portion of genes and mechanisms identified in CHD patients. More recently identified candidate genes may have unknown functions or unpredictable roles in development prior to or in conjunction with cardiac morphogenesis.

16.2.2. Identifying Additional Candidate CHD Genes

Analysis of large CHD/HTX patient cohorts using genotyping microarrays and whole exome sequencing (WES) has identified many new candidate genes. We can efficiently identify copy-number variants (CNVs) with genotyping microarrays (Alkan, Coe, and Eichler 2011). CNVs are a type of genetic perturbation that involves duplications or deletions that can span multiple genes. In such cases, it can be difficult to pinpoint disease causality to one or more genes.

An early study of CNVs from HTX patients discovered 36 relatively small CNVs affecting 61 genes. Morpholino knockdown of a subset of these genes in *Xenopus tropicalis* recapitulated HTX phenotypes. All of these genes also showed perturbations of LR patterning prior to cardiac morphogenesis (Fakhro et al. 2011). This early analysis of genetic contributions to HTX supported the need to cast a wider net to identify a greater number of rare HTX and CHD candidate genes to better understand CHD/HTX etiology.

As costs of next-generation whole genome and exome sequencing have decreased dramatically over the past decade, it has become more feasible than ever to use these approaches for identifying gene candidates. The National Heart, Lung and Blood Institute (NHLBI)'s Pediatric Cardiac Genetic Consortium (PCGC) exploited whole exome sequencing to do just that. The PCGC performed *de novo* mutation analysis with WES in cases in which patient-parent trios were available. Researchers looked for high-probability single base pair substitutions and small INDELs present in patient samples that were not present in parent samples that are rare and deleterious (nonsense, frameshift, splice-site, or damaging amino-acid altering missense mutations). Additional single-nucleotide polymorphism (SNP) array analysis has helped identify rare CNVs potentially contributing to patient CHD (Pediatric Cardiac Genomics Consortium et al. 2013). Multiple alleles for candidate genes began to emerge over the duration of the study from 2010 to 2015: 66 genes had two or more damaging *de novo* mutations, and 19 had two or more loss-of-function *de novo* mutations. Although this is a good starting point, much work remains to identify a more comprehensive list of CHD candidate genes. The most recent analysis of the PCGC study found that larger cohort sizes (10,000 patients) predict capture of 38% of CHD gene candidates (Jin et al. 2017). Therefore, patient cohort recruitment, genetic analysis, and mechanistic studies in model organisms will further improve our understanding of CHD/HTX.

16.3. APPROACHES TO STUDYING CHD/HTX IN *XENOPUS*

Following the identification of genes from CHD/HTX patients, it is critical to identify plausible mechanisms that lead to pathogenesis. *Xenopus* is a great model organism to study CHD/HTX candidate genes in early development.

16.3.1. Morphological and Developmental Benefits of Studying CHD/HTX in *Xenopus*

Xenopus are particularly useful for applying high-throughput genetic screening techniques for analyzing CHD/HTX candidate genes for a variety of reasons. First, due to their external development and transparency during organogenesis, researchers can easily view the developing heart and internal organs without dissection. Additionally, with optical coherence tomography (OCT), rapid high resolution internal anatomy of the heart is easily defined (Deniz et al. 2017). The *Xenopus* heart is structurally similar to that of humans: it has two atria separated by a septum, a single trabeculated ventricle, and an outflow tract. Because the fate map of the early blastula embryos is well mapped, it is easy to target manipulations to specific subsets of cells via microinjection of mRNAs, antisense morpholino oligonucleotides (MOs), or CRISPR/Cas9-mediated genome editing. For example, an investigator can target one cell of the two-cell embryo and subsequently select embryos in which the right or the left side of the embryo is manipulated. This is unique to *Xenopus* and is especially powerful for analyzing LR patterning (Blum et al. 2009).

16.3.2. Using CRISPR/Cas9 Genome Editing as a Screening Tool in *Xenopus*

CRISPR/Cas9 genome editing is perhaps the most important recent technology for using *X. tropicalis* as a tool for high-throughput screening. *X. tropicalis* has a diploid genome (Offield, Hirsch, and Grainger 2000), making it ideal for CRISPR/Cas9 genome editing and establishment of mutant lines. There are multiple tools for designing sgRNAs and making sgRNAs efficiently that work very well for *Xenopus*. In addition, commercially available Cas9 protein makes generating F0 knockouts quick, easy, and affordable, thus enabling genetic screens. Once researchers identify candidate genes from patients, researchers can use CRISPR/Cas9 for preliminary screening of developmental defects and continue using it as a tool to probe the cell and molecular impacts of the gene in development (Bhattacharya et al. 2015).

16.3.3. Left-Right Patterning: A Beautiful Pathway to Guide CHD/HTX Disease Mechanism Analysis

The interrelation between CHD and HTX highlights the LR patterning pathway as a powerful guide to begin analyzing CHD/HTX gene candidates on a developmental timeline. Briefly, LR patterning begins early in embryonic development at the left-right organizer (LRO), where beating of motile cilia create a unidirectional leftward extracellular fluid flow (Schweickert et al. 2010; Minegishi et al. 2017). When the flow is sensed by immotile cilia at the periphery of the LRO, the signal results in downregulation of the initially symmetric Nodal antagonist, *dand5*, on the left side of the LRO (Blum et al. 2009). Repression of *dand5* allows Nodal signaling on the left side. This left-sided Nodal signaling at the LRO is communicated to the left lateral plate mesoderm, which results in left-sided expression of *pitx2c* (Blum et al. 2009; Schweickert et al. 2010; Desgrange, Le Garrec, and Meilhac 2018). Lateral *pitx2c* expression then defines asymmetric LR heart and organ orientation (Ryan et al. 1998). Conveniently for study, each stage of the LR cascade has an informative marker: cardiac looping, *pitx2* expression, and *dand5* expression. Pinpointing the candidate gene's role in a characterized developmental pathway is particularly important for studying genes with no known function in early development.

Several developmental signaling pathways are critical for proper establishment of the LRO and for relaying signals later in this LRO pathway. Canonical Wnt is necessary for LRO tissue patterning and ciliogenesis (Zhang et al. 2012) and *pitx2c* expression in the cardiac neural crest (Kioussi et al. 2002). Non-canonical Wnt/planar cell polarity (PCP) pathways contribute to cilia polarization in the LRO (Minegishi et al. 2017). Notch signaling establishes the immotile:motile cilia ratio in the LRO (Boskovski et al. 2013) and is important for *nodal* expression (Raya et al. 2003; Krebs et al. 2003). BMP signaling regulates right-side identity establishment (Ocaña et al. 2017), whereas Nodal signaling is important for left-side identity establishment (Schweickert et al. 2010; Desgrange, Le Garrec, and Meilhac 2018). Disruption of the cellular mechanisms underlying any of these pathways could result in disastrous LR patterning and subsequent disruptions in visceral organ and cardiac structure and organization.

16.4. PATIENT-DERIVED CHD DISEASE MECHANISMS UNCOVERED IN *XENOPUS*

Researchers have used the strategies described previously to identify and study the roles of CHD candidate genes in early embryonic development. This section will highlight several studies of the roles of several CHD candidate genes.

16.4.1. Connecting Known Pathways in a New Context: Glycosylation in NOTCH Signaling and Cilia Cell Fate Determination

CHD candidate genes may be involved in well-defined pathways, such as Notch signaling or glycosylation. However, the role of these pathways in early embryonic development and CHD/HTX is often unexplored or poorly understood.

One such CHD candidate, *galnt11*, was identified as a CNV deletion in a HTX patient (Fakhro et al. 2011). *galnt11* is a member of the protein family necessary for the *O*-glycosylation of N-acetylgalactosamines (GalNAc) (Bennett et al. 2012). The role of Galnt11 or GalNAc *O*-glycosylation had not yet been studied in the context of embryonic development, and at the time, no targets of *galnt11* were known. In order to investigate its role in CHD/HTX, investigators depleted *galnt11* in *X. tropicalis* embryos. These embryos had abnormal gut and heart looping (Fakhro et al. 2011; Boskovski et al. 2013), indicating *galnt11*'s role in LR organ *situs*. Investigating earlier steps in the LR-pathway revealed that both *pitx2c* and *dand5* expression were abnormal, suggesting a defect in cilia at the LRO.

As a proxy to cilia in the LRO, the authors investigated the multiciliated cells of the embryonic epidermis. They did not find any abnormalities in the cilia structure in these cells. However, careful inspection revealed a dramatic increase in number of multiciliated cells after *galnt11* depletion. In contrast, overexpressing *galnt11* reduced the number of multiciliated cells. This phenotype was also seen when the Notch pathway was manipulated (Liu et al. 2007; Ma and Jiang 2007; Tsao et al. 2009). Thus, the authors proposed that *galnt11* may work through Notch signaling to affect multiciliated cell number.

In Notch signaling extracellular ligands, Delta or Jagged, bind to the Notch receptor, causing physical changes to the receptor that make it amenable to extracellular cleavage by the ADAM metalloprotease; this initiates an intracellular cleavage by γ-secretase and releases the Notch Intracellular Domain (NICD). NICD then translocates into the nucleus and binds to target genes via an interaction with the CSL transcription complex. NICD and a constitutively active CSL, but not Delta, rescued LR phenotypes in *galnt11*-depleted embryos. This suggested that *galnt11* impacts LR patterning through the Notch pathway at a step between binding of the extracellular ligand and NICD release into the cytoplasm.

Could the Notch receptor be a target of Galnt11? Previous studies had demonstrated that Notch is glycosylated, but not by a GalNac type glycosylation enzyme. The authors used mass spectrometry on the Notch extracellular domain peptides and an *in vitro* glycosylation assay to identify potential GalNac *O*-glycosylation sites. One of the three identified target sites is adjacent to the extracellular ADAM cleavage site. In previous work, GalNac glycosylation near a cleavage site generally prevented cleavage of the pro-peptide. However, in this case, Galnt11 appeared to enhance Notch signaling (loss of Galnt11 mimicked a loss of Notch phenotype, and gain of Notch rescued loss of galnt11 phenotypes). Unlike previous studies, in the case of the Notch receptor, Galnt11 glycosylation enhanced cleavage. This supports the interpretation that Galnt11 O-glycosylation of this adjacent site is important for proper Notch receptor processing, signaling, and LR patterning.

An important question is: How does Galnt11 and Notch signaling impact LR patterning? Previously, a two-cilia model had been proposed for vertebrate LR axis determination (McGrath et al. 2003; Tabin and Vogan 2003). The model proposed that one class of cilia in the LRO are responsible for generating flow, while another is required for sensing and translating that flow-based signal. However, how the LRO establishes both motile and immotile cilia was simply unknown. Because *galnt11*/Notch loss-of-function produced phenotypes that suggested a loss of cilia sensing, whereas *galnt11*/Notch gain-of-function led to phenotypes suggesting loss of cilia motility, the authors hypothesized that Notch may act as a switch between motile and immotile cilia types, which could solve a long-standing mystery in the two-cilia model. Using live cell imaging, the authors found that Notch/Galnt11 signaling altered the balance between motile and immotile cilia. Ultimately, the authors not only identified the role of *galnt11* in CHD/HTX pathogenesis but also uncovered a previously unknown pathway for regulating Notch signaling and the subsequent cilia distribution required for regulation of the LR patterning (Boskovski et al. 2013).

16.4.2. Uncovering Novel Roles for Well-Studied Genes: Nucleoporins in Cilia and Centrosomal Biology

Some CHD candidate genes have well-studied cellular functions that researchers may not have looked at in the context of early embryonic development. When development is disturbed, it may reveal unexpected roles for these genes. One such well-studied family is the Nucleoporins (Nups). Nups are proteins that make up the approximately 100 Megadalton (MDa) nuclear pore complexes (NPCs), which are the gatekeepers for transport in and out of the nucleus (Viso et al. 2016). The PCGC identified patient mutations in several Nups (Jin et al. 2017), found in various NPC substructures. However, since nuclear transport is fundamental for all cells, how can altering it be associated with CHD/HTX, as opposed to organism survival?

The first CHD/HTX candidate Nup studied in depth for its role in early development is Nup188, which was identified via CNV analysis (Fakhro et al. 2011). The investigators first explored the role of Nup188 in organ *situs*. Its knockdown via translation-blocking morpholino led to abnormal heart looping, gut looping, and *pitx2c* expression (Fakhro et al. 2011). However, how does Nup188 regulate LR patterning? Nup188 is a component of the inner ring of the nuclear pore complex and binds directly to Nup93. Therefore, the investigators depleted Nup93 to see if it recapitulated the phenotypes in Nup188 depletion, and indeed, it did (Viso et al. 2016). This appeared specific to the inner ring, as depletion of components of the outer ring and the central transport channel did not recapitulate Nup188 depletion phenotypes. Thus, the inner ring Nups specifically play a role in heart looping.

To look more deeply at this mechanism, the authors made a surprising observation. Embryos depleted of Nup93 and Nup188 did not glide along the surface of a Petri dish, unlike their wild-type siblings. Multiciliated cells on the embryonic epidermis beat and create extracellular fluid flow that drives this gliding. By immunohistochemistry, most cilia in these cells were depleted. This was also true of the cilia in the LRO. The authors therefore demonstrated that inner-ring Nup188 and Nup93 are uniquely important for cilia structure, as their depletion led to dramatic changes in cilia structure. However, gross nuclear pore number, structure, and function appeared intact. While the authors could not eliminate the possibility of a subtle defect, the authors began to look for mechanisms of how inner-ring Nups might affect cilia structure independently of nucleocytoplasmic transport and NPC structure.

If inner-ring Nups do not affect cilia structure via nucleocytoplasmic transport, what else could they affect? Previous studies showed that NPC substructures can have functions outside of the NPC (Belgareh et al. 2001; Loïodice et al. 2004; Hashizume et al. 2013; Itoh et al. 2013). One study proposed a model for a "ciliary pore complex" similar in function to the NPC that provides a size-selective barrier for entry of proteins into the ciliary axoneme (Kee et al. 2012). However, the evidence supporting this model has been disputed (Breslow et al. 2013). With this in mind, the authors investigated whether inner-ring Nups localize at the base of cilia. Immunofluorescence imaging of endogenous Nup93 and Nup188 in retinal pigmented epithelial (RPE) cells revealed their accumulation near the basal-bodies, while other non-inner ring Nups did not. This suggests that Nup188 and Nup93 have a role independent of the NPC at cilia and that NPC components normally needed to create the NPC diffusion barrier do not localize at the base of cilia, contradicting the "ciliary pore complex" model.

To visualize the structure of the inner-ring Nups at the base of cilia, the authors used super-resolution microscopy to visualize Nup93 and Nup188 structures at the ciliary base. Nup188 and Nup93 clustered in much smaller puncta (about 50 nm), contradicting the 100-nm ring prediction for a ciliary pore. In fact, these small clusters distributed in a pattern around the basal bodies in two barrel-like structures, a pattern congruent with the pericentriolar material (PCM) surrounding mother and daughter centrioles (Viso et al. 2016). This surprising discovery indicated that depletion of Nup188 contributes to CHD/HTX by disrupting the formation of cilia needed to establish proper LR patterning through a novel role in ciliary and centrosomal biology. This study that began with a CHD patient opened a new avenue to explore connections between Nups, centrioles, and cilia (Vishnoi et al. 2020).

16.4.3. Exploring Old Mysteries in Developmental Cell Signaling

Linking CHD/HTX variants to a developmental phenotype is often the starting point for achieving a better understanding of basic biological questions. Sometimes these questions can lead to a novel perspective on long-standing questions, for example, the nuclear transport of β-Catenin, a key effector of the Wnt signaling pathway. During Wnt signaling, Wnt ligand binding leads to sequestration of the β-Catenin degradation complex so that β-Catenin is stabilized and can accumulate and enter the nucleus to activate Wnt target genes. In fact, the tight regulation of β-Catenin is essential, and its overexpression contributes to context-dependent diseases: in the context of the early embryo, excessive ectopic ventral Wnt signaling can lead to a double axis (a conjoined-twin embryo) and in adult cells to cancers.

β-Catenin uses facilitated transport to enter the nucleus independent of Importin α/β1. Remarkably, previous studies indicated that β-Catenin nuclear transport was independent of Ran, the master regulator of facilitated nuclear transport that employs nuclear transport receptors (Fagotto, Glück, and Gumbiner 1998; Yokoya et al. 1999). However, nuclear transport was energy dependent, which led to the hypothesis that either (1) β-Catenin binds directly to the NPC via several Armadillo (ARM) repeats and acts as its own nuclear transport receptor (NTR) (Fagotto, Glück, and Gumbiner 1998), or (2) β-Catenin employs another GTPase that is not Ran. However, evidence for a direct interaction between β-Catenin and the NPC is disputed (Sharma et al. 2014; Suh and Gumbiner 2003). While the negative data mounted, a fundamental question with major impact for a host of Wnt-related diseases remained: How does β-Catenin enter the nucleus?

A guanine nuclear exchange factor, *RAPGEF5*, was identified as a CHD/HTX candidate (Fakhro et al. 2011), but no role for it in embryonic development was known (Griffin et al. 2018). Depleting *rapgef5* in *X. tropicalis* embryos led to abnormal cardiac looping similar to the patient phenotype, suggesting a role in early development. Indeed, *rapgef5* depletion also affected earlier markers of LR patterning, *pitx2* and *dand5*. However, unlike the two previous examples of CHD candidate genes, *rapgef5* depletion reduced *dand5* expression even before the commencement of cilia-driven flow. This suggested that *rapgef5* depletion affected the specification of the LRO prior to cilia-driven signaling.

The LRO is a transient mesodermal structure that is overlaid by the endoderm as gastrulation is completed. Therefore, any defects in gastrulation or formation of the mesoderm can affect the formation of the LRO and LR signaling. As depletion of *rapgef5* affected the formation of the LRO, investigators examined other markers of the mesoderm and found that most were normal, except the known Wnt responsive genes *foxj1* and *nodal3.1*. These results indicated that *rapgef5* likely disrupts LRO patterning and subsequent LR patterning by impacting Wnt signaling. Consistent with this finding, *rapgef5* depletion reduced embryonic β-Catenin levels. Given the many genes involved in transducing the Wnt signal into the cell and the degradation of β-Catenin, the investigators sought to determine if Rapgef5 affects one of these steps. To do so, they took advantage of the fact that ectopic ventral expression of β-Catenin induces a secondary axis in *Xenopus* embryos. Depleting *rapgef5* in embryos in which wild-type β-Catenin was ectopically expressed reduced the number of embryos with secondary axes. Repeating this experiment with a mutated β-Catenin such that it could not be phosphorylated by the β-Catenin degradation complex and then degraded showed that RAPGEF5 blocked the formation of secondary axes, suggesting that *rapgef5* functions downstream of β-Catenin degradation. Repeating this experiment with a β-Catenin containing the Importin α/β1-dependent nuclear localization signal prevented *rapgef5* from blocking secondary axis formation. These experiments indicate that *rapgef5* plays a role in β-Catenin nuclear transport.

Based on these results, a model for a Rap-based nuclear transport system was proposed. In support of this model, endogenous Rapgef5 protein and active Raps were found to localize to the nucleus, and Rap1 physically interacted with β-Catenin. Overexpressing constitutively active Rap1b rescued reduced *foxj1* expression in *rapgef5*-depleted gastrulae, whereas a dominant-negative permanently GDP-bound Rap1b did not. Taken together, these results support the existence and utilization of an alternate energy-dependent nuclear transport pathway by which β-Catenin can access the nucleus independently of the classical Importin α/β1-mediated Ran-dependent nuclear transport pathway (Griffin et al. 2018). Though this began with a CHD/HTX candidate gene, the developmental study of *rapgef5* opened the door to re-evaluating the β-Catenin nuclear transport pathway, with implications for other signaling effectors.

16.4.4. Shared Molecular Pathways for Multiple Diseases: Neurodevelopmental Disorders, Craniofacial Abnormalities, and CHD

An important finding from these studies inspired by patients is that genes thought to have fundamental functions in cell biology can have remarkably tissue-specific effects in human congenital malformations. For example, ribosomopathies can lead to craniofacial malformations, a far cry from the expectation that ribosomes are essential for protein production in all cells and therefore cell viability. Here we describe the role of protein folding that affects many tissues but due to a specific effect on a remarkable population of cells called the neural crest that affects cardiac development.

Variants in the gene encoding the endoplasmic reticulum membrane protein complex subunit 1 (*EMC1*) were initially identified in a patient with retinitis pigmentosa (Abu-Safieh et al. 2013) and later expanded to include patients with neurodevelopmental defects, visual impairments, craniofacial abnormalities, and CHD (Harel et al. 2016; Geetha et al. 2018; Homsy et al. 2015; Jin et al. 2017). The variants for this broad cohort appear to be randomly dispersed across the gene. Given that EMC1 is critical for the folding of multi-pass transmembrane proteins, how do we explain these phenotypes? This was first investigated by depleting *emc1* in *X. tropicalis* and cataloging any developmental defects in the resulting embryos. Defects in craniofacial cartilage development, cardiac outflow tract diameter, and embryo motility were seen, recapitulating many of the patient phenotypes. While wild-type human *EMC1* mRNA rescued each of these phenotypes in depleted embryos, mRNAs containing patient variants did not, supporting the role of *EMC1* defects in each of these patient's disease manifestations (Marquez et al. 2020).

Craniofacial, neural, and cardiac progenitors are derived from the neural crest cells (NCCs). Indeed, *emc1* morphants and embryos expressing disease variants had defects in NCC gene expression and migration.

Label-free quantitative mass spectrometry (LFQMS) of *emc1* morphants compared to WT embryos revealed a loss of proteins in the Wnt pathway, which is a critical pathway for NCC specification (Dorsky, Moon, and Raible 1998; Maj et al. 2016). As EMC1 plays a role in regulating multi-pass membrane proteins, an obvious candidate was the Frizzled2 receptor (Fzd2). Immunofluorescence microscopy in RPE cells after siRNA-mediated *emc1* depletion revealed a loss of membrane bound Fzd2 and Fzd7 and an increase in an intracellular punctate signal characteristic of misfolded proteins. These results suggested that the NCC defects in *emc1*-depleted embryos may be due to defects in proper folding and membrane integration of Fzd receptors. In support of this, when β-Catenin was expressed in NCCs, the loss of neural crest specification was rescued.

In addition to the NCC-related defects, several patients with *EMC1* mutations also had visual impairments and neurodevelopmental defects. Rhodopsin and the nicotinic acetylcholine receptor (nAChR) are also multi-pass membrane-bound proteins important for vision and neural function, respectively. As predicted, *emc1* depletion in RPE cells led to mis-localized rhodopsin and nAChR. *EMC1* patient phenotypes can therefore be explained by the defects in proper localization of these functionally important multi-pass transmembrane proteins during development (Marquez et al. 2020).

16.5. WHERE TO GO FROM HERE? THE FUTURE OF STUDYING CHD IN *XENOPUS*

With estimates of 10,000 patient-parent trios needed to achieve 80% saturation of genes associated with syndromic CHD cases (Jin et al. 2017), there is much more work to be done identifying new candidate genes, connecting their contributions to CHD/HTX, and understanding the role each gene plays in CHD/HTX etiology. Both clinicians and scientists benefit as they learn more about the genetic, molecular, and cellular mechanisms underlying the etiology of CHD/HTX. Clinicians will have access to an increasing list of studied CHD/HTX genes they can refer to when admitting, diagnosing, and treating new CHD patients. The list, along with basic science research of underlying molecular mechanisms, can help improve targeted patient care, potentially predict whether a patient may have undiagnosed extracardiac issues, and guide preventative care for CHD comorbidities/shared etiologies with other anomalies and neurodevelopmental disorders. These combined efforts also will aid genetic counseling of couples for their family planning.

Continued recruitment and genomic analysis of patients will continue to provide researchers with new genes to gain better insights into human health, development, and basic cell biology. These developmental phenotype and mechanistic studies provide scaffolding for asking new lines of questions understudied in developmental biology. Such questions shed light on shared molecular pathways between CHD/HTX and associated diseases, new molecular functions of understudied genes, and unexpected roles of known molecules in development.

REFERENCES

Abbot, Maude E. 1915. "Congenital Cardiac Disease." In *Osler's Modern Medicine*, 4th ed., 323–448. Philadelphia: Lea & Febiger.

Abu-Safieh, Leen, May Alrashed, Shamsa Anazi, Hisham Alkuraya, Arif O. Khan, Mohammed Al-Owain, Jawahir Al-Zahrani, et al. 2013. "Autozygome-Guided Exome Sequencing in Retinal Dystrophy Patients Reveals Pathogenetic Mutations and Novel Candidate Disease Genes." *Genome Research* 23 (2): 236–247. https://doi.org/10.1101/gr.144105.112.

Afzelius, Björn A. 1977. "A Human Syndrome Caused by Immotile Cilia." *Obstetrical and Gynecological Survey* 32 (1): 49–50. https://doi.org/10.1097/00006254-197701000-00024.

Alkan, Can, Bradley P. Coe, and Evan E. Eichler. 2011. "Genome Structural Variation Discovery and Genotyping." *Nature Reviews Genetics* 12 (5): 363–376. https://doi.org/10.1038/nrg2958.

Bartlett, Heather L., Lillian Sutherland, Sandra J. Kolker, Chelsea Welp, Urszula Tajchman, Vera Desmarais, and Daniel L. Weeks. 2007. "Transient Early Embryonic Expression of Nkx2–5 Mutations Linked to Congenital Heart Defects in Human Causes Heart Defects in *Xenopus laevis*." *Developmental Dynamics* 236 (9): 2475–2484. https://doi.org/10.1002/dvdy.21244.

Barton, Kevin, Natarajan Muthusamy, Christopher Fischer, Chao Nan Ting, Theresa L. Walunas, Lewis L. Lanier, and Jeffrey M. Leiden. 1998. "The Ets-1 Transcription Factor Is Required for the Development of Natural Killer Cells in Mice." *Immunity* 9 (4): 555–563. https://doi.org/10.1016/S1074-7613(00)80638-X.

Basson, Craig T., David R. Bachinsky, Robert C. Lin, Tatjana Levi, Jacob A. Elkins, Johann Soults, David Grayzel, et al. 1997. "Mutations in Human Cause Limb and Cardiac Malformation in Holt-Oram Syndrome." *Nature Genetics* 15 (1): 30–35. https://doi.org/10.1038/ng0197-30.

Belgareh, Naïma, Gwénaël Rabut, Siau Wei Baï, Megan Van Overbeek, Joël Beaudouin, Nathalie Daigle, Olga V. Zatsepina, et al. 2001. "An Evolutionarily Conserved NPC Subcomplex, Which Redistributes in Part to Kinetochores in Mammalian Cells." *Journal of Cell Biology* 154 (6): 1147–1160. https://doi.org/10.1083/jcb.200101081.

Bennett, Eric P., Ulla Mandel, Henrik Clausen, Thomas A. Gerken, Timothy A. Fritz, and Lawrence A. Tabak. 2012. "Control of Mucin-Type O-Glycosylation: A Classification of the Polypeptide GalNAc-Transferase Gene Family." *Glycobiology* 22 (6): 736–756. https://doi.org/10.1093/glycob/cwr182.

Benson, D. Woodrow, G. Michael Silberbach, Ann Kavanaugh-McHugh, Carol Cottrill, Yizhong Zhang, Steve Riggs, Octavia Smalls, et al. 1999. "Mutations in the Cardiac Transcription Factor NKX2.5 Affect Diverse Cardiac Developmental Pathways." *Journal of Clinical Investigation* 104 (11): 1567–73. https://doi.org/10.1172/JCI8154.

Bhattacharya, Dipankan, Chris A. Marfo, Davis Li, Maura Lane, and Mustafa K. Khokha. 2015. "CRISPR/Cas9: An Inexpensive, Efficient Loss of Function Tool to Screen Human Disease Genes in Xenopus." *Developmental Biology* 408 (2): 196–204. https://doi.org/10.1016/j.ydbio.2015.11.003.

Blalock, Alfred, and Helen B. Taussig. 1945. "The Surgical Treatment of Maformations of the Heart." *Journal of the American Medical Association* 128 (3): 189. https://doi.org/10.1001/jama.1945.02860200029009.

Blum, Martin, Tina Beyer, Thomas Weber, Philipp Vick, Philipp Andre, Eva Bitzer, and Axel Schweickert. 2009. "Xenopus, an Ideal Model System to Study Vertebrate Left-Right Asymmetry." *Developmental Dynamics* 238 (6): 1215–1225. https://doi.org/10.1002/dvdy.21855.

Bodmer, R. 1993. "The Gene Tinman Is Required for Specification of the Heart and Visceral Muscles in Drosophila." *Development* 118 (3): 719–729.

Bodmer, R., L. Y. Jan, and Y. N. Jan. 1990. "A New Homeobox-Containing Gene, Msh-2, Is Transiently Expressed Early during Mesoderm Formation of Drosophila." *Development* 110 (3): 661–669.

Boskovski, Marko T., Shiaulou Yuan, Nis Borbye Pedersen, Christoffer Knak Goth, Svetlana Makova, Henrik Clausen, Martina Brueckner, and Mustafa K. Khokha. 2013. "The Heterotaxy Gene GALNT11 Glycosylates Notch to Orchestrate Cilia Type and Laterality." *Nature* 504 (7480): 456–459. https://doi.org/10.1038/nature12723.

Breslow, David K., Elena F. Koslover, Federica Seydel, Andrew J. Spakowitz, and Maxence V. Nachury. 2013. "An In Vitro

Assay for Entry into Cilia Reveals Unique Properties of the Soluble Diffusion Barrier." *Journal of Cell Biology* 203 (1): 129–147. https://doi.org/10.1083/jcb.201212024.

Brown, Daniel D., Shauna N. Martz, Olav Binder, Sarah C. Goetz, Brenda M.J. Price, Jim C. Smith, and Frank L. Conlon. 2005. "Tbx5 and Tbx20 Act Synergistically to Control Vertebrate Heart Morphogenesis." *Development* 132 (3): 553–563. https://doi.org/10.1242/dev.01596.

Buckingham, Margaret, Sigolène Meilhac, and Stéphane Zaffran. 2005. "Building the Mammalian Heart from Two Sources of Myocardial Cells." *Nature Reviews Genetics* 6 (11): 826–835. https://doi.org/10.1038/nrg1710.

Castañeda, Aldo. 2005. "Congenital Heart Disease: A Surgical-Historical Perspective." *Annals of Thoracic Surgery* 79 (6): 0–3. https://doi.org/10.1016/j.athoracsur.2005.03.031.

Castaneda, Aldo R., William I. Norwood, Richard A. Jonas, Steve D. Colon, Stephen P. Sanders, and Peter Lang. 1984. "Transposition of the Great Arteries and Intact Ventricular Septum: Anatomical Repair in the Neonate." *Annals of Thoracic Surgery* 38 (5): 438–443. https://doi.org/10.1016/S0003-4975(10)64181-1.

Chapman, Deborah L., Nancy Garvey, Sarah Hancock, Maria Alexiou, Sergei I. Agulnik, Jeremy J. Gibson-Brown, Judith Cebra-Thomas, Roni J. Bollag, Lee M. Silver, and Virginia E. Papaioannou. 1996. "Expression of the T-Box Family Genes, Tbx1-Tbx5, during Early Mouse Development." *Developmental Dynamics* 206 (4): 379–390. https://doi.org/10.1002/(SICI)1097-0177(199608)206:4<379::AID-AJA4>3.0.CO;2-F.

Christianson, Arnold, Christopher P. Howson, and Bernadette Modell. 2006. "March of Dimes: Global Report on Birth Defects." In *March of Dimes Birth Defects Foundation*. White Plains, New York.

Cleaver, Ondine B., Kristin D. Patterson, and Paul A. Krieg. 1996. "Overexpression of the Tinman-Related Genes XNkx-2.5 and XNkx-2.3 in Xenopus Embryos Results in Myocardial Hyperplasia." *Development* 122 (11): 3549–3556.

Crafoord, Clarence, and G. Nylin. 1945. "Congenital Coarctation of the Aorta and Its Surgical Treatment." *Journal of Thoracic Surgery* 14 (5): 347–361. https://doi.org/10.1016/s0096-5588(20)31801-8.

Deniz, Engin, Stephan Jonas, Michael Hooper, John N. Griffin, Michael A. Choma, and Mustafa K. Khokha. 2017. "Analysis of Craniocardiac Malformations in Xenopus Using Optical Coherence Tomography." *Scientific Reports* 7: 1–13. https://doi.org/10.1038/srep42506.

Desgrange, Audrey, Jean-François Le Garrec, and Sigolène M. Meilhac. 2018. "Left-Right Asymmetry in Heart Development and Disease: Forming the Right Loop." *Development* 145 (22): dev162776. https://doi.org/10.1242/dev.162776.

Dorsky, Richard I., Randall T. Moon, and David W. Raible. 1998. "Control of Neural Crest Cell Fate by the Wnt Signalling Pathway." *Nature* 396 (6709): 370–372. https://doi.org/10.1038/24620.

Durocher, Daniel, Robert J Schwartz, and Mona Nemer. 1997. "The Cardiac Transcription Factors Nkx2–5 and GATA-4." *The EMBO Journal* 16 (18): 5687–5696. https://doi.org/10.1093/emboj/16.18.5687

Fagotto, François, Ursula Glück, and Barry M. Gumbiner. 1998. "Nuclear Localization Signal-Independent and Importin/Karyopherin-Independent Nuclear Import of β-Catenin." *Current Biology* 8 (4): 181–190. https://doi.org/10.1016/s0960-9822(98)70082-x.

Fakhro, K. A., M. Choi, S. M. Ware, J. W. Belmont, J. A. Towbin, R. P. Lifton, M. K. Khokha, and M. Brueckner. 2011. "Rare Copy Number Variations in Congenital Heart Disease Patients Identify Unique Genes in Left-Right Patterning." *Proceedings of the National Academy of Sciences* 108 (7): 2915–2920. https://doi.org/10.1073/pnas.1019645108.

Fenderson, Lewis H., and Don Miller. 1971. *Daniel Hale Williams: Open Heart Doctor*. New York: McGraw-Hill and Rutledge Books.

Geetha, Thenral S., Lokesh Lingappa, Abhishek Ravindra Jain, Hridya Govindan, Nitin Mandloi, Sakthivel Murugan, Ravi Gupta, and Ramprasad Vedam. 2018. "A Novel Splice Variant in EMC1 Is Associated with Cerebellar Atrophy, Visual Impairment, Psychomotor Retardation with Epilepsy." *Molecular Genetics and Genomic Medicine* 6 (2): 282–287. https://doi.org/10.1002/mgg3.352.

Gelb, Bruce D. 2015. "History of Our Understanding of the Causes of Congenital Heart Disease." *Circulation: Cardiovascular Genetics* 8 (3): 529–536. https://doi.org/10.1161/CIRCGENETICS.115.001058.

Gessert, Susanne, and Michael Kühl. 2009. "Comparative Gene Expression Analysis and Fate Mapping Studies Suggest an Early Segregation of Cardiogenic Lineages in *Xenopus laevis*." *Developmental Biology* 334 (2): 395–408. https://doi.org/10.1016/j.ydbio.2009.07.037.

Gibson-Brown, Jeremy J., Sergei I. Agulnik, Deborah L. Chapman, Maria Alexiou, Nancy Garvey, Silver M. Lee, and Virginia E. Papaioannou. 1996. "Evidence of a Role for T-Box Genes in the Evolution of Limb Morphogenesis and the Specification of Forelimb/Hindlimb Identity." *Mechanisms of Development* 56 (1–2): 93–101. https://doi.org/10.1016/0925-4773(96)00514-X.

Griffin, John N., Florencia del Viso, Anna R. Duncan, Andrew Robson, Woong Hwang, Saurabh Kulkarni, Karen J. Liu, and Mustafa K. Khokha. 2018. "RAPGEF5 Regulates Nuclear Translocation of β-Catenin." *Developmental Cell* 44 (2): 248–260.e4. https://doi.org/10.1016/j.devcel.2017.12.001.

Griffin, Kevin J.P., Sharon L. Amacher, Charles B. Kimmel, and David Kimelman. 1998. "Molecular Identification of Spadetail: Regulation of Zebrafish Trunk and Tail Mesoderm Formation by T-Box Genes." *Development* 125 (17): 3379–3388.

Griffin, K.J.P., J. Stoller, M. Gibson, S. Chen, D. Yelon, D.Y.R. Stainier, and D. Kimelman. 2000. "A Conserved Role for H15-Related T-Box Transcription Factors in Zebrafish and Drosophila Heart Formation." *Developmental Biology* 218 (2): 235–247. https://doi.org/10.1006/dbio.1999.9571.

Gross, Robert E., and John P. Hubbard. 1939. "Surgical Ligation of a Patent Ductus Arteriosus." *Journal of the American Medical Association* 112 (8): 729. https://doi.org/10.1001/jama.1939.02800080049011.

Grossfeld, Paul D., Teresa Mattina, Zona Lai, Remi Favier, Ken Lyons Jones, Finbarr Cotter, and Christopher Jones. 2004. "The 11q Terminal Deletion Disorder: A Prospective Study of 110 Cases." *American Journal of Medical Genetics* 129 A (1): 51–61. https://doi.org/10.1002/ajmg.a.30090.

Harel, Tamar, Gozde Yesil, Yavuz Bayram, Zeynep Coban-Akdemir, Wu Lin Charng, Ender Karaca, Ali Al Asmari, et al. 2016. "Monoallelic and Biallelic Variants in EMC1 Identified in Individuals with Global Developmental Delay, Hypotonia, Scoliosis, and Cerebellar Atrophy." *American Journal of Human Genetics* 98 (3): 562–570. https://doi.org/10.1016/j.ajhg.2016.01.011.

Hashizume, Chieko, Akane Moyori, Akiko Kobayashi, Nana Yamakoshi, Aoi Endo, and Richard W. Wong. 2013. "Nucleoporin Nup62 Maintains Centrosome Homeostasis."

Cell Cycle 12 (24): 3804–3816. https://doi.org/10.4161/cc.26671.

Homsy, Jason, Samir Zaidi, Yufeng Shen, James S. Ware, Kaitlin E. Samocha, Konrad J. Karczewski, Steven R. Depalma, et al. 2015. "De Novo Mutations in Congenital Heart Disease with Neurodevelopmental and Other Congenital Anomalies." *Science* 350 (6265): 1262–1266. https://doi.org/10.1126/science.aac9396.

Horb, Marko E., and Gerald H. Thomsen. 1999. "Tbx5 Is Essential for Heart Development." *Development* 126 (8): 1739–1751.

Itoh, Go, Shiro Sugino, Masanori Ikeda, Mayumi Mizuguchi, Shin Ichiro Kanno, Mohammed A. Amin, Kenji Iemura, Akira Yasui, Toru Hirota, and Kozo Tanaka. 2013. "Nucleoporin Nup188 Is Required for Chromosome Alignment in Mitosis." *Cancer Science* 104 (7): 871–879. https://doi.org/10.1111/cas.12159.

Jin, Sheng Chih, Jason Homsy, Samir Zaidi, Qiongshi Lu, Sarah Morton, Steven R. Depalma, Xue Zeng, et al. 2017. "Contribution of Rare Inherited and De Novo Variants in 2,871 Congenital Heart Disease Probands." *Nature Genetics* 49 (11): 1593–1601. https://doi.org/10.1038/ng.3970.

Kee, Hooi Lynn, John F. Dishinger, T. Lynne Blasius, Chia Jen Liu, Ben Margolis, and Kristen J. Verhey. 2012. "A Size-Exclusion Permeability Barrier and Nucleoporins Characterize a Ciliary Pore Complex That Regulates Transport into Cilia." *Nature Cell Biology* 14 (4): 431–437. https://doi.org/10.1038/ncb2450.

Kelley, C., H. Blumberg, L. I. Zon, and T. Evans. 1993. "GATA-4 Is a Novel Transcription Factor Expressed in Endocardium of the Developing Heart." *Development* 118 (3): 817–827.

Kioussi, Chrissa, Paola Briata, Sung Hee Baek, David W. Rose, Natasha S. Hamblet, Thomas Herman, Kenneth A. Ohgi, et al. 2002. "Identification of a Wnt/Dvl/β-Catenin → Pitx2 Pathway Mediating Cell-Type-Specific Proliferation during Development." *Cell* 111 (5): 673–685. https://doi.org/10.1016/S0092-8674(02)01084-X.

Kirk, Edwin P., Margaret Sunde, Mauro W. Costa, Scott A. Rankin, Orit Wolstein, M. Leticia Castro, Tanya L. Butler, et al. 2007. "Mutations in Crdiac T-Box Factor Gene TBX20 Are Associated with Diverse Cardiac Pathologies, Including Defects of Septation and Valvulogenesis and Cardiomyopathy." *American Journal of Human Genetics* 81 (2): 280–291. https://doi.org/10.1086/519530.

Krebs, Luke T., Naomi Iwai, Shigenori Nonaka, Ian C. Welsh, Yu Lan, Rulang Jiang, Yukio Saijoh, Timothy P. O'Brien, Hiroshi Hamada, and Thomas Gridley. 2003. "Notch Signaling Regulates Left-Right Asymmetry Determination by Inducing Nodal Expression." *Genes & Development* 17 (10): 1207–1212. https://doi.org/10.1101/gad.1084703.

Leprince, D., A. Gegonne, J. Coll, C. De Taisne, A. Schneeberger, C. Lagrou, and D. Stehelin. 1983. "A Putative Second Cell-Derived Oncogene of the Avian Leukaemia Retrovirus E26." *Nature* 306 (5941): 395–397. https://doi.org/10.1038/306395a0.

Lin, Angela E., Sergey Krikov, Tiffany Riehle-Colarusso, Jaime L. Frías, John Belmont, Marlene Anderka, Tal Geva, Kelly D. Getz, and Lorenzo D. Botto. 2014. "Laterality Defects in the National Birth Defects Prevention Study (1998–2007): Birth Prevalence and Descriptive Epidemiology." *American Journal of Medical Genetics, Part A* 164 (10): 2581–2591. https://doi.org/10.1002/ajmg.a.36695.

Liu, Yan, Narendra Pathak, Albrecht Kramer-Zucker, and Iain A. Drummond. 2007. "Notch Signaling Controls the Differentiation of Transporting Epithelia and Multiciliated Cells in the Zebrafish Pronephros." *Development* 134 (6): 1111–1122. https://doi.org/10.1242/dev.02806.

Loïodice, Isabelle, Annabelle Alves, Gwénaël Rabut, Megan van Overbeek, Jan Ellenberg, Jean-Baptiste Sibarita, and Valérie Doye. 2004. "The Entire Nup107–160 Complex, Including Three New Members, Is Targeted as One Entity to Kinetochores in Mitosis." *Molecular Biology of the Cell* 15 (7): 3333–3344. https://doi.org/10.1091/mbc.e03-12-0878.

Ma, Ming, and Yun Jin Jiang. 2007. "Jagged2a-Notch Signaling Mediates Cell Fate Choice in the Zebrafish Pronephric Duct." *PLoS Genetics* 3 (1): 0133–145. https://doi.org/10.1371/journal.pgen.0030018.

Mai, Cara T., Jennifer L. Isenburg, Mark A. Canfield, Robert E. Meyer, Adolfo Correa, Clinton J. Alverson, Philip J. Lupo, et al. 2019. "National Population-Based Estimates for Major Birth Defects, 2010–2014." *Birth Defects Research* 111 (18): 1420–1435. https://doi.org/10.1002/bdr2.1589.

Maj, Ewa, Lutz Künneke, Elisabeth Loresch, Anita Grund, Juliane Melchert, Tomas Pieler, Timo Aspelmeier, and Annette Borchers. 2016. "Controlled Levels of Canonical Wnt Signaling Are Required for Neural Crest Migration." *Developmental Biology* 417 (1): 77–90. https://doi.org/10.1016/j.ydbio.2016.06.022.

Marino, Bradley S., P. H. Lipkin, Jane W. Newburger, Georgina Peacock, Marsha Gerdes, J. William Gaynor, Kathleen A. Mussatto, et al. 2012. "Neurodevelopmental Outcomes in Children with Congenital Heart Disease: Evaluation and Management a Scientific Statement from the American Heart Association." *Circulation* 126 (9): 1143–1172. https://doi.org/10.1161/CIR.0b013e318265ee8a.

Marquez, Jonathan, June Criscione, Rebekah M. Charney, Maneeshi S. Prasad, Woong Y. Hwang, Emily K. Mis, Martín I. García-Castro, and Mustafa K. Khokha. 2020. "Disrupted ER Membrane Protein Complex-Mediated Topogenesis Drives Congenital Neural Crest Defects." *Journal of Clinical Investigation* 130 (2): 813–826. https://doi.org/10.1172/JCI129308.

Mcdermott, Deborah A., Michael C. Bressan, Jie He, Joseph S. Lee, Salim Aftimos, Martina Brueckner, Fred Gilbert, et al. 2005. "TBX5 Genetic Testing Validates Strict Clinical Criteria for Holt-Oram Syndrome." *Pediatric Research* 58 (5): 981–986. https://doi.org/10.1203/01.PDR.0000182593.95441.64.

McGrath, James, Stefan Somlo, Svetlana Makova, Xin Tian, and Martina Brueckner. 2003. "Two Populations of Node Monocilia Initiate Left-Right Asymmetry in the Mouse." *Cell* 114 (1): 61–73. https://doi.org/10.1016/S0092-8674(03)00511-7.

Meyer, D., M. Durliat, F. Senan, M. Wolff, M. Andre, J. Hourdry, and P. Remy. 1997. "Ets-1 and Ets-2 Proto-Oncogenes Exhibit Differential and Restricted Expression Patterns during *Xenopus laevis* Oogenesis and Embryogenesis." Edited by Derek Stemple. *The International Journal of Developmental Biology* 41 (4): 607–620. https://doi.org/10.1371/journal.pbio.0040010.

Minegishi, Katsura, Masakazu Hashimoto, Rieko Ajima, Katsuyoshi Takaoka, Kyosuke Shinohara, Yayoi Ikawa, Hiromi Nishimura, et al. 2017. "A Wnt5 Activity Asymmetry and Intercellular Signaling via PCP Proteins Polarize Node Cells for Left-Right Symmetry Breaking." *Developmental Cell* 40 (5): 439–452.e4. https://doi.org/10.1016/j.devcel.2017.02.010.

Mussatto, Kathleen A., Raymond G. Hoffmann, George M. Hoffman, James S. Tweddell, Laurel Bear, Yumei Cao, and Cheryl Brosig. 2014. "Risk and Prevalence of Developmental

Delay in Young Children with Congenital Heart Disease." *Pediatrics* 133 (3). https://doi.org/10.1542/peds.2013-2309.

Nie, Shuyi, and Marianne E. Bronner. 2015. "Dual Developmental Role of Transcriptional Regulator Ets1 in Xenopus Cardiac Neural Crest vs. Heart Mesoderm." *Cardiovascular Research* 106 (1): 67–75. https://doi.org/10.1093/cvr/cvv043.

Ocaña, Oscar H., Hakan Coskun, Carolina Minguillón, Prayag Murawala, Elly M. Tanaka, Joan Galcerán, Ramón Muñoz-Chápuli, and M. Angela Nieto. 2017. "A Right-Handed Signalling Pathway Drives Heart Looping in Vertebrates." *Nature* 549 (7670): 86–90. https://doi.org/10.1038/nature23454.

Offield, Martin F., Nicolas Hirsch, and Robert M. Grainger. 2000. "The Development of *Xenopus tropicalis* Transgenic Lines and Their Use in Studying Lens Developmental Timing in Living Embryos." *Development* 127 (9): 1789–1797.

Peacock, Thomas B. 1858. *On Malformation of the Human Heart with Original Cases*. London: John Churchill.

Pediatric Cardiac Genomics Consortium, Bruce Gelb, Martina Brueckner, Wendy Chung, Elizabeth Goldmuntz, Jonathan Kaltman, Juan Pablo Kaski, et al. 2013. "The Congenital Heart Disease Genetic Network Study: Rationale, Design, and Early Results." *Circulation Research* 112 (4): 698–706. https://doi.org/10.1161/CIRCRESAHA.111.300297.

Penny, L. A., M. Dell'Aquila, M. C. Jones, J. Bergoffen, C. Cunniff, J. P. Fryns, E. Grace, et al. 1995. "Clinical and Molecular Characterization of Patients with Distal 11q Deletions." *American Journal of Human Genetics* 56 (3): 676–683.

Raya, Angel, Yasuhiko Kawakami, Concepcion Rodriguez-Esteban, Dirk Buscher, Christopher M Koth, Tohru Itoh, Masanobu Morita, et al. 2003. "Notch Activity Induces Nodal Expression and Mediates the Establishment of Left-Right Asymmetry in Vertebrate Embryos." *Genes & Development* 17 (10): 1213–1218. https://doi.org/10.1101/gad.1084403.

Ryan, Aimee K., Bruce Blumberg, Concepción Rodriguez-Esteban, Sayuri Yonei-Tamura, Koji Tamura, Tohru Tsukui, J. de la Peña, et al. 1998. "Pitx2 Determines Left-Right Asymmetry of Internal Organs in Vertebrates." *Nature* 394 (6693): 545–551. https://doi.org/10.1038/29004.

Sater, A. K., and A. G. Jacobson. 1989. "The Specification of Heart Mesoderm Occurs during Gastrulation in *Xenopus laevis*." *Development* 105 (4): 821–830.

Schott, Jean Jacques, D. Woodrow Benson, Craig T. Basson, William Pease, G. Michael Silberbach, Jeffrey P. Moak, Barry J. Maron, Christine E. Seidman, and J. G. Seidman. 1998. "Congenital Heart Disease Caused by Mutations in the Transcription Factor NKX2–5." *Science* 281 (5373): 108–111. https://doi.org/10.1126/science.281.5373.108.

Schweickert, Axel, Philipp Vick, Maike Getwan, Thomas Weber, Isabelle Schneider, Melanie Eberhardt, Tina Beyer, Anke Pachur, and Martin Blum. 2010. "The Nodal Inhibitor Coco Is a Critical Target of Leftward Flow in Xenopus." *Current Biology* 20 (8): 738–743. https://doi.org/10.1016/j.cub.2010.02.061.

Sharma, Manisha, Michael Johnson, Mariana Brocardo, Cara Jamieson, and Beric R. Henderson. 2014. "Wnt Signaling Proteins Associate with the Nuclear Pore Complex: Implications for Cancer." In *Cancer Biology and the Nuclear Envelope: Recent Advances May Elucidate Past Paradoxes*, edited by Eric C. Schirmer and Jose I. de las Heras, 353–372. New York, NY: Springer New York. https://doi.org/10.1007/978-1-4899-8032-8_16.

Suh, Eun Kyung, and Barry M. Gumbiner. 2003. "Translocation of β-Catenin into the Nucleus Independent of Interactions with FG-Rich Nucleoporins." *Experimental Cell Research* 290 (2): 447–456. https://doi.org/10.1016/S0014-4827(03)00370-7.

Sumanas, Saulius, and Shuo Lin. 2006. "Ets1-Related Protein Is a Key Regulator of Vasculogenesis in Zebrafish." *PLoS Biology* 4 (1): 60–69. https://doi.org/10.1371/journal.pbio.0040010.

Sutherland, Mardi J., and Stephanie M. Ware. 2009. "Disorders of Left-Right Asymmetry: Heterotaxy and Situs Inversus." *American Journal of Medical Genetics, Part C: Seminars in Medical Genetics* 151 (4): 307–317. https://doi.org/10.1002/ajmg.c.30228.

Szeto, Daniel P., Kevin J.P. Griffin, and David Kimelman. 2002. "HrT Is Required for Cardiovascular Development in Zebrafish." *Development* 129 (21): 5093–5101.

Tabin, Clifford J., and Kyle J. Vogan. 2003. "A Two-Cilia Model for Vertebrate Left-Right Axis Specification." *Genes and Development* 17 (1): 1–6. https://doi.org/10.1101/gad.1053803.

Tahtakran, Siroun A. and Mark A.J. Selleck. 2003. "Ets-1 Expression Is Associated with Cranial Neural Crest Migration and Vasculogenesis in the Chick Embryo." *Gene Expression Patterns* 3 (4): 455–458. https://doi.org/10.1016/S1567-133X(03)00065-6.

Taussig, Helen B. 1947. *Congenital Malformations of the Heart*. New York: The Commonwealth Fund.

Tonissen, Kathryn F., Thomas A. Drysdale, Thierry J. Lints, Richard P. Harvey, and Paul A. Krieg. 1994. "XNkx-2.5, a Xenopus Gene Related to Nkx-2.5 and Tinman: Evidence for a Conserved Role in Cardiac Development." *Developmental Biology* 162 (1): 325–328. https://doi.org/10.1006/dbio.1994.1089.

Tsao, Po Nien, Michelle Vasconcelos, Konstantin I. Izvolsky, Jun Qian, Jining Lu and Wellington V. Cardoso. 2009. "Notch Signaling Controls the Balance of Ciliated and Secretory Cell Fates in Developing Airways." *Development* 136 (13): 2297–2307. https://doi.org/10.1242/dev.034884.

Virani, Salim S., Alvaro Alonso, Emelia J. Benjamin, Marcio S. Bittencourt, Clifton W. Callaway, April P. Carson, Alanna M. Chamberlain, et al. 2020. "Heart Disease and Stroke Statistics—2020 Update: A Report From the American Heart Association." *Circulation* 141 (9): e1–458. https://doi.org/10.1161/CIR.0000000000000757.

Vishnoi, Nidhi, Karthigeyan Dhanasekeran, Madeleine Chalfant, Ivan Surovstev, Mustafa K. Khokha, and C. Patrick Lusk. 2020. "Differential Turnover of Nup188 Controls Its Levels at Centrosomes and Role in Centriole Duplication." *The Journal of Cell Biology* 219 (3). https://doi.org/10.1083/jcb.201906031.

Viso, Florencia del, Fang Huang, Jordan Myers, Madeleine Chalfant, Zhang Yongdeng, Nooreen Reza, Joerg Bewersdorf, C. Patrick Lusk, and Mustafa Khokha. 2016. "Congenital Heart Disease Genetics Uncovers Context-Dependent Organization and Function of Nucleoporins at Cilia." *Developmental Cell* 38 (5): 478–492. https://doi.org/10.1158/1940-6207.CAPR-14-0359.Nrf2-dependent.

Wang, Duncheng, Shinu A. John, James L. Clements, Dean H. Percy, Kevin P. Barton, and Lee Ann Garrett-Sinha. 2005. "Ets-1 Deficiency Leads to Altered B Cell Differentiation, Hyperresponsiveness to TLR9 and Autoimmune Disease." *International Immunology* 17 (9): 1179–1191. https://doi.org/10.1093/intimm/dxh295.

Ye, Maoqing, Chris Coldren, Xingqun Liang, Teresa Mattina, Elizabeth Goldmuntz, D. Woodrow Benson, Dunbar Ivy, M. B. Perryman, Lee Ann Garrett-Sinha, and Paul Grossfeld. 2009. "Deletion of ETS-1, a Gene in the Jacobsen Syndrome Critical Region, Causes Ventricular Septal Defects and Abnormal Ventricular Morphology in Mice." *Human Molecular Genetics* 19 (4): 648–56. https://doi.org/10.1093/hmg/ddp532.

Yokoya, Fumihiko, Naoko Imamoto, Taro Tachibana, and Yoshihiro Yoneda. 1999. "β-Catenin Can Be Transported into the Nucleus in a Ran-Unassisted Manner." *Molecular Biology of the Cell* 10 (4): 1119–1131. https://doi.org/10.1091/mbc.10.4.1119.

Zaidi, Samir, Murim Choi, Hiroko Wakimoto, Lijiang Ma, Jianming Jiang, John D. Overton, Angela Romano-Adesman, et al. 2013. "De Novo Mutations in Histone-Modifying Genes in Congenital Heart Disease." *Nature* 498 (7453): 220–223. https://doi.org/10.1038/nature12141.

Zhang, Min, Junfeng Zhang, Sheng-Cai Lin, and Anming Meng. 2012. "β-Catenin 1 and β-Catenin 2 Play Similar and Distinct Roles in Left-Right Asymmetric Development of Zebrafish Embryos." *Development (Cambridge, England)* 139 (11): 2009–2019. https://doi.org/10.1242/dev.074435.

17 Craniofacial Development and Disorders—Contributions of *Xenopus*

Ashwin Lokapally and Hazel Sive

CONTENTS

17.1. What Is Craniofacial Development? ... 245
17.2. *Xenopus* as a Model for Craniofacial Development ... 245
17.3. Cells That Make Up the Craniofacial Skeleton: The Neural Crest ... 246
17.3.1. Neural Crest Cells—Contributions from *Xenopus* ... 246
17.3.2. Cranial Neural Crest Cells ... 246
17.4. The Mouth Is an Essential Facial Structure ... 246
17.4.1. Mouth Formation Is Conserved ... 246
17.4.2. Cellular Organization of the Mouth ... 247
17.4.3. Mouth Formation—*Xenopus* as a Model ... 247
17.5. The Extreme Anterior Domain Is a Craniofacial Signaling Center ... 249
17.5.1. Identification of the Extreme Anterior Domain ... 249
17.5.2. Craniofacial Organizer Function of the Extreme Anterior Domain ... 249
17.6. Landscape of Craniofacial Anomalies ... 249
17.6.1. Craniofacial Anomalies Are Associated with Multiple Factors ... 249
17.6.2. Classes of Craniofacial Anomalies ... 250
17.7. Treatment of Craniofacial Anomalies ... 251
17.7.1. Surgical Approaches ... 251
17.7.2. Tissue Engineering Approaches ... 251
17.7.3. Cell-Based Approaches ... 252
17.8. Contribution of *Xenopus* to Understanding Craniofacial Anomalies ... 252
17.9. Conclusion ... 252
References ... 253

17.1. WHAT IS CRANIOFACIAL DEVELOPMENT?

The term craniofacial is used to define the bones of the cranium (head) and the face, and the abnormalities associated with it are known as craniofacial anomalies (CFAs). The craniofacial complex is a three-dimensional structure consisting of the cranium, sense organs (i.e. eyes, ears, nose, tongue), mouth, facial bones, connective tissue, and the peripheral nerves (Wilkie and Morriss-Kay 2001; Chai and Maxson 2006). In humans, initial facial development extends from week 3 to week 12 of gestation (Som and Naidich 2013). The formation of the head and face involves a series of coordinated events that include the primary embryonic tissues, ectoderm, endoderm, and mesoderm, using multiple signaling cues to regulate proliferation, migration, and differentiation. A population of cells termed neural crest gives rise to the skull, face, jaw, and cartilages of ears and nose. Craniofacial development is susceptible to genetic and environmental perturbations, with orofacial anomalies appearing in 1 of 700 live births (Yoon, Pham, and Dipple 2016), while craniofacial anomalies result in approximately one-third of all congenital birth defects of the head and face (Twigg and Wilkie 2015), and more than 700 distinct craniofacial anomalies have been identified thus far (Terrazas et al. 2017).

17.2. *XENOPUS* AS A MODEL FOR CRANIOFACIAL DEVELOPMENT

In *Xenopus*, embryos develop externally to the mother, and the facial primordium is readily visible, facilitating live morphological observations and imaging. Since frog genes are conserved with those of mammals, application of molecular gain and loss-of-function assays and the rapid development of embryos make assays rapid and accessible to imaging. Large clutches of at least 500 eggs make for significant sample sizes. Loss-of-Function (LOF) can be elicited by injecting antisense, morpholino-oligonucleotides (MOs) into the one- to two-cell embryos (Nutt et al. 2001), using RNAi (Nakano et al. 2000), and inserting mutations obtained through CRISPR-mediated gene editing, including analysis of F0 animals (Nakayama et al. 2013; Guo et al. 2014; Willsey et al. 2018). We have developed a technique to limit LOF or gain-of-function (GOF) to

DOI: 10.1201/9781003050230-20

the extreme anterior domain (EAD) through facial transplants (Jacox, Dickinson, and Sive 2014). GOF can be temporally restricted using heat shock constructs in transient transgenic embryos (Dickinson and Sive 2009). These features make *Xenopus* an excellent model system.

One key question is whether the frog is a faithful model for human face development. Overall, it appears that vertebrate craniofacial development is well conserved (Brugmann et al. 2007; Blum, Schweickert et al. 2014; Blum, Feistel et al. 2014). The steps in *Xenopus* face formation appear similar to those described in mouse, chicken, and zebrafish systems (Szabo-Rogers et al. 2010; Mork and Crump 2015; Chen et al. 2017; Fish 2019). In frogs and mammals, craniofacial development involves migratory neural crest in conjunction with endoderm and mesoderm. These tissues form the pharyngeal or branchial arches, and a series of prominences originate from these to form the different regions of the face (Gilbert 2010). The mouth forms at the center of the prominence originating from the first pharyngeal arch but does not include tissue from the neural crest (Jacox, Sindelka et al. 2014). The early stages of face formation that can be readily studied in frogs are difficult to study in mammals, as the developing face is obscured by the large forebrain. Although it is true that every animal develops somewhat differently, the frog system allows for greater insight than any other system and is one of the most accessible vertebrate models for analysis of craniofacial development. Assays, ideas, and new signaling connections can suggest productive approaches in amniote models and contribute new information to the craniofacial community. As we will explore, *Xenopus laevis* and *Xenopus tropicalis* are yielding insights into craniofacial anomalies, informing underlying mechanisms and diagnosis of affected people.

17.3. CELLS THAT MAKE UP THE CRANIOFACIAL SKELETON: THE NEURAL CREST

We present here a concise view of neural crest cell contribution to craniofacial development.

17.3.1. Neural Crest Cells—Contributions from *Xenopus*

Much of the craniofacial skeleton arises from neural crest cells (NCCs). These are multipotent cells that arise at the border between the neural and the non-neural ectoderm during neural tube formation. Formation of the NCCs is initiated at gastrulation by induction of neural crest progenitors at the neural plate border (NPB) (LaBonne 1998; Aybar and Mayor 2002). As the neural plate closes to form the neural tube, neural crest progenitors delaminate, lose their epithelial nature, and become migratory mesenchymal cells. Migratory NCCs divide into four major populations based on their position, the cranial, cardiac, vagal, and trunk neural crests, with each contributing to distinct cell and tissue populations (Trainor 2014). Much understanding of molecular mechanisms leading to NCC formation and differentiation has come from studies in *Xenopus* (reviewed in (Meulemans and Bronner-Fraser 2004; Betancur, Bronner-Fraser, and Sauka-Spengler 2010; Klymkowsky, Rossi, and Artinger 2010; Barriga et al. 2015)).

17.3.2. Cranial Neural Crest Cells

The cranial neural crest (CNC) population gives rise to craniofacial structures under the action of multiple signaling pathways and transcription factors (Figure 17.1A). Subsequent to initial induction of CNC, migrating cranial neural crest cells arising from the first arch and frontonasal ectodermal zone give rise to the bones of the head and face (Figure 17.1B). Signaling events between the neural crest, ectoderm, and endoderm regulate cranial neural crest migration (Olesnicky Killian, Birkholz, and Artinger 2009; Theveneau and Mayor 2010; Theveneau et al. 2010; Kalcheim 2018; Duband 2006), proliferation, and differentiation into cartilage and bone (Jandzik et al. 2014; Monsoro-Burq 2015; Green, Simoes-Costa, and Bronner 2015; Shao et al. 2015; da Costa, Trentin, and Calloni 2018). Defects in signaling pathways between the ectoderm, endoderm, and neural crest are associated with craniofacial anomalies in humans (Trainor 2010; Huh and Ornitz 2010; Curtin et al. 2011). Cranial neural crest cells are further subdivided into forebrain, midbrain, and hindbrain sub-populations. Gradients of FGF, BMP, and WNT proteins specify these cells, and a *hox* gene expression gradient along the antero-posterior axis divides the CNC into two different domains (Gavalas et al. 2001; Trainor and Krumlauf 2001; Couly et al. 2002; Steventon and Mayor 2012; Raible and Ragland 2005; da Costa, Trentin, and Calloni 2018). Interestingly, a *hox*-negative CNC population at the rostral end produces the entire facial skeleton (Couly, Creuzet et al. 2002; Creuzet, Couly, and Le Douarin 2005). Although both *hox*-negative and *hox*-positive CNC domains are able to generate cartilage, only the anterior region forms the bones of the facial skeleton and palate (Vieux-Rochas et al. 2013; Creuzet, Couly, and Le Douarin 2005; Dickinson and Sive 2007). Defects in CNC development are associated with human birth disorders such as cleft lip, cleft palate, craniosynostosis, craniofacial-microsomias, and ciliopathies, as will be discussed in Section 7.

17.4. THE MOUTH IS AN ESSENTIAL FACIAL STRUCTURE

17.4.1. Mouth Formation Is Conserved

In vertebrates, the mouth is an integral part of the craniofacial system and is essential for eating and life. All multicellular animals have a mouth (Chen et al. 2017), and mouth development appears to have arisen once during evolution (Chen et al. 2017). Our group identified a cohort of genes during *Xenopus* mouth development (Dickinson and Sive 2009), many of which are expressed in similar regions in other species (Chen et al. 2017). *Xenopus* is proving an

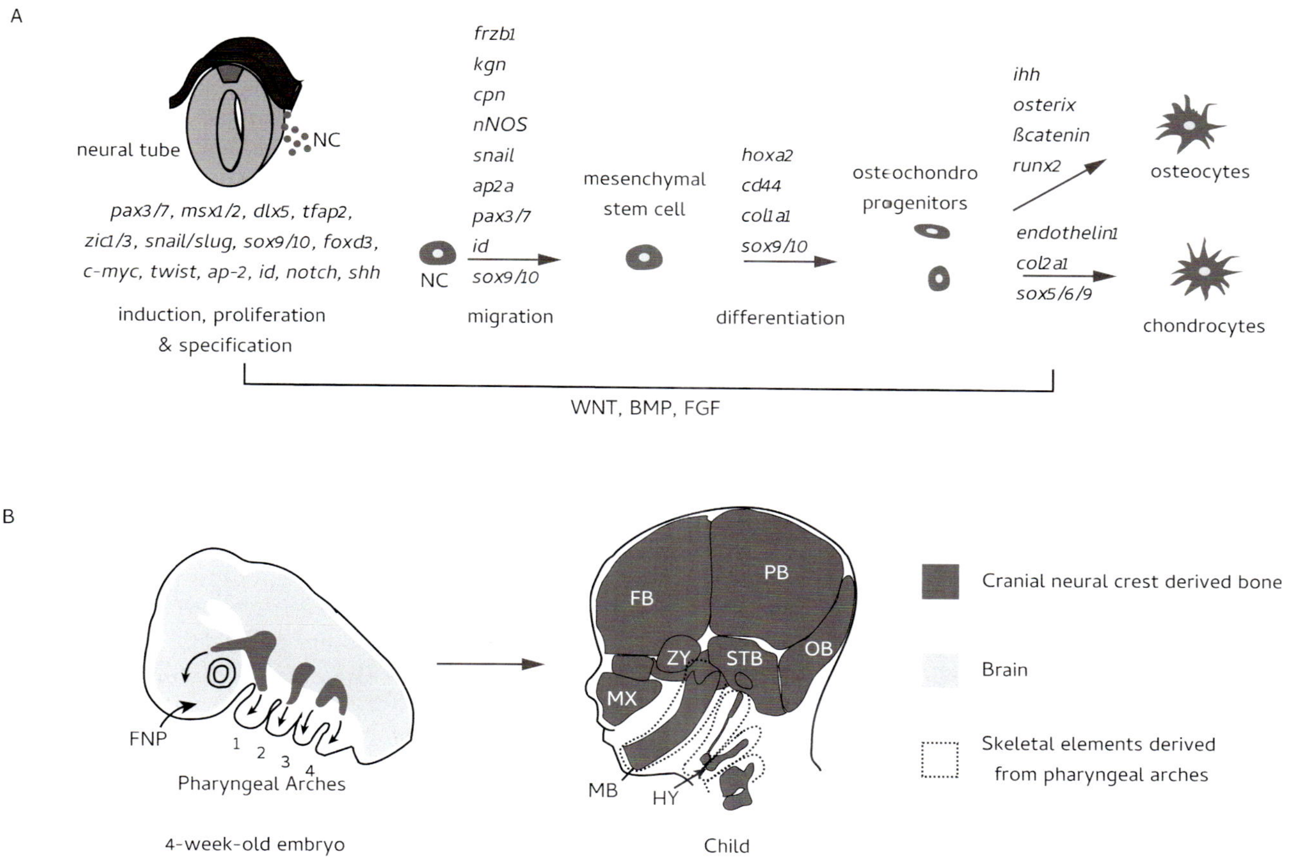

FIGURE 17.1 (A) Schematic representation of the gene regulatory network (GRN) from induction through to differentiation of NCC. WNT, BMP, and FGF are involved throughout the process from induction of the neural crest to differentiation into chondrocytes (cartilaginous cells) and osteocytes (bone cells). (B) Four-week-old human embryo and craniofacial skeleton of a child. Neural crest cells arising from distinct regions migrate into pharyngeal arches and differentiate into the bones of the face and head. Neural crest cells migrating over the eye into the frontonasal prominence (FNP) become the skeletal elements of the forehead, frontal bone (FB); upper jaw, maxillary bone (MX); and nasal bone, while neural crest migrating in the pharyngeal arches become skeletal elements of the lower jaw, mandible (MB) and throat, hyoid (HY). Parietal bone (PB), Occipital bone (OB), Squamous temporal bone (STB), and Zygoma (ZY).

Source: Figure 17.1B adapted and modified from Chen 2018.

excellent model for vertebrate mouth formation, with relevance to human birth disorders.

17.4.2. Cellular Organization of the Mouth

In chordates, mouth formation takes place in a region devoid of mesoderm where ectoderm and endoderm are directly juxtaposed (Dickinson and Sive 2007). The Sive group named this region the extreme anterior domain (Jacox, Sindelka et al. 2014). EAD cells contribute to the mouth opening, roof of the mouth, anterior pituitary, and nostrils (Jacox et al. 2016). Removal of the EAD or the EAD endoderm in *Xenopus* resulted in a smaller or unperforated mouth (Dickinson and Sive 2006; Chen 2018). Similar morphology was observed upon ablation of rostral non-neural ectoderm and endoderm in chick, mouse, and salamander embryos (Adams 1931; Couly, Creuzet et al. 2002; Cajal et al. 2014; Withington, Beddington, and Cooke 2001).

17.4.3. Mouth Formation—*Xenopus* as a Model

Mouth formation is initiated during neurulation and takes place at the anterior-most region where the ectoderm and endoderm are directly juxtaposed, without intervening mesoderm (Figure 17.2A). Mouth formation in vertebrates includes formation of a stomodeal invagination, and later, a thin buccopharyngeal membrane ruptures to form the mouth opening (Dickinson and Sive 2006). During *Xenopus* mouth formation, a multilayered ectoderm is directly juxtaposed onto endoderm and separated by a basement membrane (Figure 17.2A) (Dickinson and Sive 2006; Jacox et al. 2016). The multilayered ectoderm undergoes convergent extension to form a "pre-mouth array" two cells wide and ~ten cells deep, lying on the endodermal layer. This process is under control of the WNT-PCP pathway and signaling from the incoming neural crest (Jacox, Chen et al. 2016) (Figure 17.2B). The basement membrane between

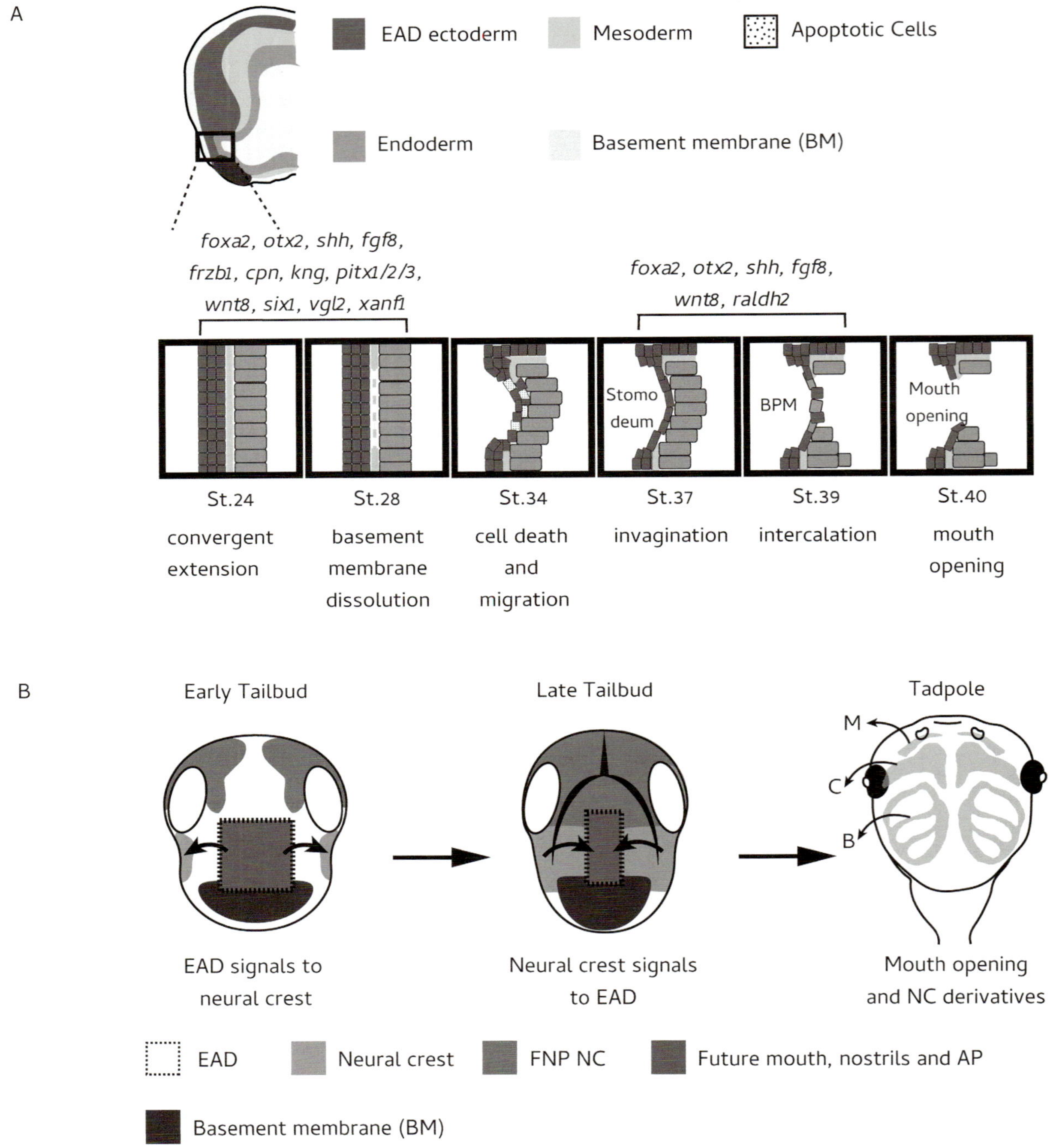

FIGURE 17.2 (A) Sagittal views of steps involved in mouth opening. Expression of extreme anterior domain (EAD) showing different germ layers at st. 24. Lower panel consists of enlarged schematics of the EAD and the subsequent steps of mouth opening. At st. 28, the basement membrane (BM) between EAD ectoderm and endoderm disintegrates activating cell death by st. 34. Stomodeal invagination is formed with concurrent bursts of apoptosis and migration of ectoderm out of the region at st. 34–37. Intercalation of ectoderm and endoderm produces the buccopharyngeal membrane (BPM), which perforates to open the mouth at tadpole stages (st. 39–40). (B) Schematic showing reciprocal signaling between EAD and cranial neural crest (CNC). At early tailbud stages, EAD secretes signals that guide the NC into the face. At late tailbud, as NCs migrate into the face, they secrete factors that stimulate EAD-convergent extension to form the "pre-mouth array." The pre-mouth array later opens down the midline to form the stomodeum and edges of the mouth, while the NCs give rise to Meckel's (M), Ceratohyal (C), and Branchial arches (B) at the tadpole stage.

Source: Figure 17.2 adapted and modified from Chen, Jacox et al. 2017.

ectoderm and endoderm disappears, requiring activation of the *β-catenin* WNT antagonist *frzb1* that is expressed in the EAD (Dickinson and Sive 2009; Tabler et al. 2014). *frzb1* expression in turn requires SHH expression (Tabler et al. 2014). At this stage, the stomodeum forms, as described in all vertebrates (Waterman 1977; Waterman and Schoenwolf 1980; Som and Naidich 2013). The cell layers thin through cell death and migration, until a single layer of ectoderm and endoderm remains, the buccopharyngeal membrane (BPM) (Dickinson and Sive 2006; Jacox et al. 2016), which perforates, leading to the mouth opening (Figure 17.2A) (Dickinson and Sive 2006). Bucopharyngeal membrane perforation is mediated by JNK signaling acting through *β-catenin* phosphorylation and E-cadherin endocytosis at adherens junctions (Houssin et al. 2017).

17.5. THE EXTREME ANTERIOR DOMAIN IS A CRANIOFACIAL SIGNALING CENTER

17.5.1. Identification of the Extreme Anterior Domain

Work from our group identified the *Xenopus* EAD as a craniofacial organizer (Jacox, Sindelka et al. 2014). The EAD comprises approximately 500 cells at the anterior of the embryo and is present from the late neurula through tailbud stages. As noted, the EAD comprises directly juxtaposed ectoderm and endoderm without the intervening mesoderm. This type of tissue arrangement has been observed across deuterostomes and is present in mammals (Chen, Jacox et al. 2017; Chen 2018; Soukup, Horacek, and Cerny 2013). The EAD contributes to similar structures in multiple species, including birds (Couly, Coltey, and Le Douarin 1993; Couly and Le Douarin 1987), mice (Osumi-Yamashita et al. 1994), zebrafish (Eberhart et al. 2006; Chapman et al. 2005), and frog (Eagleson, Ferreiro, and Harris 1995; Dickinson and Sive 2006), which include the mouth, nostrils, and anterior pituitary. Although this region is present in all chordates, including humans, work from our group in *Xenopus* has provided pivotal functional analysis.

17.5.2. Craniofacial Organizer Function of the Extreme Anterior Domain

Our group demonstrated that the *Xenopus* EAD has organizer function using a facial transplant protocol (Dickinson and Sive 2009; Jacox, Dickinson, and Sive 2014; Jacox, Sindelka et al. 2014). In particular, we observed that after transplantation of an EAD that lacked function of the WNT antagonists *frzb1* and *crescent* into a control embryo, not only did the mouth fail to form, but the rest of the face was thin and undeveloped (Dickinson and Sive 2009). We further demonstrated that the EAD acted on first arch migratory neural crest to promote its ingress into the developing face (Figure 17.2B) (Jacox, Sindelka et al. 2014). This influence required the kinin-kallikrein signaling system, shown for the first time to be involved in early craniofacial development (Jacox, Sindelka et al. 2014). The pathway includes Kininogen encoded by the *kng* gene, a precursor for the signaling peptide Bradykinin (Bdk), and the Bdk processing enzyme Carboxypeptidase N, encoded by the *cpn* gene. *cpn* RNA is localized in the EAD, while *kng* RNA is more broadly distributed. Kinin-kallikrein signaling culminates in nitric oxide (NO) production, synthesized by Nitric Oxide Synthase. We showed that NO was produced around the EAD, was absent after local *cpn* LOF, and was enhanced after implantation of beads coated with BDK peptide. This study demonstrated that the EAD signals to the neural crest through kinin-kallikrein pathway signaling to guide its development. As noted in Section 4.3, in older embryos, the first arch crest signals back to the EAD to promote its convergent extension and formation of a pre-mouth array that later opens into the stomodeum (Figure 17.2B) (Jacox, Chen et al. 2016). Chen (2018) demonstrated that the secreted WNT antagonist *fzb1* that is expressed locally in the EAD also influences neural crest development, including promoting proliferation, and importantly, also impacts brain development. These data demonstrate a global signaling role and organizer function for EAD during craniofacial development. EAD perturbation would be associated with mouth anomalies, and due to its signaling role, may contribute to other craniofacial anomalies.

17.6. LANDSCAPE OF CRANIOFACIAL ANOMALIES

Craniofacial disorders or anomalies occur during formation of the skull and facial bones and are generally congenital or present at birth. Some CFA are syndromic, associated with defined genetic changes (Buchanan, Xue, and Hollier 2014; Rice 2005; Nagy and Demke 2014).

17.6.1. Craniofacial Anomalies Are Associated with Multiple Factors

The causes of CFA are often unknown or complex but are considered to include the following.

1. *Genetic.* Gene variants associated with CFA include genes that regulate DNA replication, cell proliferation, cell signaling, or transcription (Roosenboom et al. 2016; Kobayashi et al. 2013; Alappat, Zhang, and Chen 2003; Gebuijs et al. 2019; Merkuri and Fish 2019; Twigg and Wilkie 2015; Bartzela, Carels, and Maltha 2017).
2. *Environmental.* Medication, recreational drugs, and alcohol during pregnancy have also been linked with certain craniofacial anomalies (Thompson, Levitt, and Stanwood 2009; Moiseiwitsch 2000; Seda et al. 2019; Muggli et al. 2017; Sulik 2005). Similarly, exposure to viruses such as Zika (Yan et al. 2019; Wheeler 2018, 2020) and cytomegalovirus (CMV) (Cheeran, Lokensgard, and Schleiss 2009; Jaskoll et al. 2008; Weichert et al. 2010)

during pregnancy have been implicated in craniofacial anomalies.

3. *Vitamin deficiency.* A maternal diet lacking Vitamin A, Vitamin D, folic acid (Vitamin B9), or Vitamin B12 is associated with higher risk of CFAs such as cleft lip and cleft palate (Maldonado et al. 2021; Wahl et al. 2015; Finkelstein, Layden, and Stover 2015; Pannia et al. 2016; Mulligan et al. 2010; Clagett-Dame and Knutson 2011).

17.6.2. Classes of Craniofacial Anomalies

1. *Cleft lip or cleft palate.* Cleft lip and cleft palate are the most common congenital craniofacial anomalies seen at birth. In cleft lip, the lip does not form properly. The degree of the cleft lip varies greatly, from a mild notch to a severe large opening up to the base of the nose. In cleft palate, the palatal shelves do not close completely, leaving an opening extending into the nasal cavity. The cleft can extend from the front of the mouth to the throat and may also include the lip.
2. *Craniofacial microsomia.* In these cases, one side of the face is smaller than the other, giving rise to facial asymmetry. Typically, only the lower part of the face is affected, with a flat cheek due to anomalous bone growth and an underdeveloped jaw (maxillary or mandibular hypoplasia). The external ears are malformed or absent. Craniofacial microsomia with maxillary or mandibular hypoplasia may be referred to as hemifacial microsomia and includes Goldenhar syndrome, branchial arch syndrome, and lateral facial dysplasia (Birgfeld and Heike 2012; Bogusiak, Puch, and Arkuszewski 2017; Monahan et al. 2001; Brandstetter and Patel 2016).
3. *Neurocristopathies.* Neurocristopathies are perturbations in the formation, migration, and/or differentiation of neural crest cells (Watt and Trainor 2014; Vega-Lopez et al. 2018). Cranial neural crest cells migrate into the first and second pharyngeal arches of a developing embryo. In humans, the cells in the first pharyngeal arch give rise to the maxilla, zygoma, palate, mandible, malleus, incus, muscle of mastication, and parts of the trigeminal sensory ganglion. Those of the second pharyngeal arch give rise to the hyoid cartilage, stapes, facial muscles, and parts of the facial sensory ganglia (Johnson et al. 2011; Richany, Bast, and Anson 1955; Bast, Anson, and Richany 1955). Disruptions to the first and second pharyngeal arches are associated with craniofacial anomalies known as facial dysostoses (Trainor and Andrews 2013). These include Treacher Collins syndrome, Nager syndrome, Miller syndrome, and Goldenhar syndrome (Trainor and Andrews 2013; Sato et al. 2019). The vast amount of data on neural crest development in *Xenopus* provide an important model system in which to study these syndromes (Table 17.4) (Gouignard et al. 2016; Schwenty-Lara, Pauli, and Borchers 2020; Devotta et al. 2016).
4. *Craniofacial Dysmorphism.* Craniofacial dysmorphism is defined as an abnormally formed craniofacial structure. This can include brachycephaly (flattened back of the head), highly arched eyebrows, monobrow, low-set ears, microdontism (smaller teeth), and generalized gingival hyperplasia (overgrowth of gums around the teeth). Some of the common craniofacial dysmorphisms are listed in Table 17.1.
5. *Craniosynostosis.* In the infant skull, the bones are separated by sutures or joints. These joints allow growth of skull bones in concert with the brain, and when skull growth is complete, the sutures fuse and bone growth stops. In craniosynostosis, the skull sutures fuse prematurely to inhibit skull growth, impacting brain development (Yilmaz et al. 2019; Sawh-Martinez and Steinbacher 2019). Cranial sutures consist of non-ossified mesenchymal stem cells that play an important signaling role in the development of craniofacial structure (Zhao et al. 2015; Maruyama et al. 2016). Craniosynostosis is divided into syndromic and nonsyndromic, with more than 70% of the patients diagnosed as nonsyndromic (Greenwood et al. 2014; Flaherty, Singh, and Richtsmeier 2016; Wilkie, Johnson, and Wall 2017). The most common syndromic forms of craniosynostosis and the genes involved are listed in Table 17.2.

TABLE 17.1

Description of Known Dysmorphisms, Associated Genes, Malformations, Mendelian Inheritance, and Gene-Specific Studies in *Xenopus*.

Syndrome	Associated Gene/s	Craniofacial Anomalies	OMIM
Holoprosencephaly	SHH, ZIC2, SIX3, and TGIF	Pronounced microcephaly, cyclopia (single centrally placed eye) Cebocephaly (single-nostril nose) Ethmocephaly (proboscis) Hypotelorism (increased facial width)	147250
Dandy-walker	ZIC1 and ZIC4	Macrocephaly (large brain) Hydrocephaly (dilatation of brain ventricles)	220200
Lissencephaly	RELN and TUBA1A	Microcephaly with smooth brain Cerebellar hypoplasia (small cerebellum)	607432

Source: Dubourg et al. 2007; Yamasaki and Kanemura 2015; Fry, Cushion and Pilz 2014

TABLE 17.2
Description of Known Syndromic Craniosynostosis, Associated Genes, Malformations, Mendelian Inheritance and Gene-Specific Studies in *Xenopus*.

Syndrome	Associated Gene/s	Craniofacial Anomalies	OMIM
Crouzon	FGFR2	Bicoronal craniosynostosis, lambdoid suture fusion, hypertelorism, shallow orbits, ocular proptosis, high arched palate, midface hypoplasia, low-set ears, psittichorhina (beak-like nose)	123500
Aperts	FGFR2	Tturribrachycephaly (bicoronal synostoses), high forehead, steep, flat and associated with transverse frontal skin furrow, exorbitism, proptosis, short anterior cranial fossa, cleft palate, pseudo-prognathic mandible, septal deviation	101200
Saethre-Chotzen	TWIST1	Acrocephaly (coronal 1 lambdoid), unicoronal/brachycephaly (can be bicoronal), facial asymmetry, low-set hairline, ptosis, hypertelorism, strabismus, epicanthal folds. beaked nose, nasal septal deviation cleft palate with high arch	101400
Pfeiffer	FGFR1; FGFR2	Type I—turribrachycephaly, Types II/III—Kleeblattscha del (multisuture synostosis), maxillary hypoplasia, proptosis, strabismus hypertelorism, cleft palate	101600
Carpenter	RAB23	Hypertelorism, downward sloping palpebral fissures, epicanthal folds, flat/wide nose with large nostrils	201000

Source: Yilmaz et al. 2019; Sawh-Martinez and Steinbacher 2019

TABLE 17.3
Description of Known Craniofacial Ciliopathies, Associated Genes, Malformations, Mendelian Inheritance and Gene-Specific Studies in *Xenopus*.

Syndrome	Associated Gene/S	Craniofacial Anomalies	OMIM
Bardet-Biedl	BBS 1–14	Prominent forehead, deep-set eyes, hypertelorism, downward-slanting palpebral fissures, flat nasal bridge anteverted nares, prominent nasolabial folds, long philirum, thin upper lip	209900
Joubert	INPP5E	Large head and frontal prominence, prominent forehead and nasal bridge, bitemporal narrowing, epicanthal folds, ptosis, prognathism, eyebrow abnormalities, thick earlobes, trapezoid-shaped mouth, lower lip eversion, upturned nose	213300
Meckel-Gruber	MKS1	Microcephaly, sloping forehead, occipital meningoencephalocele, cleft lip/palate, micrognathia, macrostomia, various glossal malformations	249000
Oro-facial-digital	OFD1	Malformations of the face, oral cavity, thickened alveolar ridges, abnormal dentition, absent lateral incisors, clefts of the jaw and tongue	311200
Ellis-van Creveld	EVC1–2	Cleft lip and palate gingivo, labial muscuiofibrous fraenula, premature eruption of teeth, hypodontia, small cranial base, micrognathia, increased gonial angle, malocclusion	225500

Source: Cortes et al. 2015; Schock and Brugmann 2017

6. *Craniofacial Ciliopathies.* Craniofacial ciliopathies are associated with the altered structure or function of cilia, which are essential cellular structures (Zaghloul and Brugmann 2011). Affected people characteristically display cleft lip/palate, hypertelorism (increased facial width), micrognathia (small lower jaw), and hypotelorism (decreased facial width). Human ciliopathies have been extensively reviewed by (Cortes, Metzis, and Wicking 2015; Schock and Brugmann 2017). Common syndromes affecting the genetics of the ciliary function are listed in Table 17.3.

17.7. TREATMENT OF CRANIOFACIAL ANOMALIES

Children affected by craniofacial disorders face enormous developmental challenges, and treatment options are generally limited to surgical intervention. Recent advances in tissue engineering approaches are promising (Mao et al. 2007; Petrovic et al. 2012; Velasquillo et al. 2020; Salinas and Anseth 2009; Zhang and Yelick 2018).

17.7.1. Surgical Approaches

For syndromes such as cleft lip, cleft palate, and craniosynostosis, surgery tries to correct the physical formation of the skull and facial bones. In infants with minimal deformities, laparoscopic minimally invasive surgery may be possible.

17.7.2. Tissue Engineering Approaches

Tissue engineering has promise to rebuild craniofacial structures through the combined use of cells, factors that can promote cellular differentiation into appropriate cell types needed to correct craniofacial structures, and scaffolds that can contribute structures on which cells can develop (Tevlin et al. 2014; Kim, Kim, and Kim 2020; Tollemar et al. 2016; Hollister et al. 2005; Emara and Shah 2021). Three-dimensional (3D) printing has some promise for personalized approaches to bone reconstruction (Bauermeister, Zuriarrain, and Newman 2016; Obregon et al. 2015; Flores et al. 2017; Shen et al. 2020; Chung et al. 2020).

17.7.3. Cell-Based Approaches

The idea here is to use precursor cells (osteoblasts, chondrocytes, fibroblasts) to produce cartilage and bone tissue that can repair craniofacial anomalies (Meijer et al. 2007; Teven et al. 2012; Teven et al. 2015). Use of stem cells for craniofacial repair may be extremely useful in the future, particularly autologous cells derived from the affected person, such as mesenchymal, adipogenic, skeletal stem cells, or induced pluripotential stem cells (Zuk 2008; Perez et al. 2018; Griffin et al. 2014; Tevlin, Longaker, and Wan 2020; Borrelli et al. 2020; Velasquillo, Madrazo-Ibarra et al. 2020).

17.8. CONTRIBUTION OF *XENOPUS* TO UNDERSTANDING CRANIOFACIAL ANOMALIES

Craniofacial anomalies are prevalent and severe, and there is an unmet need to define the genetic or environmental perturbations that are associated with and cause these. Contributions from *Xenopus* are providing insight into genes associated with human craniofacial disorders by two approaches. The first is understanding from fundamental research as to what genes govern craniofacial development, as explored in Sections 3 and 4 of this chapter. If such genes are later identified in human genome wide association screens (GWASs) as associated with a craniofacial anomaly, there will already be information available to help understand mechanisms that may contribute to the anomaly. In an example from our group, we identified *zic1/opl* as a gene required for neural determination and patterning (Kuo et al. 1998). Recently, heterozygous mutations in the third exon of *ZIC1* encoding the C-terminus of the protein are associated with craniosynostosis involving the coronal sutures (Twigg et al. 2015).

The second approach is to screen GWAS hits for their effects on craniofacial development (Khandelwal et al. 2013; Sanchez-Lara 2015; Yu et al. 2017; Saleem et al. 2019) and to use *Xenopus* as an assay system for gene activity (Abu-Daya, Khokha, and Zimmerman 2012; Hwang, Marquez, and Khokha 2019). This can be done by loss-of-function using antisense techniques or CRISPR-mediated genome editing in the F0 embryo. For gene variants that appear to be gain-of-function, RNA injection or (less successful) DNA expression constructs injected into the embryo can be used. Tadpoles can be assayed after the craniofacial cartilages have formed for anomalies that may be similar to affected people. Where multiple genes are implicated through human genetic analysis, *Xenopus* assays can help sort out which are the key genes contributing to such an anomaly (Table 17.4). These approaches are proving useful diagnostic tools (Devotta, Juraver-Geslin et al. 2016; Lasser et al. 2019; Schweickert and Feistel 2015).

17.9. CONCLUSION

In this chapter, we have summarized studies investigating development and disorders of the craniofacial region,

TABLE 17.4
List of Craniofacial Anomalies Modeled and Studied in *Xenopus*.

Syndrome	Affected Gene/s	Craniofacial Anomalies	Xenopus Studies	OMIM
Nager	*sf3bp4*	Downslanting palpebral fissures, malar hypoplasia (underdeveloped cheek bones), micrognathia, atresia, ear defects, and cleft palate	Devotta et al. 2016	154400
Smith-Magenis	*ra1*	Cleft lip/palate, midface hypoplasia, flat nasal bridge, brachycephaly (flattened head), and prognathia (misaligned maxilla and mandible)	Tahir et al. 2014	182290
Wolf-Hirschhorn	*wsch1*, *wsch2*, and *letm1*	Prominent forehead, widely spaced eyes (hypertelorism), wide and protrusive nasal bridge, undersized jaw (micrognathia), a short philtrum, cleft lip, and microcephaly	Mills et al. 2019	194190
Pilarowski-Bjornsson	*chd1*	Macrocephaly, depressed midface, pointed chin, translucent skin, almond-shaped eyes, downslanting palpebral fissures, periorbital fullness, and flared eyebrow	Wyatt et al. 2021	617682
CHARGE	*chd7*	Orofacial clefts, facial nerve palsy, and choanal atresia	Bajpai et al. 2010	214800
Andersen-Tawil	*kcnj2*	Low-set ears, cleft palate, mandibular hypoplasia, hypertelorism, micrognathia, a broad forehead, and dental anomalies	Adams et al. 2016	170390
Musculocontractural Ehlers-Danlos	*dse*	Micrognathia, cleft palate, brachycephaly, hypertelorism, downslanting palpebral fissures, and low-set ears	Gouignard et al. 2016	601776
Orofacial clefts	*raldh2*	Cleft lip and cleft palate	Kennedy and Dickinson 2012	
Craniosynostosis	multiple genes	Premature cranial fusion	Slater et al. 2009	
Brainbridge-Ropers	*aslx3*	Microcephaly, hypotonia, arched eyebrows, downslanting palpebral fissures, broad nasal bridge with short nose, anteverted nares, low-set ears, and small chin	Lichtig et al. 2020	615485

Source: Devotta, Juraver-Geslin et al. 2016; Tahir et al. 2014; Mills et al. 2019; Wyatt et al. 2021; Bajpai et al. 2010; Adams et al. 2016; Gouignard, Maccarana et al. 2016; Kennedy and Dickinson 2012; Slater et al. 2009; Lichtig et al. 2020; Dubey and Saint-Jeannet 2017

with emphasis on *Xenopus* as a model system. Formation of craniofacial structures is a complex process that includes neural crest cells that give rise to the cartilages or bones of the face and skull, while the extreme anterior domain forms the mouth. Gene regulatory networks drive craniofacial development, while the EAD is a major early facial signaling center or organizer. Analyses in *Xenopus* have been pivotal in uncovering processes directing craniofacial development and in defining the EAD as an organizer. Craniofacial anomalies are frequent and associated with genetic, environmental, and nutritional factors. Defining genes and processes underlying human disorders is critical for proper diagnosis and new treatments. This is being facilitated using the *Xenopus* system—either through knowledge from fundamental research or by direct assay of gene variants or environmental perturbations. *Xenopus* remains one of the most accessible systems for uncovering connections to human craniofacial disorders.

REFERENCES

Abu-Daya, A., M. K. Khokha, and L. B. Zimmerman. 2012. "The hitchhiker's guide to Xenopus genetics." *Genesis* 50 (3):164–75. doi: 10.1002/dvg.22007.

Adams, A. E. 1931. "Some effects of removal of endoderm from the mouth region of early *Amblystoma punctatum* embryos." *The Journal of Experimental Zoology* 58. doi: doi.org/10.1002/jez.1400580108.

Adams, D. S., S. G. Uzel, J. Akagi, et al. 2016. "Bioelectric signalling via potassium channels: A mechanism for craniofacial dysmorphogenesis in KCNJ2-associated Andersen-Tawil Syndrome." *J Physiol* 594 (12):3245–3270. doi: 10.1113/JP271930.

Alappat, S., Z. Y. Zhang, and Y. P. Chen. 2003. "Msx homeobox gene family and craniofacial development." *Cell Res* 13 (6):429–42. doi: doi.org/10.1038/sj.cr.7290185.

Aybar, M. J., and R. Mayor. 2002. "Early induction of neural crest cells: lessons learned from frog, fish and chick." *Current Opinion in Genetics & Development* 12 (4):452–8.

Bajpai, R., D. A. Chen, A. Rada-Iglesias, et al. 2010. "CHD7 cooperates with PBAF to control multipotent neural crest formation." *Nature* 463 (7283):958–962. doi: 10.1038/nature08733.

Barriga, E. H., P. A. Trainor, M. Bronner, et al. 2015. "Animal models for studying neural crest development: Is the mouse different?" *Development* 142 (9):1555–60. doi: 10.1242/dev.121590.

Bartzela, T. N., C. Carels, and J. C. Maltha. 2017. "Update on 13 syndromes affecting craniofacial and dental structures." *Front Physiol* 8:1038. doi: 10.3389/fphys.2017.01038.

Bast, T. H., B. J. Anson, and S. F. Richany. 1955. "The development of the second branchial arch (Reichert's cartilage), facial canal and associated structures in man." *Annals of Otology, Rhinology and Laryngology* 64 (3):802–24.

Bauermeister, A. J., A. Zuriarrain, and M. I. Newman. 2016. "Three-dimensional printing in plastic and reconstructive surgery: A systematic review." *Ann Plast Surg* 77 (5):569–76. doi: 10.1097/SAP.0000000000000671.

Betancur, P., M. Bronner-Fraser, and T. Sauka-Spengler. 2010. "Assembling neural crest regulatory circuits into a gene regulatory network." *Annu Rev Cell Dev Biol* 26:581–603. doi: 10.1146/annurev.cellbio.042308.113245.

Birgfeld, C. B., and C. Heike. 2012. "Craniofacial microsomia." *Semin Plast Surg* 26 (2):91–104. doi: 10.1055/s-0032–1320067.

Blum, M., K. Feistel, T. Thumberger, et al. 2014. "The evolution and conservation of left-right patterning mechanisms." *Development* 141 (8):1603–13. doi: 10.1242/dev.100560.

Blum, M., A. Schweickert, P. Vick, et al. 2014. "Symmetry breakage in the vertebrate embryo: When does it happen and how does it work?" *Dev Biol* 393 (1):109–23. doi: 10.1016/j.ydbio.2014.06.014.

Bogusiak, K., A. Puch, and P. Arkuszewski. 2017. "Goldenhar syndrome: current perspectives." *World J Pediatr* 13 (5):405–15. doi: 10.1007/s12519–017–0048-z.

Borrelli, M. R., M. S. Hu, M. T. Longaker, et al. 2020. "Tissue engineering and regenerative medicine in craniofacial reconstruction and facial aesthetics." *J Craniofac Surg* 31 (1):15–27. doi: 10.1097/SCS.0000000000005840.

Brandstetter, K. A., and K. G. Patel. 2016. "Craniofacial microsomia." *Facial Plast Surg Clin North Am* 24 (4):495–515. doi: 10.1016/j.fsc.2016.06.006.

Brugmann, S. A., L. H. Goodnough, A. Gregorieff, et al. 2007. "Wnt signaling mediates regional specification in the vertebrate face." *Development* 134 (18):3283–95. doi: 10.1242/dev.005132.

Buchanan, E. P., A. S. Xue, and L. H. Hollier, Jr. 2014. "Craniofacial syndromes." *Plast Reconstr Surg* 134 (1):128e–53e. doi: 10.1097/PRS.0000000000000308.

Cajal, M., S. E. Creuzet, C. Papanayotou, et al. 2014. "A conserved role for non-neural ectoderm cells in early neural development." *Development* 141 (21):4127–38. doi: 10.1242/dev.107425.

Chai, Y., and R. E. Maxson, Jr. 2006. "Recent advances in craniofacial morphogenesis." *Dev Dyn* 235 (9):2353–75. doi: 10.1002/dvdy.20833.

Chapman, S. C., A. L. Sawitzke, D. S. Campbell, et al. 2005. "A three-dimensional atlas of pituitary gland development in the zebrafish." *J Comp Neurol* 487 (4):428–40. doi: 10.1002/cne.20568.

Cheeran, M. C., J. R. Lokensgard, and M. R. Schleiss. 2009. "Neuropathogenesis of congenital cytomegalovirus infection: Disease mechanisms and prospects for intervention." *Clin Microbiol Rev* 22 (1):99–126, Table of Contents. doi: 10.1128/CMR.00023–08.

Chen, J. 2018. "The extreme anterior domain coordinates brain and craniofacial development." Thesis: Ph. D., Department of Biology, MIT, http://hdl.handle.net/1721.1/117870

Chen, J., L. A. Jacox, F. Saldanha, et al. 2017. "Mouth development." *Wiley Interdiscip Rev Dev Biol* 6 (5). doi: 10.1002/wdev.275.

Chung, J. J., H. Im, S. H. Kim, et al. 2020. "Toward biomimetic scaffolds for tissue engineering: 3D printing techniques in regenerative medicine." *Front Bioeng Biotechnol* 8:586406. doi: 10.3389/fbioe.2020.586406.

Clagett-Dame, M., and D. Knutson. 2011. "Vitamin A in reproduction and development." *Nutrients* 3 (4):385–428. doi: 10.3390/nu3040385.

Cortes, C. R., V. Metzis, and C. Wicking. 2015. "Unmasking the ciliopathies: Craniofacial defects and the primary cilium." *Wiley Interdiscip Rev Dev Biol* 4 (6):637–53. doi: 10.1002/wdev.199.

Couly, G. F., P. M. Coltey, and N. M. Le Douarin. 1993. "The triple origin of skull in higher vertebrates: A study in quail-chick chimeras." *Development* 117:409–29. doi: doi.org/10.1242/dev.117.2.409.

Couly, G. F., S. Creuzet, S. Bennaceur, et al. 2002. "Interactions between Hox-negative cephalic neural crest cells and the foregutendoderm in patterning the facial skeleton in the vertebrate head." *Development* 129:1071–73. doi: doi.org/10.1242/dev.129.4.1061.

Couly, G. F., and N. M. Le Douarin. 1987. "Mapping of the early neural primordium in quail-chick chimeras: II. The prosencephalic neural plate and neural folds: Implications for the genesis of cephalic human congenital abnormalities." *Developmental Biology* 120:198–214. doi: doi.org/10.1016/0012–1606(87)90118–7.

Creuzet, S., G. Couly, and N. M. Le Douarin. 2005. "Patterning the neural crest derivatives during development of the vertebrate head: Insights from avian studies." *Journal of Anatomy* 207:447–59. doi: 10.1111/j.1469–7580.2005.00485.x.

Curtin, E., G. Hickey, G. Kamel, et al. 2011. "Zebrafish wnt9a is expressed in pharyngeal ectoderm and is required for palate and lower jaw development." *Mech Dev* 128 (1–2):104–15. doi: 10.1016/j.mod.2010.11.003.

da Costa, M. C., A. G. Trentin, and G. W. Calloni. 2018. "FGF8 and Shh promote the survival and maintenance of multipotent neural crest progenitors." *Mech Dev* 154:251–8. doi: 10.1016/j.mod.2018.07.012.

Devotta, A., H. Juraver-Geslin, J. A. Gonzalez, et al. 2016. "Sf3b4-depleted Xenopus embryos: A model to study the pathogenesis of craniofacial defects in Nager syndrome." *Dev Biol* 415 (2):371–82. doi: 10.1016/j.ydbio.2016.02.010.

Dickinson, A. J., and H. L. Sive. 2006. "Development of the primary mouth in *Xenopus laevis*." *Dev Biol* 295 (2):700–13. doi: 10.1016/j.ydbio.2006.03.054.

Dickinson, A., and H. L. Sive. 2007. "Positioning the extreme anterior in Xenopus: Cement gland, primary mouth and anterior pituitary." *Semin Cell Dev Biol* 18 (4):525–33. doi: 10.1016/j.semcdb.2007.04.002.

Dickinson, A. J., and H. L. Sive. 2009. "The Wnt antagonists Frzb-1 and Crescent locally regulate basement membrane dissolution in the developing primary mouth." *Development* 136 (7):1071–81. doi: 10.1242/dev.032912.

Duband, J. L. 2006. "Neural-crest-delamination-and-migration." *Advances in Experimental Medicine and Biology* 589. doi: doi.org/10.1007/978-0-387-46954-6_4.

Dubey, A., and J. P. Saint-Jeannet. 2017. "Modeling human craniofacial disorders in Xenopus." *Curr Pathobiol Rep* 5 (1):79–92. doi: 10.1007/s40139-017-0128-8.

Dubourg, C., C. Bendavid, L. Pasquier, et al. 2007. "Holoprosencephaly." *Orphanet J Rare Dis* 2:8. doi: 10.1186/1750-1172-2-8.

Eagleson, G., B. Ferreiro, and W. A. Harris. 1995. "Fate of the anterior neural ridge and the morphogenesis of the xenopus forebrain." *Developmental Neurobiology* 28 (2):146–58. doi: doi.org/10.1002/neu.480280203.

Eberhart, J. K., M. E. Swartz, J. G. Crump, et al. 2006. "Early Hedgehog signaling from neural to oral epithelium organizes anterior craniofacial development." *Development* 133 (6):1069–77. doi: 10.1242/dev.02281.

Emara, A., and R. Shah. 2021. "Recent update on craniofacial tissue engineering." *Journal of Tissue Engineering* 12:1–25. doi: doi.org/10.1177/20417314211003735.

Finkelstein, J. L., A. J. Layden, and P. J. Stover. 2015. "Vitamin B-12 and perinatal health." *Adv Nutr* 6 (5):552–63. doi: 10.3945/an.115.008201.

Fish, J. L. 2019. "Evolvability of the vertebrate craniofacial skeleton." *Semin Cell Dev Biol* 91:13–22. doi: 10.1016/j.semcdb.2017.12.004.

Flaherty, K., N. Singh, and J. T. Richtsmeier. 2016. "Understanding craniosynostosis as a growth disorder." *Wiley Interdiscip Rev Dev Biol* 5 (4):429–59. doi: 10.1002/wdev.227.

Flores, R. L., H. Liss, S. Raffaelli, et al. 2017. "The technique for 3D printing patient-specific models for auricular reconstruction." *J Craniomaxillofac Surg* 45 (6):937–43. doi: 10.1016/j.jcms.2017.03.022.

Fry, A. E., T. D. Cushion, and D. T. Pilz. 2014. "The genetics of lissencephaly." *Am J Med Genet C Semin Med Genet* 166C (2):198–210. doi: 10.1002/ajmg.c.31402.

Gavalas, A, P. A. Trainor, L. Ariza-MaNaughton, et al. 2001. "Synergy betweeen Hoxa1 and Hoxb1: The relationship between arch patterning and the generation of cranial neurla crest." *Development* 128:3017–27.

Gebuijs, I. G. E., S. T. Raterman, J. R. Metz, et al. 2019. "Fgf8a mutation affects craniofacial development and skeletal gene expression in zebrafish larvae." *Biol Open* 8 (9). doi: 10.1242/bio.039834.

Gilbert, S. F. 2010. *Developmental biology*. 9th ed. Sunderland, MA: Sinauer Associates.

Gouignard, N., M. Maccarana, I. Strate, et al. 2016. "Musculocontractural Ehlers-Danlos syndrome and neurocristopathies: Dermatan sulfate is required for Xenopus neural crest cells to migrate and adhere to fibronectin." *Dis Model Mech* 9 (6):607–20. doi: 10.1242/dmm.024661.

Green, S. A., M. Simoes-Costa, and M. E. Bronner. 2015. "Evolution of vertebrates as viewed from the crest." *Nature* 520 (7548):474–82. doi: 10.1038/nature14436.

Greenwood, J., P. Flodman, K. Osann, et al. 2014. "Familial incidence and associated symptoms in a population of individuals with nonsyndromic craniosynostosis." *Genet Med* 16 (4):302–10. doi: 10.1038/gim.2013.134.

Griffin, M., D. M. Kalaskar, P. E. Butler, et al. 2014. "The use of adipose stem cells in cranial facial surgery." *Stem Cell Rev Rep* 10 (5):671–85. doi: 10.1007/s12015-014-9522-3.

Guo, X., T. Zhang, Z. Hu, et al. 2014. "Efficient RNA/Cas9-mediated genome editing in *Xenopus tropicalis*." *Development* 141 (3):707–14. doi: 10.1242/dev.099853.

Hollister, S. J., C. Y. Lin, E. Saito, et al. 2005. "Engineering craniofacial scaffolds." *Orthod Craniofac Res* 8 (3). doi: 10.1111/j.1601–6343.2005.00329.x.

Houssin, N. S., N. K. Bharathan, S. D. Turner, et al. 2017. "Role of JNK during buccopharyngeal membrane perforation, the last step of embryonic mouth formation." *Dev Dyn* 246 (2):100–15. doi: 10.1002/dvdy.24470.

Huh, S. H., and D. M. Ornitz. 2010. "Beta-catenin deficiency causes DiGeorge syndrome-like phenotypes through regulation of Tbx1." *Development* 137 (7):1137–47. doi: 10.1242/dev.045534.

Hwang, W. Y., J. Marquez, and M. K. Khokha. 2019. "Xenopus: Driving the discovery of novel genes in patient disease and their underlying pathological mechanisms relevant for organogenesis." *Front Physiol* 10:953. doi: 10.3389/fphys.2019.00953.

Jacox, L. A., J. Chen, A. Rothman, et al. 2016. "Formation of a "pre-mouth array" from the extreme anterior domain is directed by neural crest and Wnt/PCP signaling." *Cell Rep* 16 (5):1445–55. doi: 10.1016/j.celrep.2016.06.073.

Jacox, L. A., A. J. Dickinson, and H. Sive. 2014. "Facial transplants in *Xenopus laevis* embryos." *J Vis Exp* (85). doi: 10.3791/50697.

Jacox, L. A., R. Sindelka, J. Chen, et al. 2014. "The extreme anterior domain is an essential craniofacial organizer acting through Kinin-Kallikrein signaling." *Cell Rep* 8 (2):596–609. doi: 10.1016/j.celrep.2014.06.026.

Jandzik, D., M. B. Hawkins, M. V. Cattell, et al. 2014. "Roles for FGF in lamprey pharyngeal pouch formation and skeletogenesis highlight ancestral functions in the vertebrate head." *Development* 141 (3):629–38. doi: 10.1242/dev.097261.

Jaskoll, T., G. Abichaker, P. P. Sedghizadeh, et al. 2008. "Cytomegalovirus induces abnormal chondrogenesis and osteogenesis during embryonic mandibular development." *BMC Dev Biol* 8:33. doi: 10.1186/1471–213X-8–33.

Johnson, J. M., G. Moonis, G. E. Green, et al. 2011. "Syndromes of the first and second branchial arches, part 1: Embryology and characteristic defects." *AJNR Am J Neuroradiol* 32 (1):14–19. doi: 10.3174/ajnr.A2072.

Kalcheim, C. 2018. "Neural crest emigration: From start to stop." *Genesis* 56 (6–7):e23090. doi: 10.1002/dvg.23090.

Kennedy, A. E., and A. J. Dickinson. 2012. "Median facial clefts in Xenopus laevis: Roles of retinoic acid signaling and homeobox genes." *Dev Biol* 365 (1):229–240. doi: 10.1016/j.ydbio.2012.02.033.

Khandelwal, K. D., H. van Bokhoven, T. Roscioli, et al. 2013. "Genomic approaches for studying craniofacial disorders." *Am J Med Genet C Semin Med Genet* 163C (4):218–31. doi: 10.1002/ajmg.c.31379.

Kim, M. S., H. K. Kim, and D. W. Kim. 2020. "Cartilage tissue engineering for craniofacial reconstruction." *Arch Plast Surg* 47 (5):392–403. doi: 10.5999/aps.2020.01095.

Klymkowsky, M. W., C. C. Rossi, and K. B. Artinger. 2010. "Mechanisms driving neural crest induction and migration in the zebrafish and *Xenopus laevis*." *Cell Adh Migr* 4 (4):595–608. doi: 10.4161/cam.4.4.12962.

Kobayashi, G. S., L. Alvizi, D. Y. Sunaga, et al. 2013. "Susceptibility to DNA damage as a molecular mechanism for non-syndromic cleft lip and palate." *PLoS One* 8 (6):e65677. doi: 10.1371/journal.pone.0065677.

Kuo, J. S., M. Patel, J. Gamse, et al. 1998. "Opl: A zinc finger protein that regulates neural determination and patterning in Xenopus." *Development* 125:2867–82.

LaBonne, C., and M. Bronner-Fraser. 1998. "Induction and patterning of the neural crest, a stem cell-like precursor population." *Journal of Neurobiology* 36 (2):175–89.

Lasser, M., B. Pratt, C. Monahan, et al. 2019. "The many faces of Xenopus: *Xenopus laevis* as a model system to study Wolf-Hirschhorn syndrome." *Front Physiol* 10:817. doi: 10.3389/fphys.2019.00817.

Lichtig, H., A. Artamonov, H. Polevoy, et al. 2020. "Modeling bainbridge-ropers syndrome in Xenopus laevis embryos." *Front Physiol* 11:75. doi: 10.3389/fphys.2020.00075.

Maldonado, E., E. Martinez-Sanz, T. Partearroyo, et al. 2021. "Maternal folic acid deficiency is associated to developing nasal and palate malformations in mice." *Nutrients* 13 (1). doi: 10.3390/nu13010251.

Mao, J. J., W. V. Giannobile, J. A. Helms, et al. 2007. "Craniofacial tissue engineering by stem cells." *J Dent Res* 85 (11):966–79. doi: 10.1177/154405910608501101.

Maruyama, T., J. Jeong, T. J. Sheu, et al. 2016. "Stem cells of the suture mesenchyme in craniofacial bone development, repair and regeneration." *Nat Commun* 7:10526. doi: 10.1038/ncomms10526.

Meijer, G. J., J. D. de Bruijn, R. Koole, et al. 2007. "Cell-based bone tissue engineering." *PLoS Medicine* 4 (2). doi: doi.org/10.1371/journal.pmed.0040009.

Merkuri, F., and J. L. Fish. 2019. "Developmental processes regulate craniofacial variation in disease and evolution." *Genesis* 57 (1):e23249. doi: 10.1002/dvg.23249.

Meulemans, D., and M. Bronner-Fraser. 2004. "Gene-regulatory interactions in neural crest evolution and development." *Dev Cell* 7 (3):291–9. doi: 10.1016/j.devcel.2004.08.007.

Mills, A., E. Bearce, R. Cella, et al. 2019. "Wolf-Hirschhorn syndrome-associated genes are enriched in motile neural crest cells and affect craniofacial development in Xenopus laevis." *Front Physiol* 10:431. doi: 10.3389/fphys.2019.00431.

Moiseiwitsch, J. R. D. 2000. "The role of serotonin and neurotransmitters during craniofacial development." *Crit Rev Oral Biol Med* 11 (2):230–9. doi: 10.1177/10454411000110020601

Monahan, R., K. Seder P. Patel, et al. 2001. "Hemifacial microsomia: Etiology, diagnosis and treatment." *J Am Dent Assoc* 132 (10):1402–8. doi: 10.14219/jada.archive.2001.0055.

Monsoro-Burq, A. H. 2015. "PAX transcription factors in neural crest development." *Semin Cell Dev Biol* 44:87–96. doi: 10.1016/j.semcdb.2015.09.015.

Mork, L., and G. Crump. 2015. "Zebrafish craniofacial development: A window into early patterning." *Curr Top Dev Biol* 115:235–69. doi: 10.1016/bs.ctdb.2015.07.001.

Muggli, E., H. Matthews, A. Penington, et al. 2017. "Association between prenatal alcohol exposure and craniofacial shape of children at 12 months of age." *JAMA Pediatr* 171 (8):771–80. doi: 10.1001/jamapediatrics.2017.0778.

Mulligan, M. L., S. K. Felton, A. E. Riek, et al. 2010. "Implications of vitamin D deficiency in pregnancy and lactation." *Am J Obstet Gynecol* 202 (5):429 e1–9. doi: 10.1016/j.ajog.2009.09.002.

Nagy, L., and J. C. Demke. 2014. "Craniofacial anomalies." *Facial Plast Surg Clin North Am* 22 (4):523–48. doi: 10.1016/j.fsc.2014.08.002.

Nakano, H., S. Amemiya, K. Shiokawa, et al. 2000. "RNA interference for the organizer-specific gene Xlim-1 in Xenopus embryos." *Biochem Biophys Res Commun* 274 (2):434–9. doi: 10.1006/bbrc.2000.3178.

Nakayama, T., M. B. Fish, M. Fisher, et al. 2013. "Simple and efficient CRISPR/Cas9-mediated targeted mutagenesis in *Xenopus tropicalis*." *Genesis* 51 (12):835–43. doi: 10.1002/dvg.22720.

Nutt, S. L., O. J. Bronchain, K. O. Hartley, et al. 2001. "Comparison of morpholino based translational inhibition during the development of *Xenopus laevis* and *Xenopus tropicalis*." *Genesis* 30 (3):110–13. doi: 10.1002/gene.1042.

Obregon, F., C. Vaquette, S. Ivanovski, et al. 2015. "Three-dimensional bioprinting for regenerative dentistry and craniofacial tissue engineering." *J Dent Res* 94 (9 Suppl):143S–52S. doi: 10.1177/0022034515588885.

Olesnicky Killian, E. C., D. A. Birkholz, and K. B. Artinger. 2009. "A role for chemokine signaling in neural crest cell migration and craniofacial development." *Dev Biol* 333 (1):161–72. doi: 10.1016/j.ydbio.2009.06.031.

Osumi-Yamashita, N., Y. Ninomiya, K. Eto, et al. 1994. "The contribution of both forebrain and midbrain crest cells to the mesenchyme in the frontonasal mass of mouse embryos." *Developmental Biology* 164:409–19. doi: doi.org/10.1006/dbio.1994.1211

Pannia, E., C. E. Cho, R. Kubant, et al. 2016. "Role of maternal vitamins in programming health and chronic disease." *Nutr Rev* 74 (3):166–80. doi: 10.1093/nutrit/nuv103.

Perez, J. R., D. Kouroupis, D. J. Li, et al. 2018. "Tissue engineering and cell-based therapies for fractures and bone defects." *Front Bioeng Biotechnol* 6:105. doi: 10.3389/fbioe.2018.00105.

Petrovic, V., P. Zivkovic, D. Petrovic, et al. 2012. "Craniofacial bone tissue engineering." *Oral Surg Oral Med Oral Pathol Oral Radiol* 114 (3):e1–9. doi: 10.1016/j.oooo.2012.02.030.

Raible, D. W., and J. W. Ragland. 2005. "Reiterated Wnt and BMP signals in neural crest development." *Semin Cell Dev Biol* 16 (6):673–82. doi: 10.1016/j.semcdb.2005.06.008.

Rice, D. P. C. 2005. "Craniofacial anomalies: From development to molecular pathogenesis." *Current Molecular Medicine* 5:699–722. doi: 10.2174/156652405774641043.

Richany, S. F., T. H. Bast, and B. J. Anson. 1955. "The development of the first branchial arch in man and the fate of Meckel's cartilage." *Annal of Otology, Rhino logy and Laryngology* 64 (3):802–24.

Roosenboom, J., G. Hens, B. C. Mattern, et al. 2016. "Exploring the underlying genetics of craniofacial morphology through various sources of knowledge." *Biomed Res Int* 2016:3054578. doi: 10.1155/2016/3054578.

Saleem, K., T. Zaib, W. Sun, et al. 2019. "Assessment of candidate genes and genetic heterogeneity in human non syndromic orofacial clefts specifically non syndromic cleft lip with or without palate." *Heliyon* 5 (12):e03019. doi: 10.1016/j.heliyon.2019.e03019.

Salinas, C. N., and K. S. Anseth. 2009. "Mesenchymal stem cells for craniofacial tissue regeneration: Designing hydrogel delivery vehicles." *J Dent Res* 88 (8):681–92. doi: 10.1177/0022034509341553.

Sanchez-Lara, P. A. 2015. "Clinical and genomic approaches for the diagnosis of craniofacial disorders." *Curr Top Dev Biol* 115:543–59. doi: 10.1016/bs.ctdb.2015.09.004.

Sato, T. S., A. Handa, S. Priya, et al. 2019. "Neurocristopathies: Enigmatic appearances of neural crest cell-derived abnormalities." *Radiographics* 39 (7):2085–102. doi: 10.1148/rg.2019190086.

Sawh-Martinez, R., and D. M. Steinbacher. 2019. "Syndromic craniosynostosis." *Clin Plast Surg* 46 (2):141–55. doi: 10.1016/j.cps.2018.11.009.

Schock, E. N., and S. A. Brugmann. 2017. "Discovery, diagnosis, and etiology of craniofacial ciliopathies." *Cold Spring Harb Perspect Biol* 9 (9). doi: 10.1101/cshperspect.a028258.

Schweickert, A., and K. Feistel. 2015. "The Xenopus embryo: An ideal model system to study human ciliopathies." *Current Pathobiology Reports* 3 (2):115–27. doi: 10.1007/s40139-015-0074-2.

Schwenty-Lara, J., S. Pauli, and A. Borchers. 2020. "Using Xenopus to analyze neurocristopathies like Kabuki syndrome." *Genesis* 59 (1–2):1–14. doi: doi.org/10.1002/dvg.23404.

Seda, M., M. Geerlings, P. Lim, et al. 2019. "An FDA-approved drug screen for compounds influencing craniofacial skeletal development and craniosynostosis." *Mol Syndromol* 10 (1–2):98–114. doi: 10.1159/000491567.

Shao, M., C. Liu, Y. Song, et al. 2015. "FGF8 signaling sustains progenitor status and multipotency of cranial neural crest-derived mesenchymal cells in vivo and in vitro." *J Mol Cell Biol* 7 (5):441–54. doi: 10.1093/jmcb/mjv052.

Shen, C., L. Witek, R. L. Flores, et al. 2020. "Three-dimensional printing for craniofacial bone tissue engineering." *Tissue Eng Part A* 26 (23–24):1303–11. doi: 10.1089/ten.TEA.2020.0186.

Slater, B. J., K. J. Liu, M. D. Kwan, et al. 2009. "Cranial osteogenesis and suture morphology in Xenopus laevis: A unique model system for studying craniofacial development." *PLoS One* 4 (1):e3914. doi: 10.1371/journal.pone.0003914.

Som, P. M., and T. P. Naidich. 2013. "Illustrated review of the embryology and development of the facial region, part 1: Early face and lateral nasal cavities." *AJNR Am J Neuroradiol* 34 (12):2233–40. doi: 10.3174/ajnr.A3415.

Soukup, V., I. Horacek, and R. Cerny. 2013. "Development and evolution of the vertebrate primary mouth." *J Anat* 222 (1):79–99. doi: 10.1111/j.1469–7580.2012.01540.x.

Steventon, B., and R. Mayor. 2012. "Early neural crest induction requires an initial inhibition of Wnt signals." *Dev Biol* 365 (1):196–207. doi: 10.1016/j.ydbio.2012.02.029.

Sulik, K. K. 2005. "Genesis of alcohol-induced craniofacial dysmorphism." *Experimental Biology and Medicine* 230 (6):366–75. doi: doi.org/10.1177/15353702–0323006–04.

Szabo-Rogers, H. L., L. E. Smithers, W. Yakob, et al. 2010. "New directions in craniofacial morphogenesis." *Dev Biol* 341 (1):84–94. doi: 10.1016/j.ydbio.2009.11.021.

Tabler, J. M., T. G. Bolger, J. Wallingford, et al. 2014. "Hedgehog activity controls opening of the primary mouth." *Dev Biol* 396 (1):1–7. doi: 10.1016/j.ydbio.2014.09.029.

Tahir, R., A. Kennedy, S. H. Elsea, et al. 2014. "Retinoic acid induced-1 (Rai1) regulates craniofacial and brain development in Xenopus." *Mech Dev* 133:91–104. doi: 10.1016/j.mod.2014.05.004.

Terrazas, K., J. Dixon, P. A. Trainor, et al. 2017. "Rare syndromes of the head and face: Mandibulofacial and acrofacial dysostoses." *Wiley Interdiscip Rev Dev Biol* 6 (3). doi: 10.1002/wdev.263.

Teven, C. M., S. Fisher, G. A. Ameer, et al. 2015. "Biomimetic approaches to complex craniofacial defects." *Ann Maxillofac Surg* 5 (1):4–13. doi: 10.4103/2231–0746.161044.

Teven, C. M., M. Greives, R. B. Natale, et al. 2012. "Differentiation of osteoprogenitor cells is induced by high-frequency pulsed electromagnetic fields." *J Craniofac Surg* 23 (2):586–93. doi: 10.1097/SCS.0b013e31824cd6de.

Tevlin, R., M. T. Longaker, and D. C. Wan. 2020. "Skeletal stem cells: A paradigm shift in the field of craniofacial bone tissue engineering." *Frontiers in Dental Medicine* 1. doi: 10.3389/fdmed.2020.596706.

Tevlin, R., A. McArdle, D. Atashroo, et al. 2014. "Biomaterials for craniofacial bone engineering." *J Dent Res* 93 (12):1187–95. doi: 10.1177/0022034514547271.

Theveneau, E., L. Marchant, S. Kuriyama, et al. 2010. "Collective chemotaxis requires contact-dependent cell polarity." *Dev Cell* 19 (1):39–53. doi: 10.1016/j.devcel.2010.06.012.

Theveneau, E., and R. Mayor. 2010. "Integrating chemotaxis and contact-inhibition during collective cell migration: Small GTPases at work." *Small GTPases* 1 (2):113–17. doi: 10.4161/sgtp.1.2.13673.

Thompson, B. L., P. Levitt, and G. D. Stanwood. 2009. "Prenatal exposure to drugs: Effects on brain development and implications for policy and education." *Nat Rev Neurosci* 10 (4):303–12. doi: 10.1038/nrn2598.

Tollemar, V., Z. J. Collier, M. K. Mohammed, et al. 2016. "Stem cells, growth factors and scaffolds in craniofacial regenerative medicine." *Genes Dis* 3 (1):56–71. doi: 10.1016/j.gendis.2015.09.004.

Trainor, P. A. 2010. "Craniofacial birth defects: The role of neural crest cells in the etiology and pathogenesis of Treacher Collins syndrome and the potential for prevention." *Am J Med Genet A* 152A (12):2984–94. doi: 10.1002/ajmg.a.33454.

Trainor, P. A. 2014. "Neural crest cells: Evolution, development and disease." *Academic Press-Elsevier*:458.

Trainor, P. A., and B. T. Andrews. 2013. "Facial dysostoses: Etiology, pathogenesis and management." *Am J Med Genet C Semin Med Genet* 163C (4):283–94. doi: 10.1002/ajmg.c.31375.

Trainor, P. A., and R. Krumlauf. 2001. "Hox genes, neural crest cells and branchial arch patterning." *Current Opinion in Cell Biology* 13 (6):698–705.

Twigg, S. R., J. Forecki, J. A. Goos, et al. 2015. "Gain-of-function mutations in ZIC1 are associated with coronal

craniosynostosis and learning disability." *Am J Hum Genet* 97 (3):378–88. doi: 10.1016/j.ajhg.2015.07.007.
Twigg, S. R., and A. O. Wilkie. 2015. "New insights into craniofacial malformations." *Hum Mol Genet* 24 (R1):R50–9. doi: 10.1093/hmg/ddv228.
Vega-Lopez, G. A., S. Cerrizuela, C. Tribulo, et al. 2018. "Neurocristopathies: New insights 150 years after the neural crest discovery." *Dev Biol* 444 (Suppl 1):S110–S143. doi: 10.1016/j.ydbio.2018.05.013.
Velasquillo, C., A. Madrazo-Ibarra, C. Gutiérrez Gómez, et al. 2020. "Stem cells and tissue engineering: An alternative treatment for craniofacial congenital malformations and articular degenerative diseases." *Plastic and Aesthetic Research* 2020. doi: 10.20517/2347–9264.2020.30.
Vieux-Rochas, M., B. Mascrez, R. Krumlauf, et al. 2013. "Combined function of HoxA and HoxB clusters in neural crest cells." *Dev Biol* 382 (1):293–301. doi: 10.1016/j.ydbio.2013.06.027.
Wahl, S. E., A. E. Kennedy, B. H. Wyatt, et al. 2015. "The role of folate metabolism in orofacial development and clefting." *Dev Biol* 405 (1):108–22. doi: 10.1016/j.ydbio.2015.07.001.
Waterman, R. E. 1977. "Ultrastructure of oral (buccopharyngeal) membrane formation and rupture in the hamster embryo." *Developmental Biology* 58:219–29. doi: doi.org/10.1016/0012–1606(77)90088–4.
Waterman, R. E., and G. C. Schoenwolf. 1980. "The ultrastructure of oral (buccopharyngeal) membrane formation and rupture in the chick embryo." *The Anatomical Record* 197:441–70. doi: doi.org/10.1002/ar.1091970408.
Watt, K. E. N., and P. A. Trainor. 2014. Chapter 17 – Neurocristopathies: The etiology and pathogenesis of disorders arising from defects in neural crest cell development. Paul A. Trainor, *Neural Crest Cells* (eds.). Academic Press, 361–94. San Diego, CA. ISBN 9780124017306, https://doi.org/10.1016/B978-0-12-401730-6.00018-1.
Weichert, A., M. Vogt, J. W. Dudenhausen, et al. 2010. "Evidence in a human fetus of micrognathia and cleft lip as potential effects of early cytomegalovirus infection." *Fetal Diagn Ther* 28 (4):225–8. doi: 10.1159/000320203.
Wheeler, A. C. 2018. "Development of infants with congenital Zika syndrome: What do we know and what can we expect?" *Pediatrics* 141 (Suppl 2):S154–60. doi: 10.1542/peds.2017–2038D.
Wheeler, A. C., D. Toth, T. Ridenour, et al. 2020. "Developmental outcomes among young children with congenital Zika syndrome in Brazil." *JAMA Netw Open* 3 (5):e204096. doi: 10.1001/jamanetworkopen.2020.4096.
Wilkie, A. O. M., D. Johnson, and S. A. Wall. 2017. "Clinical genetics of craniosynostosis." *Curr Opin Pediatr* 29 (6):622–8. doi: 10.1097/MOP.0000000000000542.
Wilkie, A. O. M., and G. M. Morriss-Kay. 2001. "Genetics of craniofacila development and malformation." *Nature Reviews Genetics* 2.
Willsey, H. R., P. Walentek, C. R. T. Exner, et al. 2018. "Katanin-like protein Katnal2 is required for ciliogenesis and brain development in Xenopus embryos." *Dev Biol* 442 (2):276–87. doi: 10.1016/j.ydbio.2018.08.002.
Withington, S., R. Beddington, and J. Cooke. 2001. "Foregut endoderm is required at head process stages for anteriormost neural patterning in chick." *Development* 128:309–20. doi: doi.org/10.1242/dev.128.3.309.
Wyatt, B. H., T. O. Raymond, L. A. Lansdon, et al. 2021. "Using an aquatic model, Xenopus laevis, to uncover the role of chromodomain 1 in craniofacial disorders." *Genesis* 59 (1–2):e23394. doi: 10.1002/dvg.23394.
Yamasaki, M. and Y. Kanemura. 2015. "Molecular biology of pediatric hydrocephalus and hydrocephalus-related diseases." *Neurol Med Chir (Tokyo)* 55(8):640–646. doi: 10.2176/nmc.ra.2015-0075.
Yan, Y., X. T. Zhang, G. Wang, et al. 2019. "Zika virus induces abnormal cranial osteogenesis by negatively affecting cranial neural crest development." *Infect Genet Evol* 69:176–89. doi: 10.1016/j.meegid.2019.01.023.
Yilmaz, E., E. Mihci, B. Nur, et al. 2019. "Recent advances in craniosynostosis." *Pediatr Neurol* 99:7–15. doi: 10.1016/j.pediatrneurol.2019.01.018.
Yoon, A. J., B. N. Pham, and K. M. Dipple. 2016. "Genetic screening in patients with craniofacial malformations." *J Pediatr Genet* 5 (4):220–4. doi: 10.1055/s-0036–1592423.
Yu, Y., X. Zuo, M. He, et al. 2017. "Genome-wide analyses of non-syndromic cleft lip with palate identify 14 novel loci and genetic heterogeneity." *Nat Commun* 8:14364. doi: 10.1038/ncomms14364.
Zaghloul, N. A., and S. A. Brugmann. 2011. "The emerging face of primary cilia." *Genesis* 49 (4):231–46. doi: 10.1002/dvg.20728.
Zhang, W., and P. C. Yelick. 2018. "Craniofacial tissue engineering." *Cold Spring Harb Perspect Med* 8 (1). doi: 10.1101/cshperspect.a025775.
Zhao, H., J. Feng, T. V. Ho, et al. 2015. "The suture provides a niche for mesenchymal stem cells of craniofacial bones." *Nat Cell Biol* 17 (4):386–96. doi: 10.1038/ncb3139.
Zuk, P.A. 2008. "Tissue engineering craniofacial defects with adult stem cells? Are we ready yet?" *Pediatric Research* 63:478–86. doi: doi.org/10.1203/PDR.0b013e31816bdf36.

18 Modeling Digestive and Respiratory System Development and Disease in *Xenopus*

Scott A. Rankin and Aaron M. Zorn

CONTENTS

18.1. Introduction ... 259
18.2. Temporal Overview of Xenopus Endoderm Organogenesis ... 260
18.3. The Molecular Basis of Endoderm Formation ... 261
18.3.1. Maternal Pre-Pattern Phase ... 262
18.3.2. Endoderm Induction Phase ... 263
18.3.3. Endoderm Commitment Phase ... 263
18.4. Endoderm Patterning ... 263
18.4.1. Gastrula-Stage Endoderm Patterning ... 263
18.4.2. Neurula-Stage Endoderm Patterning ... 265
18.5. Induction of Endoderm Organ Fate ... 266
18.5.1. Pharyngeal Endoderm and Thyroid ... 268
18.5.2. Lung, Trachea, and Esophagus ... 268
18.5.3. Pancreas and Liver ... 268
18.5.4. Stomach ... 270
18.5.5. Intestine ... 270
18.6. Concluding Remarks ... 271
Acknowledgments ... 271
References ... 271

18.1. INTRODUCTION

The endoderm is the innermost germ layer of the vertebrate embryo that gives rise to the epithelial lining of the respiratory and digestive tracts as well as associated organs such as the stomach, liver, and pancreas. Research over the past 50 years has shown that the molecular basis of endoderm organogenesis is largely conserved from animals to human (Zorn and Wells 2009). Among vertebrate model organisms used for biomedical research, the experimental advantages of *Xenopus* embryos have proven particularly useful in elucidating key steps in endoderm formation, patterning, and early organogenesis. Large, externally developing *Xenopus* embryos allow microsurgery, explant culture, and targeted microinjection for tissue-restricted analysis of gene or pathway function, and pharmacological manipulation of pathway/gene activity is easily performed at any desired stage of development. In addition, transgenics, CRISPR-mediated genome editing, and cutting-edge genomics are widely used in *Xenopus* research. The combination of all these strategies readily permits the determination of epistatic relationships and temporal requirements of developmental events and has accelerated the utility of *Xenopus* for interrogating molecular mechanisms of endoderm organogenesis and congenital disease in ways that are much more challenging in mammals.

Xenopus experiments in the 1990s were the first to reveal the growth factor signaling pathways that induce the endoderm and mesoderm lineages in the gastrula embryo (Smith et al. 1990; Asashima et al. 1990); these findings directly led to the development of protocols enabling the directed differentiation of mouse and human endoderm tissue from pluripotent stem cells (PSCs) (Kubo et al. 2004; D'Amour et al. 2005; Loh et al. 2014). Experimental embryology in *Xenopus* has also been instrumental in defining conserved signaling events between the endoderm and adjacent splanchnic mesoderm that govern induction of cardiac progenitor cells (Nascone and Mercola 1995). Indeed, bi-directional signaling between the endoderm and mesoderm is a conserved feature of vertebrate gut tube patterning and organogenesis; this knowledge, determined in large part from *Xenopus* embryos, has helped facilitate the generation of human PSC-derived organoids useful for disease modeling and hopefully one day for regenerative medicine (McCauley and Wells 2017).

Xenopus is increasingly used to study how disruptions in developmental pathways and genes can lead to birth defects and pediatric disease of endodermal organs such as diabetes/

DOI: 10.1201/9781003050230-21

maturity onset diabetes of the young (MODY), tracheal-esophageal fistula/esophageal atresia, and intestinal malrotation (Pearl et al. 2011; Nasr et al. 2019; Grzymkowski et al. 2020; Edwards and Zorn 2021). It is estimated that about 80% of human disease-associated genes have *Xenopus* orthologs. The online biomedical knowledgebase Xenbase.org (Nenni et al. 2019) facilitates human disease modeling by curating *Xenopus* genomic, expression, genotype, and phenotype data from the published literature and integrating this with orthologous human gene, anatomy, and disease with links to National Center for Biotechnology Information (NCBI), Online Mendelian Iheritance in Man (OMIM), and human disease ontology (DO) resources.

In this chapter, we summarize the current understanding of *Xenopus* endoderm organogenesis. We organize our discussion in four major sections: (1) a temporal overview of *Xenopus* endoderm organogenesis; (2) endoderm germ layer specification, (3) progressive pattering of the endoderm gut tube, and (4) organ induction. In each section, we highlight conserved molecular mechanisms, pointing out both historical and recent contributions of *Xenopus* research. We also emphasize how the use of emerging technologies in *Xenopus* continues to inform our molecular understanding of human congenital birth defects and disease.

18.2. TEMPORAL OVERVIEW OF XENOPUS ENDODERM ORGANOGENESIS

The molecular mechanisms of endoderm formation, patterning, organ induction, and early morphogenesis are largely conserved amongst vertebrate species (Zorn and Wells 2009) and are largely indistinguishable between the two commonly used species *Xenopus laevis* and *Xenopus tropicalis*. Figure 18.1 summarizes key phases in *Xenopus* endoderm development, showing a developmental fate map (Figure 18.1A), example gene expression (Figure 18.1B), and a schematic of cell lineages (Figure 18.1C); in addition, we present the approximate timing of similar developmental transitions in mouse embryos and during the directed differentiation of human PSCs (Figure 18.1D).

Endoderm organogenesis begins with the segregation of pluripotent embryonic cells into ectoderm, mesoderm, and endoderm germ layers during the blastula and gastrula stages. Pioneering fate-mapping and transplantation studies in the 1980s demonstrated that the yolk-rich vegetal pole cells of the early cleavage stage embryo give rise to endoderm (Figure 18.1A; Moody 1987; Dale and Slack 1987), and by early gastrula stage, these vegetal cells are committed to their endoderm fate (Heasman et al. 1984; Wylie et al. 1987). A critical milestone in our understanding of endoderm germ layer specification was the discovery that the TGFβ/Activin/Nodal family of growth factors could induce mesoderm and endoderm fate in pluripotent blastula ectoderm animal cap explants in a concentration-dependent manner (Smith et al. 1990; Asashima et al. 1990). Based on these studies, the use of Activin has since become the standard approach to differentiate mouse and human PSCs into endoderm (Kubo et al. 2004; D'Amour et al. 2005; Loh et al. 2014). This, along with the discovery of the transcription factor (TF) Sox17 in *Xenopus* (Hudson et al. 1997), the first unambiguous molecular marker of vertebrate definitive endoderm, opened the door to studying the molecular mechanisms controlling endoderm formation. As detailed in the subsequent section, TGFβ/Nodal signaling activates the expression of *sox17* and other key TF encoding genes such as *foxa1* in the endoderm lineage as early as the blastula stage (approximately five hours post-fertilization [hpf], Nieuwkoop and Faber [NF] stage 8 [NF8]; Nieuwkoop and Faber 1994) (Figure 18.1A,B).

During blastula and neurula stages of development (NF9–NF20), the endoderm is progressively patterned along the anterior and posterior (A-P) and dorsal-ventral (D-V) axes into progenitor domains characterized by restricted expression of different lineage promoting TFs. Initial endoderm patterning is coincident with endoderm formation (NF9–11) and is coordinated with global D-V axial patterning of all three germ layers by Wnt and BMP signals. This initial regionalization is further elaborated during neurula and early somite stages (NF12–25) by posteriorizing WNT, BMP, FGF, and retinoic acid (RA) signals from the surrounding splanchnic mesoderm, which further subdivide the endoderm into foregut, midgut, and hindgut domains that have different developmental potential (Rankin et al. 2018): foregut-derived organs include the pharynx, thyroid, thymus, parathyroid, esophagus, trachea, lungs, liver, pancreas, and stomach; midgut endoderm gives rise to the small intestine; and hindgut endoderm gives rise to the large intestine and cloaca (Figure 18.1C) (Chalmers and Slack 2000).

Between NF15 and NF35 (approximately two days), the broad progenitor domains are progressively restricted, and organ lineages are induced by continued combinatorial signaling between the splanchnic mesoderm and endoderm epithelium. This process involves many of the same mesoderm-derived WNT, BMP, RA, and FGF signals that control early patterning, which act reiteratively in various combinations and with additional signals such as endodermal-derived Hedgehog (Shifley et al. 2012; Rankin et al. 2016; Rankin et al. 2018). The bidirectional nature of this signaling coordinates organ identity in the germ layers so that the correct mesoderm develops in contact with the appropriate endoderm. Embryological experiments show that by NF35, organ fates are mostly specified and a number of lineage-specific markers are expressed (Horb and Slack 2001; Shifley et al. 2012; Rankin et al. 2018).

Between NF35-NF43 early endoderm organ morphogenesis occurs, which include the emergence of lung, liver, and pancreatic organ buds that evaginate from the gut tube and intermingle with the adjacent mesoderm. The anterior foregut tube separates into a distinct trachea and esophagus, while the intestine elongates and undergoes a stereotypical coiling process similar to mammals (Rankin et al. 2015; Chalmers and Slack 2000; Grzymkowski et al. 2020). By NF43 (3.5 dpf) the entire gut tube can be readily microdissected, and distinct organ buds are obvious under the stereomicroscope (Figure 18.1A,B). *Xenopus* tadpoles begin

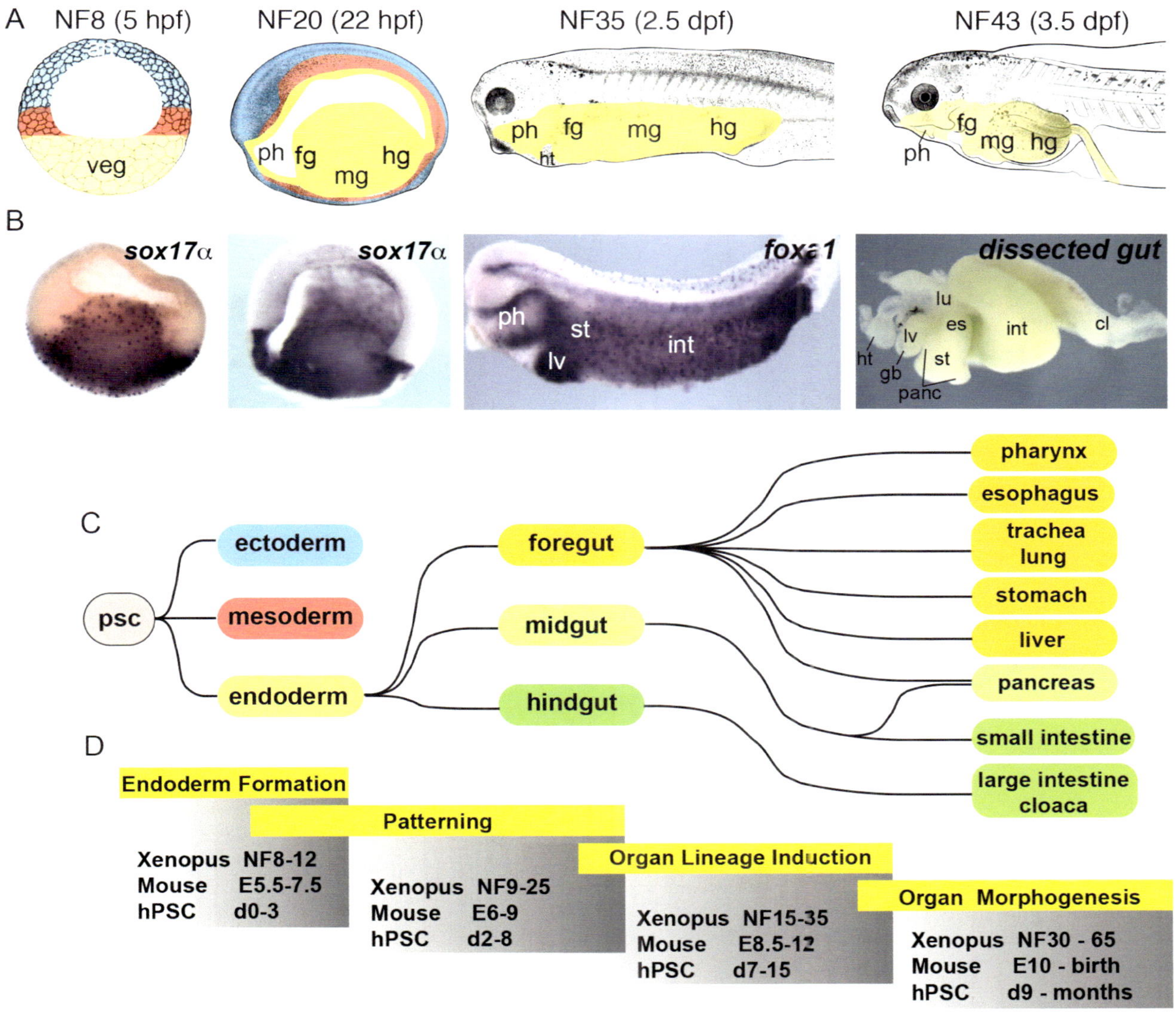

FIGURE 18.1 Temporal overview of *Xenopus* endoderm organogenesis. (A) Schematic of *Xenopus* embryos during early development. Endoderm is shaded yellow, mesoderm red, and ectoderm blue. Abbreviations: NF, Nieuwkoop and Faber; an, animal; veg, vegetal; ph, pharynx; fg, foregut; mg, midgut; hg, hindgut; ht, heart; hpf, hours post fertilization and dpf, days post fertilization, both when embryos are cultured at 23 degrees Celsius. (B) Images of actual *X. laevis* embryos stained by *in situ* hybridization and a dissected gut tube, at the stage of development indicated by the previous schematic. *In-situ* hybridization for *sox17a* or *foxa1* reveals the endoderm. Abbreviations: cl, cloaca; es, esophagus; gb, gall bladder; ht, heart; int, intestine; lu, lung; lv, liver; panc, pancreas; ph, pharynx; st; stomach. (C) Endoderm lineage tree. psc, pluripotent stem cell. (D) Approximate timing of similar phases of endoderm developmental in *Xenopus*, mouse, and during the directed differentiation of human pluripotent stem cells (hPSCs).

Source: (A) Drawings of the Xenopus embryos in the top panel reproduced with permission, © 202_ Xenbase and Natalya Zahn, licensed under CC BY-NC 4.0.

feeding around two weeks of age, and although the animals do not fully transition from gills to lungs for breathing until metamorphosis, air breathing is observable by NF47 (Rose and James 2013). During metamorphosis (stages NF56–66), the respiratory and GI organs undergo a substantial remodeling process in response to hormone signaling; these events share molecular and physiological similarities to the perinatal period of mammalian gestation around birth. As several excellent reviews describe *Xenopus* metamorphosis, we will not address it here (Brown and Cai 2007; Buchholz 2015).

18.3. THE MOLECULAR BASIS OF ENDODERM FORMATION

In *Xenopus*, endoderm formation can be conceptually summarized in 3 phases: (1) a maternal "pre-pattern" phase, from the 32-cell stage (NF6) to early blastula (NF8); (2) an "induction" phase, from NF8 to NF10; and (3) a "commitment" phase, during gastrulation (NF10 to NF12). Figure 18.2 summarizes these steps and presents an overview of the gene regulatory network (GRN) controlling endoderm formation.

18.3.1. Maternal Pre-Pattern Phase

In *Xenopus*, germ layer formation and endoderm specification are initiated by maternal factors, and transcripts encoding these factors are spatially localized along the animal-vegetal axis of eggs and early embryos. Maternal TFs transcripts localized to the animal pole include *ascl1, foxi2*, and *sox3*; these genes function to promote ectoderm fate and inhibit mesoderm and endoderm gene expression (Reich and Weinstein 2019). Vegetally localized maternal TF transcripts include *vegT, otx1*, and *sox7*; these factors are necessary, in combination with the ubiquitously expressed maternal TFs *foxh1* and *pou5f3.2/pou5f3.3*, for endoderm formation (Xanthos et al. 2001; Paraiso et al. 2020) (Figure 18.2A,G). These TFs co-bind endodermal gene enhancers prior to the onset of zygotic transcription (Paraiso et al. 2019;

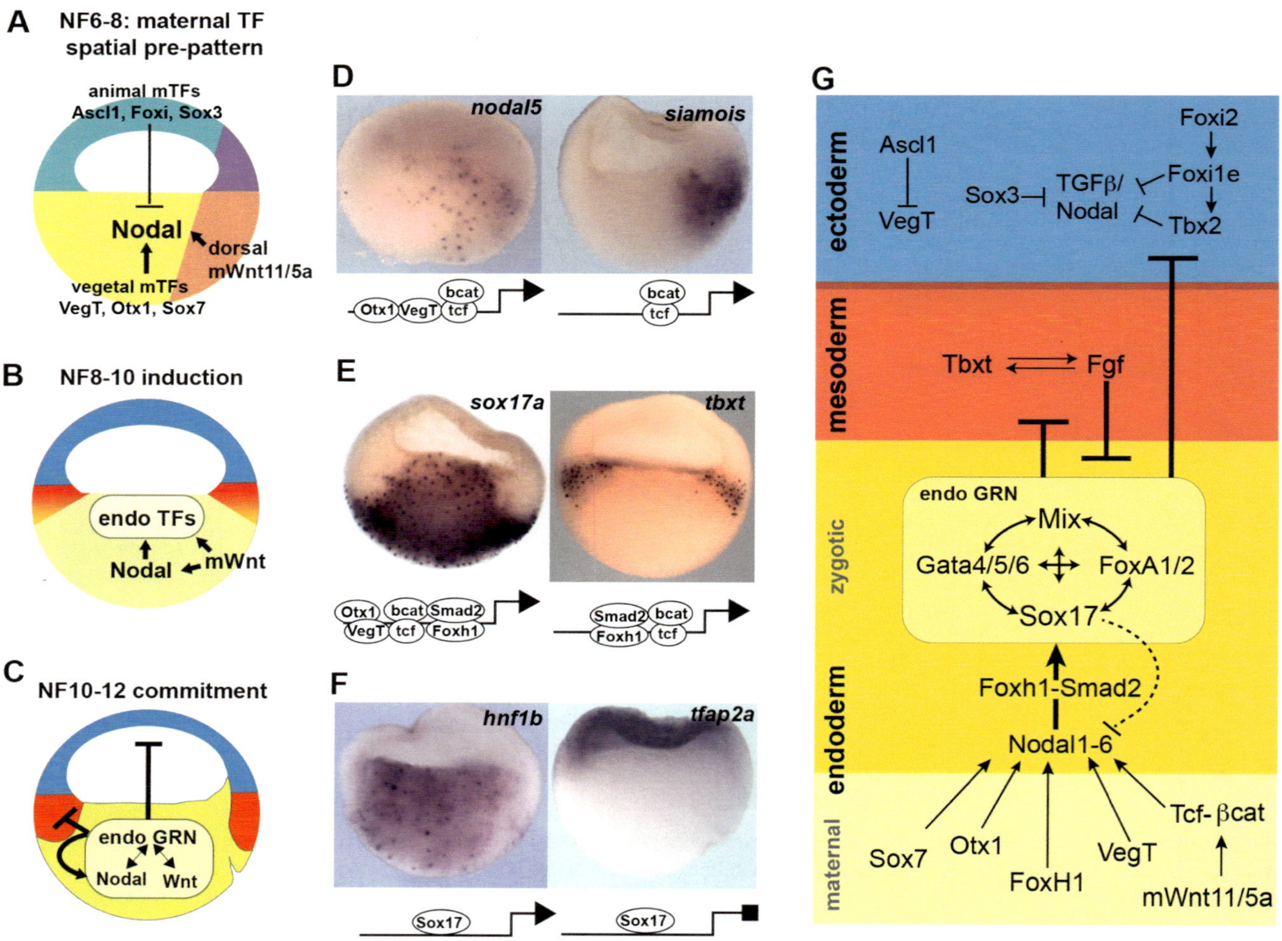

FIGURE 18.2 Three conceptual phases of *Xenopus* endoderm germ layer formation. (A,D). Maternal pre-pattern phase from the 32-cell stage (NF6) to early blastula (NF8). Transcripts of maternal TFs including *vegT, otx1*, and *sox7* are localized to the vegetal pole, and these proteins bind endoderm gene cis-regulatory elements, marking them for subsequent expression in vegetal endoderm cells. A maternal, dorsal Wnt11/5a signal results in accumulation of β-Catenin in dorsal cells, and robust *nodal* expression depends on both maternal vegetal TFs and Wnt/11/5a-dependent TCF/β-Catenin complexes. Animal-pole localized maternal TF transcripts include *ascl1, foxi2*, and *sox3*; these function to suppress *nodal* expression and mesendoderm induction in the prospective ectoderm. In panel D, *in situ* hybridization of NF9 embryos shows *nodal5* expression enriched in dorsal-vegetal cells and *siamois* (*sia*) in dorsal cells. In panels D,E,F, the gene diagrams below the *in situ* hybridization images depict binding of factors from published ChIP-seq studies described in the text. (B,E) Endoderm induction phase from NF8 to NF10. High levels of Nodal signaling, in co-operation with maternal Wnt11/5a/β-Catenin/TCF, stimulate the vegetal expression of an evolutionarily conserved group of core endodermal TFs including Sox17, Gata4–6, Foxa1–4, and Mix/Bix family members. In the equatorial marginal zone region, lower levels of Nodal signaling activate the TF *brachyury* (*tbxt*), which promotes mesoderm fate. *In situ* hybridization of *sox17a* in vegetal endoderm cells and *tbxt* in marginal zone mesoderm cells. (C,F) Endoderm commitment phase during gastrulation from NF10–NF12. During the commitment phase, core endoderm TFs function collectively with Nodal and Wnt signaling in a series of feed-forward loops to promote each other's expression and maintain endoderm identity. Sox17 reinforces Nodal-induced endoderm fate by promoting expression of endoderm genes and suppressing expression of mesoderm and ectoderm genes. Shown by *in situ* hybridization is the expression of *hnf1b*, a vegetal endoderm TF activated by Sox17, and *tfap2a*, an ectoderm TF repressed by Sox17 in vegetal cells. (G) Summary of gene regulatory networks governing *Xenopus* germ layer formation. Spatially distinct regulatory networks active along the animal-vegetal embryonic axis promote their respective germ layer identity while simultaneously repressing alternative germ layer networks.

Charney et al. 2017a, 2017b; Gentsch et al. 2019), where they appear to act as pioneering factors to decondense the chromatin prior to genome activation and inductive signaling.

The vegetally localized TFs cooperate with a maternal Wnt pathway active on the dorsal side of the embryo (Tao et al. 2005; Cha et al. 2008a). In dorsal vegetal cells, β-Catenin translocates to the nucleus, interacts with Tcf/Lef family TFs, and directly regulates expression of genes such as *siamois* (Figure 18.2A,D). β-Catenin also contributes to epigenetic priming of enhancers, recruiting the arginine methyltransferase Prmt2 to modify histones and establish a poised chromatin architecture (Blythe et al. 2010). The key result of this pre-pattern by maternal Wnt and maternal TFs is the initiation of zygotic *nodal1–6* gene expression at the mid-blastula transition in vegetal cells fated to become endoderm (Xanthos et al. 2001, 2002; Agius et al. 2000; Rex et al. 2002).

18.3.2. Endoderm Induction Phase

During the "induction phase" (NF8–NF10), high levels of Nodal signaling promote endoderm fate, while the overlying equatorial cells experience lower Nodal concentrations and are induced to adopt a *tbxt*-expressing mesoderm fate (Figure 18.2B,E) (Cha et al. 2004, 2008b; Gentsch et al. 2013; Charney et al. 2017b). Nodal-induced Smad2/Foxh1 complexes co-bind enhancers along with maternal TFs and βCatenin/TCF complexes to cooperatively stimulate the expression of an evolutionarily conserved group of zygotic endodermal TFs including Sox17, Gata4–6, Foxa1–4, and Mix/Bix family members, which collectively promote endoderm fate (Figure 18.2E) (Zorn and Wells 2009; Charney et al. 2017b). Recent genomic data suggest Smad2/Foxh1 act on super enhancers bound by maternal VegT/Oxt1/Sox7, thus linking pre-patterning to the onset of zygotic gene expression (Paraiso et al. 2019; Gentsch et al. 2019).

18.3.3. Endoderm Commitment Phase

During the "commitment" phase (NF10–NF12), the zygotic endoderm TFs function collectively with Nodal and Wnt signaling in a series of feed-forward loops to promote each other's expression and establish endoderm identity (Figure 18.2C; Sinner et al. 2006; Charney et al. 2017b). Interactions between the mutually repressive endoderm GRN and a Tbxt-FGF mesodermal GRN refine the boundary between the endoderm and mesoderm cells (Figure 18.2G). Genomic analysis recently demonstrated that Sox17 promotes endoderm commitment via multiple mechanisms: (1) Sox17 activates the expression of many key endodermal genes such as *hnf1b*; (2) Sox17 and β-Catenin co-bind about 30% of all endoderm enhancers to synergistically activate endoderm genes, illustrating a cooperation with the Wnt pathway; (3) Sox17 directly binds and suppresses mesoderm/ectoderm genes such as *tfap2a*; and (4) Sox17 acts as a negative feedback regulator to restrain *nodal* ligand expression (Figure 18.2F,G; Mukherjee et al. 2020). These studies provide molecular insights into the pioneering studies of Chris Wylie, Janet Heasman, Jim Smith, and colleagues in the 1980s that demonstrated vegetal cells become committed to endoderm fate during gastrulation (Heasman et al. 1984; Wylie et al. 1987). A spatial-temporal catalog of TFs expressed in the gastrula endoderm has been generated (Blitz et al 2017), and many TFs remain to be incorporated into the endoderm GRN. Additional details on the GRNs of germ layer formation can be found in Cho and Blitz (Chapter 12 of this book).

18.4. ENDODERM PATTERNING

In the following section, we discuss the molecular control of endoderm patterning during gastrulation (Figure 18.3) and during neurula/somitogenesis stages (Figure 18.4), which define the foregut, midgut, and hindgut progenitor domains.

18.4.1. Gastrula-Stage Endoderm Patterning

Gastrula endoderm pattern is intimately linked to its initial formation. In the 1990s and early 2000s, pioneering work by Eddy De Robertis, Richard Harland, Christof Niehrs, and colleagues defined the molecular nature of the vertebrate organizer wherein secreted Wnt/BMP-antagonists, expressed in the dorsal-anterior organizer mesendoderm, regulate pattern of all three germ layers during gastrulation; this work in *Xenopus* identified principles that turned out to be conserved in all vertebrates (De Robertis and Kuroda 2004; see also Chapter 4 of this book).

In Figure 18.3, we highlight molecular players driving distinct dorsal-anterior and ventral-posterior GRNs, which result in anterior and posterior endoderm domains identifiable by expression of the TFs *hhex* and *ventx1/2/3*, respectively (Zorn et al. 1999; Rankin et al. 2011). Interestingly, these gastrula endoderm domains have distinct organ-forming competence due to BMP and Wnt-dependent early patterning (Rankin et al. 2018).

In the dorsal-anterior GRN, maternal dorsal Wnt signaling co-operates with maternal vegetal TFs to drive high levels of zygotic Nodal signaling and induce a number of organizer-specific homedomain (HD) TFs, including Siamois (Figure 18.2D), Twin/Sia2, Otx2, Lim1/Lhx1, and Goosecoid (Figure 18.3E; Zorn et al. 1999; Rankin et al. 2011). These HD TFs in turn activate expression of BMP/Wnt-antagonists, including Chordin, Noggin, Cerberus, Dkk1, and Sfrps (secreted frizzled related proteins, Sfrp1, Sfrp2, and Frzb/Sfrp3) (Figure 18.3C,E); the HD TFs also promote transcription of *hhex*. We have performed extensive cis-regulatory analyses of the *hhex* locus, characterizing Wnt, Nodal, and HD TF responsive promoter and enhancer elements as well as Ventx-mediated repressive elements (Figure 18.3E; Rankin et al. 2011). Recent genome-wide analyses have expanded these findings and defined super-enhancers that are thought to act as transcriptional hubs that integrate combinations of the HD TFs as well as Foxh1/Smad2 and Tcf/β-Catenin complexes to control transcription of regionally expressed endoderm genes throughout

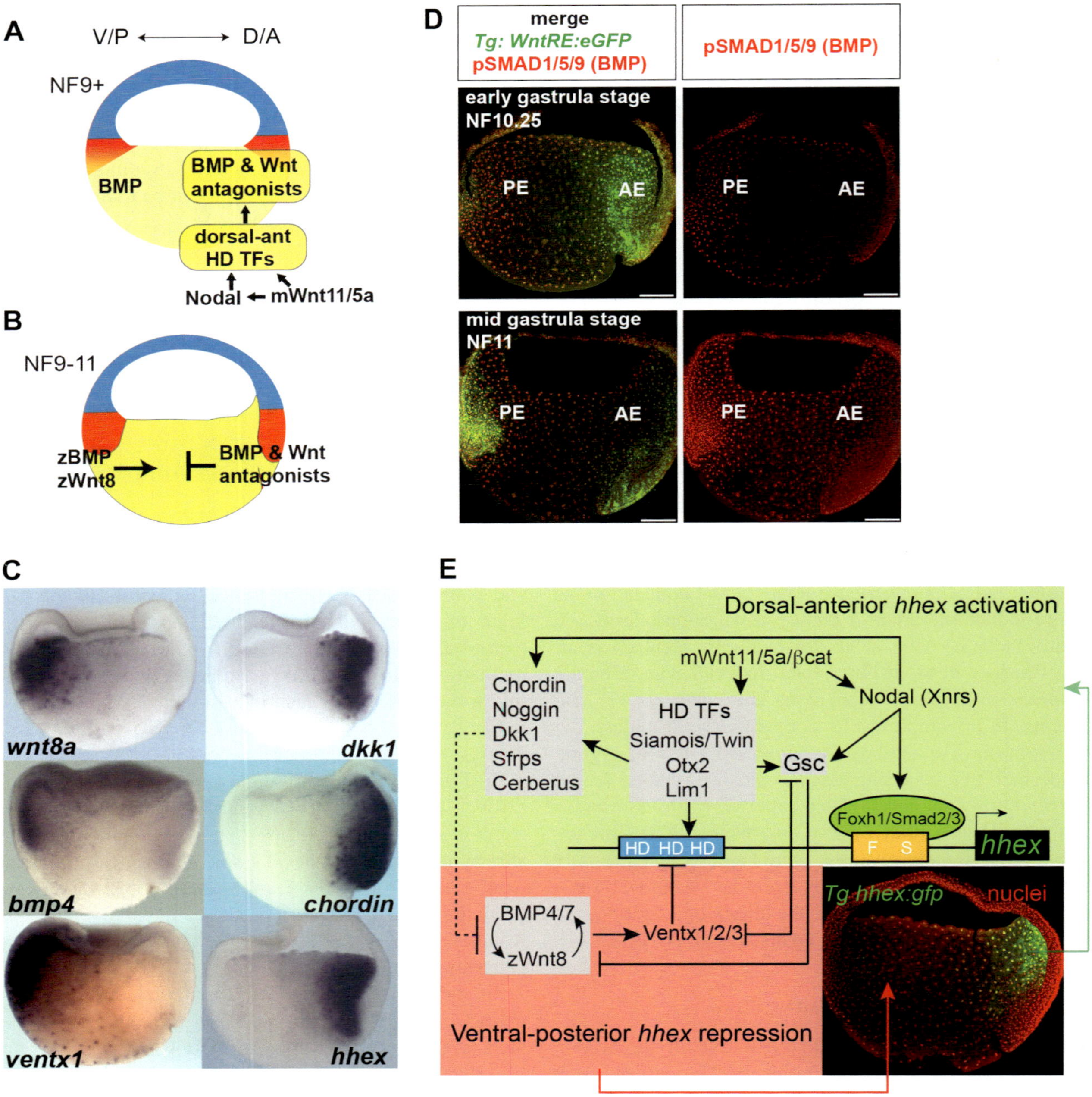

FIGURE 18.3 Blastula and gastrula stage endoderm patterning in *Xenopus*. (A-B) Initial endoderm pattern during blastula (NF9) and early gastrula (NF10–11) stages is dependent on ventral-posterior BMP and Wnt8 signaling that is suppressed in dorsal-anterior organizer territory by BMP and Wnt antagonists. Nodal and maternal Wnt11/5a signaling cooperatively induce expression of a number of Homeodomain transcription factors (HD TFs, including Siamois, Twin/Sia2, Otx2, Lim1/Lhx1, and Goosecoid), which are required to activate expression of the BMP and Wnt antagonists. Abbreviations: V/P, ventral/posterior; D/A, dorsal-anterior. (C) *In situ* hybridization of *X. laevis* bisected early gastrula NF10.5 embryos. *wnt8a, bmp4*, and *ventx1* are expressed in ventral-posterior mesendoderm, whereas anterior mesendoderm expresses *dkk1*, *chordin*, and *hhex*. (D) Transgenic *X. laevis* Wnt/β-Catenin reporter embryos *Tg(WntREs:dEGFP)*^{Vlemx} at stage NF10.25 or NF11 immunostained for eGFP (green) and phospho-Smad1/5/9 (red). At the early gastrula stage NF10.25, eGFP expression in the anterior endoderm is due to maternal Wnt11/5a activity; in contrast, in NF11 embryos, organizer-expressed Wnt antagonists suppressed this signal and zygotic Wnt8a signaling activates the eGFP in ventral-posterior mesendoderm. Abbreviations: AE, anterior endoderm; PE, posterior endoderm. (E) Anterior and posterior gene regulatory networks regulating *hhex* expression. A bisected early gastrula NF10.5 transgenic *Tg(−5kbhhex:gfp)*^{Zorn} embryo, immunostained for GFP, is shown.

Source: Panel (E) is reproduced in part from Rankin et al. (2011).

blastula and gastrula stages (Paraiso et al. 2019; Gentsch et al. 2019; Afouda et al. 2020). Additional ChIP-seq studies suggest unique combinations of HD TFs can either promote or repress transcription: for example, Oxt2/Lim1 complexes stimulate transcription of dorsal-anterior endoderm genes like *hhex*, whereas Otx2/Goosecoid complexes repress ventral-posterior genes such as *ventx1/2/3* and *wnt8* (Yasuoka et al. 2014; Yasuoka et al. 2019).

During gastrulation, there is a dramatic switch in the region of active Wnt signaling in the embryo, which can be visualized with transgenic Wnt reporter embryos *Tg(WRE:dGFP)*, wherein canonical Tcf/Lef sites drive destabilized GFP expression (Figure 18.3D; Tran et al. 2010). Maternal Wnt signaling is high in the dorsal mesendoderm and low in the ventral side of the blastula and early gastrula, but as gastrulation proceeds, the pattern is switched as zygotic *wnt8* becomes expressed in ventral posterior mesendoderm and Wnt-antagonists are expressed in the organizer.

In the ventral-posterior GRN, zygotic Wnt8/β-Catenin and Bmp4/Smad1 act in a feedforward loop (Hoppler and Moon 1998; Fuentealba et al. 2007; Kjolby et al. 2019) and directly activate expression of the Ventx family of HD transcriptional repressors (Ventx1/2/3) (Figure 18.3C; Onichtchouk et al. 1996; Karaulanov et al. 2004; Hikasa et al. 2010). Ventx factors repress transcription of dorsal-anterior genes, including *goosecoid* and *hhex* (Sander et al. 2007; Rankin et al. 2011); thus, a mutually antagonistic cross-repressive loop exists between the ventral-posterior and dorsal-anterior GRNs in the early gastrula embryo (Figure 18.3E).

18.4.2. Neurula-Stage Endoderm Patterning

During gastrulation, the *hhex*-expressing anterior endoderm migrates to the future ventral foregut adjacent to the heart as the blastopore closes and the archenteron opens. Immediately behind the migrating anterior endoderm, a thin layer of dorsal endoderm forms the roof of the archenteron, whereas the *ventx*-expressing endodermal mass makes up the bulk of the ventral gut tissue. During the neurula and early somitogenesis stages (NF12–NF25), the endoderm continues to be patterned along its D-V and A-P axes by combinatorial BMP, Wnt/β-Catenin, RA, and FGF signaling (Figure 18.4A), and distinct domains of regional endoderm TF expression are obvious by NF20 (Figure 18.4B). This post-gastrula patterning in *Xenopus* is comparable to events that occur during E7-E9 in mouse gestation and 3–8 days of many human PSC differentiation protocols.

During this post-gastrula period, *bmp2/4/7* are expressed in the ventral splanchnic mesoderm, while BMP antagonists *noggin* and *chordin* are produced dorsally from the notochord, generating a D-V gradient of phosphorylated Smad1/5/9 (pSMAD1) activity in the embryo (Figure 18.4A,B). High BMP promotes ventral gene expression in both the foregut and hindgut, whereas low BMP permits a dorsal endoderm identity (Stevens et al. 2017). BMP from the ventral mesoderm induces expression of a number of secreted Wnt antagonists in the foregut including *sfrp1/2/5* (Kenny et al. 2012; Stevens et al. 2017), which are required to maintain a low level of Wnt11-Fzd7 signaling essential for proliferation, morphogenesis, and identity of *hhex*+ progenitors (McLin et al. 2007; Li et al. 2008; Zhang et al. 2013a). In the posterior domain, BMP and Wnt ligands secreted from the mesoderm cooperate to maintain *ventx*+ hindgut fate and induce the expression of *cdx2*, a HD TF essential for intestinal development in all vertebrates (Stevens et al. 2017).

RNA-seq and ChIP-seq analyses of stage NF20 embryos have uncovered how foregut and hindgut progenitor cells transcriptomes are regulated by BMP/pSmad1 and Wnt/β-Catenin at a genomic level (Stevens et al. 2017). These studies identified hundreds of enhancers co-occupied by Smad1 and β-Catenin and revealed the unexpected observation that their chromatin occupancy was associated with both transcriptional activation and repression (Stevens et al. 2017). Moreover, Wnt-inhibited enhancers often lacking canonical TCF DNA-binding motifs, suggesting a novel mode of repression, which remains to be resolved.

RA signaling regulates both D-V and A-P pattern, with RA production being spatially controlled by restricted expression of the RA-synthesizing enzyme Aldh1a2 (Raldh2) in anterior lateral plate mesoderm and RA-degrading Cyp26 enzymes in pharyngeal and hindgut territories (Figure 18.4A,B). This is thought to generate a gradient of RA activity highest in dorsal-posterior foregut and lower in the pharynx and hindgut. Genomic studies have uncovered RA-target genes and identified a number of feedback mechanisms between BMP, Wnt, RA, and FGF. RA promotes expression of Wnt-antagonists such as *ndrg1a* and *sfrps* (Zhang et al. 2013b; Damianitsch et al. 2009) and high levels of ventral BMP restrict *aldh1a2* so that it is more robustly expressed dorsally (Stevens et al. 2017). FGFs from the posterior mesoderm co-operate with Wnt to promote the caudal expression of both *cdx2* and *cyp26*, which limits RA activity in the hindgut (Shiotsugu et al. 2004; Deimling and Drysdale 2011). RA in turn restricts expression of *fgf8* and *fgf4* to the pharynx and hindgut (Shiotsugu et al. 2004; Arima et al. 2005).

The result of this combinatorial BMP, Wnt, RA, and FGF signaling is the restricted expression of lineage-promoting TFs, a number of which are shown in Figure 18.4B. Mutations in many of these TFs are associated with human congenital syndromes affecting the digestive or respiratory systems: for example, *tbx1* is expressed in the pharyngeal endoderm and is lost in 22q11.2 Deletion syndrome (DiGeorge syndrome, OMIM# 188400); *rfx6* marks posterior foregut and mutations cause Mitchell-Riley syndrome (OMIM# 615710; Smith et al. 2010); *mnx1* is expressed in dorsal endoderm lining the archenteron roof, and mutations are associated with Currarino syndrome (OMIM#176450; Han et al. 2020);

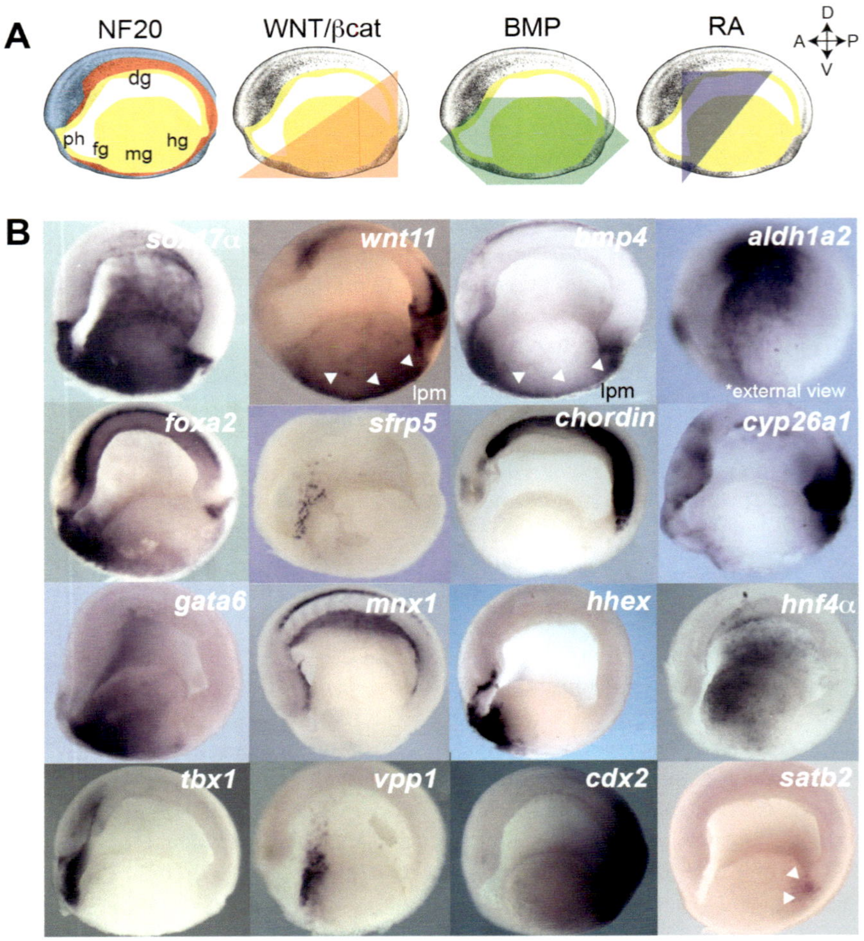

FIGURE 18.4 Neurula-stage endoderm patterning in *Xenopus*. (A) Schematic of stage NF20 *Xenopus* embryo with the signaling domains of Wnt/β-Catenin (orange), BMP (green), and RA (blue) highlighted. The combinatorial nature of these signals along the anterior-posterior (A-P) and dorsal-ventral (D-V) axes patterns the developing endoderm, resulting in expression patterns of the transcription factors shown in (B). (B) *In situ* hybridization of the indicated genes in stage NF20 *X. laevis* embryos. Expression domains are as follows: *sox17a,* pan-endoderm; *wnt11*, ventral lateral plate mesoderm; *bmp4*, ventral lateral plate mesoderm; *aldh1a2*, anterior lateral plate mesoderm (external view); *foxa2*, pharyngeal and foregut endoderm; *sfrp5*, anterior foregut endoderm; *chordin*, dorsal notochord mesoderm; *cyp26a1*, pharyngeal and hindgut mesendoderm; *gata6*, ventral foregut endoderm and ventral foregut mesoderm; *mnx1*, dorsal endoderm (archenteron roof); *hhex*, ventral foregut/hepatic endoderm; *hnf4*α, foregut and midgut endoderm; *tbx1*, pharyngeal endoderm; *vpp1*, anterior foregut/pancreatic endoderm; *cdx2*, midgut and hindgut endoderm; *satb2*, distal-most hindgut endoderm.

and *cdx2* mutations are associated with persistent cloaca and abnormal *cdx2* expression with Barrett's esophagus (Hsu et al. 2018; Colleypriest et al. 2010).

The molecular mechanisms that link these progenitor-restricted TFs to organ specification are still poorly understood. One hypothesis is these TFs regulate the epigenetic status of chromatin, which imparts developmental competence to subsequent signaling events. This may explain how the same growth factors, reiteratively active during patterning and organ induction, can regulate distinct transcriptional programs. For example, Wnt/BMP promote hindgut and repress foregut fate in the neurula stage, yet only hours later, Wnt/BMP then induce lung fate in foregut progenitors. Indeed, investigations into the molecular basis of developmental competence connecting lineage-promoting TFs to chromatin dynamics and epigenetic status of enhancers is an exciting and intensive area of endoderm organogenesis research (Wang et al. 2015; Vinckier et al. 2020). With the advent of single-cell transcriptomics in *Xenopus* embryos (Briggs et al. 2018), we predict that epigenetic analysis of experimental perturbations will reveal insight into the mechanisms of dynamic developmental competence.

18.5. INDUCTION OF ENDODERM ORGAN FATE

In Figure 18.5, we summarize the combinations of signals required for induction of different endoderm lineages. By NF35, most organ lineages can be visualized by the regional expression of TFs and signaling molecules along the embryo's A-P axis (Figure 18.5B), and in the following

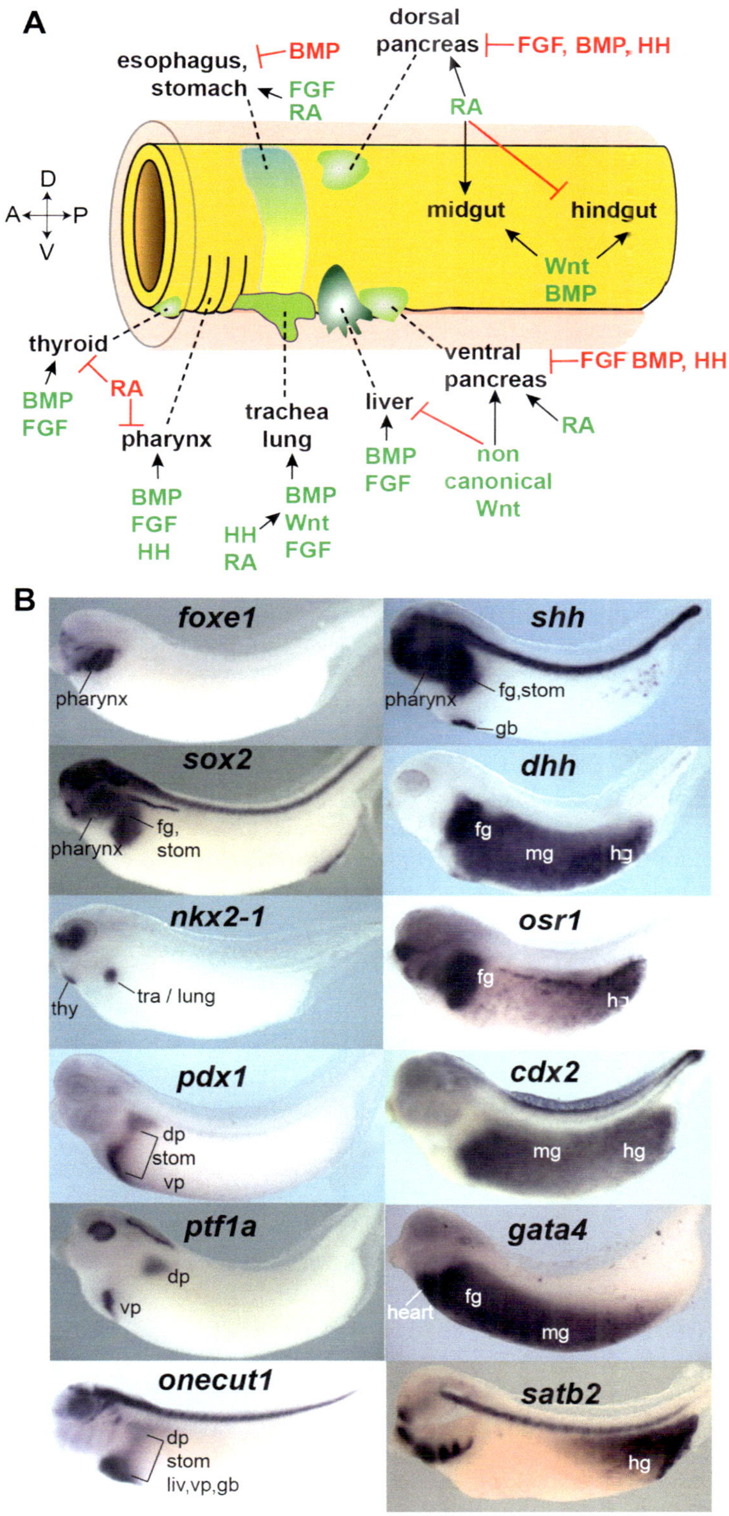

FIGURE 18.5 Combinatorial signals regulating *Xenopus* organ fate induction and expression of selected lineage-promoting transcription factors. (A) Schematic of stage NF35 *Xenopus* gut tube (yellow). Pink shading represents splanchnic mesoderm, the source of most of the indicated signaling molecules (exception of HH, which is produced by the endoderm). Organ primordia indicated in green along the endoderm gut tube. (B) *In situ* hybridization of the indicated genes in stage NF35 *X. laevis* embryos. Expression domains are as follows: *foxe1*, pharyngeal endoderm; *sox2*, pharyngeal, foregut, stomach endoderm; *nkx2–1*, thyroid, lung/trachea endoderm; *pdx1*, dorsal and ventral pancreatic, stomach endoderm; *ptf1a*, dorsal and ventral pancreatic endoderm; *onecut1*, dorsal and ventral pancreatic, stomach, liver, gall bladder endoderm; *shh*, pharyngeal, foregut, stomach, gall bladder endoderm; *dhh*, foregut, midgut, hindgut endoderm; *osr1*, foregut, midgut, hindgut endoderm; *cdx2*, midgut and hindgut endoderm; *gata4*, heart, foregut, midgut endoderm; *satb2*, hindgut endoderm.

section, we discuss how combinatorial mesenchymal signals, some of which are dependent on paracrine cross-talk with the endoderm itself, induce organ lineages.

18.5.1. Pharyngeal Endoderm and Thyroid

The pharyngeal endoderm contributes to the thymus, thyroid, and parathyroid. While the molecular markers of the thymus and parathyroid are conserved between mammals and *Xenopus* (Lee et al. 2013; Manley and Condie 2010), the molecular mechanism of their development has not been extensively investigated in *Xenopus*. We therefore focus on conserved mechanisms of pharyngeal endoderm formation and thyroid induction.

Pharyngeal endoderm cells express TFs such as Foxe1, Tbx1, and Sox2 (Figure 18.5B), and FGF signaling during NF20–NF34 promotes their expression (Rankin et al. 2012; Shifley et al. 2012; Kurmann et al. 2015). BMP signaling has temporally dynamic effects on pharyngeal endoderm development. During neurula stages, high levels of BMP in the foregut inhibit pharyngeal identity and directly repress *tbx1* (Stevens et al. 2017); however, subsequent BMP signaling, from NF20–NF34, is required for pharyngeal fate and thyroid gene expression (Kurmann et al. 2015). Thyroid progenitors, which co-express the TF genes *nkx2–1*, *foxe1*, *hhex*, and *pax2* (*pax2* in *Xenopus* is analogous to *pax8* in mammals), are induced in response to combinatorial FGF and BMP signals in a region of low RA activity (Wang et al. 2011; Kurmann et al. 2015). RA is an evolutionarily ancient patterning signal that defines the posterior boundary of the pharynx in both invertebrates and vertebrates (Kelly and Drysdale 2015). Studies in *Xenopus* have shown that in addition to suppressing *foxe1*+ pharyngeal fate (Rankin et al. 2018), RA acts as regulatory switch between thyroid and lung: exogenous RA is sufficient to drive ectopic, lung-specific surfactant gene expression in the thyroid/pharyngeal domain, simultaneously suppressing thyroid and pharyngeal markers (Wang et al. 2011; Rankin et al. 2018).

18.5.2. Lung, Trachea, and Esophagus

Reciprocal mesoderm-endoderm paracrine signaling between the foregut lateral plate mesoderm and underlying foregut endoderm is necessary for induction of Nkx2–1+ respiratory progenitors. Studies in *Xenopus* helped define a conserved GRN regulating pulmonary induction that involves RA, Hedgehog (HH), Wnt, and BMP signals, as well as the TF Tbx5 (Figure18.6A–F). Mesoderm-produced RA is required for HH ligand expression in the endoderm, and HH ligands then signal back to the mesoderm via Gli TFs, which act together with Tbx5 to promote expression of *wnt2/2b* and *bmp2/4/7* ligands in the ventral foregut mesoderm (Rankin et al. 2016; Steimle et al. 2018). Canonical Wnt/β-Catenin signaling is necessary and sufficient for *nkx2–1* induction in the ventral foregut epithelium (Steimle et al. 2018; Rankin et al. 2016). Ventral BMP signaling restricts expression of the key esophageal TF Sox2 to the dorsal foregut epithelium (Domyan et al. 2011; Rankin et al. 2015). In addition to its repressive effect on Sox2, BMP also plays an additional, yet mechanistically unknown, function in *nkx2–1* induction or early maintenance, as the experimental removal of Sox2 cannot substitute for the role of BMP during either *Xenopus* respiratory induction or during the directed differentiation of human PSCs into respiratory progenitors *in vitro* (Trisno et al. 2018).

After D-V patterning, between NF37–NF44, the foregut separates into a distinct ventral trachea and dorsal esophagus (Rankin et al. 2015; Nasr et al. 2019). In humans, disruptions in this process result in life-threatening congenital defects such as esophageal atresia (a partial absence of the esophagus), tracheoesophageal fistulas, or laryngo-tracheoesophageal clefts, which occur in about 1 out of every 3500 live births (Brosens et al. 2014). Comparative studies in *Xenopus* and mouse have identified the conserved cellular mechanism that control tracheal-esophageal (T-E) morphogenesis (Nasr et al. 2019). These include the formation of a transient epithelial septum at the Sox2/Nkx2–1 dorsal/ventral boundary as well as the resolution of this septum, a process involving endosome-mediated epithelial remodeling and localized extracellular matrix degradation (Figure 18.6G). Additional studies in *Xenopus* and mouse also found a conserved requirement of the homeodomain TF Islet1 in T-E separation (Kim et al. 2019), wherein Islet1 is thought to help maintain *nkx2–1* expression in boundary/midline epithelial cells. Interestingly, in humans, chromosomal deletions at 5q11.2, which encompass the *ISLET1* gene (and others), are found in patients with tracheal agenesis (de Jong et al. 2010).

Experiments in *Xenopus* and mouse indicate that T-E morphogenesis is regulated by HH, although the precise HH/Gli-target genes are unclear. Indeed CRISPR-mediated mutation of *gli3* in *Xenopus* to mimic heterozygous truncating mutations observed in human GLI3 in Pallister-Hall syndrome (OMIM #146510) resulted in defective epithelial remodeling and laryngo-tracheoesophageal clefts, similar to human Pallister-Hall syndrome patients (Nasr et al. 2019). The genetic etiology of trachea-esophageal birth defects is poorly understood, and although genome sequencing of patients identifies many *de novo* variants, determining which mutations are causative remains a major challenge; groups such as the CLEARconsrtium.org are using *Xenopus* CRISPR screens to validate candidate genes and place them into developmental pathways involved in T-E development.

18.5.3. Pancreas and Liver

Xenopus pancreas development has been extensively reviewed, and important insights into evolutionarily conserved mechanisms governing pancreas induction have been made in *Xenopus* (Kelly and Melton 2000; Pearl et al. 2009; Kofent and Spagnoli 2016), including the initial cloning of the master pancreatic TF gene *pdx1* (Wright et al. 1989) (Figure 18.5B). The use of *Xenopus* to study congenital pancreatic disorders and diseases such as diabetes, MODY, and

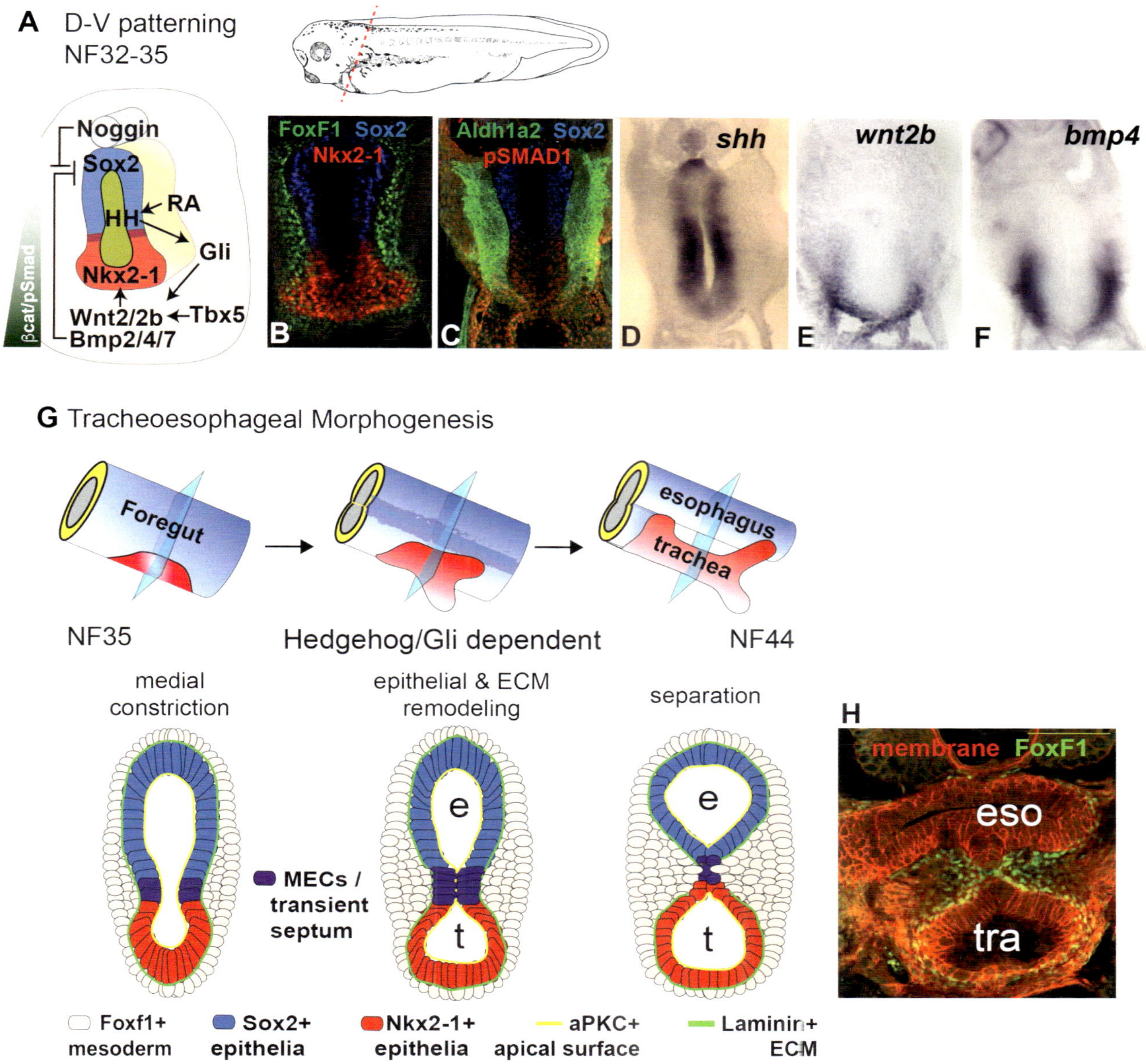

FIGURE 18.6 Regulation of respiratory progenitor induction and tracheoesophageal separation in *Xenopus*. (A) Schematic of a cross-section through the NF32-NF35 *Xenopus* foregut. Bi-directional signaling between the lateral plate mesoderm (tan) and the endoderm (blue, red) involving RA, HH, Wnt, and BMP governs induction of Nkx2–1 respiratory progenitors (red) in the ventral foregut. Notochord indicated in gray, which expresses the BMP antagonist Noggin. A ventral (high) to dorsal (low) gradient of Wnt/β-Catenin and BMP/pSmad1/5/9 activity exists along the foregut dorsal-ventral axis. (B-C) Immunostaining of NF34/35 *X. laevis* foregut sections for the indicated color-coded factors. In B, FoxF1 (green) is expressed in lateral plate mesoderm; Sox2 (blue) in dorsal esophageal progenitors; Nkx2–1 (red) in ventral respiratory progenitors. In C, Aldh1a2 (green) is expressed in lateral plate mesoderm; phosphorylated Smad1/5/9 (pSmad, red) in ventral endoderm and mesoderm. (D-F) *In situ* hybridization of NF34/35 *X. laevis* foregut sections of *shh* (expressed in both dorsal and ventral endoderm) and *wnt2b*, *bmp4* (both expressed in ventral lateral plate mesoderm). (G) Schematic of tracheal-esophageal separation and morphogenesis in *Xenopus*, from stages NF35 to NF44. Key events are diagrammed in the section schematics below the separating foregut tube cartoons. By NF35, medial constriction is progressing; subsequently, from NF36–NF40, a transient epithelial septum (purple) forms wherein midline epithelial cells (MECs) touch and down-regulate their apical membrane surface proteins. Separation of the dorsal esophagus (e) and ventral trachea (t) involves epithelial as well as extracellular matrix (ECM) remodeling and laminin basement membrane breakdown adjacent to the epithelial septum. (H) Immunostaining of NF41 *X. laevis* foregut section, showing Foxf1+ mesoderm cells (green) between the separating dorsal esophagus (e) and ventral trachea (t).

pancreatic hypoplasia/agenesis has also been thoroughly reviewed (Salanga and Horb 2015; Kofent and Spagnoli 2016). We therefore highlight studies in *Xenopus* that have defined mechanisms governing a pancreas versus liver fate choice that occurs in multipotent ventral foregut endoderm cells.

During neurula patterning stages, canonical Wnt/β-Catenin signaling must be suppressed in foregut endoderm to allow for subsequent liver and pancreas induction (McLin et al. 2007; Li et al. 2008). Studies in *Xenopus*, mouse, zebrafish, and human PSC-derived endoderm have all demonstrated that ventral pancreas and liver progenitors arise from multipotent ventral foregut endoderm cells, and prolonged FGF and BMP signaling promotes liver over pancreatic fate (Deutsch et al. 2001; Wandzioch and Zaret 2009; Shifley et al. 2012; Kenny et al. 2012; Loh et al. 2014; Twaroski et al. 2015). BMP signaling must be suppressed in a subset of these foregut progenitors to allow for pancreatic induction. This involves the TALE-family homeodomain protein Tgif2, which is initially expressed in the multipotent foregut progenitors and then selectively maintained in the pancreatic lineage. Tgif2, identified in *Xenopus* as direct target of the TF Gata5 in foregut endoderm, is required for *pdx1* expression, can physically interact with SMAD1, and functionally suppresses BMP signaling to permit pancreas fate (Spagnoli and Brivanlou 2008; Cerdá-Esteban et al. 2017). As RNA-seq and ChIP-seq studies identified Gata5 as a direct BMP/pSMAD1 target (Stevens et al. 2017), a BMP>pSMAD1>Gata5>Tgif2—| BMP negative feedback cascade thus regulates in pancreas induction. It is unclear if Tgif2 is actively repressed from the Gata5-expressing hepatic lineage or if additional signals, such as non-canonical Wnt and RA signaling, both of which also promote pancreatic fate (Stafford et al. 2004; Rodríguez-Seguel et al. 2013), selectively maintain *tgif2* expression in pancreatic progenitors.

Non-canonical Wnt signaling promotes pancreatic fate. *Xenopus* studies demonstrated that Wnt5a/Atf2 signaling can expand the expression domains of the master pancreatic TFs Pdx1 and Ptf1a at the expense of *hhex-positive/nr1h5-positive* liver progenitors (Rodríguez-Seguel et al. 2013). Similar effects of WNT5A and the expression of non-canonical WNT pathway components have been observed during the directed differentiation of both mouse and human PSCs into pancreatic progenitors (Rodríguez-Seguel et al. 2013; Cebola et al. 2015).

RA is an essential, conserved signal necessary for patterning the posterior foregut and for induction of Pdx1-expressing pancreas progenitors (Stafford et al. 2004; Chen et al. 2004). Transcriptional profiling in *Xenopus* has identified direct RA targets, which include the TF *hnf1b* and the Wnt-receptor *frizzled 4* (*fzd4*) (Gere-Becker et al. 2018). Hnf1b is a conserved regulator of liver and pancreatic development in vertebrates (Gere-Becker et al. 2018; Lokmane et al. 2008; Poll et al. 2006; Kotalova et al. 2015) and heterozygous mutations in human *HNF1B* resulting in MODY (OMIM #606391). RNA-seq studies in mouse also identified *Fzd4* expression to be enriched in pancreatic progenitors (Rodríguez-Seguel et al. 2013), but it remains to be tested if the non-canonical Wnt signal that promotes pancreatic fate utilizes Fzd4. Finally, it is still unclear how the FGF, BMP, RA, and non-canonical Wnt pathways interact in an epistatic and combinatorial manner to regulate a pancreatic versus hepatic GRN.

Repression of alternative fate by the master pancreatic TFs Pdx1 and Ptf1a is also important during pancreas induction. Experimental overexpression of *pdx1/ptf1a* messenger RNAs is sufficient to change the fate of *Xenopus* foregut endoderm into pancreatic tissue (Afelik et al. 2006; Jarikji et al. 2007); consistent with these observations, ChIP-seq studies in human pancreatic progenitors revealed that PDX1 binds and represses hepatic genes to ensure pancreatic lineage commitment (Teo et al. 2015; Wang et al. 2018). Additional *Xenopus* studies have shown that the histone methyltransferase Setd7 is necessary for *pdx1* expression and for establishment of active histone marks at pancreatic promoters (Kofent et al. 2016). How Setd7 is specifically recruited to *pdx1* and other pancreatic loci is currently unclear, but an intriguing hypothesis is that FOX TF activity and/or RA signaling, both known to regulate enhancer status, could be involved (Wang et al. 2015; Vinckier et al. 2020).

18.5.4. Stomach

RA and FGF signaling during NF25–NF35 is required for the formation of *sox2*-expressing stomach progenitors, and these signals act in part via inducing expression of the zinc-finger transcriptional repressors Osr1 and Osr2, which restrain BMP signaling (Rankin et al. 2012). Experimental BMP/Smad1 gain-of-function during this developmental window suppresses stomach progenitor induction (Rankin et al. 2012). Conserved mechanisms that drive the curvature of the stomach have also been identified in *Xenopus* (Davis et al. 2017). During stomach morphogenesis from NF34–NF39, the left stomach epithelium becomes polarized and undergoes radial rearrangement; these asymmetries are observed in both mouse and *Xenopus* and are dependent on the left-right patterning genes *foxj1*, *nodal1*, and *pitx2* (Davis et al. 2017).

18.5.5. Intestine

Intestinal fate is promoted by posteriorizing Wnt/BMP/FGF signals, which cooperate to induce and maintain expression of the master intestinal TF Cdx2 (Zorn and Wells 2009; Stevens et al. 2017; Rankin et al. 2018). RA signaling during NF14-NF25 acts as a molecular toggle between *cdx2*-positive/*gata4*-positive midgut and *cdx2*-positive/*satb2*-positive hindgut fate (Figure 18.5B; Rankin et al. 2018; Múnera et al. 2017). Although *Xenopus* has a cloaca and not a colon, the existence of *gata4*-positive midgut and *satb2*-positive distal hindgut progenitors demonstrates conserved molecular pattern of the intestine amongst frog, mouse, and human;

indeed, *SATB2* expression has proven to be a useful early patterning marker to assess successful directed differentiation of human PSCs into colonic organoids (Múnera et al. 2017).

The developmental effects resulting from simultaneous functional depletion of the three *cdx* genes (*cdx1/2/4*) typically present in vertebrate genomes was first determined in *Xenopus*; this work provided evidence that Cdx and Wnt3a function in a positive feedback loop operating in the hindgut (Faas and Isaacs 2009; Keenan et al. 2006). More recent RNA-seq studies of individual and triple Cdx loss-of-function embryos revealed further cross-talk between Cdx and FGF, identifying unappreciated negative feedback loops wherein Cdx restrains *fgf4/8* levels as well as a unique restraint on HH signaling by Cdx4 (Marlétaz et al. 2015). Mutations in human *CDX2* are associated with the congenital defect persistent cloaca (Hsu et al. 2018), and imbalances in posterior WNT, FGF, and HH signaling result in defective hindgut development in mice and humans (Runck et al. 2014). As the GRNs downstream of these signaling pathways and Cdx activity are still only partially understood, the rich resource of Cdx-regulated genes in *Xenopus*, combined with the robust ability to test epistatic relationships among signaling pathways and screen the effect of mutations in potentially causative factors, has the potential for *Xenopus* to provide needed insight into congenital syndromes affecting the intestine.

Extensive work in *Xenopus* has also identified conserved Wnt/PCP-dependent mechanisms regulating intestinal morphogenesis. During NF32-NF46, dynamic changes in intestinal endoderm cell properties, including cell-cell adhesion and microtubule architecture, are regulated by Wnt/PCP signaling mediators (Rho, ROCK, non-muscle Myosin, JNK); loss of function of these factors disrupts endoderm cell intercalation, epithelial morphogenesis, and gut tube elongation (Reed et al. 2009; Dush and Nascone-Yoder 2019). Such cell biological insights into intestinal development are also relevant to cancer, as many parallels exist between tumor pathogenesis and early embryo development. Indeed, *Xenopus* has been a useful model for studying cancer (Hardwick and Philpott 2018; see Chapter 21 of this book).

18.6. CONCLUDING REMARKS

Xenopus continues to be a powerful system to study fundamental mechanisms of digestive and respiratory system organogenesis and disease. As high-throughput genomics, genome editing, proteomics, and pharmacological screening are now routine in *Xenopus*, functional genomics and human disease modeling at a systems level is accelerating. We feel confident that future *Xenopus* research will continue to provide additional significant insights into both normal human organogenesis and congenital malformations.

ACKNOWLEDGMENTS

This work is supported in part by grants P41HD054556, P01HD093363, and R01 DK123092 to AMZ, and we acknowledge Xenbase, the *Xenopus* model organism knowledge base (RRID:SCR_003280), and the National *Xenopus* Resource (RRID:SCR_013731) for their support. Drawings of *Xenopus* embryos are reproduced with permission from Xenbase and Natalya Zahn, licensed under CC BY-NC 4.0. To view a copy of this license, visit http://creativecommons.org/licenses/by-nc/4.0/.

REFERENCES

Afelik, S., Chen, Y., & Pieler, T. (2006). Combined ectopic expression of Pdx1 and Ptf1a/p48 results in the stable conversion of posterior endoderm into endocrine and exocrine pancreatic tissue. *Genes & Development*, *20*(11), 1441–1446.

Afouda, B. A., Nakamura, Y., Shaw, S., Charney, R. M., Paraiso, K. D., Blitz, I. L., Cho, K., & Hoppler, S. (2020). Foxh1/Nodal defines context-specific direct maternal Wnt/β-Catenin target gene regulation in early development. *iScience*, *23*(7), 101314.

Agius, E., Oelgeschläger, M., Wessely, O., Kemp, C., & De Robertis, E. M. (2000). Endodermal Nodal-related signals and mesoderm induction in *Xenopus*. *Development*, *127*(6), 1173–1183.

Arima, K., Shiotsugu, J., Niu, R., Khandpur, R., Martinez, M., Shin, Y., Koide, T., Cho, K. W., Kitayama, A., Ueno, N., Chandraratna, R. A., & Blumberg, B. (2005). Global analysis of RAR-responsive genes in the *Xenopus* neurula using cDNA microarrays. *Developmental Dynamics*, *232*(2), 414–431.

Asashima, M., Nakano, H., Shimada, K., Kinoshita, K., & Ishii, K. (1990). Mesodermal induction in early amphibian embryos by activin A (erythroid differentiation factor). *Roux's archives of developmental biology: the official organ of the EDBO*, *198*(6), 330–335. https://doi.org/10.1007/BF00383771

Blitz, I. L., Paraiso, K. D., Patrushev, I., Chiu, W., Cho, K., & Gilchrist, M. J. (2017). A catalog of *Xenopus tropicalis* transcription factors and their regional expression in the early gastrula stage embryo. *Developmental Biology*, *426*(2), 409–417.

Blythe, S. A., Cha, S. W., Tadjuidje, E., Heasman, J., & Klein, P. S. (2010). Beta-Catenin primes organizer gene expression by recruiting a histone H3 arginine 8 methyltransferase, Prmt2. *Developmental Cell*, *19*(2), 220–231.

Briggs, J. A., Weinreb, C., Wagner, D. E., Megason, S., Peshkin, L., Kirschner, M. W., & Klein, A. M. (2018). The dynamics of gene expression in vertebrate embryogenesis at single-cell resolution. *Science*, *360*(6392), eaar5780.

Brosens, E., Ploeg, M., van Bever, Y., Koopmans, A. E., IJsselstijn, H., Rottier, R. J., Wijnen, R., Tibboel, D., & de Klein, A. (2014). Clinical and etiological heterogeneity in patients with tracheo-esophageal malformations and associated anomalies. *European Journal of Medical Genetics*, *57*(8), 440–452.

Brown, D. D., & Cai, L. (2007). Amphibian metamorphosis. *Developmental Biology*, *306*(1), 20–33.

Buchholz, D. R. (2015). More similar than you think: Frog metamorphosis as a model of human perinatal endocrinology. *Developmental Biology*, *408*(2), 188–195.

Cebola, I., Rodríguez-Seguí, S. A., Cho, C. H., Bessa, J., Rovira, M., Luengo, M., Chhatriwala, M., Berry, A., Ponsa-Cobas, J., Maestro, M. A., Jennings, R. E., Pasquali, L., Morán, I., Castro, N., Hanley, N. A., Gomez-Skarmeta, J. L., Vallier, L., & Ferrer, J. (2015). TEAD and YAP regulate the enhancer network of human embryonic pancreatic progenitors. *Nature Cell Biology*, *17*(5), 615–626.

Cerdá-Esteban, N., Naumann, H., Ruzittu, S., Mah, N., Pongrac, I. M., Cozzitorto, C., Hommel, A., Andrade-Navarro, M. A., Bonifacio, E., & Spagnoli, F. M. (2017). Stepwise reprogramming of liver cells to a pancreas progenitor state by the transcriptional regulator Tgif2. *Nature Communications*, *8*, 14127.

Cha, S. W., Hwang, Y. S., Chae, J. P., Lee, S. Y., Lee, H. S., Daar, I., Park, M. J., & Kim, J. (2004). Inhibition of FGF signaling causes expansion of the endoderm in *Xenopus*. *Biochemical and Biophysical Research Communications*, *315*(1), 100–106.

Cha, S. W., Tadjuidje, E., Tao, Q., Wylie, C., & Heasman, J. (2008a). Wnt5a and Wnt11 interact in a maternal Dkk1-regulated fashion to activate both canonical and non-canonical signaling in *Xenopus* axis formation. *Development*, *135*(22), 3719–3729.

Cha, S. W., Lee, J. W., Hwang, Y. S., Chae, J. P., Park, K. M., Cho, H. J., Kim, D. S., Bae, Y. C., & Park, M. J. (2008b). Spatiotemporal regulation of fibroblast growth factor signal blocking for endoderm formation in *Xenopus laevis*. *Experimental & Molecular Medicine*, *40*(5), 550–557.

Chalmers, A. D., & Slack, J. M. (2000). The *Xenopus* tadpole gut: Fate maps and morphogenetic movements. *Development*, *127*(2), 381–392.

Charney, R. M., Forouzmand, E., Cho, J. S., Cheung, J., Paraiso, K. D., Yasuoka, Y., Takahashi, S., Taira, M., Blitz, I. L., Xie, X., & Cho, K. W. (2017a). Foxh1 occupies cis-regulatory modules prior to dynamic transcription factor interactions controlling the mesendoderm gene program. *Developmental Cell*, *40*(6), 595–607.e4.

Charney, R. M., Paraiso, K. D., Blitz, I. L., & Cho, K. (2017b). A gene regulatory program controlling early *Xenopus* mesendoderm formation: Network conservation and motifs. *Seminars in Cell & Developmental Biology*, *66*, 12–24.

Chen, Y., Pan, F. C., Brandes, N., Afelik, S., Sölter, M., & Pieler, T. (2004). Retinoic acid signaling is essential for pancreas development and promotes endocrine at the expense of exocrine cell differentiation in *Xenopus*. *Developmental Biology*, *271*(1), 144–160.

Colleypriest, B. J., Farrant, J. M., Slack, J. M., & Tosh, D. (2010). The role of Cdx2 in Barrett's metaplasia. *Biochemical Society Transactions*, *38*(2), 364–369.

Dale, L., & Slack, J. M. (1987). Fate map for the 32-cell stage of *Xenopus laevis*. *Development*, *99*(4), 527–551.

Damianitsch, K., Melchert, J., & Pieler, T. (2009). XsFRP5 modulates endodermal organogenesis in *Xenopus laevis*. *Developmental Biology*, *329*(2), 327–337.

D'Amour, K. A., Agulnick, A. D., Eliazer, S., Kelly, O. G., Kroon, E., & Baetge, E. E. (2005). Efficient differentiation of human embryonic stem cells to definitive endoderm. *Nature Biotechnology*, *23*(12), 1534–1541.

Davis, A., Amin, N. M., Johnson, C., Bagley, K., Ghashghaei, H. T., & Nascone-Yoder, N. (2017). Stomach curvature is generated by left-right asymmetric gut morphogenesis. *Development*, *144*(8), 1477–1483.

Deimling, S. J., & Drysdale, T. A. (2011). Fgf is required to regulate anterior-posterior patterning in the *Xenopus* lateral plate mesoderm. *Mechanisms of Development*, *128*(7–10), 327–341.

de Jong, E. M., Douben, H., Eussen, B. H., Felix, J. F., Wessels, M. W., Poddighe, P. J., Nikkels, P. G., de Krijger, R. R., Tibboel, D., & de Klein, A. (2010). 5q11.2 deletion in a patient with tracheal agenesis. *European Journal of Human Genetics: EJHG*, *18*(11), 1265–1268.

De Robertis, E. M., & Kuroda, H. (2004). Dorsal-ventral patterning and neural induction in Xenopus embryos. *Annual Review of Cell and Developmental Biology*, *20*, 285–308.

Deutsch, G., Jung, J., Zheng, M., Lóra, J., & Zaret, K. S. (2001). A bipotential precursor population for pancreas and liver within the embryonic endoderm. *Development*, *128*(6), 871–881.

Domyan, E. T., Ferretti, E., Throckmorton, K., Mishina, Y., Nicolis, S. K., & Sun, X. (2011). Signaling through BMP receptors promotes respiratory identity in the foregut via repression of Sox2. *Development*, *138*(5), 971–981.

Dush, M. K., & Nascone-Yoder, N. M. (2019). Vangl2 coordinates cell rearrangements during gut elongation. *Developmental Dynamics*, *248*(7), 569–582.

Edwards, N. A., & Zorn, A. M. (2021). Modeling endoderm development and disease in *Xenopus*. *Current Topics in Developmental Biology*. Academic Press, ISSN 0070–2153. https://doi.org/10.1016/bs.ctdb.2021.01.001.

Faas, L., & Isaacs, H. V. (2009). Overlapping functions of Cdx1, Cdx2, and Cdx4 in the development of the amphibian *Xenopus tropicalis*. *Developmental Dynamics*, *238*(4), 835–852.

Fuentealba, L. C., Eivers, E., Ikeda, A., Hurtado, C., Kuroda, H., Pera, E. M., & De Robertis, E. M. (2007). Integrating patterning signals: Wnt/GSK3 regulates the duration of the BMP/Smad1 signal. *Cell*, *131*(5), 980–993.

Gentsch, G. E., Owens, N. D., Martin, S. R., Piccinelli, P., Faial, T., Trotter, M. W., Gilchrist, M. J., & Smith, J. C. (2013). In vivo T-box transcription factor profiling reveals joint regulation of embryonic neuromesodermal bipotency. *Cell Reports*, *4*(6), 1185–1196.

Gentsch, G. E., Spruce, T., Owens, N., & Smith, J. C. (2019). Maternal pluripotency factors initiate extensive chromatin remodelling to predefine first response to inductive signals. *Nature Communications*, *10*(1), 4269.

Gere-Becker, M. B., Pommerenke, C., Lingner, T., & Pieler, T. (2018). Retinoic acid-induced expression of *Hnf1b* and *Fzd4* is required for pancreas development in *Xenopus laevis*. *Development*, *145*(12), dev161372.

Grzymkowski, J., Wyatt, B., & Nascone-Yoder, N. (2020). The twists and turns of left-right asymmetric gut morphogenesis. *Development*, *147*(19), dev187583.

Han, L., Zhang, Z., Wang, H. et al. (2020). Novel *MNX1* mutations and genotype-phenotype analysis of patients with Currarino syndrome. *Orphanet Journal of Rare Diseases*, *15*, 155.

Hardwick, L., & Philpott, A. (2018). *Xenopus* models of cancer: Expanding the oncologist's toolbox. *Frontiers in Physiology*, *9*, 1660.

Heasman, J., Wylie, C. C., Hausen, P., & Smith, J. C. (1984). Fates and states of determination of single vegetal pole blastomeres of *X. laevis*. *Cell*, *37*(1), 185–194.

Hikasa, H., Ezan, J., Itoh, K., Li, X., Klymkowsky, M. W., & Sokol, S. Y. (2010). Regulation of TCF3 by Wnt-dependent phosphorylation during vertebrate axis specification. *Developmental Cell*, *19*(4), 521–532.

Hoppler, S., & Moon, R. T. (1998). BMP-2/-4 and Wnt-8 cooperatively pattern the *Xenopus* mesoderm. *Mechanisms of Development*, *71*(1–2), 119–129.

Horb, M. E., & Slack, J. M. (2001). Endoderm specification and differentiation in *Xenopus* embryos. *Developmental Biology*, *236*(2), 330–343.

Hsu, J., So, M., Tang, C., Karim, A., Porsch, R. M., Wong, C., Yu, M., Yeung, F., Xia, H., Zhang, R., Cherny, S. S., Chung, P., Wong, K., Sham, P. C., Ngo, N. D., Li, M., Tam, P., Lui, V., & Garcia-Barcelo, M. M. (2018). De novo mutations in caudal type homeo box transcription factor 2 (CDX2) in patients with persistent cloaca. *Human Molecular Genetics*, *27*(2), 351–358.

Hudson, C., Clements, D., Friday, R. V., Stott, D., & Woodland, H. R. (1997). Xsox17alpha and -beta mediate endoderm formation in *Xenopus*. *Cell*, *91*(3), 397–405.

Jarikji, Z. H., Vanamala, S., Beck, C. W., Wright, C. V., Leach, S. D., & Horb, M. E. (2007). Differential ability of Ptf1a and Ptf1a-VP16 to convert stomach, duodenum and liver to pancreas. *Developmental Biology*, *304*(2), 786–799.

Karaulanov, E., Knöchel, W., & Niehrs, C. (2004). Transcriptional regulation of BMP4 synexpression in transgenic *Xenopus*. *The EMBO Journal*, *23*(4), 844–856.

Keenan, I. D., Sharrard, R. M., & Isaacs, H. V. (2006). FGF signal transduction and the regulation of Cdx gene expression. *Developmental Biology*, *299*(2), 478–488.

Kelly, G. M., & Drysdale, T. A. (2015). Retinoic acid and the development of the endoderm. *Journal of Developmental Biology*, *3*(2), 25–56.

Kelly, O. G., & Melton, D. A. (2000). Development of the pancreas in *Xenopus laevis*. *Developmental Dynamics*, *218*(4), 615–627.

Kenny, A. P., Rankin, S. A., Allbee, A. W., Prewitt, A. R., Zhang, Z., Tabangin, M. E., Shifley, E. T., Louza, M. P., & Zorn, A. M. (2012). Sizzled-tolloid interactions maintain foregut progenitors by regulating fibronectin-dependent BMP signaling. *Developmental Cell*, *23*(2), 292–304.

Kim, E., Jiang, M., Huang, H., Zhang, Y., Tjota, N., Gao, X., Robert, J., Gilmore, N., Gan, L., & Que, J. (2019). Isl1 regulation of Nkx2.1 in the early foregut epithelium is required for trachea-esophageal separation and lung lobation. *Developmental Cell*, *51*(6), 675–683.e4.

Kjolby, R., Truchado-Garcia, M., Iruvanti, S., & Harland, R. M. (2019). Integration of Wnt and FGF signaling in the *Xenopus* gastrula at TCF and Ets binding sites shows the importance of short-range repression by TCF in patterning the marginal zone. *Development*, *146*(15), dev179580.

Kofent, J., & Spagnoli, F. M. (2016). *Xenopus* as a model system for studying pancreatic development and diabetes. *Seminars in Cell & Developmental Biology*, *51*, 106–116.

Kofent, J., Zhang, J., & Spagnoli, F. M. (2016). The histone methyltransferase Setd7 promotes pancreatic progenitor identity. *Development*, *143*(19), 3573–3581.

Kotalova, R., Dusatkova, P., Cinek, O., Dusatkova, L., Dedic, T., Seeman, T., Lebl, J., & Pruhova, S. (2015). Hepatic phenotypes of HNF1B gene mutations: A case of neonatal cholestasis requiring portoenterostomy and literature review. *World Journal of gastroenterology*, *21*(8), 2550–2557.

Kubo, A., Shinozaki, K., Shannon, J. M., Kouskoff, V., Kennedy, M., Woo, S., Fehling, H. J., & Keller, G. (2004). Development of definitive endoderm from embryonic stem cells in culture. *Development*, *131*, 1651–1662.

Kurmann, A. A., Serra, M., Hawkins, F., Rankin, S. A., Mori, M., Astapova, I., Ullas, S., Lin, S., Bilodeau, M., Rossant, J., Jean, J. C., Ikonomou, L., Deterding, R. R., Shannon, J. M., Zorn, A. M., Hollenberg, A. N., & Kotton, D. N. (2015). Regeneration of thyroid function by transplantation of differentiated pluripotent stem cells. *Cell Stem Cell*, *17*(5), 527–542.

Lee, Y. H., Williams, A., Hong, C. S., You, Y., Senoo, M., & Saint-Jeannet, J. P. (2013). Early development of the thymus in *Xenopus laevis*. *Developmental Dynamics*, *242*(2), 164–178.

Li, Y., Rankin, S. A., Sinner, D., Kenny, A. P., Krieg, P. A., & Zorn, A. M. (2008). Sfrp5 coordinates foregut specification and morphogenesis by antagonizing both canonical and noncanonical Wnt11 signaling. *Genes & Development*, *22*(21), 3050–3063.

Loh, K. M., Ang, L. T., Zhang, J., Kumar, V., Ang, J., Auyeong, J. Q., Lee, K. L., Choo, S. H., Lim, C. Y., Nichane, M., Tan, J., Noghabi, M. S., Azzola, L., Ng, E. S., Durruthy-Durruthy, J., Sebastiano, V., Poellinger, L., Elefanty, A. G., Stanley, E. G., Chen, Q., Prabhakar, S., Weissman, I. L., & Lim, B. (2014). Efficient endoderm induction from human pluripotent stem cells by logically directing signals controlling lineage bifurcations. *Cell Stem Cell*, *14*(2), 237–252.

Lokmane, L., Haumaitre, C., Garcia-Villalba, P., Anselme, I., Schneider-Maunoury, S., & Cereghini, S. (2008). Crucial role of vHNF1 in vertebrate hepatic specification. *Development*, *135*(16), 2777–2786.

Manley, N. R., & Condie, B. G. (2010). Transcriptional regulation of thymus organogenesis and thymic epithelial cell differentiation. *Progress in Molecular Biology and Translational Science*, *92*, 103–120.

Marlétaz, F., Maeso, I., Faas, L., Isaacs, H. V., & Holland, P. W. (2015). Cdx ParaHox genes acquired distinct developmental roles after gene duplication in vertebrate evolution. *BMC Biology*, *13*, 56.

McCauley, H. A., & Wells, J. M. (2017). Pluripotent stem cell-derived organoids: Using principles of developmental biology to grow human tissues in a dish. *Development*, *144*(6), 958–962.

McLin, V. A., Rankin, S. A., & Zorn, A. M. (2007). Repression of Wnt/beta-catenin signaling in the anterior endoderm is essential for liver and pancreas development. *Development*, *134*(12), 2207–2217.

Moody, S. A. (1987). Fates of the blastomeres of the 32-cell-stage *Xenopus* embryo. *Developmental Biology*, *122*(2), 300–319.

Mukherjee, S., Chaturvedi, P., Rankin, S. A., Fish, M. B., Wlizla, M., Paraiso, K. D., MacDonald, M., Chen, X., Weirauch, M. T., Blitz, I. L., Cho, K. W., & Zorn, A. M. (2020). Sox17 and β-catenin co-occupy Wnt-responsive enhancers to govern the endoderm gene regulatory network. *eLife*, *9*, e58029.

Múnera, J. O., Sundaram, N., Rankin, S. A., Hill, D., Watson, C., Mahe, M., Vallance, J. E., Shroyer, N. F., Sinagoga, K. L., Zarzoso-Lacoste, A., Hudson, J. R., Howell, J. C., Chatuvedi, P., Spence, J. R., Shannon, J. M., Zorn, A. M., Helmrath, M. A., & Wells, J. M. (2017). Differentiation of human pluripotent stem cells into colonic organoids via transient activation of BMP signaling. *Cell Stem Cell*, *21*(1), 51–64.e6.

Nascone, N., & Mercola, M. (1995). An inductive role for the endoderm in *Xenopus* cardiogenesis. *Development*, *121*, 515–523.

Nasr, T., Mancini, P., Rankin, S. A., Edwards, N. A., Agricola, Z. N., Kenny, A. P. Kinney, J. L., Daniels, K., Vardanyan, J., Han, L., Trisno, S. L., Cha, S. W., Wells, J. M., Kofron, M. J., & Zorn, A. M. (2019). Endosome-mediated epithelial remodeling downstream of Hedgehog-Gli is required for tracheoesophageal separation. *Developmental Cell*, *51*(6), 665–674.e6.

Nenni, M. J., Fisher, M. E., James-Zorn, C., Pells, T. J., Ponferrada, V., Chu, S., Fortriede, J. D., Burns, K. A., Wang, Y., Lotay, V. S., Wang, D. Z., Segerdell, E., Chaturvedi, P., Karimi, K., Vize, P. D., & Zorn, A. M. (2019). Xenbase: Facilitating the use of *Xenopus* to model human disease. *Frontiers in Physiology*, *10*, 154.

Nieuwkoop, P. D., & Faber, J. (1994). *Normal table of Xenopus laevis (Daudin)*. Garland Publishing Inc, New York. ISBN 0-8153-1896-0.

Onichtchouk, D., Gawantka, V., Dosch, R., Delius, H., Hirschfeld, K., Blumenstock, C., & Niehrs, C. (1996). The Xvent-2 homeobox gene is part of the BMP-4 signalling pathway controlling dorsoventral patterning of *Xenopus* mesoderm. *Development*, *122*(10), 3045–3053.

Paraiso, K. D., Blitz, I. L., Coley, M., Cheung, J., Sudou, N., Taira, M., & Cho, K. (2019). Endodermal maternal transcription factors establish super-enhancers during zygotic genome activation. *Cell Reports*, *27*(10), 2962.e5–2977.e5.

Paraiso, K. D., Cho, J. S., Yong, J., & Cho, K. (2020). Early *Xenopus* gene regulatory programs, chromatin states, and the role of maternal transcription factors. *Current Topics in Developmental Biology*, *139*, 35–60.

Pearl, E. J., Bilogan, C. K., Mukhi, S., Brown, D. D., & Horb, M. E. (2009). *Xenopus* pancreas development. *Developmental Dynamics*, *238*(6), 1271–1286.

Pearl, E. J., Jarikji, Z., & Horb, M. E. (2011). Functional analysis of Rfx6 and mutant variants associated with neonatal diabetes. *Developmental Biology*, *351*(1), 135–145.

Poll, A. V., Pierreux, C. E., Lokmane, L., Haumaitre, C., Achouri, Y., Jacquemin, P., Rousseau, G. G., Cereghini, S., & Lemaigre, F. P. (2006). A vHNF1/TCF2-HNF6 cascade regulates the transcription factor network that controls generation of pancreatic precursor cells. *Diabetes*, *55*(1), 61–69.

Rankin, S. A., Gallas, A. L., Neto, A., Gómez-Skarmeta, J. L., & Zorn, A. M. (2012). Suppression of Bmp4 signaling by the zinc-finger repressors Osr1 and Osr2 is required for Wnt/β-catenin-mediated lung specification in *Xenopus*. *Development*, *139*(16), 3010–3020.

Rankin, S. A., Han, L., McCracken, K. W., Kenny, A. P., Anglin, C. T., Grigg, E. A., Crawford, C. M., Wells, J. M., Shannon, J. M., & Zorn, A. M. (2016). A retinoic acid-Hedgehog cascade coordinates mesoderm-inducing signals and endoderm competence during lung specification. *Cell Reports*, *16*(1), 66–78.

Rankin, S. A., Kormish, J., Kofron, M., Jegga, A., & Zorn, A. M. (2011). A gene regulatory network controlling hhex transcription in the anterior endoderm of the organizer. *Developmental Biology*, *351*(2), 297–310.

Rankin, S. A., McCracken, K. W., Luedeke, D. M., Han, L., Wells, J. M., Shannon, J. M., & Zorn, A. M. (2018). Timing is everything: Reiterative Wnt, BMP and RA signaling regulate developmental competence during endoderm organogenesis. *Developmental Biology*, *434*(1), 121–132.

Rankin, S. A., Thi Tran, H., Wlizla, M., Mancini, P., Shifley, E. T., Bloor, S. D., Han, L., Vleminckx, K., Wert, S. E., & Zorn, A. M. (2015). A molecular atlas of *Xenopus* respiratory system development. *Developmental Dynamics*, *244*(1), 69–85.

Reed, R. A., Womble, M. A., Dush, M. K., Tull, R. R., Bloom, S. K., Morckel, A. R., Devlin, E. W., & Nascone-Yoder, N. M. (2009). Morphogenesis of the primitive gut tube is generated by Rho/ROCK/myosin II-mediated endoderm rearrangements. *Developmental Dynamics*, *238*(12), 3111–3125.

Reich, S., & Weinstein, D. C. (2019). Repression of inappropriate gene expression in the vertebrate embryonic ectoderm. *Genes*, *10*(11), 895.

Rex, M., Hilton, E., & Old, R. (2002). Multiple interactions between maternally-activated signalling pathways control *Xenopus* nodal-related genes. *The International Journal of Developmental Biology*, *46*(2), 217–226.

Rodríguez-Seguel, E., Mah, N., Naumann, H., Pongrac, I. M., Cerdá-Esteban, N., Fontaine, J. F., Wang, Y., Chen, W., Andrade-Navarro, M. A., & Spagnoli, F. M. (2013). Mutually exclusive signaling signatures define the hepatic and pancreatic progenitor cell lineage divergence. *Genes & Development*, *27*(17), 1932–1946.

Rose, C. S., & James, B. (2013). Plasticity of lung development in the amphibian, *Xenopus laevis*. *Biology Open*, *2*(12), 1324–1335.

Runck, L. A., Method, A., Bischoff, A., Levitt, M., Peña, A., Collins, M. H., Gupta, A., Shanmukhappa, S., Wells, J. M., & Guasch, G. (2014). Defining the molecular pathologies in cloaca malformation: similarities between mouse and human. *Disease Models & Mechanisms*, *7*(4), 483–493.

Salanga, M. C., & Horb, M. E. (2015). *Xenopus* as a model for GI/pancreas disease. *Current Pathobiology Reports*, *3*(2), 137–145.

Sander, V., Reversade, B., & De Robertis, E. M. (2007). The opposing homeobox genes Goosecoid and Vent1/2 self-regulate *Xenopus* patterning. *The EMBO Journal*, *26*(12), 2955–2965.

Shifley, E. T., Kenny, A. P., Rankin, S. A., & Zorn, A. M. (2012). Prolonged FGF signaling is necessary for lung and liver induction in *Xenopus*. *BMC Developmental Biology*, *12*, 27.

Shiotsugu, J., Katsuyama, Y., Arima, K., Baxter, A., Koide, T., Song, J., Chandraratna, R. A., & Blumberg, B. (2004). Multiple points of interaction between retinoic acid and FGF signaling during embryonic axis formation. *Development*, *131*(11), 2653–2667.

Sinner, D., Kirilenko, P., Rankin, S., Wei, E., Howard, L., Kofron, M., Heasman, J., Woodland, H. R., & Zorn, A. M. (2006). Global analysis of the transcriptional network controlling *Xenopus* endoderm formation. *Development*, *133*(10), 1955–1966.

Smith, J. C., Price, B. M., Van Nimmen, K., & Huylebroeck, D. (1990). Identification of a potent *Xenopus* mesoderm-inducing factor as a homologue of activin A. *Nature*, *345*(6277), 729–731.

Smith, S. B., Qu, H. Q., Taleb, N., Kishimoto, N. Y., Scheel, D. W., Lu, Y., Patch, A. M., Grabs, R., Wang, J., Lynn, F. C., Miyatsuka, T., Mitchell, J., Seerke, R., Désir, J., Vanden Eijnden, S., Abramowicz, M., Kacet, N., Weill, J., Renard, M. E., Gentile, M., Hansen, I., Dewar, K., Hattersley, A. T., Wang, R., Wilson, M. E., Johnson, J. D., Polychronakos, C., & German, M. S. (2010). Rfx6 directs islet formation and insulin production in mice and humans. *Nature*, *463*(7282), 775–780.

Spagnoli, F. M., & Brivanlou, A. H. (2008). The Gata5 target, TGIF2, defines the pancreatic region by modulating BMP signals within the endoderm. *Development*, *135*(3), 451–461.

Stafford, D., Hornbruch, A., Mueller, P. R., & Prince, V. E. (2004). A conserved role for retinoid signaling in vertebrate pancreas development. *Development Genes and Evolution*, *214*(9), 432–441.

Steimle, J. D., Rankin, S. A., Slagle, C. E., Bekeny, J., Rydeen, A. B., Chan, S. S., Kweon, J., Yang, X. H., Ikegami, K., Nadadur, R. D., Rowton, M., Hoffmann, A. D., Lazarevic, S., Thomas, W., Boyle Anderson, E., Horb, M. E., Luna-Zurita, L., Ho, R. K., Kyba, M., Jensen, B., Zorn, A. M., Conlon, F. L., & Moskowitz, I. P. (2018). Evolutionarily conserved *Tbx5-Wnt2/2b* pathway orchestrates cardiopulmonary development. *Proceedings of the National Academy of Sciences of the United States of America*, *115*(45), E10615–E10624.

Stevens, M. L., Chaturvedi, P., Rankin, S. A., Macdonald, M., Jagannathan, S., Yukawa, M., Barski, A., & Zorn, A. M. (2017). Genomic integration of Wnt/β-catenin and BMP/Smad1 signaling coordinates foregut and hindgut transcriptional programs. *Development*, *144*(7), 1283–1295.

Tao, Q., Yokota, C., Puck, H., Kofron, M., Birsoy, B., Yan, D., Asashima, M., Wylie, C. C., Lin, X., & Heasman, J. (2005). Maternal wnt11 activates the canonical wnt signaling pathway required for axis formation in *Xenopus* embryos. *Cell*, *120*(6), 857–871.

Teo, A. K., Tsuneyoshi, N., Hoon, S., Tan, E. K., Stanton, L. W., Wright, C. V., & Dunn, N. R. (2015). PDX1 binds and represses hepatic genes to ensure robust pancreatic commitment in differentiating human embryonic stem cells. *Stem Cell Reports*, *4*(4), 578–590.

Tran, H. T., Sekkali, B., Van Imschoot, G., Janssens, S., & Vleminckx, K. (2010). Wnt/beta-catenin signaling is involved in the induction and maintenance of primitive hematopoiesis in the vertebrate embryo. *Proceedings of the National Academy of Sciences of the United States of America*, *107*(37), 16160–16165.

Trisno, S. L., Philo, K., McCracken, K. W., Catá, E. M., Ruiz-Torres, S., Rankin, S. A., Han, L., Nasr, T., Chaturvedi, P., Rothenberg, M. E., Mandegar, M. A., Wells, S. I., Zorn, A. M., & Wells, J. M. (2018). Esophageal organoids from human pluripotent stem cells delineate Sox2 functions during esophageal specification. *Cell Stem Cell*, *23*(4), 501–515. e7.

Twaroski, K., Mallanna, S. K., Jing, R., DiFurio, F., Urick, A., & Duncan, S. A. (2015). FGF2 mediates hepatic progenitor cell formation during human pluripotent stem cell differentiation by inducing the WNT antagonist NKD1. *Genes & Development*, *29*(23), 2463–2474.

Vinckier, N. K., Patel, N. A., Geusz, R. J., Wang, A., Wang, J., Matta, I., Harrington, A. R., Wortham, M., Wetton, N., Wang, J., Jhala, U. S., Rosenfeld, M. G., Benner, C. W., Shih, H. P., & Sander, M. (2020). LSD1-mediated enhancer silencing attenuates retinoic acid signalling during pancreatic endocrine cell development. *Nature Communications*, *11*(1), 2082.

Wandzioch, E., & Zaret, K. S. (2009). Dynamic signaling network for the specification of embryonic pancreas and liver progenitors. *Science*, *324*(5935), 1707–1710.

Wang, A., Yue, F., Li, Y., Xie, R., Harper, T., Patel, N. A., Muth, K., Palmer, J., Qiu, Y., Wang, J., Lam, D. K., Raum, J. C., Stoffers, D. A., Ren, B., & Sander, M. (2015). Epigenetic priming of enhancers predicts developmental competence of hESC-derived endodermal lineage intermediates. *Cell Stem Cell*, *16*(4), 386–399.

Wang, J. H., Deimling, S. J., D'Alessandro, N. E., Zhao, L., Possmayer, F., & Drysdale, T. A. (2011). Retinoic acid is a key regulatory switch determining the difference between lung and thyroid fates in *Xenopus laevis*. *BMC Developmental Biology*, *11*, 75.

Wang, X., Sterr, M., Burtscher, I., Chen, S., Hieronimus, A., Machicao, F., Staiger, H., Häring, H. U., Lederer, G., Meitinger, T., Cernilogar, F. M., Schotta, G., Irmler, M., Beckers, J., Hrabě de Angelis, M., Ray, M., Wright, C., Bakhti, M., & Lickert, H. (2018). Genome-wide analysis of PDX1 target genes in human pancreatic progenitors. *Molecular Metabolism*, *9*, 57–68.

Wright, C. V., Schnegelsberg, P., De Robertis, E. M. (1989, April). XlHbox 8: A novel *Xenopus* homeo protein restricted to a narrow band of endoderm. *Development*, *105*(4), 787–794.

Wylie, C. C., Snape, A., Heasman, J., & Smith, J. C. (1987). Vegetal pole cells and commitment to form endoderm in *Xenopus laevis*. *Developmental Biology*, *119*(2), 496–502.

Xanthos, J. B., Kofron, M., Wylie, C., & Heasman, J. (2001). Maternal VegT is the initiator of a molecular network specifying endoderm in Xenopus laevis. *Development*, *128*(2), 167–180.

Xanthos, J. B., Kofron, M., Tao, Q., Schaible, K., Wylie, C., & Heasman, J. (2002). The roles of three signaling pathways in the formation and function of the Spemann organizer. *Development*, *129*(17), 4027–4043.

Yasuoka, Y., Suzuki, Y. Takahashi, S., Someya, H., Sudou, N., Haramoto, Y., Cho, K. W., Asashima, M., Sugano, S., & Taira, M. (2014). Occupancy of tissue-specific cis-regulatory modules by Otx2 and TLE/Groucho for embryonic head specification. *Nature Communications*, *5*, 4322.

Yasuoka, Y., Tando, Y., Kubokawa, K., & Taira, M. (2019). Evolution of cis-regulatory modules for the head organizer gene *goosecoid* in chordates: Comparisons between *Branchiostoma* and *Xenopus*. *Zoological Letters*, *5*, 27.

Zhang, Z., Rankin, S. A., & Zorn, A. M. (2013a). Different thresholds of Wnt-Frizzled 7 signaling coordinate proliferation, morphogenesis and fate of endoderm progenitor cells. *Developmental Biology*, *378*(1), 1–12.

Zhang, T., Guo, X., & Chen, Y. (2013b). Retinoic acid-activated Ndrg1a represses Wnt/β-catenin signaling to allow *Xenopus* pancreas, oesophagus, stomach, and duodenum specification. *PLoS One*, *8*(5), e65058.

Zorn, A. M., Butler, K., & Gurdon, J. B. (1999). Anterior endomesoderm specification in *Xenopus* by Wnt/beta-catenin and TGF-beta signalling pathways. *Developmental Biology*, *209*(2), 282–297

Zorn, A. M., & Wells, J. M. (2009). Vertebrate endoderm development and organ formation. *Annual Review of Cell and Developmental Biology*, *25*, 221–251.

19 Functional Neurobiology in *Xenopus* Provides Insights into Health and Disease

Clayton Gordy, Michael Forsthofer, Parthena Soupiadou, Suzan Özugur, and Hans Straka

CONTENTS

19.1. Historical Background ... 277
19.2. Past Observations in Neurobiology Using *Xenopus* ... 278
19.3. Present Status of the Field—Experimental Approaches ... 279
19.4. Functional Neurobiology Using *In Vitro* Preparations ... 282
19.5. Future Directions ... 285
Acknowledgments ... 286
References ... 286

19.1. HISTORICAL BACKGROUND

Frogs were among the first species used in biomedical studies, including experimental approaches related to impairments of brain function. Early discoveries focused on the identification of general functional principles of the nervous system, which eventually led towards insights into the origin, progression, and potential treatment of neurological diseases in humans (e.g. Galvani, 1791; Aubert, 1881; Ewald, 1892). For a long time, these studies were almost exclusively conducted on various species of adult *ranid* frogs, such as *Rana esculenta*, *temporaria*, and *catesbeiana*, or on different species of toads, such as *bufo* (Ewald, 1892; Rubin, 1936; Guardabassi, 1955). *Xenopus laevis* as a model species appeared only between the 1930s and 1950s, in part driven by the increasing popularity of developmental biology (Blum and Ott, 2018). The first documented use of *Xenopus* in neurobiological research commenced with exploring the ability to regenerate tadpole tails (Jurand et al., 1954) and by exploring neurosecretory and regulatory mechanisms, including the pituitary gland (e.g. Charles, 1931; Dodd and Landgrebe, 1953). The subsequent worldwide radiation of *Xenopus* as a model organism for neuroscience benefitted considerably from the prior use of *ranid* frog species. These latter frogs provided a large body of morpho-physiological information on the peripheral (PNS) and central nervous system (CNS), facilitated by the qualitative and quantitative phylogenetic conservation of many features of the brain and spinal cord in anurans (Llinás and Precht, 1976). The gradual but constant transition from ranid frogs to *Xenopus* as an anuran model was also driven by the progressive classification of increasing numbers of frogs as endangered species over the past 50 years, impacting the use of the former as experimental animals (Stuart et al., 2004). The reduced availability of many ranid frog and other anuran species has in the meantime been more than compensated for by *Xenopus* species, with the predominance of *laevis* and *tropicalis* (Pearl et al., 2012). These species became highly suitable for experimental settings, as they can be made available in abundance at any time of the year due to their easy maintenance in captivity (Nieuwkoop and Faber, 1994) across all developmental stages, ranging from embryos to tadpoles, and adults (McNamara et al., 2018).

The systematically increasing use of *Xenopus* vastly expanded anuran experimental research into diverse biological fields (Cline and Kelley, 2012). Such an expansion is credited to the large spectrum of modern technical, molecular, and genetic innovations that have emerged over the past decades, which can be easily applied to *Xenopus* (e.g. Pratt and Khakhalin, 2013). These latter advancements have facilitated the probing and manipulating of behaviorally relevant neuronal circuits with considerable implications for the understanding of basic neuro-computational and neuro-developmental aspects (Constantine-Paton and Cline, 1998; Ruthazer and Cline, 2004), as well as clinically relevant patho-physiological conditions (Lambert and Straka, 2012). While the overall sensory-motor capacity and behavioral performance of *Xenopus* adhere to general vertebrate principles, the particular eco-physiology, such as the permanent aquatic lifestyle, makes this species excellently suited to study specific sensory/motor adaptations. As one of the major hallmark features, and in contrast to other aquatic model species such as zebrafish, *Xenopus* allows studying the morpho-physiological transition from an animal with a fish-like swim style into a tetrapod with limb-based propulsion with direct relevance for the functional organization of quadrupedal locomotion in terrestrial vertebrates (Combes

DOI: 10.1201/9781003050230-22

et al., 2004). The remarkable shapeshifting capacity during metamorphosis, even though inferior and irreversible compared to the extraterrestrial species of the changelings (Carey, 1999), is accompanied by an equally extensive neuronal restructuration and adjustment of motor proficiency. This reorganization thus offers unique insight into the extent of adaptive plasticity of the vertebrate nervous system, making *Xenopus* an outstanding model to study major neurobiological questions. *Xenopus* is therefore well suited to identify functionally relevant aspects of neurobiological systems, which is fundamental for the identification and understanding of vertebrate diseases, including those present in humans.

19.2. PAST OBSERVATIONS IN NEUROBIOLOGY USING *XENOPUS*

Most neuroscientific discoveries in frogs resulted from complementary work on *ranid* species and *Xenopus*. From early on, these studies focused on basic principles, such as anatomical connectivity, signaling patterns, neuronal plasticity, and behavior, which are summarized in the comprehensive masterpiece of Nieuwenhuys et al. (1998). Collectively, these studies provided the foundational understanding for functional aspects of the anuran nervous system (Figure 19.1A1–5), though often without particular focus on topics related to health and disease. Novel tract tracing and intracellular recording techniques along with manipulations of embryonic tissue allowed the exploration of the organization and capacity of inducing plastic adaptations in a variety of sensory systems. While not necessarily exhaustive, this section highlights major discoveries of neurobiological principles obtained in anurans, with emphasis on the erstwhile available state-of-the-art methodologies and outcomes.

Early research on cranial sensory organs and associated central targets produced multiple ground-breaking findings on topics of neuroanatomy and connectivity, largely based on cut-and-paste tissue experiments (Fritzsch et al., 2019). Hallmark studies included the discovery of visual field representations in the central nervous system (Sperry, 1944, 1956). Combining optic nerve transections with behavioral assessment in adult frogs, a remarkable reconnection of optic nerve fibers with proper areas of the optic tectum was observed, which permitted restoration of visual capabilities. This culminated in sets of experiments in which eyes were rotated by 180°, effectively flipping and mirroring the retina. At that time, most behavioral experiments were limited to simple categorical evaluations, such as observing the direction of a guided behavior in response to an external stimulus. Eye-rotated animals were presented with prey in a specific part of their visual field. This elicited capture responses which were oriented in the opposite direction to the location of the prey, confirming that the visual field had been mirrored along with the retina (Sperry, 1944). Although initial experiments were performed in urodeles, the outcome was validated in frogs (Sperry, 1944). Combining such targeted perturbations and behavioral assays therefore demonstrated the presence of peripheral subdivisions and tectal topographies based on the identity of retinal subregions (Sperry, 1956) (Figure 19.1A1). Further exploration of visual system connectivity continued with the milestone experiments by Constantine-Paton and Law (1978), which combined embryonic manipulations and neuronal tract/nerve tracings. Embryonic transplantation of an optic anlage onto *Rana* embryos generated three-eyed animals and allowed the evaluation of the connectivity pattern and neural targeting strategy of the supernumerary eye in the midbrain. Available techniques at that time, such as tracing with radioactive amino acids, revealed that the third eye generated discrete alternating columns innervated by either one of the other eyes along the tectal surface (Constantine-Paton and Law, 1978). By expanding the findings on principles of visual connectivity specification, these results not only uplifted frogs into the realm of animal models with broad neurobiological relevance, but also demonstrated the suitability of these animals for determining key concepts of development and function of sensory systems (Constantine-Paton and Law, 1978; Fritzsch et al., 2019).

The early contributions of *Xenopus* were no less prolific than those obtained in *ranid* frogs and further illustrated the ease of assessing sensory system networks and function in these species. The simplicity of embryonic manipulations in frogs, particularly in *Xenopus laevis*, allowed generating animals with all-nasal or all-temporal retinas by grafting together halves of eye primordia in embryos before the initial innervation of the tectum (Gaze et al., 1963). With the advent of robust and reliable electrophysiological recordings, it was possible to probe single-cell neuronal activity of tectal neurons in adult frogs after metamorphosis, providing an objective functional assessment of sensory inputs into the midbrain. Visual field representation across the tectal surface in such animals revealed that each half-retina was mapped onto the tectum in a mirror-like fashion (Gaze et al., 1963), leading to the conclusion that each half-retina reorganizes itself into a small full retina, with respective nasal and temporal identities (Meyer and Sperry, 1976). The biological implications of these results, particularly those obtained in *Xenopus*, emphasized the remarkable amount of plasticity and reorganizational capacity of developmental processes as opposed to regeneration in adult animals (Meyer and Sperry, 1976). Moreover, these studies also showed the analytical power of electrophysiological recordings in the quest to decipher central representations of the sensory periphery.

Combining early embryonic manipulations and tract tracing in *Xenopus* also helped exploring other sensory and motor systems as well as axonal pathfinding mechanisms, which are at the origin of the ontogenetic assembly of circuits. Accordingly, eye primordia grafted onto the dorsal trunk (Giorgi and van der Loos, 1978) demonstrated correct development of corresponding connections through the spinal cord by visualizing retinal ganglion cell neurites in the brain. Despite the caudally displaced origin of optic nerve fibers, the grafted eye innervated the correct tectal target site, provided that both innate eyes were ablated beforehand (Giorgi and van der Loos, 1978). This method was further refined by employing xenografts from different *Xenopus* species and different sensory tissue origins (Koo and Graziadei, 1995). Eye primordia in *Xenopus borealis* were successfully replaced with olfactory primordia of *Xenopus laevis*, demonstrating that grafted olfactory tissue, identified by fluorescent chromosome staining, innervated the brain at the level of the missing optic nerve (Koo and Graziadei, 1995). This replacement approach was complemented by target-specific ablations, which, in combination with visualization of axonal pathways, revealed the extent of *Xenopus* CNS plasticity for rewiring within brain regions. Specifically, removal of the otic placode caused an expansion of adjacent cranial nerve projections into areas of the hindbrain that would normally receive inner ear afferent input (Fritzsch, 1990) (Figure 19.1A3).

Embryonic manipulation-based approaches to assess neuronal circuit formation during *Xenopus* development were paralleled by more descriptive approaches, excessively exploiting horseradish-peroxidase placement and tracing techniques. The vast amount of resultant anatomical connection schemes, such as cerebellar projections (Gonzalez et al., 1984; van der Linden et al., 1990) (Figure 19.1A2) or descending pathways to the spinal cord (ten Donkelaar et al., 1981, 1991), formed the basis of the present knowledge about connectivity and circuit roadmaps of the *Xenopus* CNS (Nieuwenhuys et al., 1998). These anatomical descriptions guided corresponding physiological studies, thereby complementing the understanding of how anatomically defined neural circuits acquire and execute neuronal computations based on afferent and efferent connections, although many of these electrophysiological studies were made on *ranid* frogs (Llinás and Precht, 1976). The discovery of neurophysiological principles particularly benefitted from studies on the frog spinal cord (Eccles, 1944). This included fundamental aspects of sensory signal processing as well as the morphological arrangement of spinal motoneurons (Frank and Westerfield, 1982). Initiated by Eccles' use of frogs to study signal propagation and synaptic transmission in spinal motoneurons utilizing ortho- and antidromic electrical stimulation of spinal roots (Eccles, 1944), the spinal cord became for a long time a dominating model for studying motor control principles. Facilitated by their large cell size, single-cell recordings characterized how motoneurons interact with each other (Grinnell, 1966) and how muscle-specific functional neuronal populations innervate and control muscle fibers (Frank and Westerfield, 1982). This ultimately led to formulating hypotheses about physiological principles of limb motion control by the spinal cord in the framework of "force-fields" (Giszter et al., 1993; Tresch et al., 1999).

Meanwhile, physiological studies on *Xenopus* utilized the specific eco-physiology of this species, which in contrast to most other frogs includes a retention of the mechanosensory lateral line system in adults (Shelton, 1970). Following the anatomical description of this sensory system throughout development (Shelton, 1970; Winklbauer, 1989), subsequent functional analyses characterized the properties of neuromast sensory afferents in response to mechanical stimuli (Figure 19.1A4, top) (Kroese et al., 1978). The easy access to this sensory system at all hierarchical neuronal levels, in contrast to other aquatic anamniotes, made *Xenopus* an excellent model to investigate the spatio-temporal resolution of sensory encoding of water motion and the subsequent neuronal representation and computation, as well as the range and performance of induced motor responses such as turning behavior (Elepfandt et al., 1985) (Figure 19.1A4, bottom). However, the assessment of function by either orienting behaviors in adult *Xenopus* or of motoneuron activity during the ontogenetic implementation of spinal circuits (Sillar et al., 1992) (Figure 19.1A5) was at that time limited to qualitative descriptions without the possibility for a more fine-tuned differentiation.

In summary, *Xenopus* and *Rana* greatly contributed to the understanding of how sensory systems are peripherally structured, centrally represented, and involved in general vertebrate behavioral repertoires (Figure 19.1A). Various anuran species individually assisted in establishing the comprehension of neuronal connectivity in the CNS with growing relevance of *Xenopus* for investigating embryonic development. In particular, grafting methods followed by subsequent anatomical and functional assessments became instrumental in determining these principles. However, the lack of appropriate experimental tools often prevented quantification at cellular and sub-cellular levels, and due to limited recording techniques, behavioral analyses were largely qualitative and categorical. Nonetheless, such detailed analyses are required for a more profound understanding of basic neurophysiological concepts and thus also for pathophysiological principles related to diseases.

19.3. PRESENT STATUS OF THE FIELD—EXPERIMENTAL APPROACHES

Recent decades have witnessed a considerable advancement in technical and analytical approaches in neuroscientific research. While many of these innovations were initiated in individual model systems, such as the development of the patch-clamp technique in *Rana* muscle fibers (Neher and Sakmann, 1976), a transfer to other organisms and cell types with respective species-specific adaptations has nonetheless expanded steadily (e.g. Liu et al., 2018). Such methods, however, are only as useful as they are applicable to a particular model system. In *Xenopus*, functional neurobiological studies have profited considerably from characteristics that permit the use of a wide variety of analytical techniques, which allow a sizable depth of probing and experimental manipulations (Figure 19.1B). In the following, these attributes will be featured to highlight the increasing ease with which *Xenopus* is currently used to answer fundamental neurobiological questions, driven to a considerable extent by more modern methodologies and conceptually innovative approaches.

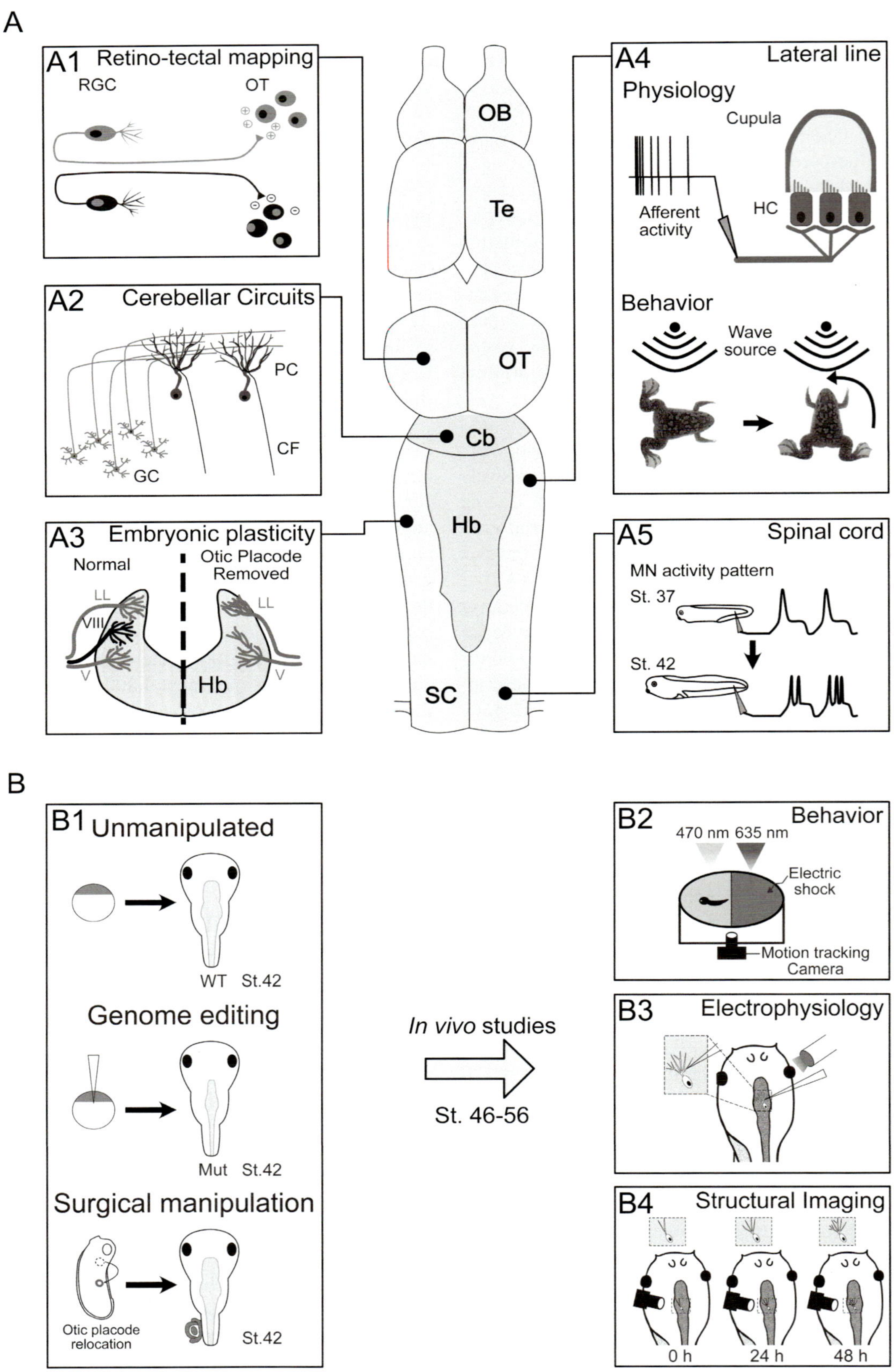

FIGURE 19.1 Examples of *Xenopus* neural systems and analysis techniques demonstrating morpho-physiological aspects of the nervous system. (A) Selection of sensory (A1,3,4) and motor systems (A5), as well as central circuits (A2) to identify developmental and computational principles; optic nerve and eye manipulations combined with behavioral observation revealed rules of retino-tectal

FIGURE 19.1 (Continued)
pathfinding (A1); tract-tracing and electrophysiological recordings outlined cerebellar anatomy and function (A2); embryonic removal of the inner ear demonstrated plasticity-driven reorganization during hindbrain development (A3); the use of mechanosensory stimuli (water waves) characterized the neuronal activity of lateral line nerve afferents (A4, top) and the discrimination capabilities underlying turning behavior (A4, bottom); the ontogeny of motoneuron activity revealed the network maturation dynamics across developmental stages (A5). (B) Examples of modern approaches in neurobiology using *Xenopus*. Embryos can be raised to unmanipulated, wild-type tadpoles (B1, top) or can be altered by genetic (B1, middle) and/or surgical manipulations (B1, bottom) for subsequent functional profiling; quantitative behavioral tracking, such as of avoidance responses, permits measurements of learned behavior (B2); single cells can be recorded in response to sensory stimulation (B3); the transparency of *Xenopus* tadpoles allows live imaging of selected neurons, demonstrating the progressive maturation of dendritic arborizations over consecutive days (B4). Cb, cerebellum; CF, climbing fiber; GC, granule cell; Hb, hindbrain; HC, hair cell; LL, lateral line; MN, motoneuron; Mut, Mutation; OB, olfactory bulb; OT, optic tectum; PC, Purkinje cell; RGC, retinal ganglion cell; SC, spinal cord; St., stage (Nieuwkoop and Faber, 1994); Te, telencephalon; V, 5th cranial nerve (trigeminal nerve); VIII, 8th cranial nerve; WT, wild-type.

Source: Panels A3, A4 (bottom), A4 (top), A5, B3, B2, B4 are based on data by Fritzsch, 1990; Elepfand et al., 1985; Kroese et al., 1978; Sillar et al., 1992; Munz et al., 2014; Blackiston and Levin, 2013; Santos et al., 2018, respectively.

Appreciation of the transformative anuran life cycle, particularly in *Xenopus laevis*, is evident from the early use of this model organism to establish neuroanatomical principles (e.g. Figure 19.1A1,2; Nieuwenhuys et al., 1998) and to assess questions of plasticity in embryonic circuit formation, as mentioned previously (e.g. Figure 19.1A3; Fritzsch, 1990). However, even greater advancements have been achieved by the use of developmental stage-specific analytical instruments and approaches (Figure 19.1B). *Xenopus* embryos in particular are highly suited for a variety of manipulations that influence the formation of neuronal circuits (Figure 19.1B1). Techniques such as cutting and pasting of embryonic tissue have continued their use in modern studies (Figure 19.1B1, lower; Elliott et al., 2015; Blackiston et al., 2017), with a progressive introduction of combinatorial genetic and surgical manipulations (Duncan et al., 2019). Indeed, the ease with which gene expression can be modulated (Figure 19.1B1, middle; Tandon et al., 2017; Naert et al., 2020) or molecular pathways perturbed nowadays (James et al., 2015) offers appealing methods for targeted disruption of morpho-physiological characteristics that classify specific neuronal circuits and brain regions (Figure 19.1B1). These abilities, coupled with the abundance and size of *Xenopus* embryos, permit accessibility for various techniques. Such features are also favorable for reporting on the phenotypic success of a manipulation, such as genetically encoded fluorescently tagged proteins in combination with confocal microscopy (Bestman et al., 2006). Extensive work using these methods has established *Xenopus* as a representative model for neurodevelopmental disorders, where disease modeling is readily approached by targeting clinically relevant genes and pathways (e.g. Lee et al., 2010; Pratt and Khakhalin, 2013; Willsey et al., 2020, 2021). Given the effectiveness of modifications of the early developing nervous system, functional insight emerging from such manipulations have been best provided through the use of tadpoles, where a post-embryonic nervous system can be readily profiled (Figure 19.1B1).

Xenopus tadpoles express many advantageous features which facilitate the evaluation of neurobiological principles. In particular, *in vivo* profiling of tadpole behavior (Figure 19.1B2) forms an active way of approximating neuronal function by assessing the execution of motor commands (Dong et al., 2009; Blackiston and Levin, 2013). Current methodologies in behavioral assessments are strengthened by equipment with high-resolution motion tracking abilities (Viczian and Zuber, 2014), such as high-speed cameras that enable frame-by-frame comparisons of, for example, swimming-related tail undulations (Lambert et al., 2020) and head/body turn directions (Zarei et al., 2017; Gambrill et al., 2018; Hänzi and Straka, 2017). Such measurements are often accompanied by automated tracking algorithms, exploiting, for example, the contrast between the relatively opaque eyes and the translucent body to demonstrate the developmental progression of eye movement performance (Lambert et al., 2020). Behavioral studies in this manner offer the possibility for non-invasive measurements of functional consequences of CNS manipulations, such as those following alteration of gene expression in embryos (Falk et al., 2007; Tandon et al., 2017; Duncan et al., 2019) or tadpoles (Liu and Haas, 2011; Bestman and Cline, 2020) (Figure 19.1B1). Thus, conserved vertebrate genes and molecular pathways can be readily targeted either early in development or after a particular brain region has fully formed and followed by subsequent evaluation of corresponding phenotypes (e.g. Tosa et al., 2015). Such assessments thus interrelate to genes and pathways that are commonly disrupted in disease models.

The ability to profile and quantify tadpole behavior permits assessment of cognitively derived responses (Figure 19.1B2). Learned associations of a visual stimulus with noxious shocking punishments provides a method to determine the extent of recall performance in *Xenopus* tadpoles (Blackiston and Levin, 2013). When limited to visual input exclusively from a singular grafted eye on the trunk, this methodology revealed the ability of ectopic sensory input to be successfully integrated, despite the absence of clearly defined sensory pathways into the brain (Blackiston and Levin, 2013). This latter result is particularly striking given the ability to augment successful integration with pharmacological agents (Blackiston et al., 2017). Such results are promising for insight into regenerative therapies in humans and highlight the successful use of *Xenopus* in this field. While this review has so far made no distinction between tadpole stages, general behavioral tracking is applicable to animals of all age groups, including post-metamorphic froglets (Combes et al., 2004; De Vidts et al., 2019). This further presents as a method to identify differences in kinematic profiles between *Xenopus* stages (Hänzi and Straka, 2017), allowing inferences across developmental periods.

The computational ability of cells and neuronal circuits, while often approximated by behavior, are more precisely investigated by electrophysiological recordings (Figure 19.1B3). *Xenopus* species are suitable for such studies due to the accessibility of the CNS. Patch-clamping and sharp electrode intracellular recordings can be performed at various levels, depending on the feature of interest. The resolution of patch-clamping (Figure 19.1B3) can infer ion channel current dynamics and synaptic properties and thus yield insight into the integrative capacity of cell membranes (Engert et al., 2002; Pratt and Aizenman, 2007). Extracellular single- or multi-unit recordings, while lacking this specificity, nonetheless approximate neuronal activity and are experimentally more rapid, effectively increasing the number of recorded cells. In *Xenopus* research, such techniques were extensively used to characterize neuronal response characteristics in the midbrain to a range of stimuli following mechanosensory lateral line stimulation (Behrend et al., 2006) or to measure receptive fields in the optic tectum (Gaze et al., 1963). On a broader level, low-impedance electrodes record the field potential produced by larger numbers of neurons, providing the activity status across an entire group of neurons (Bibikov and Elepfandt, 2005). Collectively, these methods are ideally suited for linking a particular behavior to the underlying neuronal activity. For example, neuronal correlates of avoidance behavior in *Xenopus* were approximated by examining receptive fields of tectal neurons following visual scene motion (Dong et al., 2009). After quantifying receptive field size, a correlation of the strength of the avoidance response with sharper receptive fields was discovered. Chemical and physical manipulations of the tectum or training of the visual system further explored the dependency of this behavior on receptive field properties. While certain electrode recordings are still unparalleled in their level of detail on the electrical signature of neurons (see also subsequently), non-invasive alternatives, such as calcium-imaging, have now all but replaced extracellular recordings. For these optical methods, *Xenopus* tadpoles excel due to their transparency and small size and have been used in determining computations across entire brain regions. Calcium-imaging of cellular activity is particularly compelling in its capacity to assess larger brain regions (Podgorski et al., 2012), an ability that is often constrained by size and transparency in mammalian models. In line with such optical measurements, time-lapse imaging of neurons and their associated neurites are readily accomplished in *Xenopus* (Figure 19.1B4), and are suitable for imaging of, for example, dendritic growth of tectal cells in real time (Munz et al., 2014). While most physiological studies focus predominantly on visual pathways (e.g. Liu et al., 2018), investigations of other systems have been just as illuminating, including motion-sensitive circuits during locomotion (Lambert et al., 2020) and mechanosensation (Behrend et al., 2006). Given the conserved nature of these sensory and motor control systems, emerging conclusions find themselves applicable to other vertebrates.

While embryos and tadpoles represent the stages most commonly used with modern technical applications, post-metamorphic stages are just as suited to a variety of functional studies. A particularly useful and unique experimental feature of *Xenopus* is the gradual change in body plan during metamorphosis. *Xenopus* possess most advantages present also in zebrafish, such as transparency of the brain and body, small numbers of neurons, and almost unlimited accessibility for experimentation. However, *Xenopus* has additional benefits provided by the metamorphic transition into a quadrupedal vertebrate, with comparable motor control principles as present in mammalian species. This duality in lifestyle makes *Xenopus* ideally suited for studies aiming at the identification of functional features that undergo modification during metamorphosis, such as spinal motor patterns (Combes et al., 2004), which provides standing precedence for the plasticity of spinal network function in vertebrates.

19.4. FUNCTIONAL NEUROBIOLOGY USING *IN VITRO* PREPARATIONS

Disorders of the nervous system often manifest as functional impairments, which can involve a variety of brain regions and modalities (Raichle, 2015). While an *in vivo* approach is usually necessary for systemic assessment of biomedical questions, it is often associated with considerable experimental constraints. Aspects such as level and type of anesthesia, bleeding, accessibility for surgery, and unexpected movements form considerable challenges that can render neurobiological studies on intact animals extremely difficult, if not impossible. *In vitro* models, such as slice preparations (e.g. Götz et al., 2021) or organotypic cell cultures (e.g. Koehler et al., 2017) have historically been employed as suitable alternatives to circumvent these constraints. However, while extremely beneficial for the discovery of many cellular, subcellular, and molecular aspects of brain function, questions concerning systemic neurobiological principles or even behavioral consequences of CNS manipulations or impairments remain out of reach with *in vitro* approaches such as these.

While the benefits of experimental accessibility limit the range of addressable scientific questions for most *in vitro* vertebrate models, amphibians form a remarkable exception. In particular, *Xenopus* species are able to almost entirely bypass such *in vitro* restrictions. This is related to the fact that larvae and adults of *Xenopus* allow the generation of isolated, semi-intact *in vitro* preparations with various levels of surgical reductions of tissues (Straka and Simmers, 2012). Preparations, such as isolated whole heads, maintained in simple frog Ringer solution remain viable for several days and provide access to all CNS circuits between the olfactory bulb and the caudal end of the spinal cord. This plain visibility and accessibility of the CNS in isolated *Xenopus* preparations (Figure 19.2) represent unique advantages that allow unimpaired, µm-precise impalements of specific brain

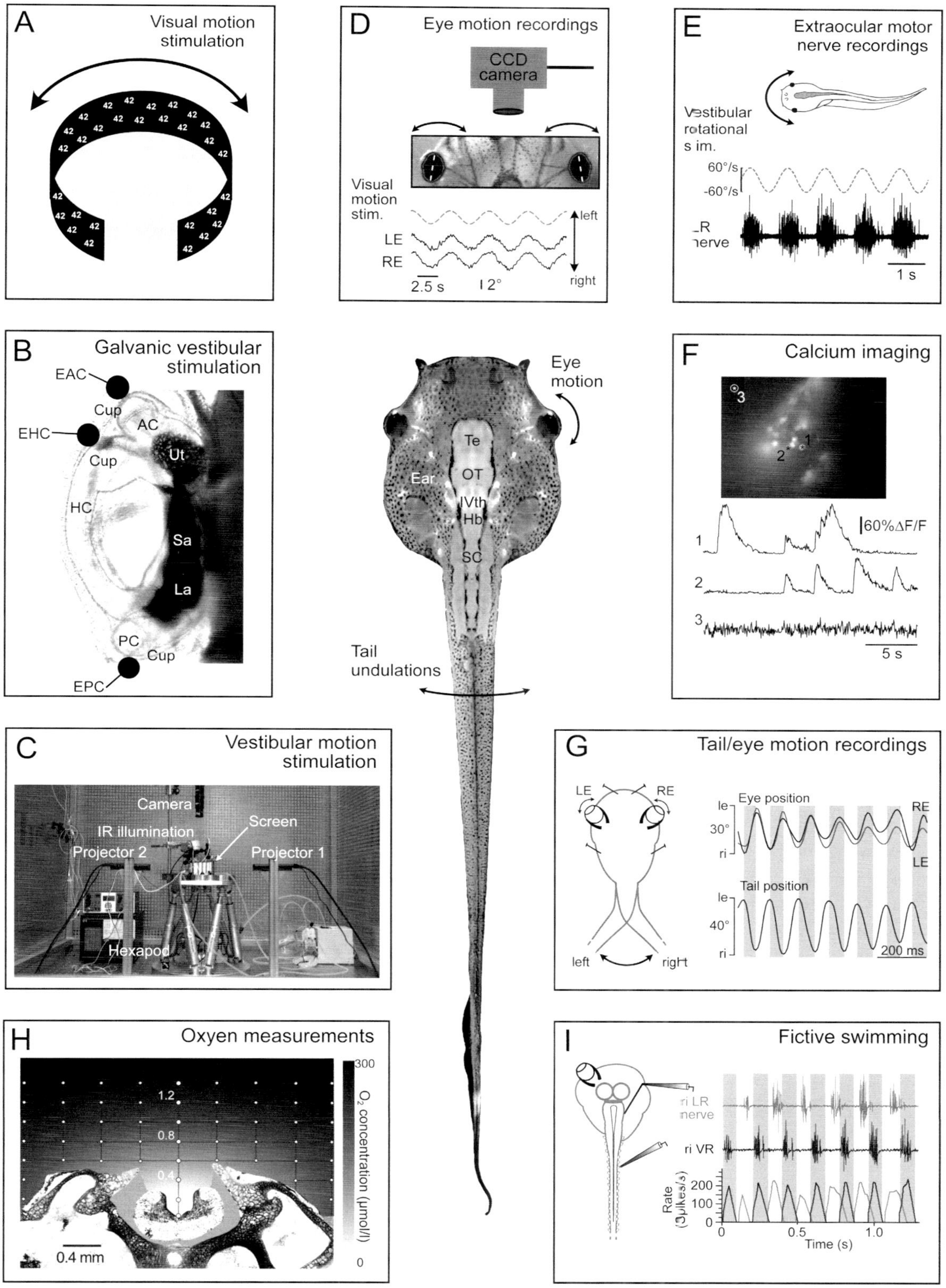

FIGURE 19.2 Selected spectrum of applicable stimulation and recording techniques in isolated *Xenopus in vitro* preparations. (A–C) Visual and vestibular circuits can be activated by large-field visual motion patterns (A), by galvanic vestibular stimulation of inner ear endorgans (B), or by natural motion with a 6D stimulator (Hexapod; C). (D–I) Motor behaviors in such preparations can be recorded as movements of the eyes (D), the tail (G) or appendages (not illustrated) during swimming; following further isolation of the tissue, motor nerve spike discharge can be recorded, representing fictive eye movements (E) or fictive axial- (I) or limb-based swimming; neuronal activity of cells and circuits in the central nervous system can be recorded by calcium-imaging (F) or by evaluating the oxygen

FIGURE 19.2 (Continued)
consumption in specific brain regions in relation to sensory or motor computations (H). Anterior, horizontal, posterior semicircular canal; Cup, cupula; EAC, EHC, EPC, electrode stimulating the AC, HC, PC sensory epithelium; Hb, hindbrain; IR, infrared; IVth, IVth ventricle; LE, RE, left eye, right eye; La, lagena; LR, lateral rectus; OT, optic tectum; Sa, saccule; SC, spinal cord; Te, telencephalon; Ut, utricle; VR, spinal ventral root.

Source: Panels B, D, H, G, I are adapted from Gensberger et al., 2016; Soupiadou et al., 2020; Özugur et al., 2020; Lambert et al., 2012, respectively.

compartments with any type of recording electrode, as well as imaging of intracellular calcium gradients in identifiable neuronal populations at cellular resolution (e.g. Gensberger et al., 2016; Lambert et al., 2018; Özugur et al., 2020). Moreover, the vitality and functionality of isolated amphibian whole-head preparations represents an unprecedented experimental condition, comparable to the circumstances of the isolated brain of the fictional character William living in Ringer solution (Dahl, 1960). The integrity of practically all sense organs such as eyes or inner ear endorgans in these preparations (Lambert et al., 2008; Gravot et al., 2017) allows application of natural stimuli in an experimentally reconstructed sensory environment, such as provided by virtual reality setups (Figure 19.2A). In addition, simple motor behaviors like swim-related tail-oscillations or eye movements (Lambert et al., 2020) can be recorded at high resolution with spatio-temporal characteristics that match those expected from intact animals (Hänzi and Straka, 2017). The virtually unlimited access to the CNS, the use of multi-methodological approaches, and the possibility to study animals at all developmental stages make *Xenopus* a unique *in vitro* animal model (Straka and Simmers, 2012).

The introduction of isolated frog *in vitro* preparations began during the early period of electrophysiological studies. Electrode recordings in the frog cerebellum were among the first to benefit from isolated brains (Hackett, 1972). Such preparations, however, were only used by relatively few scientists over the subsequent decades, mostly for pharmacological studies of synaptic connections (e.g. Cochran et al., 1987) or to demonstrate the spatio-temporal specificity of defined neuronal circuits (Straka and Dieringer, 1993). Beyond functional aspects, neuroanatomical studies benefited from the use of *in vitro* preparations, particularly due to visually guided application of retro- and anterogradely transported neuronal tracers to brain areas that are inaccessible in intact animals (Birinyi et al., 2000; Straka et al., 2001). Isolated anuran whole-brain or head preparations were traditionally used for combined anatomical and physiological exploration of small ensembles of neurons with relatively defined synaptic connections (Straka and Dieringer, 1993). Subsequent use of this approach has since expanded into more distributed circuits and entire systems predominantly using *Xenopus laevis* (Lambert et al., 2008, 2020). This extension was possible due to several characteristic features, which are favorable toward systemic studies under sustained *in vitro* conditions. This includes retention of those parts of the tissue that are relevant for a particular question, such as the brain and the eyes (Gravot et al., 2017), when profiling central circuits that control visuo-motor responses (Figure 19.2A,D). Furthermore, targeted surgical manipulations can be performed with ease at any level of the circuit or system, without drawbacks typical for *in vivo* studies, such as bleeding, pulsations induced by respiration, and blood circulation or consequences of the anesthesia, which are all absent *in vitro* (Soupiadou et al., 2020). Several neuroscientific fields have thus benefited considerably from the use of *in vitro Xenopus* preparations, particularly over the past 15 years, due to the applicability of modern analytical tools. Collectively, these applications have yielded in-depth protocols for generating preparations and applying experimental regimes to answer specific questions on, for example, the retinotectal circuitry (Pratt, 2021), arguably one of the most heavily studied fields in *Xenopus* neuroscience. Olfactory research has also taken advantage of *in vitro* preparations, using the spatial proximity of the olfactory epithelium and olfactory bulb to generate slices and explants that effectively consist only of the sensory epithelia and the first central neuronal processing center. Due to the aquatic lifestyle of *Xenopus*, naturalistic yet highly controlled stimulation of the mucosa is achieved by simply flushing odorants into the Ringer solution, with the olfactory bulb available for any manner of functional imaging or electrophysiological recording (Manzini et al., 2002; Offner et al., 2020).

Another field which utilizes *in vitro Xenopus* preparations aims at deciphering general principles of self-motion processing and has succeeded in discovering a wide spectrum of developmental and computational mechanisms involved in implementing the respective CNS circuitry. Self-motion perception derives from a combination of visual, vestibular, and predictive motor signals, which converge centrally from independent afferent pathways. Dissociation of the relative influences of each modality was accomplished with such preparations by natural stimulation paradigms coupled with measurements of neuronal activity (Lambert et al., 2012; von Uckermann et al., 2016). Motion-sensitive visual input can be activated either in isolation (Figure 19.2A) or in tandem with galvanically evoked activation of vestibular endorgans in the inner ear (Figure 19.2B). Activation of the vestibular system can in addition be achieved by motion of a multi-axis turntable (Figure 19.2C) with or without corresponding visual scene motion (Lambert et al., 2012; Soupiadou et al., 2020). The tractability of sensory stimuli in these paradigms recapitulates *in vivo*-like conditions while simultaneously granting a high degree of flexibility to selectively activate desired sensory pathways. As in *in vivo* studies, traditional methods of profiling neuronal activity range from behavior (Soupiadou et al., 2020), to electrophysiological recordings of selected cranial nerves (Lambert et al., 2008), to calcium-imaging of identified neurons (Gensberger et al., 2016). In the case of neuronal activity during self-motion, appropriate functional proxies include motion of the eyes and tail, which report on the extent and pattern of sensorimotor transformations (Lambert et al., 2020). The kinematics of eye motion can be captured non-invasively (Figure 19.2D) in such preparations with infrared video-recordings. Extracellular recordings of extraocular motor nerves (Figure 19.2E), calcium-imaging of corresponding neurons (Figure 19.2F), and motion recording of tail oscillations (Figure 19.2G) contributed toward the understanding of the spatio-temporal processing of self-motion. A particular advantage of *in vitro* methods in *Xenopus* is in the ability to acquire neuronal activity during fictive behaviors, such as swimming (Figure 19.2I). Fictive *in vitro* behaviors, which can be elicited experimentally at all developmental stages (Combes et al., 2004), allow simultaneous recordings of the activity carried by, for example, spinal or extraocular motor nerves (Figure 19.2G) in the absence of muscle contractions.

In addition to discoveries of computational features of cells and circuits (Gensberger et al., 2016; Dietrich et al., 2017), *in vitro* approaches in *Xenopus* are utilized in studies with a focus on translative neurobiological concepts with direct relevance for human pathologies (I Gusti Bagus et al., 2019; Soupiadou et al., 2020). Such studies often make use of targeted manipulations, either in the embryo (Gordy et al., 2018) or tadpole (Lambert et al., 2009), and are followed by functional profiling to gauge neuronal response mechanisms, which are often the result of conserved plasticity processes. Embryonic addition of ectopic inner ears, for example, challenged the classical understanding of developmental processes by incorporating additional sensory signals from a novel source and served to identify features which drive circuit formation in the brainstem (Gordy et al., 2018). Such studies have therapeutic consequences for aging populations or those with inner ear impairments. Similar perturbations of inner ear sensory organs, such as targeted lesions of the eighth cranial nerve or the entire ensemble of inner ear endorgans in *Xenopus* tadpoles, revealed an adaptive plasticity with considerable developmental consequences, which recapitulate features commonly observed in patients with vestibular impairments (Branoner et al., 2016; Lambert et al., 2009; Soupiadou et al., 2020). More clinically relevant approaches, such as drug application (I Gusti Bagus et al., 2019), imitated biological principles resulting from therapeutic targeting. Furthermore, *in vitro Xenopus* preparations have helped to identify cellular substrates and response patterns of neurons following galvanic vestibular stimulation, a common diagnostic tool used in the clinic (Gensberger et al., 2016). This latter finding is particularly relevant given the need to reclassify the outcome of clinical tests with respect to the interpretation of underlying vestibular pathologies.

In vitro methodologies in *Xenopus laevis* offer additional benefits beyond insights into functional neurobiology. *In vivo* studies usually require institutional animal protocol approvals for the use of *Xenopus* older than stage 46, a developmental time point at which the intracellular yolk has been consumed and self-feeding commences (Nieuwkoop and Faber, 1994). In contrast, euthanasia prior to the isolation of the tissue to obtain an *in vitro* preparation renders the legal requirements and compliance documentation considerably simpler and usually puts such preparations at the same legal level as the generation of mammalian brain slice preparations. This is useful for researchers opting to study the nervous system in older tadpoles and adults. In addition, longitudinal studies are approachable when using *Xenopus in vitro* preparations, given the period of days that they remain functionally viable in Ringer solution (Lambert et al., 2008). Collectively, *in vitro* approaches employing isolated brain/body *Xenopus laevis* preparations at any developmental stage can be used to study neurobiological questions and offer a wide variety of methods which can supplement *in vivo* approaches. This approach is therefore beneficial for a considerably larger number of members of the *Xenopus* scientific community. The experimental and administrative simplicity renders normally invasive technical approaches readily feasible and brings scientific questions that are only addressable in developmentally advanced stages into reach for meaningful answers.

19.5. FUTURE DIRECTIONS

Xenopus as a model system has considerably advanced our understanding of biological principles of the nervous system. As technical and theoretical innovations have ushered in an era of neurobiological research with declining numbers of model systems, *Xenopus* continues to advance in stride. This is largely due to its suitability for many fields of research rather than for a single, specialized application. The comprehensive toolkit available in these animals, which includes tractable genetic and molecular manipulations as well as analytical techniques, allows a sizable depth of neuronal assessments. Innovative next steps in neuroscientific discoveries will continue to capitalize on these existing methods, particularly those which influence functional characteristics in both developing and mature neuronal circuits. These approaches offer valuable insight into defined neuronal disease phenotypes, such as those which manifest with computational impairments. Such cases often arise as the result of congenital miswiring, after CNS injury, or due to age-related deteriorations. The latter is highly relevant in modern neurobiological practice given the increase in age-related impairments in elderly populations. Continued manipulations in this manner will advance our current understanding of basic biological principles of the nervous system both during states of health and disease. More provocatively, the use of *Xenopus* offers possible exploration into avenues which are only just emerging. Neurobiological implications of spaceflight, a steadily evolving enterprise, are of particular interest due to the influence of microgravity on sensory signal processing and the retention of navigational skills. Such research could benefit from *Xenopus* due to the variety of ways to manipulate microgravity while simultaneously permitting analytical profiling of all sensorimotor circuits. In addition, the ease of neuronal accessibility in *Xenopus* can also be used to explore metabolic processes of neurons, shedding light on energetic demands during computational tasks. In summary, *Xenopus* constitutes a well-suited experimental system for neuroscientific practice and will continue to emerge as a source for a constant proliferation of our knowledge about the nervous system.

ACKNOWLEDGMENTS

The authors acknowledge financial support from the Research Training Group 2175 and the Collaborative Research Center 870 of the German Science Foundation (CRC 870, RTG 2175).

REFERENCES

Aubert H. 1881. Über den Einfluß der Temperatur auf die Kohlensäureausscheidung und die Lebensfähigkeit der Frösche in sauerstoffloser Luft. *Pflügers Arch Ges Physiol* 26:293–323.

Behrend O., Branoner F., Zhivkov Z., Ziehm U. 2006. Neural responses to water surface waves in the midbrain of the aquatic predator *Xenopus laevis*. *Eur J Neurosci* 23:729–744.

Bestman J.E., Cline H.T. 2020. Morpholino studies in *Xenopus* brain development. *Methods Mol Biol* 2047:377–395.

Bestman J.E., Ewald R.C., Chiu S.L., Cline H.T. 2006. In vivo single-cell electroporation for transfer of DNA and macromolecules. *Nat Protoc* 1:1267–1272.

Bibikov N.G., Elepfandt A. 2005. Auditory evoked potentials from medulla and midbrain in the clawed frog, *Xenopus laevis*. *Hear Res 204*:29–36.

Birinyi A., Straka H., Matesz C., Dieringer N. 2000. Location of dye-coupled second order and of efferent vestibular neurons labeled from individual semicircular canal or otolith organs in the frog. *Brain Res* 921:44–59.

Blackiston D.J., Levin M. 2013. Ectopic eyes outside the head in *Xenopus* tadpoles provide sensory data for light-mediated learning. *J Exp Biol* 216:1031–1040.

Blackiston D.J., Vien K., Levin M. 2017. Serotonergic stimulation induces nerve growth and promotes visual learning via posterior eye grafts in a vertebrate model of induced sensory plasticity. *NPJ Regen Med* 2:8.

Blum M., Ott T. 2018. *Xenopus*: An undervalued model organism to study and model human genetic disease. *Cells Tissues Organs* 205:303–313.

Branoner F., Chagnaud B.P., Straka H. 2016. Ontogenetic development of vestibulo-ocular reflexes in amphibians. *Front Neural Circuits* 10:91.

Carey D. 1999. *The Dominion Wars: Book 2: Call to Arms*. New York: Pocket Books.

Charles E. 1931. Metabolic changes associated with pigmentary effector activity and pituitary removal in *Xenopus laevis*. I. Respiratory exchange. *Proc Roy Soc* 107:486–503.

Cline H.T., Kelly D. 2012. *Xenopus* as an experimental system for developmental neuroscience: introduction to a special issue. *Dev Neurobiol* 72:463–464.

Cochran S.L., Kasik P., Precht W. 1987. Pharmacological aspects of excitatory synaptic transmission to second-order vestibular neurons in the frog. *Synapse* 1:102–123.

Combes D., Merrywest S.D., Simmers J., Sillar K.T. 2004. Developmental segregation of spinal networks driving axial- and hindlimb-based locomotion in metamorphosing *Xenopus laevis*. *J Physiol* 559:17–24.

Constantine-Paton M., Cline H.T. 1998. LTP and activity-dependent synaptogenesis: the more alike they are, the more different they become. *Curr Opin Neurobiol* 8:139–148.

Constantine-Paton M., Law M.I. 1978. Eye-specific termination bands in tecta of three-eyed frogs. *Science* 202: 639–641.

Dahl, R. 1960. William and Mary. In *Kiss Kiss*. New York: Alfred A. Knopf.

De Vidts S., Méndez-Olivos E., Palacios M., Larraín J., Mery D. 2019. Characterization of spinal cord damage based on automatic video analysis of froglet swimming. *Biol Open*, 8(12).

Dietrich H., Glasauer S., Straka H. 2017. Functional organization of vestibulo-ocular responses in abducens motoneurons. *J Neurosci* 37:4032–4045.

Dodd J.M., Landgrebe F.W. 1953. Assay of thyroid-stimulating hormone, thyroid and thyroid-like substances on *Xenopus* tadpoles. *Nature* 172:121–122.

Dong W., Lee R.H., Xu H., Yang S., Pratt K.G., Cao V., Song Y.K., Nurmikko A., Aizenman C.D. 2009. Visual avoidance in *Xenopus* tadpoles is correlated with the maturation of visual responses in the optic tectum. *J Neurophysiol* 101:803–815.

Duncan J.S., Fritzsch B., Houston D.W., Ketchum E.M., Kersigo J., Deans M.R., Elliott K.L. 2019. Topologically correct central projections of tetrapod inner ear afferents require Fzd3. *Sci Rep* 9:10298.

Eccles J.C. 1944. Synaptic transmission in the spinal cord. *Nature* 153:432.

Elepfandt A., Seiler B., Aicher B. 1985. Water wave frequency discrimination in the clawed frog, *Xenopus laevis*. *J Comp Physiol A* 157:255–261.

Elliott K.L., Houston D.W., Fritzsch B. 2015 Sensory afferent segregation in three-eared frogs resemble the dominance columns observed in three-eyed frogs. *Sci Rep* 5:8338.

Engert F., Tao H.W., Zhang L.I., Poo M.M. 2002. Moving visual stimuli rapidly induce direction sensitivity of developing tectal neurons. *Nature* 419:470–475.

Ewald J.R. 1892. *Physiologische Untersuchungen über das Endorgan des N. Oktavus*. Wiesbaden: Bergmann.

Falk J., Drinjakovic J., Leung K.M., Dwivedy A., Regan A.G., Piper M., Holt C.E. 2007. Electroporation of cDNA/morpholinos to targeted areas of embryonic CNS in *Xenopus*. *BMC Dev Biol* 7:107.

Frank E., Westerfield M. 1982. Synaptic organization of sensory and motor neurones innervating triceps brachii muscles in the bullfrog. *J Physiol* 324:479–494.

Fritzsch B. 1990. Experimental reorganization in the alar plate of the clawed toad, *Xenopus laevis*. I. Quantitative and qualitative effects of embryonic otocyst extirpation. *Dev Brain Res* 51:113–122.

Fritzsch B., Elliott K.L., Pavlinkova G. 2019. Primary sensory map formations reflect unique needs and molecular cues specific to each sensory system. *F1000Res* 8:F1000 Faculty Rev-345.

Gambrill A.C., Faulkner R.L., Cline H.T. 2018. Direct intertectal inputs are an integral component of the bilateral sensorimotor circuit for behavior in *Xenopus* tadpoles. *J Neurophysiol*. 119:1947–1961.

Galvani L. 1791. *De viribus electricitatis in motu musculari commentarius*. Bologna. Translated by Foley, M.G. 1953. *Luigi Galvani: Commentary on the Effects of Electricity on Muscular Motion*. Norwalk CT: Burndy Library.

Gaze R.M., Jacobson M., Szekely C. 1963. The retino-tectal projection in *Xenopus* with compound eyes. *J Physiol* 165:484–499.

Gensberger K.D., Kaufmann A.K., Dietrich H., Branoner F., Banchi R., Chagnaud B.P., Straka H. 2016. Galvanic vestibular stimulation: Cellular substrates and response patterns of neurons in the vestibulo-ocular network. *J Neurosci* 36:9097–9110.

Giorgi P.P., van der Loos H. 1978. Axons from eyes grafted in Xenopus can grow into the spinal cord and reach the optic tectum. *Nature* 275:746–748.

Giszter S.F., Mussa-Ivaldi F.A., Bizzi E. 1993. Convergent force fields organized in the frog's spinal cord. *J Neurosci* 13:467–491.

Gonzalez A., ten Donkelaar H.J., de Boer-Van Huizen R. 1984. Cerebellar connections in *Xenopus laevis*: An HRP study. *Anat Embryol* 169:167–176.

Gordy C., Straka H., Houston D.W., Fritzsch B., Elliott K.L. 2018. Transplantation of ears provides insights into inner ear afferent pathfinding properties. *Dev Neurobiol* 78:1064–1080.

Götz S., Bribian A., López-Mascaraque L., Götz M., Grothe B., Kunz L. 2021. Heterogeneity of astrocytes: Electrophysiological properties of juxtavascular astrocytes before and after brain injury. *Glia* 69:346–361.

Gravot C.M., Knorr A.G., Glasauer S., Straka H. 2017. It's not all black and white: Visual scene parameters influence optokinetic reflex performance in *Xenopus laevis* tadpoles. *J Exp Biol* 220:4213–4224.

Grinnell A.D. 1966. A study of the interaction between motoneurones in the frog spinal cord. *J Physiol* 182:612–648.

Guardabassi A. 1955. Experiments on exchange of ocular anlagen in *Bufo vulgaris* embryos; connections of the nerves with brain centers. *Arch Ital Anat Embriol* 60:488–514.

Hackett J.T. 1972. Electrophysiological properties of neuronal circuits in the frog cerebellum in vitro. *Brain Res* 48:385–389.

Hänzi S., Straka H. 2017. Developmental changes in head movement kinematics during swimming in *Xenopus laevis* tadpoles. *J Exp Biol* 220:227–236.

I Gusti Bagus M., Gordy C., Sanchez-Gonzalez R., Strupp M., Straka H. 2019. Impact of 4-aminopyridine on vestibulo-ocular reflex performance. *J Neurol* 266(Suppl 1):93–100.

James E.J., Gu J., Ramirez-Vizcarrondo C.M., Hasan M., Truszkowski T.L., Tan Y., Oupravanh P.M., Khakhalin A.S., Aizenman C.D. 2015. Valproate-induced neurodevelopmental deficits in *Xenopus laevis* tadpoles. *J Neurosci* 35: 3218–3229.

Jurand A., Maron K., Olekiewicz M., Skowron S. 1954. Effect of excision of the telencephalon on regeneration rate in the tail in *Xenopus laevis* tadpoles. *Folia Biol (Krakow)* 2:3–29.

Koehler K.R., Nie J., Longworth-Mills E., Liu X.P., Lee J., Holt J.R., Hashino E. 2017. Generation of inner ear organoids containing functional hair cells from human pluripotent stem cells. *Nat Biotechnol* 35:583–589.

Koo H., Graziadei P.P. 1995. Cell migration from the transplanted olfactory placode in *Xenopus*. *Anat Embryol* 191:171–181.

Kroese A.B.A., van der Zalm J.M., van der Bercken J. 1978. Frequency response of the lateral-line organ of *Xenopus laevis*. *Pflügers Arch* 375:167–175.

Lambert F.M., Bacqué-Cazenave J., Le Seach A., Arama J., Courtand G., Tagliabue M., Eskiizmirliler S., Straka H., Beraneck M. 2020. Stabilization of gaze during early Xenopus development by swimming-related utricular signals. *Curr Biol* 30:1–8.

Lambert F.M., Beck J.C., Baker R., Straka H. 2008. Semicircular canal size determines the developmental onset of angular vestibuloocular reflexes in larval *Xenopus*. *J Neurosci* 28:8086–8095.

Lambert F.M., Cardoit L., Courty E., Bougerol M., Thoby-Brisson M., Simmers J., Tostivint H., Le Ray D. 2018. Functional limb muscle innervation prior to cholinergic transmitter specification during early metamorphosis in *Xenopus*. *Elife* 7:e30693.

Lambert F.M., Combes D., Simmers J., Straka H. 2012. Gaze stabilization by efference copy signaling without sensory feedback during vertebrate locomotion. *Curr Biol* 22: 1649–1658.

Lambert F.M., Malinvaud D., Glaunès J., Bergot C., Straka H., Vidal P.P. 2009. Vestibular asymmetry as the cause of idiopathic scoliosis: A possible answer from *Xenopus*. *J Neurosci* 29:12477–12483.

Lambert F.M., Straka H. 2012. The frog vestibular system as a model for lesion-induced plasticity: Basic neural principles and implications for posture control. *Front Neurol* 3:42.

Lee R.H., Mills E.A., Schwartz N., Bell M.R., Deeg K.E., Ruthazer E.S., Marsh-Armstrong N., Aizenman C.D. 2010. Neurodevelopmental effects of chronic exposure to elevated levels of pro-inflammatory cytokines in a developing visual system. *Neural Dev* 5:2.

Liu X.F., Haas K. 2011. Single-cell electroporation in *Xenopus*. *Cold Spring Harb Protoc* 9:pdb.top065607.

Liu Z., Donnelly K.B., Pratt K.G. 2018. Preparations and protocols for whole cell patch clamp recording of *Xenopus laevis* tectal neurons. *J Vis Exp* 133:57465.

Llinás R., Precht W. 1976. *Frog Neurobiology*. Berlin, Heidelberg, New York: Springer-Verlag.

Manzini I., Rössler W., Schild D. 2002. cAMP-independent responses of olfactory neurons in *Xenopus laevis* tadpoles and their projection onto olfactory bulb neurons. *J Physiol* 545:475–484.

McNamara S., Wlizla M., Horb M.E. 2018. Husbandry, general care, and transportation of *Xenopus laevis* and *Xenopus tropicalis*. *Methods Mol Biol* 1865:1–17.

Meyer R.L., Sperry R.W. 1976. Retinotectal specificity: Chemoaffinity theory. *Studies on the Development of Behavior and the Nervous System*. G. Gottlieb, Elsevier. 3:111–149.

Munz M., Gobert D., Schohl A., Poquérusse J., Podgorski K., Spratt P., Ruthazer E.S. 2014. Rapid Hebbian axonal remodeling mediated by visual stimulation. *Science* 344:904–909.

Naert T., Tulkens D., Edwards N.A., Carron M., Shaidani N.I., Wlizla M., Boel A., Demuynck S., Horb M.E., Coucke P., Willaert A., Zorn A.M., Vleminckx K. 2020. Maximizing CRISPR/Cas9 phenotype penetrance applying predictive modeling of editing outcomes in Xenopus and zebrafish embryos. *Sci Rep* 10:14662.

Neher E., Sakmann B. 1976. Single-channel currents recorded from membrane of denervated frog muscle fibres. *Nature* 260:799–802.

Nieuwenhuys R., ten Donkelaar H.J., Nicholson C. 1998. *The Central Nervous System of Vertebrates*. Berlin, Heidelberg: Springer.

Nieuwkoop P.D., Faber J. 1994. *Normal Table of Xenopus laevis (Daudin): A Systematical and Chronological Survey of the Development from the Fertilized Egg Till the End of Metamorphosis*. New York: Garland Publisher.

Offner T., Daume D., Weiss L., Hassenklöver T., Manzini I. 2020. Whole-brain calcium imaging in larval *Xenopus*. *Cold Spring Harb Protoc* 12:pdb.prot106815.

Özugur S., Kunz L., Straka H. 2020. Relationship between oxygen consumption and neuronal activity in a defined neural circuit. *BMC Biol*. 18:76.

Pearl E.J., Grainger R.M., Guille M., Horb M.E. 2012. Development of *Xenopus* resource centers: The national *Xenopus* resource and the European *Xenopus* resource center. *Genesis* 50: 155–163.

Podgorski K., Dunfield D., Haas K. 2012. Functional clustering drives encoding improvement in a developing brain network during awake visual learning. *PLoS Biol* 10:e1001236.

Pratt K.G. 2021. Electrophysiological approaches to studying normal and abnormal retinotectal circuit development in the *Xenopus* tadpole. *Cold Spring Harb Protoc* 2021(2):pdb.prot106898.

Pratt K.G., Aizenman C.D. 2007. Homeostatic regulation of intrinsic excitability and synaptic transmission in a developing visual circuit. *J Neurosci* 27:8268–8277.

Pratt K.G., Khakhalin A.S. 2013. Modeling human neurodevelopmental disorders in the *Xenopus* tadpole: From mechanisms to therapeutic targets. *Dis Model Mech* 6:1057–1065.

Raichle M.E. 2015. The brain's default mode network. *Annu Rev Neurosci* 8:433–447.

Rubin M.A. 1936. Adaptation of cutaneous tactile receptors: IV. Electrolyte content of frog skin. *J Gen Physiol* 19:935–937.

Ruthazer E.S., Cline H.T. 2004. Insights into activity-dependent map formation from the retinotectal system: A middle-of-the-brain perspective. *J Neurobiol* 59:134–146.

Santos R.A., Fuertes A.J.C., Short G., Donohue K.C., Shao H., Quintanilla J., Malakzadeh P., Cohen-Cory S. 2018. DSCAM differentially modulates pre- and postsynaptic structural and functional central connectivity during visual system wiring. *Neural Dev* 13:22.

Shelton P.M.J. 1970. The lateral line system at metamorphosis in *Xenopus laevis* (Daudin). *J Embryol Exp Morphol* 24:511–524.

Sillar K.T., Simmers A.J., Wedderburn J.F. 1992. The post-embryonic development of cell properties and synaptic drive underlying locomotor rhythm generation in *Xenopus* larvae. *Proc Roy Soc Lond* 249:65–70.

Soupiadou P., Gordy C., Forsthofer M., Sanchez-Gonzalez R., Straka H. 2020. Acute consequences of a unilateral VIIIth nerve transection on vestibulo-ocular and optokinetic reflexes in *Xenopus laevis* tadpoles. *J Neurol* 267(Suppl 1):62–75.

Sperry R.W. 1944. Optic nerve regeneration with return of vision in anurans. *J Neurophysiol* 7:57–69.

Sperry R.W. 1956. The eye and the brain. *Scientific American* 194:48–53.

Straka H., Baker R., Gilland E. 2001. Rhombomeric organization of vestibular pathways in larval frogs. *J Comp Neurol* 437:42–55.

Straka H., Dieringer N. 1993. Electrophysiological and pharmacological characterization of vestibular inputs to identified frog abducens motoneurons and internuclear neurons in vitro. *Eur J Neurosci* 5:251–260.

Straka H., Simmers J. 2012. *Xenopus laevis*: An ideal experimental model for studying the developmental dynamics of neural assembly and sensory motor computations. *Dev Neurobiol* 72:649–663.

Stuart S.N., Chanson J.S., Cox N.A., Young B.E., Rodrigues A.S., Fischman D.L., Waller R.W. 2004. Status and trends of amphibian declines and extinctions worldwide. *Science*. 306:1783–1786.

Tandon P., Conlon F., Furlow J.D., Horb M.E. 2017. Expanding the genetic toolkit in *Xenopus*: Approaches and opportunities for human disease modeling. *Dev Biol* 426:325–335.

ten Donkelaar H.J., de Boer-van Huizen R., Schouten F.T.M., Eggen S.J.H. 1981. Cells of origin of descending pathways to the spinal cord in the clawed toad (*Xenopus laevis*). *Neuroscience* 6:2297–2312.

ten Donkelaar H.J., de Boer-van Huizen R., van der Linden J.A.M. 1991. Early development of rubrospinal and cerebellorubral projections in *Xenopus laevis*. *Dev Brain Res* 58:297–300.

Tosa Y., Tsukano K., Itoyama T., Fukagawa M., Nii Y., Ishikawa R., Suzuki K.T., Fukui M., Kawaguchi M., Murakami Y. 2015. Involvement of Slit-Robo signaling in the development of the posterior commissure and concomitant swimming behavior in *Xenopus laevis*. *Zoological Lett* 1:28.

Tresch M.C., Saltiel P., Bizzi E. 1999. The construction of movement by the spinal cord. *Nat Neurosci* 2:162–167.

van der Linden J.A.M., ten Donkelaar H.J.T, de Boer-van Huizen R. 1990. Development of olivocerebellar fibers in the clawed toad, *Xenopus laevis*: A light and electron microscopical HRP study. *J Comp Neurol* 293:236–252.

Viczian A.S., Zuber M.E. 2014. A simple behavioral assay for testing visual function in *Xenopus laevis*. *J Vis Exp* 88:51726.

von Uckermann G., Lambert F.M., Combes D., Straka H., Simmers J. 2016. Adaptive plasticity of spino-extraocular motor coupling during locomotion in metamorphosing *Xenopus laevis*. *J Exp Biol* 219:1110–1121.

Willsey H.R., Exner C.R.T., Xu Y., Everitt A., Sun N., Wang B., Dea J., Schmunk G., Zaltsman Y., Teerikorpi N., Kim A., Anderson A.S., Shin D., Seyler M., Nowakowski T.J., Harland R.M., Willsey A.J., State M.W. 2021. Parallel in vivo analysis of large-effect autism genes implicates cortical neurogenesis and estrogen in risk and resilience. *Neuron* 109:788–804.

Willsey H.R., Xu Y., Everitt A., Dea J., Exner C.R.T., Willsey A.J., State M.W., Harland R.M. 2020. The neurodevelopmental disorder risk gene *DYRK1A* is required for ciliogenesis and control of brain size in *Xenopus* embryos. *Development* 147:dev189290.

Winklbauer R. 1989. Development of the lateral line system in *Xenopus*. *Prog Neurobiol* 32:181–206.

Zarei K., Elliott K.L., Zarei S., Fritzsch B., Buchholz J.H.J. 2017. A method for detailed movement pattern analysis of tadpole startle response. *J Exp Anal Behav* 108:113–124.

20 Leaping toward Understanding of Spinal Cord Regeneration

Paula G Slater, Gabriela Edwards-Faret, and Juan Larraín

CONTENTS

20.1. Introduction ... 289
20.2. Historical Background and Past Observations ... 289
20.3. Present Status of the Field ... 290
20.3.1. *X. laevis:* A Model Organism to Study Spinal Cord Regeneration ... 290
20.3.2. Cellular Response to Spinal Cord Injury ... 291
20.3.3. The Role of Neural Stem and Progenitor Cells and Neurogenesis ... 293
20.3.4. Axon Regeneration ... 294
20.3.5. Identification of Biological Processes Involved in Spinal Cord Regeneration and Comparison to Other Models ... 294
20.4. Future Directions and Important Questions ... 296
Acknowledgments ... 297
References ... 297

20.1. INTRODUCTION

The spinal cord is composed of neurons that receive sensory information and control the motor response (Watson et al. 2009). Therefore, spinal cord injury (SCI) generates paralysis caudal to the injury site, and internal organs are disconnected from central nervous system regulation. Due to the limited regenerative capacity of humans and mammals in general, and the absence of therapies allowing functional and efficient recovery, this affliction is permanent (Organization and Society 2013; Thuret et al. 2006).

In mammals, the damage produced by a SCI is composed of two main phases. The primary injury starts with the initial mechanical insult and generates a hemostatic response, damage of axons and death of oligodendrocytes, resulting in tissue structural changes and functional loss. This is followed by a secondary phase, leading to further damage both rostral and caudal to the injury site (Grossman et al. 2001; Quadri et al. 2020). This second response is composed of three different cellular phases: cell death and inflammation, cell proliferation and tissue replacement, and tissue remodeling (Burda and Sofroniew 2014). In mammals, this response protects the spinal cord from further damage (Sabelström et al. 2013) but blocks proper regeneration.

In contrast to mammals, non-mammals, including urodele amphibians (e.g. salamanders) and teleost fish (e.g. zebrafish), excel in regenerative capacities, and thus, they have been extensively used to study regeneration (Chernoff et al. 2003; Zupanc and Sîrbulescu 2011; Diaz Quiroz and Echeverri 2013; Lee-Liu et al. 2013). Unlike these animal models that retain regenerative capacities throughout their life spans, *Xenopus laevis (X. laevis)* tadpoles are able to regenerate many tissues, including spinal cord, at Nieuwkoop and Faber (NF) pre-metamorphic developmental stages 46–54 (Nieuwkoop and Faber 1994), but this ability decreases through the pro-metamorphic phase (NF stage 54–58) and metamorphosis climax (NF stage 58–66) and is almost completely lost after metamorphosis (NF stage 66) (Hooker 1925; Beattie et al. 1990). This changing capacity to regenerate allows a comparison between regenerative (R-) and non-regenerative (NR-) mechanisms in the same species and placed *Xenopus* in a phylogenetic position between mammals and urodeles and fish (Phipps et al. 2020), making it an ideal model to study spinal cord regeneration.

This chapter provides an historical background and past observations about spinal cord regeneration in amphibians, followed by a description of the current status of knowledge in this area. A particular focus is given to findings that describe the cellular response to injury, the genetic networks involved in spinal cord regeneration, and the role of neural progenitor stem cells (NSPCs). Finally, a discussion about the pitfalls on this area and the future directions in spinal cord regeneration research in *Xenopus* is presented.

20.2. HISTORICAL BACKGROUND AND PAST OBSERVATIONS

Understanding organ and tissue regeneration has been a question driving human curiosity since the beginning of scientific inquiry. Aristotle, already around 350 BC, in his book about the history of animals, commented, "if the tails of serpents or lizards be cut off, they will be reproduced" (Duncan and Sánchez Alvarado 2019). Amphibians, including the

DOI: 10.1201/9781003050230-23

order of urodeles and anurans (e.g. *Xenopus laevis*), have been among the favorite models of choice to study regeneration since the beginnings of experimental biology (Freitas et al. 2019). Lorenzo Spallanzani demonstrated that the tail of salamanders, including its spinal cord, grows back after tail amputation (Spallanzani 1768). A century later, George H. Lewes demonstrated that anurans also were able to regenerate the spinal cord (Lewes 1859), followed by the important finding that the regenerative capability decreased with tadpole development (Hooker 1925), which was later corroborated with modern approaches (Sims 1962; Filoni et al. 1984).

During the second half of the 20th century, taking advantage of the progress of microscopic techniques, a more detailed description of spinal cord regeneration process was attained in salamanders, axolotl, and frogs, mainly *Xenopus*. One set of observations demonstrated that, in contrast to mammals, axonal outgrowth is an important component of spinal cord reconstitution. Axonal growth from severed neurons rostral and caudal to the injury site was demonstrated in salamanders (Piatt 1955), although it can take as long as 23 months to recover the number of axons observed in the uninjured animals (Clarke et al. 1988; Davis et al. 1989). Similarly, in *X. laevis* tadpoles, axons growing through the injury site was demonstrated (Sims 1962; Filoni et al. 1984). Studies in *Rana catesbiana* and *X. laevis* tadpoles showed descending axonal projections that regenerate and cross the injury site, but regeneration of ascending (sensory) axons was not observed at any stage (Forehand and Farel 1982; Beattie et al. 1990). The absence of regeneration of ascending axons was also reported in newts (Zukor et al. 2011). In line with the absence of functional recovery, the axonal regenerative capacity in *X. laevis* was no longer observed in animals transected after tail resorption (Beattie et al. 1990).

A second group of responses triggered by SCI, revealed by descriptive studies in salamanders, is the activation of ependymoglial cell proliferation. Ependymoglial cells are probably the equivalent to mammalian radial glial cells and neural stem cells (NSCs) and are found lining the central canal of the spinal cord of salamanders (Freitas et al. 2019). These proliferative ependymoglial cells (Stefanelli 1951) lead to neurogenesis, which seems to be a necessary process for the reconstruction of the nervous tissue (Butler and Ward 1965; Benraiss et al. 1999). In addition, the basal processes of these cells build an ependymoglial tube or bridge in the injury site, providing a pathway and permissive substrate for axonal regeneration (Egar and Singer 1972). Similarly, after spinal cord transection in *X. laevis*, axons grow through the ablation gap in association with a bridge of ependymal processes (Michel and Reier 1979; Filoni et al. 1984). These observations show that the environment is permissive for axon growth, probably due to the absence of a glial scar, in salamanders and in *X. laevis* R-stages.

The advances from the last century established a very detailed description of the response to SCI in urodeles and anurans but failed to provide a mechanistic explanation of the process. In summary, they described the presence of active axon regeneration, proliferation of putative NSPCs that give rise to new neurons, and presence of a regenerative-permissive environment, a response that is lost with metamorphosis in the case of *X. laevis*.

20.3. PRESENT STATUS OF THE FIELD

20.3.1. *X. laevis*: A Model Organism to Study Spinal Cord Regeneration

As indicated previously, one of the great advantages of *X. laevis* as a model organism to study spinal cord regeneration is the possibility to perform experiments in R- and NR-stages. This comparison allows the identification of the cellular, molecular, and genetic mechanisms involved in regeneration and those that are responsible for the loss of the regenerative capacities (Gaete et al. 2012; Lee-Liu et al. 2014; Muñoz et al. 2015; Lee-Liu et al. 2018). In addition, the NR-stages provide an experimental paradigm to test new therapies such as genetic manipulation and pharmacological treatments that could enhance the regenerative capacities (Phipps et al. 2020).

Many methodological developments make *X. laevis* an amenable system to perform functional studies (Harland and Grainger 2011), including the generation of transgenic lines (Amaya and Kroll 1999); mutagenesis using CRISPR and TALEN systems (Nakajima and Yaoita 2015; Nakayama et al. 2020); injection and electroporation of DNA, mRNA, or morpholinos (Blum et al. 2015; Gomez et al. 2003; Bestman et al. 2006; Eide et al. 2000); and intracoelomic injection of drugs (Edwards-Faret et al. 2017). Additionally, standardized protocols for optimized husbandry have been established, allowing the acquisition of a large number of regenerative tadpoles in three weeks and non-regenerative froglets in two months, making them easily accessible for the isolation of material and to have enough animals for statistically robust experimental procedures (Edwards-Faret et al. 2017).

The study of spinal cord regeneration in frogs can be approached using two main types of injury methods: tail amputation and spinal cord transection or resection. Tail amputation removes the entire tail, including tissues such as muscle, notochord, and spinal cord, followed by their regeneration (Beck et al. 2003). The finding that there is a refractory period (NF stage 46–48) during which tail regeneration does not occur, and instead it only heals, has been very useful to understand tail, including spinal cord, regeneration (Slack et al. 2008). Using this injury method, it has been demonstrated that cellular dedifferentiation does not occur; instead, the spinal cord and notochord are regenerated from their corresponding pre-existing tissue in the tail stump, indicating that cell-lineage restriction is maintained during this process (Gargioli and Slack 2004). Similarly, the regenerated muscle originates from skeletal muscle satellite cells (Gargioli and Slack 2004), and melanophores arise from melanophore precursors present in the tail (Lin et al. 2007). In addition, tail amputation has been a fertile ground to identify the signals and genetic mechanisms involved in tissue regeneration. BMP, FGF, Wnt, and Notch signaling play an important role in tail regeneration (Beck et al. 2003). Furthermore, the Amaya lab has reported a key role for reactive oxygen species (ROSs) in the initiation of the regenerative response that seems to be evolutionarily conserved among many species (Love et al. 2011; Love et al. 2013; Phipps et al. 2020).

Although fruitful to understand the basic and general mechanism of tissue regeneration, the tail amputation paradigm presents caveats for SCI studies. On one hand, it involves regeneration of multiple tissues, not just of the spinal cord, and on the other hand, it is different from the situation that occurs in humans who suffer SCI. For these reasons, the second model of injury, spinal cord transection and resection, seems more relevant for the study of SCI (Filoni et al. 1984; Lee-Liu et al. 2013). Spinal cord transection consists of a transverse cut that completely severs the spinal cord at the thoracic level, and spinal cord resection consists of two transverse cuts followed by the removal of the spinal cord segment, leaving a gap between the rostral and caudal stumps (Edwards-Faret et al. 2017; Slater and Larraín 2021). Of note, these injury paradigms are surgically very simple, can be performed in R- and NR-stages, and are very reproducible across animals. Although these experimental approaches are still not identical to SCI in humans, they resemble the models of injury used in rodents, one of the favorite models to study SCI, allowing the comparative investigation of the mechanisms involved in this process in *X. laevis* and other model organisms.

Furthermore, sensory and motor functional recovery can be evaluated using simple behavioral tests. For this, tadpole swimming behavior can be classified qualitatively into paraplegia, partial locomotor recovery, and coordinated swimming phenotypes (Gaete et al. 2012). Additionally, the free swimming distance can be quantified using a semi-automatized video-tracking system (Edwards-Faret et al. 2017; Muñoz et al. 2015). More recently, a new method was developed using kinematic technology to allow the detection of slight improvements in NR-stage swimming recovery that could be helpful to identify compounds that improve spinal cord regeneration (De Vidts et al. 2019). This method consists of determining kinematic features during swimming, including synchronization and symmetry between the right and left hindlimb and the right and left foot range movement.

20.3.2. Cellular Response to Spinal Cord Injury

1. *Histological differences in R- and NR-stage spinal cord*: R-stage animals have a spinal cord of 400–600 μm diameter, with a cell rich ventricular layer lacking a complex stratification. In contrast, NR-stage animals have a spinal cord with a diameter of 800–1600 μm, with a more complex stratification and cellular organization in the ventricular zone (Edwards-Faret et al. 2018; Muñoz et al. 2015). The cells lining the central canal are heterogeneous, encompassing five different cells types, with fundamental differences between R- and NR-stages (Edwards-Faret et al. 2018). R-stage cells are uniciliated, exhibit a radial morphology and elongated nuclei with lax chromatin, and thus resemble mammalian radial glial cells. Nevertheless, in NR-stages, the most abundant cells are multiciliated and reveal extensive changes in the maturation and differentiation state, including round-oval shaped nucleus, with deep invaginations and clumped chromatin (Edwards-Faret et al. 2018). R-stages contain a significant number of proliferative cells, whereas NR-stages have a lower proportion of these cells (Edwards-Faret et al. 2018; Thuret et al. 2015), which correlates with the more differentiated profile of their cells.
2. *Early cellular response:* In R-stage animals, a rapid sealing of the injured stumps is observed at 2 days post-transection (dpt). The cells that accomplish sealing resemble the ones lining the central canal but with a few new features such as a fusiform shape, lack of cilia, and lack of cell junctions between them, suggesting they are migratory. Additionally, macrophage infiltration in the ablated gap leads to an early immune response by phagocytosing cell debris (Figure 20.1, Table 20.1) (Edwards-Faret et al. 2021). In contrast, in NR-stages, the cells lining the central canal appear necrotic and therefore are not able to seal the ablated stumps, leading to a disorganized central canal. There is also an abundant infiltration of red blood cells (RBCs) that fill the ablated gap (Figure 20.1, Table 20.1) (Edwards-Faret et al. 2021).
3. *Intermediate cellular response:* In R-stages, at 6 dpt, Sox2/3 expressing cells self-organize into rosettes or neural tube-like structures in the ablation gap (Figure 20.1), and cells lining the central canal differentiate into neurons, which extend their axons into the ablation gap (Edwards-Faret et al. 2018; Muñoz et al. 2015). Abundant unmyelinated bundles of axons populate the lumen of the central canal caudal to the gap at 10 dpt (Figure 20.1, Table 20.1) and are in close contact with neuronal somas, postsynaptic densities, and synaptic vesicles, suggesting that active synaptic processes occurred in the lumen of the caudal central canal (Edwards-Faret et al. 2021). Importantly, a transient increase of Fibronectin and Collagen is detected, probably serving as supportive extracellular matrix (ECM) for the advancing axons to cross through the ablation gap. In contrast, in the NR-stage, massive disorganization of the central canal persists at 6 dpt, with extracellular spaces and vacuolated cells lining the central canal. Glial processes are observed surrounding the borders of both stumps, and RBC are accompanied by a massive macrophage infiltration in the injury site. Moreover, abundant ECM

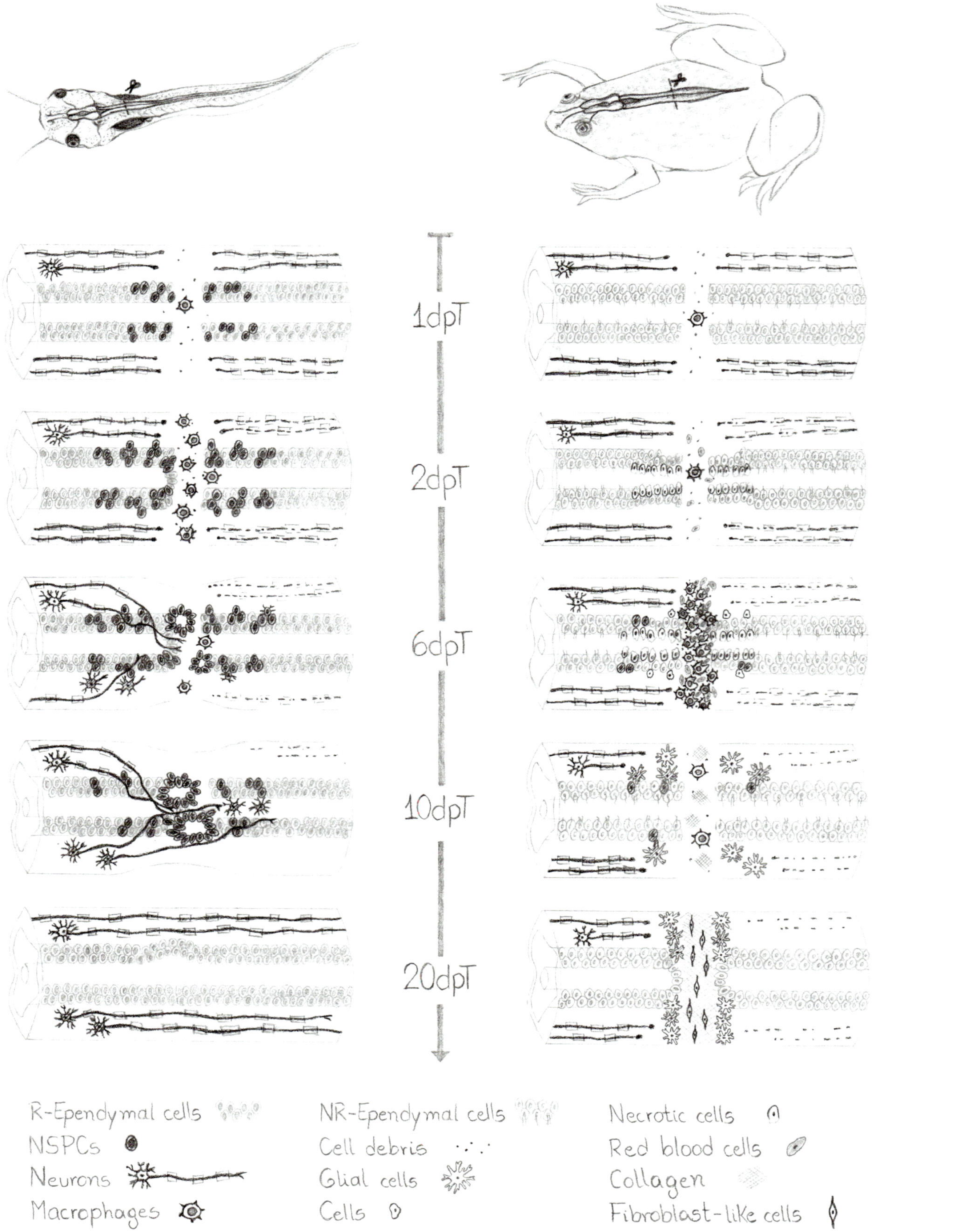

FIGURE 20.1 Comparative cellular response to SCI in R-stage and NR-stage. Drawing representing the diversity of the cellular responses between R-stage (left; NF stage 50) and NR-stage (right; NF-stage 66) at different time points after SCI. *1 dpt:* R-stage shows initial NSPC

FIGURE 20.1 (Continued)
proliferation and restricted macrophage infiltration; NR-stage shows limited macrophage infiltration. At *2dpt*, the R-stage animals show abundant NSPC proliferation and the peak of macrophage invasion, and NR-stage shows disorganized and necrotic cells lining the central canal and abundant infiltration by red blood cells (RBCs). At *6 dpt*, the R-stage animals have NSPC rosettes, NSPC differentiation into neurons, and neurons elongating axons through the ablation gap, and NR-stage have abundant RBC and macrophage infiltration into the ablation gap. At *10 dpt*, the R-stage animals still have NSPC rosettes and now also axons crossing the ablation gap, and the NR-stage showed glial processes surrounding the stumps and macrophages at the ablation gap. Finally, at *20 dpt*, the R-stage exhibits an almost complete regenerated spinal cord, and, on the contrary, the NR-stage shows the presence of a glial scar and accumulation of ECM and glial processes.

TABLE 20.1
Cellular Response to Spinal Cord Injury in R- and NR-Stages.

Cellular Response to SCI	R-Stages	NR-Stages
Morphology of cells lining the central canal	Almost healthy and normal	Damaged cells. Rupture of cell membrane and organelle loss (2–6 dpt)
Cell death	Reduced and controlled (1dpt)	Massive. Necrotic cells (2–6 dpt)
Stump closure	Early (2 dpt) By cells lining the central canal	Late (20 dpt) By cells lining the central canal
Immune response	Rapid infiltration, debris clearance by macrophages (2 dpt)	Slow infiltration, prolonged red blood cell clearance by macrophages (6–10 dpt)
Proliferation	Early (2 dpt) by NSPCs	Late (6–10 dpt), mainly other cells
Cells in the ablation gap	NSPCs forming neural tube-like structures (rosettes) (6–10 dpt)	Red blood/immune cells (6 dpt) Fibroblast-like cells (20 dpt) ECM deposition (10–20 dpt)
Neurogenesis	Differentiation of NSPCs into neurons and astrocytes	Not clearly defined
Axon regeneration	Axon tips in ablation gap (6 dpt) Bundle of axons crossing the ablation gap (10 dpt)	Not achieved
Continuity of ependymal canal	Achieved (20 dpt)	Not achieved Border of stumps surrounded by glial processes (10–20 dpt)

Source: Muñoz et al. 2015; Edwards-Faret et al. 2018, 2021

dpt: days post transection, NSPCs: neural stem and progenitor cells; ECM: extracellular matrix.

such as Fibronectin and Collagen fill the injury site (Table 20.1) (Edwards-Faret et al. 2021).

4. *Late cellular response:* In the R-stage at 20 dpt, cells lining the central canal complete the reconstruction of the pseudo-stratified epithelium, the central canal is continuous across the injury site, and axons cross it (Figure 20.1; Table 20.1) (Edwards-Faret et al. 2021), correlating with the recovery of swimming capabilities (Muñoz et al. 2015). On the other hand, in the NR-stage, although the cells lining the central canal have recovered a pseudo-stratified epithelial structure and the central canal of both stumps is finally sealed, a glial-scar like structure remains in the ablation gap, and no reconnection between the stumps is observed. This glial-scar like structure is characterized by both stumps being surrounded by glial processes and the injury gap being filled by fibroblast-like cells and a dense ECM containing Collagen and Chondroitin Sulfate Proteoglycans (Edwards-Faret et al. 2021). The lack of axonal connections between both stumps persists after 40 days, leading to the complete lack of swimming capacities in NR-stage (Edwards-Faret et al. 2021; Muñoz et al. 2015).

20.3.3. The Role of Neural Stem and Progenitor Cells and Neurogenesis

The cells lining the central canal of the spinal cord have a fundamental role during spinal cord regeneration. *sox2*, which is a marker of stem cells and also neural stem cells (Ellis et al. 2004; Pevny and Nicolis 2010), is expressed in cells lining the central canal during R-stages, and a progressive decrease in the number of Sox2-expressing cells takes place during metamorphosis (Gaete et al. 2012; Muñoz et al. 2015). Sox2 protein and mRNA expression are upregulated after tail amputation, concomitant with a proliferative response of Sox2-expressing cells (Gaete et al. 2012), which is also observed in R-stages after spinal cord transection (Muñoz et al. 2015). *In vivo* time-lapse imaging has shown, for the first time, their capacity to generate new neurons in response to injury (Figure 20.1; Table 20.1), which correlates with an increase in neurogenic markers (Muñoz et al. 2015). Consistent with this, Sox2 functional knockdown using morpholino oligonucleotides or overexpressing a dominant negative form of Sox2 results in spinal cord regeneration impairment (Muñoz et al. 2015) and in defective tail regeneration (Gaete et al. 2012). The neural stem progenitor cell identity of some *sox2*-expressing cells lining the central canal was recently corroborated (Edwards-Faret et al. 2021). Taking advantage of a reporter transgenic line (Xla.Tg(Dre.gfap::EGFP)^{Larra}), which expresses EGFP in spinal cord cells with radial glial cell morphology (Edwards-Faret et al. 2018, 2021), it was possible to perform a transcriptomic (RNA-seq) analysis of these cells in isolation and show the enrichment of transcripts mainly related to neural precursor cell identities and stem cell proliferation genes (Edwards-Faret et al. 2021). Even more exciting, the dependency of proper spinal cord regeneration on the presence of NSPCs was demonstrated by ablation of NSPCs in R-stages using the nitroreductase/metronidazole system in the transgenic line (Xla. Tg(Dre.gfap::mCherry-Nitroreductase)^{Larra}, which led to the loss of functional recovery, determined by the decrease in swimming capacity (Edwards-Faret et al. 2021).

In contrast, NR-stages have reduced levels of Sox2 and a delayed and poor proliferative response of Sox2/3$^+$ cells, concomitant with no increase in neurogenic markers (Muñoz et al. 2015). Interestingly, transplantation of R-stage Sox2/3+ cells into the NR-stage injury site provided further evidence that Sox2/3$^+$ cells are NSPCs (Méndez-Olivos et al. 2017). Transplanted cells were able to self-organize, proliferate, and differentiate into neurons in the host NR-environment, with axons growing in the grafted tissue and into the host spinal cord. Surprisingly, regeneration of axons coming from the host was also observed, suggesting that transplanted cells are able to provide a permissive environment for axon outgrowth (Méndez-Olivos et al. 2017). It remains to be elucidated which intrinsic and extrinsic factors are provided by the donor cells that are responsible for this regeneration in NR-stages.

20.3.4. Axon Regeneration

For successful spinal cord regeneration and functional recovery, nerve regeneration and reconstitution of lost connections is necessary. As early as the beginning of the 1960s, *X. laevis* axons traversing the injury gap were seen, but their origin was not possible to identify (Sims 1962). In the mid-1980s, it was determined that some of the growing axons were from serotonergic neurons (Beattie et al. 1990), whose projections are formed early during *X. laevis* development (Van Mier et al. 1986), indicating the axons were regenerating. Later, it was demonstrated that metamorphosis is preceded by a new wave of neurogenesis and the generation of new neuronal connections to allow for locomotor changes (Kollros 1981; Sillar et al. 2008), showing the possibility that these neurons could compensate for the interrupted spinal cord tracts following SCI.

It was not until 2006 that Gibbs and Szaro were able to demonstrate that axonal regeneration occurs and that the axons originate in the hindbrain, mainly in the reticular and raphe nuclei (Gibbs and Szaro 2006). Additionally, they showed that axon regeneration capacity depends on metamorphosis progression; inhibiting metamorphosis allows animals to regenerate axons and reestablish swimming, while induction of metamorphosis prematurely in R-stage animals by thyroid hormone 3,3′,5′-Triiodo-L-thyronine (T3) treatment results in impaired axon regeneration and promotes a differential transcriptional response in the hindbrain compared to untreated animals. Therefore, intrinsic transcriptional changes in the neuron soma accompany axon R- and NR-responses in the spinal cord (Gibbs et al. 2011). The loss of regenerative capacity after metamorphosis correlates with the progressive changes in the cells lining the central canal of the spinal cord (Edwards-Faret et al. 2018), concomitant with the increase in thyroid hormone (TH) levels (Brown and Cai 2007) and the down-regulation of *sox2* (Gaete et al. 2012; Muñoz et al. 2015) and *lin28*, which is an heterochronic gene that controls developmental timing (Faunes et al. 2017). Interestingly, overexpression of *lin28* regulates *X. laevis* metamorphosis by disturbing TH function, resulting in improved digit regeneration (Faunes et al. 2017). These observations add to the interest of studying genetic interactions during metamorphosis.

20.3.5. Identification of Biological Processes Involved in Spinal Cord Regeneration and Comparison to Other Models

The different regenerative abilities of *X. laevis* prior to and during metamorphosis, and its genome availability (Session et al. 2016), have made it possible to perform high-throughput experiments using RNA-seq at 1, 2, and 6 dpt, and Proteomics at 1 dpt, to compare the response to SCI in R- versus NR-stages (Lee-Liu et al. 2014, 2018). Even though both stages display a massive transcriptomic change response, they differ in terms of timing and levels of gene expression. R-stages show a rapid response, with most of the transcript changes observed at 1 dpt, whereas NR-stages present a later response, with most of the changes observed at 6 dpt (Table 20.2) (Lee-Liu et al. 2014).

Gene ontology enrichment analysis shows that most of the genes that change their expression levels in response to injury, which are mostly differentially regulated between R- and NR-stages, participate in biological processes such as metabolism, response to stress, cell cycle, development, immune response and inflammation, neurogenesis, and axonal regeneration (Table 20.2). These biological processes are related to the SCI secondary injury cellular phases mentioned previously, cell death and inflammation, cell proliferation and tissue replacement, and tissue remodeling (Burda and Sofroniew 2014). Therefore, the different transcriptional responses of R- and NR-stages could lead to dissimilar cellular responses and be responsible for the regenerative capacity at the R-stages and its loss at the NR-stage. Even though the contribution of transcript and protein-level changes has been only partially tested in *X. laevis* during SCI, their comparison with other animal models or paradigms is important for guiding future directions in the understanding of R- and NR-stage cell response after SCI. Here we interpret these findings in the context of what is known about spinal cord regeneration in other experimental paradigms.

1. *Cell death*: In *X. laevis* tail amputation experiments, an early and controlled process of cellular apoptosis takes place. Apoptosis is required during the first 24 h, as its inhibition during the first day abolishes tail regeneration, whereas later inhibition

has no effect (Tseng et al. 2007). Even more interesting, tail amputation during the refractory period results in broader apoptosis, which could be related to the absence of regeneration (Tseng et al. 2007). Concordantly, in the transection paradigm, R-stages showed downregulation of cell death transcripts (Lee-Liu et al. 2014) and upregulation of proteins that negatively regulate programmed cell death (Lee-Liu et al. 2018) at 1 dpt, whereas NR-stages show no down-regulation of cell death at any analyzed time point (Table 20.2).

2. *Immune cell response and inflammation*: The immune response in the central nervous system has a dual effect, defined by the different microglia/macrophage phenotypes: the M1 or pro-inflammatory and the M2 or anti-inflammatory (Jha et al. 2016). The pro-inflammatory response is needed to clear cellular debris and facilitate repair, but the release of pro-inflammatory cues also could exacerbate cellular and extracellular matrix damage and increase immune cell infiltration. On the other hand, an anti-inflammatory response is needed for inflammatory resolution, tissue remodeling, and repair (Hamilton et al. 1999). In regenerative model animals, this dual immune response is observed during regeneration. Zebrafish respond to SCI with an initial M1 phenotype, which is necessary for induction of axon regeneration, as inhibition of TNF-α impairs axon re-growth. This is followed by a M2 polarization within 1 dpt, which is necessary for reducing pro-inflammatory cytokines, as excessive IL-1β also impairs axon re-growth (Tsarouchas et al. 2018). Mice show a completely different response. Initially, the genes involved in M1 and M2 polarization are activated. This activation is predominant for the M1-related genes, leading to M1 polarization lasting for at least 1 month, which results in chronic inflammation (Kigerl et al. 2009). In agreement with these results, both transcriptomic (Lee-Liu et al. 2014) and proteomic (Lee-Liu et al. 2018) analyses in *X. laevis* show that NR-stages present a positive regulation of immune response related genes and proteins, while in R-stages, a negative regulation is mainly observed (Table 20.2).
3. *Neurogenesis*: After SCI, extensive cell death and tissue damage occur, and cellular proliferation and differentiation capacities correlate with the ability of certain animals, like urodele and anuran amphibians and teleost fish, to regenerate their spinal cord (Ferretti et al. 2003; Diaz Quiroz and Echeverri 2013). R-stage animals respond to SCI with an early up-regulation of transcripts related to cell cycle (1–2 dpt) and cell division (2 dpt), whereas NR-stage animals present a late up-regulation of transcripts related to cell cycle (6 dpt) (Table 20.2) (Lee-Liu et al. 2014). Following cell proliferation, cell fate commitment is a decisive aspect for spinal cord regeneration. Regenerative animal models are able to generate new neurons after SCI (Benraiss et al. 1999; Ghosh and Hui 2016; Muñoz et al. 2015), whereas mammals generate only new astrocytes and oligodendrocytes (Barnabé-Heider et al. 2010; Meletis et al. 2008). Transcript and protein analyses in *X. laevis* have shown that this different capacity is in part intrinsically regulated within the cells. R-stage animals show an up-regulation of neurogenesis-related transcripts (Lee-Liu et al. 2014; Muñoz et al. 2015) and proteins (Muñoz et al. 2015), while they are absent or down-regulated in NR-stage animals (Lee-Liu et al. 2014; Muñoz et al. 2015) (Table 20.2).
4. *Axon regeneration*: Regenerative animal models, like zebrafish, regenerate axons and recover lost connections after SCI (Ghosh and Hui 2018), while mammals present a very limited capacity (Kerschensteiner et al. 2005). In the same way, in R-stage animals, it is possible to see axon bundles crossing the injury gap, which are absent in the NR-stage animals (Figure 20.1) (Edwards-Faret et al. 2021; Muñoz et al. 2015). These axons may be derived from new neurons (Muñoz et al. 2015), regenerating axons (Gibbs and Szaro 2006), or both. Consistent with these observations, transcripts related to axon growth, including growth cone and axonal guidance, are differentially regulated in R- versus NR-stages, being down-regulated only in NR-stages (Table 20.2) (Lee-Liu et al. 2014).
5. *Metabolic regulation*: More than 50% of regulated transcripts in R-stage animals are metabolic related, being highly predominant in both up- and down-regulated genes starting at 1 dpt and continuing until 6 dpt, while NR-stage animals present a later regulation of metabolic related transcripts, observed only at 6 dpt (Table 20.2). The preponderance of metabolic genes changing their expression during the early phases of the regenerative response can be explained by many reasons. Among others, we propose the following:

   i. The regulation of microglia/macrophage activation and M1 to M2 polarization by cellular metabolic changes has been extensively studied in murine cell lines. Under normal conditions, immune cells are quiescent, and when they are exposed to a pro-inflammatory environment, a metabolic switch from an oxidative to a glycolytic metabolism occurs, resulting in their activation and polarization to a M1 phenotype, while a new metabolic change favoring oxidative phosphorylation directs towards an M2 phenotype (Afridi et al. 2020; Jha et al. 2016).
   ii. Cellular proliferation is an energy demanding process, which needs the production of the cellular building blocks like lipids, proteins, and

nucleic acids. Rapidly dividing cells, like the ones present in tumors and embryonic tissue, present metabolic changes favoring aerobic glycolysis, and the carbons provided by glycolysis enter the pentose phosphate pathway for macromolecular biosynthesis (Vander Heiden et al. 2009; Krisher and Prather 2012; Love et al. 2014).

iii. NSPC maintenance and differentiation are regulated by metabolism. A glycolytic metabolism is necessary for *in vitro* stem cell maintenance and self-renewal, whereas an oxidative metabolism is associated with differentiation (Rafalski et al. 2012; Khacho et al. 2019). Additionally, the expression profile and epigenetic state of the stem cells regulate stemness and differentiation, and many mitochondrial metabolic intermediates are used as co-factors by many histone-modifying proteins (Khacho et al. 2019; Zhang et al. 2018).

iv. Axon growth and the generation of collateral projections to re-establish lost connections are also energy-demanding processes (Bradke et al. 2012). Enhancing energetic metabolism by administration of creatine, which is processed by creatine kinase for ATP generation, as well as enhancing axonal mitochondrial transport, enhance axonal regeneration after SCI in a murine model (Han et al. 2020).

6. *Tissue remodeling*: The lesion site in mammals is characterized by the presence of a glial scar formed by a non-neural lesion core containing fibroblasts, pericytes, and ECM components, among others, surrounded by astrocytes that function to contain the damage in order to prevent its spread onto adjacent viable neural tissue (O'Shea et al. 2017). In *X. laevis*, R-stage animals present continuity of the central canal and axon tracts at 20 dpt. In contrast, NR-stage animals are not able to reconstitute the spinal cord, and the ablation gap is filled with fibroblast-like cells and ECM components (Figure 20.1; Table 20.1) (Edwards-Faret et al. 2021; Muñoz et al. 2015). In agreement with these observations, transcriptomic analyses showed that ECM-related transcripts are up-regulated only in NR-stage animals at 6 dpt, supporting that NR-stage, but not R-stage, develop a glial scar (Table 20.2) (Edwards-Faret et al. 2021; Lee-Liu et al. 2014).

TABLE 20.2
Biological Process Regulated at the Transcriptomic and/or Proteomic Level in R- and NR-Stages after Spinal Cord Injury.

Biological Process	R-Stages	NR-Stages
Cell death	Early down-regulation (1 dpt)	Early up-regulation (1 dpt)
Blood coagulation	Mild early up-regulation (1 dpt)	Robust early up-regulation (1 dpt)
Cell cycle	Early up-regulation (1, 2 dpt)	Late up-regulation (6 dpt)
Development	Early down-regulation (1, 2 dpt)	Late down-regulation (6 dpt)
Response to stress	Early and transient up-regulation (1, 2 dpt)	Sustained up-regulation (1, 2, 6 dpt)
Metabolism	Sustained up-regulation (1, 2, 6 dpt)	Late up-regulation (6 dpt)
Immune response Inflammation	Mainly down-regulation (1, 2 dpt)	Mainly up-regulation (1, 2 dpt)
Neurogenesis	Early up-regulation (1, 2 dpt)	Late up-regulation (6 dpt)
Axonal growth cone (axonal regeneration)	No change	Late down-regulation (6 dpt)
Glial scar	No change	Late up-regulation (6 dpt)

Source: Lee-Liu et al. 2014; Muñoz et al. 2015; Lee-Liu et al. 2018; Edwards-Faret et al. 2021

dpt: days post transection

20.4. FUTURE DIRECTIONS AND IMPORTANT QUESTIONS

This chapter presented an overview of the different cellular and genetic responses associated with SCI and regeneration, in addition to evidence about how proper regulation of them may explain the difference between R- and NR-capacities. Even though progress has been made in understanding the cellular and molecular mechanisms underlying these cellular responses, there is still much to elucidate.

The questions for future directions that now arise are about the signaling pathways that are activated or repressed in a differential manner between R- and NR-stages, allowing protection against further cell death in the R-stage and induction of cellular proliferation. Some studies in other model organisms have spotlighted the anti-apoptotic Bcl family, which has been demonstrated to play a role in planarian regeneration (Pellettieri et al. 2010). Additionally, the Wnt signaling pathway in planaria (Chera et al. 2009), and the JNK and Wingless signaling pathways in *Drosophila* (Ryoo et al. 2004), are activated during cell death after injury, subsequently leading to proliferation and regeneration. It will be important to corroborate if these responses are conserved across organisms and identify how they are regulated. As cell death and proliferation transcripts are already regulated at 1 dpt in R-stage animals, a starting point would be the study of the transcript and protein responses prior to 24 hours after SCI.

Another important aspect to study is the differentiation of the proliferative NSPCs into postmitotic neurons. As mentioned previously, regenerative organisms generate new neurons, while mammals do not. Interestingly, the mammalian cells lining the central canal are able to generate new neurons *in vitro* (Barnabé-Heider et al. 2010; Meletis et al. 2008) and when transplanted to a neurogenic-permissive environment (Shihabuddin et al. 2000). These observations raise the question about the signals that trigger and the signaling cascades that should be activated to induce, and in the future direct, differentiation towards neurons. In addition, it will be important to determine if it is possible to improve regeneration in NR-stage by activating NSPC proliferation and forcing their differentiation to neurons.

Additionally, as mentioned previously, the cell bodies of the damaged axons that are located in the hindbrain present a differential transcriptional response in R- versus NR-stages. It is very important to determine the role of these transcripts to understand the molecular and cellular mechanisms underlying the intrinsic regenerative capacity of these neurons and to determine if modifications in the expression levels of these genes is necessary and sufficient for induction of axon regeneration at NR-stages.

Finally, if we consider all the aforementioned evidence about metabolic regulation of the cellular phases involved in SCI response, some open questions remain: Which metabolic regulation, in which cell type, and at what time after SCI are needed for regeneration? And, is it possible to improve regeneration in NR-stage animals by mimicking R-stage metabolic responses?

As *X. laevis* presents R- and NR-stages during its development, it is an outstanding model animal to answer all these questions and to evaluate new potential therapies that would help deepen the understanding of the mechanisms involved in a successful regenerative process and how to improve regeneration in non-regenerative organisms.

ACKNOWLEDGMENTS

This work was supported by FONDECYT 1180429 for JL and 3190820 for PGS, and a post-doctoral fellow of the Fonds Wetenschappelijk Onderzoek (FWO) Grant 1214420N for GE-F. We thank Mimí Palacios for the Figure 20.1 drawings. The authors declare no conflict of interest.

REFERENCES

Afridi, Ruqayya, Jong-Heon Kim, Md Habibur Rahman, and Kyoungho Suk. 2020. Metabolic regulation of glial phenotypes: Implications in neuron-glia interactions and neurological disorders. *Frontiers in Cellular Neuroscience* 14:20.

Amaya, Enrique, and Kristen L. Kroll. 1999. A method for generating transgenic frog embryos. In *Molecular Embryology*. Totowa, NJ: Humana Press. pp. 393–414.

Barnabé-Heider, Fanie, Christian Göritz, Hanna Sabelström, et al. 2010. Origin of new glial cells in intact and injured adult spinal cord. *Cell Stem Cell* 7 (4):470–482.

Beattie, Michael S., Jacqueline C. Bresnahan, and Glenn Lopate. 1990. Metamorphosis alters the response to spinal cord transection in *Xenopus laevis* frogs. *Journal of Neurobiology* 21 (7):1108–1122.

Beck, Caroline W., Bea Christen, and Jonathan M.W. Slack. 2003. Molecular pathways needed for regeneration of spinal cord and muscle in a vertebrate. *Developmental Cell* 5 (3):429–439.

Benraiss, A., J.P. Arsanto, J. Coulon, and Y. Thouveny. 1999. Neurogenesis during caudal spinal cord regeneration in adult newts. *Development Genes and Evolution* 209 (6):363–369.

Bestman, Jennifer E., Rebecca C. Ewald, Shu-Ling Chiu, and Hollis T. Cline. 2006. In vivo single-cell electroporation for transfer of DNA and macromolecules. *Nature Protocols* 1 (3):1267.

Blum, Martin, Edward M. De Robertis, John B. Wallingford, and Christof Niehrs. 2015. Morpholinos: Antisense and sensibility. *Developmental Cell* 35 (2):145–149.

Bradke, Frank, James W. Fawcett, and Micha E. Spira. 2012. Assembly of a new growth cone after axotomy: The precursor to axon regeneration. *Nature Reviews Neuroscience* 13 (3):183–193.

Brown, Donald D., and Liquan Cai. 2007. Amphibian metamorphosis. *Developmental Biology* 306 (1):20.

Burda, Joshua E., and Michael V. Sofroniew. 2014. Reactive gliosis and the multicellular response to CNS damage and disease. *Neuron* 81 (2):229–248.

Butler, Elmer G., and Margery B. Ward. 1965. Reconstitution of the spinal cord following ablation in urodele larvae. *Journal of Experimental Zoology* 160 (1):47–65.

Chera, Simona, Luiza Ghila, Kevin Dobretz, et al. 2009. Apoptotic cells provide an unexpected source of Wnt3 signaling to drive hydra head regeneration. *Developmental Cell* 17 (2):279–289.

Chernoff, Ellen A.G., David L. Stocum, Holly L.D. Nye, and Jo Ann Cameron. 2003. Urodele spinal cord regeneration and related processes. *Developmental Dynamics: An Official Publication of the American Association of Anatomists* 226 (2):295–307.

Clarke, J.D.W., Ruth Alexander, and Nigel Holder. 1988. Regeneration of descending axons in the spinal cord of the axolotl. *Neuroscience Letters* 89 (1):1–6.

Davis, Brian M., Mark T. Duffy, and Sidney B. Simpson Jr. 1989. Bulbospinal and intraspinal connections in normal and regenerated salamander spinal cord. *Experimental Neurology* 103 (1):41–51.

De Vidts, Sebastián, Emilio Méndez-Olivos, Miriam Palacios, Juan Larraín, and Domingo Mery. 2019. Characterization of spinal cord damage based on automatic video analysis of froglet swimming. *Biology Open* 8 (12).

Diaz Quiroz, Juan Felipe, and Karen Echeverri. 2013. Spinal cord regeneration: Where fish, frogs and salamanders lead the way, can we follow? *Biochemical Journal* 451 (3):353–364.

Duncan, Elizabeth M., and Alejandro Sánchez Alvarado. 2019. Regulation of genomic output and (Pluri) potency in regeneration. *Annual Review of Genetics* 53:327–346.

Edwards-Faret, Gabriela, Arantxa Cebrián-Silla, Emilio E. Méndez-Olivos, Karina González-Pinto, José Manuel García-Verdugo, and Juan Larraín. 2018. Cellular composition and organization of the spinal cord central canal during metamorphosis of the frog *Xenopus laevis*. *Journal of Comparative Neurology* 526 (10):1712–1732.

Edwards-Faret, Gabriela, Karina González-Pinto, Arantxa Cebrián-Silla, Johany Peñailillo, José Manuel García-Verdugo, and Juan Larraín. 2021. Cellular response to spinal cord injury in regenerative and non-regenerative stages in *Xenopus laevis*. *Neural Development* 16 (1):2.

Edwards-Faret, Gabriela, Rosana Muñoz, Emilio E. Méndez-Olivos, Dasfne Lee-Liu, Victor S. Tapia, and Juan Larraín.

2017. Spinal cord regeneration in *Xenopus laevis*. *Nature Protocols* 12 (2):372–389.

Egar, Margaret, and Marcus Singer. 1972. The role of ependyma in spinal cord regeneration in the urodele, Triturus. *Experimental Neurology* 37 (2):422–430.

Eide, Fernette F., Staci R. Eisenberg, and Timothy A. Sanders. 2000. Electroporation-mediated gene transfer in free-swimming embryonic *Xenopus laevis*. *FEBS Letters* 486 (1):29–32.

Ellis, Pam, B. Matthew Fagan, Scott T. Magness, et al. 2004. SOX2, a persistent marker for multipotential neural stem cells derived from embryonic stem cells, the embryo or the adult. *Developmental Neuroscience* 26 (2–4):148–165.

Faunes, Fernando, Daniel G. Gundermann, Rosana Muñoz, Renzo Bruno, and Juan Larraín. 2017. The heterochronic gene Lin28 regulates amphibian metamorphosis through disturbance of thyroid hormone function. *Developmental Biology* 425 (2):142–151.

Ferretti, Patrizia, Fang Zhang, and Paul O'Neill. 2003. Changes in spinal cord regenerative ability through phylogenesis and development: Lessons to be learnt. *Developmental Dynamics: An Official Publication of the American Association of Anatomists* 226 (2):245–256.

Filoni, S., L. Bosco, and C. Cioni. 1984. Reconstitution of the spinal cord after ablation in larval *Xenopus laevis*. *Acta embryologiae et morphologiae experimentalis (1980)* 5 (2):109–129.

Forehand, Cynthia J., and Paul B. Farel. 1982. Anatomical and behavioral recovery from the effects of spinal cord transection: Dependence on metamorphosis in anuran larvae. *Journal of Neuroscience* 2 (5):654–662.

Freitas, Polina D., Anastasia S. Yandulskaya, and James R. Monaghan. 2019. Spinal cord regeneration in amphibians: A historical perspective. *Developmental Neurobiology* 79 (5):437–452.

Gaete, Marcia, Rosana Muñoz, Natalia Sánchez, Ricardo Tampe, Mauricio Moreno, Esteban G. Contreras, Dasfne Lee-Liu, and Juan Larraín. 2012. Spinal cord regeneration in *Xenopus* tadpoles proceeds through activation of Sox2-positive cells. *Neural Development* 7(1): 1–17.

Gargioli, Cesare, and Jonathan M.W. Slack. 2004. Cell lineage tracing during Xenopus tail regeneration. *Development* 131 (11):2669–2679.

Ghosh, Sukla, and Subhra Prakash Hui. 2016. Regeneration of zebrafish CNS: Adult neurogenesis. *Neural Plasticity* 2016.

Ghosh, Sukla, and Subhra Prakash Hui. 2018. Axonal regeneration in zebrafish spinal cord. *Regeneration* 5 (1):43–60.

Gibbs, Kurt M., Sridar V. Chittur, and Ben G. Szaro. 2011. Metamorphosis and the regenerative capacity of spinal cord axons in *Xenopus laevis*. *European Journal of Neuroscience* 33 (1):9–25.

Gibbs, Kurt M., and Ben G. Szaro. 2006. Regeneration of descending projections in *Xenopus laevis* tadpole spinal cord demonstrated by retrograde double labeling. *Brain Research* 1088 (1):68–72.

Gomez, Timothy M., Dan Harrigan, John Henley, and Estuardo Robles. 2003. Working with Xenopus spinal neurons in live cell culture. *Methods Cell Biol* 71:129–156.

Grossman, S.D., L.J. Rosenberg, and J.R. Wrathall. 2001. Temporal-spatial pattern of acute neuronal and glial loss after spinal cord contusion. *Experimental Neurology* 168 (2):273–282.

Hamilton, T.A., Y. Ohmori, J.M. Tebo, and R. Kishore. 1999. Regulation of macrophage gene expression by pro-and anti-inflammatory cytokines. *Pathobiology* 67 (5–6):241–244.

Han, Qi, Yuxiang Xie, Josue D. Ordaz, et al. 2020. Restoring cellular energetics promotes axonal regeneration and functional recovery after spinal cord injury. *Cell Metabolism* 31 (3):623–641. e8.

Harland, Richard M., and Robert M. Grainger. 2011. Xenopus research: metamorphosed by genetics and genomics. *Trends in Genetics* 27 (12):507–515.

Hooker, Davenport. 1925. Studies on regeneration in the spinal cord. III. Reestablishment of anatomical and physiological continuity after transection in frog tadpoles. *Journal of Comparative Neurology* 38 (3):315–347.

Jha, Mithilesh Kumar, Won-Ha Lee, and Kyoungho Suk. 2016. Functional polarization of neuroglia: Implications in neuroinflammation and neurological disorders. *Biochemical Pharmacology* 103:1–16.

Kerschensteiner, Martin, Martin E. Schwab, Jeff W. Lichtman, and Thomas Misgeld. 2005. In vivo imaging of axonal degeneration and regeneration in the injured spinal cord. *Nature Medicine* 11 (5):572–577.

Khacho, Mireille, Richard Harris, and Ruth S. Slack. 2019. Mitochondria as central regulators of neural stem cell fate and cognitive function. *Nature Reviews Neuroscience* 20 (1):34–48.

Kigerl, Kristina A., John C. Gensel, Daniel P. Ankeny, Jessica K. Alexander, Dustin J. Donnelly, and Phillip G. Popovich. 2009. Identification of two distinct macrophage subsets with divergent effects causing either neurotoxicity or regeneration in the injured mouse spinal cord. *Journal of Neuroscience* 29 (43):13435–13444.

Kollros, Jerry J. 1981. Transitions in the nervous system during amphibian metamorphosis. In *Metamorphosis*. Boston, MA: Springer. pp. 445–459.

Krisher, Rebecca L., and Randall S. Prather. 2012. *A role for the Warburg effect in preimplantation embryo development: Metabolic modification to support rapid cell proliferation*. Hoboken, NJ: Wiley Online Library. pp. 311–320.

Lee-Liu, Dasfne, Gabriela Edwards-Faret, Víctor S. Tapia, and Juan Larraín. 2013. Spinal cord regeneration: Lessons for mammals from non-mammalian vertebrates. *Genesis* 51 (8):529–544.

Lee-Liu, Dasfne, Mauricio Moreno, Leonardo I. Almonacid, et al. 2014. Genome-wide expression profile of the response to spinal cord injury in *Xenopus laevis* reveals extensive differences between regenerative and non-regenerative stages. *Neural Development* 9 (1):1–20.

Lee-Liu, Dasfne, Emilio E. Méndez-Olivos, Rosana Muñoz, and Juan Larraín. 2017. The African clawed frog *Xenopus laevis*: A model organism to study regeneration of the central nervous system. *Neuroscience Letters* 652:82–93.

Lee-Liu, Dasfne, Liangliang Sun, Norman J. Dovichi, and Juan Larraín. 2018. Quantitative proteomics after spinal cord injury (SCI) in a regenerative and a nonregenerative stage in the frog *Xenopus laevis*. *Molecular & Cellular Proteomics* 17 (4):592–606.

Lewes, G.H. 1859. Part second: American and Foreign intelligence. *The Ohio Medical and Surgical Journal (1848–1878)* 11 (5):378.

Lin, Gufa, Ying Chen, and Jonathan M.W. Slack. 2007. Regeneration of neural crest derivatives in the Xenopus tadpole tail. *BMC Developmental Biology* 7 (1):1–14.

Love, Nick R., Yaoyao Chen, Boyan Bonev, et al. 2011. Genome-wide analysis of gene expression during *Xenopus tropicalis* tadpole tail regeneration. *BMC Developmental Biology* 11 (1):1–15.

Love, Nick R., Yaoyao Chen, Shoko Ishibashi, et al. 2013. Amputation-induced reactive oxygen species are required for successful Xenopus tadpole tail regeneration. *Nature Cell Biology* 15 (2):222–228.

Love, Nick R., Mathias Ziegler, Yaoyao Chen, and Enrique Amaya. 2014. Carbohydrate metabolism during vertebrate appendage regeneration: What is its role? How is it regulated? A postulation that regenerating vertebrate appendages facilitate glycolytic and pentose phosphate pathways to fuel macromolecule biosynthesis. *Bioessays* 36 (1):27–33.

Meletis, Konstantinos, Fanie Barnabé-Heider, Marie Carlén, et al. 2008. Spinal cord injury reveals multilineage differentiation of ependymal cells. *PLoS Biol* 6 (7):e182.

Méndez-Olivos, Emilio E., Rosana Muñoz, and Juan Larraín. 2017. Spinal cord cells from pre-metamorphic stages differentiate into neurons and promote axon growth and regeneration after transplantation into the injured spinal cord of non-regenerative *Xenopus laevis* froglets. *Frontiers in Cellular Neuroscience* 11:398.

Michel, Mary Ellen, and Paul J. Reier. 1979. Axonal-ependymal associations during early regeneration of the transected spinal cord in *Xenopus laevis* tadpoles. *Journal of Neurocytology* 8 (5):529–548.

Muñoz, Rosana, Gabriela Edwards-Faret, Mauricio Moreno, Nikole Zuñiga, Hollis Cline, and Juan Larraín. 2015. Regeneration of *Xenopus laevis* spinal cord requires Sox2/3 expressing cells. *Developmental Biology* 408 (2):229–243.

Nakajima, Keisuke, and Yoshio Yaoita. 2015. Highly efficient gene knockout by injection of TALEN mRNAs into oocytes and host transfer in *Xenopus laevis*. *Biology Open* 4 (2):180–185.

Nakayama, Takuya, Robert M. Grainger, and Sang-Wook Cha. 2020. Simple embryo injection of long single-stranded donor templates with the CRISPR/Cas9 system leads to homology-directed repair in *Xenopus tropicalis* and *Xenopus laevis*. *Genesis* 58 (6):e23366.

Nieuwkoop, Pieter D., and Jacob Faber. 1994. *Normal table of Xenopus laevis (Daudin)*. New York: Garland Publishing, Inc. ISBN 0-8153-1896-0.

Organization, World Health, and International Spinal Cord Society. 2013. *International Perspectives on Spinal Cord Injury*. Geneva: World Health Organization.

O'Shea, Timothy M., Joshua E. Burda, and Michael V. Sofroniew. 2017. Cell biology of spinal cord injury and repair. *The Journal of Clinical Investigation* 127 (9):3259–3270.

Pellettieri, Jason, Patrick Fitzgerald, Shigeki Watanabe, Joel Mancuso, Douglas R. Green, and Alejandro Sánchez Alvarado. 2010. Cell death and tissue remodeling in planarian regeneration. *Developmental Biology* 338 (1):76–85.

Pevny, Larysa H., and Silvia K. Nicolis. 2010. Sox2 roles in neural stem cells. *The International Journal of Biochemistry & Cell Biology* 42 (3):421–424.

Phipps, Lauren S., Lindsey Marshall, Karel Dorey, and Enrique Amaya. 2020. Model systems for regeneration: Xenopus. *Development* 147 (6).

Piatt, Jean. 1955. Regeneration of the spinal cord in the salamander. *Journal of Experimental Zoology* 129 (1):177–207.

Quadri, Syed A., Mudassir Farooqui, Asad Ikram, et al. 2020. Recent update on basic mechanisms of spinal cord injury. *Neurosurgical Review* 43 (2):425–441.

Rafalski, Victoria A., Elena Mancini, and Anne Brunet. 2012. Energy metabolism and energy-sensing pathways in mammalian embryonic and adult stem cell fate. *Journal of Cell Science* 125 (23):5597–5608.

Ryoo, Hyung Don, Travis Gorenc, and Hermann Steller. 2004. Apoptotic cells can induce compensatory cell proliferation through the JNK and the Wingless signaling pathways. *Developmental Cell* 7 (4):491–501.

Sabelström, Hanna, Moa Stenudd, Pedro Réu, et al. 2013. Resident neural stem cells restrict tissue damage and neuronal loss after spinal cord injury in mice. *Science* 342 (6158): 637–640.

Session, Adam M., Yoshinobu Uno, Taejoon Kwon, et al. 2016. Genome evolution in the allotetraploid frog *Xenopus laevis*. *Nature* 538 (7625):336–343.

Shihabuddin, Lamya S., Philip J. Horner, Jasodhara Ray, and Fred H. Gage. 2000. Adult spinal cord stem cells generate neurons after transplantation in the adult dentate gyrus. *Journal of Neuroscience* 20 (23):8727–8735.

Sillar, Keith T., Denis Combes, Sankari Ramanathan, Micol Molinari, and John Simmers. 2008. Neuromodulation and developmental plasticity in the locomotor system of anuran amphibians during metamorphosis. *Brain Research Reviews* 57 (1):94–102.

Sims, R.T. 1962. Transection of the spinal cord in developing *Xenopus laevis*. *Development* 10 (2):115–126.

Slack, J.M., G. Lin, and Y. Chen. 2008. The Xenopus tadpole: A new model for regeneration research. *Cellular and Molecular Life Sciences: CMLS* 65 (1):54–63.

Slater, Paula G, and Larrain, Juan. 2021. Spinal Cord Transection In Xenopus laevis Tadpoles. *Journal of Visualized Experiments* (178), e63276.

Spallanzani, Lazzaro. 1768. *Prodromo di un'opera da imprimersi sopra le riproduzioni animali*. In Modena: Nella Stamperia di Giovanni Montanari. Italy.

Stefanelli, Alberto. 1951. I fenomeni rigenerativi e degenerativi del midollo spinale caudale degli Anfibi e dei Rettili. *Italian Journal of Zoology* 18 (4–6):279–290.

Thuret, Raphaël, Hélène Auger, and Nancy Papalopulu. 2015. Analysis of neural progenitors from embryogenesis to juvenile adult in *Xenopus laevis* reveals biphasic neurogenesis and continuous lengthening of the cell cycle. *Biology Open* 4 (12):1772–1781.

Thuret, Sandrine, Lawrence D.F. Moon, and Fred H. Gage. 2006. Therapeutic interventions after spinal cord injury. *Nature Reviews Neuroscience* 7 (8):628.

Tsarouchas, Themistoklis M., Daniel Wehner, Leonardo Cavone, et al. 2018. Dynamic control of proinflammatory cytokines Il-1β and Tnf-α by macrophages in zebrafish spinal cord regeneration. *Nature Communications* 9 (1):1–17.

Tseng, Ai-Sun, Dany S. Adams, Dayong Qiu, Punita Koustubhan, and Michael Levin. 2007. Apoptosis is required during early stages of tail regeneration in *Xenopus laevis*. *Developmental Biology* 301 (1):62–69.

Vander Heiden, Matthew G., Lewis C. Cantley, and Craig B. Thompson. 2009. Understanding the Warburg effect: The metabolic requirements of cell proliferation. *Science* 324 (5930):1029–1033.

Van Mier, P., H.W.J. Joosten, R. Van Rheden, and H.J. Ten Donkelaar. 1986. The development of serotonergic raphespinal projections in *Xenopus laevis*. *International Journal of Developmental Neuroscience* 4 (5):465–475.

Watson, Charles, George Paxinos, and Gulgun Kayalioglu. 2009. *The spinal cord: A Christopher and Dana Reeve Foundation text and atlas*. Cambridge, MA: Academic Press.

Zhang, Hongbo, Keir J. Menzies, and Johan Auwerx. 2018. The role of mitochondria in stem cell fate and aging. *Development* 145 (8).

Zukor, Katherine A., David T. Kent, and Shannon J. Odelberg. 2011. Meningeal cells and glia establish a permissive environment for axon regeneration after spinal cord injury in newts. *Neural Development* 6 (1):1–23.

Zupanc, Günther K.H., and Ruxandra F. Sîrbulescu. 2011. Adult neurogenesis and neuronal regeneration in the central nervous system of teleost fish. *European Journal of Neuroscience* 34 (6):917–929.

21 Studying Tumor Formation and Regulation in *Xenopus*

Dieter Tulkens and Kris Vleminckx

CONTENTS

21.1. Historical Background 301
21.1.1. *Xenopus*, from Research Tool in Cell and Developmental Biology to Genetically Engineered Cancer Model Organism 301
21.1.2. A Diploid Genome Favors Genetic Research in *Xenopus tropicalis* 302
21.2. Past Observations 302
21.2.1. Early *Xenopus* Embryogenesis as a Source of Information for Studying Pathways and Cellular Processes Involved in Cancer Initiation and Progression 302
21.2.2. Naturally Occurring Tumors in *Xenopus* 302
21.2.3. Induced Tumor–Like Structures 302
21.2.4. Tumor–Immune System Interactions 303
21.2.5. Genetically Engineered *Xenopus* Models 303
21.3. Challenges to Overcome 303
21.4. Genetically Engineered *Xenopus* Models for Cancer Research 303
21.4.1. Generation of Clinically Relevant GEXMs 303
21.4.2. TALEN-Mediated TSG Disruption as a First Genetic *Xenopus tropicalis* Cancer Model 304
21.4.3. CRISPR/Cas9 Cancer Modeling in *Xenopus tropicalis* 305
21.5. Application Potential of *Xenopus tropicalis* Cancer Modeling 305
21.5.1. GEXM for Identification of Novel Tumor Dependencies 305
21.5.2. GEXMs as Tool for Novel Anti-Cancer Compound Validation 306
21.6. Current and Future *Xenopus tropicalis* Cancer Modeling Methodologies 307
21.6.1. Tools for *In Vivo* Monitoring of Tumor Progression 307
21.6.2. Overview of Tumor Cell Transplantation Possibilities 307
21.6.3. Future Prospects for Precise *Xenopus* Gene Editing 308
Acknowledgments 308
References 309

21.1. HISTORICAL BACKGROUND

21.1.1. *Xenopus*, from Research Tool in Cell and Developmental Biology to Genetically Engineered Cancer Model Organism

The first reported experimental animal cancer model was published at the beginning of the previous century by Yamagiwa and Ichikawa, in which the authors treated rabbit ears with coal tar, yielding the first animal model for squamous cell carcinoma (1918). Since then, cancer modeling using laboratory animals evolved as a tremendously broad field accompanied by an ever-evolving assortment of tools and techniques that have aided the establishment and monitoring of these models. In 1988, the first transgenic mouse cancer model was patented, called "Oncomouse" by a group of Harvard researchers, based on their paper in which they generated mice expressing oncogenic fusion genes, resulting in the induction of mammary adenocarcinomas (Stewart, Pattengale and Leder 1984). Cancer modeling in *Xenopus* is a more recent development, especially aided with the invention of genome engineering techniques such as zinc finger nucleases (ZFNs), transcription activator-like effector nucleases (TALENs), and in particular CRISPR/Cas9 (Guo et al., 2014; Nakayama et al., 2014). As already extensively described in other chapters, *Xenopus* earned significant credit for its valuable contributions to the exploration of early developmental processes and the molecular dissection of developmental signaling pathways that, in the majority of cases, appear to be highly conserved between frog and human. However, for studying human genetic diseases, mice and zebrafish tend to be preferred by researchers as genetic vertebrate models. This is evidently because of the size of the research field, the many genetically mutant lines, transgenic reporter animals, standardized experimental protocols, and (for the mouse) a wide range of verified antibodies, as well as well-established genomic data integration networks for these

DOI: 10.1201/9781003050230-24

species (Sprague et al., 2008; Noy et al., 2009). Nevertheless, over the last decades, *Xenopus* has received increasing interest and recognition for its valuable use as a model for human disease (reviewed by Blum & Ott, 2018), especially strengthened by its unique features (further described subsequently). Very recently, *Xenopus* (in particular the true diploid *X. tropicalis*) entered the cancer modeling field as the first robust genetically engineered *Xenopus* model (GEXM) for familial adenomatous polyposis (Van Nieuwenhuysen et al., 2015). With the further expansion of the current genome engineering revolution, we can expect rapid establishment of novel cancer models in *Xenopus* that can serve for exploration of cooperative gene mutations involved in cancer induction or progression and the identification of cancer cell vulnerabilities, which can offer therapeutic opportunities for targeted cancer treatments (Naert et al., 2021).

21.1.2. A Diploid Genome Favors Genetic Research in *Xenopus tropicalis*

With its high fecundity, large externally developing embryos, simple housing demands, and highly conserved developmental pathways, *Xenopus* has proven to be extremely well suited for biomedical research (Tandon et al., 2017). However, since *Xenopus laevis* harbors an allotetraploid genome, employing genetic engineering to mimic human genetic diseases or cancer can be challenging for certain genomic regions in this species (Session et al., 2016). The same applies to some extent to zebrafish, which has at least 20% of its genome duplicated (Postlethwait et al., 2000). In contrast, *Xenopus tropicalis* harbors, just like humans and mice, a true diploid genome displaying high synteny with the human genome (Hellsten et al., 2010). In addition, despite the obvious evolutionary distance between frogs and humans, from what is known today, 79% of human disease genes have a *Xenopus tropicalis* orthologue (Hellsten et al., 2010; Khokha, 2012). In conclusion, these special features place *X. tropicalis* in a unique position for employing nuclease-based techniques with regard to disease/cancer modeling.

21.2. PAST OBSERVATIONS

21.2.1. Early *Xenopus* Embryogenesis as a Source of Information for Studying Pathways and Cellular Processes Involved in Cancer Initiation and Progression

Many mechanisms and hallmarks underlying cancer initiation and progression (Hanahan and Weinberg, 2000, 2011) are comparable between a wide range of species, including *Xenopus* (Hardwick and Philpott, 2018). As cancer development and progression often involve the re-use of particular embryonic pathways and processes (Pennisi, 1998), *Xenopus* serves as an important tool for elucidating these normal developmental processes in order to understand the parallel mechanisms and molecular players underlying malignant transformation. A considerable amount of research has indeed exposed high similarities between normal embryonic developmental processes and particular tumor cell behavior, especially at the level of gene and protein expression, epigenetic regulation, and cell invasion and migration (Ma et al., 2010). As an example, Wnt signaling, which is known to play a key role in embryonic development (e.g. during body axis formation, cell migration, cell fate determination, and others), is deregulated in the vast majority of human colon cancers (Hardwick and Philpott, 2015). Interestingly, *Xenopus* researchers have exploited a functional biological assay based on duplication of the dorsal axis upon the ventral injection of mRNA-encoding mediators of the Wnt pathway, such as β-catenin, to develop chemical screens (Kühl and Pandur, 2008). Regarding embryonic morphogenetic processes re-activated during tumor progression, epithelial to mesenchymal transition (EMT), occurring for instance during neural crest cell delamination, is a key process during tumor cell invasion (Yang et al., 2020). The molecular processes underlying EMT during neural crest cell delamination have been intensively investigated by several *Xenopus* research groups (Pegoraro and Monsoro-Burq, 2013). Interestingly, *Xenopus* anti-EMT compound screens have been deployed in embryos, which identified chemicals that also affect cancer cell invasion (Tanaka et al., 2016) (see section 5.2).

21.2.2. Naturally Occurring Tumors in *Xenopus*

Spontaneous occurrence of tumors is rather uncommon in *Xenopus*, leading to the now-contested belief that frogs are rather recalcitrant to carcinogenesis. This was further strengthened by the fact that potent human carcinogens have shown limited to no effect in *Xenopus laevis* (Hardwick and Philpott, 2018). For example, attempts with N-methyl-N-nitrosourea (NMU), a known human carcinogen, failed to induce tumors in *Xenopus laevis* (Goyos and Robert, 2009). However, cases of spontaneous adenocarcinoma, fibroma, lipoma, lymphosarcoma, and leukemia have been occasionally documented (Balls, 1962; Stacy and Parker, 2004; Suzuki et al., 2020). Finally, a number of naturally occurring thymic tumors, from which several lymphoid tumor cell lines could be established, are now being used for elucidating key mechanisms for tumor versus immune system interactions, especially via transplantation experiments (see subsequently) (Robert, Guiet and du Pasquier, 1994; Goyos and Robert, 2009).

21.2.3. Induced Tumor-Like Structures

The very first report of genetically induced tumors in *Xenopus laevis* embryos dates from 1997, where mRNA encoding human dominant-negative p53 was injected in early embryos, inducing the formation of embryonic tumors, so-called induced tumor-like-structures (ITLSs) (Wallingford et al., 1997; Wallingford, 1999). Closely thereafter, similar approaches were undertaken to model tumors in embryos upon injection of *gli1* or *rel* (*Xrel3*) mRNA (Dahmane et al.,

1997; Yang et al., 1998). Relatively recently, the technique for inducing tumors in embryos was used to study novel mechanisms in tumor formation. By investigating ITLSs, it was found that transmembrane voltage potential is a crucial cellular parameter for tumor detection and control, highlighting opportunities for novel therapeutic intervention (Chernet and Levin, 2013). The same research group also identified the importance of long-range gap junctional signaling in regulating tumorigenesis (Chernet, Fields and Levin, 2015).

21.2.4. Tumor–Immune System Interactions

Xenopus appears to be well positioned for investigating the role of the immune system against cancer. The majority of the mechanisms and cell types governing both the innate and the adaptive immune system are highly conserved between *Xenopus* and humans (Robert and Ohta, 2009), and experiments in *Xenopus laevis* have exposed crucial roles for CD8+, NK, and NKT-like cells in anti-tumor defenses (Goyos and Robert, 2009; Banach and Robert, 2017; Banach et al., 2019). In addition, an elegant strategy for modeling tumors in *Xenopus* is used based on the subcutaneous transplantation of a lymphoid tumor cell/collagen mixture matrix in isogenic tadpoles, allowing the study of the tumor microenvironment, neovascularization, and immune cell interactions (Haynes-Gimore et al., 2015). Interestingly, in light of the recent promising developments in the use of immune checkpoint inhibition for treating a range of aggressive cancers in humans (Taube et al., 2014), the *Xenopus* genome contains the orthologues for both the human receptors *PDCD1* (*PD1*) and *CTLA4* as well as their respective ligands *CD274* (*PDL1*) and *CD86*.

21.2.5. Genetically Engineered *Xenopus* Models

With the invention of nuclease-based genome editing systems like ZFNs, TALENs, and CRISPR/Cas9, cancer research experienced an unprecedented push forward. Next to mice and zebrafish, numerous genetically engineered organisms have emerged for cancer modeling, ranging from yeast (reviewed by Guaragnella et al., 2014), *Drosophila* (reviewed by Mirzoyan et al., 2019), and swine (reviewed by Watson et al., 2016) to *Xenopus tropicalis* (Van Nieuwenhuysen et al., 2015; Naert et al., 2016; Naert, Dimitrakopoulou, et al., 2020). Especially favored by its diploid genome, genetically engineered *Xenopus tropicalis* might become extremely useful in deciphering factors that drive cancer or expose genetic cancer vulnerabilities.

21.3. CHALLENGES TO OVERCOME

As with other animal models, GEXMs can experience some criticism from the scientific and medical communities. Therefore, in this section, we will highlight some concerns that might arise and how to answer potential reviewers' critique regarding the use of GEXM in cancer research.

An obvious criticism that may arise is that of the greater evolutionary distance that exists between amphibia and mammals. Without any doubt, mice models currently better reflect the patient situation. However, *Xenopus*—like zebrafish—can perfectly serve as complementary model, since generating disease models (even in F0) is extremely straightforward, and findings can be eventually extrapolated to mammalian organisms such as mice. Additionally, it is worth mentioning that in contrast to its teleost counterpart, *Xenopus* does have limbs, lungs, and urinary bladder, allowing the modeling of diseases and or cancer in these organ systems. An additional major shortcoming of current GEXMs is that the genetic insults are not introduced in a tissue-, organ-, or cell-type-specific manner. While inducible Cre-Lox technology would be applicable in *Xenopus*, there is limited reason or benefit to introduce this technology for cancer modeling in *Xenopus*, given the extensive genetic tools currently available in the mouse. However, the ease with which targeted injections can be performed in the external developing *Xenopus* embryos, especially with the existence of a detailed fate map (Dale and Slack, 1987; Moody, 1987), makes tissue-restricted genome engineering possible (Van Nieuwenhuysen et al., 2015). This in turn also lowers the danger of abrogating embryonic processes leading to lethality. Of note, we want to point out that even for efficient guide RNAs injected in one- or two-cell-stage embryos, the significant occurrence of small functional in-frame mutations, as well as the possibility to carefully dose the concentration of the genome-editing reagents, might be sufficient to retain normal functionality in order to avoid perturbing embryonic development (Naert, Tulkens, et al., 2020).

Another major shortcoming is the fact that cancer formation in GEXMs generated via multiplexing injections results from the simultaneous introduction of mutations in genes, while in normal oncogenesis, this is a sequential process, with intermediate selection of certain cell sub-populations. A final drawback that one might notice is the limitation of only being able to generate loss-of-function mutations, as knock-in engineering remains challenging (Aslan et al., 2017). For that, we refer to recently published technologies that may be used in the future to induce expression of oncogenes (described in section 6.3).

Of note, an issue that might arise when using CRISPR- or TALEN-mediated genome editing is the phenomenon of off-target effects. However, when targeting cancer-related genes, the genetic insult will be under positive selection. Therefore, if a developing tumor does not show the intended biallelic inactivation of the TSG or monoallelic inactivation combined with loss of heterozygosity (LOH), tumor induction may have been the result of an off-target effect. Similarly, when introducing activating mutations in a proto-oncogene, this should be apparent in at least one of the two targeted alleles of the sampled tumor.

21.4. GENETICALLY ENGINEERED *XENOPUS* MODELS FOR CANCER RESEARCH

21.4.1. Generation of Clinically Relevant GEXMs

From genome-wide studies, it is becoming clear that specific genetic alterations found in human cancers are often associated with activation of discrete signaling pathways. Interestingly, activation of these cancer pathways can often be mediated by either the aberrant activation of oncogenes or by the inactivation of tumor suppressor genes. As an example, the Wnt pathway can be activated by either a gain-of-function (GOF) mutation in the proto-oncogene *CTNNB1* or a loss-of-function (LOF) mutation in the tumor suppressor gene *APC*. Similarly, GOF mutation in *PI3KCA* or LOF of the TSG *PTEN* in essence have a similar functional outcome. As described earlier, the inactivation of TSGs is most convenient to achieve via CRISPR. In addition, it is also very straightforward to introduce LOF mutations simultaneously in multiple tumor suppressor genes by means of multiplexed injections.

Modeling human cancer in *Xenopus* can encompass the generation of heterozygote or homozygote lines harboring mutations in key TSGs (e.g. *tp53-/-, tp53+/-, apc+/-*) that consequently are prone to accelerated cancer development during their life span (Naert, Dimitrakopoulou, et al., 2020). However, in these models, cancer initiation is dependent on the stochastic acquisition of second hit mutations or LOH, which may still take several months or even years. Alternatively, since cells carrying biallelic hits in TSGs encounter strong positive selective pressure, rapid F0 modeling is possible even with low-efficiency nucleases and in a multiplexed fashion (Van Nieuwenhuysen et al., 2015; Zuckermann et al., 2015; Naert et al., 2016; Naert, Dimitrakopoulou, et al., 2020). Since loss of biallelic WT alleles and/or second hit mutations are experimentally induced at an early developmental time point, these models in general are very fast, mostly requiring less than three months until tumor formation, and are very penetrant (Figure 21.1).

21.4.2. TALEN-Mediated TSG Disruption as a First Genetic *Xenopus tropicalis* Cancer Model

The very first genuine genetic *Xenopus tropicalis* tumor model was established in 2015 using TALEN nucleases guided to introduce truncating mutations in the hotspot mutation region of the TSG *apc*. The resulting tadpoles presented with a tumor spectrum resembling the familial adenomatous polyposis (FAP) and Gardners' syndrome, including intestinal neoplasia (but no intestinal adenomas), desmoid tumors, medulloblastomas, retinal hyperproliferation, and epidermoid

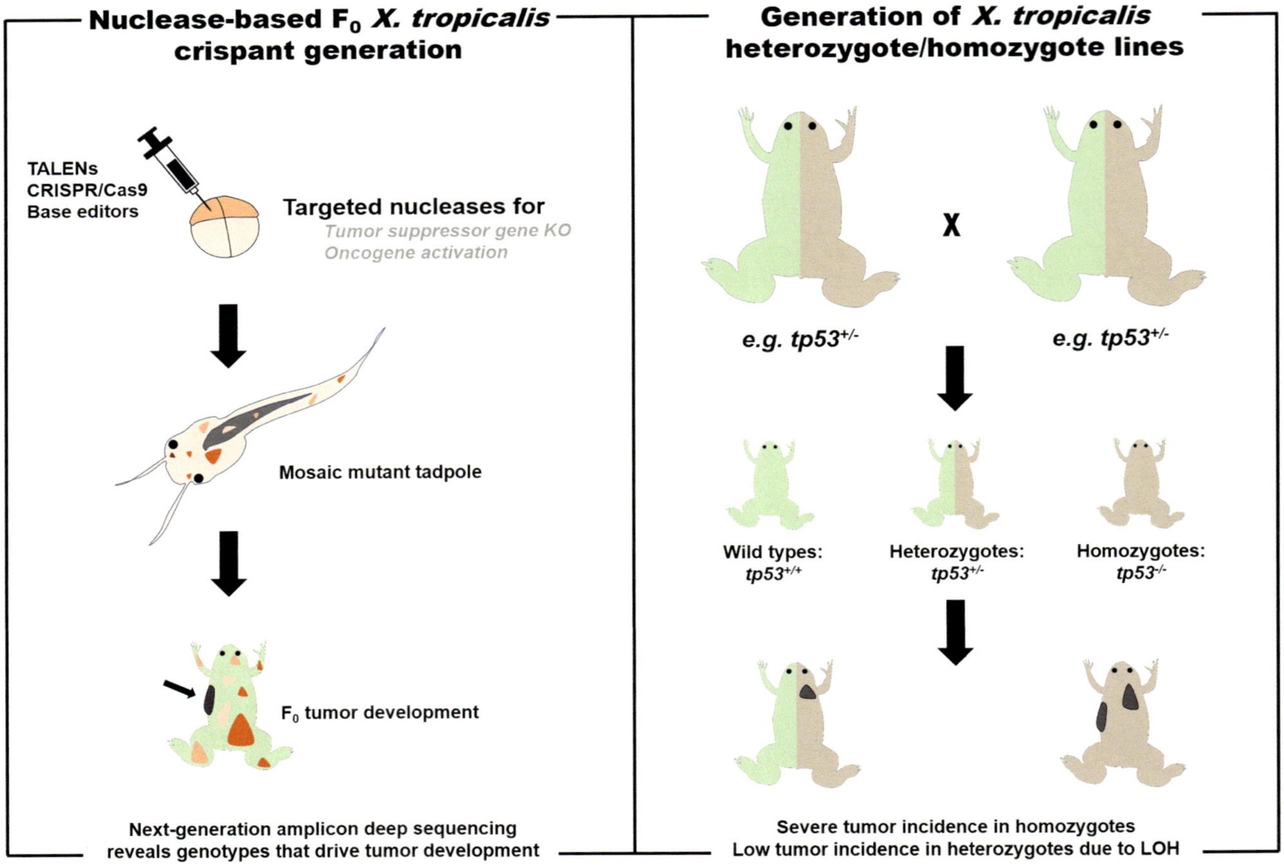

FIGURE 21.1 (Left) Pipeline for the generation of F_0 *X. tropicalis* tumor bearing crispants by targeted blastomere injections using nucleases. Tumor formation relies on the principle of positive selection, in which mutant cells can obtain a proliferative advantage, further validated via next-generation amplicon deep sequencing. (Right) Generation of *X. tropicalis* cancer models via heterozygote intercrossing of animals carrying a loss-of-function mutation in a tumor suppressor gene (e.g. $tp53^{+/-}$). In heterozygote animals tumor formation in general occurs upon loss-of heterozygosity (LOH).

cysts. Already four to five weeks after injection, tadpoles showed organization abnormalities in the intestine, and desmoid tumors became apparent around metamorphosis (Van Nieuwenhuysen et al., 2015). Of note, by using targeted injection of the *apc* TALEN mRNAs in a single blastomere of an eight-cell embryo, it was found that the occurrence of specific tumor phenotypes and survival rates highly depended on the blastomere that was injected, further demonstrating the virtue of targeted injections. For example, desmoid tumors were not observed when embryos were injected animal-dorsally (blastomere assignation) (Van Nieuwenhuysen et al., 2015). As a side note, these TALEN injections eventually led to the generation of a heterozygote *apc* line.

21.4.3. CRISPR/Cas9 Cancer Modeling in *Xenopus tropicalis*

TALENs can be extremely useful as genome editing reagents and are by their design and molecular mode of action less prone to off-target effects. However, CRISPR, although being PAM sequence restricted, has several interesting features over TALENs, such as higher *in vivo* effectiveness, ease of use, shorter time required for reagent assembly, and higher multiplexing potential (Nemudryi et al., 2014). These unique characteristics were translated into the generation of the first CRISPR/Cas9 cancer models in *Xenopus tropicalis*.

Shortly after the first *Xenopus tropicalis apc* TALEN tumor model, the first *Xenopus tropicalis* CRISPR/Cas9 tumor model was generated, closely mimicking human retinoblastoma by the concomitant biallelic inactivation of *rb1* and *rbl1* (Naert et al., 2016). Interestingly, while in humans, biallelic disruption of the *RB1* gene is sufficient to induce retinoblastoma (Lohmann, 1999), the additional inactivation of the retinoblastoma-like *rbl1* gene was required for retinoblastoma formation in *Xenopus*. In fact, *Xenopus tropicalis* thereby parallels the situation in the mouse, where the simultaneous inactivation of *Rb1* and *Rbl1/p107* is also required (Robanus-Maandag et al., 1998). Importantly, retinoblastoma induction in the *rb1/rbl1* crispant model showed very high penetrance (up to 73%) and short latency (median time of 69 days). In addition to the prevalent retinoblastoma, occasional appearance of a brain tumor type (probably pinealoblastoma), as well as choroid plexus neoplasms and small cell lung cancer, was observed (Naert et al., 2016).

Apart from the *apc* heterozygote line, generated via TALENs, two novel cancer-prone *Xenopus tropicalis* lines (*tp53+/−* and *tp53−/−*) were recently generated using CRISPR/Cas9 (Naert, Dimitrakopoulou, et al., 2020). These animals resemble the human Li-Fraumeni syndrome (LFS), with manifestation of hematologic malignancies and sarcomas in direct accordance with previously published *Tp53* mutant mouse models (Donehower et al., 1992; Jacks et al., 1994). While *tp53+/− Xenopus* animals showed normal survival in the first two years, *tp53−/−* animals already demonstrated clear signs of disease at the age of one year, and only 34% of the animals were alive after two years (Naert, Dimitrakopoulou, et al., 2020). It is important to notice that the time until morbidity was substantially longer than is the case for homozygous *Tp53* knockout mice (4.5 months) (Donehower, 1996), which we believe may be related to the longer life span of *Xenopus* compared to the mouse.

Given the involvement of *TP53* mutations as a cooperative event in the vast majority of human cancers, the availability of *tp53* mutant lines also offers opportunities for new applications for a range of additional cancer models in *Xenopus*. In the cancer models obtained by simultaneous disruption of *rb1* and *rbl1*, it was found that additional *tp53* disruption aggravated tumor malignancy for the choroid plexus carcinomas and the gliomas (Konukiewitz et al., 2017; Naert, Dimitrakopoulou, et al., 2020). In a pilot multiplexing experiment combining four sgRNAs (*rb1, rbl1, pten*, and *tp53*), it could be determined that *pten* disruption further aggrevated the glioblastoma phenotype (Naert, Dimitrakopoulou, et al., 2020), which is in line with clinical data (Xiao et al., 2002; Chow et al., 2011).

21.5. Application Potential of *Xenopus tropicalis* Cancer Modeling

21.5.1. GEXM for Identification of Novel Tumor Dependencies

Currently, there is an increased interest in the identification of druggable gene products that are crucial for tumor growth. These targets, further called dependency factors, are nowadays being identified employing so-called "negative selection dependency screens" using techniques such as RNAi, shRNA, and CRISPR/Cas9 (Chen et al., 2015, 2018; Tzelepis et al., 2016; Tsherniak et al., 2017). Interestingly, recently, a method for *in vivo* CRISPR/Cas9 Selection-mediated Identification of Dependencies (CRISPR-SID) was developed using the *apc*-based *Xenopus tropicalis* desmoid tumor model (Naert et al., 2021). Using a multiplexing strategy, a guide RNA for the tumor suppressor gene *apc* is co-injected with a guide RNA against a putative dependency gene, and animals are grown until they develop desmoid tumors. Due to their pure monoclonal and non-metastasizing growth, the desmoid tumors are ideally suited for performing these targeted dependency screening approaches. Genotyping of the isolated desmoid tumors allows the identification of dependency genes. More specifically, one can quantitatively assess selection mechanisms by comparing for a particular guide RNA, the induced double strand break (DSB) repair outcomes in normal embryonic tissue in the absence of negative selection (determined by deep amplicon sequencing) with the repair outcomes observed in experimental tumors. Specifically, the tumor will select for out-of-frame insertion-deletion (INDEL) mutations in tumor suppressor genes (= positive selection) while favoring absence of gene editing or the presence of in-frame INDELs in essential genes or dependency genes. Hence, for a genuine dependency gene, in all the tumors sampled, at least one of the two alleles of the targeted dependency gene will be either wild type or have an in-frame mutation that maintains protein function to allow sustained tumor growth (Figure 21.2).

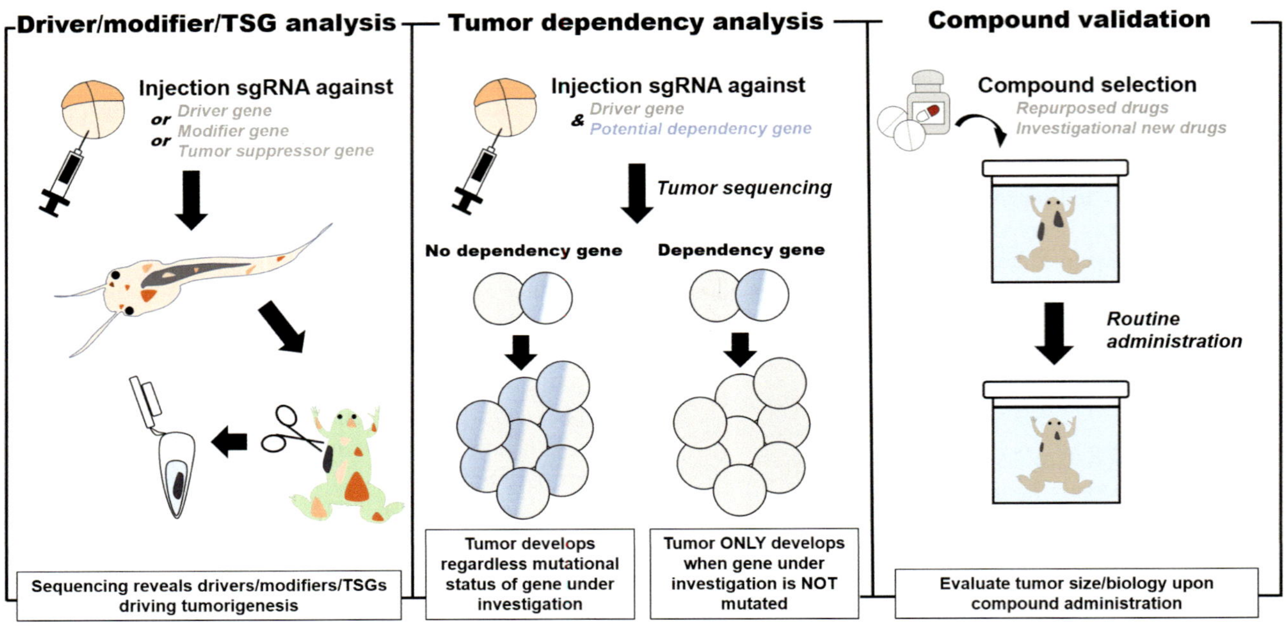

FIGURE 21.2 Overview of applications in *Xenopus tropicalis* cancer models. (Left) F0 *X. tropicalis* cancer modeling using targeted nucleases combined with next-generation amplicon deep sequencing can provide rapid (two to three months) genetic information exposing cancer drivers/modifiers/TSGs. (Middle) Robust *X. tropicalis* cancer models can be used for the identification of novel tumor dependency genes. (Right) GEXMs can be used for validating potential compounds (e.g. repurposed or investigational new drugs).

This CRISPR-SID methodology can now be implemented in other *Xenopus tropicalis* solid tumor cancer models (Naert, Dimitrakopoulou, et al., 2020) or even hematological cancer models (reviewed by Dimitrakopoulou et al., 2019). *Xenopus tropicalis* can thus serve as a rapid and efficient preclinical model for the validation of dependency genes, which in turn opens the road for novel molecular therapies.

21.5.2. GEXMs as Tool for Novel Anti-Cancer Compound Validation

Similar to zebrafish (for comprehensive review, see Letrado et al., 2018), *Xenopus* can be used for identification and validation of novel anti-cancer compounds. In zebrafish compound administration can be done simply by addition to the water (Lieschke and Currie, 2007), via direct intraperitoneal (Kinkel et al., 2010) and retro-orbital injection (Pugach et al., 2010), or even via oral gavage (Dang, Fogley and Zon, 2016). In addition, rapid zebrafish embryonal anti-cancer compound screens, in which embryos with a "proxy-cancer phenotype" (embryos that show tumor-like structures in early embryonal stages) are subjected to hundreds of compounds, have proven very effective (White, Rose and Zon, 2013). When the compound is administered to the zebrafish water, it can be taken up via the skin or via direct swallowing, as demonstrated by a study where uptake was monitored using liquid chromatography-tandem mass spectrometry (LC-MS/MS) (Zhang et al., 2015). *Xenopus* has also been used in screens to identify compounds for the treatment of rare human inherited diseases or to uncover novel agonists and antagonist of key embryonic pathways (reviewed by Schmitt et al., 2014; Wheeler & Liu, 2012). As was already described, early *Xenopus* embryos have also been subjected to a range of potential anti-EMT compounds and screened for possible perturbations of gastrulation and migration of cranial neural crest cells (Tanaka et al., 2016). The identified potent compounds interfering with embryonic EMT processes were subsequentially investigated in mouse cancer models (Tanaka et al., 2016). While screening of early embryonic phenotypes can be performed on a large scale in a small multi-well format, tumor-bearing GEXMs in general are too large for such an application. However, compound validation experiments can be performed in these GEXMs (Figure 21.2). In a proof-of-concept experiment, our research group recently applied a *Xenopus tropicalis* drug treatment approach in which desmoid tumor-bearing adult animals were treated with Tazemetostat (EZP-6438), a known human EZH2 inhibitor, by simple administration of the compound to the rearing water. Interestingly, magnetic resonance imaging (MRI) of treated animals showed a significant shrinkage of the tumors (Naert et al., 2021). We truly believe that these findings might accelerate the process of novel anti-cancer compound identification. Also, with the advances in *Xenopus* tumor cell transplantations and availability of reporter lines, we foresee that novel experiments will become feasible combining allotransplantations with compound administration.

It should be highlighted that for compound treatments, parameters like absorption, bioavailability, dosage, water solubility, and degradability are to be taken into account when small molecules are administered to the water. This is already the case for embryos and larvae where compounds are taken up via the skin or gills or via direct swallowing (Wheeler and Brändli, 2009) but is likely even more important when treating post-metamorphic animals. A critical question that arises when adding compounds to the water is the determination of the dose and the compound refreshment scheme. There is no "gold standard" for this, but it may evidently be possible to determine the concentration of the compound in the liver or in targeted organs by performing mass spectrometry. However, the best test to determine the required concentration is to look at a biological read-out for the compound. As an example, we documented H3K27me3 levels in the liver to determine the required concentration of the previously mentioned EZH2 inhibitor Tazemetostat. Of note, the required concentration in the frog water was 100 times of what was needed in cell culture experiments. We refreshed half of the water (with compound) daily or every second day. For compounds that are not taken up via the rearing water or that are very expensive, intraperitoneal injection is relatively easy to perform, and a protocol for oral gavage has been described for *Xenopus laevis* (Du, Mashoof and Criscitiello, 2012).

21.6. CURRENT AND FUTURE *XENOPUS TROPICALIS* CANCER MODELING METHODOLOGIES

21.6.1. Tools for *In Vivo* Monitoring of Tumor Progression

Correct follow-up of tumor progression and quantitative phenotypic analysis are essential to cancer research studies. While external signs of morbidity associated with cancer development such as lethargy, cachexia, or a swollen abdomen may be straightforward to observe, they only provide limited information on the actual pathological status of the underlying malignancy. Furthermore, when investigating responses on treatments, a longitudinal follow-up of disease status is paramount to assess the effectiveness of the treatment. Currently, micro CT scanning has proven its usefulness for uncovering ectopic calcified outgrowths, such as osteosarcoma, in *tp53* heterozygote animals (Naert, Dimitrakopoulou, et al., 2020). Likewise, magnetic resonance imaging (MRI) of adult *Xenopus tropicalis* tumor-bearing animals is a useful tool to follow efficacy of compound treatments (Naert et al., 2021). In addition, optical coherence tomography may serve as alternative tool for monitoring cancer development, which was nicely demonstrated in mice (reviewed by Vakoc et al., 2012). While the latter method should be applicable to *Xenopus* (Boppart et al., 1997), its use may be primarily restricted to tumors in early tadpoles, since the tissue penetrance depth is limited to approximately 2 mm. As an alternative, a penetrance depth of centimeters can be obtained in *Xenopus* via ultrasound imaging (Bartlett et al., 2010; Slater et al., 2019) and could be applicable for monitoring tumor progression, as is done in mouse studies (Snyder et al., 2009). Finally, the generation of stable transgenic reporter lines can be a very valuable tool for following tumor progression. Of note for the detection of tumors in the internal organs, and given the size of the frogs and the opacity of the post-metamorphic skin, bioluminescent reporters may be better suited than fluorescent reporter genes to detect tumor masses.

21.6.2. Overview of Tumor Cell Transplantation Possibilities

Transplantation experiments are useful tools for investigating the role of the host immune system upon subcutaneous tumor cell injection in *Xenopus* tadpoles (Haynes-Gimore et al., 2015) and are also useful for propagating tumor cells. In addition, with regard to cancer research, investigation of tumor engraftment potential has already shown its usefulness in validating zebrafish leukemia models (Smith et al., 2010; Borga et al., 2019). Currently, multiple possibilities exist that are suitable for performing *Xenopus* tumor transplantation experiments. First, transplantation of tumor cells in thymectomized *Xenopus* animals, thus lacking the main functional T-cell compartment, offers a possibility to avoid graft rejection (Robert et al., 1997). Besides, syngeneic lines (e.g. LG-6, LG-15 . . .) being MHC identical have proven their usefulness for tumor transplantation purposes (Hadji-Azimi and Fischberg, 1971; Rau et al., 2002; Haynes-Gilmore et al., 2014, 2015). Additionally, sublethal gamma irradiation has been used to allow transplantation of spontaneous lymphoid tumors (Robert, Guiet and du Pasquier, 1995; Rau, Cohen and Robert, 2001) or even for transplantation of leukemic cells derived from GEXMs (Tulkens et al., 2021). We would also like to mention that a *rag2* homozygote knock-out line (lacking mature T- and B-cells), which has already proved its usefulness in zebrafish transplantation experiments (Tang et al., 2014), has recently been generated in *Xenopus tropicalis* and shows engraftment potential for multiple tumor types (Tulkens et al., 2021). Finally, while in *rag2* knockout animals, rejection of allografted tumors may still occur by natural killer cells or other populations of the innate immune system, this potential problem can be circumvented by the use of fully immunocompetent *X. tropicalis* inbred lines (Sato et al., 2018).

Recently, zebrafish larvae and immunocompromised adults have been used as recipients for xenotransplantation of human cancer cells and even patient-derived tumor biopsies. These so-called "avatars" can be used for phenotypic testing of drug responses with the ultimate goal of finding patient-tailored molecular therapies (Fazio et al., 2020). Recent work by the Langenau laboratory described the generation of a semi-transparent *prkdc−/−*, *il2rgα−/−* line that lacks adaptive immune cells and natural killer cells and can be adapted to 37°C before being engrafted with human cancer cells, after which candidate therapeutics were administered via oral gavage (Yan et al., 2019). While this may seem an attractive model to introduce in *Xenopus*, the need to adapt the animals to 37°C may not be achievable, and one may also wonder what the added value over the zebrafish model would be.

21.6.3. Future Prospects for Precise *Xenopus* Gene Editing

Current GEXMs only encompass inactivating mutations in tumor suppressor genes, as generating activating mutations in proto-oncogenes is currently challenging and inefficient. However, a major advantageous characteristic is that tumor formation is under positive selection, thereby requiring the activating mutation(s) to be present in a limited number of cells, and alternative methods exist for the efficient induction of oncogenic gene activation.

1. For example, in the context of leukemia, a hyperactive Notch1 protein can be obtained simply by targeting a specific region in the last exon of the *notch1* gene. Introduction of a frameshift INDEL mutation and subsequent translational termination will thus result in the expression of a truncated Notch1 protein that lacks the C-terminal PEST sequence that targets the protein for proteolytic degradation. This in fact recapitulates the mutations found in a substantial fraction of T-ALL patients and results in increased Notch signaling, ultimately driving oncogenic transformation in the T-cell lineage (Baldus et al., 2009). Of note, while the inactivation of tumor suppressor genes requires the disruption of the two alleles, oncogenic mutations are dominant, allowing the maintenance of one wild type allele.
2. Many cancers are driven by oncogenic fusion proteins. *In vivo* viral delivery of CRISPR/Cas9 reagents in mice has been used for efficient generation of oncogenic fusion proteins showing penetrant lung tumor formation (Blasco et al., 2014; Maddalo et al., 2014). This strategy builds on the simple injection of two guide RNAs targeting intron regions of two genes on the same chromosome, being separated by more than 10 Mb from each other, resulting in a fusion of both loci mimicking the patient situation (Blasco et al., 2014; Maddalo et al., 2014). Considering the concept of positive selection, this technique might be extrapolated to generate F0 mosaic mutant GEXMs harboring fusion oncogenes, whether or not in combination with cooperating loss-of-function event(s) in one or more tumor suppressor gene(s).
3. The generation of knock-ins in embryos using CRISPR/HDR (homology directed repair)-mediated targeted gene editing in *Xenopus* remains challenging, as low homologous repair rates (<3%) are observed in *Xenopus* F0 mosaic mutant embryos (Aslan et al., 2017). This could hamper the generation of additional cancer predisposition models such as Li-Fraumeni syndrome, where particular missense mutations may be associated with dissimilar tumor spectra or latency for disease onset (Bougeard et al., 2008). However, Aslan et al. (2017) described that *Xenopus* oocytes show significant higher homology-directed repair activity. For this methodology, CRISPR/Cas9 reagents and a single-strand DNA template were co-injected in oocytes, and subsequent host-transfer yielded editing efficiencies up to 25.7% (Aslan et al., 2017). Therefore, this method would allow the modeling of variants of uncertain significance (VUS) in the contexts of human cancer predisposition (Eggington et al., 2014).
4. Additionally, we would like to mention that, very recently, we validated for the first time that *in vivo* genome editing outcomes using CRISPR/Cas9 can be accurately predicted using *in silico* prediction software like InDelphi both in *Xenopus tropicalis* and *laevis* as well as in zebrafish (Naert, Tulkens, et al., 2020). This interesting finding offers unique opportunities for selecting guide RNAs favoring frameshift mutations, thus maximizing F0 phenotype penetrance or even for selecting guides that give a predominance for a specific small INDEL which can facilitate modeling patient specific mutations in a rapid manner (Naert, Tulkens, et al., 2020).
5. Finally, we would like to mention that with the generation of CRISPR base editors (Komor et al., 2016), the PITCh (precise integration into target chromosome) system (Sakuma et al., 2016), and CRISPR prime editing (Anzalone et al., 2019), precise cancer modeling will be significantly enhanced in the (near) future, offering novel opportunities for more GEXMs.

In conclusion, in the *Xenopus* field, major steps have already been taken in unraveling genetic factors that drive cancer formation and maintenance. However, we believe that this might be only an initial step stone to more extensive applications, as with the expansion and fine-tuning of genome engineering techniques, this field is rapidly evolving. Considering thereby the unique benefits of performing cancer research in *Xenopus tropicalis*, primarily fast and efficient GEXM generation with concomitant targeting of multiple genes, this organism can eventually be nicely complementary to mice and zebrafish and aid in the development of novel precise cancer therapies.

ACKNOWLEDGMENTS

Research in the Vleminckx laboratory is supported by the Research Foundation–Flanders (FWO-Vlaanderen) and by the Concerted Research Actions from Ghent University (BOF15/GOA/011 and BOF20/GOA/23). Further support was obtained by the Desmoid Tumor Research Foundation, the Desmoid Tumor Foundation of Canada, and SOS Desmoïde.

REFERENCES

Anzalone, A. V. et al. (2019) 'Search-and-replace genome editing without double-strand breaks or donor DNA', *Nature*, 576(7785), pp. 149–157. doi: 10.1038/s41586-019-1711-4.

Aslan, Y. et al. (2017) 'High-efficiency non-mosaic CRISPR-mediated knock-in and indel mutation in F0 Xenopus', *Development (Cambridge)*, 144(15), pp. 2852–2858. doi: 10.1242/dev.152967.

Baldus, C. D. et al. (2009) 'Prognostic implications of NOTCH1 and FBXW7 mutations in adult acute T-lymphoblastic leukemia', *Haematologica*, 94(10), pp. 1383–1390. doi: 10.3324/haematol.2008.005272.

Balls, M. (1962) 'Spontaneous neoplasms in amphibia: A review and descriptions of six new cases', *Cancer Research*, 22(10).

Banach, M. and Robert, J. (2017) 'Tumor immunology viewed from alternative animal models—the Xenopus story', *Current Pathobiology Reports*, 5(1), pp. 49–56. doi: 10.1007/s40139-017-0125-y.

Banach, M. et al. (2019) 'Impacts of the MHC class I-like XNC10 and innate-like T cells on tumor tolerance and rejection in the amphibian Xenopus', *Carcinogenesis*, 40(7), pp. 924–935. doi: 10.1093/carcin/bgz100.

Bartlett, H. L. et al. (2010) 'Echocardiographic assessment of cardiac morphology and function in Xenopus', *Comparative Medicine*, 60(2), pp. 107–113. Available at: /pmc/articles/PMC2855036/?report=abstract (Accessed: 12 November 2020).

Blasco, R. B. et al. (2014) 'Simple and rapid in vivo generation of chromosomal rearrangements using CRISPR/Cas9 technology', *Cell Reports*, 9(4), pp. 1219–1227. doi: 10.1016/j.celrep.2014.10.051.

Blum, M. and Ott, T. (2018) 'Xenopus: An undervalued model organism to study and model human genetic disease', *Cells Tissues Organs*, 205(5–6), pp. 303–313. doi: 10.1159/000490898.

Boppart, S. A. et al. (1997) 'Noninvasive assessment of the developing Xenopus cardiovascular system using optical coherence tomography', *Proceedings of the National Academy of Sciences of the United States of America*, 94(9), pp. 4256–4261. doi: 10.1073/pnas.94.9.4256.

Borga, C. et al. (2019) 'Simultaneous B and T cell acute lymphoblastic leukemias in zebrafish driven by transgenic MYC: Implications for oncogenesis and lymphopoiesis', *Leukemia*, 33(2), pp. 333–347. doi: 10.1038/s41375-018-0226-6.

Bougeard, G. et al. (2008) 'Molecular basis of the Li-Fraumeni syndrome: An update from the French LFS families', *Journal of Medical Genetics*, 45(8), pp. 535–538. doi: 10.1136/jmg.2008.057570.

Chen, L. et al. (2018) 'CRISPR-Cas9 screen reveals a MYCN-amplified neuroblastoma dependency on EZH2', *Journal of Clinical Investigation*, 128(1), pp. 446–462. doi: 10.1172/JCI90793.

Chen, S. et al. (2015) 'Genome-wide CRISPR screen in a mouse model of tumor growth and metastasis', *Cell*, 160(6), pp. 1246–1260. doi: 10.1016/j.cell.2015.02.038.

Chernet, B. T., Fields, C. and Levin, M. (2015) 'Long-range gap junctional signaling controls oncogene-mediated tumorigenesis in *Xenopus laevis* embryos', *Frontiers in Physiology*, 5, p. 519. doi: 10.3389/fphys.2014.00519.

Chernet, B. T. and Levin, M. (2013) 'Transmembrane voltage potential is an essential cellular parameter for the detection and control of tumor development in a Xenopus model', *DMM Disease Models and Mechanisms*, 6(3), pp. 595–607. doi: 10.1242/dmm.010835.

Chow, L. M. L. et al. (2011) 'Cooperativity within and among Pten, p53, and Rb pathways induces high-grade astrocytoma in adult brain', *Cancer Cell*, 19(3), pp. 305–316. doi: 10.1016/j.ccr.2011.01.039.

Dahmane, N. et al. (1997) 'Activation of the transcription factor Gli1 and the sonic Hedgehog signalling pathway in skin tumours', *Nature*, 389(6653), pp. 876–881. doi: 10.1038/39918.

Dale, L. and Slack, J. M. (1987) 'Fate map for the 32-cell stage of *Xenopus laevis*', *Development*, 99(4), pp. 527–551.

Dang, M., Fogley, R. and Zon, L. I. (2016) 'Identifying novel cancer therapies using chemical genetics and zebrafish', in *Advances in Experimental Medicine and Biology*, 916, pp. 103–124. doi: 10.1007/978-3-319-30654-4_5.

Dimitrakopoulou, D. et al. (2019) '*Xenopus tropicalis*: Joining the Armada in the fight against blood cancer', *Frontiers in Physiology*, 10, p. 48. doi: 10.3389/fphys.2019.00210.

Donehower, L. A. (1996) 'The p53-deficient mouse: A model for basic and applied cancer studies', in *Seminars in Cancer Biology*, 7(5), pp. 269–278. doi: 10.1006/scbi.1996.0035.

Donehower, L. A. et al. (1992) 'Mice deficient for p53 are developmentally normal but susceptible to spontaneous tumours', *Nature*, 356(6366), pp. 215–221. doi: 10.1038/356215a0.

Du, C. C., Mashoof, S. M. and Criscitiello, M. F. (2012) 'Oral immunization of the African clawed frog (*Xenopus laevis*) upregulates the mucosal immunoglobulin IgX', *Veterinary Immunology and Immunopathology*, 145(1–2), pp. 493–498. doi: 10.1016/j.vetimm.2011.10.019.

Eggington, J. M. et al. (2014) 'A comprehensive laboratory-based program for classification of variants of uncertain significance in hereditary cancer genes', *Clinical Genetics*, 86(3), pp. 229–237. doi: 10.1111/cge.12315.

Fazio, M. et al. (2020) 'Zebrafish patient avatars in cancer biology and precision cancer therapy', *Nature Reviews Cancer*. Nature Research, pp. 263–273. doi: 10.1038/s41568-020-0252-3.

Goyos, A. and Robert, J. (2009) 'Tumorigenesis and anti-tumor immune responses in Xenopus', *Frontiers in Bioscience*, 14(1), pp. 167–176. doi: 10.2741/3238.

Guaragnella, N. et al. (2014) 'The expanding role of yeast in cancer research and diagnosis: Insights into the function of the oncosuppressors p53 and BRCA1/2', *FEMS Yeast Research*, 14(1), pp. 2–16. doi: 10.1111/1567–1364.12094.

Guo, X. et al. (2014) 'Efficient RNA/Cas9-mediated genome editing in *Xenopus tropicalis*', *Development (Cambridge)*, 141(3), pp. 707–714. doi: 10.1242/dev.099853.

Hadji-Azimi, I. and Fischberg, M. (1971) 'Normal and cancerous tissue transplantation in allogeneic and syngeneic *Xenopus laevis*', *Cancer Research*, 31(11), pp. 118–132.

Hanahan, D. and Weinberg, R. A. (2000) 'The hallmarks of cancer', *Cell*, 100(1), pp. 57–70. doi: 10.1016/S0092-8674(00)81683-9.

Hanahan, D. and Weinberg, R. A. (2011) 'Hallmarks of cancer: The next generation', *Cell*, 144(5), pp. 646–674. doi: 10.1016/j.cell.2011.02.013.

Hardwick, L. J. A. and Philpott, A. (2015) 'An oncologist's friend: How Xenopus contributes to cancer research', *Developmental Biology*, 408(2), pp. 180–187. doi: 10.1016/j.ydbio.2015.02.003.

Hardwick, L. J. A. and Philpott, A. (2018) 'Xenopus models of cancer: Expanding the oncologist's toolbox', *Frontiers in Physiology*, 9, p. 27. doi: 10.3389/fphys.2018.01660.

Haynes-Gilmore, N. et al. (2014) 'A critical role of nonclassical MHC in tumor immune evasion in the amphibian Xenopus model', *Carcinogenesis*, 35(8), p. 1807. doi: 10.1093/carcin/bgu100.

Haynes-Gimore, N. et al. (2015) 'Semi-solid tumor model in *Xenopus laevis/gilli* cloned tadpoles for intravital study of neovascularization, immune cells and melanophore infiltration', *Developmental Biology*, 408(2), pp. 205–212. doi: 10.1016/j.ydbio.2015.01.003.

Hellsten, U. et al. (2010) 'The genome of the western clawed frog *Xenopus tropicalis*', *Science*, 328(5978), pp. 633–636. doi: 10.1126/science.1183670.

Jacks, T. et al. (1994) 'Tumor spectrum analysis in p53-mutant mice', *Current Biology*, 4(1), pp. 1–7. doi: 10.1016/S0960-9822(00)00002-6.

Khokha, M. K. (2012) '*Xenopus* white papers and resources: Folding functional genomics and genetics into the frog', *genesis*, 50(3), pp. 133–142. doi: 10.1002/dvg.22015.

Kinkel, M. D. et al. (2010) 'Intraperitoneal injection into adult zebrafish', *Journal of Visualized Experiments*, 42, p. 2126. doi: 10.3791/2126.

Komor, A. C. et al. (2016) 'Programmable editing of a target base in genomic DNA without double-stranded DNA cleavage', *Nature*, 533(7603), pp. 420–424. doi: 10.1038/nature17946.

Konukiewitz, B. et al. (2017) 'Somatostatin receptor expression related to TP53 and RB1 alterations in pancreatic and extrapancreatic neuroendocrine neoplasms with a Ki67-index above 20%', *Modern Pathology*, 30(4), pp. 587–598. doi: 10.1038/modpathol.2016.217.

Kühl, M. and Pandur, P. (2008) 'Dorsal axis duplication as a functional readout for Wnt activity', *Methods in Molecular Biology, 469*, pp. 467–476. doi: 10.1007/978-1-60327-469-29.

Letrado, P. et al. (2018) 'Zebrafish: Speeding up the cancer drug discovery process', *Cancer Research*, 78(21), pp. 6048–6058. doi: 10.1158/0008-5472.CAN-18-1029.

Lieschke, G. J. and Currie, P. D. (2007) 'Animal models of human disease: Zebrafish swim into view', *Nature Reviews Genetics*, 8(5), pp. 353–367. doi: 10.1038/nrg2091.

Lohmann, D. R. (1999) 'RB1 gene mutations in retinoblastoma', *Human Mutation*, 14(4), pp. 283–288. doi: 10.1002/(SICI)1098-1004(199910)14:4<283::AID-HUMU2>3.0.CO;2-J.

Ma, Y. et al. (2010) 'The relationship between early embryo development and tumourigenesis', *Journal of Cellular and Molecular Medicine*, 14(12), pp. 2697–2701. doi: 10.1111/j.1582-4934.2010.01191.x.

Maddalo, D. et al. (2014) 'In vivo engineering of oncogenic chromosomal rearrangements with the CRISPR/Cas9 system', *Nature*, 516(7531), pp. 423–428. doi: 10.1038/nature13902.

Mirzoyan, Z. et al. (2019) '*Drosophila melanogaster*: A model organism to study cancer', *Frontiers in Genetics*, 10, p. 51. doi: 10.3389/fgene.2019.00051.

Moody, S. A. (1987) 'Fates of the blastomeres of the 16-cell stage Xenopus embryo', *Developmental Biology*, 119(2), pp. 560–578. doi: 10.1016/0012-1606(87)90059-5.

Naert, T. et al. (2016) 'CRISPR/Cas9 mediated knockout of rb1 and rbl1 leads to rapid and penetrant retinoblastoma development in *Xenopus tropicalis*', *Scientific Reports*, 6(1), pp. 1–10. doi: 10.1038/srep35264.

Naert, T., Dimitrakopoulou, D., et al. (2020) 'RBL1 (p107) functions as tumor suppressor in glioblastoma and small-cell pancreatic neuroendocrine carcinoma in *Xenopus tropicalis*', *Oncogene*, 39(13), pp. 2692–2706. doi: 10.1038/s41388-020-1173-z.

Naert, T., Tulkens, D., et al. (2020) 'Maximizing CRISPR/Cas9 phenotype penetrance applying predictive modeling of editing outcomes in Xenopus and zebrafish embryos', *Scientific Reports*, 10(1), p. 14662. doi: 10.1038/s41598-020-71412-0.

Naert, T. et al. (2021) 'CRISPR-SID: Identifying EZH2 as a druggable target for desmoid tumors via in vivo dependency mapping', *Proceedings of the National Academy of Sciences of the United States of America*. in press. doi: 10.1073/pnas.2115116118.

Nakayama, T. et al. (2014) 'Cas9-based genome editing in *Xenopus tropicalis*', *The Use of CRISPR/cas9, ZFNs, TALENs in Generating Site Specific Genome Alterations*, 546, pp. 355–375. doi: 10.1016/B978-0-12-801185-0.00017-9.

Nemudryi, A. A. et al. (2014) 'TALEN and CRISPR/Cas genome editing systems: Tools of discovery', *Acta Naturae*, 6(3), pp. 19–40. doi: 10.32607/20758251-2014-6-3-19-40.

Noy, N. F. et al. (2009) 'BioPortal: Ontologies and integrated data resources at the click of a mouse', *Nucleic Acids Research*, 37(Suppl. 2), pp. W170–W173. doi: 10.1093/nar/gkp440.

Pegoraro, C. and Monsoro-Burq, A. H. (2013) 'Signaling and transcriptional regulation in neural crest specification and migration: Lessons from xenopus embryos', *Wiley Interdisciplinary Reviews: Developmental Biology*, 2(2), pp. 247–259. doi: 10.1002/wdev.76.

Pennisi, E. (1998) 'How a growth control path takes a wrong turn to cancer', *Science*, 281, pp. 1438–1441. doi: 10.1126/science.281.5382.1438.

Postlethwait, J. H. et al. (2000) 'Zebrafish comparative genomics and the origins of vertebrate chromosomes', *Genome Research*, 10(12), pp. 1890–1902. doi: 10.1101/gr.164800.

Pugach, E. K. et al. (2010) 'Retro-orbital injection in adult zebrafish', *Journal of Visualized Experiments*, 34, p. e1645. doi: 10.3791/1645.

Rau, L., Cohen, N. and Robert, J. (2001) 'MHC-restricted and -unrestricted CD8 T cells: An evolutionary perspective', *Transplantation*, 72(11), pp. 1830–1835. doi: 10.1097/00007890-200112150-00020.

Rau, L. et al. (2002) 'Identification and characterization of Xenopus CD8+ T cells expressing an NK cell-associated molecule', *European Journal of Immunology*, 32(6), pp. 1574–1583. doi:10.1002/1521-4141(200206)32:6<1574::AID-IMMU1574>3.0.CO;2-4.

Robanus-Maandag, E. et al. (1998) 'p107 is a suppressor of retinoblastoma development in pRB-deficient mice', *Genes and Development*, 12(11), pp. 1599–1609. doi: 10.1101/gad.12.11.1599.

Robert, J., Guiet, C. and du Pasquier, L. (1994) 'Lymphoid tumors of *Xenopus laevis* with different capacities for growth in larvae and adults', *Developmental Immunology*, 3(4), pp. 297–307. doi: 10.1155/1994/37392.

Robert, J., Guiet, C. and du Pasquier, L. (1995) 'Ontogeny of the alloimmune response against a transplanted tumor in *Xenopus laevis*', *Differentiation*, 59(3), pp. 135–144. doi: 10.1046/j.1432-0436.1995.5930135.x.

Robert, J. et al. (1997) 'Effects of thymectomy and tolerance induction on tumor immunity in adult *Xenopus laevis*', *International Journal of Cancer*, 70(3), pp. 330–334. doi: 10.1002/(SICI)1097-0215(19970127)70:3<330::AID-IJC14>3.0.CO;2-J.

Robert, J. and Ohta, Y. (2009) 'Comparative and developmental study of the immune system in Xenopus', in *Developmental Dynamics*, 238(6), pp. 1249–1270. doi: 10.1002/dvdy.21891.

Sakuma, T. et al. (2016) 'MMEJ-assisted gene knock-in using TALENs and CRISPR-Cas9 with the PITCh systems', *Nature Protocols*, 11(1), pp. 118–133. doi: 10.1038/nprot.2015.140.

Sato, K. et al. (2018) 'Flow cytometric analysis of *Xenopus laevis* and *X. tropicalis* blood cells using acridine orange', *Scientific Reports*, 8(1). doi: 10.1038/s41598-018-34631-0.

Schmitt, S. M., Gull, M. and Brändli, A. W. (2014) 'Engineering Xenopus embryos for phenotypic drug discovery screening', in *Advanced Drug Delivery Reviews*, 69–70, pp. 225–246. doi: 10.1016/j.addr.2014.02.004.

Session, A. M. et al. (2016) 'Genome evolution in the allotetraploid frog *Xenopus laevis*', *Nature*, 538(7625), pp. 336–343. doi: 10.1038/nature19840.

Slater, R. T. et al. (2019) 'Radiographic and ultrasonographic appearance of pneumonia in a frog', *Veterinary Radiology and Ultrasound*, 62(4), pp. 35–39. doi: 10.1111/vru.12796.

Smith, A. C. H. et al. (2010) 'High-throughput cell transplantation establishes that tumor-initiating cells are abundant in zebrafish T-cell acute lymphoblastic leukemia', *Blood*, 115(16), pp. 3296–3303. doi: 10.1182/blood-2009-10-246488.

Snyder, C. S. et al. (2009) 'Complementarity of ultrasound and fluorescence imaging in an orthotopic mouse model of pancreatic cancer', *BMC Cancer*, 9(1), p. 106. doi: 10.1186/1471-2407-9-106.

Sprague, J. et al. (2008) 'The Zebrafish information network: The zebrafish model organism database provides expanded support for genotypes and phenotypes', *Nucleic Acids Research*, 36(Suppl. 1), pp. D768–D772. doi: 10.1093/nar/gkm956.

Stacy, B. A. and Parker, J. M. (2004) 'Amphibian oncology', in *Veterinary Clinics of North America: Exotic Animal Practice*, 7(3), pp. 673–695. doi: 10.1016/j.cvex.2004.04.001.

Stewart, T. A., Pattengale, P. K. and Leder, P. (1984) 'Spontaneous mammary adenocarcinomas in transgenic mice that carry and express MTV/myc fusion genes', *Cell*, 38(3), pp. 627–637. doi: 10.1016/0092-8674(84)90257-5.

Suzuki, M. et al. (2020) 'Spontaneous neoplasia in the western clawed frog *Xenopus tropicalis*', *microPublication biology*, 2020. doi: 10.17912/micropub.biology.000294.

Tanaka, M. et al. (2016) 'Identification of anti-cancer chemical compounds using Xenopus embryos', *Cancer Science*, 107(6), pp. 803–811. doi: 10.1111/cas.12940.

Tandon, P. et al. (2017) 'Expanding the genetic toolkit in Xenopus: Approaches and opportunities for human disease modeling', in *Developmental Biology*, 426(2), pp. 325–335. doi: 10.1016/j.ydbio.2016.04.009.

Tang, Q. et al. (2014) 'Optimized cell transplantation using adult rag2 mutant zebrafsh', *Nature Methods*, 11(8), pp. 821–824. doi: 10.1038/nmeth.3031.

Taube, J. M. et al. (2014) 'Association of PD-1, PD-1 ligands, and other features of the tumor immune microenvironment with response to anti-PD-1 therapy', *Clinical Cancer Research*, 20(19), pp. 5064–5074. doi: 10.1158/1078-0432.CCR-13-3271.

Tsherniak, A. et al. (2017) 'Defining a cancer dependency map', *Cell*, 170(3), pp. 564–576.e16. doi: 10.1016/j.cell.2017.06.010.

Tulkens, D. et al. (2021) 'Engraftment of allotransplanted tumour cells in adult rag2 mutant Xenopus tropicalis', *BioRxiv*, doi:10.1101/2021.11.15.468684

Tzelepis, K. et al. (2016) 'A CRISPR dropout screen identifies genetic vulnerabilities and therapeutic targets in acute myeloid leukemia', *Cell Reports*, 17(4), pp. 1193–1205. doi: 10.1016/j.celrep.2016.09.079.

Vakoc, B. J. et al. (2012) 'Cancer imaging by optical coherence tomography: Preclinical progress and clinical potential', *Nature Reviews Cancer*, 12(5), pp. 363–368. doi: 10.1038/nrc3235.

Van Nieuwenhuysen, T. et al. (2015) 'TALEN-mediated apc mutation in *Xenopus tropicalis* phenocopies familial adenomatous polyposis', *Oncoscience*, 2(5), pp. 555–566. doi: 10.18632/oncoscience.166.

Wallingford, J. B. (1999) 'Tumors in tadpoles: The Xenopus embryo as a model system for the study of tumorigenesis', *Trends in Genetics* 15(10), pp. 385–388. doi: 10.1016/S0168-9525(99)01800-4.

Wallingford, J. B. et al. (1997) 'p53 activity is essential for normal development in Xenopus', *Current Biology*, 7(10), pp. 747–757. doi: 10.1016/S0960-9822(06)00333-2.

Watson, A. L. et al. (2016) 'Engineered swine models of cancer', *Frontiers in Genetics*, 7, p. 78. doi: 10.3389/fgene.2016.00078.

Wheeler, G. N. and Brändli, A. W. (2009) 'Simple vertebrate models for chemical genetics and drug discovery screens: Lessons from zebrafish and *Xenopus*', *Developmental Dynamics*, 238(6), pp. 1287–1308. doi: 10.1002/dvdy.21967.

Wheeler, G. N. and Liu, K. J. (2012) '*Xenopus* : An ideal system for chemical genetics', *Genesis*, 50(3), pp. 207–218. doi: 10.1002/dvg.22009

White, R., Rose, K. and Zon, L. (2013) 'Zebrafish cancer: The state of the art and the path forward', *Nature Reviews Cancer*, 13(9), pp. 624–636. doi: 10.1038/nrc3589.

Xiao, A. et al. (2002) 'Astrocyte inactivation of the pRb pathway predisposes mice to malignant astrocytoma development that is accelerated by PTEN mutation', *Cancer Cell*, 1(2), pp. 157–168. doi: 10.1016/S1535-6108(02)00029-6.

Yamagiwa, K. and Ichikawa, K. (1918) 'Experimental study of the pathogenesis of carcinoma', *Journal of Cancer Research*, 3(1), pp. 1–29. doi: 10.1158/jcr.1918.1.

Yan, C. et al. (2019) 'Visualizing engrafted human cancer and therapy responses in immunodeficient Zebrafish', *Cell*, 177(7), pp. 1903–1914.e14. doi: 10.1016/j.cell.2019.04.004.

Yang, J. et al. (2020) 'Guidelines and definitions for research on epithelial-mesenchymal transition', *Nature Reviews in Molecular and Cellular Biology*, 21(6), pp. 341–352. doi: 10.1038/s41580-020-0237-9.

Yang, S. et al. (1998) 'Overexpression of a novel Xenopus Rel mRNA induces tumors in early embryos', *Journal of Biological Chemistry*, 273(22), pp. 13746–13752. doi: 10.1074/jbc.273.22.13746.

Zhang, F. et al. (2015) 'Antibiotic toxicity and absorption in Zebrafish using liquid chromatography-tandem mass spectrometry', *PLoS One*, 10(5), p. e0124805. doi: 10.1371/journal.pone.0124805.

Zuckermann, M. et al. (2015) 'Somatic CRISPR/Cas9-mediated tumour suppressor disruption enables versatile brain tumour modelling', *Nature Communications*, 6, p. 7391. doi: 10.1038/ncomms8391.

22 *Xenopus*
A Model to Study Natural Genetic Variation and Its Disease Implications

Avi Leibovich, Sally A. Moody, Steven L. Klein, and Abraham Fainsod

CONTENTS

22.1. Historical Background 313
22.1.1. Natural Polymorphisms in the Human Genome 313
22.1.2. The Contribution of Genetic Polymorphisms to Human Disease 314
22.2. Summary of the Field 314
22.2.1. Mouse Inbred Lines 315
22.2.2. Common Fish Lines 315
22.2.3. Available *Xenopus* Lines 315
22.3. Present State of the Field 316
22.3.1. The Advantages of Outbred Lines 316
22.3.2. Using *Xenopus* to Study the Impact of Genetic Variation on Development and Disease 316
22.4. Future Outlook 320
Acknowledgments 320
Funding 320
References 320

Polymorphisms in protein-coding exons can result in nonfunctional, hypomorphic, or hyperactive proteins, whereas those in regulatory elements can influence penetrance, dominance, expressivity, and pleiotropy. For nearly 40 years, research to understand the function of individual genes has focused on utilizing inbred animals to eliminate the confounding influences of polymorphisms, second-site mutations, and modifier genes. However, it has become increasingly clear that these genetic polymorphisms in the human population are very prevalent and contribute to variable phenotypes, disease susceptibility, and responses to environmental factors and therapies. This chapter summarizes the relationship between human genetic variation and disease, discusses why inbred experimental models are inadequate for examining this relationship, and summarizes the important role of the frog, *Xenopus*, as an excellent outbred experimental system to study the effects of genetic polymorphisms in human disease.

22.1. HISTORICAL BACKGROUND

22.1.1. Natural Polymorphisms in the Human Genome

Even though the first draft of the human genome sequence was released nearly 20 years ago, the current human reference genome is still predominantly derived from a very limited number of individuals (International Human Genome Sequencing Consortium, 2004; Lander et al., 2001; Venter et al., 2001). Accordingly, it does not represent the considerable genetic variation that exists across the world's human population (Ballouz et al., 2019; Wong et al., 2020). We now appreciate the significance of this deficiency because this genetic variation helps to explain why different populations are differentially susceptible to certain diseases and exhibit different responses to environmental factors and medical treatments (Choudhury et al., 2014; Posey, 2019; Zerbino et al., 2020). Accordingly, there is growing interest in understanding these genetic variations and in constructing databases that catalogue them.

Genetic diversity among ethnicities, called genetic polymorphisms (Kruglyak and Nickerson, 2001), make up only about 1% of the genome and are responsible for common differences between humans such as eye, skin, and hair color (Pavan and Sturm, 2019; White and Rabago-Smith, 2011) but can also affect susceptibility to disease. A "genetic polymorphism" is a variation in a given gene locus that occurs with a frequency of 1% or more in a given population. Less common variations are referred to as "mutations" (Karki et al., 2015; Kruglyak and Nickerson, 2001; Stenson et al., 2017). The most common form is the substitution of a single base pair, known as a single-nucleotide polymorphism (SNP) (Kruglyak and Nickerson, 2001). This small change can have different consequences depending on the location or degree of change within the gene: (1) Synonymous

DOI: 10.1201/9781003050230-25

SNPs appear inside the coding region and have no effect on the amino acid sequence or the protein function. However, some synonymous SNPs have been linked to human disease (Sauna and Kimchi-Sarfaty, 2011). (2) SNPs within the coding region that cause protein sequence changes are classified as non-synonymous. These can be missense SNPs that result in amino acid changes or nonsense (stop codon) SNPs that cause premature termination of the protein. (3) SNPs located in non-protein coding regions can affect the sequence and structure of the encoded RNA, regulatory elements or promoters, splicing properties, or RNA stability.

Other common polymorphisms in the human genome include short tandem repeats (STRs), insertions/deletions (indels), transposable elements (TEs) or Alu repeats, structural variations (SVs), and copy number variations (CNVs). STRs (microsatellites) are tandem 1–6 bp repeats that make up ~3% of the human genome. Most are polymorphic in nature and thus are used in forensic DNA typing (Novroski et al., 2018; Saini et al., 2018). STRs can be located within promoters, exons, introns, or intergenic regions; they can modulate gene expression and alter proteins by coding for repeated amino acids, as in some nervous system diseases (Saini et al., 2018). Other types of repeats include minisatellites (10–100 bp) and macrosatellites (>100 bp) (Richard et al., 2008). Indels can range from 1 to several hundred bps in length, are widespread across the genome, and are the second most common type of genetic variability. Indels cause extensive variation in human genes; those located in functionally important sites are likely to affect traits and disease susceptibility (Barton and Zeng, 2018; Montgomery et al., 2013; Mullaney et al., 2010).

SVs refer to architectural and quantitative chromosomal rearrangements, usually involving DNA segments of 1Kb or more. SVs include deletions, duplications, insertions, and translocations (Spielmann et al., 2018). If the SV causes changes in the diploid status of a genomic region, it is defined as a copy number variation (Spielmann et al., 2018; Zarrei et al., 2015). SVs contribute to the genetic diversity of the human genome and play important roles in cancer genetics, rare diseases, and evolutionary genetics. SVs can affect either coding or non-coding regions or the three-dimensional organization of the DNA by disrupting higher-order chromatin structure influencing the expression of distant genes, thereby causing disease (Spielmann et al., 2018).

Transposable elements are segments of DNA that can move around the genome and were previously referred to as "junk DNA," but today we know they are important for gene regulation and evolution (Biémont and Vieira, 2006). Sometimes TE insertion disrupts a gene's function or expression pattern or triggers chromosomal rearrangements that are involved in cancer and other diseases (Jönsson et al., 2020).

22.1.2. The Contribution of Genetic Polymorphisms to Human Disease

Numerous studies have analyzed the involvement of genetic polymorphisms in human disease. For example, drug metabolism performed by cytochrome P450 (CYP) enzymes can be affected by genetic polymorphisms, resulting in differences in drug response between individuals and even adverse drug reactions (Manikandan and Nagini, 2018; Zhou et al., 2009, 2017). Depending on the CYP polymorphism, individuals can metabolize drugs either poorly, extensively, or ultrarapidly. Many polymorphisms are ethnic group dependent, leading to differences in drug responsiveness between world populations (Bachtiar et al., 2019; McGraw and Waller, 2012).

Polymorphisms also affect the severity of defects seen in fetal alcohol spectrum disorder (FASD), a complex set of congenital malformations, neurobehavioral anomalies, and intellectual disabilities resulting from maternal alcohol consumption during pregnancy. The biochemical similarity between ethanol metabolism and retinoic acid (RA) biosynthesis led to the suggestion that ethanol clearance competes with the metabolism of retinol (vitamin A), thus reducing RA signaling (reviewed in Fainsod et al., 2020; Shabtai and Fainsod, 2018). Epidemiological studies indicate that the incidence and severity of FASD phenotypes have a genetic component (Eberhart and Parnell, 2016; Garic et al., 2014; Green et al., 2007). Studies focusing on the *ADH1B* gene found "protective" polymorphisms. *ADH1B* has 3 common alleles encoding for isozymes with different kinetic characteristics. The ADH1B * 2 and ADH1B * 3 isozymes have a turnover rate over 80 times greater than the ADH1B * 1 variant. Expression of the "fast" alleles results in rapid acetaldehyde accumulation at the same blood alcohol concentration compared to individuals carrying the *ADH1B * 1* allele (Hurley and Edenberg, 2012). Except for one study, all studies found that *ADH1B*3* lowers the risk of FASD (Green and Stoler, 2007).

The pathologies of many infectious, autoimmune, and malignant diseases are influenced by the profiles of cytokine production in pro-inflammatory (TH1) and anti-inflammatory (TH2) T cells. Individual differences in cytokine profiles appear to be due, at least in part, to genetic polymorphisms within regulatory regions of cytokine genes (Bidwell et al., 1999). Toll-like receptor (TLR) genes, whose proteins are an important link between innate and adaptive immunity, exhibit SNPs, small-scale indels, polymorphic repetitive elements, and microsatellite variation. Polymorphic variants of TLRs are associated with several inflammatory disorders, including a higher risk of prostate cancer (Sun et al., 2006), protection against leprosy (Johnson et al., 2007), higher risk of developing tuberculosis (Wu and Yang, 2015), susceptibility to inflammatory bowel disease (Török et al., 2017), increased risk for gram-negative bacteremia and sepsis, and increased prevalence of hepatitis B virus infection (Vijay, 2018).

22.2. SUMMARY OF THE FIELD

Although genetic polymorphisms impact human health and disease, research has commonly been performed in animals that share extensive genotypic composition to minimize the effects of polymorphisms. Accordingly, genetic experiments are usually performed in inbred animal lines, defined as the result of at least 20 sequential generations of sister-brother mating, to increase homozygosity.

22.2.1. Mouse Inbred Lines

Inbred strains are the foundation of mouse developmental genetics because they reduce genetic variability within and across experiments performed in different laboratories. Although many mouse strains used in current research are close to homozygosity at any one locus, each strain has unique sets of polymorphisms and modifiers that affect their responsiveness to genetic manipulations and treatments. As examples: C57BL/6J is considered optimal for genetic engineering approaches, strain 129 is commonly used to produce targeted mutations due to the availability of many embryonic stem cell lines, and FVB is favored for transgenic microinjections due to very large pronuclei and large litter sizes (Blair et al., 2011; Bryant, 2011; Taketo et al., 1991). The advantage of each strain, however, is accompanied by different disease susceptibilities: C57BL/6 tends to develop age-related hearing loss, type 2 diabetes, and atherosclerosis; 129 is susceptible to testicular teratomas; and FVB carries a mutation that results in blindness (Bryant, 2011; Rashid et al., 2019; Stevens and Hummel, 1957). Thus, although the fixed genetic composition of inbred lines makes them extremely useful for specific experimental manipulations, their underlying genetic constraints render them less than optimal for determining how health-related manipulations will affect genetically variable human populations.

An additional caution is that there are sublines of the common inbred lines that can be distinctly different at the genomic level, such as multiple 129 substrains (Simpson et al., 1997; Threadgill et al., 1997) or the two C57BL/6 substrains (6J, 6N), which have been separated for about 220 generations and differ by numerous SNPs, indels, and SVs (Simon et al., 2013). In addition to substrain variability, there can be significant differences between mice originating from the same substrain but housed within different laboratory colonies (Justice and Dhillon, 2016). Thus, an inbred mouse strain may not be as "isogenic" as presumed.

22.2.2. Common Fish Lines

Medaka (*Oryzias latipes*) and zebrafish (*Danio rerio*) are the most common fish used in biomedical research. Several inbred medaka lines have been generated by sibling mating for over 20 generations, and several of these inbred lines have heterozygosity levels that are nearly 100-fold lower than in wild-caught fish (Spivakov et al., 2014). In contrast, zebrafish are prone to "inbreeding depression," in which during the inbreeding process they exhibit a decline in successful matings and clutch size and an increase in offspring sterility (Monson and Sadler, 2010; Shinya and Sakai, 2011). The two available lines considered inbred (C32, SJD) were derived by early-pressure egg parthenogenesis (Johnson et al., 1995; Streisinger et al., 1981). Similar to mouse, different zebrafish strains are genetically divergent from each other, from wild-caught fish from different localities, and from isolates of the same strain kept in different laboratories (Balik-Meisner et al., 2018; Franek et al., 2020; Suurväli et al., 2020). The inter-lab genetic diversity is thought to result from using a breeding laboratory population that is too small (Suurväli et al., 2020).

22.2.3. Available *Xenopus* Lines

Xenopus is the most common amphibian model used for biomedical research, and there are a few inbred lines. The *Xenopus laevis* J strain originated from a laboratory in Switzerland and was sent to the United States and then to Japan (hence "J" strain). In 1992, the 21st generation no longer exhibited long-term skin rejection, indicating that they were extensively homozygous. A descendant from the 30th generation was used for genome sequencing (Session et al., 2016), and subsequent generations were used to create BAC libraries (32nd and 33rd), RNAseq data (33rd and 34th), and chromosomal fluorescent *in situ* hybridization mapping (33rd) (Session et al., 2016). Like other inbred animals, the more uniform genetic background of J strain frogs enables improved accuracy for genome editing techniques (e.g. Ratzan et al., 2017), and the homozygosity of the MHC locus provides an important resource for immunological research (Gantress et al., 2003). The *Xenopus laevis* B strain originated at the Institute of Developmental Biology in Moscow as a spontaneous albino mutation, *periodic albinism* (*ap*) (Hoperskaya, 1975). Individuals were imported to Berkeley in the 1980s, and as a result of edemas, they were occasionally outbred to pigmented frogs and then crossed back to obtain the albino phenotype without edema (Savova et al., 2017). Since 1994, B strain frogs have been interbred for approximately ten generations (Savova et al., 2017).

Xenopus tropicalis was introduced to biomedical research because it has a simpler genome (diploid vs. allotetraploid) and a shorter generation time (5–8 months vs. 6–12 months) than *Xenopus laevis*. Beginning in the early 1990s, several laboratories obtained wild-caught *Xenopus tropicalis* from Nigeria (N) and the Ivory Coast (IC) (Grainger, 2012). The Golden strain was produced by selecting N frogs for rapid growth rate and early sexual maturity, resulting in a shorter generation time (Horb et al., 2019). Three stock centers (European Xenopus Resource Centre [EXRC, England], National BioResource Center [NBRC, Japan], National Xenopus Resource [NXR, USA]) produced eight *Xenopus tropicalis* lines (5 N and 3 IC), and by 2015, three N lines had achieved a high degree of homozygosity (Igawa et al., 2015). Currently, N lines are at the 10th (NH), 11th (Golden, BH), and 18th (NA) generations, and the IC strains are at least at the 5th generation.

22.3. PRESENT STATE OF THE FIELD

22.3.1. The Advantages of Outbred Lines

Another complication with inbred lines is the different phenotypes observed for the same mutation on different genetic backgrounds (Montagutelli, 2000). For example, deletion of the EGF receptor is peri-implantation lethal on a CF-1 background, but in CD-1 mice, the pups live for up to three weeks postnatally (Threadgill et al., 1995). Differences in the severity of phenotypes linked to which inbred line was used have been reported for gene mutations involved in polycystic kidney disease, tooth development, spermatogenesis, and craniofacial development (Li et al., 2013; Percival et al., 2017; Sakai et al., 2019; Sommardahl et al., 2001). Likewise, different lines show different susceptibility to neural tube defects in mice (Leduc et al., 2017), and to prenatal alcohol exposure, resulting in FASD in mice, rats, chickens, and zebrafish (Eberhart and Parnell, 2016).

For several decades, our reasoning has focused on the benefits of using inbred lines to minimize the influence of polymorphisms when examining single gene function. Now we recognize that there are many advantages to using outbred models. First and foremost, the lack of genetic diversity in inbred strains may not adequately inform studies in patient populations, which are genetically diverse (Justice and Dhillon, 2016). Recently, there has been a call for including biological variability in animal-based biomedical research, including the use of genetic polymorphisms (Voelkl et al., 2020). In support of using genetically heterogeneous animal models, it was recently shown that for the majority of phenotypes assessed, outbred mice were as phenotypically stable as inbred mice and were more resistant to housing differences between laboratories (Tuttle et al., 2018). The advantages of outbred strains were acknowledged by developing genetically diverse strains, including the Collaborative Cross mice, the Diversity Outbred mice (Harrill and McAllister, 2017; Saul et al., 2019; Threadgill et al., 2011), and the T5D zebrafish line (Balik-Meisner et al., 2018).

22.3.2. Using *Xenopus* to Study the Impact of Genetic Variation on Development and Disease

22.3.2.1. Most *Xenopus* Laboratory Populations Are Genetically Diverse

Genetically diverse, outbred *Xenopus* populations have been the traditional source of a rich history of biomedical research. Their genetic diversity derives from three primary factors: outbred commercial stocks, random matings to obtain experimental offspring, and design strategies to increase diversity within an experimental group.

Most laboratories using *Xenopus* do not establish their breeding colony by growing up the embryos produced by the colony. Instead, they house colonies of up to several hundred sexually mature individuals purchased from commercial frog breeders, such as Xenopus 1, Xenopus Express, and Nasco. These commercial suppliers maintain large populations of outbred adults that are sold to research laboratories. Xenopus 1 maintains both a wild-caught and a closed laboratory-bred colony, which are housed in separate facilities. For their laboratory-bred population, which numbers approximately 20,000 including tadpoles, they introduce about 1000 embryos from the wild-caught colony every two to three years. These embryos are grown and bred into the existing laboratory-bred colony to maintain genetic diversity. Their wild-caught colony is a steady-state population of about 4500 adults replenished with five imports from Chile per year; Xenopus were apparently introduced into Chile in the 1950s (R. Weymouth, Xenopus 1, personal communication). The Nasco laboratory-bred colony consisted of about 30,000 adults that were derived from wild-caught founding adults imported from South Africa between the early 1970s and 1996 (D. Brattlie, Nasco, personal communication). Although the Xenopus 1 and Nasco populations are "closed," their enormous sizes minimize inbreeding and maximize genetic diversity. In both facilities, mating is random, producing laboratory-bred progeny that are sold to research laboratories as well as added to the breeding colony when they mature. Wild-caught animals can be purchased from Xenopus 1 and Xenopus Express. These animals are particularly favored by the research community that uses oocytes for expression studies and electrophysiology. Wild-caught animals are considered more robust and disease resistant due to growing up in a natural environment and are quarantined and tested for pathogens before sale.

Typically, an individual laboratory maintains its breeding colony of commercial and/or stock center acquired-adults to produce offspring for experiments. Because the adults reproduce for up to 15 years, the colony within a single laboratory often is composed of frogs that were purchased over many years from more than one source. Accordingly, the adult frogs in a given laboratory breeding colony are expected to be unrelated and to display a high degree of genetic diversity, as shown by extensive EST and cDNA sequencing (Blackshear et al., 2001; Fierro et al., 2007; Gilchrist et al., 2004; Gilchrist and Pollet, 2012; Hellsten et al., 2007; Klein et al., 2006). Even the comparison of RNAseq and genome data between the inbred J and B strains revealed a degree of sequence variation that resembled the situation determined from comparing human genomes (Savova et al., 2017). Since the typical commercial and laboratory breeding schemes maintain a high level of genetic diversity in the offspring, *Xenopus* has been proposed to be an ideal model to study the functional relevance of specific gene variants (Savova et al., 2017).

The most common practices for generating offspring also add genetic diversity to individual experiments. Typically, embryos are produced by either natural mating of randomly selected male and female pairs or by artificial insemination using the sperm of one randomly selected male to fertilize eggs from several randomly selected females. Because embryos from different clutches might respond differently to treatments (Figure 22.1A), in a "typical" experiment, embryos from different clutches are combined (clutch mixing) to make up one biological replicate of the experiment (Figure 22.1B). This experimental design "averages" differences between clutches due to genetic polymorphisms, minimizing differences between replicates and focusing on the experimental manipulation. However, since *Xenopus* females can lay thousands of eggs in one day, making it possible to reach a statistically acceptable experimental sample size from a single cross, an alternative approach is to obtain each experimental replicate from a single, isolated clutch (Figure 22.1C). In this approach, genetic polymorphisms can be identified that play an important role in responsiveness to experimental treatments (Figure 22.1A). Although the polymorphisms might hinder the efficacy of knock-down approaches due to sequence variation in binding sites for antisense oligonucleotides or CRISPR-Cas9 guide RNAs, it can be an asset to study how genetic variability contributes to human disease.

22.3.2.2. The Unique Ability to Make Clutch Comparisons

Can the variable results of a manipulation across clutches that are genetically divergent be similar to studying human populations? We first noticed significant clutch differences by analyzing the size distribution of *Xenopus laevis* embryos (Leibovich et al., 2020). Analysis of over 2200 embryos from 33 different females revealed that each female lays

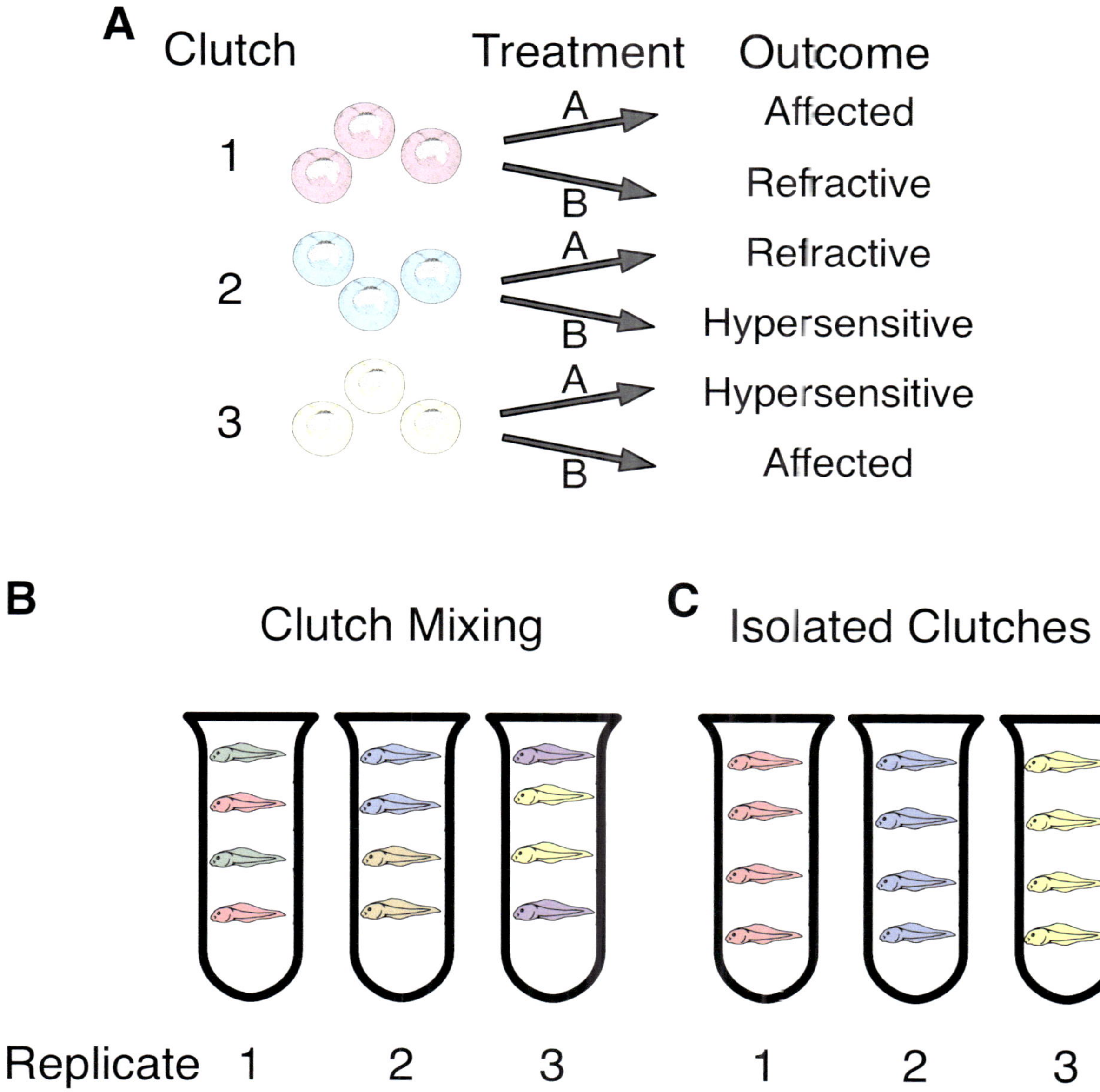

FIGURE 22.1 Clutch variability and experimental design in *Xenopus*. (A) Schematic representation of the experimental outcome of genetic variation between *Xenopus* clutches. (B, C) Approaches to the experimental design of biological replicates by mixing (B) or keeping clutches separate (C).

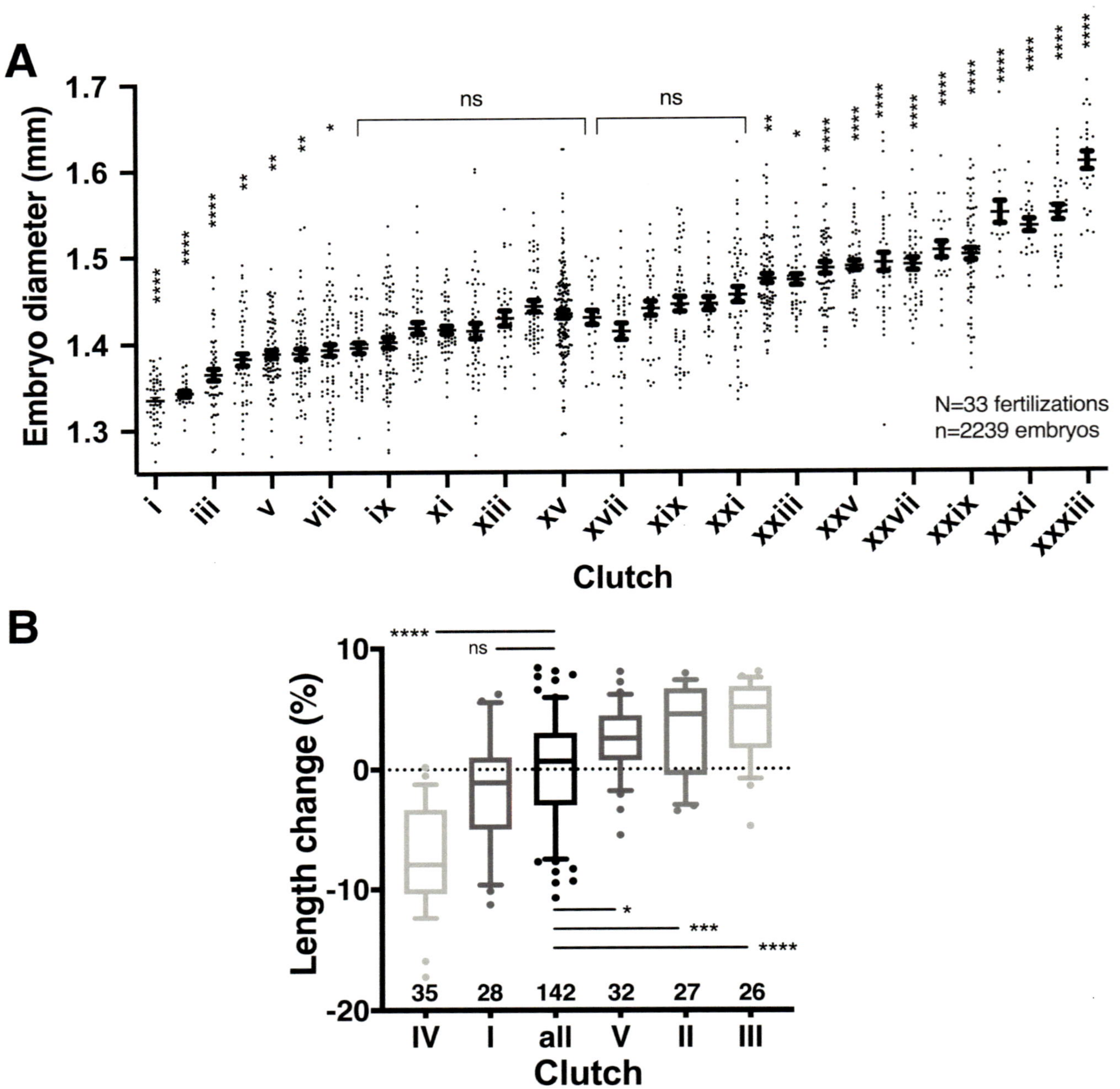

FIGURE 22.2 Natural size polymorphism associated with genetic variation in *Xenopus laevis*. (A) Distribution of embryo size between 33 embryo clutches. Diameters were measured before the onset of gastrulation. For each clutch on the scatter dot plot the mean and the SEM is shown. (B) Individual plotting of the five embryo clutches comprising five biological replicates. Embryos were measured at early tailbud stage (st. 30). The length change was calculated by comparing each clutch to the overall average length of all five clutches together. For each clutch, a boxplot of the interquartile range and the median diameter for the clutch is shown. Whiskers mark the range from the 10th to the 90th percentile for each clutch.

Note: ****, $p < 0.0001$; **, $p < 0.01$; *, $p < 0.05$; ns, not significant.

eggs within a restricted size range (Figure 22.2A). There also was a size-dependent effect on the RNA content of the embryos, quantitative and spatial adaptation of gene expression patterns, the size of tissues generated, and even the pattern of cell division. Because many of the clutches were generated in parallel and multiple embryos were analyzed from each clutch, we ruled out technical issues as the source of the variability and proposed that genetic polymorphisms might be responsible for these differences between clutches (Leibovich et al., 2020).

The outbred composition of most laboratory colonies raises the possibility that when a disease model is established in *Xenopus*, the effects of genetic polymorphisms on the severity and penetrance can be effectively explored using the "isolated" clutch approach (Figure 22.1C). For example, we used this approach to analyze the induction of fetal growth restriction (FGR) resulting from ethanol exposure or inhibition of RA biosynthesis (Shukrun et al., 2019). FGR is the *Xenopus* equivalent of intrauterine growth restriction (IUGR), commonly observed in human fetuses (Cox and Marton, 2009; Gurugubelli Krishna and Vishnu Bhat, 2018). IUGR can be subdivided into two types: asymmetric, in which only the trunk displays growth restriction, or symmetric, in which both the trunk and the head are restricted (Cox and Marton, 2009). Analysis of multiple isolated clutches revealed that some exhibited asymmetric and others symmetric FGR, recapitulating the human disease in *Xenopus* embryos (Shukrun et al., 2019). IUGR, sometimes also called small for gestational age (SGA), is defined as fetuses failing to reach their growth potential during embryogenesis (Mandruzzato et al., 2008; Visser et al., 2014). Extreme IUGR cases can lead to disease and even lethality (Bukowski, 2010; Gascoin and Flamant, 2013; Mandruzzato et al., 2008). Similarly, analysis of the length of larvae generated from multiple isolated clutches revealed a statistically significant dependence on their maternal origin (Figure 22.2B), and the natural size variability observed in *Xenopus* embryos paralleled the size variability described for human fetuses (Imdad et al., 2011; Sharma et al., 2016). This use of the isolated clutch approach is likely to continue to contribute to understanding the induction of IUGR and the genetic contribution to this defect (Sharma et al., 2017). This example demonstrates the usefulness of the outbred nature of *Xenopus* frogs to explore the genetic impact in important human health conditions.

22.3.2.3. Exploiting the Genetic Variability of *Xenopus* for Disease Modeling

The *Xenopus laevis* genome provides an additional level of genetic variation that can be exploited to further understand and model disease-causing genetic polymorphisms in humans. *Xenopus laevis* arose around 17–18 million years ago as an allotetraploid hybrid that carries distinct chromosomes derived from the two original species (Session et al., 2016). Of the original gene pairs, one copy has been lost in about 43% of protein-coding genes; the surviving singletons behave as in a diploid species. However, the remaining protein-coding genes are present as homoeologue pairs with different levels of subfunctionalization as a result of evolution (Hellsten et al., 2007; Session et al., 2016). While some homoeologue pairs overlap extensively in expression pattern and level, sharing conserved protein product functions (Kondo et al., 2017; Tour et al., 2001; Watanabe et al., 2017), in other homoeologue pairs, these features have diverged, often as the result of mutations in a regulatory element, similar to congenital-defect-causing mutations (Kondo et al., 2017; Ochi et al., 2017a, 2017b; Watanabe et al., 2017). Can we take advantage of the "enhanced" genetic diversification of homoeologues to identify changes that resemble disease conditions in humans?

Xenopus genetic variability can also provide insights into the variability of human responses to environmental factors and therapies. High-throughput chemical screens have been performed to study the toxicity of environmental toxins and drugs, as well as to identify small inhibitory molecules that target specific proteins or processes for functional studies and drug discovery (Blay et al., 2020; Tomlinson et al., 2012). Several chemical screens have been performed in *Xenopus* to study signaling pathway components, angiogenesis, and lymphangiogenesis (Kälin et al., 2009; Peterson et al., 2006). Taking advantage of the natural genetic variation in *Xenopus* laboratory colonies and the isolated clutch experimental design, variants exhibiting enhanced sensitivity or resistance to a specific small molecule can be identified. These variants then can be compared to the polymorphisms known in the homologous human protein to further inform drug screening, development, and medical treatments.

The *Xenopus* oocyte also serves as an *in vivo* "test-tube" to express proteins and analyze their function during normal or disease conditions (see Chapter 9). The large size of the oocyte and its efficiency in translating exogenous, injected mRNAs have made it an exemplary experimental system in which to study secreted and membrane-bound proteins, including channels and transporters (Marchant, 2018; Mowry, 2020). Also, cellular processes, developmental mechanisms, physiological events, molecular biology, viral infection, and drug discovery have been studied in *Xenopus* oocytes (Au et al., 2010; Lin-Moshier and Marchant, 2013; Zeng et al., 2020). These studies have been expanded to include proteins involved in abnormal physiological processes and neuropathological conditions (Baker et al., 2020; Meyer et al., 2020; Nashimoto et al., 2020; Singh et al., 2020). Importantly, injecting mRNA-encoding human proteins opens the possibility to directly study the effect of genetic polymorphisms on protein function. Studies using this approach have analyzed the effect of protein variants on cholesterol transport, GABAp receptor function, and Na/K pump function and on individuals suffering from Bartter Syndrome type 3 (Baker et al., 2020; Meyer et al., 2020; Nashimoto et al., 2020; Seys et al., 2017). *Xenopus* oocytes also were utilized to transplant membranes from Alzheimer's patients to study their neuropathology and how they differ (Singh et al., 2020).

RNAs encoding human protein variants can also be injected into *Xenopus* embryos with the advantage that specific tissues or organs can be targeted. For example, embryos have been employed to study the role of RAD21 variants in the induction of sclerocornea (Zhang et al., 2019) and the effects of SIX1 variants on craniofacial development (Shah et al., 2020; Mehdizadeh et al., 2021). We recently used this approach to identify polymorphisms in *ALDH1a2* (*RADLH2*) that affect the level of enzymatic activity (Shabtai et al., 2016). Analysis of embryos during late blastula, before endogenous RA signaling begins, allows analysis of the injected enzymatic activity with minimal endogenous background, whereas analysis during gastrula stages explores the interaction of the human variants with the frog network components. This experimental design can be further manipulated by partially removing endogenous components to simplify the analysis using CRISPR/Cas9 or morpholino antisense oligonucleotides (Blum et al., 2015; Tandon et al., 2017). These have already demonstrated the usefulness of *Xenopus* embryos and oocytes to explore the contribution of genetic polymorphisms to human disease and the spectrum of severities encountered among patients.

22.4. FUTURE OUTLOOK

It is well established that different human ethnic groups display differential disease susceptibility and variable responses to environmental factors and medical therapies. Likewise, mutations in a given gene produce highly variable dysmorphologies in several congenital syndromes. A significant factor in these differential responses is the underlying genetic diversity of the population. Inbred lines of various animal experimental models have been an invaluable tool for examining the specific role of single genes in development, cell biology, and physiology; they also are commonly used to explore medical treatments and drug responses. However, the variable responses common in human populations may not be observed in a given inbred line because it has a fixed set of polymorphisms. Accordingly, these responses are best examined in animal models that remain outbred and continue to be genetically polymorphic. We posit that an ideal animal model for these studies is *Xenopus*. Although inbred *Xenopus* lines are being used for specialized purposes, outbred *Xenopus* remains one of the common research models and offers many experimental advantages to establish disease models and study the effects of genetic variability. These advantages enable outbred *Xenopus* to elucidate the important relationship between genetic variation and disease and provide disease models to study possible treatments.

ACKNOWLEDGMENTS

We thank Dan Brattlie from Nasco and Rob Weymouth from Xenopus 1 for sharing the history of their breeding colonies and their maintenance practices. We also thank Martin Blum, Dale Frank, and Graciela Pillemer for critically reading this chapter.

FUNDING

This work was funded in part by grants from the United States-Israel Binational Science Foundation (2013422 and 2017199) to SM and AF, The Israel Science Foundation (668/17) to AF, and the Wolfson Family Chair in Genetics to AF.

REFERENCES

Au, S., Cohen, S., Panté, N., 2010. Microinjection of *Xenopus laevis* oocytes as a system for studying nuclear transport of viruses. *Methods* 51, 114–120. doi:10.1016/j.ymeth.2010.02.001

Bachtiar, M., Ooi, B.N.S., Wang, J., Jin, Y., Tan, T.W., Chong, S.S., Lee, C.G.L., 2019. Towards precision medicine: interrogating the human genome to identify drug pathways associated with potentially functional, population-differentiated polymorphisms. *Pharmacogenomics J.* 19, 516–527. doi:10.1038/s41397-019-0096-y

Baker, B.W., Dougherty, D.A., Lummis, S.C.R., 2020. Proline residues contribute to efficient gabap receptor function. *ACS Chem. Neurosci.* doi:10.1021/acschemneuro.0c00483

Balik-Meisner, M., Truong, L., Scholl, E.H., Tanguay, R.L., Reif, D.M., 2018. Population genetic diversity in zebrafish lines. *Mamm. Genome* 29, 90–100. doi:10.1007/s00335-018-9735-x

Ballouz, S., Dobin, A., Gillis, J.A., 2019. Is it time to change the reference genome? *Genome Biol.* 20, 159. doi:10.1186/s13059-019-1774-4

Barton, H.J., Zeng, K., 2018. New methods for inferring the distribution of fitness effects for indels and snps. *Mol. Biol. Evol.* 35, 1536–1546. doi:10.1093/molbev/msy054

Bidwell, J., Keen, L., Gallagher, G., Kimberly, R., Huizinga, T., McDermott, M.F., Oksenberg, J., McNicholl, J., Pociot, F., Hardt, C., D'Alfonso, S., 1999. Cytokine gene polymorphism in human disease: on-line databases. *Genes Immun.* 1, 3–19. doi:10.1038/sj.gene.6363645

Biémont, C., Vieira, C., 2006. Genetics: Junk DNA as an evolutionary force. *Nature* 443, 521–524. doi:10.1038/443521a

Blackshear, P.J., Lai, W.S., Thorn, J.M., Kennington, E.A., Staffa, N.G., Moore, D.T., Bouffard, G.G., Beckstrom-Sternberg, S.M., Touchman, J.W., Bonaldo, M.F., Soares, M.B., 2001. The NIEHS Xenopus maternal EST project: Interim analysis of the first 13,879 ESTs from unfertilized eggs. *Gene* 267, 71–87. doi:10.1016/s0378-1119(01)00383-3

Blair, K., Wray, J., Smith, A., 2011. The liberation of embryonic stem cells. *PLoS Genet.* 7, e1002019. doi:10.1371/journal.pgen.1002019

Blay, V., Tolani, B., Ho, S.P., Arkin, M.R., 2020. High-throughput screening: Today's biochemical and cell-based approaches. *Drug Discov. Today*. doi:10.1016/j.drudis.2020.07.024

Blum, M., De Robertis, E.M., Wallingford, J.B., Niehrs, C., 2015. Morpholinos: Antisense and sensibility. *Dev. Cell* 35, 145–149. doi:10.1016/j.devcel.2015.09.017

Bryant, C.D., 2011. The blessings and curses of C57BL/6 substrains in mouse genetic studies. *Ann. N. Y. Acad. Sci.* 1245, 31–33. doi:10.1111/j.1749-6632.2011.06325.x

Bukowski, R., 2010. Stillbirth and fetal growth restriction. *Clin. Obstet. Gynecol.* 53, 673–680. doi:10.1097/GRF.0b013e3181eba0db

Choudhury, A., Hazelhurst, S., Meintjes, A., Achinike-Oduaran, O., Aron, S., Gamieldien, J., Jalali Sefid Dashti, M., Mulder, N., Tiffin, N., Ramsay, M., 2014. Population-specific common SNPs reflect demographic histories and highlight regions of genomic plasticity with functional relevance. *BMC Genomics* 15, 437. doi:10.1186/1471-2164-15-437

Cox, P., Marton, T., 2009. Pathological assessment of intrauterine growth restriction. *Best Pract. Res. Clin. Obstet. Gynaecol.* 23, 751–764. doi:10.1016/j.bpobgyn.2009.06.006

Eberhart, J.K., Parnell, S.E., 2016. The genetics of fetal alcohol spectrum disorders. *Alcohol. Clin. Exp. Res.* 40, 1154–1165. doi:10.1111/acer.13066

Fainsod, A., Bendelac-Kapon, L., Shabtai, Y., 2020. Fetal alcohol spectrum disorder: embryogenesis under reduced retinoic acid signaling conditions. *Subcell Biochem* 95, 197–225. doi:10.1007/978-3-030-42282-0_8

Fierro, A.C., Thuret, R., Coen, L., Perron, M., Demeneix, B.A., Wegnez, M., Gyapay, G., Weissenbach, J., Wincker, P.,

Mazabraud, A., Pollet, N., 2007. Exploring nervous system transcriptomes during embryogenesis and metamorphosis in *Xenopus tropicalis* using EST analysis. *BMC Genomics* 8, 118. doi:10.1186/1471-2164-8-118

Franek, R., Baloch, A.R., Kaspar, V., Saito, T., Fujimoto, T., Arai, K., Psenicka, M., 2020. Isogenic lines in fish: A critical review. *Reviews in Aquaculture* 12, 1412–1434.

Gantress, J., Maniero, G.D., Cohen, N., Robert, J., 2003. Development and characterization of a model system to study amphibian immune responses to iridoviruses. *Virology* 311, 254–262. doi:10.1016/S0042-6822(03)00151-X

Garic, A., Berres, M.E., Smith, S.M., 2014. High-throughput transcriptome sequencing identifies candidate genetic modifiers of vulnerability to fetal alcohol spectrum disorders. *Alcohol. Clin. Exp. Res.* 38, 1874–1882. doi:10.1111/acer.12457

Gascoin, G., Flamant, C., 2013. Long-term outcome in context of intra uterine growth restriction and/or small for gestational age newborns. *J Gynecol Obstet Biol Reprod (Paris)* 42, 911–920. doi:10.1016/j.jgyn.2013.09.014

Gilchrist, M.J., Pollet, N., 2012. Databases of gene expression in Xenopus development. *Methods Mol. Biol.* 917, 319–345. doi:10.1007/978-1-61779-992-1_19

Gilchrist, M.J., Zorn, A.M., Voigt, J., Smith, J.C., Papalopulu, N., Amaya, E., 2004. Defining a large set of full-length clones from a *Xenopus tropicalis* EST project. *Dev. Biol.* 271, 498–516. doi:10.1016/j.ydbio.2004.04.023

Grainger, R.M., 2012. *Xenopus tropicalis* as a model organism for genetics and genomics: Past, present, and future. *Methods Mol. Biol.* 917, 3–15. doi:10.1007/978-1-61779-992-1_1

Green, M.L., Singh, A.V., Zhang, Y., Nemeth, K.A., Sulik, K.K., Knudsen, T.B., 2007. Reprogramming of genetic networks during initiation of the fetal alcohol syndrome. *Dev. Dyn.* 236, 613–631. doi:10.1002/dvdy.21048

Green, R.F., Stoler, J.M., 2007. Alcohol dehydrogenase 1B genotype and fetal alcohol syndrome: a HuGE minireview. *Am. J. Obstet. Gynecol.* 197, 12–25. doi:10.1016/j.ajog.2007.02.028

Gurugubelli Krishna, R., Vishnu Bhat, B., 2018. Molecular mechanisms of intrauterine growth restriction. *J. Matern. Fetal Neonatal Med.* 31, 2634–2640. doi:10.1080/14767058.2017.1347922

Harrill, A.H., McAllister, K.A., 2017. New rodent population models may inform human health risk assessment and identification of genetic susceptibility to environmental exposures. *Environ. Health Perspect.* 125, 086002. doi:10.1289/EHP1274

Hellsten, U., Khokha, M.K., Grammer, T.C., Harland, R.M., Richardson, P., Rokhsar, D.S., 2007. Accelerated gene evolution and subfunctionalization in the pseudotetraploid frog *Xenopus laevis*. *BMC Biol.* 5, 31. doi:10.1186/1741-7007-5-31

Hoperskaya, O.A., 1975. The development of animals homozygous for a mutation causing periodic albinism (ap) in *Xenopus laevis*. *J. Embryol. Exp. Morphol.* 34, 253–264.

Horb, M., Wlizla, M., Abu-Daya, A., McNamara, S., Gajdasik, D., Igawa, T., Suzuki, A., Ogino, H., Noble, A., Centre de Ressource Biologique Xenope team in France, Robert, J., James-Zorn, C., Guille, M., 2019. Xenopus resources: Transgenic, inbred and mutant animals, training opportunities, and web-based support. *Front. Physiol.* 10, 387. doi:10.3389/fphys.2019.00387

Hurley, T.D., Edenberg, H.J., 2012. Genes encoding enzymes involved in ethanol metabolism. *Alcohol Res.* 34, 339–344.

Igawa, T., Watanabe, A., Suzuki, A., Kashiwagi, A., Kashiwagi, K., Noble, A., Guille, M., Simpson, D.E., Horb, M.E., Fujii, T., Sumida, M., 2015. Inbreeding ratio and genetic relationships among strains of the Western clawed frog, *Xenopus tropicalis*. *PLoS One* 10, e0133963. doi:10.1371/journal.pone.0133963

Imdad, A., Yakoob, M.Y., Siddiqui, S., Bhutta, Z.A., 2011. Screening and triage of intrauterine growth restriction (IUGR) in general population and high risk pregnancies: A systematic review with a focus on reduction of IUGR related stillbirths. *BMC Public Health* 11, Suppl 3, S1. doi:10.1186/1471-2458-11-S3-S1

International Human Genome Sequencing Consortium, 2004. Finishing the euchromatic sequence of the human genome. *Nature* 431, 931–945. doi:10.1038/nature03001

Johnson, C.M., Lyle, E.A., Omueti, K.O., Stepensky, V.A., Yegin, O., Alpsoy, E., Hamann, L., Schumann, R.R., Tapping, R.I., 2007. Cutting edge: A common polymorphism impairs cell surface trafficking and functional responses of TLR1 but protects against leprosy. *J. Immunol.* 178, 7520–7524. doi:10.4049/jimmunol.178.12.7520

Johnson, S.L., Africa, D., Horne, S., Postlethwait, J.H., 1995. Half-tetrad analysis in zebrafish: Mapping the ros mutation and the centromere of linkage group I. *Genetics* 139, 1727–1735.

Jönsson, M.E., Garza, R., Johansson, P.A., Jakobsson, J., 2020. Transposable elements: A common feature of neurodevelopmental and neurodegenerative disorders. *Trends Genet.* 36, 610–623. doi:10.1016/j.tig.2020.05.004

Justice, M.J., Dhillon, P., 2016. Using the mouse to model human disease: Increasing validity and reproducibility. *Dis. Model. Mech.* 9, 101–103. doi:10.1242/dmm.024547

Kälin, R.E., Bänziger-Tobler, N.E., Detmar, M., Brändli, A.W., 2009. An in vivo chemical library screen in Xenopus tadpoles reveals novel pathways involved in angiogenesis and lymphangiogenesis. *Blood* 114, 1110–1122. doi:10.1182/blood-2009-03-211771

Karki, R., Pandya, D., Elston, R.C., Ferlini, C., 2015. Defining "mutation" and "polymorphism" in the era of personal genomics. *BMC Med. Genomics* 8, 37. doi:10.1186/s12920-015-0115-z

Klein, S.L., Gerhard, D.S., Wagner, L., Richardson, P., Schriml, L.M., Sater, A.K., Warren, W.C., McPherson, J.D., 2006. Resources for genetic and genomic studies of Xenopus. Methods Mol. *Biol.* 322, 1–16. doi:10.1007/978-1-59745-000-3_1

Kondo, M., Yamamoto, T., Takahashi, S., Taira, M., 2017. Comprehensive analyses of hox gene expression in *Xenopus laevis* embryos and adult tissues. *Dev. Growth Differ.* 59, 526–539. doi:10.1111/dgd.12382

Kruglyak, L., Nickerson, D.A., 2001. Variation is the spice of life. *Nat. Genet.* 27, 234–236. doi:10.1038/85776

Lander, E.S., Linton, L.M., Birren, B., Nusbaum, C., Zody, M.C., Baldwin, J., Devon, K., Dewar, K., Doyle, M., FitzHugh, W., Funke, R., Gage, D., Harris, K., Heaford, A., Howland, J., Kann, L., Lehoczky, J., LeVine, R., McEwan, P., McKernan, K., et al., 2001. Initial sequencing and analysis of the human genome. *Nature* 409, 860–921. doi:10.1038/35057062

Leduc, R.Y.M., Singh, P., McDermid, H.E., 2017. Genetic backgrounds and modifier genes of NTD mouse models: An opportunity for greater understanding of the multifactorial etiology of neural tube defects. *Birth Defects Res.* 109, 140–152. doi:10.1002/bdra.23554

Leibovich, A., Edri, T., Klein, S.L., Moody, S.A., Fainsod, A., 2020. Natural size variation among embryos leads to the corresponding scaling in gene expression. *Dev. Biol.* 462, 165–179. doi:10.1016/j.ydbio.2020.03.014

Li, Y., Konicki, W.S., Wright, J.T., Suggs, C., Xue, H., Kuehl, M.A., Kulkarni, A.B., Gibson, C.W., 2013. Mouse genetic background influences the dental phenotype. *Cells Tissues Organs* (Print) 198, 448–456. doi:10.1159/000360157

Lin-Moshier, Y., Marchant, J.S., 2013. The Xenopus oocyte: A single-cell model for studying Ca2+ signaling. *Cold Spring Harb. Protoc*. 2013. doi:10.1101/pdb.top066308

Mandruzzato, G., Antsaklis, A., Botet, F., Chervenak, F.A., Figueras, F., Grunebaum, A., Puerto, B., Skupski, D., Stanojevic, M., WAPM, 2008. Intrauterine restriction (IUGR). *J. Perinat. Med*. 36, 277–281. doi:10.1515/JPM.2008.050

Manikandan, P., Nagini, S., 2018. Cytochrome P450 structure, function and clinical significance: A review. *Curr. Drug Targets* 19, 38–54. doi:10.2174/13894501186661701251 44557

Marchant, J.S., 2018. Heterologous protein expression in the Xenopus oocyte. *Cold Spring Harb. Protoc*. 2018. doi:10.1101/pdb.prot096990

McGraw, J., Waller, D., 2012. Cytochrome P450 variations in different ethnic populations. *Expert Opin. Drug Metab. Toxicol*. 8, 371–382. doi:10.1517/17425255.2012.657626

Mehdizadeh, T., Majumdar, H.D., Ahsan, S., Tavares, A.L.P., Moody, S.A., 2021. Mutations in SIX1 associated with Branchio-oto-renal Syndrome (BOR) differentially affect otic expression of putative target genes. *J. Devel. Biol*. 9, 25. doi: 10.3390/jdb9030025.

Meyer, D.J., Bijlani, S., de Sautu, M., Spontarelli, K., Young, V.C., Gatto, C., Artigas, P., 2020. FXYD protein isoforms differentially modulate human Na/K pump function. *J. Gen. Physiol*. 152. doi:10.1085/jgp.202012660

Monson, C.A., Sadler, K.C., 2010. Inbreeding depression and outbreeding depression are evident in wild-type zebrafish lines. *Zebrafish* 7, 189–197. doi:10.1089/zeb.2009.0648

Montagutelli, X., 2000. Effect of the genetic background on the phenotype of mouse mutations. *J. Am. Soc. Nephrol*. 11, Suppl 16, S101–S105.

Montgomery, S.B., Goode, D.L., Kvikstad, E., Albers, C.A., Zhang, Z.D., Mu, X.J., Ananda, G., Howie, B., Karczewski, K.J., Smith, K.S., Anaya, V., Richardson, R., Davis, J., 1000 Genomes Project Consortium, MacArthur, D.G., Sidow, A., Duret, L., Gerstein, M., Makova, K.D., Marchini, J., Lunter, G., 2013. The origin, evolution, and functional impact of short insertion-deletion variants identified in 179 human genomes. *Genome Res*. 23, 749–761. doi:10.1101/gr.148718.112

Mowry, K.L., 2020. Using the Xenopus oocyte toolbox. *Cold Spring Harb. Protoc*. 2020, 095844. doi:10.1101/pdb.top095844

Mullaney, J.M., Mills, R.E., Pittard, W.S., Devine, S.E., 2010. Small insertions and deletions (INDELs) in human genomes. *Hum. Mol. Genet*. 19, R131–R136. doi:10.1093/hmg/ddq400

Nashimoto, S., Yagi, S., Takeda, N., Nonaka, M., Takekuma, Y., Sugawara, M., Sato, Y., 2020. A new system to evaluate characteristics of Niemann-Pick C1 Like 1-mediated cholesterol transport using *Xenopus laevis* oocytes. *Biochim. Biophys. Acta Biomembr*. 183508. doi:10.1016/j.bbamem.2020.183508

Novroski, N.M.M., Woerner, A.E., Budowle, B., 2018. Potential highly polymorphic short tandem repeat markers for enhanced forensic identity testing. *Forensic Sci. Int. Genet*. 37, 162–171. doi:10.1016/j.fsigen.2018.08.011

Ochi, H., Kawaguchi, A., Tanouchi, M., Suzuki, N., Kumada, T., Iwata, Y., Ogino, H., 2017a. Co-accumulation of cis-regulatory and coding mutations during the pseudogenization of the *Xenopus laevis* homoeologs six6.L and six6.S. *Dev. Biol*. 427, 84–92. doi:10.1016/j.ydbio.2017.05.004

Ochi, H., Suzuki, N., Kawaguchi, A., Ogino, H., 2017b. Asymmetrically reduced expression of hand1 homeologs involving a single nucleotide substitution in a cis-regulatory element. *Dev. Biol*. 425, 152–160. doi:10.1016/j.ydbio.2017.03.021

Pavan, W.J., Sturm, R.A., 2019. The genetics of human skin and hair pigmentation. *Annu. Rev. Genomics Hum. Genet*. 20, 41–72. doi:10.1146/annurev-genom-083118-015230

Percival, C.J., Marangoni, P., Tapaltsyan, V., Klein, O., Hallgrímsson, B., 2017. The interaction of genetic background and mutational effects in regulation of mouse craniofacial shape. *G3* (Bethesda) 7, 1439–1450. doi:10.1534/g3.117.040659

Peterson, J.R., Lebensohn, A.M., Pelish, H.E., Kirschner, M.W., 2006. Biochemical suppression of small-molecule inhibitors: A strategy to identify inhibitor targets and signaling pathway components. *Chem. Biol*. 13, 443–452. doi:10.1016/j.chembiol.2006.02.009

Posey, J.E., 2019. Genome sequencing and implications for rare disorders. *Orphanet J. Rare Dis*. 14, 153. doi:10.1186/s13023-019-1127-0

Rashid, K., Dannhausen, K., Langmann, T., 2019. Testing for known retinal degeneration mutants in mouse strains. *Methods Mol. Biol*. 1834, 45–58. doi:10.1007/978-1-4939-8669-9_3

Ratzan, W., Falco, R., Salanga, C., Salanga, M., Horb, M.E., 2017. Generation of a *Xenopus laevis* F1 albino J strain by genome editing and oocyte host-transfer. *Dev. Biol*. 426, 188–193. doi:10.1016/j.ydbio.2016.03.006

Richard, G.-F., Kerrest, A., Dujon, B., 2008. Comparative genomics and molecular dynamics of DNA repeats in eukaryotes. *Microbiol. Mol. Biol. Rev*. 72, 686–727. doi:10.1128/MMBR.00011-08

Saini, S., Mitra, I., Mousavi, N., Fotsing, S.F., Gymrek, M., 2018. A reference haplotype panel for genome-wide imputation of short tandem repeats. *Nat. Commun*. 9, 4397. doi:10.1038/s41467-018-06694-0

Sakai, K., Ito, C., Wakabayashi, M., Kanzaki, S., Ito, T., Takada, S., Toshimori, K., Sekita, Y., Kimura, T., 2019. Usp26 mutation in mice leads to defective spermatogenesis depending on genetic background. *Sci. Rep*. 9, 13757. doi:10.1038/s41598-019-50318-6

Saul, M.C., Philip, V.M., Reinholdt, L.G., Center for Systems Neurogenetics of Addiction, Chesler, E.J., 2019. High-diversity mouse populations for complex traits. *Trends Genet*. 35, 501–514. doi:10.1016/j.tig.2019.04.003

Sauna, Z.E., Kimchi-Sarfaty, C., 2011. Understanding the contribution of synonymous mutations to human disease. *Nat. Rev. Genet*. 12, 683–691. doi:10.1038/nrg3051

Savova, V., Pearl, E.J., Boke, E., Nag, A., Adzhubei, I., Horb, M.E., Peshkin, L., 2017. Transcriptomic insights into genetic diversity of protein-coding genes in X. *laevis*. *Dev. Biol*. 424, 181–188. doi:10.1016/j.ydbio.2017.02.019

Session, A.M., Uno, Y., Kwon, T., Chapman, J.A., Toyoda, A., Takahashi, S., Fukui, A., Hikosaka, A., Suzuki, A., Kondo, M., van Heeringen, S.J., Quigley, I., Heinz, S., Ogino, H., Ochi, H., Hellsten, U., Lyons, J.B., Simakov, O., Putnam, N., Stites, J., Rokhsar, D.S., 2016. Genome evolution in the allotetraploid frog *Xenopus laevis*. *Nature* 538, 336–343. doi:10.1038/nature19840

Seys, E., Andrini, O., Keck, M., Mansour-Hendili, L., Courand, P.-Y., Simian, C., Deschenes, G., Kwon, T., Bertholet-Thomas, A., Bobrie, G., Borde, J.S., Bourdat-Michel, G., Decramer, S., Cailliez, M., Krug, P., Cozette, P., Delbet, J.D., Dubourg, L., Chaveau, D., Fila, M., Vargas-Poussou, R., 2017. Clinical and genetic spectrum of Bartter syndrome type 3. *J. Am. Soc. Nephrol*. 28, 2540–2552. doi:10.1681/ASN.2016101057

Shabtai, Y., Fainsod, A., 2018. Competition between ethanol clearance and retinoic acid biosynthesis in the induction of fetal alcohol syndrome. *Biochem. Cell Biol.* 96, 148–160. doi:10.1139/bcb-2017-0132

Shabtai, Y., Jubran, H., Nassar, T., Hirschberg, J., Fainsod, A., 2016. Kinetic characterization and regulation of the human retinaldehyde dehydrogenase 2 enzyme during production of retinoic acid. *Biochem. J.* 473, 1423–1431. doi:10.1042/BCJ20160101

Shah, A.M., Krohn, P., Baxi, A.B., Tavares, A.L.P., Sullivan, C.H., Chillakuru, Y.R., Majumdar, H.D., Neilson, K.M., Moody, S.A., 2020. Six1 proteins with human branchio-oto-renal mutations differentially affect cranial gene expression and otic development. *Dis. Model. Mech.* doi:10.1242/dmm.043489

Sharma, D., Shastri, S., Farahbakhsh, N., Sharma, P., 2016. Intrauterine growth restriction—part 1. *J. Matern. Fetal Neonatal Med.* 29, 3977–3987. doi:10.3109/14767058.2016.1152249

Sharma, D., Sharma, P., Shastri, S., 2017. Genetic, metabolic and endocrine aspect of intrauterine growth restriction: an update. *J. Matern. Fetal Neonatal Med.* 30, 2263–2275. doi:10.1080/14767058.2016.1245285

Shinya, M., Sakai, N., 2011. Generation of highly homogeneous strains of zebrafish through full sib-pair mating. *G3* (Bethesda) 1, 377–386. doi:10.1534/g3.111.000851

Shukrun, N., Shabtai, Y., Pillemer, G., Fainsod, A., 2019. Retinoic acid signaling reduction recapitulates the effects of alcohol on embryo size. *Genesis* 57, e23284. doi:10.1002/dvg.23284

Simon, M.M., Greenaway, S., White, J.K., Fuchs, H., Gailus-Durner, V., Wells, S., Sorg, T., Wong, K., Bedu, E., Cartwright, E.J., Dacquin, R., Djebali, S., Estabel, J., Graw, J., Ingham, N.J., Jackson, I.J., Lengeling, A., Mandillo, S., Marvel, J., Meziane, H., Brown, S.D.M., 2013. A comparative phenotypic and genomic analysis of C57BL/6J and C57BL/6N mouse strains. *Genome Biol.* 14, R82. doi:10.1186/gb-2013-14-7-r82

Simpson, E.M., Linder, C.C., Sargent, E.E., Davisson, M.T., Mobraaten, L.E., Sharp, J.J., 1997. Genetic variation among 129 substrains and its importance for targeted mutagenesis in mice. *Nat. Genet.* 16, 19–27. doi:10.1038/ng0597-19

Singh, A., Allen, D., Fracassi, A., Tumurbaatar, B., Natarajan, C., Scaduto, P., Woltjer, R., Kayed, R., Limon, A., Krishnan, B., Taglialatela, G., 2020. Functional integrity of synapses in the central nervous system of cognitively intact individuals with high Alzheimer's disease neuropathology is associated with absence of synaptic tau oligomers. *J Alzheimers Dis.* doi:10.3233/JAD-200716

Sommardahl, C., Cottrell, M., Wilkinson, J.E., Woychik, R.P., Johnson, D.K., 2001. Phenotypic variations of orpk mutation and chromosomal localization of modifiers influencing kidney phenotype. *Physiol. Genomics* 7, 127–134. doi:10.1152/physiolgenomics.00089.2001

Spielmann, M., Lupiáñez, D.G., Mundlos, S., 2018. Structural variation in the 3D genome. *Nat. Rev. Genet.* 19, 453–467. doi:10.1038/s41576-018-0007-0

Spivakov, M., Auer, T.O., Peravali, R., Dunham, I., Dolle, D., Fujiyama, A., Toyoda, A., Aizu, T., Minakuchi, Y., Loosli, F., Naruse, K., Birney, E., Wittbrodt, J., 2014. Genomic and phenotypic characterization of a wild medaka population: Towards the establishment of an isogenic population genetic resource in fish. *G3* (Bethesda) 4, 433–445. doi:10.1534/g3.113.008722

Stenson, P.D., Mort, M., Ball, E.V., Evans, K., Hayden, M., Heywood, S., Hussain, M., Phillips, A.D., Cooper, D.N., 2017. The human gene mutation database: Towards a comprehensive repository of inherited mutation data for medical research, genetic diagnosis and next-generation sequencing studies. *Hum. Genet.* 136, 665–677. doi:10.1007/s00439-017-1779-6

Stevens, L.C., Hummel, K.P., 1957. A description of spontaneous congenital testicular teratomas in strain 129 mice. *J Natl Cancer Inst* 18, 719–747.

Streisinger, G., Walker, C., Dower, N., Knauber, D., Singer, F., 1981. Production of clones of homozygous diploid zebra fish (*Brachydanio rerio*). *Nature* 291, 293–296. doi:10.1038/291293a0

Sun, J., Wiklund, F., Hsu, F.-C., Bälter, K., Zheng, S.L., Johansson, J.-E., Chang, B., Liu, W., Li, T., Turner, A.R., Li, L., Li, G., Adami, H.-O., Isaacs, W.B., Xu, J., Grönberg, H., 2006. Interactions of sequence variants in interleukin-1 receptor-associated kinase4 and the toll-like receptor 6–1–10 gene cluster increase prostate cancer risk. *Cancer Epidemiol. Biomarkers Prev.* 15, 480–485. doi:10.1158/1055–9965.EPI-05-0645

Suurväli, J., Whiteley, A.R., Zheng, Y., Gharbi, K., Leptin, M., Wiehe, T., 2020. The laboratory domestication of zebrafish: From diverse populations to inbred substrains. *Mol. Biol. Evol.* 37, 1056–1069. doi:10.1093/molbev/msz289

Taketo, M., Schroeder, A.C., Mobraaten, L.E., Gunning, K.B., Hanten, G., Fox, R.R., Roderick, T.H., Stewart, C.L., Lilly, F., Hansen, C.T., 1991. FVB/N: An inbred mouse strain preferable for transgenic analyses. *Proc Natl Acad Sci USA* 88, 2065–2069. doi:10.1073/pnas.88.6.2065

Tandon, P., Conlon, F., Furlow, J.D., Horb, M.E., 2017. Expanding the genetic toolkit in Xenopus: Approaches and opportunities for human disease modeling. *Dev. Biol.* 426, 325–335. doi:10.1016/j.ydbio.2016.04.009

Threadgill, D.W., Dlugosz, A.A., Hansen, L.A., Tennenbaum, T., Lichti, U., Yee, D., LaMantia, C., Mourton, T., Herrup, K., Harris, R.C., 1995. Targeted disruption of mouse EGF receptor: Effect of genetic background on mutant phenotype. *Science* 269, 230–234. doi:10.1126/science.7618084

Threadgill, D.W., Miller, D.R., Churchill, G.A., de Villena, F.P.-M., 2011. The collaborative cross: A recombinant inbred mouse population for the systems genetic era. *ILAR J.* 52, 24–31. doi:10.1093/ilar.52.1.24

Threadgill, D.W., Yee, D., Matin, A., Nadeau, J.H., Magnuson, T., 1997. Genealogy of the 129 inbred strains: 129/SvJ is a contaminated inbred strain. *Mamm. Genome* 8, 390–393. doi:10.1007/s003359900453

Tomlinson, M.L., Hendry, A.E., Wheeler, G.N., 2012. Chemical genetics and drug discovery in Xenopus. *Methods Mol. Biol.* 917, 155–166. doi:10.1007/978-1-61779-992-1_9

Török, H.P., Bellon, V., Konrad, A., Lacher, M., Tonenchi, L., Siebeck, M., Brand, S., De Toni, E.N., 2017. Functional toll-like receptor (TLR)2 polymorphisms in the susceptibility to inflammatory bowel disease. *PLoS One* 12, e0175180. doi:10.1371/journal.pone.0175180

Tour, E., Pillemer, G., Gruenbaum, Y., Fainsod, A., 2001. The two Xenopus Gbx2 genes exhibit similar, but not identical expression patterns and can affect head formation. *FEBS Lett.* 507, 205–209.

Tuttle, A.H., Philip, V.M., Chesler, E.J., Mogil, J.S., 2018. Comparing phenotypic variation between inbred and outbred mice. *Nat. Methods* 15, 994–996. doi:10.1038/s41592-018-0224-7

Venter, J.C., Adams, M.D., Myers, E.W., Li, P.W., Mural, R.J., Sutton, G.G., Smith, H.O., Yandell, M., Evans, C.A., Holt, R.A., Gocayne, J.D., Amanatides, P., Ballew, R.M., Huson, D.H., Wortman, J.R., Zhang, Q., Kodira, C.D., Zheng, X.H., Chen, L., Skupski, M., et al., 2001. The sequence of the human genome. *Science* 291, 1304–1351. doi:10.1126/science.1058040

Vijay, K., 2018. Toll-like receptors in immunity and inflammatory diseases: Past, present, and future. *Int. Immunopharmacol.* 59, 391–412. doi:10.1016/j.intimp.2018.03.002

Visser, G.H.A., Bilardo, C.M., Lees, C., 2014. Fetal growth restriction at the limits of viability. *Fetal Diagn. Ther.* 36, 162–165. doi:10.1159/000358058

Voelkl, B., Altman, N.S., Forsman, A., Forstmeier, W., Gurevitch, J., Jaric, I., Karp, N.A., Kas, M.J., Schielzeth, H., Van de Casteele, T., Würbel, H., 2020. Reproducibility of animal research in light of biological variation. *Nat. Rev. Neurosci.* 21, 384–393. doi:10.1038/s41583-020-0313-3

Watanabe, M., Yasuoka, Y., Mawaribuchi, S., Kuretani, A., Ito, M., Kondo, M., Ochi, H., Ogino, H., Fukui, A., Taira, M., Kinoshita, T., 2017. Conservatism and variability of gene expression profiles among homeologous transcription factors in *Xenopus laevis*. *Dev. Biol.* 426, 301–324. doi:10.1016/j.ydbio.2016.09.017

White, D., Rabago-Smith, M., 2011. Genotype-phenotype associations and human eye color. *J. Hum. Genet.* 56, 5–7. doi:10.1038/jhg.2010.126

Wong, K.H.Y., Ma, W., Wei, C.-Y., Yeh, E.-C., Lin, W.-J., Wang, E.H.F., Su, J.-P., Hsieh, F.-J., Kao, H.-J., Chen, H.-H., Chow, S.K., Young, E., Chu, C., Poon, A., Yang, C.-F., Lin, D.-S., Hu, Y.-F., Wu, J.-Y., Lee, N.-C., Hwu, W.-L., Kwok, P.-Y., 2020. Towards a reference genome that captures global genetic diversity. *Nat. Commun.* 11, 5482. doi:10.1038/s41467-020-19311-w

Wu, H., Yang, L., 2015. Arg753Gln polymorphisms in toll-like receptor 2 gene are associated with tuberculosis risk: A meta-analysis. *Med. Sci. Monit.* 21, 2196–2202. doi:10.12659/MSM.893214

Zarrei, M., MacDonald, J.R., Merico, D., Scherer, S.W., 2015. A copy number variation map of the human genome. *Nat. Rev. Genet.* 16, 172–183. doi:10.1038/nrg3871

Zeng, S.L., Sudlow, L.C., Berezin, M.Y., 2020. Using Xenopus oocytes in neurological disease drug discovery. *Expert Opin. Drug Discov.* 15, 39–52. doi:10.1080/17460441.2020.1682993

Zerbino, D.R., Frankish, A., Flicek, P., 2020. Progress, challenges, and surprises in annotating the human genome. *Annu. Rev. Genomics Hum. Genet.* 21, 55–79. doi:10.1146/annurev-genom-121119-083418

Zhang, B.N., Wong, T.C.B., Yip, Y.W.Y., Liu, Z., Wang, C., Wong, J.S.C., He, J.N., Chan, T.C.Y., Jhanji, V., Pang, C.P., Zhao, H., Chu, W.K., 2019. A sclerocornea-associated RAD21 variant induces corneal stroma disorganization. *Exp. Eye Res.* 185, 107687. doi:10.1016/j.exer.2019.06.001

Zhou, S.-F., Liu, J.-P., Chowbay, B., 2009. Polymorphism of human cytochrome P450 enzymes and its clinical impact. *Drug Metab. Rev.* 41, 89–295. doi:10.1080/03602530902843483

Zhou, Y., Ingelman-Sundberg, M., Lauschke, V.M., 2017. Worldwide distribution of cytochrome P450 alleles: A meta-analysis of population-scale sequencing projects. *Clin. Pharmacol. Ther.* 102, 688–700. doi:10.1002/cpt.690

23 Using *Xenopus* to Understand Pluripotency and to Reprogram Cells for Therapeutic Use

Meghana S. Oak and Eva Hörmanseder

CONTENTS

23.1. Introduction ... 325
23.2. Inception of Nuclear Reprogramming ... 325
23.2.1. Early SCNT Experiments Establish the Reprogramming Field ... 326
23.2.2. Nuclear Reprogramming Using *Xenopus laevis* Oocytes, Eggs, and Egg Extracts ... 327
23.3. Nuclear Reprogramming Takes a Leap from Frog to Mammals and Clinical Application ... 327
23.3.1. Mammalian Nuclear Transfer and Generation of Pluripotent Stem Cell Lines ... 327
23.3.2. Use of Alternative Methods for Nuclear Reprogramming ... 328
23.4. Molecular Insights into the Process of Nuclear Reprogramming Gained in *Xenopus laevis* ... 329
23.4.1. Reprogramming and DNA Replication ... 329
23.4.2. Change of Gene Expression Patterns during Reprogramming ... 330
23.4.3. Oocyte Factors Promote Nuclear Reprogramming ... 330
23.4.4. DNA Methylation as a Barrier to Nuclear Reprogramming ... 332
23.4.5. Histone Modifications and Resistance to Nuclear Reprogramming ... 332
23.5. Conclusion and Future Perspectives ... 332
References ... 333

23.1. INTRODUCTION

Every multicellular organism possesses a variety of cells of different types. Differentiated cells are extremely stable and, once defined, do not ordinarily change their fate. We now know that differentiation of cells takes place through distinct transcription patterns and is maintained with the help of epigenetic modifications and chromatin structure, even though the exact mechanistic details are still not fully understood. It has been shown using different techniques that the nucleus of a differentiated cell can be reverted to a state mimicking that of an undifferentiated, totipotent cell, which can itself develop into an entire organism, a process commonly referred to as "nuclear reprogramming."

One method that achieves reprogramming of nuclei is a technique called somatic cell nuclear transfer (SCNT), in which a somatic nucleus is transplanted into the enucleated egg of an organism. The early attempts at SCNT were performed on amphibians, and much of the seminal work that proved for the first time that nuclear reprogramming to totipotency is possible in differentiated somatic nuclei was done on *Xenopus laevis*. The use of *Xenopus* in developmental biology, although fortuitous, is tremendously advantageous. The early accomplishments seen using this model organism eventually led to successful cloning in mammalian systems and new reprogramming techniques being developed using *in vitro* models. The clinical applications of nuclear reprogramming and cellular plasticity are many, including therapeutical cloning, cell replacement therapies, and modeling of human diseases *in vitro*. The establishment of nuclear reprogramming in more advanced model systems has allowed this technology to make its way into clinical trials. The relevance of nuclear reprogramming studies using *Xenopus* as a model system was recognized in 2012, when the Nobel Prize in Physiology or Medicine was awarded to Sir John B. Gurdon, together with Shinya Yamanaka, for their discovery that mature cells can be reprogrammed to become pluripotent.

23.2. INCEPTION OF NUCLEAR REPROGRAMMING

Do all cell types house the same set of genes?

The totipotent zygote divides and differentiates into a number of different cell types to give rise to a live, fully developed organism. Differentiated cells rarely, if ever, reverse or change their state of terminal differentiation. This begs the question—do all cell types house the same set of genes? The question of how cell differentiation is established and maintained paved the way to the cell fate reprogramming

DOI: 10.1201/9781003050230-26

field. In 1893, Weismann put forth the germ-plasm hypothesis, in which he suggested that as development proceeds, genetic determinants are segregated into cells based on their type. It was quite intuitively supposed that genes that are not required for the intended fate of the cell are either permanently deactivated or, more likely, disposed of. An experiment in which the totipotent nucleus of the zygote is replaced by a differentiated, somatic nucleus was considered a definitive approach to solving this question, because proper development of the transplanted embryo would confirm that the genetic material remains intact and that differentiation does not irreversibly affect the plasticity of the nucleus.

23.2.1. Early SCNT Experiments Establish the Reprogramming Field

Between 1952 and 1957, Briggs and King revisited this crucial biological question and performed nuclear transfer (NT) experiments using the frog *Rana pipiens* as a model system. They introduced a blastula nucleus into an enucleated egg and obtained normal, living tadpoles (Briggs and King, 1952). This demonstrated that, at least until the blastula stage, the derived nuclei were able to support normal development and could give rise to complete organisms. Following this experiment, Briggs and King attempted to introduce nuclei of further differentiated endoderm cells into enucleated *Rana pipiens* eggs. The embryos resulting from this NT experiment did not survive, due to which they concluded that the ability of a nucleus to support normal development either decreases with, or is permanently discontinued at some point during their own differentiation. However, the curiosity of many scientists was not satisfied by these results, and one group in particular decided to tackle this problem once more.

John Gurdon began his Ph.D. in 1956 in the lab of Michail Fischberg with the same question in mind: Do all cell types have the same set of genes? For this, it was crucial to pick up where Briggs and King left off—with a repetition of the SCNT experiment. Gurdon performed a series of NTs using donor nuclei isolated from successive developmental stages of *Xenopus laevis*. The donor nuclei were derived from the endoderm, taken either from the vegetal cell mass at blastula or gastrula stages or the floor of gut-lumen in hatched tadpoles. He showed that, contrary to the earlier results obtained by Briggs and King, donor nuclei from more advanced stages of development can also support normal growth via NT (Gurdon, 1962). In fact, the organisms resulting from similar experiments most strikingly developed into sexually mature adults that could produce normally developing embryos themselves (Gurdon and Uehlinger, 1966).

The discrepancy between these findings and those of the widely respected and recognized scientists Briggs and King led to a certain amount of incredulity among the scientific population. However, there were certain noteworthy differences between the two studies. First, choosing the optimal model organism is extremely important. The use of *Rana pipiens* brings with it a number of technical disadvantages, one of which is its seasonal reproductive cycle. *Xenopus laevis* can be induced to lay eggs at any point of the year using mammalian hormones such as gonadotropic hormone. Furthermore, *Xenopus* has high disease resistance, and the growth of *Xenopus laevis* to sexual maturity occurs frequently in laboratory conditions and within one year. Second, a key difficulty of NT experiments at the time was the inability to safely penetrate the protective jelly surrounding eggs, which was necessary for enucleation. Gurdon used UV irradiation to destroy the chromosomes inside the egg, also making the egg jelly more penetrable in the process for microinjection during NT. Third, experimental outcomes were corrected for variations caused by technical damage to the oocyte during the NT procedure and for varying egg quality. This left innate properties of the donor nuclei as the only significant variable that might hinder efficiency of reprogramming in NT experiments (Gurdon, 1960). The fourth reason, which ultimately ensured the irrefutability of the study, was the use of marked donor nuclei. Nuclei obtained from a single-nucleolus strain of *Xenopus* were used as donors for NT, and the resulting successively transferred embryos were easily distinguishable from wild-type 2-nucleolated strains (Elsdale et al., 1960). This marker was crucial to validate Gurdon's NT experiments against all doubt.

The most puzzling outcome of these experiments was that the efficiency of survival and appropriate development of the NT embryo decreased with an increase in the differentiation state of their respective donor nuclei. For example, the efficiency of NT from a donor of an early developmental stage, that is, the blastula stage, was rather high (60% of cleaving NT-embryos), whereas the efficiency of NT from a donor of highly differentiated state was extremely low (~1–2% of cleaving NT-embryos) (Gurdon, 1962). This observation pointed to an existing resistance of cells to reprogramming that becomes more prominent with differentiation. This can now be considered one of the earliest experimental indicators of epigenetics—that a separate regulatory mechanism might exist on top of DNA sequence to specifically supervise the expression of the genome based on its cell type. The implications of this finding for the mechanism's underlying cell-fate stability and reprogramming resistance will be discussed in the following sections. Regardless, his work answered a long-standing open question in developmental biology—it showed that genes unnecessary for the intended cell type are not lost or irreversibly inactivated through the course of cellular differentiation. In fact, differentiated nuclei retain the complete genome throughout development, including the genes important for forming functional germ cells, as evident from the fertile frogs resulting from NT (Gurdon and Uehlinger, 1966). Importantly, this new possibility of reprogramming the nucleus of a cell to totipotency or even a different cell fate has major implications for the field of regenerative medicine. It suggests that, in theory, any cell of our bodies can be changed to any other type to replace damaged or irreversibly lost cells.

23.2.2. Nuclear Reprogramming Using *Xenopus laevis* Oocytes, Eggs, and Egg Extracts

In amphibians, eggs develop from germ cells called oocytes, which harbor a nucleus called a germinal vesicle (GV). The GV is densely packed with components essential for development. Oocytes are arrested at prophase of meiosis I, and hormones such as progesterone induce their progression past meiosis I and into metaphase of meiosis II, at which they are stalled again. These cells are then ready for fertilization—in mammals, this stage is termed the "MII oocyte"; in *Xenopus*, it is called the egg. All early SCNT experiments were performed at this second stage of arrest, as the closest means of mimicking natural fertilization, and hence this method is also termed "egg-NT." In an egg-NT experiment (Figure 23.1), microinjection of the somatic nucleus causes activation of the egg, which, like in fertilized embryos, is then followed by several rapid, transcriptionally quiescent cell divisions that are accompanied with DNA replication until the zygotic genome is ultimately activated. In an optimal NT experiment, cell-type specific gene expression patterns are established and the cell types are correctly formed, allowing the embryo to further develop successfully. *Xenopus* egg-NT as an experimental system provides several advantages, as it represents a unique model to study successful reprogramming but also to investigate why reprogramming so often fails due to unsuccessful erasure of the previous somatic cell identity. For example, it can help to identify the barriers present in somatic cells that confer cellular memory of its differentiated state and that prevent reprogramming. It can help to reveal the propagation mechanisms of cellular memory and the associated chromatin factors from the differentiated donor nucleus to all cells of the NT embryo throughout several cell divisions, independently of transcription. Furthermore, it offers the advantage that reprogramming efficiencies can be directly measured by monitoring the establishment of cell-type-specific transcriptional networks and the formation of functional cell types in the developing NT embryo. However, the rapid early cleavage cycles contrast with the typical slow division rate of the somatic nucleus, and this transition can take a toll on the nucleus, exposing the genome to chromosomal damage. Thus, despite its many advantages, this can make it more challenging to use this method for studying the alterations in transcription throughout the reprogramming procedure and the chromatin modifications driving resistance.

The oocyte GV, when arrested in prophase I, does not undergo cell division or DNA replication and can be a highly transcriptionally active region. It acts as a reserve for the generation and accumulation of large numbers of maternal transcripts that will drive early development of the embryo after fertilization. Due to this quality, it also serves as a prime opportunity for the study of nuclear reprogramming. The transplantation of donor nuclei directly into the GV of an amphibian oocyte—a method also called "oocyte-NT" (Figure 23.1)—induces rapid de-differentiation of the somatic nuclei along with reactivation and increased expression of previously repressed genes and pluripotency genes (Halley-Stott et al., 2010; Pasque et al., 2011). Several hundred donor nuclei can be transplanted at a time into the GV of a single oocyte. The high number of transcripts resulting from this technique facilitates both the analysis of gene expression trends during reprogramming and the resistance that the donor nuclei experience by means of epigenetic modifications. Although this technique generates oocytes arrested in meiotic prophase I and cannot produce embryos that support normal growth through specification of cell lineages, it is a useful system for studying reprogramming in a purely transcriptional context. *Xenopus* oocytes also have the handy ability to specifically transcribe nuclei from different species, due to which it is possible to transfer mammalian or cultured donor cells and study transcription distinctly induced by reprogramming without the noise of endogenous maternal transcripts (De Robertis and Gurdon, 1977).

In addition to the use of oocytes as NT recipients to study nuclear reprogramming *in vitro*, the use of *Xenopus* egg extracts and oocyte extracts has also proved surprisingly beneficial (Hansis et al., 2004). Initially used to study the fertilization process more closely (Lokha and Masui, 1983), the use of this cell-free system has since been adapted as an excellent tool to mimic the reprogramming environment. Cultured cells display striking reprogramming activity when introduced to metaphasic *Xenopus* egg extracts, including alterations in their replicative nature, epigenetic signature, and chromatin organization (Figure 23.1; Ganier et al., 2011). This occurs initially in the absence of transcription, which is mostly inhibited in eggs and embryos until zygotic genome activation at mid-blastula transition. It has been shown that transient exposure of donor nuclei to egg extracts followed by typical NT to an enucleated egg primes them for significantly more efficient reprogramming (Ganier et al., 2011). Similar experiments have been carried out on oocyte extracts, and this cell-free system has also been developed for use in mammals (Miyamoto et al., 2009). The ability to easily manipulate the composition of egg extracts through biochemical methods and the ease with which the resulting effects can be analyzed and transferred to an *in vivo* setting make the use of egg extracts for reprogramming invaluable.

23.3. NUCLEAR REPROGRAMMING TAKES A LEAP FROM FROG TO MAMMALS AND CLINICAL APPLICATION

23.3.1. Mammalian Nuclear Transfer and Generation of Pluripotent Stem Cell Lines

It took a little over three decades for the early results in *Xenopus* to be reproduced in mammals, but finally, the generation of the first clones in a number of different mammals succeeded: cloning of mouse embryos (Tsunoda et al., 1987; Wakayama et al., 1998), bovine embryos (Prather et al., 1987), sheep (Campbell et al., 1996; Wilmut et al., 1997), and pigs (Polejaeva et al., 2000). Not

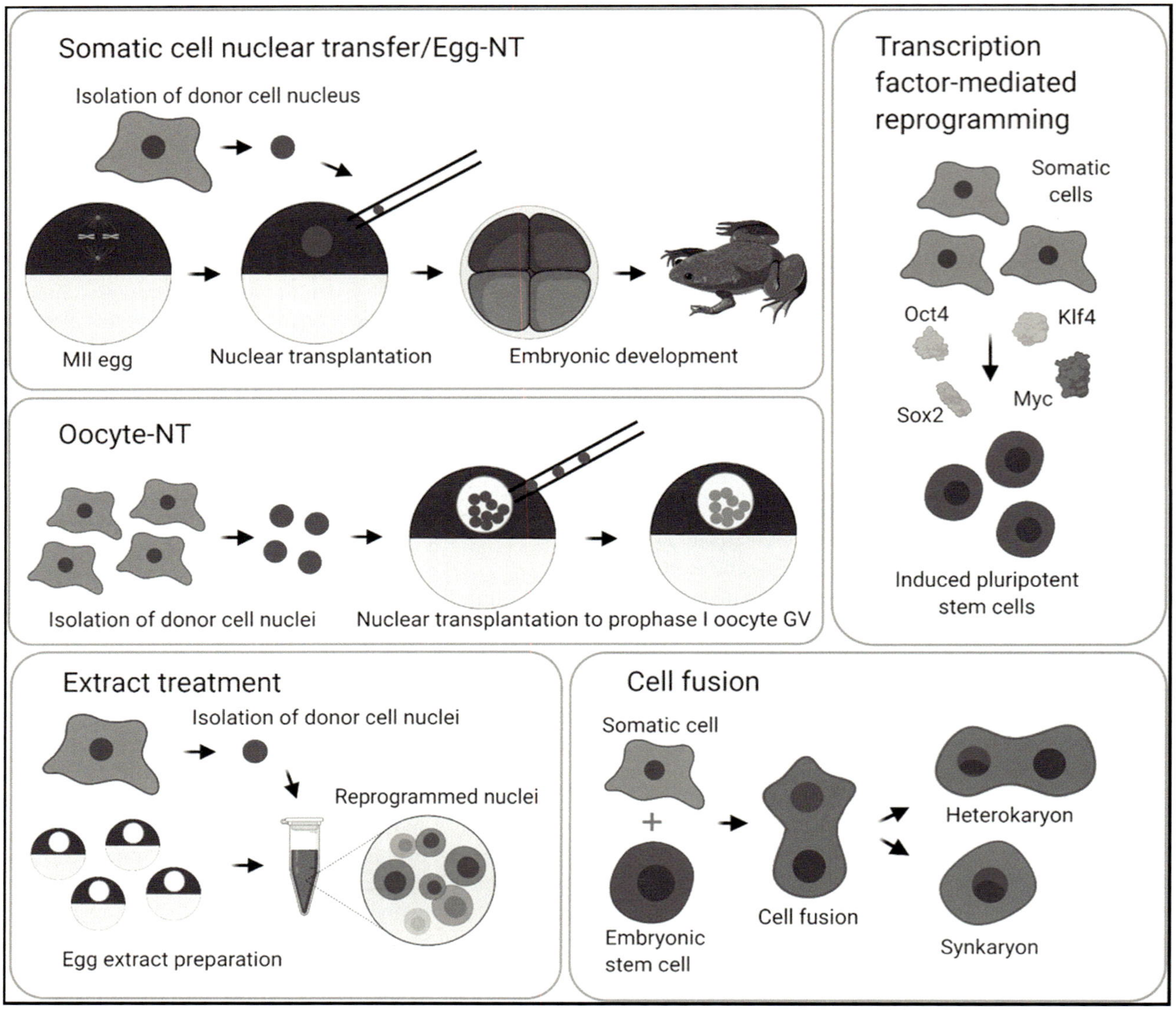

FIGURE 23.1 Schematic of the different methods used for nuclear reprogramming. (*Clockwise from top left*) Somatic cell nuclear transfer/egg-NT: Nucleus of a somatic cell is isolated and transferred to an enucleated egg; the formed zygote then can develop into a cloned organism. Transcription factor-mediated reprogramming: Cultured cells are induced to pluripotency by overexpression of indicated transcription factors. Cell fusion: Somatic cells are fused with embryonic stem cells. The embryonic components of the ES cell induce reprogramming of the differentiated cell. Extract treatment: Somatic cell nuclei are introduced to *Xenopus* egg extracts, which induce their reprogramming. Oocyte-NT: Multiple somatic cell nuclei are transplanted into the germinal vesicle (nucleus) of an oocyte. Oocyte factors reprogram the nuclei.

too long before NT was proven possible on mammalian systems, another prominent study published for the first time the establishment of progressively growing cultures of pluripotent embryonic cells derived from mice (Evans and Kaufman, 1981). Another breakthrough followed almost two decades later, with the establishment of human embryonic stem cell lines (Thomson et al., 1998). These studies opened up the possibility of isolating embryonic cells from cloned embryos for clinical applications. In conjunction with one another, both these techniques can have far-reaching effects on therapeutical cloning and human disease modeling.

23.3.2. Use of Alternative Methods for Nuclear Reprogramming

As the popularity of the initial SCNT results grew, other techniques for reprogramming cells to pluripotency emerged (Figure 23.1). Reprogramming by cell fusion, for example, functions through the fusion of plasma membranes of an ES-cell with a differentiated cell type (Figure 23.1; Blau et al., 1983). In the resulting fused cell, the differentiated nucleus is exposed to embryonic components of the ES cell, leading to expression of stem cell-specific genes, and is thus reprogrammed. The beginning of the 21st century witnessed another formative step in the path to manipulation of pluripotency and cell fate reprogramming, with the generation of induced pluripotent stem cells (iPSCs) (Takahashi and Yamanaka, 2006). In a strictly *in vitro* setting, the over-expression of a certain set of pioneer transcription factors can reprogram differentiated cells to an induced pluripotent state (Figure 23.1). iPS cells generated in this way are in many aspects functionally equivalent to ES cells. Due to its ease of implementation, the iPS procedure has advanced the journey towards cell replacement to a great extent. Nevertheless, the generation of iPSCs via transcription factor over-expression remains inefficient and stochastic. In addition, it is argued that the quality of ES-cells derived from NT-embryos is higher than of iPSCs, as measured by chromosomal abnormalities and differentiation potential (Matoba and Zhang, 2018).

The early advances in nuclear reprogramming procedures were further complemented by the discovery of transdifferentiation—a technique that aims at reprogramming the cell from its inherent differentiated type directly to another differentiated cell type, all while surpassing the dedifferentiation process that is key in NT. For example, in one of the first transdifferentiation experiments, ectopic expression of the pioneer transcription factor *MyoD* by transfection of its cDNA into mouse fibroblasts converted the cells into myoblasts (Davis et al., 1987).

While the mechanisms of cell fate reprogramming may vary among these different techniques, what they all have in common is that differentiated cells subjected to these reprogramming methods show resistance, and reprogramming efficiencies are low. In the past, using NT and *Xenopus laevis* as model system, key mechanisms of nuclear reprogramming were revealed, and initial insights into the mechanistic basis of resistance to reprogramming were gained. Importantly, the obtained results were confirmed in mammalian reprogramming systems and hence significantly advanced the entire field of nuclear reprogramming, as summarized next.

23.4. MOLECULAR INSIGHTS INTO THE PROCESS OF NUCLEAR REPROGRAMMING GAINED IN *XENOPUS LAEVIS*

The last few decades of the 1900s witnessed a lag in the use of SCNT to study the resistance of differentiated cell types against reprogramming. Spanning amphibians to mammals, the efficiency of cloning was consistently low, and absence of the high-throughput technology available today limited further work on this problem. Developments in molecular cloning, refined tissue culture methods, and other molecular biology techniques were used hand in hand with the alternative reprogramming technologies discussed previously. However, while these methods broadened the scope of the field, the issue of resistance against reprogramming in differentiated cells was universal, with little headway into the reasons causing this. Whether *in vivo* enucleated oocytes or *in vitro* egg extracts, there existed a pervading struggle between the reprogramming activities of the egg and the mechanisms maintaining stable cell fates of the differentiated nucleus—a "battle for supremacy," as termed by John Gurdon. In the 21st century, the reprogramming field was revived with new vigor with the onset of genomic and proteomic techniques. The development of transcriptional profiling techniques, histone modification-specific proteomic tools, and, later, sequencing technology was key to driving the current era of the reprogramming field and led to important discoveries concerning resistance to reprogramming.

23.4.1. Reprogramming and DNA Replication

One evident incongruity in the development of a NT embryo is the frequency of cell division of an adult cell versus that of the donor nucleus in SCNT experiments. Due to the rather quiescent cell division of an adult cell, the extremely rapid cell divisions enforced on the donor nucleus during NT may cause severe replication defects, limiting the probability of such a cell to give rise to a fully developed, fertile organism. During SCNT, microinjection causes activation of the oocyte. As the egg divides 90 minutes after activation, the DNA synthesis within the transplanted nucleus—a process which normally requires approximately six hours in somatic cells—must be acquired within this time frame upon NT to the egg. This transition from slow to rapid DNA replication can take a toll on the nucleus, causing DNA damage or a lag in DNA replication. Thus, occasionally, the entire replicating genome may move into only one of the dividing blastomeres, leading to a partially cleaved blastula. Previous attempts to increase reprogramming efficiency of adult cells using "serial nuclear transfers"—NT is repeated using a partial or complete blastula cell nucleus derived from a developing NT embryo—have proved substantially successful (King and Briggs, 1956; Gurdon et al., 1958; Laskey and Gurdon, 1970). It is thought that a subsequent NT event provides the embryo a second opportunity to complete DNA replication, which allows it to develop much further. Studies in *Xenopus* egg extracts confirmed that differentiated erythrocyte nuclei indeed replicate inefficiently in interphase *Xenopus* egg extracts when compared to sperm nuclei, and this difference in replication efficiency decreases when differentiated nuclei are allowed to progress through mitosis first (Lemaitre et al., 2005). Each additional round of mitosis furthers the capability of differentiated nuclei to undergo DNA replication upon NT. Thus, remodeling of the chromatin, which occurs at each embryonic metaphase, may support reprogramming of replicon organization in NT embryos to an early embryonic state and so permit successful development in some instances. However, some somatic nuclei, such as spermatid nuclei, show efficient reprogramming of replication origins comparable to those of sperm nuclei (Teperek et al., 2016). While sperm nuclei almost always support successful embryonic development, spermatids do not and show just as low developmental potential as other somatic nuclei when transplanted to eggs to generate NT embryos. In summary, this suggests that an adjustment of the replication machinery to an embryonic state is likely beneficial for nuclear reprogramming and also that additional layers of resistance exist on differentiated chromatin that further prevent complete cell-fate reprogramming to pluripotency.

23.4.2. Change of Gene Expression Patterns during Reprogramming

Specific genes required during early development are typically repressed within the genome of a differentiated cell, whereas genes characteristic of the cell type are highly expressed. For reprogramming to be successful, it appears necessary that the gene expression pattern of the differentiated donor cell be fully changed to that of a pluripotent cell and further to that of a specialized cell in the developing embryo. Thus, when reprogramming fails, it was hypothesized that this could be the result of inefficient silencing of genes that were expressed in the differentiated donor cell, together with unsuccessful activation of genes typically expressed in the reprogrammed cell type.

Indeed, *Xenopus* cell lineages isolated from an NT-embryo continue to express genes that are strictly characteristic of the donor cell type and should not be active in these reprogrammed cell lineages (Hörmanseder et al., 2017). The first evidence for a failure in inactivating genes during reprogramming was obtained when muscle nuclei were used as donor cells. Neurectoderm and endoderm lineages derived from roughly half of the resulting NT embryo continued to inappropriately express certain muscle genes (Ng and Gurdon, 2005). A serial NT study showed that the expression of muscle genes persisted even in approximately 50% of the second-generation NT embryos, addressing the stability and potential transmission of the donor-like expression state (Ng and Gurdon, 2008). The transcriptional quiescence of *Xenopus* embryos for the first 12 divisions until ZGA suggests that the "memory" of the past active transcriptional state is transmitted through subsequent embryonic cell divisions, independently of ongoing gene expression and the signal that induced the state. However, the molecular basis that maintains this memory of past active states remains to be elucidated. Inefficient reactivation of certain pluripotency genes, such as *Oct4* (*Pou5f1*), is observed in mouse NT-embryos (Matoba and Zhang, 2018). Interestingly, such decreased expression of pluripotency genes was found to be associated with a subsequent failure in embryonic development in mouse-derived SCNT-embryos (Boiani et al., 2002; Bortvin et al., 2003) and was later confirmed in human NT-embryos (Chung et al., 2015). Similar anomalous expression patterns are observed in iPSCs in which differentiation bias exists towards the donor cell type used (Polo et al, 2010). This bias can be attenuated upon persistent cell divisions, suggesting only transient influence levied by the original cell type in iPSCs. It is, however, clear that persistence of active and inactive past transcriptional states points to conserved mechanisms that restrict reprogramming of these genes. There exists a conflict between persistence of the memory state and imposition of the reprogramming process. It has been observed that the factors present in the oocyte strongly favor reprogramming, whereas the epigenetic state of the donor cell type contributes to the resistance against it.

23.4.3. Oocyte Factors Promote Nuclear Reprogramming

NT provides a unique experimental advantage to study the mechanisms of nuclear reprogramming with the hope to apply the obtained knowledge to improve reprogramming efficiencies in other reprogramming systems. A majority of the transcriptional reprogramming that a somatic nucleus undergoes takes place within the first 48 hours of its transfer to an egg, as compared to the two to eight weeks required to generate an iPSC line. In particular, the transfer of multiple somatic nuclei to the GV of an oocyte produces a rapid transcriptional reprogramming system of high quality as well as high efficiency.

The introduction of somatic nuclei to GV of *Xenopus* oocytes or to egg/oocyte extracts immediately triggers a cascade of events within the system, enabling the process of transcriptional reprogramming (Figure 23.2). When differentiated nuclei are injected into *Xenopus* egg extracts, decondensation of the somatic chromatin is accompanied by certain modifications to histone marks typically associated with an open chromatin structure (such as H3K14 acetylation) as well as the mobilization of the heterochromatin proteins HP1β and TIF1β (Trim28) from the nuclei (Tamada et al., 2006). Rapid exchange of somatic linker histones with oocyte linker histones occurs almost immediately post-transfer. The somatic linker histone H1 is replaced by the *Xenopus* oocyte-specific B4, which is bound on somatic chromatin within 24 hours post-transfer (Jullien et al., 2010). In mammals, the orthologous linker histone is H1foo, which acts in a similar manner (Teranishi et al., 2004). This exchange is closely followed by the recruitment of the Pol II subunit—oocyte RPB1 in its hypophosphorylated form. Within 48 hours post-transfer, the Pol IIA is phosphorylated on the C-terminal domain (CTD) at Ser5 (for initiation of transcription) and Ser2 (for elongation of transcription) (Jullien et al., 2014). A genome-wide shift in transcription follows, in which genes responsible for transcription and development are upregulated, while those involved in signaling pathways (probably enriched according to the properties of the somatic cell) are downregulated. Here, the *Xenopus* orthologs of highly expressed genes in MEF-transplanted oocytes closely resemble the expression pattern of wild-type *Xenopus* oocytes, much more so than the expression pattern of mouse ESCs. Due to this, it is evident that NT to oocytes induces a shift from the somatic to oocyte cell type rather than the pluripotent stem cell pattern (Jullien et al., 2014).

Besides the exchange of linker histones between the somatic nucleus and the oocyte GV, deposition of other histone factors is also affected. The recruitment of Pol IIA coincides with a drop in somatic RBP1 as well as an increase in the deposition of oocyte histone H2B (Jullien et al., 2014). Histone chaperone HIRA-dependent deposition of oocyte-specific H3.3 occurs on transplanted somatic nuclear chromatin within the first 24 hours (Jullien et al., 2012). ASF1A, a histone variant highly enriched in MII stage oocytes that cooperates with histone chaperone HIRA, is also involved in acquisition of pluripotency in humans (Gonzalez-Muñoz et al., 2014). The macroH2A histone variant is replaced from somatic nuclei following NT (Chang et al., 2010; Pasque et al, 2011). This variant has been widely characterized as a repressive factor that hinders chromatin remodeling and binding of transcription factors by recruiting HDACs and interfering with the binding of the SWI/SNF remodeling complex (Angelov et al., 2003). Furthermore, the somatic TATA-binding protein (TBP) involved in the formation of the pre initiation complex of transcription is replaced by its oocyte counterpart (Jullien et al., 2014). The chromatin remodeler Brg1 (Smarca4) helps to activate the expression of pluripotency factor Oct4 during NT of human somatic nuclei to *Xenopus* egg extracts (Hansis et al, 2004). Similarly, in induction of pluripotency in mouse embryonic fibroblasts (MEFs), the chromatin remodeler Baf155 causes an increase in Oct4 expression (Singhal et al, 2010).

Nuclear actin, which is commonly associated with several different remodeling complexes, undergoes polymerization upon SCNT. It has been shown that this polymerization is also essential for the reactivation of *Oct4*, with the help of the actin signaling protein Toca-1 (Fnbp1l) and its downstream target Wave1 (Wasf1), both of which are enriched in the oocyte GV (Miyamoto et al., 2011; Miyamoto et al., 2013). The ability of Wave1 to interact with Ser2P Pol II helps to increase the efficiency of transcriptional reprogramming in *Xenopus* oocytes (Miyamoto et al., 2013). In this manner, the oocyte rapidly employs several mechanisms to promote reprogramming of the donor nucleus, irrespective of the cell type of origin.

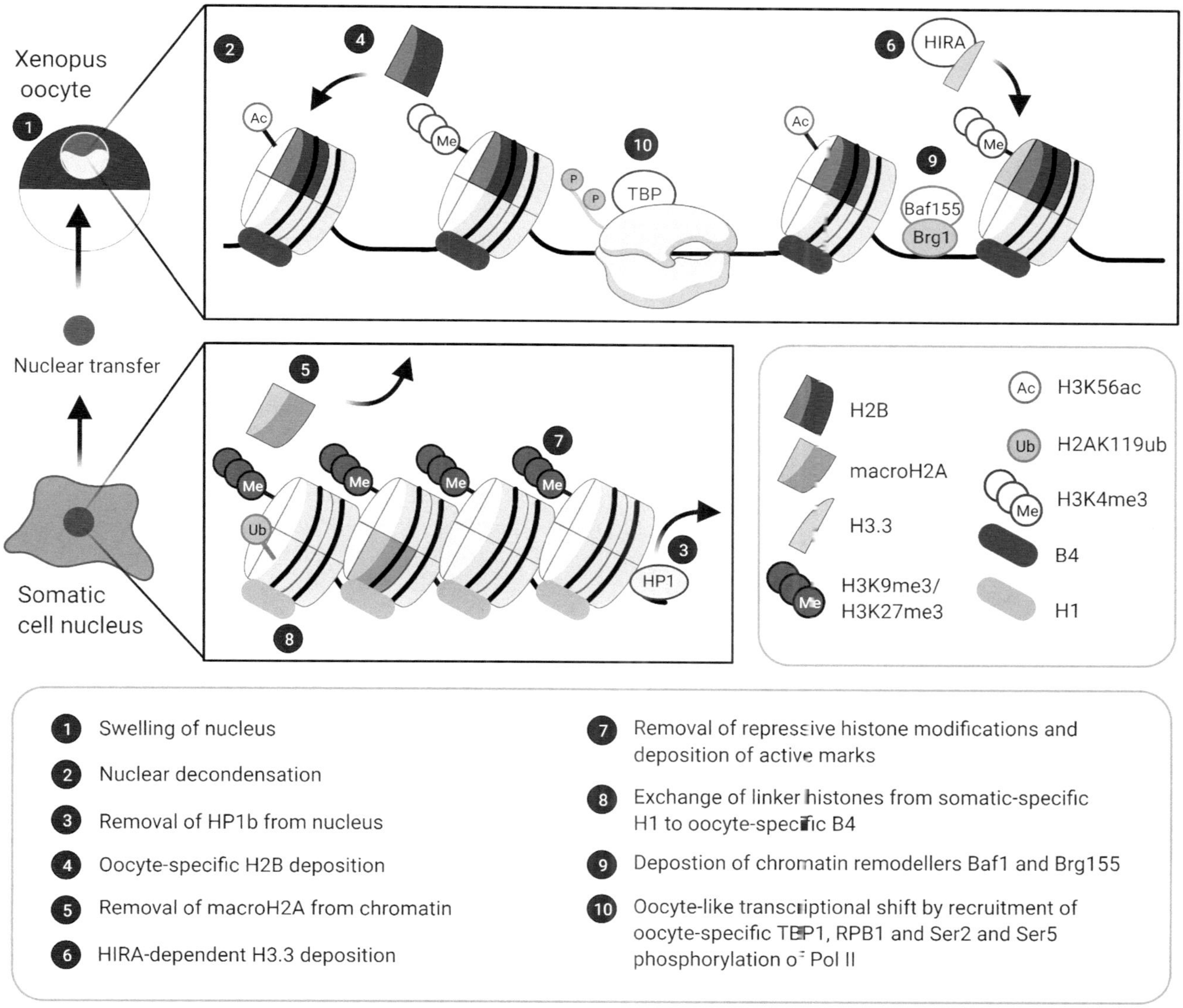

FIGURE 23.2 Schematic model highlighting roles of oocyte factors in driving reprogramming of the somatic nucleus to the oocyte after nuclear transfer. Oocyte factors incorporated into the transplanted somatic nuclei (top), as well as chromatin factors that are removed from the somatic nuclei by oocyte components (bottom) during reprogramming, are highlighted.

23.4.4. DNA Methylation as a Barrier to Nuclear Reprogramming

DNA methylation, specifically 5-methylcytosine (5mC) methylation, is a fairly stable epigenetic mark that is widely associated with the repressive state of gene expression (Greenberg and Bourc'his, 2019). Nuclear reprogramming depends on the ability to activate expression of several stably repressed genes in the donor nucleus in order to attain totipotency. It has been shown using multiple reprogramming systems that the repressive nature of DNA methylation poses a roadblock to efficient transcriptional reprogramming and that DNA demethylation is necessary for reprogramming somatic cell nuclei in *Xenopus* and in mammals (Simonsson and Gurdon, 2004; Mikkelsen et al., 2008). Aberrant erasure of DNA methylation marks in the donor nucleus causes inefficient cloning and may also lead to anomalies in development, as demonstrated in mammalian systems, including mouse and bovine embryos (Dean et al., 2001; Kang et al., 2001). Similar incomplete DNA methylation reversal has been reported in human iPSCs (Lister et al., 2011), showing that DNA methylation is a definite cause of resistance to reprogramming, regardless of the approach. However, the dependence of DNA demethylation during successful reprogramming in the *Xenopus* system remains to be fully elucidated.

23.4.5. Histone Modifications and Resistance to Nuclear Reprogramming

Chromatin organization and the epigenetic signature of the donor nucleus contribute to the resistance against reprogramming. Histone modifications affect both, contribute to maintenance of cellular identity, and can influence the ability of genes to be successfully reprogrammed (Kouzarides, 2007). Several studies have been conducted on the role of repressive histone marks as reprogramming barriers in *Xenopus*. In mouse SCNT-embryos, these barriers are thought to prevent early developmental genes from being reprogrammed into active expression. The histone marks H3K27me3, H3K9me3, and H2AK119 monoubiquitylation and the histone variant macro H2A, for example, are often associated with gene repression. In *Xenopus* and in mammals, H3K9me3 is enriched in reprogramming-resistant regions, and removal of this mark using either ectopic expression of its specific demethylase Kdm4d or knock-down of the H3K9 methyltransferase Suv39h in *Xenopus* and mouse NT experiments has been shown to be promising for reprogramming (Jullien et al., 2017; Matoba et al., 2014; Liu et al., 2016). H3K27me3 is enriched at genes important for embryonic development and also acts as an epigenetic barrier to reprogramming (Zhang et al., 2009; Jullien et al., 2017) in *Xenopus* and in mouse. Overexpression of the H3K27 demethylase KDM6A improved overall transcriptional reprogramming in *Xenopus* and mouse, as well as the developmental potential of mouse NT embryos. (Jullien et al., 2017; Zhou et al., 2019). H2AK119 monoubiquitylation is a repressive mark that confers reprogramming resistance (Jullien et al., 2017), as its removal via USP21 overexpression improved transcriptional reprogramming efficiencies in *Xenopus* oocyte NT. The histone variant macroH2A is involved in the stability of repressed states during *Xenopus* oocyte-NT, as seen by its role in resisting the reversal of X chromosome inactivation in mouse donor cells (Pasque et al., 2011). A role of macroH2A as reprogramming barrier was confirmed in the iPSC reprogramming system (Pasque et al., 2012).

However, aberrant gene expression patterns in the NT embryo cannot be exclusively attributed to repressive modifications resisting reprogramming—they may be also due to the transmission of active gene states. Intriguingly, the low efficiency of cell-fate reprogramming *Xenopus* NT embryos has been linked to the retention of donor cell type-specific memory of an active chromatin state stabilized by histone 3 lysine 4 (H3K4) methylation. The number of genes with persisting active and inactive gene states in *Xenopus* embryos is surprisingly similar (Hörmanseder et al., 2017). The genes resistant to reprogramming that are thus inappropriately active are implicated in functions related to the donor cell type, such as cell type-specific transcription factors, and showed increased H3K4me3 intensities and domain breadth when compared to genes that were properly reprogrammed, that is, downregulated (Hörmanseder et al., 2017). When nuclei with experimentally reduced H3K4 methylation were used to generate NT-embryos, a reduction of ON-memory and an improvement of cell-fate conversion was observed. This phenomenon was found to be conserved in mouse NT-embryos (Hörmanseder et al., 2017; Liu et al., 2016). These studies suggest that H3K4me3 stabilizes donor cell-fate memory in *Xenopus* NT-embryos and that active histone marks could contribute, in combination with repressive histone marks, to the epigenetic mechanisms of cellular memory.

The previously described chromatin modifications are largely observed in resistant genes regardless of the means of reprogramming—oocyte-NT, egg-NT, iPS, and cell fusion alike, suggesting similar mechanisms of epigenetic resistance (Jullien et al., 2017). Although it is well accepted that silent chromatin states characterized by modifications such as H3K9me3 contribute to cellular memory, it is still under debate whether active histone marks like H3K4me3 also have the potential for this (Stewart-Morgan et al., 2020) and whether other active histone marks impose barriers to successful nuclear reprogramming. Furthermore, it is not well understood how the different chromatin marks interact with each other, potentially in combination with transcription factors, to safeguard cellular identities and to prevent efficient nuclear reprogramming.

23.5. CONCLUSION AND FUTURE PERSPECTIVES

The search for an answer to the fundamental question of whether all cells in an organism have an identical set of genes opened the door to the untapped field of nuclear reprogramming. This field holds endless possibilities and immense potential for therapeutical use. During the 70 years since the first NT experiments, we have uncovered a vast reserve of knowledge that has aided the development of reprogramming technology. SCNT can be applied to agriculture, namely to aid in the preservation of endangered species. In hand with the recently lauded gene-editing technology CRISPR, SCNT can enable the production of cloned organisms with useful traits such as disease resistance. Development of SCNT has led to the ability to generate patient-derived NT embryonic cells (ntESCs). The isogeneity of ntESCs eliminates the possibility of immune rejection, marking an exciting feat in the therapeutic progress of this field (Tachibana et al., 2013). ntESCs could greatly aid in the development of cell replacement therapies and disease modeling. A decade after cloning the first mammal, we have been successful in cloning the first non-human primate (Liu et al., 2018).

We have taken a significant leap from the initial *in vivo* SCNT experiments carried out on *Xenopus* embryos to the successful cloning of primates. SCNT has several advantages in comparison to other reprogramming techniques, the chief advantage being its rapidity and the high quality of stem cells that can be produced. However, there are still several roadblocks to reprogramming that limit its translation into the clinic. Reprogramming efficiency continues to be low, which is the primary disadvantage in this technique, followed closely by the ethical and practical limitations that arise with theacquirement of human eggs. To improve efficiency, SCNT experiments into the *Xenopus* oocyte or using *Xenopus* oocyte/egg extracts continue to be extremely valuable model systems to study reprogramming factors and factors causing resistance. In addition, understanding the factors causing resistance of differentiated cells to nuclear reprogramming will also reveal important safeguarding mechanisms of cells, which help to maintain differentiated cell identities in healthy organisms and that are impaired in disease. The value of *Xenopus* as a model system for studying reprogramming continues to be noteworthy.

REFERENCES

Angelov, D., Molla, A., Perche, P.-Y., Hans, F., Côté, J., Khochbin, S., Bouvet, P., & Dimitrov, S. (2003). The histone variant macroH2A interferes with transcription factor binding and SWI/SNF nucleosome remodeling. *Molecular Cell*, *11*(4), 1033–1041. https://doi.org/10.1016/s1097-2765(03)00100-x

Blau, H. M., Chiu, C.-P., & Webster, C. (1983). Cytoplasmic activation of human nuclear genes in stable heterocaryons. *Cell*, *32*(4), 1171–1180. https://doi.org/10.1016/0092-8674(83)90300-8

Boiani, M., Eckardt, S., Schöler, H. R., & McLaughlin, K. J. (2002). Oct4 distribution and level in mouse clones: Consequences for pluripotency. *Genes & Development*, *16*(10), 1209–1219. https://doi.org/10.1101/gad.966002

Bortvin, A., Eggan, K., Skaletsky, H., Akutsu, H., Berry, D. L., Yanagimachi, R., Page, D. C., & Jaenisch, R. (2003). Incomplete reactivation of Oct4-related genes in mouse embryos cloned from somatic nuclei. *Development*, *130*(8), 1673–1680. https://doi.org/10.1242/dev.00366

Briggs, R., & King, T. J. (1952). Transplantation of living nuclei from blastula cells into enucleated frogs' eggs. *Proceedings of the National Academy of Sciences of the United States of America*, *38*(5), 455–463.

Campbell, K. H., McWhir, J., Ritchie, W. A., & Wilmut, I. (1996). Sheep cloned by nuclear transfer from a cultured cell line. *Nature*, *380*(6569), 64–66. https://doi.org/10.1038/380064a0

Chang, C.-C., Gao, S., Sung, L.-Y., Corry, G. N., Ma, Y., Nagy, Z. P., Tian, X. C., & Rasmussen, T. P. (2010). Rapid elimination of the histone variant MacroH2A from somatic cell heterochromatin after nuclear transfer. *Cellular Reprogramming*, *12*(1), 43–53. https://doi.org/10.1089/cell.2009.0043

Chung, Y. G., Matoba, S., Liu, Y., Eum, J. H., Lu, F., Jiang, W., Lee, J. E., Sepilian, V., Cha, K. Y., Lee, D. R., & Zhang, Y. (2015). Histone demethylase expression enhances human somatic cell nuclear transfer efficiency and promotes derivation of pluripotent stem cells. *Cell Stem Cell*, *17*(6), 758–766. https://doi.org/10.1016/j.stem.2015.10.001

Davis, R. L., Weintraub, H., & Lassar, A. B. (1987). Expression of a single transfected cDNA converts fibroblasts to myoblasts. *Cell*, *51*(6), 987–1000. https://doi.org/10.1016/0092-8674(87)90585-x

Dean, W., Santos, F., Stojkovic, M., Zakhartchenko, V., Walter, J., Wolf, E., & Reik, W. (2001). Conservation of methylation reprogramming in mammalian development: Aberrant reprogramming in cloned embryos. *Proceedings of the National Academy of Sciences*, *98*(24), 13734–13738. https://doi.org/10.1073/pnas.241522698

De Robertis, E. M. D., & Gurdon, J. B. (1977). Gene activation in somatic nuclei after injection into amphibian oocytes. *Proceedings of the National Academy of Sciences*, *74*(6), 2470–2474. https://doi.org/10.1073/pnas.74.6.2470

Elsdale, T. R., Gurdon, J. B., & Fischberg, M. (1960). A description of the technique for nuclear transplantation in *Xenopus laevis*. *Development*, *8*(4), 437–444.

Evans, M. J., & Kaufman, M. H. (1981). Establishment in culture of pluripotential cells from mouse embryos. *Nature*, *292*(5819), 154–156. https://doi.org/10.1038/292154a0

Ganier, O., Bocquet, S., Peiffer, I., Brochard, V., Arnaud, P., Puy, A., Jouneau, A., Feil, R., Renard, J.-P., & Méchali, M. (2011). Synergic reprogramming of mammalian cells by combined exposure to mitotic Xenopus egg extracts and transcription factors. *Proceedings of the National Academy of Sciences*, *108*(42), 17331–17336. https://doi.org/10.1073/pnas.1100733108

Gonzalez-Muñoz, E., Arboleda-Estudillo, Y., Otu, H. H., & Cibelli, J. B. (2014). Histone chaperone ASF1A is required for maintenance of pluripotency and cellular reprogramming. *Science*, *345*(6198), 822–825. https://doi.org/10.1126/science.1254745

Greenberg, M., & Bourc'his, D. (2019). The diverse roles of DNA methylation in mammalian development and disease: Nature reviews. *Molecular Cell Biology*, *20*(10), 590–607. https://doi.org/10.1038/s41580-019-0159-6

Gurdon, J. B. (1960). The developmental capacity of nuclei taken from differentiating endoderm cells of *Xenopus laevis*. *Journal of Embryology and Experimental Morphology*, *8*, 505–526.

Gurdon, J. B. (1962). The developmental capacity of nuclei taken from intestinal epithelium cells of feeding tadpoles. *Development*, *10*(4), 622–640.

Gurdon, J. B., Elsdale, T. R., & Fischberg, M. (1958). Sexually mature individuals of *Xenopus laevis* from the transplantation of single somatic nuclei. *Nature*, *182*(4627), 64–65. https://doi.org/10.1038/182064a0

Gurdon, J. B., & Uehlinger, V. (1966). "Fertile" intestine nuclei. *Nature*, *210*(5042), 1240–1241. https://doi.org/10.1038/2101240a0

Halley-Stott, R. P., Pasque, V., Astrand, C., Miyamoto, K., Simeoni, I., Jullien, J., & Gurdon, J. B. (2010). Mammalian nuclear transplantation to germinal vesicle stage Xenopus oocytes: A method for quantitative transcriptional reprogramming. *Methods (San Diego, Calif.)*, *51*(1), 56–65. https://doi.org/10.1016/j.ymeth.2010.01.035

Hansis, C., Barreto, G., Maltry, N., & Niehrs, C. (2004). Nuclear reprogramming of human somatic cells by Xenopus egg extract requires BRG1. *Current Biology*, *14*(16), 1475–1480. https://doi.org/10.1016/j.cub.2004.08.031

Hörmanseder, E., Simeone, A., Allen, G. E., Bradshaw, C. R., Figlmüller, M., Gurdon, J., & Jullien, J. (2017). H3K4 methylation-dependent memory of somatic cell identity inhibits reprogramming and development of nuclear transfer embryos. *Cell Stem Cell*, 21(1), 135–143.e6. https://doi.org/10.1016/j.stem.2017.03.003

Jullien, J., Astrand, C., Halley-Stott, R. P., Garrett, N., & Gurdon, J. B. (2010). Characterization of somatic cell nuclear reprogramming by oocytes in which a linker histone is required for pluripotency gene reactivation. *Proceedings of the National Academy of Sciences of the United States of America*, *107*(12), 5483–5488. https://doi.org/10.1073/pnas.1000599107

Jullien, J., Astrand, C., Szenker, E., Garrett, N., Almouzni, G., & Gurdon, J. B. (2012). HIRA dependent H3.3 deposition is required for transcriptional reprogramming following nuclear transfer to Xenopus oocytes. *Epigenetics & Chromatin*, *5*(1), 17. https://doi.org/10.1186/1756-8935-5-17

Jullien, J., Miyamoto, K., Pasque, V., Allen, G. E., Bradshaw, C. R., Garrett, N. J., Halley-Stott, R. P., Kimura, H., Ohsumi, K., & Gurdon, J. B. (2014). Hierarchical molecular events driven by oocyte-specific factors lead to rapid and extensive reprogramming. *Molecular Cell*, *55*(4), 524–536. https://doi.org/10.1016/j.molcel.2014.06.024

Jullien, J., Vodnala, M., Pasque, V., Oikawa, M., Miyamoto, K., Allen, G., David, S. A., Brochard, V., Wang, S., Bradshaw, C., Koseki, H., Sartorelli, V., Beaujean, N., & Gurdon, J. (2017). Gene resistance to transcriptional reprogramming following nuclear transfer is directly mediated by multiple chromatin-repressive pathways. *Molecular Cell*, *65*(5), 873–884.e8. https://doi.org/10.1016/j.molcel.2017.01.030

Kang, Y.-K., Koo, D.-B., Park, J.-S., Choi, Y.-H., Chung, A.-S., Lee, K.-K., & Han, Y.-M. (2001). Aberrant methylation of donor genome in cloned bovine embryos. *Nature Genetics*, *28*(2), 173–177. https://doi.org/10.1038/88903

King, T. J., & Briggs, R. (1956). Serial transplantation of embryonic nuclei. *Cold Spring Harbor Symposia on Quantitative Biology*, *21*, 271–290. https://doi.org/10.1101/sqb.1956.021.01.022

Kouzarides, T. (2007). Chromatin modifications and their function. *Cell*, *128*(4), 693–705. https://doi.org/10.1016/j.cell.2007.02.005

Laskey, R. A., & Gurdon, J. B. (1970). Genetic content of adult somatic cells tested by nuclear transplantation from cultured cells. *Nature*, *228*(5278), 1332–1334. https://doi.org/10.1038/2281332a0

Lemaitre, J. M., Danis, E., Pasero, P., Vassetzky, Y., & Méchali, M. (2005). Mitotic remodeling of the replicon and chromosome structure. *Cell*, *123*(5), 787–801. https://doi.org/10.1016/j.cell.2005.08.045

Lister, R., Pelizzola, M., Kida, Y. S., Hawkins, R. D., Nery, J. R., Hon, G., Antosiewicz-Bourget, J., O'Malley, R., Castanon, R., Klugman, S., Downes, M., Yu, R., Stewart, R., Ren, B., Thomson, J. A., Evans, R. M., & Ecker, J. R. (2011). Hotspots of aberrant epigenomic reprogramming in human induced pluripotent stem cells. *Nature*, *471*(7336), 68–73. https://doi.org/10.1038/nature09798

Liu, W., Liu, X., Wang, C., Gao, Y., Gao, R., Kou, X., Zhao, Y., Li, J., Wu, Y., Xiu, W., Wang, S., Yin, J., Liu, W., Cai, T., Wang, H., Zhang, Y., & Gao, S. (2016). Identification of key factors conquering developmental arrest of somatic cell cloned embryos by combining embryo biopsy and single-cell sequencing. *Cell Discovery*, *2*(1), 1–15. https://doi.org/10.1038/celldisc.2016.10

Liu, Z., Cai, Y., Wang, Y., Nie, Y., Zhang, C., Xu, Y., Zhang, X., Lu, Y., Wang, Z., Poo, M., & Sun, Q. (2018). Cloning of macaque monkeys by somatic cell nuclear transfer. *Cell*, *172*(4), 881–887.e7. https://doi.org/10.1016/j.cell.2018.01.020

Lohka, M. J., & Masui, Y. (1983). Formation in vitro of sperm pronuclei and mitotic chromosomes induced by amphibian ooplasmic components. *Science*, *220*(4598), 719–721. https://doi.org/10.1126/science.6601299

Matoba, S., Liu, Y., Lu, F., Iwabuchi, K. A., Shen, L., Inoue, A., & Zhang, Y. (2014). Embryonic development following somatic cell nuclear transfer impeded by persisting histone methylation. *Cell*, *159*(4), 884–895. https://doi.org/10.1016/j.cell.2014.09.055

Matoba, S., & Zhang, Y. (2018). Somatic cell nuclear transfer reprogramming: Mechanisms and applications. *Cell Stem Cell*, *23*(4), 471–485. https://doi.org/10.1016/j.stem.2018.06.018

Mikkelsen, T. S., Hanna, J., Zhang, X., Ku, M., Wernig, M., Schorderet, P., Bernstein, B. E., Jaenisch, R., Lander, E. S., & Meissner, A. (2008). Dissecting direct reprogramming through integrative genomic analysis. *Nature*, *454*(7200), 49–55. https://doi.org/10.1038/nature07056

Miyamoto, K., Pasque, V., Jullien, J., & Gurdon, J. B. (2011). Nuclear actin polymerization is required for transcriptional reprogramming of Oct4 by oocytes. *Genes & Development*, *25*(9), 946–958. https://doi.org/10.1101/gad.615211

Miyamoto, K., Teperek, M., Yusa, K., Allen, G. E., Bradshaw, C. R., & Gurdon, J. B. (2013). Nuclear wave1 is required for reprogramming transcription in oocytes and for normal development. *Science (New York, N.Y.)*, *341*(6149), 1002–1005. https://doi.org/10.1126/science.1240376

Miyamoto, K., Tsukiyama, T., Yang, Y., Li, N., Minami, N., Yamada, M., & Imai, H. (2009). Cell-free extracts from mammalian oocytes partially induce nuclear reprogramming in somatic cells. *Biology of Reproduction*, *80*(5), 935–943. https://doi.org/10.1095/biolreprod.108.073676

Ng, R. K., & Gurdon, J. B. (2005). Epigenetic memory of active gene transcription is inherited through somatic cell nuclear transfer. *Proceedings of the National Academy of Sciences of the United States of America*, *102*(6), 1957–1962. https://doi.org/10.1073/pnas.0409813102

Ng, R. K., & Gurdon, J. B. (2008). Epigenetic memory of an active gene state depends on histone H3.3 incorporation into chromatin in the absence of transcription. *Nature Cell Biology*, *10*(1), 102–109. https://doi.org/10.1038/ncb1674

Pasque, V., Gillich, A., Garrett, N., & Gurdon, J. B. (2011). Histone variant macroH2A confers resistance to nuclear reprogramming. *The EMBO Journal*, *30*(12), 2373–2387. https://doi.org/10.1038/emboj.2011.144

Pasque, V., Radzisheuskaya, A., Gillich, A., Halley-Stott, R. P., Panamarova, M., Zernicka-Goetz, M., Surani, M. A., & Silva, J. C. (2012). Histone variant macroH2A marks embryonic differentiation in vivo and acts as an epigenetic barrier to induced pluripotency. *Journal of Cell Science*, *125*(Pt 24), 6094–6104. https://doi.org/10.1242/jcs.113019

Polejaeva, I. A., Chen, S. H., Vaught, T. D., Page, R. L., Mullins, J., Ball, S., Dai, Y., Boone, J., Walker, S., Ayares, D. L., Colman,

A., & Campbell, K. H. (2000). Cloned pigs produced by nuclear transfer from adult somatic cells. *Nature*, *407*(6800), 86–90. https://doi.org/10.1038/35024082

Polo, J. M., Liu, S., Figueroa, M. E., Kulalert, W., Eminli, S., Tan, K. Y., Apostolou, E., Stadtfeld, M., Li, Y., Shioda, T., Natesan, S., Wagers, A. J., Melnick, A., Evans, T., & Hochedlinger, K. (2010). Cell type of origin influences the molecular and functional properties of mouse induced pluripotent stem cells. *Nature Biotechnology*, *28*(8), 848–855. https://doi.org/10.1038/nbt.1667

Prather, R. S., Barnes, F. L., Sims, M. M., Robl, J. M., Eyestone, W. H., & First, N. L. (1987). Nuclear transplantation in the bovine embryo: Assessment of donor nuclei and recipient oocyte. *Biology of Reproduction*, *37*(4), 859–866. https://doi.org/10.1095/biolreprod37.4.859

Simonsson, S., & Gurdon, J. (2004). DNA demethylation is necessary for the epigenetic reprogramming of somatic cell nuclei. *Nature Cell Biology*, *6*(10), 984–990. https://doi.org/10.1038/ncb1176

Singhal, N., Graumann, J., Wu, G., Araúzo-Bravo, M. J., Han, D. W., Greber, B., Gentile, L., Mann, M., & Schöler, H. R. (2010). Chromatin-remodeling components of the BAF complex facilitate reprogramming. *Cell*, *141*(6), 943–955. https://doi.org/10.1016/j.cell.2010.04.037

Stewart-Morgan, K. R., Petryk, N., & Groth, A. (2020). Chromatin replication and epigenetic cell memory. *Nature Cell Biology*, *22*(4), 361–371. https://doi.org/10.1038/s41556-020-0487-y

Tachibana, M., Amato, P., Sparman, M., Gutierrez, N. M., Tippner-Hedges, R., Ma, H., Kang, E., Fulati, A., Lee, H.-S., Sritanaudomchai, H., Masterson, K., Larson, J., Eaton, D., Sadler-Fredd, K., Battaglia, D., Lee, D., Wu, D., Jensen, J., Patton, P., . . . Mitalipov, S. (2013). Human embryonic stem cells derived by somatic cell nuclear transfer. *Cell*, *153*(6), 1228–1238. https://doi.org/10.1016/j.cell.2013.05.006

Takahashi, K., & Yamanaka, S. (2006). Induction of pluripotent stem cells from mouse embryonic and adult fibroblast cultures by defined factors. *Cell*, *126*(4), 663–676. https://doi.org/10.1016/j.cell.2006.07.024

Tamada, H., Van Thuan, N., Reed, P., Nelson, D., Katoku-Kikyo, N., Wudel, J., Wakayama, T., & Kikyo, N. (2006). Chromatin decondensation and nuclear reprogramming by nucleoplasmin. *Molecular and Cellular Biology*, *26*(4), 1259–1271. https://doi.org/10.1128/MCB.26.4.1259-1271.2006

Teperek, M., Simeone, A., Gaggioli, V., Miyamoto, K., Allen, G. E., Erkek, S., Kwon, T., Marcotte, E. M., Zegerman, P., Bradshaw, C. R., Peters, A. H., Gurdon, J. B., & Jullien, J. (2016). Sperm is epigenetically programmed to regulate gene transcription in embryos. *Genome Research*, *26*(8), 1034–1046. https://doi.org/10.1101/gr.201541.115

Teranishi, T., Tanaka, M. Kimoto, S., Ono, Y., Miyakoshi, K., Kono, T., & Yoshimura, Y. (2004). Rapid replacement of somatic linker histones with the oocyte-specific linker histone H1foo in nuclear transfer. *Developmental Biology*, *266*(1), 76–86. https://doi.org/10.1016/j.ydbio.2003.10.004

Thomson, J. A., Itskovitz-Eldor, J., Shapiro, S. S., Waknitz, M. A., Swiergiel, J. J., Marshall, V. S., & Jones, J. M. (1998). Embryonic stem cell lines derived from human blastocysts. *Science*, *282*(5391), 1145–1147. https://doi.org/10.1126/science.282.5391.1145

Tsunoda, Y., Yasui, T., Shioda, Y., Nakamura, K., Uchida, T., & Sugie, T. (1987). Full-term development of mouse blastomere nuclei transplanted into enucleated two-cell embryos. *Journal of Experimental Zoology*, *242*(2), 147–151. https://doi.org/10.1002/jez.1402420205

Wakayama, T., Perry, A. C., Zuccotti, M., Johnson, K. R., & Yanagimachi, R. (1998). Full-term development of mice from enucleated oocytes injected with cumulus cell nuclei. *Nature*, *394*(6691), 369–374. https://doi.org/10.1038/28615

Weismann, A. (1893). *The germ-plasm: A theory of heredity*. Translated by W. Newton Parker and Harriet Rönnfeldt. Scribner, New York.

Wilmut, I., Schnieke, A. E., McWhir, J., Kind, A. J., & Campbell, K. H. S. (1997). Viable offspring derived from fetal and adult mammalian cells. *Nature*, *385*(6619), 810–813. https://doi.org/10.1038/385810a0

Zhang, M., Wang, F., Kou, Z., Zhang, Y., & Gao, S. (2009). Defective chromatin structure in somatic cell cloned mouse embryos. *The Journal of Biological Chemistry*, *284*(37), 24981–24987. https://doi.org/10.1074/jbc.M109.011973

Zhou, C., Wang, Y., Zhang, J., Su, J., An, Q., Liu, X., Zhang, M., Wang, Y., Liu, J., & Zhang, Y. (2019). H3K27me3 is an epigenetic barrier while KDM6A overexpression improves nuclear reprogramming efficiency. *FASEB Journal: Official Publication of the Federation of American Societies for Experimental Biology*, *33*(3), 4638–4652. https://doi.org/10.1096/fj.201801887R

Index

Note: Page numbers in *italic* indicate a figure and page numbers in **bold** indicate a table.

45S pre-ribosomal RNA gene, 159

A

Abbot, Maude E., 233
Activation-transformation model (Nieuwkoop), *53*
Activators, absence, 59
ADAM metalloprotease, usage, 78, 237
Adenomatous Polyposis Coli (APC) protein, overexpression, 31
ADH1B gene, alleles (commonality), 314
Agonist binding sites, change, 144
Allo-alleles, presence, 199
Allopolyploidy, 155–156
Allotetraploid genomes, 155–156
Allotetraploidization, usage, 161
Allotetraploidy, recognition, 156
AMPA/GABA ratio, differences, 147
AMPA receptors, characteristics, 147
Amphibia (color changes), hormonal control (Huxley study), 7
Amphibians
- egg/embryonic patterning, organization, *26*
- embryos, axial pattern (organizer identification), 25
- embryos, fate map, *53*
- localization problem, 25–26
- models, importance, 126

Anaphase promoting complex (APC), 15
Anaphase promoting complex/Cyclosome (APC/C), activation, 15
Animal cap cells, Pax3/Zic1-mediated reprogramming, 133
Antagonistic H3K4ME3/H3K27ME3 marks, acquisition, 178–180
Anterior-posterior (AP) embryonic patterning
- classical/modern studies/issues, synthesis, 58–60
- control, signaling pathways (impact), 54–58

Anterior-posterior (AP) neural induction
- models, *53*
- molecular analyses, 54
- principles, elucidation, 52

Anterior-posterior (AP) neural patterning models, experimental embryology studies, 51–54
Anterior-posterior (AP) patterning
- control, 58
- signaling pathways, 51

Anti-cancer compound validation, GEXM (usage), 306–307
Anti-EMT compounds, range, 306
Anti-inflammatory (TH2) cells, cytokine production, 314
Antisense DNA oligonucleotide technology, *Xenopus* oocytes (usage), 29
Antisense MOs, usage, 45, 236
Anti-sense MO, usage, 226, 228
Antisense mRNA depletion, usage, 33
Antisense oligonucleotides, usage, 317
APC (canonical signal transduction component), 66
apc TALEN mRNAs, targeted injection (usage), 305
Aquaporins 1/9, identification, 146
Armadillo (ARM) repeats, usage, 238
Artificial allopolyploid analysis, 166–167
ATAC-seq data sets, training, 191
Atrio-septal defects (ASDs), 235
Autoregulation subcircuits, 188–189
Autoregulatory loop, Pax3 creation, 128
Axin (canonical signal transduction component), 66
Axin1 dorsal degradation (promotion), Hwa protein (impact), 32
Axis-inducing activity (displacement), cortical rotation (impact), 31
Axon growth, 296
Axon regeneration, 294, 295

B

Bacterial artificial chromosomes (BACs)
- location, 159
- sequencing, 157

Batrachian, 3
Beddard, Frank, 4
Bellerby, Charles, 7
Beta-catenin (β-catenin)
- degradation, promotion, 57
- dorsal β-catenin, cytoplasmic activation, 31–32
- epigenetic priming, 263
- independence, 68
- independent pathways, addition, 71
- mutant, destabilization, 90
- mutation, 239
- protein, N-terminal phosphorylation, 66
- protein, nuclear localization, *67*
- regulation function, 31
- secreted ligand activation, 32

bHLH proteins, impact, 99
bHLH transcription factors, mesodermal development role, 129–130
Bicaudal C (Bicc1), post-transcriptional regulator, 228
Bighead, head formation usage, 57
Bioinformatics, challenge, 198
Biological processes, identification, 294–296
Biophysics, *Xenopus* oocytes (usage), 143
Birnstiel, Max, 9
Blackler, Antoine, 8
Blastomeres, profiling, 203
Blastula
- blastula-stage embryo, fate map (production), 8–9
- blastula-stage embryo, prospective dorsal side, *67*
- mesoderm/endoderm, *187*
- stage endoderm patterning, *264*

Blastula Chordin and Noggin-expressing (NCNE) center, 90
Blastula stem cells, NCCs (shared features), 133
Bles, Edward, 4
Blum, Martin, 45
BMP4 signaling, impact, 45
Bmp4/Smad1, action, 265
BMP antagonists
- depletions, impact, 46–47
- expression, activation, 263
- tissue differentiation, 45–47

BMP cell specification, 246
BMP pathways, genes (identification), *44*
BMP signaling, 30
- pathways, reactivation, 209

Boie, Heinrich (letter), Wagler footnote, *6*
Bone reconstruction, approaches, 251
Bouncing eggs, detection, 15
Branchial arch syndrome, 250
Branchiostoma floridae, Hox cluster, 162
Brown, Don, 9

C

C57BL/6J, genetic engineering usage, 315
C57MG mouse mammary epithelial cells, transformation, 65–66
Cadherin switching, promotion, 128
Cadherin switch, neural crest migration requirement, 130
Calcineurin (non-canonical Wnt signaling pathway component), 71
Callan, Harold, 8
Call lineage gene co-expression profiles network, 191
CamKII (non-canonical Wnt signaling pathway component), 71
Cancer
- CRISPR/Cas9 cancer modeling, 305
- genetic *Xenopus tropicalis* cancer model, TALEN-mediated TSG disruption, 304–305
- initiation/progression, pathways/cellular processes (study), 302
- research, genetically engineered *Xenopus* models (usage), 303–305
- *Xenopus* cancer modeling, application potential, 305–307

Candidate CHD genes, identification, 235–236
Canonical notch signaling, 78
- device, *79*
- pathway, scheme, *79*

Canonical Wnt/b-catenin signaling, *67*
Canonical Wnt, requirement, 236–237
Canonical Wnt signaling, 66, 68
importance, 127
- link, 65–66
- mediator, 177–178
- non-canonical Wnt signaling, integration, 68

Cardiac organogenesis, Wnt signaling, 69
Casein Kinase I (CKI), canonical signal transduction component, 66
Cdc25 activation, prevention, 20
Cdk2
- Cyclin E binding, 19
- identification, 19

Cdt1 degradation, 18
Cell biology, proteomics (relationship), 200
Cell cycle
early development, relationship, *14*
events, *in vitro* recapitulation, 13–15
oscillatory nature, impact, 17
pausing, 16
progression, cytoplasmic factors (impact), 15
protein degradation, impact, 15–16
regulation, 102
systems biology, 16–17
transitions, control, 16–17
Cell death, 294–295
Cell developmental biology, *Xenopus* (research tool), 301–302
Cell division
control, study, 13
cycle, existence (demonstration), 13
Cell-fate reprogramming *Xenopus* NT embryos, efficiency, 332
Cell-free translation assays, usage, 26
Cell lineages, creation, 25
Cell nuclei, beta-catenin (β-catenin) localization, *67*
Cells, reprogramming, 325
Cellular activity, calcium-imaging, 282
Cellular proliferation, 295–296
Central left-right organizer (cLRO), characterization, *227*
Central nervous system (CNS)
accessibility, 282
circuitry, implementation, 284
manipulations, functional consequences, 281
morpho-physiological information, 277
visual field representations, 278
Centrosomal biology, nucleoporins (presence), 237–238
Cerberus (secreted factor)
identification, 226
impact, 57
CerS, impact, 30
CG (CpG) dinucleotides, DNA methylation (occurrence), 180
Channels (biophysics/pharmacological properties), *Xenopus* oocytes (usage), 143
Charles, Enid, 7
Child, craniofacial skeleton, *247*
ChIP-re-ChIP experiments, usage, 178
ChIP-seq analysis, 186
ChIP-seq data sets, training, 191
ChIP-sequencing experiments, 179
Chordamesoderm, formation, 188
Chordin/Tolloid/Twisted gastrulation/Crossveinless-2/BMP ancestrally conserved D-V patterning system, 47
Chromatin
accessibility, 176–178
accessibility, transcription factor binding (relationship), 177–178
assembly, replication-coupled chromatin assembly dynamics, 178
assembly, repressiveness, 177
assembly, usage, 177
chromatin-associated proteins, role, 173
heterochromatin marks, 180
immunoprecipitation, **174**
remodeling, dynamics, 173
state, DNA packaging/regulatory substrate, 173–174
state maps, genome browser view, *175*
Chromosomes
bouquet, formation, 34
comparison, 158
fusion sites, identification, 158
identification, 158–159
sex chromosomes, W-specific/Z-specific regions, 161
Cilia-based flow reception, 229
Cilia biology, nucleoporins (presence), 237–238
Cilia cell fate determination, glycosylation, 237
Ciliary pore complex model, contradiction, 238
Cis-regulatory modules (CRMs)
binding, 188–189
regulatory effector interactions, display, 185
CLEAR consortium, 268
Cleavage-stage embryos
arrest, 16
Cleavage-stage *Xenopus* embryos
cleavage, occurrence (timing), 17
H3K4me3/H3K27me3, absence, 179
Cleft lip/palate, 250
Clinically relevant GEXMs, generation, 303–304
Clock hypothesis, postulation, 107, 111
Cluster-specific gene expansions, *164*
Clutch comparisons, making, 317–319
Clutch variability, *317*
CMB:GFP neurula stage embryo, grafting, 210
CMV promoter, impact, 177
c-Myc promoter, usage, 177
Coherent feedforward loop, *189*
Collateral projections, generation, 296
Collinearity, 163
Comparative genome analysis, 166
Complementary deoxyribonucleic acid (cDNA)
clones, synthesis, 144
data, usage, 157
functional expression library screening, 57
isolation, 143
libraries, preparation, 44
reverse transcription, 203
screen, *gata4* identification, 235
sequencing, 316
transfection, 329
translating, 199
usage, 158
Congenital heart disease (CHD), 239–240
candidate CHD genes, identification, 235–236
cilia/centrosomal biology, nucleoporins (presence), 237–238
clinical therapies, 234
developmental cell signaling, mysteries, 238–239
diagnosis/studying, difficulties, 234
disease mechanism analysis, left-right patterning (usage), 236–237
genes, function (discovery), 233
historical/medical background, 233–234
patient cohorts, analysis, 235–236
roots, 233–234
shared molecular pathways, 239–240
study, future, 240
studying, approaches (*Xenopus*), 236–237
studying, morphological/developmental benefits, 236
surgical interventions, 234
understanding, molecular genetic studies (usage), 234–236
Congenital Malformations of the Heart (Taussig), 234
Connexins, gap junction component, 146
Context-specific Wnt signaling, 70
Copy-number variant (CNV)
analysis, 238
deletion, 237
identification, 235–236
Cortical rotation, 25–26
model, *26*
molecular control, 31
Cranial neural crest (CNC)
cells, differentiation (requirements), 131
EAD reciprocal signaling, *248*
population, impact, 246
Cranial placode development, notch pathway role, 107
Craniofacial abnormalities (CFAs), 239–240
Craniofacial anomalies (CFAs), 245
causes, 249–250
cell-based approaches, 252
classes, 250–251
environmental causes, 249–250
factors, association, 249–250
genetic causes, 249
landscape, 249–251
list, **252**
surgical approaches, 251
tissue engineering approaches, 251
treatment, 251–272
understanding, 252
vitamin deficiency, impact, 250
Craniofacial ciliopathies, 251, **251**
Craniofacial development
defining, 245
Xenopus, contributions, 245
Xenopus model, 245–246
Craniofacial disorders, 245
Craniofacial microsomia, 250
Craniofacial organizer, function, 249
Craniofacial signaling center, extreme anterior domain (comparison), 249
Craniofacial skeleton, *247*
cells, presence, 246
Craniofacial structure, development, 250
Craniosynostosis, 250
Crew, Frank, 7
CRISPR
CRISPR-mediated genome editing, usage, 252, 303
knock-ins, presence, 203–204
screens, 268
systems, usage, 290
CRISPR-based genome editing, usage, 200
CRISPR-based mutations, isolation, 8
CRISPR/Cas9 genome editing, usage, 236
CRISPR/Cas9-mediated genome editing, 127
CRISPR/Cas9 Selection-mediated Identification of Dependencies (CRISPR-SID)
development, 305
methodology, 306
Crosstalk, histone modifications (usage), 181
C-terminal domain (CTD), 330
CUB domains, 47
Cuvier, Georges, 4
Cyclin/Cdk complex, impact, 19
Cyclin, identification, 14–15
Cys-loop family, oocyte expression, 147
Cytoplasm, canonical Wnt signaling, 66, 68
Cytoplasmic activation, 31–32
Cytoplasmic determinants, examples, 25–26

Cytoplasmic dorsalizing activity, GBP proteins (impact), 31–32
Cytoskeleton, control, 127–128
Cytostatic factor (CSF)
 activity, impact, *15*
 arrest dependence, 16

D

Dactylethra (genus name, usage), 4
Dand5
 flow-induced post-transcriptional repression, *229*
 flow target, evolution, 229
 identification, 226, 228
 nodal-DAND5 module, high sequence diversity, **230**
 repression, 236
Dand5 mRNA, post-transcriptional regulation, 228
Daudin, François Marie, 3–4
 Xenopus description, *5*
DaVinci, Leonardo, 233
Dawid, Igor, 9
Degradation-ensuring sequence motif (degon), presence, 16
Deleted in AZoospermia-like (Dazl) family, germ cell development, 33
De novo methylation, involvement, 179
Deoxyribonucleic acid (DNA)
 overreplication, prevention, 18
 packaging, 173–174
 polymerase inhibitors, addition, 20
 recovery, *179*
 re-replication, prevention, 18
 segments, involvement, 314
 SOM, metacluster linkage, 191
Deoxyribonucleic acid (DNA) methylation
 barrier, 332
 dynamics, 180–181
 histone modifications, repression/modulation (contrast), 180–181
 repression/crosstalk, histone modifications (usage), 181
Deoxyribonucleic acid (DNA) replication, 176
 control, 17–18
 DNA replication-coupled chromatin assembly, impact, 178
 feedback, 15
 induction, *Xenopus* egg cytoplasm (impact), 13
 initiation, control, 18–19
 low-speed *Xenopus* egg extract, usage (advantages), 17
 ordering events, 18–20
 origins/timing, 20
 study, 13
De Robertis, Eddy, 263
Desmoid tumors, 304
Deuchar, Elizabeth, 3, 8
Developmental biology
 Gurdon, impact, 8
 researchers, impact, 8–9
Developmental cell signaling, mysteries, 238–239
Development/disease, genetic variation (impact), 316–320
Dexter, John S., 77
Diabetes, 260, 268
Dickkopf (Dkk1)
 experiments, loss, 69
 induction, 57
Dickkopf (Dkk2) functions, 69
Dickkopf (DKK) protein family, binding, 66, 68
DIF-FRAC, 200
Diploid genome, impact, 302
Disease
 insights, *Xenopus* functional neurobiology (impact), 277
 modeling, *Xenopus* genetic variability (exploitation), 319–320
Disease ontology (DO) resources, 260
DKL1 genes, linkage, 147
Dlc/Notch/RBPG signaling, experimental perturbation (consequences), 111
dll1STU/dll1 knock-down, impact, 95
Dominant-negative (DN) receptors, introduction, 54–55
Dorsal blastomeres, β-catenin (nuclear localization), *67*
Dorsal center development, prevention, *91*
Dorsal gene activation, β-catenin regulation function, 31
Dorsal involuting marginal zone (DIMZ), *91*
Dorsal midline (DML) development
 hes/hey genes, impact, **96–97**
 notch pathway, core components, **92–94**
Dorsal midline (DML) precursors, Notch pathway differential expression, 98
Dorsal midline (DML) tissues, 98–99
Dorsal-ventral (D-V) axis
 establishment, 90
 patterning, notch (role), *91*
Dorsal-ventral (D-V) patterning, signaling components, 43, *44*
Dorsal-ventral (D-V) tissues, origins (diagram), *46*
Dorsal β-catenin, cytoplasmic activation, 31–32
Dorsal midline (DML)
 germ layers development, notch pathway (core components), **92–94**
 segregation, *91*
Double strand break (DSB) repair, 305
Driesch, Hans, 43
Drosophila fruit fly
 chordin, homologue, 47
 mutant phenotype, discovery, 77
 Trithorax/Polycobm group gene identification, 178
 usage, 43
Drosophila melanogaster, homeobox gene screen, 235
dsBreaks, usage, 212, 214
D-V gradient, generation, 265
D-V histotypic cell differentiation, analysis, 45
D-V patterning, 268
Dynein Axonemal Particles (DynAPs), usage, 200
Dysmorphisms, description, **250**

E

EAF 2 (transcriptional-level regulation), 71
Early blastula mesoderm/endoderm, *187*
Early embryogenesis, notch signaling functions, 77
Early gastrula mesoderm/endoderm, *187*
Early-localizing mRNAs, *27*
Early *Xenopus* embryogenesis, information source, 302
Ectoderm
 development, pluripotency-related transcription factor (Pou5fl) homologues (usage), 31
 specification, 30–31
EGFP expression, 293
EGF receptor, deletion, 316
Egg constriction experiments, 25
Egg & Ego (Slack), 9
Egg storage histones, 178
Einsteck experiment, 52
Embryological experiments, *Xenopus* (usage), 210
Embryonic development, chromatin remodeling (dynamics), 173
Embryonic manipulation-based approaches, usage, 279
Embryos
 antagonistic H3K4ME3/H3K27ME3 marks, acquisition, 178–180
 histone modification maps, **174**
 proxy-cancer phenotype, presence, 306
EMC1 variants, molecular connection, 71
Endoderm
 commitment phase, 263
 development, phases (timing), *261*
 endoderm-specific transcription factor, co-occupation, 178
 formation, molecular basis, 261–263
 gastrula-stage endoderm patterning, 263, 265
 germ layer formation, conceptual phases, *262*
 induction phase, 263
 maternal control, 29–30
 maternal pre-pattern phase, 262–263
 neurula-stage endoderm patterning, 265–266, *266*
 organ fate, induction, 266–271
 organogenesis, temporal overview, 260–261, *261*
 patterning, 263–266
 pharyngeal endoderm, thyroid (relationship), 268
 specification, gene regulatory networks (usage), *187*
Endodermal lineages, formation, 186
Endogenous ion channels, usage, **145–146**
Entwicklungsmechanic, 126
ENU amplification, usage, 211
Ependymoglial cells, comparison, 290
Epidermal differentiation, *hes/hey* genes (role), **103–104**
Epidermoid cysts, 304–305
Epigenetic poising mechanism, usage, 31
Epigenome
 H3K9 methylation/H4K20me3 decoration, 180
Epigenome maps, 174, 176
Epithelial sodium channel (ENaC), beta subunit (encoding), 146–147
Epithelium-to-mesenchyme transition (EMT), 125, 128, 302
 anti-EMT compounds, range, 306
 neural crest EMT/migration, 129–130
Erythropoiesis, GATA-1a/b function, 156
Esophageal atresia, 260
Esophagus, endoderm organ fate, 268
Eukaryotic mRNA isolation, 9
Eukaryotic transcription factor (TFIIIA), isolation, 91
European *Xenopus* Resource Centre (EXRC), transgenic line, 208
Exogenous DNA, carrying, 207
Exome sequencing, 236
Experimental embryology, 43–44, 126
 studies, 51–54
Extracellular Chordin/Tolloid/Tsg/CV2/BMP pathway, components, *44*

Extracellular environment, active remodeling (requirement), 130
Extracellular matrix (ECM)
- components, presence, 296
- degradation, 268
- role, 291, 293

Extraocular motor nerves, extracellular recordings, 285
Extreme anterior domain (EAD)
- cells, impact, 247
- craniofacial organizer function, 249
- craniofacial signaling center, comparison, 249
- expression, *248*
- identification, 249
- LOF/GOF limitation, 245–246

F

F0 *pkd1* knockout tadpole, *215*
False discovery rate (FDR), 198
Familial adenomatous polyposis (FAP), 304
Fate map, *53*
Feedforward loops (FFLs), subcircuits, 190
FERM-domain containing protein, impact, 57
Fetal alcohol spectrum disorder (FASD)
- defects, polymorphisms (impact), 314
- result, 316

Fetal growth restriction (FGR), induction, 319
FGF3/FGF8, transcription (activation), 58
FGF/MAPK/GSK3 phosphorylations, impact, 47
Fibroblast growth factor (FGF)
- caudalizing factor, 56
- gradients, cell specification, 246
- ligands, ectopic expression, 55
- pathway, role, 55

Fibroblast growth factor (FGF) signaling, 55–56
- promotion, 128
- requirement, 105
- TF expression, 268

Fischberg, Michaïl, 7–8, 326
FISH analyses, usage, 158–159
Fish lines, 315
Flow-induced post-transcriptional repression, *229*
Flow-induced symmetry, 226
Follicle-enclosed oocytes
- presence, 147
- receptor presence, 143

Follicle, native ion currents, 147
Force-fields, framework, 279
Forward genetic screens, usage, 210–211
Four-channel confocal image, merged image, *46*
Foxa1-4, 263
Foxd3 (forkhead box/winged helix transcription factor), importance, 130
FOXG1 genes, linkage, 147
Foxh1, Tle recruitment, 177
Frizzled2 receptor (Fzd2), 239
Frizzled3 responsive element, identification, 71
Frizzled-like gene Frzb-1, impact, 57
Frog egg
- fertilization, appearance, *26*
- post-fertilization, appearance, *26*

Functional cDNA expression library screening, usage, 57
Functional neurobiology, *in vitro* preparations (usage), 282–285

G

Gain-of-function analysis, 186
Gain-of-function assays, usage, 67
Gain-of-function experiments, 102
- complementing, 54–55

Gain of function (GOF), limitation, 245–246
Gain-of-function (GOF) mutation, 304
Gall, Joe, 9
Galnt11/GalNAc *O*-glycosylation, role, 237
Gap junctions, connexins (relationship), 146
Gardner's syndrome, 304
Gastrula
- embryos, transplantation, 52–53
- gastrula-stage endoderm patterning, 263, 265
- mesoderm/endoderm, *187*
- stage endoderm patterning, *264*

Gastrulating embryo, diagram, *91*
Gastrulation, 265
- assays, usage, 69
- movements, impact, 52

Gastulation, gene regulatory networks (usage), 188
GATA-1a/b, erythropoiesis function, 156
Gata4-6, 263
Gaunt, Stephen, 45
Geminin protein, impact, 18
Gene cluster, formation, 163
Gene expression
- chromatin accessibility, 176–178
- reference, 163
- regulation, 71

Gene regulatory network (GRN), 127, 261
- component set, 129
- constructs, 203–204
- impact, 186–188
- regulation, novel/non-canonical genes (importance), **129**
- schematic representation, *247*
- status, 190–192
- subcircuits, 188–190
- usage, 185, *187*, 188

Genes
- batteries (expression), TF regulation, 185
- clusters, cluster-specific gene expansions, *164*
- editing, 210–214
- gene-specific studies, **251**
- *hox* genes, analyses, 162–163
- models, refinement, 203
- near genes, heterochromatin marks, 180
- organization, detail, 163, 165
- structures, intron-exon-UTR annotation (adjustment), 198
- study, novel roles (discovery), 237–238
- symbols (high-resolution proteomic analyses), 199–200
- *Xenopus* gene editing, future, 308

Genetically altered *Xenopus* (creation), targeted nucleases (usage), **212–214**
Genetically engineered cancer model organism, *Xenopus* (research tool), 301–302
Genetically engineered *Xenopus* models (GEXMs), 302, 303
- clinically relevant GEXMs, generation, 303–304
- usage, 305–307

Genetic research, diploid genome (impact), 302
Genetics, 210–214
- reverse genetics, dsBreaks (usage), 212, 214

Genetic *Xenopus tropicalis* cancer model, TALEN-mediated TSG disruption, 304–305
Genomes
- assembly scaffolds, 199
- comparative genome analysis, 166
- editing tools, advances, 207
- human genome, natural polymorphisms, 313–314
- notch signaling responsiveness, **82**
- *Xenopus laevis* genome, sequencing/assembly, 157–158
- zygotic genome activation (onset), chromatin assembly (relationship), 177

Genome wide association screens (GWASs), usage, 252
Genomic DNA, sequencing, 157
Genomic resources, improvement, 211
Germ cell development, Dazl family involvement, 33
Germ layer
- boundaries, refinement, 91, 95
- patterning, maternal control, 29–32
- segregation, notch targets (involvement), 95, 98
- theory, 126

Germ layer development
- *hes/hey* genes, impact, **96–97**
- notch pathway, core components, **92–94**

Germ layer formation, 90–91, 95, 98
- gene regulatory network, impact, 186–188

Germ layer induction
- maternal control, 29–32
- maternal secreted molecules, 30

Germ plasm
- identification, 32–33
- maternal germ plasm, assembly, 33–34
- mRNAs, study, 33
- presence, 26

Global transitions, genome-wide analyses, 174
Glycosylation, 237
Goldenhar syndrome, 250
Goosecoid, identification, 44–45
goosecoid-related homeobox genes, PCR screen (usage), 225–226
GPCR proteins, impact, 57
Growth factor signaling pathways (revealing), *Xenopus* experiments (usage), 259
gsc CRMs, input integration, 188
GSK3 (canonical signal transduction component), 66
GTP binding, encoding, 116
Guanine nuclear exchange factor, identification, 238–239
Gunn, J.W.C., 3, 7
Gurdon, John, 7–9, 143, 325, 326

H

H2A.Z histone variant, presence, 176–177
H3K4ME3
- DNA recovery, *179*
- increase, 180
- mark, involvement, 176

H3K4 methylation, increase, 179
H3K27 acetylation mark, deposit, 176
Hamburger, Viktor, 7, 44
Haramoto, Yoshikazu, 163
Hardman, John Maurice, 234
Harland, Richard, 263
Hartwell, Leland (genetic experiments), 13
Hay, Elizabeth, 91
Head formation, Wnt pathway, 56–57
Head mesoderm, appearance, 186
Health (insights), *Xenopus* functional neurobiology (impact), 277
Heasman, Janet, 29

Heat shock-inducible HSP70 promoter, usage, 209
Hemifacial microsomia, 250
Hes1-7 genes, involvement, 102
hes1-7/hey groups, *Xenopus tropicalis* genes (notch signaling responsiveness), **82**
hes4-7/hey genes, early expression patterns, *80*
Hes4, role, 105, 107
hes genes, impact, **110**, 111
Hes genes, role, 107
Hes/Hey bHLH-O transcription factors, binding, 78
hes/hey genes
  cross-regulation, **88–89**
  impact, **96–97**
  role, **103–104**
Heterochromatinization, drivers, 180
Heterochromatin marks, 180
Heterologous mRNAs, expression, 29
Heterosis, 156
Heterotaxy (HTX)
  birth defects, 234–235
  developmental cell signaling, mysteries, 238–239
  disease mechanism analysis, left-right patterning (usage), 236–237
  internal organs, misorientation, 233
  patient cohorts, analysis, 235–236
  studying, approaches, 236–237
  studying, morphological/developmental benefits, 236
  understanding, molecular/genetic studies, 234–236
hhex-positive/nrlh5-positive liver progenitors, problem, 270
HH/Gli-target genes, 268
High-resolution proteomic analyses
  database, 198–199
  development, 197
  nomenclature/gene symbols, 199–200
High-throughput RNA-seq, possibility, 44
Hippo signaling pathway, 71
HIRA histone chaperone complex, 176
Histone deacetylases (HDACs)
  Hdac1 activity, importance, 133
  impact, 111
  recruitment, 78
Histone H3, N-terminal tails (modifications), 174
Histone modifications, 332
  acquisition/dynamics, 178–180
  comparisons, 179
  maps, **174**
  repression/modulation, contrast, 180–181
Histone resistance, 332
Histone variants, usage, 176–177
His, Wilhelm, 126, *126*
Hogben, Lancelot, 6–7
Hogben Pregnancy Test, 7
Homeologous gene loci, comparison, 159
Homeologs, 156
Homozygous *muzak* tadpoles, *myh6* nonsense mutation (impact), 211
Hornblatt, 126
Hörstadius, Sven, *126*
Hox clusters
  identification, 162
  sequence determination, 158
hox genes
  analyses, 162–163
  identification, 162
  necessity, 116
  patterning role, occurrence, 131
HPLC, usage, 54
Human ciliopathies, study, 251
Human disease
  genetic polymorphism, contribution, 314
  Xenopus model, 71–72
Human embryo, representation, *247*
Human genome, natural polymorphisms, 313–314
Human organisms, whole genome sequences (publications), **157**
Human protein variants (encoding), RNAs (usage), 319–320
Huxley, Julien, 7

I

Immotile:motile cilia ratio, establishment, 237
Immune cell response, inflammation (relationship), 295
INDELs, presence, 236
Induced double strand break (DSB) repair, 305
Induced pluripotent stem cells (iPSCs)
  expression patterns, 330
  generation, 329
Induced tumor-like structures (ITLSs), 302–303
Inducible promoters, usage, 209
Inner ear endorgans, galvanic vestibular stimulation, *283*
Insertional mutagenesis, 211–212
Insertion-deletion (INDEL) mutations, 305
Insulin-like growth factor (IGF) signaling, impact, 58
Integrin protein, information relay, 130
Intermediate-localizing mRNAs, *27*
Internal organs, misorientation, 233
Intestinal malrotation, 260
Intestinal neoplasia, 304
Intestines, endoderm organ fate, 270–271
Intrauterine growth restriction (IUGR), FGR (comparison), 319
Intron-exon-UTR annotation, adjustment, 198
In vitro preparations, usage, 282–285
In vitro Xenopus preparations, usage, 284
Involuting marginal zone (IMZ), 98
  contribution, 90–91
I-SceI recognition site, 208
Isis (Oken), 4
Isogenic tadpoles, lymphoid tumor cell/collagen mixture matrix (subcutaneous transplantation), 303
iTRAQ, usage, 197

J

Jelly coats, removal, *19*
JNK (non-canonical Wnt signaling pathway component), 71
  signaling, 249

K

KEN box, 16
Kinetochores, concentration (low level), 16
Kinin-kallikrein signaling, NO production, 249
Kremen2, mediation role, 128

L

Late-localizing mRNAs, *27*
Lateral facial dysplasia, 250
Laterality, nodal cascade (impact), 225–226
Lateral plate, appearance, 186
Lateral plate mesoderm (LPM)
  Nodal cascade, activity, 228
  Nodal signaling, propagation, 225
LEF/TCF proteins, transcriptional activation complex, 66
Left-right (LR) asymmetry, impact (absence), 228
Left-right (LR) axis
  formation, 225
  present state, 225–229
Left-right (LR) development, 225
Left-right organizer (LRO), 226
  cells, flow target Dand5 (role), *227*
  pathway, signals (relaying), 236–237
  patterning, 239
  tissue, mispositioning, 228
Left-right (LR) patterning, 236–237
Left-right (LR)-relevant Nodal signaling module components, 229
Left-right (LR) signaling module, percent identity matrix, 229
Leftward flow target, Dand5 identification, 226, 228
Le Règne Animal (Cuvier), 4
Liddle syndrome, 146
Ligand-gated ion channels, expression (rarity), 147
Lineage tracing, 203–204
Linkage groups, mapping, 157
Linker histones, usage, 176–177
Liquid chromatography-tandem mass spectrometry (LC-MS/MS), usage, 306
Liver, endoderm organ fate, 268, 270
Localization elements (LEs) identification, structure-function mutagenesis (usage), 27
Localized maternal mRNAs, 26–27
Localized mRNAs, roles, *27*
Localized RNAs, molecular characterization, 26–28
Lohka, Manfred, 9
Loss-of-function analysis, 186
Loss-of-function approaches, 209
Loss-of-function assays, usage, 57
Loss of function (LOF), elicitation, 245
Loss-of-function experiments, 31, 102
XFD/DN-Fgfr1/DN-Ras, usage, 56
Loss-of-function *in vivo* studies, 59
Loss-of-function (LOF) mutation, 304
Loss-of-function studies, 54, 70
Loss of heterozygosity (LOH), 303
LRP6 receptor endocytosis, promotion, 44
LRP intracellular domain, 66
L subgenomes
  asymmetrical evolution, 159, 161
  identification, 158–159, *160–161*
Lungs, endoderm organ fate, 268
Lymphoid tumor cell/collagen mixture matrix, subcutaneous transplantation, 303
Lysis buffer, usage, 203

M

M1/M2 polarization, regulation, 295
Magnetic resonance imaging (MRI), usage, 306, 307

Malformations, **250**, **251**
Maller, Jim, 9
Mammalian nuclear transfer, 327–328
Mammary adenocarcinomas, induction, 301
Mangold, Hilde, 43
Mangold, Otto, 52
MAPK phosphorylation, impact, 47
MAPK signaling, decrease, 133
Marginal zone (MZ)
contribution, 90–91
dorsal descendants, allocation, 98
Mass spectrometry (MS)
MS-based proteomics, 197
Xenopus MS spectrometry, usage, 198
Mastermind-like (MAML) co-activator, recruitment, 78
Masui, Yoshio, 9
Maternal chromosomes, loss, 211
Maternal cytokeratin (krt8.l) mRNA, depletion, 29
Maternal ectoderm specifying pathway, zygotic transcription factors (identification), 30
Maternal effect mutation, 28–29
Maternal gene function, analysis, 28–29
Maternal gene plasm, assembly, 33–34
Maternal mRNAs, 25
depletion studies, 32
localized maternal mRNAs, 26–27
molecular characterization, 26–28
obtaining, cell-free translation assays (usage), 26
Maternal pre-pattern phase, 261–263
Maternal protein dowry, synthesis, 203
Maternal RNA, enrichment, 187
Maternal T-box transcription factor, importance, 187
Maternal-to-zygotic transition (MZT), 29
Maturation promoting factor (MPF)
activation, prevention, 20
characterization, 14–15
definition, 14
kinase activity, change, 18
kinase activity, increase/sustaining, *14*
Maturity onset diabetes of the young (MODY), 260, 268
Maxwell, A.K., 4, 6
Mdm2 binding protein (MTBP), usage, 19
Medaka (*Oryzias latipes*)
biomedical research usage, 315
male sex-determining gene, 161
Medulloblastomas, 304
Megadalton (MDa) NPCs, 238
Melanocytes, impact, 131
Membrane, canonical Wnt signaling, 66, 68
Mendelian inheritance, **250**, **251**
Mesendoderm
induction *in vivo* (blocking), CerS (usage), 30
regulation, GRN (impact), 186
Mesoderm
maternal control, 29–30
specification, gene regulatory networks (usage), *187*
specification, *hes7.3/esr5* promotion/*hes5.1* inhibition, 95, 98
Mesodermal lineage, formation, 186
Messenger RNA (mRNA)
concentration, proportion, 201
Dand5 mRNA, post-transcriptional regulation, 228
decay, 228
expression, 293
germ plasm mRNAs, study, 33
heterologous mRNAs, expression, 29
localization mechanisms, 27–28
microinjection, 208
mRNA-encoding mediators, ventral injection, 302
preparation/selection, 144
Metabolic regulation, 295–296
Metaclusters, presence, *191*
Methylcytosine-binding protein 2 (*MECP2*) genes, mutations, 147
Microglia/macrophage activation, regulation, 295
Mid-blastula transition (MBT), 29
impact, 56
Midbrain-hindbrain boundary, 102, 105, 107
Midbrain/hindbrain boundary (MHB), 107
establishment, *hes* genes (impact), **110**
Miller syndrome, 250
Mini chromosome Maintenance 2-7 (MCM2-7) protein complex, recruitment, 18
Mitochondrial DNA, discovery, 9
Mitotic Cyclin (Cyclin Δ90), N terminus deletion (impact), 15
Mitotic entry, prevention, 16
Mix/Bix clusters, 165
Mix/Bix family clusters, 263
mix/bix gene clusters, sequence determination, 158
Model organisms, 3
whole genome sequences (publications), **157**
Modified dual-component GAL4-UAS system, usage, 209
Modified two-signal hypothesis, articulation, 53
Molecular biological studies, usage, 54–58
Molecular biology
neural crest development, insights, 127–131
Molecular biology, *Xenopus* (usage), 9
Molecular cloning technology, advances, 26–27
Monoubiquitylation, 332
Morgan, Thomas Hunt, 43, 77
Moriyama, Yuki, 45–46
Morphogenesis, Wnt signaling, 65
Morpholino-based depletion, 31
Morpholino knockdown, usage, 33
Morpholino oligonucleotide (MO), 245
anti-sense MO, usage, 226, 228
usage, 45
Mosaic development, 25
Mouse embryonic fibroblasts (MEFs), pluripotency (induction), 331
Mouse ES cells, differentiation, 178
Mouse inbred lines, 315
Mouth
cellular organization, 247
facial structure, 246–249
formation, conservation, 246–247
formation, *Xenopus* model, 247, 249
opening, sagittal views, *248*
MS1 ion-current, usage, 201
Multi-pass membrane proteins, regulation, 239
Murine transthyretin promoter, usage, 209–210
Museum d'Histoire, 4
Mutagenesis screens, *X. tropicalis* usage, 157
Mutant alleles, impact, 77
Mutate genes, cluster feature, 163
myh6 nonsense mutation, impact, 211

N

Nakamura, Osamu, 8–9
Nanorana parkeri (orthologous genes), 165
National Center for Biotechnology Information (NCBI), 260
Natural cell cycle arrests, strengths, 202
Natural genetic variation/disease implications
field, summary, 314–315
present state, 316–320
study model, 313
Naturally occurring tumors, 302
Natural size polymorphism, *Xenopus laevis* genetic variation (relationship), *318*
Natural stimulation paradigms, 284
Natürliches System der Amphibien (Wagler), 4
Near genes, heterochromatin marks, 180
Negative autoregulation subcircuits, occurrence, 189
Neural border/descendants
hes genes, role, **108–109**
notch pathway, core components, **106**
Neural crest
derivatives, examples, *132*
developmental mechanisms, knowledge, 125
development/evolution, 125
development, insights, 127–131
development, knowledge, 133–134
ectoderm, appearance, 186
EMT control, 130
EMT/migration, 129–130
GRN regulation, novel/non-canonical genes (importance), **129**
lineage diversification, 130–131
migration, extracellular environment active remodeling (requirement), 130
ontogeny, 127
pluripotency, insights, *132*
potential, origins, 131–133
research, amphibian models (importance), 126
stem cells, establishment, 129
Xenopus, contributions, 246
Neural crest cells (NCCs)
appearance, 102, 105
blastula stem cells, shared feature, 133
development, Hes4 development, 105, 107
differentiation, *247*
progenitor derivation, 239
Neural crest cell transcription factors (NCC-TFs), encoding, 105
Neural crest-derived head skeleton, formation (requirements), 131
Neural crest progenitors, 127–129
induction, 247
Neural ectoderm, induction, 9
Neural induction, 47
Neural plate border (NB)
development, Hes4 role, 105, 107
establishment, 127–129
midbrain-hindbrain boundary, 102, 105, 107
neural crest progenitor induction, 247
region, characterization, 128
Neural progenitor stem cells (NSPCs)
maintenance/differentiation, metabolism regulation, 296
presence, 294
proliferation, 290, *293*
role, 289
Neural stem/progenitor cells, role, 293–294
Neural systems, examples, *280–281*
Neural tissues
AP patterning, 56–57
induction, mesoderm (impact), 52

Neurobiology
observations, *Xenopus* (usage), 278–279
present status, experimental approaches, 279, 281–282
Neurocristopathies, 250
Neurodevelopmental disorders (NDDs), 239–240, 281
Neurogenesis, 293–295
differentiation, *hes/hey* genes (role), **103–104**
Neuronal circuit formation (assessment), embryonic manipulation-based approaches (usage), 279
Neurons, calcium-imaging, 285
Neurula patterning stages, 270
Neurula-stage endoderm patterning, 265–266, *266*
New Head hypothesis, 126
NF35-NF43 early endoderm organ morphogenesis, occurrence, 260–261
NFAT (calcium-sensitive transcription factor), activation, 68
Nicotinic acetylcholine receptor (nAChR), multi-pass membrane-bound protein, 239–240
Niehrs, Christof, 263
Nieuwkoop center (NC), 90
Nieuwkoop, Pieter, 7, 59
Nitric oxide (NO) production, 249
nkx2.5-GFP insert, presence, 212
N-methyl-N-nitrosourea (NMU), usage, 302
nodal3 clusters, 163, 165
nodal5 clusters, 163, 165
Nodal antagonist (CerS), impact, 30
Nodal cascade, impact, 225–226
Nodal-Dand5 module, high sequence diversity, **230**
Nodal-related genes, activation, 30
Nodal signaling
propagation, 225
spatiotemporal gradients, impact, 29
Non-canonical notch signaling, 78
Non-canonical Wnt/PCP signaling, *67*
Non-canonical Wnt signaling, 68
canonical Wnt signaling, integration, 68
gene expression, regulation, 71
importance, 127
Non-histone chromatin state maps, 174
Non-involuting marginal zone (NIMZ), *91*
contribution, 90–91
transcripts, accumulation, 904
Non-regenerative (NR-) mechanisms, comparison, 289, 291
Non-regenerative (NR-) stages, 295
Sox2 reduction, 294
notch1 activity, regulation, 90
Notch1/RBPJ signaling, response (change), 95
Notch/genes/pathways, interplay, 111, 116
Notch Intracellular Domain (NICD), usage, 77, 237
Notch ligands, 99, 102
impact, 111
Notch pathway, 77–78
components, **81**
components, impact, 99
core components, **106**
genes, expression, **112–115**
hes1-7/hey genes responsiveness, experimental evidence, **83–87**
involvement, 107
roles, 77, 107
Notch perturbation, time-dependent opposite responses, 105
Notch receptors/ligands, early expression patterns, *79*
Notch signaling, 78
action, modes, *79*
canonical notch signaling, 78
functions, 77
glycosylation, 237
immotile:motile cilia ratio establishment, 237
increase, 308
involvement, 116
non-canonical notch signaling, 78
pathway, 71
pathway, reactivation, 209
Notch targets, involvement, 95, 98
Notochord, appearance, 186
Nuclear pore complexes (NPCs), 238
Nuclear reprogramming
alternative methods, usage, 328
changes, 327–329
DNA methylation barrier, 332
histone modifications/resistance, 332
methods, schematic, *328*
process, molecular insights, 329–332
promotion, oocyte factors, 330–331
Xenopus laevis oocytes/eggs/egg extracts, usage, 327
Nuclear reprogramming, inception, 325–327
Nuclear transport receptor (NTR), ARM action, 238
Nuclear transport (NT), usage, 326
Nucleoplasmic extract (NPE)
creation, process, *19*
development, 19
Nucleoporins, presence, 237–238
Nucleosome
dynamics/accessibility, histone variants/linker histones (usage), 176–177
positions, importance, 173
Nucleus, canonical Wnt signaling, 66
Nup188, inner ring component, 238

O

OFF-memory, 180
Oken, Lorenz, 4
-Omics, neural crest development (insights), 127–131
Online Mendelian Inheritance in Man (OMIM), 260
ON-memory, 180
Oocyte factors, 330–331
schematic model, *331*
Oocytes
follicle-enclosed oocytes, receptor presence, 143
histones, 178
mRNA localization, mechanisms, 27
RNA polymerases, presence, 9
Optical coherence tomography (OCT), 236
Ordinary differential equations (ODEs), usage, 69–70
Organ fate induction, combinatorial signals, *267*
Origin recognition complex (ORC), 18
otx1 (endodermal identity), 34
Outbred lines, advantages, 316

P

Pancreas, endoderm organ fate, 268, 270
Pancreatic hypoplasia/agenesis, study, 270
PAPC (transcriptional-level regulation), 71
Pardue, Mary Lou, 9
Patient-derived CHD disease mechanisms, discovery, 237–240
pCMBSS, water-transporting channel sensitivity, 146
pDXTP (destination vector), usage, 209
Peacock, Bevill, 233
Pecker, Malcolm, 4
Pediatric Cardiac Genetic Consortium (PCGC), exome sequencing, 236
Peptide-spectra matching, protein reference set (obtaining), **199**
Pericentriolar material (PCM), 238
Peripheral nervous system (PNS), morpho-physiological information, 277
Perturbations, impact, 174
Perutz, Max, 4
Phantom sequences, providing, 199
Pharmacological properties, *Xenopus* oocytes (usage), 143
Pharyngeal endoderm, thyroid (relationship), 268
Phloretin, water-transporting channel sensitivity, 146
Phosphorylated Smad1/5/9 activity, D-V gradient (generation), 265
Phospho-Smad1/5/8,usage, 47
PI3K/Akt signaling, increase, 133
PIF/harbinger-type transposons, distribution, 159
Pigmentation, melanocytes (impact), 131
Placode development, notch pathway role, 107
Platt, Julia, *126*
Pluripotency
demonstration, *132*
maintenance, 128
understanding, *Xenopus* (usage), 325
Pluripotency-related transcription factor (Pou5fl), homologues (usage), 31
Pluripotent embryonic cells, segregation, 260
Pluripotent stem cells (PSCs), 259
human PSC-derived endoderm, 270
human PSCs, differentiation, 260, 265, 268
induced pluripotent stem cells (iPSCs), generation, 329
lines, generation, 327–328
Pole plasm, 26
Polycomb group genes, identification, 178
Polycomb Repression Complex 2 (PRC2), transcription function, 176
Polymerase chain reaction (PCR)
amplification, usage, 211
screen, 225–226
usage, 198
Polymorphisms
impact, 314
simple sequence length polymorphism (SSLP), 157
Polyploid organisms, whole genome sequences (publications), **157**
Polyubiquitination, 47
Positive autofeedback loop, *189*
Posterior neural markers, induction, 56
Post-fertilization embryos (arrest), CSF (impact), 16
Post-involuted mesendoderm, usage, 59
Post-translational modifications, 202
Post-translational protein modification, 202
Precise integration into target chromosome (PITCh) system, usage, 308
Pregnancy Diagnosis Center, 8
Xenopus source, 7

Pre-placodal ectoderm (PPE), formation, 105
Prepattern genes, balanced expression, 99
Pre-placodal ectoderm (PPE)
appearance, 186
development, Hes genes (role), 107
formation, 105
Primary ciliary dyskinesia (PCD), 226
Primary neurogenesis, 99, 102
Hes1-7 genes, involvement, 102
Primordial germ cell (PGC)
formation, maternal control, 32–34
induction, 32–33
specification, germ plasm mRNAs (study), 33
Progenitor-restricted TFs, molecular mechanisms, 266
Pro-inflammatory (TH1) cells, cytokine production, 314
Proliferating cell nuclear antigen-interacting protein (PIP) box, usage, 18
Proliferating cell nuclear antigen (PCNA), interaction, 18
Proneural genes, impact, 99
Pro-neural *soxb1* expression, repressing, 128–129
Protein expression
developmental atlas, 201–202
encoding, *201*
Proteins
degradation, impact, 15–16
post-translational sequence editing, 199
recruitment steps, 18
Proteoglycans, importance, 69
Proteomics
cell biology, relationship, 200
high-resolution proteomic analyses, 197–200
single-cell proteomics, 202–203
protocadherin 8 (*pcdh8*), repression, 111, 116
Proxy-cancer phenotype, presence, 306
Pseudogenes, cluster feature, 163
Pseudo-stratified epithelial structure, recovery, 2932
pTransgenesis, usage, 208–209

R

Rabbit *globin* gene, eukaryotic mRNA isolation, 9
Rana catesbiana studies, 290
Rana pipiens eggs
enucleated eggs, endoderm cells (differentiation), 326
nuclei, usage, 14
Rana pipiens sperm nuclei, usage, 14
ranid species, study, 278
RAPGEF5, blocking, 239
RAR (receptor construct), usage, 54–55
RBPJ (DNA-binding protein), participation/mediation, 78
Reactive oxygen species (ROSs), role, 291
Receptors (biophysics/pharmacological properties), *Xenopus* oocytes (usage), 143
Recombinant DNA technologies, invention, 9
Red blood cells (RBCs), infiltration, 291
Regenerative (R-) mechanisms, comparison, 289, 291
Regenerative (R-) stages, 294, 295
Regional neural inducers, model (Mangold), *53*
Region-specific microsurgeries, performing, 52
Regulative development, 25
Regulatory substrate, 173–174
Repetitive elements, heterochromatin marks, 180
Replication
dynamics (study), *Xenopus* egg extracts (usage), 20
licensing, 17–18
origin firing, 20
replication-coupled chromatin assembly dynamics, 178
timing, concept (development), 20
Repression, histone modifications (usage), 181
Reprogramming. *See* Nuclear reprogramming
field (establishment), SCNT experiments (usage), 326
gene expression patterns, change, 330
Respiratory progenitor induction, regulation, *269*
Restriction enzyme-mediate integration (REMI), 207–208
Retinal hyperproliferation, 304
Retinal pigmented epithelial (RPE) cells, accumulation, 238
Retinoic acid (RA)
biosynthesis, 314
signaling, 54–55, 270
Reverse genetics, dsBreaks (usage), 212, 214
Reverse transcription (RT) reagents, usage, 203
Rhodopsin, multi-pass membrane-bound protein, 239–240
RhoGEF TRIO, 212–213
Ribonucleic acid (RNA). *See* Messenger RNA
RNA-based SOM, generation, *191*
RNA-binding proteins, identification, 200
RNAi, usage, 246, 305
RNA SOM metacluster linkage, 191
Ribosomal DNA (rDNA), separation, 9
Ribosome profiling (RIBO-seq), 200
Right half embryo, sagittal cut, *48*
Roeder, Bob, 9
Roux, Wilhelm, 43, 126
R-stage/NR-stage, transcriptomic/proteomic level (biological process regulation), 296
RT-qPCR, usage, 163
rtTA expression, control, 209
RXR (receptor construct), usage, 54–55

S

Sanger method, usage, 157
Scaffolds
BAC location, 159
mapping, 157
Sclerostin (SOST), binding, 66, 68
Secreted frizzled-related protein (sFRP), Wnt binding ability (loss), 45
Secreted ligand activation, 32
Selenoproteins, sequence, 199
Self-organizing map (SOM)
application, 190–191
linked SOM method, 191
metaclusters, presence, *191*
RNA-based SOM, generation, 191
Sensory left-right organizer (sLRO) cells, pre-flow stages, *227*
Sex chromosomes, W-specific/Z-specific regions, 161
Sex-determining gene (*DM-W*), female specificity, 161
sfrp1 expression, confinement, 70
Shapiro, Hillel, 7
Shared molecular pathways, 239–240
shRNA, usage, 209
Siamois clusters, 165
siamois gene cluster, *sia* expansion, 165
Signaling molecules, dynamic temporal expression, 59
Signaling pathways
directions/questions, 71
importance, 132–133
usage, 54–58
Wnt/beta-catenin (β-catenin) signaling pathway, 56–57
Signal integration, 57–58
Signal transduction (mediation), NICD (usage), 77
Simple sequence length polymorphism (SSLP), 157
Single-cell proteomics, 202–203
push, 203
Single-nucleotide polymorphism (SNP), 313–314
array analysis, usage, 236
Single-stranded DNA, replication, 17
Skp1-cullin-F-box protein (CCF) ubiquitin ligase, impact, 19
Skull bones, growth, 250
Slack, Jonathan, 9
Sleeping Beauty, usage, 211–212
Slome, David, 7
Smad4-dependent Tgfβ signaling, inhibitor, 31
Smit, Pierre Jacques (*Xenopus* tadpole illustration), *5*
snai1 expression, regulation, 129
snai2 expression
induction, 127
repression, 129
Snail family transcription factors, importance, 130
Sodium channels, identification, 146–147
Somatic cell nuclear transfer (SCNT), 325
experiments, impact, 326
Somite, appearance, 186
Somitogenesis, 107, 111, 116
notch/genes/pathways, interplay, 112, 116
notch ligands/*hes* genes, involvement, 111
notch pathway genes, expression, **112–115**
Somitomeres
segmental prepattern, 111
whorls, formation, 111
Southern blotting, usage, 181
Sox2 protein
expression, 293
reduction, 294
Sox17, 263
Spatial exclusion, *189*
subcircuits, 190
Spemann, Hans
egg constriction experiments, 25, *26*
Spemann-Mangold experiment, 43, 48
Spemann organizer
formation, 186
inductive signals, *46*
molecules, search, 44–45
neural induction, 47
relocalization, self-regulation, 47–48
S phase, M phase (transitions), 17
Spinal cord injury (SCI), 289
comparative cellular response (R-stage/NR-stage), *292–293*
early cellular response, 291
intermediate cellular response, 291, 293
late cellular response, 293
R-stage/NR-stage spinal cord, histological differences, 291

R-stage/NR-stage, transcriptomic/proteomic level (biological process regulation), **296**
zebrafish response, 295
Spinal cord injury (SCI), cellular response, 291, 293
R-stages/NR-stages, **293**
Spinal cord regeneration
axon regeneration, 294
biological processes, identification, 294–296
neural stem/progenitor cells, role, 293–294
neurogenesis, 293–294
observations, 289–290
present status, 290–296
study, *Xenopus laevis* (model organism), 290–291
understanding, 289
S subgenomes
asymmetrical evolution, 159, 161
identification, 158–159, *160–161*
Stage NF20 embryos, RNA-seq/ChIP-seq analyses, 265
Stage NF20 *Xenopus* embryo, schematic, *266*
Stage-specific microsurgeries, performing, 52
Steinbeisser, Herbert, 45
Stomach, endoderm organ fate, 270
Structure-function mutagenesis, usage, 27
Subgenomes, evolution, 161
Surani, Azim, 45
Symmetrization, rotation, 25–26
Syndromic craniosynostosis, description, **251**

T

T3 treatment, 294
Tadpole behavior, profiling/quantification ability, 281
Tailbud, dorsal half embryo, *48*
Tail oscillations, motion recording, 285
Takahashi, Shuji, 163
Talin protein, information relay, 130
Targeted nucleases, usage, **212–214**
Targeting induced local lesions in genomes (TILLING), 211
screen, usage, 211
TATA-binding protein, pre-initiation complex involvement, 331
TATA box, usage, 176
Taussig, Helen Brooke, 233
TAZ (Hippo signaling pathway component), 71
TBX20 mutations, presence, 235
tbxt (pan-mesodermal marker)
inhibition, 95
regulation, 188
Tbxt-expressing cells, change, 188
Tc1/mariner-type DNA transposons, distribution, 159
TCF proteins, repressing, 31
Tdgf1.3, maternal supply, 30
Teleost lineage, WGD (occurrence), 163
Temperature-inducible *hsp70* promoter, usage, 209
Temporal collinearity, definition (ambiguity), 163
Tetracycline-inducible (TRE) promoter, usage, 209
TGFβ2 (transcriptional-level regulation), 71
Thyroid hormone (TH) levels, increase, 294
Thyroid, pharyngeal endoderm (relationship), 268
TILLING. *See* Targeting induced local lesions in genomes
Tissue differentiation, 45–47
Wnt signaling, 65
Tissue remodeling, 296
Tissue-specific promoters, usage, 209
TMEM16A (calcium-dependent chloride channel), 147
activation, 146
discovery, 144, 146
TMT-C+, usage, 197
TMT MS-3, usage, 197
Toll-like receptor (TLR) genes, impact, 314
Tolloid chordinase, 47
Tomlin, S.G., 8
Topologically associating domains (TADs), 173
Trachea, endoderm organ fate, 268
Tracheal-esophageal (T-E) fistula/esophageal atresia, 260
Tracheal-esophageal (T-E) morphogenesis, 268
Tracheoesophageal separation, regulation, *269*
Transcription activator-like effector nucleases (TALENs), 301
apc TALEN mRNAs, targeted injection (usage), 305
students, 212
systems, usage, 290
TALEN-mediated genome editing, 303
TALEN-mediated TSG disruption, 304–305
tumor model, generation, 305
Transcriptional reprogramming, usage, 330
Transcription factors (TFs)
binding, chromatin accessibility (relationship), 177–178
binding, physical sites (identification), 190
CRMs, relationship, 188
docking, 185
expression, 268
importance, 132–133
neural crest cell transcription factors (NCC-TFs), encoding, 105
regulation, Vegt/Vegt-induced Nodal signals (usage), 30
target DNA sequence binding, 78
Transforming growth factor β (Tgfβ), encoding, 26–27, 30
Transgene construction, streamlining, 208–209
Transgenic strategies, dependence, 210
Transgenic *Xenopus*
generation, 207–209
pTransgenesis, usage, 208–209
technology development, 207–208
uses, 209–210
Transient receptor potential (TRP) cation channels, 146
Transient receptor potential melastatin (TRPM) family, 146–147
Transmembrane receptors, Fizzled family receptors (binding), 65
Transposable elements, 314
localization, 180
Treacher Collins syndrome, 250
Treslin/TICRR, usage, 19
Trithorax group genes, identification, 178
Tumor dependencies (identification), GEXM (usage), 305–306
Tumor-immune system interactions, 303
Tumor induction, 303
Tumor progression, *in vivo* monitoring tools, 307–308
Twinned embryos, result, *26*
Twist1/Twist2 proteins (bHLH transcription factors), mesodermal development role, 129–130

U

Ultraviolet (UV) irradiation, usage, 326
Undiluted cell lysates, making, 200
Unmethylated CpGs, capture, **174**
UUCAC/UUUCU motifs, presence, 27

V

Vasointestinal peptide, 143
Vegetally localized mRNAs, **28**
encoding, velo1 (usage), 33
vegt
endodermal identity establishment, 34
endoderm/mesoderm, maternal control, 29–30
locus, *175*
mRNA, maternal role (assessment), 29–30
Vegt (maternal T-box transcription factor), importance, 187
Velo11, usage, 33–34
Ventral BMP signaling, expression restriction, 268
Ventral genes, BMP4 signaling (impact), 45
Ventral half embryo, *48*
Ventral mesenchyme, appearance, 186
Ventral mesoderm, formation, 188
Ventral-posterior GRN, 265
Ventral segments 3/4, progeny, *46*
Ventral signaling center, 45
Ventx clusters, 165
Vertebrates
cardiogenesis, studies, 235
development, molecules/genes (roles), 127
embryonic development, regulatory network diagrams (establishment), 185–186
embryonic development, Wnt signaling (integrated understanding), 69
gastrulation, 68
neural crest, development/evolution, 125
notch signaling pathway, scheme, *79*
somitogenesis, Clock/Wavefront hypothesis (postulation), 107, 111
vg1 clusters, 163, 165
Vinculin protein, information relay, 130
Visual stimulus, learned associations, 281–282
Voltage sensor, 144

W

Waddington, C.H., 7, 132
Wagler, Johann Georg, 4
footnote, *6*
Water-transporting channel sensitivity, 146
Wavefront hypothesis, postulation, 107, 111
Weissman, August, 126
Whole genome bisulfite sequencing, 174
Whole genome duplication (WGD), 155–156
usage, 162–163
Whole genome sequences, publications, 157
Whole genome sequencing (WGS), 157
Williams, Daniel Hale, 234
Wingless-type MMTV integration site (Wnt) genes, cloning efforts, 27
Wnt11b, classification, 32
Wnt antagonists
expression, signaling, 263
overexpression, 32

Wnt/beta-catenin (β-catenin) signaling pathway, 56
Wnt/BMP/FGF signals, posteriorization, 270–271
Wnt/calcium signaling, 68
Wnt genes, co-discovery, 65
Wnt interactions, directions/questions, 71
Wnt pathways
genes, identification, *44*
head formation, 56–57
Wnt/PCP/JNK signaling, 68
WNT-PCP pathway/signaling, control, 247, 249
WNT proteins, cell specification, 246
Wnt receptor function, 66
Wnt signaling, 29, 65
canonical Wnt signaling, 66, 68
cardiac organogenesis, 69
context-specific Wnt signaling, 70
embryonic signaling/regulatory environment, integration, 71
integrated signaling, 69
levels, functional manipulation, 56
modeling, 69–70
pathways, *67*
pathways, characterization, 65–69
potentiation, 32
requirement, 105
W-specific regions, 161, *162*
Wylie, Chris, 29

X

Xelaev1801880tm, identification, 199
Xenopus. *See* Transgenic *Xenopus*
acquisition, importance, 7
analysis techniques, examples, *280–281*
blastula stage endoderm patterning, *264*
behavioral performance, 277–278
blastula ectodermal explants, usage, 47
blastula, regional specification (model), *26*
blastula-stage embryo, fate map (production), 8–9
cancer modeling, application potential, 305–307
cell-fate reprogramming *Xenopus* NT embryos, efficiency, 332
cell lineages, isolation, 330
clutches (genetic variation), experimental outcome (schematic representation), *317*
clutch variability/experimental design, *317*
congenital heart disease, study (future), 240
craniofacial development/disorders, 245
CRISPR mutants, presence, 212
description, 3–4, 6
developmental biology researchers, involvement, 8–9
digestive/respiratory system development/disease, modeling, 259
dorsal midline (development), notch pathway (core components), **92–94**
dorsal-ventral patterning, signaling components, *44*
D-V tissues, origins (diagram), *46*
early development, histone modifications (acquisition/dynamics), 178–180
early embryos (cell lineages), maternal mRNAs (relationship), 25
egg constriction experiments, 25
egg cytoplasm, DNA replication induction, 13–14
embryonic manipulations, 278–279
endoderm germ layer formation, conceptual phases, *262*
endoderm organogenesis, temporal overview, 260–261, *261*
endoderm patterning, 69
explants, usage, 68
follicle-enclosed oocytes, electrophysiological characterization, 144
functional neurobiology, 277
gene editing, future, 308
gene-specific studies, **250**, **251**
genes study, novel roles (discovery), 237–238
genetically altered *Xenopus* (creation), targeted nucleases (usage), **212–214**
genetically engineered *Xenopus* models, 303
genetics/gene editing, 210–214
genetic variability, exploitation, 319–320
hes4-7/hey genes, early expression patterns, *80*
hes/hey genes, cross-regulation, **88–89**
high-resolution proteomic analyses, development, 197
history, 3
images, *5*
in vitro preparations, applicable stimulation/recording techniques, *283*
laboratory animal, usage (increase), 6–8
laboratory populations, genetic diversity, 316–317
large-scale genomic/genetic resource development, 211
lines, availability, 315
literature, 9–10
mass spectrometry, usage, 198
maternal inheritance, demonstration, 9
mesendodermal GRN, FFL usage, 190
mesoderm/endoderm specification, gene regulatory networks (usage), *187*
metamorphosis, 209
midbrain/hindbrain boundary (MHB) establishment, *hes* genes (impact), **110**
model, 16, 71–72, 225
model system, success, 65
Nasco laboratory-bred colony, 316
naturally occurring tumors, 302
neural border/descendants, *hes* genes (role), **108–109**
neural border/descendants, notch pathway (core components), **106**
neural crest development, 127–131
neural systems, examples, *280–281*
neurogenesis/epidermal differentiation, *hes/hey* genes (role), **103–104**
oocytes, RNA polymerases (presence), 9
oocytes, test system importance, 29
ORFeome, incompatibility, 208
organ fate induction, combinatorial signals, *267*
organizer, 43
organizer, historical background, 43–44
pathways, connection, 237
patient-derived CHD disease mechanisms, discovery, 237–240
physiological studies, 279
proteomics, 203
rDNA, eukaryotic gene isolation, 9
research, advances, 69–70
research tool, 301–302
respiratory progenitor induction, regulation, *269*
RNA-based SOM, generation, 191
sensory-motor capacity, 277–278
simple motor behaviors, 284
somitogenesis, notch pathway gene (expression), **112–115**
source (Pregnancy Diagnosis Center), 7
species, phylogenetic relationships, *160–161*
Spemann organizer molecules, search, 44–45
stage NF20 *Xenopus* embryo, schematic, *266*
system, strengths, 202
tracheoesophageal separation, regulation, *269*
tract tracing, 278–279
tumor formation/regulation, study, 301
twinned embryos, result, *26*
usage, 9, 278–279, 316–320, 325
vegetally localized mRNAs, **28**
vertebrate cardiogenesis, studies, 235
wild-caught colony, population, 316
Xenopus BMP4, expression, 45
Xenopus, CHD/HTX (studying)
approaches, 236–237
morphological/developmental benefits, 236
Xenopus development
localized mRNAs, roles, *27*
maternal gene function, analysis, 28–29
overview, 186
Xenopus dorsal midline development
hes/hey genes, impact, **96–97**
notch pathway, core components, **92–94**
Xenopus egg extracts
cell division control/DNA replication, study, 13
low-speed *Xenopus* egg extracts, advantages, 17
preparation, *19*
single-molecule approaches, usage, 20
system, development, 91
tractable system, providing, 19–20
usage, 15–16, 327
utility, increase, 19
Xenopus embryogenesis
control, gene regulatory networks (usage), 185
early *Xenopus* embryogenesis, information source, 302
notch signaling, 78
past observations, 186–190
Xenopus embryos
32-cell stage, lineage tracing, 46
cell lineage, predictability, 45–46
early development, schematic, *261*
gene regulatory networks (GRNs), usage, 188–190
histone modification maps, **174**
Xenopus genome
evolution, 155
history, 155–156
sequences, usage (studies), 161–165
sequencing, 157–161, 198
Xenopus germ layers development
hes/hey genes, impact, **96–97**
notch pathway, core components, **92–94**
Xenopus laevis (XLA)
allotetraploidy, 155
chromosomes, *164*
collection, 201
data (peptide-spectra matching), protein reference set (obtaining), **199**

genome sequencing, 158
half embryos, self-organizing properties, *48*
homeologous chromosomes, designation, 158
model organism, usage, 290–291
natural size polymorphism, relationship, *318*
nuclear reprogramming process, molecular insights, 329–332
pkd1 (CRISPR knockout), *215*
subgenomes, gene organization (details), 163, 165
Xenopus laevis (XLA) embryos
developmental stages, workflow schematic, *201*
images, *261*
research, 48
Xenopus laevis (XLA) genomes
assembly, 157–158
notch signaling responsiveness, **82**
sequencing, 44, 157–158
Xenopus laevis (XLA)/*Xenopus tropicalis*
chromosomal relationships, *160–161*
comparison, 158
contrast, 200–201
Xenopus notch pathway
components, **81**
hes1-7/hey genes, responsiveness (experimental evidence), **83–87**
Xenopus oocytes
endogenous ion channels, usage, **145–146**
usage, 143–144
vegetal pole mRNAs, cell-free translation, 26
Xenopus oocytes, usage, 143
past observations, 143–144
present status, 144–147
Xenopus tadpole
immunostaining, usage, *5*
Smit illustration, *5*
Xenopus tropicalis (XTR)
biomedical research usage, 315
cancer modeling methodologies, 307–308
cancer models, applications, *306*
chromosomes, *164*
CRISPR/Cas9 cancer modeling, 305
development, chromatin state maps (genome browser view), *175*
F0 *X. tropicalis* tumor, crispants (presence), *304*
forward genetic screens, usage, 210–211
genetic research, diploid genome (impact), 302
genome sequencing, 157
hes1-7/hey groups genes, notch signaling responsiveness, **82**
Xenopus laevis, chromosomal relationships, *160–161*
Xenopus laevis, comparison, 158
Xenopus laevis, contrast, 200–201
Xic1, modification, 19
XLA2Lq32-33, W-specific/Z-specific regions, *162*
XLA3L, 165
XLA9_10, chromosome fusion sites (identification), 158
Xvent1/2, function, 45

Y

Yamada, Tsuneo, 8
Yamanaka, Shinya, 325
YAP (Hippo signaling pathway component), 71
Yong, Ed, 6

Z

Zebrafish (*Danio rerio*)
biomedical research usage, 315
DNA methylation, oocyte/sperm patterns (differences), 181
larvae/immunocompromised adults, xenotransplantation recipient usage, 307–308
SCI response, 295
Zebrafish, histone modification dynamics, 179
Zebrafish International Resource Center (ZIRC), donor vector procurement, 208
Zinc Finger Nucleases (ZFNs), usage, 212, 301
Z-specific regions, 161, *162*
Zwarenstein, Harry, 7
Zwischenstrang, 126
Zygotic genome activation (ZGA), 187
maternal Wnt signaling, relationship, 70
onset, chromatin assembly (relationship), 177
Zygotic transcription factors, identification, 30
Zygotic Wnt8/beta-catenin, action, 265